Lecture Notes in Electrical Engineering

Volume 123

Gary Lee (Ed.)

Advances in Automation and Robotics, Vol. 2

Selected Papers from the 2011 International Conference on Automation and Robotics (ICAR 2011), Dubai, December 1–2, 2011

Springer

Prof. Gary Lee
Information Engineering Research Institute
100 Continental Dr
Newark, DE 19713
United States
E-mail: garrylee@ier-institute.org

ISBN 978-3-642-25645-5 e-ISBN 978-3-642-25646-2

DOI 10.1007/978-3-642-25646-2

Lecture Notes in Electrical Engineering ISSN 1876-1100

Library of Congress Control Number: 2011941700

Typeset by Scientific Publishing Services Pvt. Ltd., Chennai, India.

Printed on acid-free paper

9 8 7 6 5 4 3 2 1

springer.com

Preface

The international conference on Automation and Robotics-ICAR2011 is held during December 12–13, 2011 in Dubai, UAE. The proceedings of ICAR2011 have been published by Springer Lecture Notes in Electrical Engineering, which include about 163 excellent papers selected from more than 400 submitted paper(s).

The conference is intended to bring together the researchers and engineers/technologists working in different aspects of intelligent control systems and optimization, robotics and automation, signal processing, sensors, systems modeling and control, industrial engineering, production and management. In addition to the contributed papers, the conference committee has invited papers by active researchers from various countries in relevant topic areas covered at ICAR2011. Internationally known experts from several countries are invited to deliver plenary/Keynote lectures at ICAR2011.

Much of the credit of the success of the conference is due to topic coordinators who have devoted their expertise and experience in promoting and in general co-ordination of the activities for the organization and operation of the conference. The coordinators of various session topics have devoted a considerable time and energy in soliciting papers from relevant researchers for presentation at the conference. The Chairpersons of the different session played important role in conducting the proceedings of the session in a timely and efficient manner and the Committee expresses sincere appreciation for their involvement.

The reviewers of the manuscripts, those by tradition would remain anonymous, have been very helpful in efficiently reviewing the manuscripts, providing valuable comments well within the time allotted to them. The Editors express their sincere and grateful thanks to all reviewers.

The ICAR2011 International Committee is proud to dedicate this conference proceeding to the Honor Chair David Wang from IEEE Nanotechnology Council Cambodia Chapter Past Chair, Cambodia. And Mark Zhou, from Hong Kong Education Society, Hong Kong, was not only an outstanding researcher and visionary but was also a humble human being.

Finally, our sincere thanks to the supper from Information Engineering Research Institute (USA) for designing the conference web page and also spending countless days in preparing the final conference program in time for printing. Many thanks to Miss Han: ICAR2011 secretary for arranging to type a large number of the invitation letters and for establishing the vast ICAR2011 conference address data base. Sincere thanks to our organization committee for their several months' hard work in sorting out manuscripts from various authors. Also thanks to other staffs assisting in the various stage of the editorial work, especially the international committees of our conference.

Garry Lee
Information Engineering Research Institute, USA

Conference Committee

Honor Chair

Gerald Schaefer	Loughborough University, U.K.

General Chair

Garry Lee	Information Engineering Research Institute, USA
Xiang Li	IERI Circuit and System Society, USA

Publication Chair

Garry Lee	Information Engineering Research Institute, USA

Organizing Chair

Ming Ma	Singapore NUS ACM Chapter Chair, Singapore
Paris Kitsos	American Applied Sciences Research Institute, USA

International Committee

Yuanzhi Wang	Anqing Normal University, China
Xiaofang Zhu	Jiangxi Normal Univesity, China
Gerald Schaefer	Loughborough University, UK
Qihai Zhou	South Western University of Finance and Economics, China
Srinivas Aluru	ACM NUS Singapore Chapter, Singapore
Tatsuya Akutsu	ACM NUS Singapore Chapter, Singapore
Aijun An	National University of Singapore, Singapore
Minli Dai	Suzhou University, China
Yijin Wu	Asia Pacific Environmental Science Research Center, Hong Kong
Zhenghong Wu	East China Normal University, China
Biswanath Vokkarane	Society on Social Implications of Technology and Engineering
Nian Zhang	Suzhou University, China
Xiang Li	IERI Circuit and System Society, USA

ICAR 2011 Sponsored by

Information Engineering Research Institute, USA

IERI

IERI Circuits & Systems Society

IERI Circuit and System Society

Contents

Automatic Reasoning Technology Based on Secondary CBR

Haobin Shi[1], Wenjie Dong[1], Linquan Yang[2], and Zhujun Yu[1]

[1] School of Computer, Northwestern Polytechnical University, Shanxi, Xi'an 710129, China
[2] Faculty of Information Engineering, China University of Geosciences, Wuhan 430074, China

Abstract. Because of the unique knowledge representation method and natural learning ability, researchers pay more and more attention to the Case-Based Reasoning (CBR). Currently, the most CBR technology just a simple case reuse, it cannot doing the case revise or case revise didn't have good generality. Aiming at the problem of it, this paper proposed a framework of automatic reasoning technology based on secondary CBR. The framework drive the case reuse by case revise, storing the base case with primary case base, use the secondary difference case base to store the case adjust knowledge, and with the secondary CBR assist the whole automatic reasoning process completely. Practical application shows that the technology has good practicality and generality.

Keywords: CBR, case reuse, case revise, automatic reasoning.

1 Introduction

Automatic reasoning is the key technology of artificial intelligence. It could make a conclusion according to some strategies. Recently, some research of artificial intelligence technology such as the artificial intelligence technology based on qualitative reasoning, rough set, evidence theory, group hierarchical analysis and Mult-Agent, which has greatly promote the development of artificial intelligence. Although these technologies solve some the problems in the field of artificial intelligence from different angles and in different extent, it is still difficult to solve the following problems in a better way: domain knowledge lack, experts solve problems by experience, there are lots of facts not include in rules, hard to establish model, semi-structured and non-structure problems exist. To solve the problems list above is the CBR technology advantage [1], it aims to solve new problem with existing cases. CBR is an incremental learning method, it avoid the bottleneck problem of traditional artificial intelligence in knowledge acquisition, overcome defects of knowledge hard to acquire and reasoning vulnerability in the system based on rule. With the vigorous development of artificial intelligence field, the CBR technology has more and more be taken seriously and has been widely application in various areas [2].

In current CBR research, the important and complex technology is case adaptation. Case adaptation is a technology, which is when the target case has not completely matching original case in case base, how to adjust original case to meet the target case. Now, there are two types of ideas to solve this problem. One type is pre-storage

G. Lee (Ed.): Advances in Automation and Robotics, Vol.2, LNEE 123, pp. 1–8.
Springerlink.com

the adjust knowledge with the form of rule to build the adaption knowledge database. It could use to guide case adaptation by query knowledge in the base. As the rule-based reasoning, this method needs to acquire knowledge from domain experts. The process involves knowledge engineering, and it also repeated the bottleneck problem of knowledge acquisition. The other method is mainly depends on the machine learning technology, such as neural networks and decision trees. It guides case adaptation by extract knowledge from case base. In literature [3], learning rules from differential case to guide case adaptation. This method require appropriate rule learning algorithm, and the corresponding algorithm often have strong relevance. Literature [4] discusses the differential case adaptation guide method in linear field. Literature [5, 6] discusses case adaptation technology in non-linear field with local linear regression model.

Throughout these case adaptation method we could find that their common weakness is although these methods could solve case adaptation problem to some extent, it do not have good versatility. In order to solve the problem, this paper proposes an automatic reasoning method based on secondary CBR.

2 CBR Basic Principle and Constitute

CBR technology is a kind of problem solving and learning method based on knowledge in artificial intelligence. The problem solving process is similar to the process of person thinking. It is a simulation to cognitive psychological activity of human [7]. From the angle of cognitive psychology, the technology solves complex problems according to human daily thinking model.

A typical CBR problem solving process can be divided into four mainly process: retrieve, reuse, revise and retain. So, the CBR also called 4R [7-10]. It working process was as shown in figure 1.

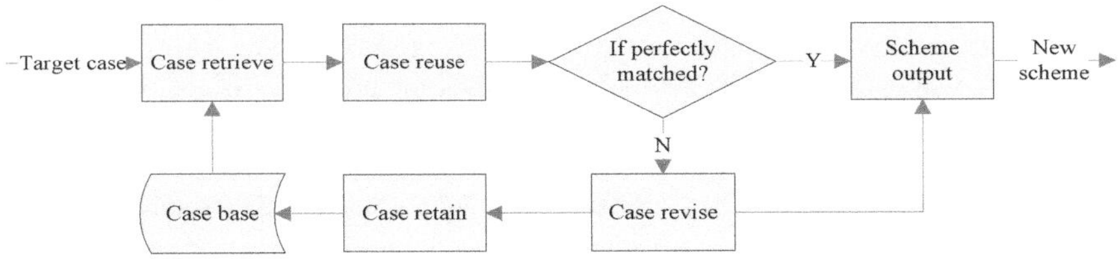

Fig. 1. Case reasoning 4R model

In CBR, the problems or situation will solved currently is called target case. The experience or historical case often called base case. Figure 1 shows that the CBR basic process of solving problem is [7] as bellow list. A new problem to be solved appearing, it is target case. Case retrieve process is query similarity case of the past with the description information of target case, and then, get the base case which is similar with target case. Therefore, gain some new problem solutions. If the base case perfectly matched with target case, the base case could output directly. If not, it will enter case re process. In order to adjust the situation of target case, the base case scheme will be modified by case revise. At the same time, the new solution scheme could be output and access into the case retain step in order to take it as reference at the next time when meet the similar problem. At the last, by analogy the effectiveness of solution should get evaluation. CBR take the cases as knowledge element, the knowledge acquisition and representation is natural and direct, and it has self-learning function. Its essence is analogical reasoning based on similarity, which is consistent with the logical of human analogical thinking. For example, when we could not understanding what is automatic reasoning, we will open the search engine firstly, and then, input "reasoning" or "automatic reasoning" to doing search. At the last, find the exact explanation and record it. We do it because our previous experience accumulation. When we meet other similar problem before, we use the same method to solve it. That is to say, when we encounter a situation, we used to recall the manner, method, strategies and solutions in previous situation to help us find solutions of current problems. The cognition thinking perspective shows the mental model of human memory, planning, learning and problem solving, it reflecting the more advanced knowledge environment. It is integrated of variety artificial intelligence.

In the process of case reasoning, case representation, case retrieval and case revise are the core issue in case reasoning research. Most of current case reasoning system is the case retrieval and case reuse system basically. Case revise is often completely by case reasoning system administrator.

3 Automatic Reasoning Based on Secondary CBR

This paper proposed the automatic reasoning technology based on secondary CBR. The system structure is shown as Figure 2. It storing the base case with primary case base, use the secondary difference case base to store the case revises knowledge, and with the secondary CBR assists the whole automatic reasoning process completely.

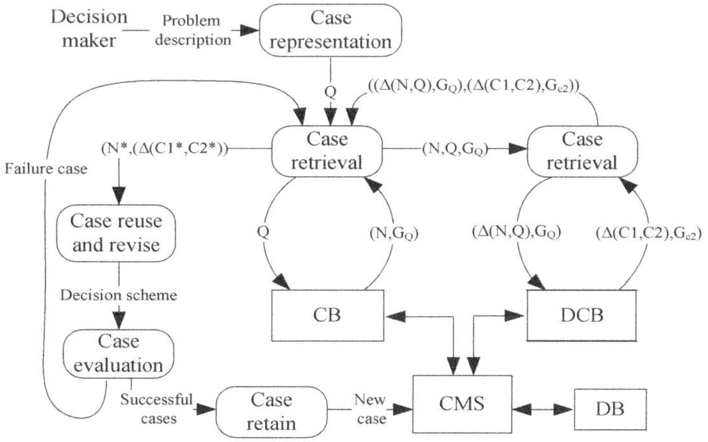

Fig. 2. Secondary CBR automatic reasoning system structure

In the reasoning process, firstly, the decision-maker will give the problem description which will be solved. The problem could be expressed as standard case form Q by the process of case representation. Secondly, in traditional case base CB, case N which is most similar with Q could be get and the curvature GQ nearby Q would be calculated with case retrieval. Case curvature is the case attribute rate of the case in it space, to the case express in multi-dimensional space vector, the curvature also is multi-dimensional vector. It is primary CBR reasoning process. Thirdly, the result (N,Q,GQ) get in the second step will be take as input of secondary CBR reasoning process. Difference case (Δ(C1,C2), GC2) which is most similar with (Δ(N,Q),GQ) would be retrieval in difference case base (DCB) by case retrieval. Fourthly, reasoning result could be transfer to primary CBR process. The reasoning result of primary CBR combine with secondary CBR could used to comprehensive evaluate the retrieval result and find the best similar case and difference case with (N*,Δ(C1*,C2*)). The process may repeat the second and third step until get the best revise scheme. Fifth, (N*,Δ(C1*,C2*)) will pass to case reuse and revise process. By reuse N* and use Δ(C1*,C2*) the doing case revise to N*, the new problem solution will be get. Finally, evaluate the new problem solution scheme, if the scheme evaluation failure, it will return to primary CBR system and change condition to redoing reasoning. If the scheme evaluation successful, it will access into the case retain process and the new case could plus into case base, at the same time, the case maintenance system (CMS) could used to update and maintain CB and DCB to achieve the purpose of case study.

The system actually use the approach of drive the case reuse by case revise, so the final reuse case not always is the old case which is most similar with new problem.

3.1 General Case and Difference Case Representation

Generally, case content has three mainly part. (1) Problem or situation description. It is the problem when the case occurred needs to solve and the state of word around. (2) Solutions. It is the problem solution. (3) Result. It is the result after implement the solution (the new state of world around). Any case reasoning system must have the solution of problem or situation description. It is an essential part. The result part may not exist in some system.

Most cases description is adopt eigenvector representation method in existing system. That is: let $(f_1, f_2, f_3, ..., f_n)$ be feature attribute set of research object, V_i is range of feature f_i $(i = 1, 2, ..., n)$, thus, $(v_1, v_2, v_3, ..., v_n)$ is a value vector of feature set, and it could be called feature vector. Where v_i is belongs to V_i $(i = 1, 2, ..., n)$, and a case could express as a feature vector.

Literature [11] mention that when involves complex task, it will be difficult to distinguish the problem and answer part of a case. So, the simple eigenvector representation is hard to meet the case express need of complex task, and it form also limits the further application of eigenvector. Literature [12] adopts frame representation, but it does not suit calculation in practical application. Literature [13] proposes a hybrid representation method, it improves the ability of case representation, but it is a variant of eigenvector representation actually.

This paper generalizes eigenvector representation by comprehensive considering above factor and gets the general case vector representation. It is $C(c_1,c_2,c_3,...,c_n,d_1,d_2,d_3,...,d_m)$, where c_i $(i=1,2,...,n)$ express the i sub-system in condition attribute of case C. d_i $(i=1,2,...,m)$ express i sub-system in decision attribute of case C. Obviously, the $(c_1,c_2,c_3,...,c_n)$ constitute the case condition space, and $(d_1,d_2,d_3,...,d_m)$ constitute the case decision space. In special circumstances, when each subsystem of case C is constitutes by a property and the decision space is one-dimensional, it variant to general eigenvector representation. It is $C(c_1,c_2,c_3,...,c_n,d)$.

To difference case, here with $\Delta(C1,C2)$ express the difference of two vectors, $C1(c_{11},c_{12},...,c_{1n},d1)$ and $C2(c_{21},c_{22},...,c_{2n},d2)$. Where $\Delta(C1,C2)=(\Delta(c_{11},c_{21}),\Delta(c_{12},c_{22}),...,\Delta(c_{1n},c_{2n}),\Delta(d_1,d_2))$. The difference between two cases is a vector constitute with the difference of each case eigenvector.

If the case feature properties are numeric attributes, the attribute difference could expressed by the corresponding attribute discrepancy of two cases. Essentially, the expression of difference case is the same with general case. If the case attributes are non-numeric, the $\Delta(\tau,\rho)$ could express with $"\tau \rightarrow \rho"$.

3.2 Case Retrieval

Case retrieval and matching is the method of retrieval and select the available source case from source case base by retrieval information, and evaluate the similarity between new case and source case. It is a key step in CBR system. Case retrieval will reach follow two objects finally: the retrieval cases should as less as possible, and it should be as much similar or match as possible with current case (target case). Case retrieval has big distinction with general retrieval (such as Web search, database retrieval). Traditional database index mechanism emphasizes keep balance of storage structure. CBR used to point out a separate case in needed situation and it care about how to divide the case base into useful segment in concept. The operation of traditional database is exact match. CBR is similarity match. In CBR retrieval, the key technology includes the design and selection of search algorithm and similar algorithm. The system structure adopts comprehensive representation of overall similarity and attributes similarity and combined with the nearest neighbor strategy to measure the case similarity.

3.3 Case Reuse and Revise

Case reuse means that the source case in case base could reused. When retrieved similar cases, the solution of the similar case could used to help solve the problem of target case. To same cases with good matching, the solution of matching case retrieved could be taken as the solution of target case directly. Case reuse could bring a good result. But in CBR system, it difficult to find a case perfectly matched with target case in case base, usually, we can only find a source case which is similar with target case. And then, revise it to make it adapt new situations, thus to solve.

Case revise also could be called case adaptation. Usually, the proposal get in case reuse process is most similar with current problem, but it still exist some differences, it causing the solutions and solution strategies in proposal are not complete suitable for current problem solution. So, according to current problems, domain knowledge would be used in specific environmental to modify the unqualified solution in order to fit the solution of current problems. After that, proven the new case, if it is feasible or correct, it could be stored in case base. The process could be called case base update.

In the automatic reasoning technology based on secondary CBR, the differences between each case in case base would be used to guide case revise. The reasoning will combine two model, local weight linear regression and KNN (K-Nearest-Neighbor). When case containing the non-numeric attributes, it could be ignore, and the premise is it at least contain a numeric attribute for build regression model.

3.4 Case Evaluation and Maintain

For the specific problem to be solved, the solutions automatic reasoning based on secondary CBR needs to be evaluation. After the evaluation successful, it would be saved as a new case into the case base. To the failure evaluation case, it will return to CBR system and re-reasoning by changing conditions.

Case base maintenance means that realize some update strategy of case base structure or content, it including domain content, describing information, implementation. It is use to ensure the future reasoning complete specific performance indicators. When the case number in case base is less, it does not appear bottleneck of retrieve efficiency generally. Because of case study is a kind of incremental learning, the CBR could solve more problems and reflecting its intelligence. With the case number increase, if retain case unconditional will cause another problem. It is can not control the growth of case number, then, dropped the operating efficiency of system and raise the cost of retrieval. In order to ensure the case retrieval speed, under the prerequisite of maintaining the case typical, it need to constantly reduce case redundancy. Case maintenance system plays the role of case maintenance and management. Each process of case reasoning will by case maintenance system to realize interaction. In addition, case maintenance system has also play the role of contact the case base and database. Case maintenance involves many contents, such as case increase, query, delete, study and maintenance.

4 Reasoning Technology Application

In this paper, the automatic reasoning technology based on secondary CBR is applied to the digital camera market price aided making system. The system mainly gives a forecast of certain case attributes on camera market price by sale case reasoning. For businesses, system could reason appropriate market price position by new product performance characteristics parameters of given camera. It will help businesses establish reasonable marketing strategy.

The whole system could divided into several functions, including human-computer interaction function, case representation function, case reasoning function, case reuse and revise function, case evaluation and case maintenance function etc.. Among them, case maintenance function contain sub-function, such as case add, delete, query, case maintenance and case study.

The various properties of new products entered into system, and use automatic reasoning module get the product price. The market commissioner evaluate price, if the market price appropriate, it will put into use and save, at the same time doing the case maintenance. If the market price considered inappropriate, then return to reason module to choice new case and difference case to reuse and revise. Application shows that the automatic reasoning technology based on secondary CBR has good practicability.

5 Conclusions

With the special knowledge representation, case reasoning successful around the difficult problem, "knowledge acquisition". Compare with the traditional rule-based reasoning, case reasoning has some advantages, such as knowledge acquisition easy, system maintenance simple and strong self-learning ability. These features make the case reasoning technology could be used in some complex field, such as knowledge is difficult to obtain and it has accumulated a wealth of cases. However, most studies are focusing on case retrieval and case reuse at present, to the respects of case representation, case revise and case maintenance is involved less. This paper proposed the automatic reasoning technology based on secondary CBR, it use case revise to drive case reuse. The technology solved the drawback of traditional CBR technology is only simple case reuse and can not revise case, it improve the flexibility and versatility of CBR system.

Acknowledgement. This work was supported by a grant from the National Natural Science Foundation of China (No. 61003129).

The Project was Supported by the Fundamental Research Funds for the Central Universities, China University of Geosciences (Wuhan), Grant CUGL100243.

References

1. Li, F.-G.: Research on Intelligent Decision Techniques Based on Optimized Case-Based Reasoning. Hefei University of Technology (2007)
2. Liu, F.: Research and application of intelligence decision support system based on CBR. Lanzhou University (2008)
3. Hanney, K., Keane, M.: Learning Adaptation Rules from a Case-Base. In: Smith, I., Faltings, B.V. (eds.) EWCBR 1996. LNCS, vol. 1168, pp. 179–192. Springer, Heidelberg (1996)
4. McSherry, D.: An Adaptation Heuristic for Case-Based Estimation. In: Smyth, B., Cunningham, P. (eds.) EWCBR 1998. LNCS (LNAI), vol. 1488, pp. 184–195. Springer, Heidelberg (1998)

5. McDonnell, N., Cunningham, P.: Using Case Differences for Regression in CBR Systems. In: Proc. 25th Annual International Conference of the BCS SGAI, pp. 219–232. Springer, Heidelberg (2005)
6. McDonnell, N., Cunningham, P.: A knowledge-Light Approach to Regression Using Case-Based Reasoning. In: Roth-Berghofer, T.R., Göker, M.H., Güvenir, H.A. (eds.) ECCBR 2006. LNCS (LNAI), vol. 4106, pp. 91–105. Springer, Heidelberg (2006)
7. Han, J.-F., Yu, X.-L.: Research and application of case reasoning technology. Science and Technology Information 18, 133–134 (2007)
8. Yang, X.-Q., Li, J.-Y., Lan, T.-C., Liu, H.-T.: Researches on application of case-based reasoning in decision science. Journal of East China Institute of Technology (Natural Science Edition) 31(2), 187–191 (2008)
9. Li, F., Wei, Y.: Case Retrieval Algorithm Based on Preference Information. Computer Engineering 34(24), 28–30 (2008)
10. Zhen, Y.-H., Kou, Y.-Z., Zhang, W.-J.: Research of case retrieval technology based on fuzzy reasoning. Sichuan Ordnance Journal 30(7), 122–124 (2009)
11. Li, J.-Y., Chen, X.-Y., Liu, H.-T., Xie, X.-Z.: Researches on Case-Representation in Case-Based Reasoning System. Journal of Hefei University (Natural Sciences) 17(3), 26–29 (2007)
12. Li, L.-J., Wang, R.-C.: Research on CBR in Rule-Based IDS. Computer Science 33(5), 117–120 (2006)
13. Wang, Y., Fan, J.-H., Tian, S.-G.: Method for Case Representation in Expert System Based on Case-based Reasoning. Journal oF Shanghai University of Engineering Science 19(1), 42–46 (2005)

Experiments with Adaptable Multimodal User Interfaces for Mobile Devices

Andreas Truar and Norbert Kuhn*

Institut für Softwaresysteme in Wirtschaft, Umwelt und Verwaltung
University of Applied Sciences Trier, Umwelt-Campus Birkenfeld,
Campusallee, D-55761 Birkenfeld
{a.truar,n.kuhn}@umwelt-campus.de

Abstract. Our daily living environment contains more and more technical systems, which need human control. Modern mobile devices offer functionalities to act as generic control devices for these technical systems. Besides classical user interactions they provide components to process new ways of human-machine interactions like speech and touch inputs. In this article, we describe a software framework to support multimodal inputs for Android based mobile devices. We exploit this framework in a case study to develop adaptable, multimodal user interfaces for two different instances of robot controls. Furthermore, we present some results from evaluating these interfaces with a couple of test persons.

Keywords: Android Platform, Multimodal User Interfaces, Design Pattern, Robot Control, Usability, Embedded Systems.

1 Introduction

Since the introduction of the iPhone mobile phone in 2007, almost all mobile platforms provide new possibilities of human-machine interactions. The massmarket availability of capacitive touch displays, cheap mobile broadband connections and powerful mobile hardware are the basic requirements for these new interactions. Today's mobile devices are highly integrated and these new interaction methods don't need additional hardware. Instead, the human body (finger, speech) is the "hardware" that triggers system actions. Evaluations showed that multimodal interfaces can speed up the task execution by 10 percent 1. This faster interaction with an anytime available device changed the way people consume and control information.

Another technical development in the recent years is the linking-up of daily used devices in our environment. Modern embedded systems enrich our environment with interactive elements and create a ubiquitous infrastructure 2 that needs control and visualization compatibilities. Modern smartphones present a generic way to control and visualize embedded systems because control elements are defined in software and are visualized via touch screen displays.

* This work is supported by the German Federal Ministry of Education and Research under Grant 17043X11.

G. Lee (Ed.): Advances in Automation and Robotics, Vol.2, LNEE 123, pp. 9–16.
Springerlink.com

Unfortunately, there is no common model that describes the integration of multimodal interactions with mobile applications. Every interaction mode has special limitations as regards to execution time, accuracy, threading and configuration. We solve these problems in the framework architecture through the use of multiple abstraction layers, that manage the parameters which are specific for the input source. Furthermore, we use the Command Pattern 4 to provide an easy to use framework for multimodal interactions.

2 Techniques and Architecture

Today, nearly all mobile platforms support touch, speech and gesture recognition to control the mobile devices. The native software development kits of these platforms provide components that use hardware and services to process user-inputs. In this article, we describe a framework to simplify the use of user-inputs, in particular multimodal interactions. The framework provides an XML-based dialect to configure the input recognizers and defines a standard to react on recognized events. Figure 1 gives an overview of the framework. It supports the dynamic adding and removing of available input sources. Furthermore, it provides functionality to configure the recognition process and offers a defined way to react on user-input events. Currently the input sources can be classic GUI elements, as well as touch inputs, speech and sensor states (gyroscope). Since the design follows the Command Pattern, future interaction modes, e.g. freehand gestures, can be included seamlessly and configured as a new input source. The dynamic binding between the abstract event and the concrete event is then automatically provided by the framework.

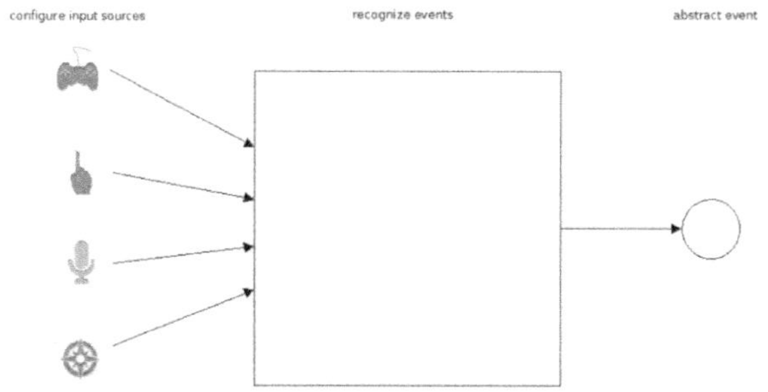

Fig. 1. Overview of the event handling for different input sources

The framework is implemented for the Android platform. We decided to use the Android framework because

- it provides good support by device manufacturers,
- it is easily adaptable for java desktop applications,
- it comes up with free development environments and emulators,
- it includes opensource components for speech and gesture recognition

2.1 Speech Recognition in Android

The Android speech recognition is a web service hosted by Google. The speech engine on the device digitalizes the user's spoken input and sends it to the cloud using an internet connection. In the cloud, the single words are compared to statistic models that Google's algorithms gathered over the years. The server-side processing of speech recognition has the advantage that the CPU/Battery of the smartphone is not used and the recognition results are better due to the huge amount of available statistical data. In addition, new recognition algorithms are directly available for all users. According to Google, 25 percent of all mobile search queries are made through speech recognition 1. The recognition is available without a user-specific training but correct recognition results increase because a personalized user speech model is created and improved over time. A disadvantage is the fact that the service needs a permanent internet connection. Long speech requests often result in error exceptions due to the high bandwidth consumption of the service.

2.2 Gesture Recognition in Android

In contrast to speech recognition, gesture recognition is usable without an internet connection. The Android framework provides three methods to react on user gesture events. They differ in the complexity of the recognizable gesture. 0 presents the three gesture types and lists its parameters.

Table 1. Types of gestures and its parameters in the Android framework

Name	**Parameter**	
	Event and data	***Event interface***
MotionEvent	single motion; low-level data	onTouchEvent(MotionEvent)
GestureDetector	single-double tab, fling, long tab, pinch; motion data, multitouch: velocity and distance	SimpleOnGestureListener
GestureOverlayView	finished gesture input; shape with multiple strokes (timed sequence of points)	onGesturePerformedListener()

We used the GestureOverlayView to identify gesture inputs in the framework. A major advantage of this gesture type is the possibility to define own gestures and use a prediction score to rate user gesture inputs. The framework simplifies the loading of gesture libraries and the invocation of concrete events. It follows the command pattern that belongs to the class of behavioral design patterns 4. With this design pattern a certain decoupling between the client, that requests an operation and the receiver, that executes the operation, can be achieved. The command object serves as an interface between the client and the receiver thus representing the invocation. Through using a command object, the framework gains flexibility because the object can be stored for later processing and previous executions can be reversed. Furthermore, new receivers, e.g. input sources, that make invocations and clients, which process invocations, can be added without changing the application (loose coupling between receiver and client).

3 Case Study: Controlling Mindstorms Robots with Multimodal Interactions

To evaluate the usability of the interaction framework, we used it to built an application to control Lego Mindstorms robots over a Bluetooth data connection. We chose a robot application because it is a good example for a technical embedded system that needs real-time control. Furthermore, the robot movement gives the user a direct feedback to the actual interaction. In addition, this setup provides a playful setting to test and evaluate new forms of human-machine interactions. Figure 2 shows a screenshot of the interface on the Android platform.

Fig. 2. Screenshot of the remote control screen for the Mindstorms application

At the top of the display, the user can choose among different input modalities (remote, speech, gesture and gyroscope). According to the selected input source, the bottom of the window shows information about the input modality. In the remote control input source a screen with classical GUI elements is presented. The Mindstorm robot connected to this interface provides several concrete activities (drive home, parking or rotate), which can be activated with different input modalities. E.g., the activity "Rotate" can be activated either by pushing the appropriate button, by speaking the word(s) "rotate (left)", by the gesture  and even by using the buttons to control the low level motion actions of the robot. During the usability study which we discuss in the following we evaluated the different input modalities and how the user rates them.

The usability study was designed according to the "discount usability engineering method" introduced by Jakob Nielsen [5] This method is based on the use of the following four techniques, namely the *user and task observation*, *scenarios*, *simplified thinking aloud*, and *heuristic evaluation*.

To evaluate the usability of the framework and the available input modalities, we developed a qualitative user survey, based on these four techniques. Every user interview is executed according to the following schema:

Composition of the Scenario ⇒ User introduction to the use of the application ⇒

Free testing of the application and the input modalities (10 minutes) ⇒

Scenario task execution (observation) ⇒ Guided interview (thinking aloud)

The guided interview comprised questions about the user, about the application's functionality (ease of use) and about personal feelings (joy of use) while using the application. During the scenario task execution, different time measurements between input modalities gave comparative values between the robot control modes (remote, speech, gesture and gyroscope).

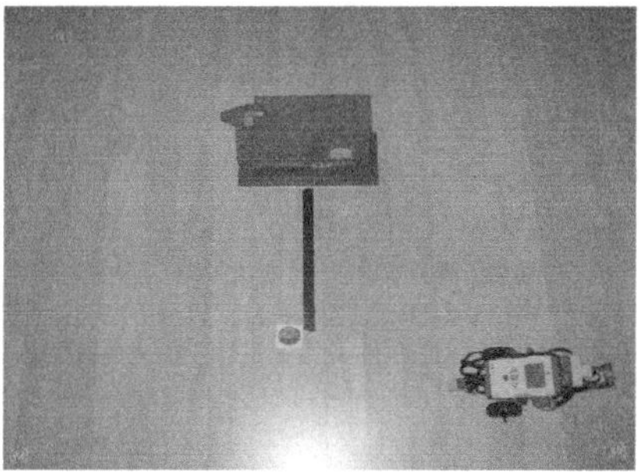

Fig. 3. The "Riding Home Scenario"

Figure 3 shows an example of a scenario which we used for the evaluation. Here, the user has to navigate the robot from a given position to its home, where he has to park it backside along the black line. The different scenarios should cover different kinds of activations composed of generic and complex actions of the robot.

Furthermore, we constructed two different robots for our study, one which can mainly be driven around and a forklift robot which can also be used to load, transport and unload goods. This enabled us to set up a great variety of scenarios. In the following section, we give a brief summary of the heuristic evaluation.

4 Results of the Usability Study

The heuristic evaluation of the usability study has been made using descriptive statistics. It follows the appearance of the questions in the guided interview.

4.1 General Information

In the first section of the interview, general information about the test person was collected. Altogether, ten users participated in the usability study. 30 percent of the participants were women. The youngest test person was 9 years old, the oldest one 58 years. The other test persons were between 25 and 35 years old with an overall average of the age of about 29 years.

Then, we asked the users about their experience with the use of modern smartphones. Most test persons who owned a smartphone estimated their experience with modern touch-based smartphones as "very good" (3) or "good" (2). Four of the test persons who did not own a smartphone estimated their experience as "low", the other one rated his experience with "very good". An interesting fact is that test persons who possessed a smartphone rated the idea of controlling technical systems with it "very good". They wanted to control almost every embedded system around them with their phone. 60 percent of the non smartphone owners rated the idea of controlling embedded devices as "good". So we observed a tendency that owners of smartphones appreciate the possibility to control embedded systems with their mobile device.

4.2 Questions Concerning the Application

In the second section, the users should rate the whole Android application that controls the robots. The functionality (ease of use) of the application was mainly rated as "good". Most of the test persons wanted more robot movements and a better gesture recognition. Thus, a new version of the application provides new gestures with better separation. In addition to the ease of use, the test persons rated the joy of use. Most of the users found it very amazing to use the application and to control the robot with multiple input modalities. The following descriptive statistics in table 2 shows the answers in detail.

Table 2. Ease of Use and Joy of Use for the Test Persons

Ease of use		
	Votes	Percent
very good	2	20,0
Good	7	70,0
no opinion	1	10,0

Joy of use		
	Votes	Percent
very good	6	60,0
good	3	30,0
satisfying	1	10,0

4.3 Questions about the Input Modalities

The next section discusses the different input modalities. In this part of the interview, the test persons had to rate how fast the robot reacts on the different input modalities and how accurate the control abilities of the modalities are. Figure 4 shows that the users rated the reaction time of the Remote control with classic GUI elements as the best. The reaction time of the speech mode is rated as the slowest form of interaction. As expected, the recognition of speech needs more execution time due to the web service invocation process. Similar picture is obtained for the field of usability (cf. fig. 5).

To compare the different input modes, we merged the given user ratings of reaction, accuracy and usability by interaction mode. As a result, the users rated the button based interaction mode with the best value of 1.3 and the gyroscope with a good value of 1.49. Although the merged value of the speech (3.43) and gesture input (2.66) is not as good, the ratings exhibit a high variance. Some of the test persons liked the speech and gesture input mode and gave them a good rating. To provide a good usability, applications should support at least one of the "classic" interaction modes based on buttons or sensor states. The speech and gesture interactions should be given as an additional way to activate functionalities often used.

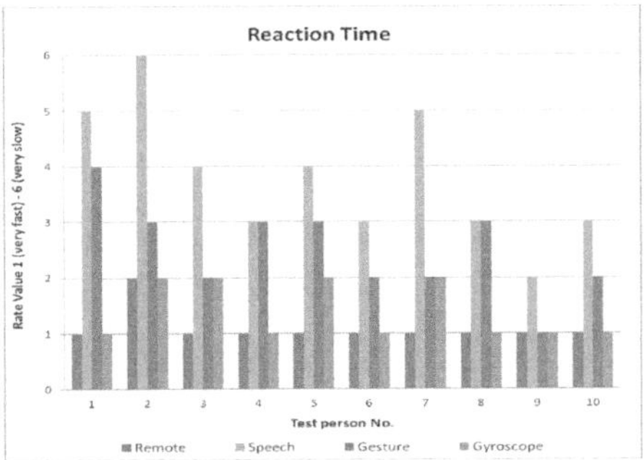

Fig. 4. Reaction time for the different input modalities

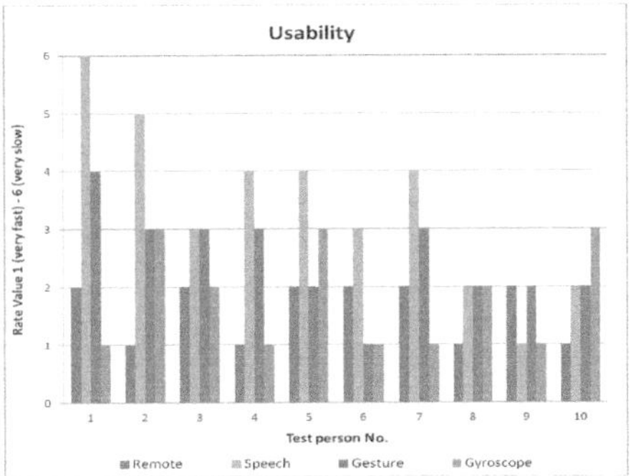

Fig. 5. Usability of the different input modalities

4.4 Performance Test

In this section, the test persons executed the scenarios. Compared to the previous questions these time measurements gave an objective value about the per-formance of the input modalities. The scenario tasks consisted of two parts. First, the drive home and the parking commands should be executed with speech, gesture and an arbitrary input mode. In the second task, the test person had to drive along four given landmarks. Fig. 6 and fig. 7 show the respective time measurements. An interesting fact is that the speech input mode activates the commands in task 1 faster than the gesture mode. So there is a discrepancy in the objective usability of speech input mode and the users' feelings about this kind of interaction. When users have the free choice of their interaction modality, they tend to use the button or gyroscope input mode.

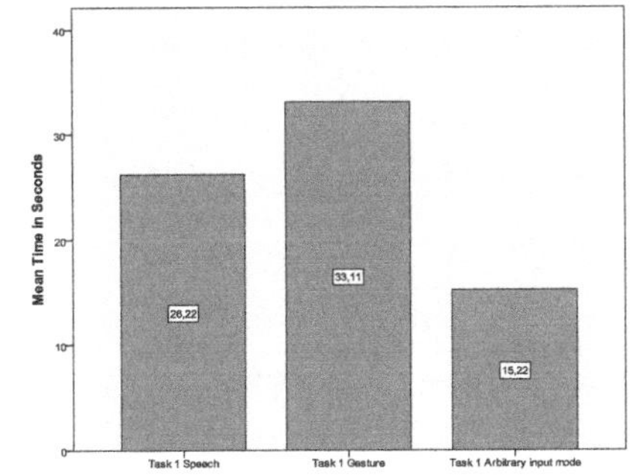

Fig. 6. Task 1, mean times of all test persons with different input modes

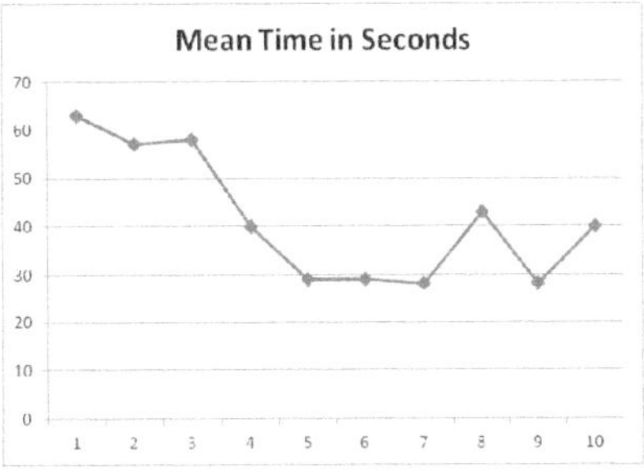

Fig. 7. Mean times of single test persons for task 2

5 Conclusion and Future Work

In this article, we discussed a software framework to provide multimodal interactions for the Android platform. The results of a usability study realized with these multimodal interactions are given. The usability study shows that multimodal interactions are suitable for controlling systems even with real-time requirements. However, there are differences between the individual users. Some have good experience with new mobile devices and are familiar with interactions based on speech and gestures. They use new interaction modes as an additional way of controlling certain functionalities. Less experienced users have problems to activate and to control multimodal interactions. Here, future versions of the interaction framework should include a user-specific profile to allow an individual configuration of the framework. With this user-specific profile, the dynamic binding of interaction modes can be user-dependent. In this scenario, e.g. a visual impaired user could use a mobile application better because the touch based input modes are blocked and speech and sensor based inputs are responsible for more tasks. Moreover, user-specific speech grammars and gestures could be bound by the framework to give a better usability experience. In addition to a better model of the user, the technical systems should be described in a detailed way to provide an optimized controlling. For instance, information about the used technical system and possible interactions could lead to a dynamic creation of user interfaces.

References

1. Oviatt, S.L.: Multimodal interactive maps: Designing for human performance. Human-Computer Interaction 12, 93–129 (1997)
2. Weiser, M.: The computer of the 21st century. Scientific American 265(3), 66–75 (1991)
3. Schalkwyk, J., Beeferman, D., Beaufays, F., Byrne, B., Chelba, C., Cohen, M., Garrett, M., Strope, B.: Google Search by Voice: A Case Study. In: Advances in Speech Recognition: Mobile Environments, Call Centers, and Clinics, 1st edn., p. 61 (2010)
4. Gamma, E., Helm, R., Johnson, R., Vlissides, J.: Design Patterns. In: Elements of Reusable Object-Oriented Software. Addison Wesley (1995)
5. Nielsen, J.: Usability Inspection Methods. In: Proceedings of the CHI 1994, pp. 413–414 (1994)

A New Receiver for the Data Traffic of the EDEG System under High Speed Environments

Leyu Zhang, Tao Wang, and Jie Zou

Department of foundational Courses Mathematic Teaching and Reach Office, Chongqing Telecommunication College, Shapingba District, Chongqing Municipality, P.R. of China
zhangleyucq@163.com

Abstract. Higher speed will cause more serious Doppler spread, meanwhile change the character of the wireless channel, which breaks suppose that the channel impulse respond (CIR) is a constant during a time slot. In order to overcome this problem, this paper proposed an algorithm which is RS-MAP-PSP (adaptive reduced state maximum a posteriori equalizer based on per-survivor processing) in conjunction with DFE (decision feedback equalizer) for the EDGE system with 8PSK modulation. To reduce the state of the equalizer, a pre-filter is followed by the MAP-PSP to reshape CIR to have approximately a minimum phase property. Our simulation results show that the proposed algorithm can improve the performance of the EDGE system, compared with general receiver algorithm especially for the high coding rate traffic and/or under the high speed environments.

Keywords: Equalizer, PSP, EDGE, MAP.

1 Introduction

Under the wireless communication environments, there usually exists inter-symbol interference (ISI) caused by time-varying channels, which are characterized by the presence of both delay and Doppler spreading. In order to cancel/reduce the ISI, there is an efficient detection technique which is MLSE implemented by the VITERBI algorithm (VA) [1, 2]. Due to the high order modulation and fast time-varying ISI channels specified in EDGE systems [3], the MLSE-PSP is becoming an attractive approach by updating the channel parameters for survivor of each state in the VA and improving the system performance via the PSP [4, 5].

Generally, much wireless channel has long CIR length in the EDGE system in which the signals are higher order modulations, and the number of the states in the VA is extremely large. In order to improve this problem, sub-optimal methods must be sought to make the MLSE applicable. Mainly, there are two categories for reducing the complexity of MLSE receiver employing the VA. The first category directly shortens the overall effective CIR length handled in the VA by DFE in order to reduce the total number of states in the trellis diagram. The second category selects a subset of total number of trellis states, as in the RSSE [6]. RSSE is an approach that reduces the complexity of VA receiver but still offer at least partial performance gains

G. Lee (Ed.): Advances in Automation and Robotics, Vol.2, LNEE 123, pp. 17–25.
Springerlink.com

of MLSE. The RSSE algorithm is suboptimum but less complex alternatives to VA. It can be consider as the combination of VA and DFE-Decided Feedback Estimation. It was suggested by Eyuboglu and Qureshi [8]. Among those various approaches, the delayed decision feedback sequence estimation (DDFSE) [7] in the first category and RSSE with the set partitioning principle in the second one are promising candidates in the sense of making good trade-off between the performance and complexity. By plenty simulations, it can be found that the performance of the RSSE-PSP algorithm [9] can't achieve the standard of the 3GPP for the high coding rate traffic.

A reduced state maximum a posteriori (RS-MAP) algorithm based on PSP is proposed. MAP equalizer is based on the theory of maximum a priority to reduce the performance loss and pass the standard mentioned in the 3GPP. It involves three variables-forward metrics, branch metrics and backward metrics. The simulations show that the performance of the RS-MAP-PSP equalizer proposed in this paper is better than the traditional RSSE-PSP scheme.

2 System Model

In this paper, a multi-path fading and AWGN channels are considered respectively. Source consisted of random bits are formatted to the EDGE burst structure and gray mapped to 8PSK symbols. Every burst includes a training sequence of 26 symbols in the middle, 3 tail symbols at either end, and 8.25 guard symbols at one end as shown in Fig.1.

3*3 tail	3*58 data	3*26 TS	3*58 data	3*3 tail	3*8 guard

Fig. 1. EDGE Burst structure

The gray coding, modulation and pulse shaping filtering of the EDGE signal, which is 3π/8-shifted 8PSK, is shown in Fig.2.

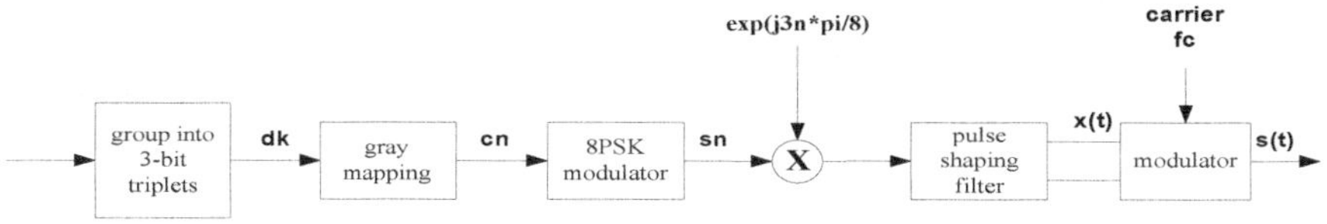

Fig. 2. Modulation of EDGE signal

After 8PSK modulation, the signal is corrupted by ISI and AWGN. Thus the received sequence y_k can be expressed as

$$y_k = \sum_{i=0}^{L} h_i a_{k-i} + \eta_k \tag{1}$$

where h_i denotes the discrete overall impulse response with length L, a_k is the transmitted 8PSK symbol and η_k is a AWGN sequence. The received signal y_k is fed to a pre-filter, in which the settings are calculated based on the estimated CIR by the least squares algorithm with mid-amble. The aim of the pre-filter is to reshape the estimated CIR to have approximately a minimum phase property.

3 Pre-filter

In order to achieve a high performance of reduced-state equalization such as RS-MAP equalizer, a discrete-time pre-filter is required in front of equalizer to transform the channel impulse response into its minimum phase or maximum phase equivalent. There are three common approaches for the pre-filter calculation: Root-finding, MMSE-DFE and Linear-Prediction (LP). Taking the practical implementation into account, the LP method is much less complicated.

The pre-filter A(z) can be viewed as a cascade of two filter A1(z) and A2(z), i.e. A(z) = A1(z)A2(z). Where A1(z) is a matched filter, and A2(z) may be viewed as a noise whitening filter. Basically, the A2(z) can be solved efficiently through a prediction-error filter:

$$1-P(z)=1-\sum_{1}^{Np} p\_f(k)z^{-k} \tag{2}$$

The prediction filter P(z) is designed to minimize the output power of the error filter. The optimum coefficients for this criterion are given by the solution of the Yule-Walker equations: $\Phi_{hh} * p\_f = \varphi_{hh}$. Where

$$\Phi hh = \begin{bmatrix} \varphi_{hh}(0) & \varphi_{hh}(-1) & .... & \varphi_{hh}(-N_p+1) \\ \varphi_{hh}(1) & \varphi_{hh}(0) & .... & \varphi_{hh}(-N_p+2) \\ . & . & . & . \\ . & . & . & . \\ \varphi_{hh}(N_p-1) & \varphi_{hh}(N_p-2) & .... & \varphi_{hh}(0) \end{bmatrix} \tag{3}$$

$$p\_f = [p\_f(1) \quad p\_f(2) \quad .... \quad p\_f(N_p)]^T \tag{4}$$

$$\varphi_{hh} = [\varphi_{hh}(1) \quad \varphi_{hh}(2) \quad .... \quad \varphi_{hh}(N_p)]^T \tag{5}$$

where $\varphi_{hh}(k)$ is the autocorrelation of lag k computed as:

$$\varphi_{hh}(k) = \sum_{m=0}^{L-1-k} h(m) * h^*(m+k) \tag{6}$$

Φ_{hh} is a Hermitian, positive-definite and Toeplitz matrix. The AR coefficients of Yule-Walker equation can be solved through Levinson-Durbin Algorithm (LDA), which is an iterative update algorithm.

Supposed that the n taps of the pre-filter based on the estimated CIR are calculated, and labelled as P_f_n. Then the CIR and the received signal y_k are fed to this filter, and their outputs are expressed as

$$pre\_y_k = filter(P\_f_n, 1, y_k) \tag{7}$$

$$pre\_h_i = filter(P\_f_n, 1, h_i) \tag{8}$$

The pre-filtered signal pre_y_k is then processed in the RS-MAP-PSP with the initial estimated reshaped CIR pre_h_i. In order to track the channel variations, the estimated reshaped CIR pre_h_i in the acquisition phase is updated with the LMS or RLS algorithm for every survivor inside the PSP in the tracking phase.

4 DFE

A schematic diagram of a DFE is shown in Fig. 3. An (n,m) DFE denotes an equalizer with n tapped delayed inputs and m feedback signals. So, m output samples are fed back to the input through a feedback filter in addition to the input samples. This feedback helps the system to de-correlate the noise that is produced by the ISI at the final output. DFEs are usually implemented using LMS or RLS algorithms.

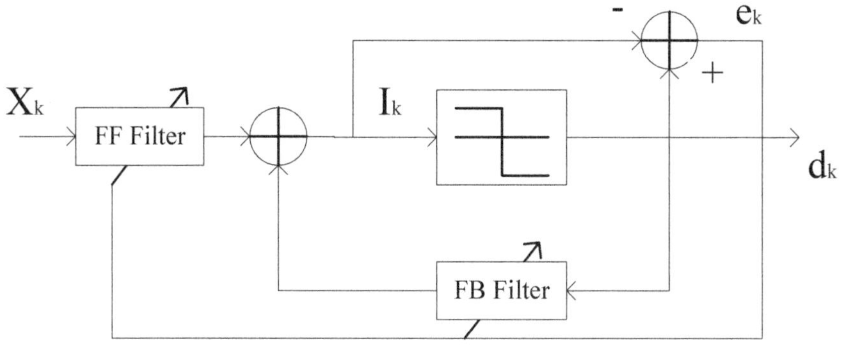

Fig. 3. DEF using two FIR filters, one as feed-forward and another as feed-backward.

Since the DFE is considered to be a nonlinear equalizer, it is used more often than linear equalizers, especially for the case of severe-ISI channels. These channels are characterized in their frequency response by the existence of frequency nulls that make them totally nonlinear and produce disturbed output.

The performance of DFE depends on the number of the delayed inputs and the number of the feedback signals from output to input. It can be improved by feeding an error signal (the difference between the expected output and the produced output) back to the input in addition to the normal feedback signals.

5 PSP and Equalizer Module

The MAP (Maximum A Posterior) equalization algorithm computes the probability of each transmitted bit given by the channel outputs and provides the optimal soft bit information that can be passed to the channel decoder. The MAP algorithm can be described by forward and backward passes over the trellis with the MAX operation to simplify the practical implementation.

$$\alpha_n(j) = \underset{i}{MAX}\{\alpha_{n-1}(i) + \gamma_n(i,j)\} \tag{9}$$

$$\beta_n(j) = \underset{i}{MAX}\{\beta_{n+1}(i) + \gamma_{n+1}(i,j)\} \tag{10}$$

$$\gamma_n(i,j) = | r(n) - \sum_{k=0}^{L-1} f(k)x(n-k) |^2 \tag{11}$$

The MAX operation is

$$MAX(x,y) = \log(e^x + e^y) = \max(x,y) + \log(1 + e^{-|x-y|}) \tag{12}$$

In the reduced-State MAP equalizer, the channel taps are divided into two groups, and then the transition metric will be expressed as

$$\gamma_n(i,j) = | r(n) - \sum_{k=0}^{\mu-1} f(k)x(n-k) - \sum_{k=\mu}^{L-1} f(k)\hat{x}(n-k) |^2 \tag{13}$$

Now we apply the PSP principle [10] to the RS-MAP. When calculating the forward factor alpha, the transition metric from the l-th state at time index k to the m-th state at time index k+1 as

$$e(t_k \to t_{k+1}) = | y_k - \langle X(t_k \to t_{k+1}), f_k \rangle |^2 \tag{14}$$

where the expression $(X(t_k \to t_{k+1}))$ is an inner product between the symbol vector at time index k and the vector f_k. The channel updating, by the LMS algorithm, is as follows:

$$f_{k+1} = f_k + \beta e(t_k \to t_{k+1}) X^*(t_k \to t_{k+1}) \quad (15)$$

where $X^*(t_k \to t_{k+1})$ is the symbol vector, e is the error signal and β is the step size.

The structure of the RS-MAP-PSP equalizer is diagramming in Fig.4

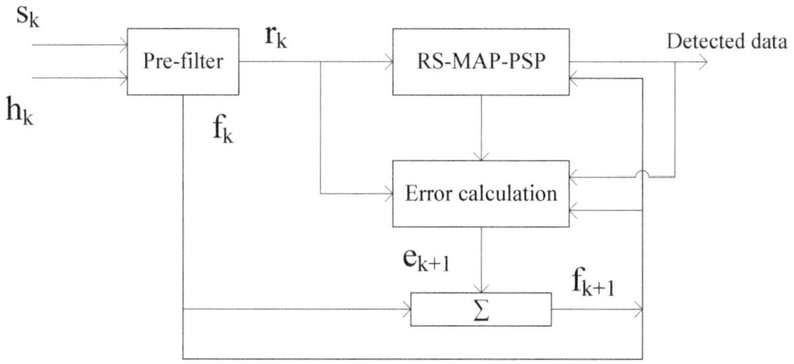

Fig. 4. The system structure of the RS-MAP-PSP

6 Normalized Soft Information

Because of the difference of the quality of each burst, a normalization model is needed to normalize the soft information from the output of the equalizer.

First, the noise energy is estimated using by the hard symbol and the received signal;

$$En = sum(| y - \hat{h} * hard\_\inf |^2) \quad (16)$$

Second, the soft information is divided by the noise energy estimated in the first step.

$$soft\_\inf = soft\_\inf / En \quad (17)$$

7 Simulation Result

The simulation results are all based on the TU50 channel environment, and the MCS5 and MCS9 are chosen as the traffic channel. The trellis state of the RS-MAP is set as 64, and the tracking step of the PSP is fixed on the 0.02.

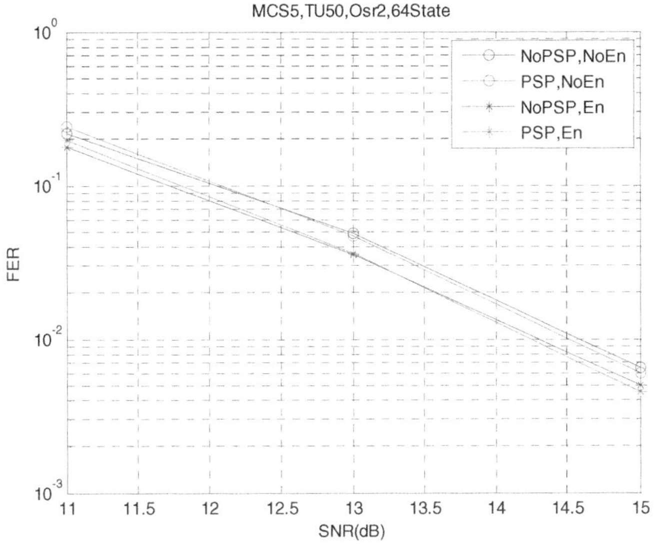

Fig. 5. The performance comparison between normalization and not

Fig.5 shows that the normalization based on noise energy can obtain about 0.5dB gain under MCS5 and TU50 conditions, whatever conjoint with PSP or not.

Fig.6 and Fig.7 show the effect of PSP on the performance under low or high coding rate. Taking the MCS5 for example as the low coding rate of traffic, whose results were shown in the Fig.6. The result shows that the PSP algorithm only obtains about 0.5 dB for low coding rate. The high coding rate, taking MCS9 for example, has more than 4 dB gains due to the PSP algorithm, which shown in the Fig.7.

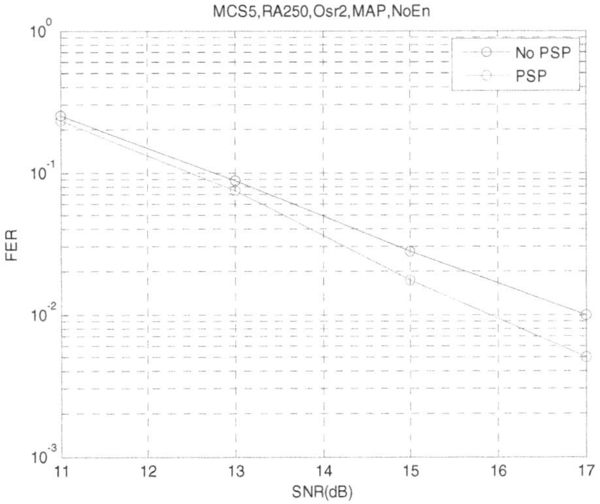

Fig. 6. The performance of MCS5 with and without PSP

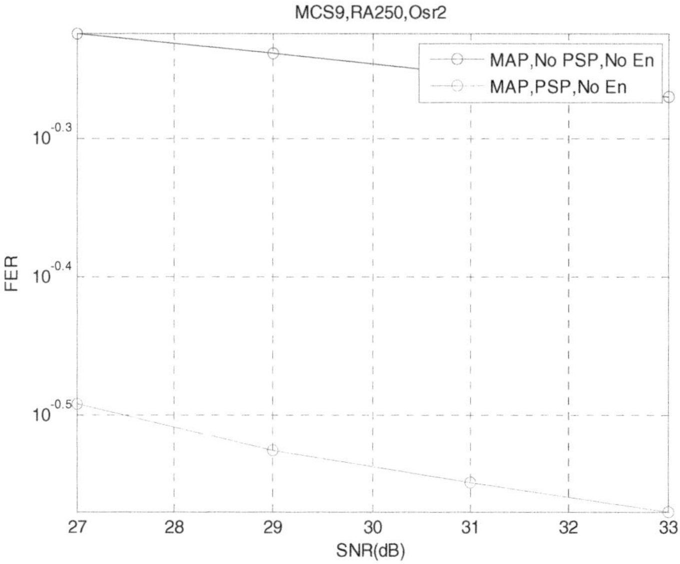

Fig. 7. The performance of MCS9 with and without PSP

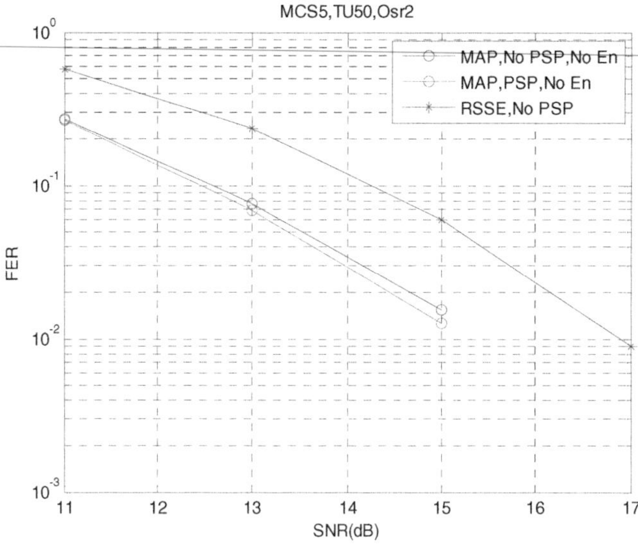

Fig. 8. The performance comparison between MAP and RSSE

Fig.8 shows the performance comparison between MAP and RSSE under conditions of MCS5,Tu50 channel type, which can be obtained that the MAP algorithm outperform about 2dB gain than RSSE while the FER is 10%.

8 Conclusion

This paper proposed a RS-MAP-PSP to improve the performance of EDGE equalizer. The simulation results show this proposed algorithm can achieve a better FER than the RSSE-PSP equalizer, and especially for high coding rate traffic. Meanwhile, the noise energy normalization is necessary for the equalizer, which can also contribute performance gain to the whole system. Since the complexity of this algorithm proposed in this paper is higher than RSSE, the schemes which can reduce the complexity are our research plan in the next step.

References

1. Duel-Hallen, A., Heegard, C.: Delayed Decision-Feedback Sequence Estimation. IEEE Transactions on Communications 37(3), 428–436 (1989)
2. Al-Dhahir, N., Cioffi, J.M.: Fast Computation of Channel-Estimate Based Equalizers in Packet Data Transmission. IEEE Transactions on Signal Processing 43(11), 2462–2473 (1995)
3. Li, Z., Piirainen, O., Mammela, A.: An adaptive RSSE-PSP receiver with a pre-filter for EDGE systems. In: IEEE International Conference on Communications, vol. 2003(5), pp. 3594–3598.
4. Raheli, R., Polydoros, A., Tzou, C.-K.: Per-survivor Processing: A General Approach to MLSE in Uncertain Environments. IEEE Transactions on Communications 43(2/3/4), 354–364 (1995)
5. ETSI, GSM 05.04 V8.1.0, Digital Cellular Telecommunications System (Phase2+); Radio Transmission reception, GSM 05.04, Version 8.1.0 (December 1999)
6. Li, Z., Mämmelä, A.: Adaptive PSP-MLSE Receiver for GSM/EDGE System in Fast Multipath Fading Channels. In: IEEE International Conference on Personal and Wireless Communications, Hyderabad, India, pp. 184–188 (December 2000)
7. Forney, G.D.: Maximum-Likelihood Sequence Estimation of Digital Sequence in the Presence of Inter-symbol Interference. IEEE Transactions on Information Theory 18(5), 363–378 (1972)
8. Simmons, S.J.: Breadth-First Trellis Decoding with Adaptive Effort. IEEE Transactions on Communications 38(1), 3–12 (1990)
9. Eyuboglu, M.V., Qureshi, S.U.H.: Reduced-State Sequence Estimation with Set Partitioning and Decision Feedback. IEEE Transactions on Communications 36(1), 13–30 (1988)
10. Anderson, J.B., Mohan, S.: Sequential Coding Algorithm: A Survey and Cost Analysis. IEEE Transactions on Communications 32(2), 169–176 (1984)

Model-Based Human Pose Estimation with Hierarchical ICP from Single Depth Images

Maoying Qiao[1,2], Jun Cheng[1,2], and Wenchuang Zhao[1]

[1] Shenzhen Institutes of Advanced Technology, Chinese Academy of Sciences, China
[2] The Chinese University of Hong Kong
{my.qiao,jun.cheng,wch.zhao}@siat.ac.cn

Abstract. In this paper, human poses which are presented by real 3D points cloud got from Kinect sensor are estimated and tracked by a hierarchical human model in ICP framework. There are several novel points in this paper. First, we compute human models' nearest points rather than points' nearest limbs as traditional methods do to make every limb have points. Second, we consider global information while hierarchically do ICP for every local limbs to conserve articulated kinematics chain. Third, by analyzing the four limbs (two legs and two arms) and enforcing joint constraints, we solve several specific problems, such as leg- or arm-crossing, etc. Experimental results including kinds of real human actions verify our method's effectiveness.

Keywords: Human poses estimation, Hierarchical ICP, 3D points cloud.

1 Introduction

Human pose estimation is an important task in computer vision for many promising applications, such as man-machine interaction, human gesture recognition, video indexing and retrieval, character animation, and so on[1], [2]. However, recovering human pose from visual observations is one of the most challenging problems in computer vision because of the complexity of the human pose [3].

The introduction of depth cameras appeals lots of researches' attention from 2D images to 3D depth images for human behavior analysis [4], [5], [6], [7]. There are three main reasons for this trend. First, 3D images provide depth information which is useful for reducing ambiguity in the 2D images. Second, the 3D depth information avoids scale of one observation which is an inevitable problem in 2D images. Third, 3D depth images are more and more accurate and robust and can be got easier than before. Recently, along with the launch of Kinect gaming platform [8], whose camera's output is dense and accurate single-view depth images, kinds of applications and algorithms on human motion based on it are emerging at a rapid rate.

Shotton et al. [9] builds a probabilistic decision tree and find an approximate pose of body parts by a local optimization step. Although their method is very successful, it relies on heavy training [10]. Another main branch of these methods is the model-based Iterative Closest Point (ICP) framework [9], [11]. In [12], soft-joint constraints are added to the ICP cost function. In [13], [14], the author projects unconstrained motions onto the linearized articulated motion space in an optimal way.

G. Lee (Ed.): Advances in Automation and Robotics, Vol.2, LNEE 123, pp. 27–35.
Springerlink.com

In [15], the author constructs three kinds joint constraint models, and they add correspondences between joint constraints model and the real data points into the ICP framework. In [16] and [17], the authors model the human body in a hierarchical manner, namely, the torso drives the motion of head and for limbs, and the up arms' and up legs' motion lead to down arms' and down legs' movement. In this hypothesis, one does not enforce joint constraints during the registration. Inspiring with this simple and natural idea, we also adopt the hierarchical model in our algorithm.

In this paper, we match data points with current human body model in the hierarchical structure by ICP algorithm. Comparing with computing each data point's closest limb model, we compute each limb model's nearest data points for several advantages. And to be simpler, we directly apply the parents' movement to the children rather than through the weight constraint to keep the articulated kinematics chain. Moreover, we apply our algorithm to the real 3D human points cloud got by Kinect sensor and solve several specific problems (limb-crossing, etc) existing in the real behavior by enforcing some extra joint constraints.

The rest of this paper is arranged as below. In section 2, we concisely introduce our human body model and review the ICP algorithm. Our method for human pose estimation is stated in section 3 and experimental results with several of poses are showed in section 4 and at the end, we briefly conclude our work.

2 Background

2.1 Human Body Model

In this paper, we adopt an articulated human body model developing from [5]'s hierarchical stick model, and it is shown in Fig. 1.

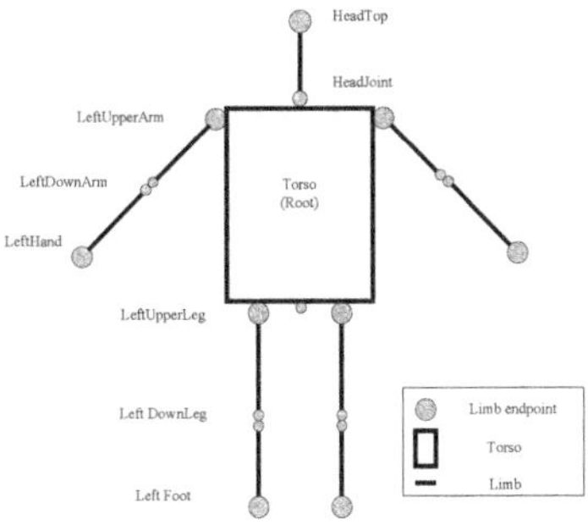

Fig. 1. Articulated Human Body Model

Demonstrated by Fig. 1, the articulated human body model comprises one rectangle for torso and nine sticks for head, arms and legs. Simply, we just use fixed-length sticks, each of whom has two endpoints, as the arms' and legs' model. And their lengths are measured by anthropometry. The torso appears as a cube and the normal direction of its front is changed while the human does some rotation actions.

So, we use a rectangle instead of stick as the torso model. And the rectangle's height and width also satisfy the anthropometry. In the hierarchical aspect, the torso is in the root position that dominates the whole body's location and orientation. The head, arms and legs just do rotate transformation around their respective connecting joint. In this manner, the articulated kinematic chain is modeled in a natural way.

2.2 Review of ICP

The ICP algorithm is very popular in registrating of 3D points sets for its simplicity and efficiency. In this paper, ICP is used to align the model of each human limb and torso with 3D data points got from Kinect sensor.

Given two sets of n corresponding 3D points of one limb (the corresponding relationship will be given in next section), the data points set$\{ d_i \}_{i=1,\dots,n}$ and the model points set $\{m_i\}_{i=1,\dots,n}$, the registration problem is modeled as a problem of minimizing the sum of squares of residual error, namely,

$$\begin{aligned} \arg \quad & \min_{R,t} \sum_{i=0}^{n} \left\| m_i R + t - d_i \right\|_2 \\ s.t. \quad & R^T R = I \end{aligned} \quad . \tag{1}$$

Where R and t are rotation matrix and translation vector respectively. In our hierarchical human model, while the torso controls the whole body's orientation and translation, other limbs just rotate around their parents to locate themselves. And the below optimization problem (2) is the fitting problem for the head and four limbs.

$$\begin{aligned} \arg \quad & \min_{R} \sum_{i=0}^{n} \left\| m_i R - d_i \right\|_2 \\ s.t. \quad & R R^T = I \end{aligned} \quad . \tag{2}$$

Again, R is the orthogonal rotation matrix for head and four limbs' models. Obviously, (2) is a special situation of (1).

In [18], the specific process of solving the problems (1) and (2) are derived. Here, we apply the results R and t directly to our problems.

Until now, the motions of torso and each limb have been calculated. In the next section, we will show how to apply the result of R and t to human body model.

3 Our Method

Our human body data points are segmented from the whole scene points cloud got from the Kinect sensor. First, we register our model to the human points cloud. Then, we track the model in the following frames.

3.1 Our Hierarchically Sequential ICP Algorithm

Initially, we use the model shown in Fig. 1 as our original pose. Aligning the model to the human data points cloud is this section's main task. Before we start the ICP algorithm, the corresponding relationship between human model and human points cloud should be calculated. In [5], [6], [7], they globally project every point onto every limb model and choose the closest foot point as corresponding point for each data point. Differently, in this paper, we gradually compute every limb's closest data points. The number of the closest data points is proportional to the area of itself which satisfies the anthropometry. The specific method is summed up in the flow chart shown in Fig. 2.

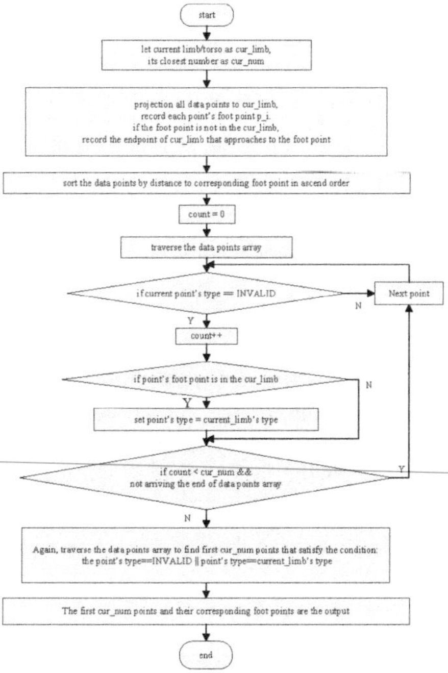

Fig. 2. Flowchart of finding corresponding relationship

From the flow chart, first, we process limbs one by one with priority ordered by hierarchical level and global constraint that sorting data points by distance other than label all the points at the same time. Second, keeping every limb have fixed number points makes the algorithm have the ability of dealing with some occluding situation. That's to say, when one limb is occluded, there is still points on the limb, regardless of whether their foot points are on the limb or not. And then, the occluded limb can still be transformed by these closest points. Moreover, it is also convenient to determine whether the current limb is occluded or not by comparing the cur_num and sum count of the foot points coinciding with the endpoints.

At present, we do prepare everything for the registering work. And the complete ICP algorithm is described in Algorithm 1.

Obviously, in Algorithm 1, we transform not only the current limb but also all of its descendants. This just takes the natural sense that the descendants' moving is partly driven by their ancestor. Transform the descendants along with their own ancestor's movement is equivalent to do motion prediction for the descendants. Comparing to that doing respective rotation and translation independently, the motion prediction makes the descendants' model register to the data points more quickly and accurately.

Algorithm 1. Hierarchically sequential ICP algorithm

Input:

Human body data points cloud D;

Initial human body model M0(m0, m1, m2, …, m9);

Output:

Registered human pose M1(m0', m1', m2',…, m9');

0: align the root with human data points' gravity;

1: traverse the hierarchical human body model and start from the torso;

2: set the current limb/torso model as mi;

3: compute the corresponding closest points sets <Mi, Di> by method shown in Fig. 3. Points set Di is a part of D and points set Mi is the projective foot point of Di;

4: compute rotation matrix R (and translation vector t, if current limb is torso) by Solving the optimization problem (2) (or (1) for torso).

5: do transformation with R (and t if current limb is torso) for current limb and all of its descendants in the hierarchical structure.

6: if the residual error does not satisfy the requiring accuracy, then go to step 3 for a loop.

7: if the residual error of current limb/torso is small enough, and there still existing limbs not be processed, then go to next limb in the hierarchical structure using a level-order traversal method and jump to step2 for a new start.

3.2 Tracking

To estimate human pose in a successive video, we apply tracking to the estimation system.

The start of the video, we use the model shown in Fig. 1 as initial model for the first frame whose pose is similar to the model. For the subsequent frames, the last frame's result model is used as initial model for the current frame. Before applying the last frame's result model to the current frame, we do some analysis work to add some constraints into the model below.

One situation, the leg models are crossed while the true postures of the legs points are not, is really hard to automatically recovery from the crossing situation for the ICP algorithm as the closest points data will always be classified wrongly. To settle this problem, we manually reinitialize the legs model by parallel sticks and apply the ICP algorithm to only the legs again to get a new residual error. If current residual error is less than the former model's, we deliver the new model to the next frame; else the former model is delivered.

We decide whether the arms are occluded or not by comparing the number of foot points that are on the limb and the cur_num. If both up and down arms are occluded, we reinitialize the arms' model using the arms' model gained from adjacent last frames. If only the up arm is occluded, we do nothing and directly deliver the former human model to next frame.

Besides, some human kinematic constraints can be added through simple angle limit, such as the angle between up and down legs cannot be less than 90 degree only when the down leg is occluded, the angle between the head and torso also cannot be less than 90 degree in the unoccluded situation. If the result model violates these constraints, we just simply reinitialize these limbs' model.

At last, every limb of the whole human body, whose residual error is larger than the predefined threshold, will be reinitialized by choosing a better one from the model shown in Fig. 1 or the last several models.

4 Experimental Results

In this section, we perform some experiments to evaluate our method.

We use real depth data got from the Kinect sensor. We get the body's joints' true position by hand-labeled for two different human actions videos: Arm-lift, akimbo walk. Each video lasts about 100 frames. The Kinect sensor output both the scene's color image and depth image. First, we box the joints' position shown obviously in

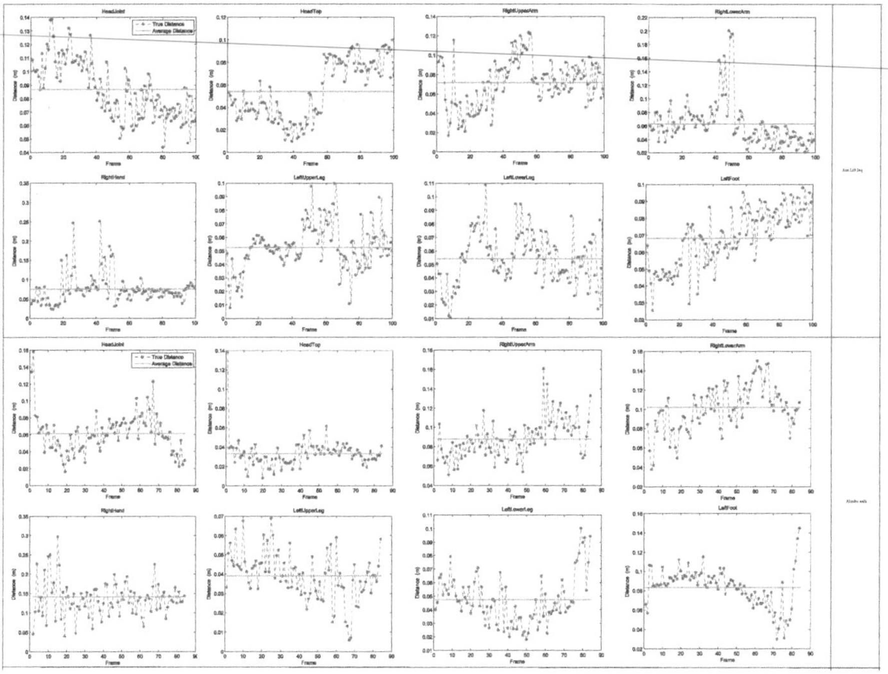

Fig. 3. Distance between the true position and the position got by our method for every joint

the color image. Then, we compute the average 3D position in the box from the depth image. The resulted position serves as the true joints' position.

We suppose that the joints position got by our method is true only when it locates into the true position' spherical area with radius = 0.1m. We just simply use the distance between the true position and the position got by our method for every joint as error to evaluate our method. The quantitative results are shown in Fig. 3. From the graph, most of the average errors are below 0.1m. Apparently, for the hierarchical process, the error of the extremes, namely hands and feet, is a little higher than other limbs.

In Fig. 4, we compare some frames of true joints model and human model got by our method. Qualitatively, most of the two models are coincident except for some limbs. The intrinsic reason of the wrong joints' position is that the points cloud is wrongly clustered for these limbs. Take the 81th frame of akimbo walk sequence as an example. The relevant clustering result got by our method is shown in Fig. 5. The points actually belong to the up arm are labeled as down arm and this is caused by that we only use distance information. In the near future, we will trial other features extracted from the points to make a better result.

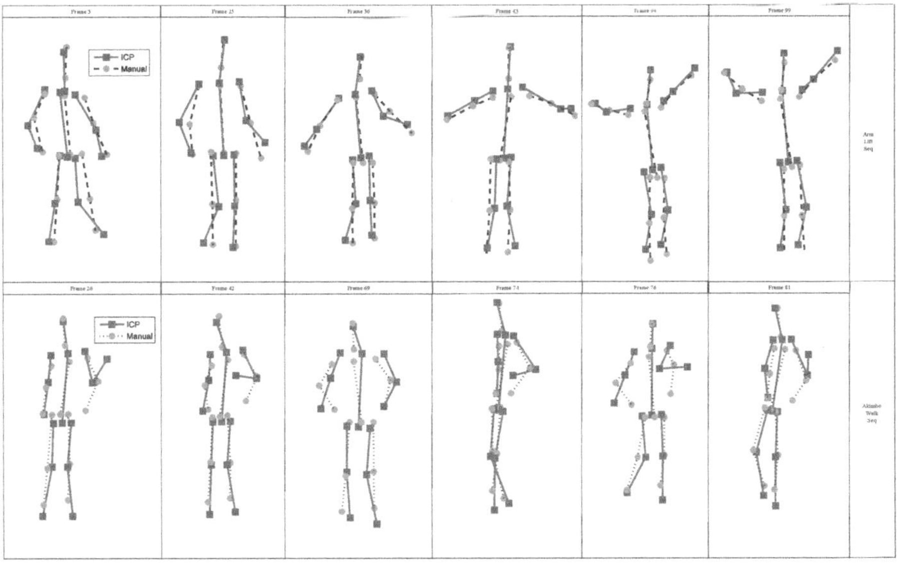

Fig. 4. Frames of true joints model and human model got by our method

Fig. 5. The 81th frame's Clustering result of akimbo walk

5 Conclusion

Summing up, generally, our method is capable of dealing with the natural poses in estimating and tracking human pose. But, as mentioned above, the distance measurement is not sufficient. In our further work, this will be the first barrier we should solve. Another aspect is that, we use the fast sorting algorithm which is the most time-consuming operation in the whole algorithm. To put our method into real time application, we will also take the running time into consider in the future.

Acknowledgments. Special thanks to Key Laboratory of Robotics and Intelligent System of Guangdong Province (2009A060800016), Shenzhen Key Laboratory of Precision Engineering (Project No: CXB201005250018A), CAS and Locality Cooperation Projects (ZNGZ-2011-012).

References

1. Wang, L., Hu, W., Tan, T.: Recent Developments in Human Moton Analysis. J. Pattern Recognition, 585–601 (2003)
2. Zhu, Y., Dariush, B., Fujimura, K.: Kinematic Self Retargeting: A Framework for Human Pose Estimation. J. Compute Vision and Image Understanding 114, 1362–1375 (2010)
3. Microsoft Corp. Redmond WA. Kinect for Xbox 360
4. Siddiqui, M., Medioni, G.: Human Pose Estimation from a Single View Point, Real-time Range Sensor. In: Computer Vision and Pattern Recognition Workshops, pp. 1–8 (2010)
5. Grest, D., Woetzel, J., Koch, R.: Nonlinear Body Pose Estimation from Depth Images. J. Pattern Recognition 3663, 285–292 (2005)
6. Knoop, S., Vacek, S., Dillmann, R.: Sensor Fusion for 3D Human Body Tracking with an articulated 3D Body Model. In: IEEE International Conference on Robotics and Automation, pp. 1686–1691 (2006)
7. Weik, S., Liedtke, C.E.: Hierarchical 3D Pose Estimation for Articulated Human Body Models from a Sequence of Volume Data. In: Klette, R., Peleg, S., Sommer, G. (eds.) RobVis 2001. LNCS, vol. 1998, pp. 27–34. Springer, Heidelberg (2001)
8. Ganapathi, V., Plagemann, C., Koller, D., Thrun, S.: Real time Motion Capture Using a Single Time-of-Flight Camera. In: IEEE Conference on Computer Vision and Pattern Recognition, pp. 755–762 (2010)
9. Shotton, J., Fitzgibbon, A., Cook, M., Sharp, T., Finocchio, M., Moore, R., Kipman, A., Blake, A.: Real Time human pose recognition in parts from a single depth image. In: IEEE Conference on Computer Vision and Pattern Recognition (2011)
10. Ly, D.L., Saxena, A., Lipson, H.: Pose Estimation from a Single Depth Image for Arbitrary Kinematic Skeletons. J. Computing Research Repository (2011)
11. Rusinkiewicz, S., Levo, M.: Efficient Variants of the ICP Algorithm. In: International Conference on 3D Imaging and Modeling, pp. 145–152 (2001)
12. Demirdjian, D.: Enforcing Constraints for Human Body Tracking. In: Computer Vision and Pattern Recognition Workshop, pp. 102–102 (2003)
13. Knoop, S., Vacek, S., Dillmann, R.: Modeling Joint Constraints for an Articulated 3D Human Body Model with Artificial Correspondences in ICP. In: 5th IEEE-RAS International Conference on Humanoid Robots, pp. 74–79 (2005)

14. Besl, P.J., McKay, N.D.: A Method for Registration of 3D shapes. J. IEEE Transaction of Pattern Analysis and Machine Intelligence 14, 239–256 (1992)
15. David, D., Trevor, D.: 3D Articulated Pose Tracking for Untethered Diectic Reference. In: 4th IEEE International Conference on Multimodal Interfaces, pp. 267–272 (2002)
16. Mündermann, L., Corazza, S., Andriacchi, T.P.: Accurately Measuring Human Movement Using Articulated ICP with Soft-joint Constraints and Repository of Articulated Model. In: IEEE International Conference on Computer Vision and Pattern Recognition, pp. 1–6 (2007)
17. Moschini, D., Fusiello, A.: Tracking Stick Figures with Hierarchical Articulated ICP. In: Proceedings THEMIS, pp. 61–68 (2008)
18. Gruen, A.W., Akca, M.D.: Generalized Procrustes Analysis and its Applications in Photogrammetry (2003)

Analysis of Motion Space for Modified Screwdrive Bending Mechanism

Chiharu Ishii

Department of Mechanical Engineering, Hosei University,
Tokyo 184-8584, Japan
c-ishii@hosei.ac.jp

Abstract. The purpose of this paper is to extend the motion space of a new screwdrive mechanism, termed Double-Screw-Drive (DSD) mechanism, which was developed by the author so far. To this end, the former DSD mechanism was improved so that the degree of freedom of motion of the tip position is extended, and new DSD mechanism is proposed. In order to control bending motion of the new DSD mechanism, inverse kinematics problem is considered. Namely, equations which give the amount of rotations of each linkage are derived. In addition, using the derived equations, the motion space of the tip position is clarified by software analysis.

Keywords: Bending mechanism, Double-Screw-Drive (DSD) mechanism, motion space, manipulators.

1 Introduction

Minimally invasive surgery (MIS), represented by laparoscopic surgery, has great advantages for patients, since trauma to the body, pain and duration of hospitalization can be reduced significantly. However, MIS is difficult operation for surgeons because degrees of freedom of the surgical instruments are limited due to a small work space. In order to compensate drawbacks of MIS, various surgical assistance devices with the application of robot technology have been proposed. Relevant technology and applications related to computer integrated surgery are listed in [1]. A survey of medical robot systems used in surgery is presented in [2]-[3].

In particular, the development of multi-DOF robotic forceps manipulators capable of reproducing complex human finger movements in laparoscopic surgery is one of the most important issues in the field of robotic surgical systems. A large number of conventional multi-DOF robotic forceps manipulators are of the wire actuation type [4]. However, the rigidity and the durability of wires are poor. In order to improve the rigidity of the forceps manipulator, we have developed a multi-DOF robotic forceps manipulator for MIS using a novel omnidirectional bending technique with screwdrive mechanism termed double-screw-drive (DSD) mechanism in [5]. The robotic forceps manipulator incorporating the DSD mechanism (DSD forceps) has two linkages, each consisting of a universal joint of the screwdrive and a spline nut, and bending motion is achieved since one side of the universal joint has a right-handed screw and the other has a left-handed screw. Therefore, the DSD forceps can

G. Lee (Ed.): Advances in Automation and Robotics, Vol.2, LNEE 123, pp. 37–45.
Springerlink.com

bend without using wires, and omnidirectional bending motion is realized by rotation of the two linkages.

In this paper, the DSD mechanism is improved so that the degree of freedom of motion of the tip position is extended. In the former DSD mechanism, the tip position could reach only on the hemisphere surface. In order to apply the DSD mechanism as a manipulator which works in various kinds of fields, such as a power shovel for civil engineering and construction and hazardous environmental robots for the nuclear reactor, seabed and space, it is necessary to extend the degree of freedom of motion of the tip position to the inside and outside of the hemisphere surface. Thus, design of the DSD mechanism is modified, and new DSD mechanism is proposed. In order to control bending motion of the new DSD mechanism, equations which give the amount of rotations of each linkage to achieve given bending posture are derived. In addition, the motion space of the tip position is analyzed using the derived equations.

2 Structure of DSD Forceps Manipulator

2.1 Former DSD Mechanism

One module of the bending mechanism, which is shown in Fig.1 left, is the former DSD mechanism.

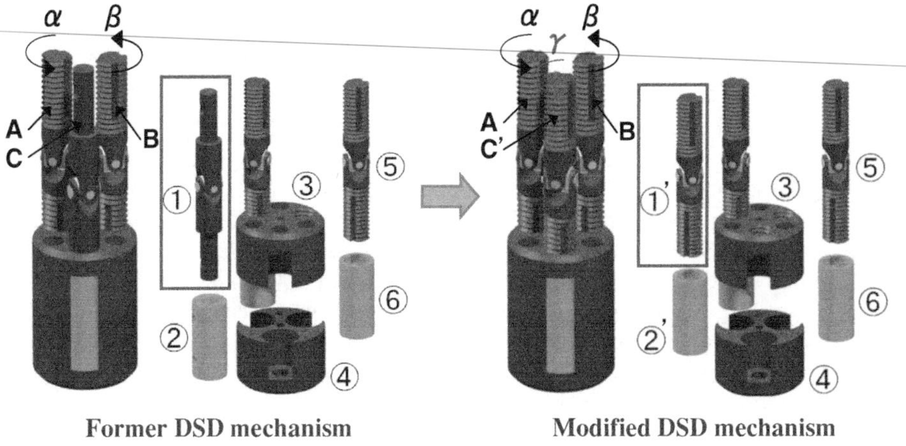

Fig. 1. One module of DSD mechanism. (A) and (B) Bending linkage, (C) Grasping linkage. ① Universal joint shaft. ② Coupling. ③ Plate with left-handed threaded hole. ④ Plate with right-handed threaded hole. ⑤ Universal joint of the screwdrive. ⑥ Spline nut.

The DSD mechanism has three linkages, and when examined in cross-sectional view, each linkage is 120 degrees apart from the other linkages and 6 mm from the center of the cross-section. Let us denote the group consisting of part ⑤ and part ⑥ as a "bending linkage" and the group consisting of part ① and part ② as a "grasping linkage". Bending motion is achieved by rotating the two bending linkages, and grasping linkage is used for actuating the gripper.

The key point of this mechanism is that one side of part ⑤ is a left-handed screw and the other side is a right-handed screw. When a DSD module is connected to another module, a joint is formed. The principle of the bending motion for such a joint is illustrated in Fig.2.

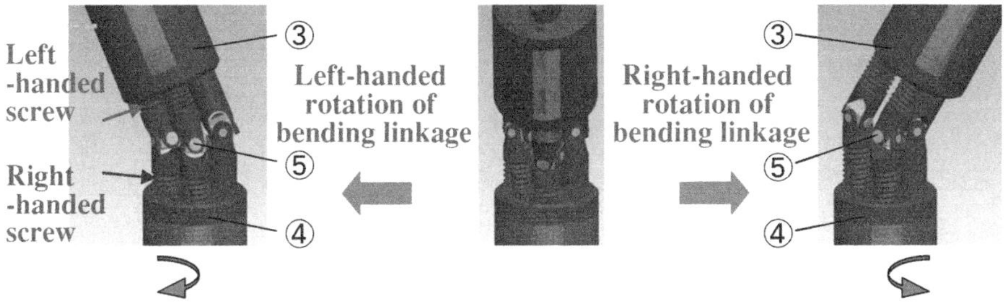

Fig. 2. Principle of bending motion

The left-handed screw of part ⑤ connects to part ③, and the right-handed screw of part ⑤ connects to part ④ of another module. The rotation of the linkage changes the connecting length of the screw and the plate at both ends of part ⑤. As a result, an angle is formed between part ③ and part ④.

For example, when the linkage rotates clockwise, part ③ and part ④ approach each other, and when the linkage rotates counterclockwise, they move away from each other. Thus, bending motion is achieved.

The maximum bending angle of one joint is between -30 and +30 degrees since this is the allowable bending angle of the universal joint. One bending linkage allows for one-DOF bending motion, and by using two bending linkages and controlling their amount of rotations, arbitrary omnidirectional bending motion can be attained.

The third linkage is similar to the other two linkages which comprise the bending mechanism, with the exception that it is not threaded. Therefore, the third linkage, the grasping linkage, only transmits the rotational force from the motor to the other end of the grasping linkage. In the DSD forceps manipulator, the grasping linkage is used for rotation of the gripper.

2.2 DSD Forceps Manipulator

The DSD mechanism is used in a multi-DOF robotic forceps manipulator. Fig.3 shows the complete assembly of the DSD forceps manipulator.

Since the allowable bending angle of a universal joint is between -30 and +30 degrees, the bending part of the DSD forceps manipulator consists of three joints, which allows for bending between -90 and +90 degrees. The total length of the bending part is 85 mm, including the gripper. The diameter from the top of the bending part to the root of the rod is 10 mm. The total length of the DSD forceps manipulator is 635 mm, and its gross weight is 1050 g.

The inside part of the rod consists of three shafts, each 2 mm in diameter and 300 mm long. Each motor in the drive unit and each linkage in the DSD mechanism are connected to each other through a shaft. Therefore, the rotation of each motor is transmitted to each respective linkage through a shaft.

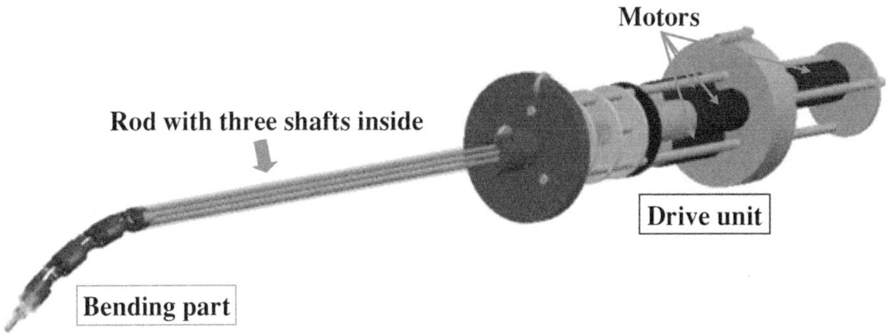

Fig. 3. DSD forceps manipulator

2.3 Kinematics of the Former DSD Mechanism

In order to realize arbitrary bending for the DSD forceps manipulator, an inverse kinematics problem where it is necessary to determine the rotation angle of the two bending linkages given the elevation and the azimuth of the forceps manipulator, was considered.

The coordinates for the DSD mechanism are shown in Fig.4, where θ represents the elevation, which is the deflection of the forceps from the vertical, and ϕ represents the azimuth, which is the angle between the forceps and the grasping linkage. The angles are referred to as θ "bending angle" and ϕ "rotation angle".

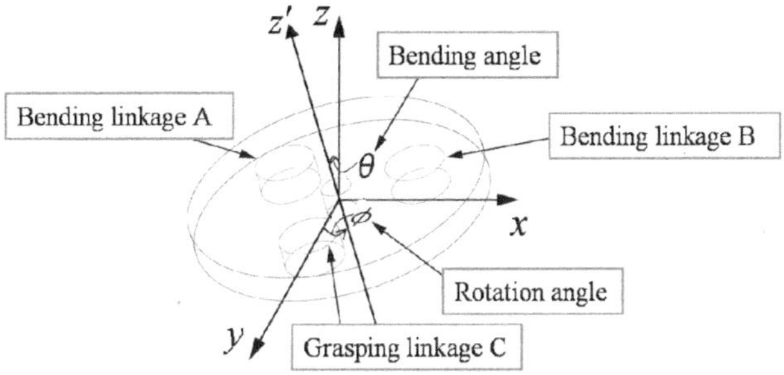

Fig. 4. Bending-rotation coordinates

Then, from geometrical analysis, the amount of rotations of the bending linkage A, denoted as α, and linkage B, denoted as β, are given as follows.

$$\alpha = 10.4\tan\frac{\theta}{6}\cos(\phi - \frac{\pi}{6}),\quad \beta = 10.4\tan\frac{\theta}{6}\cos(\phi + \frac{\pi}{6}) \tag{1}$$

Therefore, provided that the bending angle θ and the rotation angle ϕ are specified, the amount of rotations α and β of each linkage can be determined.

3 Extension of Degree of Freedom of Motion

3.1 Modification of DSD Mechanism

As shown in Fig.5, in the former DSD mechanism, the tip position could be controlled arbitrarily on the hemisphere surface by rotation of two bending linkages, which would be enough degree of freedom in the use as multi-DOF forceps.

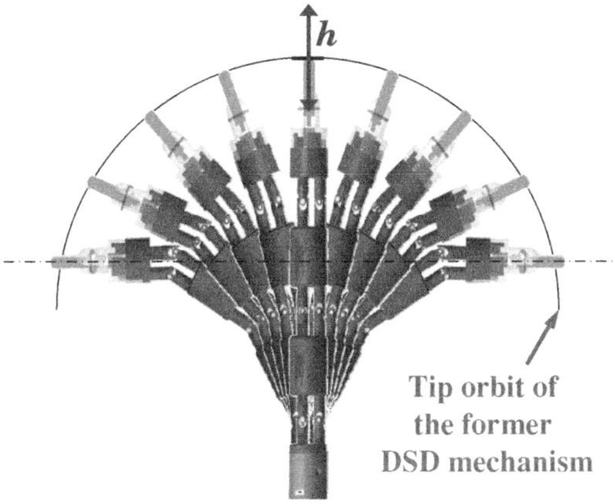

Fig. 5. Tip orbit of the former DSD mechanism

However, when applying this bending mechanism as a manipulator in other fields, it is indispensable to extend the motion space of the tip position to inside and outside of the hemisphere surface. This is achievable by replacing the third linkage currently used as the grasping linkage to the bending linkage. In order to control the tip position of the new DSD mechanism arbitrarily, it is necessary to calculate the amount of rotations of each bending linkage using inverse kinematics. Thus, it is considered that utility of the DSD mechanism in industrial field will spread.

Modification of the one module of the DSD mechanism is shown in Fig.1 right. Note that the linkage C of the former DSD mechanism, namely the grasping linkage, was exchanged to the bending linkage in the new DSD mechanism. Hence, it is necessary to also consider the amount of rotations of the bending linkage C', denoted as γ, in the inverse kinematics problem.

3.2 Kinematics of the New DSD Mechanism

For the modified DSD mechanism, consider a displacement of the tip position in vertical direction h as shown in Fig.5. h is a new parameter to specify a bending posture. In the following, equations which give the amount of rotations (α, β, γ) of each linkage to realize the given bending posture (θ, ϕ, h) are derived, where θ is "bending angle", ϕ is "rotation angle" and h is "tip height".

To this end, depending on the rotation angle ϕ, the control area is divided into three domains as shown in Fig.6, which are ① ϕ=0° to 120°, ② ϕ=120° to 240° and ③ ϕ=240° to 360°.

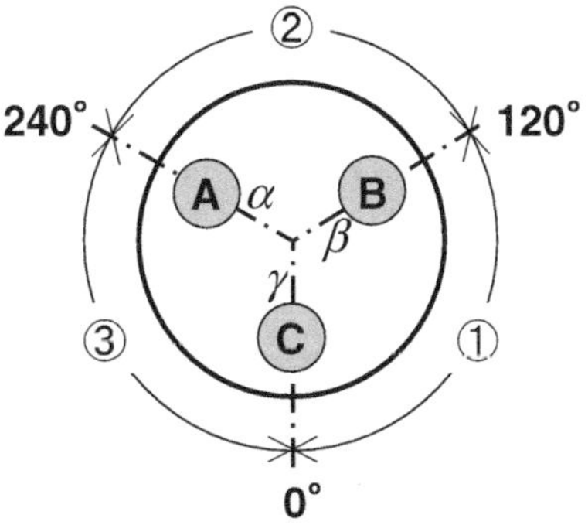

Fig. 6. Domain division by rotation angle

In order to attain a specified posture (θ, ϕ), only two bending linkages are needed. Therefore, the given bending angle and rotation angle (θ, ϕ) are realized by using the linkages B and C (β, γ) in domain ①, using the linkages A and B (α, β) in domain ② and using the linkages A and C (α, γ) in domain ③.

Then, the remaining linkage that is not used in each domain is used for adjusting the tip height h. When the three linkages are rotated by the same amount of rotations simultaneously, the height at the tip can be shifted having maintained the posture (θ, ϕ).

The permitted translation distance of the screw of the universal joint in each linkage is ±1.4mm. Each linkage has three universal joints, and both ends of the universal joint are screw. Therefore, the tip height h is ±1.4mm x 6= ±8.4mm movable at the maximum when the bending part forms upright posture $(\theta=0°, \phi=0°)$.

Since the tip height h is a function of θ and ϕ, the following inequality holds.

$$-8.4 \le h(\theta,\phi) \le 8.4 \tag{2}$$

Given θ and ϕ, $h_{\min}$ and $h_{\max}$ which satisfies

$$h_{\min} \le h(\theta,\phi) \le h_{\max} \tag{3}$$

are obtained by the following equations.

$$\begin{cases} h_{\min} = 6\{-1.4 - \min(0.5\alpha_0, 0.5\beta_0, 0.5\gamma_0)\} \\ h_{\max} = 6\{1.4 - \max(0.5\alpha_0, 0.5\beta_0, 0.5\gamma_0)\} \end{cases} \tag{4}$$

Let δ denote the amount of rotations of each linkage for shifting the tip height. Since the pitch of the screw is p=0.5mm, δ is given as follows.

$$\delta = \frac{h(\theta,\phi)/6}{p} = \frac{h(\theta,\phi)}{3} \tag{5}$$

Hence, from equation (1), the amount of rotations (α, β, γ) of each bending linkage to realize a bending posture (θ, ϕ, h) are given by the following equations.

① ϕ=0° to 120°

$$\alpha = \delta, \quad \beta = 10.4\tan\frac{\theta}{6}\cos(\phi + \frac{\pi}{2}) + \delta, \quad \gamma = 10.4\tan\frac{\theta}{6}\cos(\phi + \frac{5\pi}{6}) + \delta \tag{6}$$

② ϕ=120° to 240°

$$\alpha = 10.4\tan\frac{\theta}{6}\cos(\phi - \frac{\pi}{6}) + \delta, \quad \beta = 10.4\tan\frac{\theta}{6}\cos(\phi + \frac{\pi}{6}) + \delta, \quad \gamma = \delta \tag{7}$$

③ ϕ=240° to 360°

$$\alpha = 10.4\tan\frac{\theta}{6}\cos(\phi - \frac{\pi}{2}) + \delta, \quad \beta = \delta, \quad \gamma = 10.4\tan\frac{\theta}{6}\cos(\phi - \frac{5\pi}{6}) + \delta \tag{8}$$

Thus, the equations which give the amount of rotations of the three bending linkages so that the specified bending posture is achieved, were derived.

3.3 Analysis for Motion Space

The motion space of the tip position of the bending mechanism was analyzed under the derived equations using the numerical analysis software MATLAB/Simulink. The cross sectional view of the motion space at the tip of the bending part of the modified DSD mechanism is shown in Fig.7.

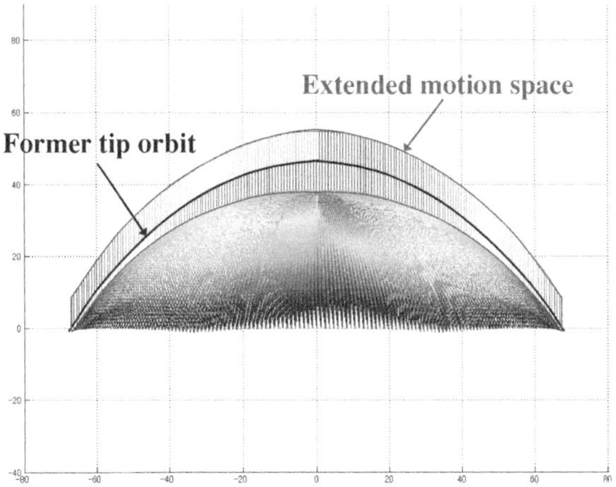

Fig. 7. Cross sectional view of the motion space at the tip

As shown in Fig.7, although the former tip orbit is hemisphere surface, the modified tip orbit was extended inside and outside of the hemisphere surface. Thus, the motion space of the tip position was extended.

From the above analysis, the control method in the use of the modified DSD mechanism as a manipulator was determined and the motion space of the tip position was clarified.

In the case where the gripper is attached at the tip of the modified DSD mechanism as an end effector, operation of extracting the sediment, which was not able to be performed by the manipulator incorporating the former DSD mechanism, can be achieved. Therefore, the modified DSD mechanism is applicable to the manipulator which works in various kinds of fields, such as a power shovel for civil engineering and construction and hazardous environmental robots for the nuclear reactor, seabed and space.

4 Conclusions

In this paper, the screwdrive bending mechanism termed DSD mechanism, which achieves omnidirectional bending motion by rotating two bending linkages, each consisting of a universal joint of the screwdrive and a spline nut, and the robotic forceps manipulator incorporating the DSD mechanism were introduced. The problem of the present DSD mechanism was mentioned, and in order to extend the degree of freedom of motion at the tip of the bending part, redesign of the DSD mechanism was performed. In order to control bending motion of the modified DSD mechanism, the equations which give the amount of rotations of the three bending linkages to achieve the specified bending posture, were derived. In addition, the motion space of the tip position was analyzed using the derived equations.

In this study, the analysis was carried out under the conditions that the number of modules in the bending part is three and the diameter of the bending part is 10mm. In the case where it is necessary to enlarge the bending angle more than the present one, every 30 degrees of increase of the bending angle is achieved by adding a module of the DSD mechanism. In addition, as for displacement of the tip height, magnitude of the displacement of the tip height can be increased by lengthening the thread part of the universal joint. Thus, the motion space at the tip position can be enlarged further. Therefore, the new DSD mechanism can be adjusted to the formation which is suitable for the use.

References

1. Taylor, R.H., Lavallèe, S., Burdea, G.S., Mösges, R. (eds.): Computer-Integrated Surgery—Technology and Clinical Applications. MIT Press, Cambridge (1995)
2. Taylor, R.H., Stoianovici, D.: Medical robotics in computer-integrated surgery. IEEE Trans. Robot. Autom. 19(5), 765–781 (2003)

3. Dario, P., Hannaford, B., Menciassi, A.: Smart surgical tools and augmenting devices. IEEE Trans. Robot. Autom. 19(5), 782–792 (2003)
4. Ikuta, K., Hasegawa, T., Daifu, S.: Hyper redundant miniature manipulator 'hyper finger' for remoteminimally invasive surgery in deep area. In: Proc. 2003 IEEE Int. Conf. Robot. Autom., Taipei, Taiwan, pp. 1098–1102 (2003)
5. Ishii, C., Kobayashi, K., Nishitani, Y.: Robotic forceps manipulator with a novel bending mechanism. IEEE/ASME Trans. Mechatronics 15(5), 671–684 (2010)

Research on General Error Modeling and Instructions Correction Method of Multi-axis CNC Machine Tools

Xiaofeng Wang, Jinwei Fan, Dongju Chen, and Yun Li

The College of Mechanical and Application Electronic Engineering
Beijing University of Technology Beijing 100024 PR China
wxf19982000@163.com

Abstract. Machine tools has become an indispensable processing tool in industrial field. Especially the appearance of multi-axis machine tools has changed the traditional processing method. This paper focuses on the research of improving the machine accuracy. Based on the general error model which has been established using multi-body theory, it has given the precise NC instructions of multi-axis machine tools. Taking C-A type machine tool for example, the precise NC instructions of circle angular has been given.

Keywords: Multi-axis machine tools, error model, multi-body theory.

1 Introduction

The progress of science and technology has promoted the development of every industry. The machine tool also has made revolutionary improvement. From the first machine tool to compound machine tool, the machine tool has experienced the leap development. Multi-axis linkage CNC machine tools, especially five and five or more axis linkage CNC machine tools play very important role in the processing of large and allotype complex parts. It represents the highest level of nation's NC technology in mechanical manufacturing industry. Precision and ultraprecision linkage CNC machine tools together with other precision and ultraprecision technology [1-2] are the important branches of current manufacturing technology. It is the important mark of nation's precision manufacturing ability and technology level. So the improvement of machining accuracy is very important. Error compensation[3-5] is one of the most effective methods. This paper established the model of multi-axis machine tools using multi-body kinematic theory[6-7]. It has given the new instructions correction method.

2 Error Modeling on Multi-axis Machine Tools

2.1 The Basic Description Method of Multi-body System

It is proved by research that multi-body kinematic theory is applied to solve the problem of precision controlling, especially in solving the complex problem of mechanic precision controlling. Therefore this article has established the model to describe the structure of machine tools using multi-body theory. As shown in Figure 1,

G. Lee (Ed.): Advances in Automation and Robotics, Vol.2, LNEE 123, pp. 47–53.
Springerlink.com

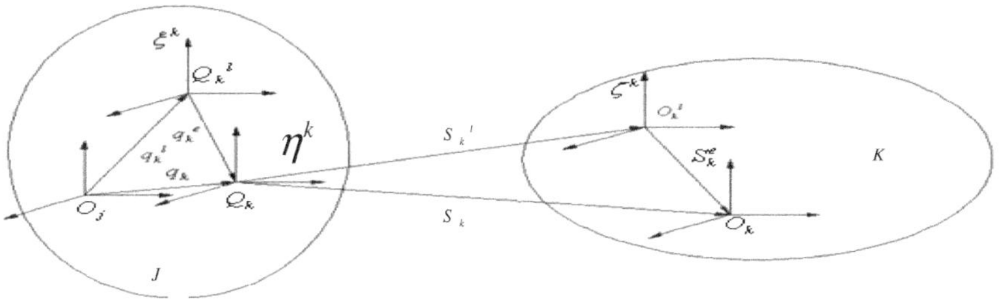

Fig. 1. Schematic diagram of relative movement between neighboring bodies in errors

J body is adjacent low-order body of K body. q_k, q_k^l, q_k^e denote respectively the practical position vector, the ideal position vector and the position error vector of K body's motion reference coordinate origin Q_k in adjacent low-order body. S_k, S_k^l S_k^e denote respectively the practical displacement vector, the ideal displacement vector and the displacement error vector of K body's reference coordinate origin O_k in adjacent low-order body. The error can be divided into two categories: one is position error, which can be described by the position error vector q_k^e and directional error structure matrix $\left[S\xi^k\eta^k\right]$,the other one is kinematic error, Which can be described by the displacement error vector S_k^e and rotation error transformation matrix $\left[S\zeta^k K\right]$. According to the above description method, we can get the following relation:

$$\vec{q}_k = \vec{q}_k^{\,l} + \vec{q}_k^{\,e} \tag{1}$$

$$\vec{S}_k = \vec{S}_k^{\,l} + \vec{S}_k^{\,e} \tag{2}$$

$$[SJK] = [SJ\xi^k][S\xi^k\eta^k][S\eta^k\zeta^k][S\varsigma^k K] \tag{3}$$

The position equation that K body's any point in J body can be described as:

$$\vec{P}_k = \overrightarrow{O_j O_k} + \vec{r}_k \tag{4}$$

Machine tools is composed of many bodies. When we need to describe the position relation of the point in end body we can use the above method, describing the position and orientation relation in adjacent low-order body, then, describing the adjacent low-order body in its the adjacent low-order body. By this way, we can describe the point which we want in the coordinate which we need.

2.2 The General Model of Multi-axis Machine Tools

There are many kinds of machine tools , but after analysizing, we can find that the machine tools is composed of two branches: one is workpiece branch which is made up of m kinematic pairs from workpiece to bed; the other one is tool branch which is made up of n kinematic pairs from tool to bed. According to Figure2, the workpiece branch and tool branch each has five bodies. This is due to that although the machine tools has five bodies at most, but the bodies' motion form and the branch which they belong to are flexible. So we have given the model in Figure 2. We can shield the redundant body which we don't need according to the structure of machine tools. Two branches in figure 2,"workpiece-bed"branch B-W is composed of following bodies: $B(O)$-W_1-W_2 $-W_3$-W_4-W_5-W, W_1,W_2, W_3, W_4, W_5 denote five moving bodies, the movement between adjacent body is translation or rotation; "tool-bed" branch B-T is composed of following bodies: $B(O)$-T_1- T_2-T_3-T_4-T_5-T, T_1, T_2, T_3, T_4, T_5 denote five moving bodies the movement between adjacent body is translation or rotation. In the condition of error, the position of typical body's point "P" in "R" can be expressed as:

$$\begin{bmatrix} \{p_o\} \\ 1 \end{bmatrix} = [SOK] \begin{bmatrix} \{p_k\} \\ 1 \end{bmatrix} \tag{5}$$

[SOK] denote the transformation matrix of typical body "B_k" coordinate system relative to inertial coordinate system "R" in error condition. It can be expressed as:

$$[SOK] = \prod_{t=u}^{0}([SSV]_p[SSV]_{pe}[SSV]_s[SSV]_{se}) \tag{6}$$

$[SSV]_p$——denote the transformation matrix of "B_v" body motion coordinate system relative to adjacent low-order "B_s" body coordinate system in ideal condition.

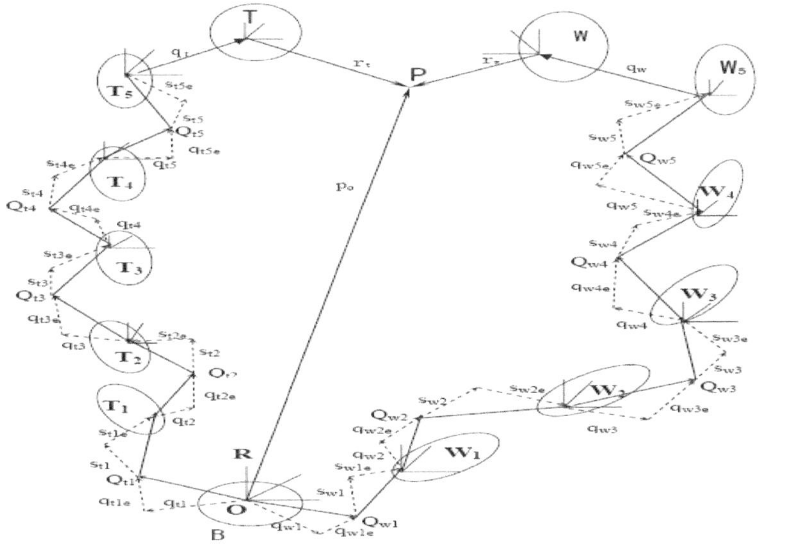

Fig. 2. General model of multi-body in geometry errors

$[SSV]_s$——denote the transformation matrix of "B_v" body coordinate system relative to its practical motion coordinate system in ideal condition.

$[SSV]_{pe}$——denote the transformation matrix of "B_v" body position error in error condition.

$[SSV]_{se}$——denote the transformation matrix of "B_v" body displacement error in error condition.

According to the above, point "P" on typical body "W" coordinate system can be expressed with the branch of "workpiece-bed" in "R" coordinate system, the position p_w can be expressed as follows:

$$\begin{bmatrix} \{p_w\} \\ 1 \end{bmatrix} = [SOW] \begin{bmatrix} \{r_w\} \\ 1 \end{bmatrix} \tag{7}$$

p_w——denote the position matrix of point "P" on typical body(workpiece) coordinate system relative to "R"

r_w——denote the position matrix of point "P" relative to workpiece coordinate system

$[SOW]$——denote the practical transformation matrix of typical body "W" coordinate system relative to "R"

we can get:

$$\begin{aligned}
[SOW] &= [SOW_1][SW_1W_2][SW_2W_3][SW_3W_4][SW_4W_5][SW_5W] \\
&= [SOW_1]_p[SOW_1]_{pe}[SOW_1]_s[SOW_1]_{se}[SW_1W_2]_p[SW_1W_2]_{pe}[SW_1W_2]_s[SW_1W_2]_{se} \\
&\quad [SW_2W_3]_p[SW_2W_3]_{pe}[SW_2W_3]_s[SW_2W_3]_{se}[SW_3W_4]_p[SW_3W_4]_{pe}[SW_3W_4]_s[SW_3W_4]_{se} \\
&\quad [SW_4W_5]_p[SW_4W_5]_{pe}[SW_4W_5]_s[SW_4W_5]_{se}[SW_5W]_p[SW_5W]_{pe}[SW_5W]_s[SW_5W]_{se}
\end{aligned}$$

We can know by the same way, point "P" on typical body "T" coordinate system can be expressed with the branch of "tool-bed" in "R" coordinate system, the position p_t can be expressed as follows:

$$\begin{bmatrix} \{p_t\} \\ 1 \end{bmatrix} = [SOT] \begin{bmatrix} \{r_t\} \\ 1 \end{bmatrix} \tag{8}$$

p_t——denote the position matrix of point "P" on typical body(tool) coordinate system relative to "R"

r_t——denote the position matrix of point "P" relative to tool coordinate system

$[SOT]$ ——denote the practical transformation matrix of typical body "T" coordinate system relative to "R"

we can get:

$$\begin{aligned}
[SOT] &= [SOT_1][ST_1T_2][ST_2T_3][ST_3T_4][ST_4T_5][ST_5T] \\
&= [SOT_1]_p[SOT_1]_{pe}[SOT_1]_s[SOT_1]_{se}[ST_1T_2]_p[ST_1T_2]_{pe}[ST_1T_2]_s[ST_1T_2]_s \\
&\quad [ST_2T_3]_{pe}[ST_2T_3]_{pe}[ST_2T_3]_s[ST_2T_3]_{se}[ST_3T_4]_p[ST_3T_4]_{pe}[ST_3T_4]_s[ST_3T_4]_{se} \\
&\quad [ST_4T_5]_p[ST_4T_5]_{pe}[ST_4T_5]_s[ST_4T_5]_{se}[ST_5T]_p[ST_5T]_{pe}[ST_5T]_s[ST_5T]_{se}
\end{aligned}$$

3 Instructions Correction Method of Multi-axis CNC Machine Tools Analysis

3.1 Mapping Relation between Ideal Normal Vector and Rotation Instructions

Because the translation along x-axis, y-axis, z-axis doesn't affect the rotation of multi-axis machine tools. We can analyze the rotation without considering the translation movement. The formula can be expressed as follows:

$$\left([SW_0W_1]_s[SW_0W_1]_{se}[SW_1W_2]_s[SW_1W_2]_{se}\right)(n_x,n_y,n_z,0)^T =$$

$$\left([ST_0T_1]_s[ST_0T_1]_{se}[ST_1T_2]_s[ST_1T_2]_{se}\right)(t_x,t_y,t_z,0)^T \tag{9}$$

In the formula, "tool branch" and "workpiece branch" both have two rotational parts. This is due to that although the multi-axis machine tools have two rotational parts at most, but the bodies' motion form and the branch which they belong to are flexible. We can shield the redundant body which we don't need according to the structure of machine tools. Taking five-axis C-A type machine tools for example, the ideal equation can be expressed as follows:

$$\begin{bmatrix} n_x \\ n_y \\ n_z \\ 0 \end{bmatrix} = \begin{bmatrix} \cos\gamma & -\sin\gamma & 0 & 0 \\ \sin\gamma & \cos\gamma & 0 & 0 \\ 0 & 0 & 1 & 0 \\ 0 & 0 & 0 & 1 \end{bmatrix} \begin{bmatrix} 1 & 0 & 0 & 0 \\ 0 & \cos\alpha & -\sin\alpha & 0 \\ 0 & \sin\alpha & \cos\alpha & 0 \\ 0 & 0 & 0 & 1 \end{bmatrix} \begin{bmatrix} 0 \\ 0 \\ 1 \\ 0 \end{bmatrix} \tag{10}$$

Solving the equation ,we can get the rotation instruction

We can get: $\cos\alpha = n_z$, $\tan\gamma = -\dfrac{n_x}{n_y}$ (11)

3.2 The Solution of Rotation Instructions' Correction Value

We can get the mapping relation between ideal normal vector and rotation instructions according to (9), as follows:

$$\begin{bmatrix} n_x \\ n_y \\ n_z \\ 0 \end{bmatrix} = \begin{bmatrix} \cos\gamma & -\sin\gamma & 0 & 0 \\ \sin\gamma & \cos\gamma & 0 & 0 \\ 0 & 0 & 1 & 0 \\ 0 & 0 & 0 & 1 \end{bmatrix} \begin{bmatrix} 1 & -\varepsilon_z(\gamma_0) & \varepsilon_y(\gamma_0) & \delta_x(\gamma_0) \\ \varepsilon_z(\gamma_0) & 1 & -\varepsilon_x(\gamma_0) & \delta_y(\gamma_0) \\ -\varepsilon_y(\gamma_0) & \varepsilon_x(\gamma_0) & 1 & \delta_z(\gamma_0) \\ 0 & 0 & 0 & 1 \end{bmatrix} \times$$
$$\begin{bmatrix} 1 & 0 & 0 & 0 \\ 0 & \cos\alpha & -\sin\alpha & 0 \\ 0 & \sin\alpha & \cos\alpha & 0 \\ 0 & 0 & 0 & 1 \end{bmatrix} \begin{bmatrix} 1 & -\varepsilon_z(\alpha_0) & \varepsilon_y(\alpha_0) & \delta_x(\alpha_0) \\ \varepsilon_z(\alpha_0) & 1 & -\varepsilon_x(\alpha_0) & \delta_y(\alpha_0) \\ -\varepsilon_y(\alpha_0) & \varepsilon_x(\alpha_0) & 1 & \delta_z(\alpha_0) \\ 0 & 0 & 0 & 1 \end{bmatrix} \begin{bmatrix} 0 \\ 0 \\ 1 \\ 0 \end{bmatrix} \tag{12}$$

In the formula, (α_0,γ_0) denote the initial rotation value, normal vector $[n_x,n_y,n_z]^T$ can be solved according to (10) using the initial rotation value, we can get:

$$\begin{bmatrix} n_x \\ n_y \\ n_z \\ 0 \end{bmatrix} = \begin{bmatrix} \cos\gamma \bullet a + \sin\gamma \bullet b \\ \sin\gamma \bullet a - \cos\gamma \bullet b \\ -\varepsilon_y(\gamma_0)\varepsilon_y(\alpha_0) + \cos\alpha \bullet c - \sin\alpha \bullet d \\ 0 \end{bmatrix} \tag{13}$$

In the formula,

$$a = \cos\alpha\,(\varepsilon_y(\gamma_0) + \varepsilon_z(\gamma_0)\varepsilon_x(\alpha_0)) + \sin\alpha\,(\varepsilon_z(\gamma_0) - \varepsilon_x(\alpha_0)\ \varepsilon_y(\gamma_0)) + \varepsilon_y(\alpha_0)$$

$$b = \cos\alpha(\ \varepsilon_x(\gamma_0) + \varepsilon_x(\alpha_0)) + \sin\alpha\,(1 - \varepsilon_x(\alpha_0)\ \varepsilon_x(\gamma_0)) - \varepsilon_y(\alpha_0)\ \varepsilon_z(\gamma_0)$$

$$c = 1 - \varepsilon_x(\alpha_0)\ \varepsilon_x(\gamma_0)$$

$$d = \varepsilon_x(\gamma_0) + \varepsilon_x(\alpha_0)$$

We can get:

$$\alpha = k_1\pi - \arcsin\frac{n_z + \varepsilon_y(\gamma_0)\varepsilon_y(\alpha_0)}{\sqrt{c^2+d^2}} - \arcsin\frac{c}{\sqrt{c^2+d^2}} \tag{14}$$

$$\gamma = k_2\pi + \arcsin\frac{n_y}{\sqrt{a^2+b^2}} + \arcsin\frac{b}{\sqrt{a^2+b^2}} \tag{15}$$

The range of α, γ depends on the specific machine tools, our paper has given the value as follows:

$$\begin{cases} -\frac{\pi}{2} \le \arcsin\frac{c}{\sqrt{c^2+d^2}} \le \frac{\pi}{2} \\ -\pi \le \alpha \le 0 \end{cases} \quad \begin{cases} -\frac{\pi}{2} \le \arcsin\frac{b}{\sqrt{a^2+b^2}} \le \frac{\pi}{2} \\ 0 \le \gamma \le 2\pi \end{cases} \tag{16}$$

We can get:

$$-\frac{\pi}{2} \le \arcsin\frac{c}{\sqrt{c^2+d^2}} - \alpha \le \frac{3\pi}{2} \qquad -\frac{\pi}{2} \le \gamma - \arcsin\frac{b}{\sqrt{a^2+b^2}} \le \frac{5\pi}{2}$$

When $-\frac{\pi}{2} \le \arcsin\frac{c}{\sqrt{c^2+d^2}} - \alpha_0 \le \frac{\pi}{2}, k_1 = 0$

When $\frac{\pi}{2} < \arcsin\frac{c}{\sqrt{c^2+d^2}} - \alpha_0 \le \frac{3\pi}{2}, k_1 = -1$

When $-\frac{\pi}{2} \le \gamma - \arcsin\frac{b}{\sqrt{a^2+b^2}} \le \frac{\pi}{2},\ k_2 = 0$

When $\frac{\pi}{2} < \gamma - \arcsin\frac{b}{\sqrt{a^2+b^2}} \le \frac{3\pi}{2}, k_2 = -1$

When $\frac{3\pi}{2} < \gamma - \arcsin\frac{b}{\sqrt{a^2+b^2}} \le \frac{5\pi}{2},\ k_2 = -2$

From the above we can get the first correction value which can be described as (α_1,γ_1), put(α_1,γ_1) into formula (12) and updated the error parameters from database, we can calculate the motion transformation matrix. Repeating the formula(12) we can get (α_2,γ_2), calculate the(α_i,γ_i) by the same way. When it achieved the terminal condition, stop calculating. The terminal condition is :

$$\theta = \arccos \frac{n_x n_{xi} + n_y n_{yi} + n_z n_{zi}}{\sqrt{n_x^2 + n_y^2 + n_z^2}\sqrt{n_{xi}^2 + n_{yi}^2 + n_{zi}^2}} \leq \Delta \tag{17}$$

In the formula, $[n_{xi},n_{yi},n_{zi}]^T$ can be calculated using (α_i,γ_i)(this time it is the known value) and formula(12). Δ is permission error.

4 Conclusions

The complex multi-axis machine tools is abstracted as two branches based on multi-body theory in this paper. The general model has been built. This paper analyzed the rotational parts and has given the mapping relation between ideal normal vector and rotation instructions. Taking C-A type five-axis machine tools for example, the solution of instructions have been given based on iterative calculation method. It has given a new idea to solve the rotation problem of multi-axis machine tools.

References

1. Yuan, Z.J., Wang, X.K.: Precision and Ultra-precision Manufacturing technology. China Machine Press (1999)
2. Li, S.Y., Dai, Y.F., Peng, X.Q.: Ultra-precision Machine Tools and Development of the Latest Technology. Journal Of National University Of Defense Technology (2), 95–100 (2000)
3. Ramesh, R., Mannan, M.A., Poo, A.N.: Error compensation in machine tool—a review part I geometric, cutting – force induced and fixture – dependent errors. International Journal of machine Tools & Manufacture, 1235–1256 (2000)
4. Error compensation in CNC turning solely from dimensional measurements of previously machined parts. Machine Tools & Manufacture, 155–164 (1998)
5. Ferreira, P.M., Liu, C.R.: A Contribution to the Analysis and Compensation of the Geometric Error of a Machine Center. Annals of CIRP 35(1), 259–263 (1986)
6. Huston, R.L., Liu, Y.W.: Multi-body System Dynamics Theory. Tianjin University Press (1987)
7. Zhang, G.X.: Error Compensation of Coordinate Measuring Machines. CIRP 37(1), 515–519 (1988)

Vehicle Routing Problem with Stochastic Demands and Simultaneous Delivery and Pickup Based on the Cross-Entropy Method

Chuansheng Wang and Yue Qiu

Information College
Capital University of Economic and Business
Beijing 100070, China
qiuyue@cueb.edu.cn

Abstract. This paper considers the problem of Vehicle Routing Problem with stochastic demands and simultaneous delivery and pickup. An effective algorithm based on cross-entropy is designed to solve the model. The cross-entropy method is a new generic approach to combinatorial and multi-extremal optimization and rare event simulation. Numerical experiments have been conducted and the results indicate that the method can effectively solve this problem.

Keywords: cross-entropy, vehicle routing problem, stochastic demands.

1 Introduction

The Vehicle Routing Problem (VRP) was motivated by Dantzig and Ramser in 1959, which refers to a whole class of problems. From a theoretical point-of-view, VRP belongs to the class of NP-hard problems [6]. The Vehicle Routing Problem (VRP) can be described as the problem of designing optimal delivery or collection routes from one or several depot(s) to a number of geographically scattered customers subject to side constraints. The objective is to find a set of routes which minimizes the total distance traveled. The VRP is categorized by HVRP, CVRP, MDVRP, VRPTW, SVRP, VRPPD, and so on with different constraints (Bodin et al., 1983). The vehicle routing problem, which is used in practice, considers these constraints simultaneously. However, only one or two constraint(s) have been examined in past research. It was also assumed that the demand at any demand point did not exceed the loading capability (Danzig and Ramser, 1959). When demand exceeded capability, another vehicle would be needed. Therefore the opportunity for optimization was often lost (Shin and Kang, 1991).

Due to the solving complexity, people have been putting much emphasis on SVRP With computer technology's development. The influence of continuously changing goods amount in the transportation process to route and cost is hardly considered, but the character of cost economic of transportation industry indicate that service cost is relation to production batch size[1], which the truck's fuel consumption and the cost of escort was include. Transportation cost will be different along with the variation of

G. Lee (Ed.): Advances in Automation and Robotics, Vol.2, LNEE 123, pp. 55–60.
Springerlink.com

vehicle load. The influence of continuously changing goods amount in the transportation process and customer stochastic demand to vehicle route is considered in this paper. In this paper a model named WVRPSD with weight coefficients and stochastic demands is proposed in order to overcome this weakness of the past researches. However, it is categorized as an NP-hard problem according its complexity as indicated by Lenstra and Rinnooy (1981). Various mathematical formulations of the VRP were given by Laporte (1992). As the size of problem gets larger, it is nearly impossible to get a solution. Therefore, different approaches, tabu search(Osman;1993, Gendreau et al.;1994, Rego and Roucairol;1996, Barbarosoglu and Ozgur;1999), simulated annealing(Osman;1993, Hiquerbran et al.;1994), genetic algorithms, and ant colony optimization have been(Bullnheimer et al.;1999, Gambardella et al.; 1999) developed so far[3-5].

The cross-entropy (CE) method is a new generic approach to rare event simulation [2]. The CE method was motivated by an adaptive algorithm for estimating probabilities of rare events in complex stochastic networks by Rubinstein in 1997, which involves variance minimization [2]. It was soon realized that a simple cross-entropy modification could be used for not only estimating probabilities of rare events but solving difficult COPs as well. Cross entropy method has been widely used recently. The rest of this paper is organized as follows: Section 2 explains the structures of model; Section 3 describes methodology; an experiment analysis is shown in section 4 and finally some concluding remarks are drawn from section 5.

2 Structure of WVRPSD

WVRPSD can be described as follows: Let $G=(V,A,C)$ be a complete graph with vertex set $V=\{v_1,\ldots,v_n\}$, the arc set $A=\{(v_i,v_j): i\neq j; v_i,v_j\in V\}$ and cost set $C=\{c_{v_iv_j}: i\neq j; v_i,v_j\in V\}$. Where v_1 represent the storehouse, other vertex be the customs.

Let $L=\{l_{v_iv_j}: i\neq j; v_i,v_j\in V\}$ be distance set satisfied Symmetry and triangle inequality. Let the cost of vehicles running unit distance denotes by c_l and the cost of vehicles with unit weight running unit distance denote by c_w. Let the customers' demand denote by independent and identically distributed discrete random variables d_{v_i}, and its real value will be known until vehicles come. Let Q be the vehicle capacity.

For any custom sequence, that is route $R=(r_1,\ldots,r_n,r_{n+1})$, where $r_1=r_{n+1}=v_1$ denote storehouse. Its cost function can be defined as follows

$$f(R)=\sum_{i=1}^{n} c_{r_ir_{i+1}} I(r_i)+\sum_{i=2}^{n} g(r_i) \tag{1}$$

Let $remain(r_i)$ be remaining goods for vehicle arrive r_i and unload, then

$$remain(r_1) = Q$$

$$remain(r_i) = \begin{cases} remain(r_{i-1}) - d_{r_i} & remain(r_{i-1}) > 0 \\ Q - d_{r_i} & remain(r_{i-1}) = 0 \\ Q + remain(r_{i-1}) - d_{r_i} & remain(r_{i-1}) < 0 \end{cases} \quad i = 2, \ldots, n$$

where

$$I(r_i) = \begin{cases} 0 & remain(r_i) \le 0 \\ 1 & remain(r_i) > 0 \end{cases} \quad i = 1, \ldots, n$$

$c_{r_i r_j}$ denote the cost between node r_i and r_j:

$$c_{r_i r_j} = l_{r_i r_j} \times c_l + l_{r_i r_j} \times \left[remain(r_i) + \sum_{k=1}^{i} pick(r_k) \right] \times c_w$$

where, $pick(r_k)$ be pickup goods for vehicle arrive r_k.

Let the extra cost along the route due to limited vehicle capacity denote by $g(r_i)$.

$$g(r_i) = \begin{cases} 0 & remain(r_{i-1}) > 0, \left[remain(r_{i-1}) + \sum_{k=1}^{i-1} pick(r_k) \right] \le Q \\ c_{r_i 0} + c_{0 r_{i+1}} & remain(r_i) = 0, \left[remain(r_{i-1}) + \sum_{k=1}^{i-1} pick(r_k) \right] \le Q \\ c_{r_i 0} + c_{0 r_i} & remain(r_i) < 0 \\ c_{r_i 0} + c_{0 r_i} & \left[remain(r_{i-1}) + \sum_{k=1}^{i-1} pick(r_k) \right] > Q \end{cases} \quad i = 2, \ldots, n$$

We generate M samples of customs' demand by using Monte Carlo sampling method; cross-entropy method is adopted to compute route cost according to stochastic demands.

3 Algorithm Scheme Based on Cross-Entropy Methods

Straightforward simulation for rare events requires a large number of trials and hard to implement because the occurrence of rare events are extremely little in a standard simulation, hence new methods are needed to be investigated and developed. IS is based on the idea to make the occurrence of rare events more frequent, or in other words, to speed up the simulation. Technically, IS aims to select a probability distribution that minimizes the variance of the IS estimate. The efficiency of IS depends on obtaining a good change of measure. Large deviation results have proven their applicability for small systems with medium complexity[2].

Let $x=(X_1,\ldots,X_n)$ be a random vector taking values in some space Ω and $f(x)$ be a probability density function on Ω. Let s be some real-valued function on Ω. Suppose we are interested in the probability that $s(x)$ is greater than or equal to some real number r and this probability can be expressed as $l=P\{s(x)\ge r\}=EI_{(s(x)\ge r)}$.

An methods is based on importance sampling: take a random sample $X_1,\ldots,X_N$ from an *importance sampling density* $g(x)$ on Ω, then

$$l=\int_{\{s(x)\ge r\}} f(x)dx=\int_{\{s(x)\ge r\}} L(x)g(x)dx$$

In estimate rare event probability: draw a random sample $X_1,\ldots,X_N$ from $g(x)$, and then get estimator

$$\hat{l}_{IS}=\sum_{i=1}^{R} I_{\{s(X_i)\ge r\}}L(X_i)$$

where $L(x)=\dfrac{f(x)}{g(x)}$ is called the *likelihood ratio*. Minimizing the Kullback-Leibler distance between $g_{opt}(x)$ and $f(\cdot;v)$ is equivalent to solving the maximization problem

$$v^*=\arg\max_v E_w I_{\{s(X)\ge r\}} W(x;U,w)\ln f(x;v)$$

We generate stochastic demand samples $D^1,\ldots,D^M$ and average cost value of this M samples denote any route R cost, that is

$$F(R)=\frac{1}{M}\sum_{i=1}^{M} f(R,D^i) \tag{2}$$

Thus, the solution of (1) may be readily obtained by solving (2). The algorithm as following:

step 1. Generate the initial transfer matrix P, that is $p_{ij}=\begin{cases}\dfrac{1}{n-1} & i\ne j\\ 0 & i=j\end{cases}$,

Define $P^{(1)}=P$,let $t=1$, $\rho=0.01$;

step 2 Generate a sample $R_1,\ldots,R_N$ from transfer matrix $P^{(t)}$, Random generate M demand samples

step 3. Compute the average cost of each route for M customs' demand samples. Denote the sample $(1-\rho)$-quantile of $F(R_1),\ldots,F(R_N)$ by γ_t

step 4 . Update transfer matrix according to:

$$p_{r_i r_j} = \frac{\sum_{k=1}^{N} I_{\{F(R_k) \le r_t\}} I_{\{(r_i, r_j) \in R_k\}}}{\sum_{k=1}^{N} I_{\{F(R_k) \le r_t\}}} \quad (3)$$

Step 5. For $t \ge d$, stop iteration when $r_t = r_{t-1} = \cdots = r_{t-d}$; else set $t = t + 1$ and reiterate from step 2.

4 Experiment Results

To illustrate the ideas set forth in the previous sections, we present now numerical results. Generate the distance between two nodes from [0, 100], generate custom demand and pickup goods demands in [0, 60]. The vehicle capacity is 50, that is custom demand can be bigger than vehicle capacity. The cost of vehicles running unit distance is 15 and vehicles with unit weight running unit distance is 25. $5n^2$ is the number of samples, n is the customers number。 Random generate 100 samples, the following datas are average value. Customer number is 5, the initial transfer matrix:

$$P^{(1)} = \begin{bmatrix} 0 & 0.25 & 0.25 & 0.25 & 0.25 \\ 0.25 & 0 & 0.25 & 0.25 & 0.25 \\ 0.25 & 0.25 & 0 & 0.25 & 0.25 \\ 0.25 & 0.25 & 0.25 & 0 & 0.25 \\ 0.25 & 0.25 & 0.25 & 0.25 & 0 \end{bmatrix}$$

The transfer matrix for t=250 as follows:

$$P^{(250)} = \begin{bmatrix} 0 & 0.9 & 0 & 0 & 0.1 \\ 0 & 0 & 0 & 1 & 0 \\ 0 & 0 & 1 & 0 & 0 \\ 0.85 & 0 & 0 & 0 & 0.15 \\ 0.05 & 0 & 0.07 & 0 & 0.88 \end{bmatrix}$$

The optimal route is (2, 4, 3, 1, 5), the cost is 63474.

5 Conclusion

A class of vehicle routing problem with stochastic demands and simultaneous delivery and pickup is discussed in this paper. Due to the characteristic that the strong correlation between transportation cost and customers' demand is exist and the customers' demand is stochastic, we describe the cost objective function using the expectation of object function instead of the original one. The influence of current vehicle capacity to transportation cost is considered when describe the expectation of object function. An algorithm is proposed based on importance sampling based on expectation to solve this complicated route cost function. Numerical experiments have shown that the method has a strong capability to solve this problem and offers a good tool for VRP.

Acknowledgment. This research was supported by the field grade scientific research project of Capital University of Economics and Business (No: 2011XJ035); project funds (No: 00791154430107); the Beijing Municipal Education Commission Foundation of China (No. KM201110038002); the Funding Project for Academic Human Resources Development in Institutions of Higher Learning Under the Jurisdiction of Beijing Municipality (No. PHR201007117).

References

1. Ballou, R.H.: Business Logistics Management: Planning, Organizing, and Controlling the Supply Chain. Pearson Education North Asia Limited (1992)
2. Chepuri, K., Homem-de-Mello, T.: Solving the vehicle routing problem with stochastic demands using the Cross-entropy method. Annals of Operation Research 134(1), 153–181 (2005)
3. Teodorovic, D., Pavkovic, G.: A simulated annealing technique approach to the vehicle routing problem in the case of stochastic demand. Transportation Planning and Technology 16, 261–270 (1992)
4. Gendreau, M., Laporte, G., Séguin, R.: A tabu search heuristic for the vehicle routing problem with stochastic demands and customers. Operations Research 44(3), 4692477 (1996)
5. Guo, Z.G., Mak, K.L.: A heuristic algorithm for the stochastic vehicle routing problems with soft time windows. In: Proc of the 2004 Congress on Evolutionary Computation (CEC 2004), Portland, pp. 1449–1456 (2004)
6. Laporte, G., Gendreau, M., Potvin, J.Y., et al.: Classical and modern heuristics for the vehicle routing problem. International Transactions in Operational Research 7, 285–300 (2000)

Modeling and Simulation of Formed Milling Cutter for Screw Based on Non-instantaneous Envelope Method

Yun Li, Jin-wei Fan, Dong-ju Chen, and Xiao-feng Wang

College of Mechanical Engineering and Applied Electronics Technology,
Beijing University of Technology, Beijing 100124, PR China
handanliyun@126.com

Abstract. According to the shape of screw, the spiral surface equation of screw was established; and through analyzing the relative motion between screw and formed milling cutter, find the contact conditions of screw and cutter, further deduce the meshing conditions of screw and cutter, and the contact line equation was established based on non-instantaneous envelope method. From the axial cross-section shapes which the contact line rotates around the axis of cutter formed rotary surface, the profile equation of formed milling cutter was established. Analyzing the causes of processing error after wear of formed milling cutter, the tooth back curve equation was established, finally got the profile equation of formed milling cutter have constant back edge. By interference checking for profile equation of formed milling cutter, ensure correctness of milling cutter profile.

Keywords: screw, non-instantaneous envelope, profile equation, formed milling cutter.

1 Introduction

Along with the development of science and technology, all kinds of screw machinery equipment widely used day by day. Eccentric screw is the key components of oil extraction progressing cavity pump. Their shape and processing quality decides the meshing status of screw and spiral cavity. Different helical camber of screw directly determines the differences of technical performance index of working pressure, rated flow, volume efficiency, vibration and noise etc.

Eccentric screw processing methods are various. In order to improve machining accuracy of the screw, the research of tool for spiral surface is very important. From the characteristic parameters of progressing cavity pump, Mimmi, G. and Pennacchi, P. had established the screw model determined the contact line between cutter and screw, got the tool profile [1], and analyzed the processing error caused by the tool wear of screw processing [2]. Chi-Jung Chiang and Zhang-Hua Fong proposes a mathematical method for finding the distribution of cutter body inserts that will result in an equal wear rate for their cutting edges [3]. Using projection principle and matrix transformation, Zheng Mei-sheng etc had realized the tool design by numerical calculation and computer programming [4]. Zhao chun-xiu etc through analyzing

G. Lee (Ed.): Advances in Automation and Robotics, Vol.2, LNEE 123, pp. 61–68.
Springerlink.com

coordinate calculation method of processing track for formed milling cutter of screw, obtained a kind of computation method for solving the profile of milling cutter [5]. Wu Qin etc had achieved the milling cutter design of complex spiral surface based on the envelop principle of milling cutter and screw [6]. Using the contact lines of screw processing forms have the characteristics of space position unchanged, Peng Xue-yuan etc had designed hob blade of screw [7]. Sun Xing-wei etc according to the approaching theory of spatial arc, achieve the optimum design of cutter parameters for screw [8]. Wang Tai-yong etc come up with high precision tool path planning algorithm of error compensation, leads to a high processing precision of complex parametric surface [9].

This paper through analyzing the shape of screw, the spiral surface equation of screw was established. And from the relative motion between cutter and screw, according to the contact principle of cutter and screw, the meshing conditions of cutter and screw was deduced, and the contact line equation was established based on non-instantaneous envelope method. From the axial cross-section shapes which the contact line rotates around the axis of cutter formed rotary surface, the profile equation of cutter was established. Analyzing the causes of processing error after wear of cutter, the tooth back curve equation was established based on spiral of Archimedes, and got the profile equation of cutter have constant back edge. The correctness of milling cutter profile is ensured by interference checking.

2 The Non-instantaneous Envelope Method

Non-instantaneous envelope method is a kind of processing method, by milling cutter and work piece do relative movement, cutting edge rotation trajectory formed surface envelope generated work piece surface profile.

Formed milling cutter is special tool for processing spiral surface; the profile of formed milling cutter is axial cross-section shapes which cutting edge around its axis rotation produced helical camber. By non-instantaneous envelope method known, formed milling cutter processing helicoids, helical camber of screw is the envelope surface which milling cutter edge relative to the screw made spiral movement. Contact line of the cutter rotary surface and screw spiral surface is a space curve. Therefore, these two surfaces relative movement in space any instantaneous must tangent to this space curve, and this space curve is contact line. When the contact line makes helical motion around axis line of screw, it formed helical camber of screw; if the contact line makes rotation motion around axis line of tool, it formed rotary surface of tool.

3 Profile Design of Formed Milling Cutter for Screw

3.1 Spiral Surface Equation of Screw

The eccentric screw surface are formed by the spiraling of eccentric circle, establish coordinate system of screw as shown in Fig.1 below, and the z axis and screw axis overlap. $o_1x_1y_1z_1$ is the dynamic coordinate system, and oxyz is the fixed coordinate system, oo_1 is the eccentricity e, T is lead of screw, R is the cross section radius of screw. The center o_1 of cross section rotates around the center line of screw with the eccentricity e, and then does uniform motion along the center line of screw to get the helical camber. Therefore, the profile equations of the screw in the coordinate system oxyz are:

$$\begin{cases} x = R\cos\phi + e\cos\theta \\ y = R\sin\phi + e\sin\theta \\ z = \dfrac{T}{2\pi}\theta \end{cases} \quad (1)$$

Where ϕ represents the intersection angle of Mo_1 and x_1. θ represents the intersection angle of oo_1 and axis x.

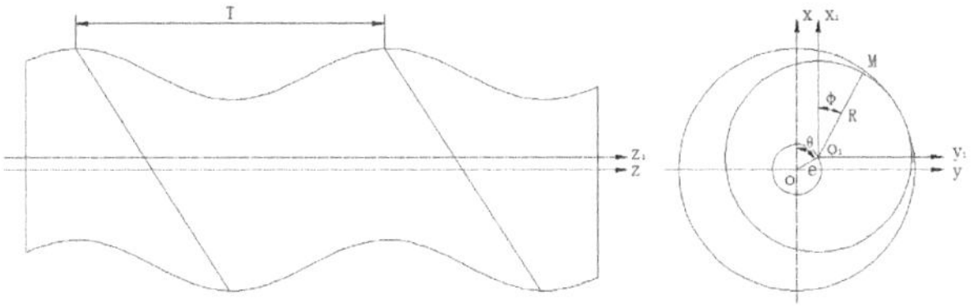

Fig. 1. The coordinate systems of eccentric screw

3.2 Coordinate Transformation of Screw and Tool

Establish coordinate system of screw and tools, coordinate system of screw is oxyz, its z axis and screw axis overlap; and coordinate system of tool is OXYZ, its z axis and tool axis overlap. Axis x in coordinate system of screw and axis X in coordinate system of tool overlap, they have same direction. The two coordinate systems in the space location are fixed, not turn with screw and tool. The shortest distance between the tool axis and the screw axis is L, the angle of two axes is α, and thus the relationship between coordinate system of screw and tool is:

$$\begin{cases} X = x - L \\ Y = y\cos\alpha + z\sin\alpha \\ Z = -y\sin\alpha + z\cos\alpha \end{cases} \quad (2)$$

3.3 Contact Conditions of Tool Rotary Surface and Screw Spiral Surface

As profile P-P parallel to the plane XOY, it perpendicular to axis of tool, and the distance of profile P-P and plane XOY is A. The intersecting line of profile P-P and tool rotary surface is a circle C, the intersecting line of profile P-P and screw spiral surface is a plane curve I, as shown in Fig.2. Because tool rotary surface and screw spiral surface is the tangent, so the circle C and the plane curve I will at some point in tangent. It can be seen, the intersection N of circle C and plane curve I is the point of tangency, namely the intersection N is a contact point on the contact line of tool rotary surface and screw spiral surface.

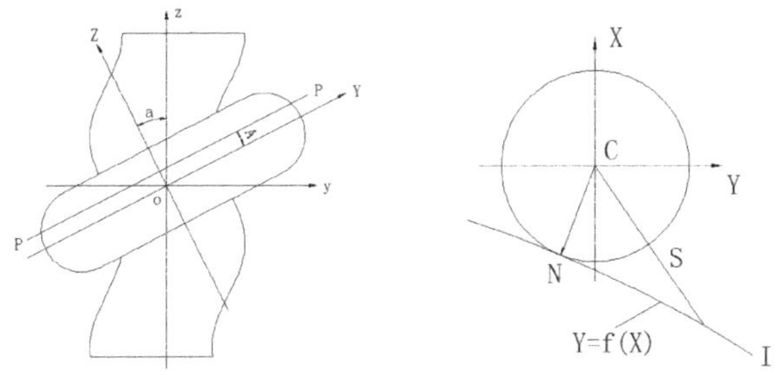

Fig. 2. The contact conditions of tool and screw

3.4 The Contact Line Equation of Tool Rotary Surface and Screw Spiral Surface

Hypothesis the space A between profile P-P and plane XOY is a constant, the third formula in formula (2) can be expressed as:

$$Z = -y\sin\alpha + z\cos\alpha = A \tag{3}$$

Hypothesis the equation of plane curve I in tool coordinate system is:

$$Y = f(X) \tag{4}$$

Hypothesis the distance from arbitrary point of the plane curve I to the axis of tool are S, S is a function of one variable X. According to the tangent conditions of the circle C and the plane curve I, for the radius of circle C is solving the minimum of function S. So to the distance S derivation and order derivatives equal to zero.

$$XdX + YdY = 0 \tag{5}$$

To get the differential of formula (3):

$$dZ = -dy\sin\alpha + dz\cos\alpha = 0 \tag{6}$$

To get the complete differential of formula (1):

$$\begin{cases} dx = -R\sin\phi d\phi - e\sin\theta d\theta \\ dy = R\cos\phi d\phi + e\cos\theta d\theta \\ dz = \dfrac{T}{2\pi}d\theta \end{cases} \tag{7}$$

Put dy and dz in formula (7) generation formula (6) to get the relation of $d\phi$ and $d\theta$:

$$d\phi = \frac{\frac{T}{2\pi}\cot\alpha - e\cos\theta}{R\cos\phi}d\theta \tag{8}$$

To get the complete differential of formula (2):

$$\begin{cases} dX = dx \\ dY = dy\cos\alpha + dz\sin\alpha \\ dZ = -dy\sin\alpha + dz\cos\alpha \end{cases} \tag{9}$$

From the formula (6) to get:

$$dy = \cot\alpha dz \tag{10}$$

Put formula (10) generation the second formula in formula (9) to get:

$$dY = (\cot\alpha\cos\alpha + \sin\alpha)dz = \frac{dz}{\sin\alpha} = \frac{T}{2\pi\sin\alpha}d\theta \tag{11}$$

Put formula (2), formula (9) and formula (11) generation formula (5) to get:

$$(x-L)(-R\sin\phi d\phi - e\sin\theta d\theta) + (y\cos\alpha + z\sin\alpha)\frac{T}{2\pi\sin\alpha}d\theta = 0 \tag{12}$$

Put formula (8) generation formula (12) to get the contact line equation of tool rotary surface and screw spiral surface:

$$(L-x)\left[\tan\phi\left(\frac{T}{2\pi}\cot\alpha - e\cos\theta\right) + e\sin\theta\right] + (y\cot\alpha + z)\frac{T}{2\pi} = 0 \tag{13}$$

In formula, L, T, e and α are constant, x, y and z are function of parametric ϕ and θ, so the contact line equation of tool rotary surface and screw spiral surface are the equation of parametric ϕ and θ.

3.5 Profile Equation of Formed Milling Cutter

From the contact line equation of tool rotary surface and screw spiral surface known, if selected a value ϕ, that is, selected a point in the cross-section shape of screw, it can work out a corresponding values θ, this means the point turned an angle θ becomes the contact point. Then put value (ϕ, θ) generation formula (1) and formula (2) to get the coordinate (X, Y, Z) of contact point in the tool coordinate system. According to the parameters scope of cross-section shape of screw spiral surface, it can get the axial cross-section shape of tool rotary surface, which is the profile equation of formed milling cutter.

3.6 Establish the Tooth Back Curve Equation

When grinding the milling cutter, the diameter of milling cutter reduced, this will cause the shortest distance L between tool axis and screw axis decrease, it can make processing error of screw spiral surface. Therefore, the profile of milling cutter should make corresponding change, establish a profile equation have constant back angle. That is establish the tooth back equation of milling cutter, make tooth back curve meet the design back angle, and after grinding the rake face of milling cutter, its back angle remain unchanged. Therefore, with spiral of Archimedes as tooth back curve.

Every point radius vector ρ of spiral of Archimedes will proportion to add or reduce follow its rotor angle φ, when $\varphi = 0°$, then $\rho = r$, r is radius of milling cutter; while $\varphi > 0°$, $\rho < r$. Therefore, normal equation of spiral of Archimedes is:

$$\rho = r - c\varphi \tag{14}$$

Where c is constant.

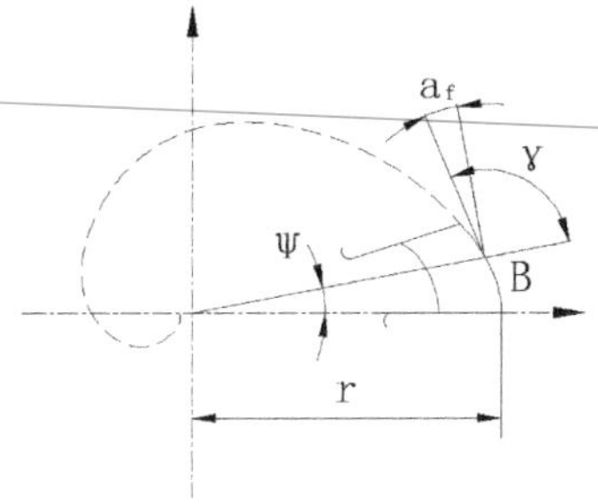

Fig. 3. Spiral of Archimedes

The included angle γ between the tangent and the radius vector which arbitrary point B of tooth back curve can be used for get:

$$\tan \gamma = \frac{\rho}{d\rho/d\varphi} = \frac{r - c\varphi}{-c} = \varphi - \frac{r}{c} \tag{15}$$

The back angle of tooth back curve is α_f, $\alpha_f = \gamma - 90°$. When $\varphi = 0°$, from formula (15) to get:

$$c = r \tan \alpha_f \tag{16}$$

The tooth back curve equation is:

$$\rho = r - \varphi r \tan \alpha_f = r(1 - \varphi \tan \alpha_f) \tag{17}$$

Finally, get profile equation of milling cutter have constant back angle:

$$\begin{cases} X = R\cos\phi + e\cos\theta - L \\ Y = (R\sin\phi + e\sin\theta)\cos\alpha + \dfrac{T}{2\pi}\theta\sin\alpha \\ Z = -(R\sin\phi + e\sin\theta)\sin\alpha + \dfrac{T}{2\pi}\theta\cos\alpha \\ -X\left[\tan\phi\left(\dfrac{T}{2\pi}\cot\alpha - e\cos\theta\right) + e\sin\theta\right] + \dfrac{T}{2\pi}\left[(R\sin\phi + e\sin\theta)\cot\alpha + \dfrac{T}{2\pi}\theta\right] = 0 \\ \rho = r(1 - \varphi\tan\alpha_f) \end{cases} \tag{18}$$

4 Interference Checking for Formed Milling Cutter of Screw

In order to ensure the correctness of design results, the processing screw was simulated by SolidWorks. The interference between the profile of milling cutter and the spiral surface of eccentric screw was observed from all angles. The results showed that the profile of milling cutter coincided with the spiral surface of eccentric screw. The simulation of interference checking was shown in Fig.4.

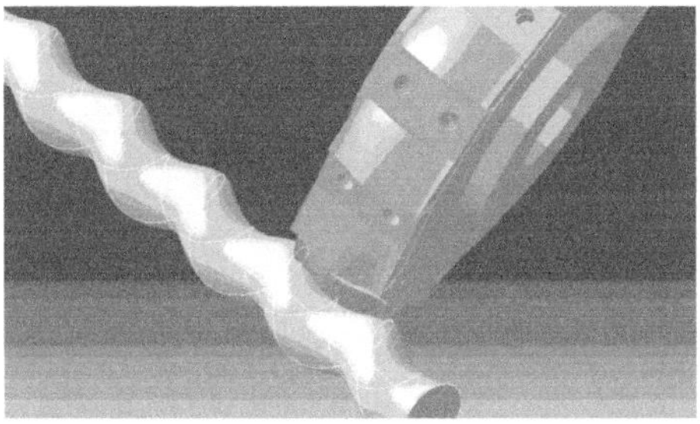

Fig. 4. The interference checking of milling cutter profile

5 Conclusions

Through analyzing the shape of screw, the spiral surface equation of screw was established; and from the relative motion between cutter and screw, according to the contact principle of cutter and screw, the meshing conditions of cutter and screw was deduced, and the contact line equation was established based on non-instantaneous envelope method. Further establish the profile equation of formed milling cutter.Analyzing the causes of processing error after wear of formed milling cutter, the tooth back curve equation was established based on spiral of Archimedes, and got the profile equation of formed milling cutter have constant back edge. Make wear of milling cutter, after grinding still has good profile, do not affect machining accuracy of eccentric screw. The correctness of milling cutter profile is ensured by interference checking for profile equation of formed milling cutter. The profile equation of formed milling cutter provides a theoretical foundation for the design and manufacture, and it would have great significance to the actual production.

Acknowledgments. This work was financed by the National Natural Science Foundation of China and Beijing Municipal Education Commission.

References

1. Mimmi, G., Pennacchi, P.: Determination of tool profile for the milling of three screw pump rotors. Meccanica 32(4) (1997)
2. Mimmi, G., Pennacchi, P.: Deviations Induced by Tool Sharpening in the Profile of Three Screw Pump Rotors. Meccanica 32, 567–576 (1997)
3. Chiang, C.-J., Fong, Z.-H.: Design of form milling cutters with multiple inserts for screw rotors. Mechanism and Machine Theory (45), 1613–1627 (2010)
4. Zheng, M., Xi, W.: A study on the shaped cutting tool's calculation method on symmetry style screw spindle. Journal of Nanjing Forestry University 20(4), 18–20 (1996)
5. Zhao, C., Liu, J.: Calculation method on the contour coordinates of moulding milling cutter in screw machining. Journal of Machine Design 25(6), 69–70 (2008)
6. Wu, Q., Rui, Z.-Y., Yang, J.-J.: Design on contour of milling cutter in machining complex spiral curved surface. Modular Machine Tool & Automatic Manufacturing Technique (7), 78–81 (2009)
7. Peng, X.-Y., Wu, H.-G., Xing, Z.-W.: Design of the hobbing tool profile for manufacturing screw rotors. Journal of Xi an Jiao Tong University 37(1), 103–104 (2003)
8. Sun, X., Zhang, P., Wang, K., Wang, J.: Optimum design of cutter parameters machining complex helical surface. Modular Machine Tool & Automatic Manufacturing Technique (2), 61–62 (2007)
9. Wang, T., Zhang, Z., Wang, T., Xu, A., Hu, S., Li, Z.: High precision tool path planning algorithm for complex parametric surface. Chinese Journal of Mechanical Engineering 43(12), 109–113 (2007)

Study and Design of an Inchworm-Like Micro-robot Walking Mechanism

Li Zu, Yi Liang, and Meili Song

School of Mechanical Engineering, Nanjing University of Science and Technology, 210094 Nangjing, China
mzmllovey@vip.sina.com

Abstract. The inchworm movement principle is studied in this paper firstly, and then by imitating the biological movement of the inchworm, the walking mechanism of inchworm-like micro-robot is studied and designed. The robot can achieve the unique rhythm of the inchworm-1ike movement based on the innovative design of the linkage mechanism. For the inchworm-like micro-robot, both walking gait and rotating gait are studied by the simulation research. The results show that the inchworm-1ike micro-robot walking mechanism is efficient, stable, and easy to be controlled.

Keywords: bionic inchworm, micro-robot, walking mechanism, design.

1 Introduction

By imitating body structure and function of biology in the nature, the biomimetic robots can be made to do the work with the biological characteristics, and the robots are expected to replace the traditional robots as well as to be the future development direction in the robotics field [1][2]. Biomimetic robots imitate the performance and behaviors of organisms. The biological structure, movement mechanism and behavior etc. are used in the robot design. So the robot system has some natural or biological function and external shape. Movement bionics is the premise of robot research and development. The key of the movement imitation lies in the modeling of the movement principle. Legged and footed robots can be used in many different areas, such as space exploration, hazardous environment work, industrial pipe inspection, medical invasive surgery etc. they have wide application potentials. One kind of bionic robot is to imitate movement principle of the inchworm, so the micro-robot can achieve the unique rhythm of the inchworm-1ike movement with the performance of small size, low power consumption, light weight and stable movement[3][4].

There are some inchworm-like movement mechanisms. Mantian Li etc. designed the mobile micro-robot adopting inchworm-like biped mechanical structure [5], and it is actuated by DC motors with four DOFs. The robot can move in two different modes of 1ocomotion such as crawling and overturn according to environments, and this robot can change its motion direction and transit between different surfaces. Dongsheng Qu etc. designed a new micro positioning bionic model based on inchworm principle [6]. It was actuated by the piezoelectric ceramic and has the

G. Lee (Ed.): Advances in Automation and Robotics, Vol.2, LNEE 123, pp. 69–78.
Springerlink.com

oriented mechanism with the flexible hinge. Based on the model the 2D micro bionic robot was realized to achieve a wide range movement and high-precision positioning. Tong Guo etc. designed an inchworm-like micro-robot used in pipe actuated by piezoelectric cymbal actuators [7]. It Can crawl into 10mm curved pipes with high positioning accuracy. The inchworm movement principle is studied in this paper, and the linkage mechanism is adopted for innovative design to achieve the unique rhythm of the robot movement like inchworm, and it is efficient, stable, and easy to be controlled.

2 Movement Principle of Inchworm

Inchworm is the common name of geometridae Larval. Inchworm moves with operation of bending and extending its body just like a bridge. As shown in the Fig.1, its forefoot and rear foot contact the surface with two attachment points while its trunk is up, and then the front foot moves towards the other attachment point while the trunk keeps close to the surface. So inchworm completes one gait and moves forward. With the cycle exercise Inchworm can perform its movement.

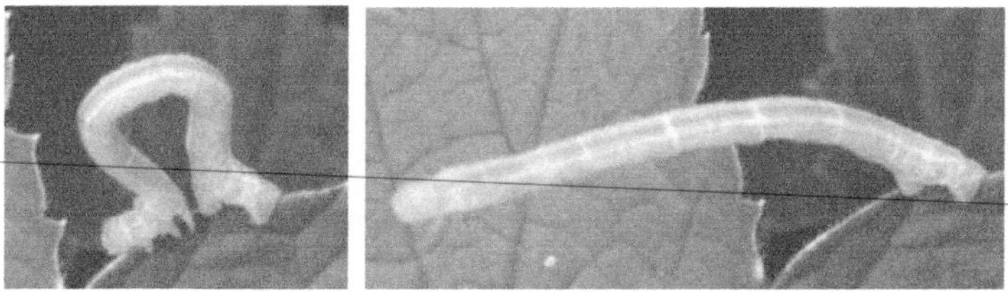

Fig. 1. Movement of inchworm

Some tentacles locating at the rear foot have strong grip to support the weight of its own, so inchworm can not only crawl on the ground but also in trees or on cliffs, it can reach a very wide field. The only defect is that the frequency and speed of its movement are too slow. In the actual design we should take full advantage of its small size, large field reach, not affected by ground conditions and so on to design a micro-robot with the fast action and steady movement.

3 Design of the Walking Mechanism

Bionic inchworm robot is one kind of Legged and footed crawling robots. The design of its structure is more crucial, and it will directly affect the robot's movement principle, walking gait and control method etc. Different structural forms have their advantages and disadvantages. The number of legs, drive and drive method, the collocation of legs, the degrees of freedom and the walking mechanism should be studied in actual design.

3.1 Number of Legs

As shown in the Table 1, for Legged and footed robot the performance of its walking mechanism has relationship with the number of legs. For achieving linear motion, legs with the even number can produce an effective gait, so the majority robots have legs with the even number. The robot can move slowly by heavy load with more legs, while legs with a small number would be more flexible. The practical legged and footed robot should be able to maintain a certain position when it pauses to complete a specific task, so at least three legs or more legs are required considering robot's ability to maintain the stable posture in any terrain. When the number of legs is 4 to 6, the static stability to walk can be greatly improved. However the number of legs increases again, the action seems to be unconspicuous. When the number of legs is greater than or equal to 3, the robot has more superior dynamic stability to walk. It is better than one leg or two legs in the view of control and application. The outstanding feature is that, even if the robot is out of balance and tumbling, the swinging legs can support the robot automatically to recover the normal position and continue moving. More legs mean that there are more degrees of freedom. For the robot designed by the linkage mechanism, it will make the drive system more complex considering the DOFs. So taking one thing with another, four legs is a better choice to satisfy requirements also with smaller size and lighter weight. In order to make the robot rotate better and faster, a rotating chassis is designed for the inchworm-like micro-robot.

Table 1. Comparisons of different number of legs

Evaluation index	One	Two	Three	Four	Five	Six	Seven	Eight
Ability to maintain stable posture	No	No	Good	Best	Best	Best	Best	Best
Static stability to walk	No	No	No	Good	Best	Best	Best	Best
Dynamic stability to walk	Yes	Yes	Best	Best	Best	Good	Good	Good
High speed stability to walk	No	No	No	Yes	Good	Best	Best	Best
Simplicity considering DOFs	Best	Best	Good	Good	Good	Yes	Yes	Yes

3.2 Walking Mechanism

The legs and feet of bionic inchworm robot designed by linkage mechanism can adopt zoom-based mechanism, four-bar linkage mechanism, articulated structure mechanism and series-wound mechanism. The gravity can be decoupled in the zoom-based mechanism, which is the significant advantage. The zoom-based Legs have the proportional feature, so the robot feet can reach longer distance with enlarging the drive extension proportionally. However the scale and weight of zoom-based mechanism will increase. For four-bar linkage mechanism, the trajectory of the link can be used as the robot feet trajectory. So the walking mechanism is simple and the motor can be set within the body, the legs of robot are correspondingly light. However this structure limits the rotation range of the joint. The articulated mechanism has compact and simple structure, larger extension and flexible movement, but it needs more degree of freedom. The series-wound mechanism can achieve the omnidirectional movement for the Legged and footed robot, but if more than four feet are designed for the robot, at least twelve degrees of freedom need to be controlled, and there are some difficulties in designing the robot control system [8]. In this paper, taking advantages of the four-bar linkage mechanism and the articulated structure mechanism, the walking mechanism of the bionic inchworm robot is designed. As Shown in Fig.2, there are two points on the inchworm robot body, point a represents the midpoint of the trunk, which is assumed to be the center of mass, point b is a point on the front foot. Assuming the point a is static, then the speed of the point b relative to the point a is shown in Fig.3 [9] [10]. By evolution the relative velocity between two points is approximately oval.

Fig. 2. Creeping posture of inchworm

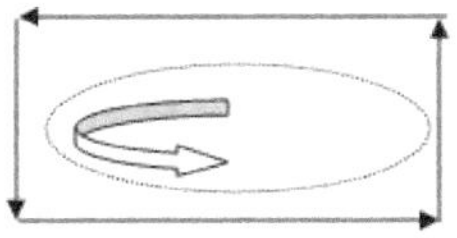

Fig. 3. Speed of the point b relative to the point a

The walking mechanism designed for the inchworm-like micro-robot should achieve the following objectives: (1) the trajectory of robot's leg end (that's the foot) should approximately be the regular arc-shaped, either oval or circle will ensure the robot to walk steadily. (2) Because the step of the bar end will affect the moving distance of the robot's barycenter, so the right walking step should be designed. (3) When the robot is walking, the ratio of time for supporting its body to robot's walk cycle is called the duty factor, forelegs and hind legs for the robot need the appropriate duty factor, and all feet's duty factor should be equal. The inchworm-like micro-robot should have the regular gait and stable movement. So the walking mechanism designed for the robot is shown in Fig.4, the symmetrical structures are designed for the forelegs and hind legs to achieve the elliptical trajectory of feet. The robot structure is compact and simple. By one motor set within the body, the robot can achieve the flexible movement and greater range of joint rotation.

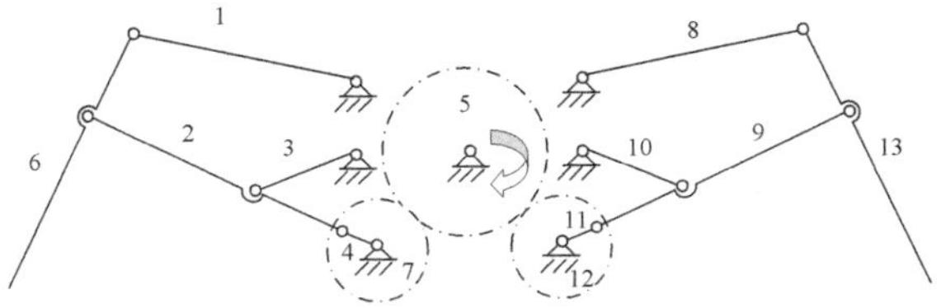

Fig. 4. Kinematic diagram of walking mechanism for the robot

In Fig.4 the long rocker for hind leg is denoted by number 1; number 2 represents the link for hind leg; number 3 represents the short rocker for hind leg; number 4 is the crank for hind leg; number 5 represents the driving gear; the hind leg is denoted by number 6; number 7 represents the driven gear for hind leg; number 8 is the long rocker for foreleg; number 9 represents the link for foreleg; number 10 is the short rocker for foreleg; the crank for foreleg is denoted by number 11; number 12 represents the driven gear for foreleg; number 13 is the foreleg.

Firstly the trajectory planning for legs is conducted, and the virtual prototype model of the inchworm-like micro-robot based on ADAMS is built. As shown in Fig.5, by imitating the biological structure, the prototype model has forelegs and hind legs; the middle part is the trunk. The link for hind leg denoted by number 2 shown in Fig.4 is designed to be 65mm long, by adapting different lengths of long rocker denoted by number 1 and short rocker denoted by number 3, the different trajectories of the leg end can be obtained as shown in Fig.6. The structure sizes for Forelegs and hind legs are same and symmetrical.

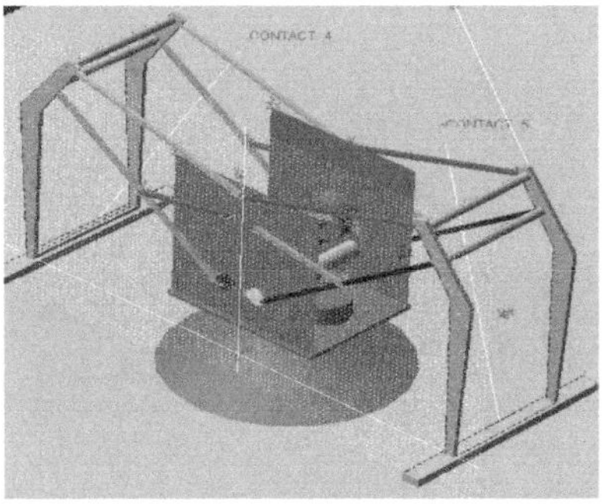

Fig. 5. Virtual prototype model of the inchworm-like micro-robot

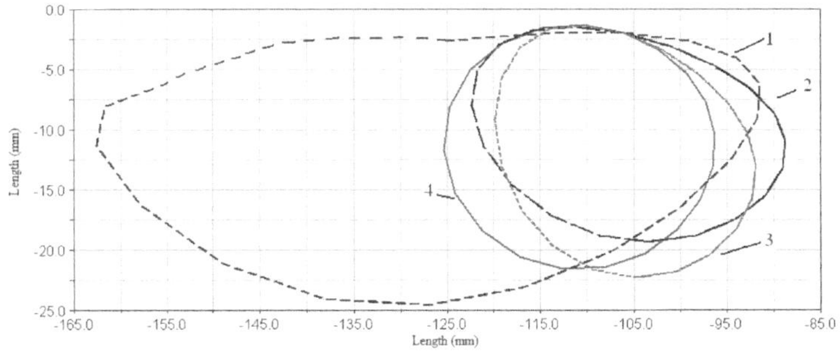

Fig. 6. Different trajectories of the leg end by adapting different lengths of long rocker 1 and short rocker 3

When long rocker 1 is equal to 51mm and short rocker 3 is equal to 15mm, the trajectory of the leg end is denoted by number 1; When long rocker 1 is equal to 48mm and short rocker 3 is equal to 27mm, the trajectory of the leg end is denoted by number 2; When long rocker 1 is equal to 42mm and short rocker 3 is equal to 24mm, the trajectory of the leg end is denoted by number 3; When long rocker 1 is equal to 51mm and short rocker 3 is equal to 24mm, the trajectory of the leg end is denoted by number 4.

As can be seen from Fig.6, for the walking mechanism of bionic inchworm, the foot trajectory varies greatly by designing different lengths of bars. Therefore to ensure the walking stability of the robot, the size of bar and its installation location should be chosen correctly. Now the length of long rocker 1 is designed to be 51mm and the length of short rocker 3 is designed to be 24mm. Trajectories of the foreleg end and hind leg end are shown in Fig.7.

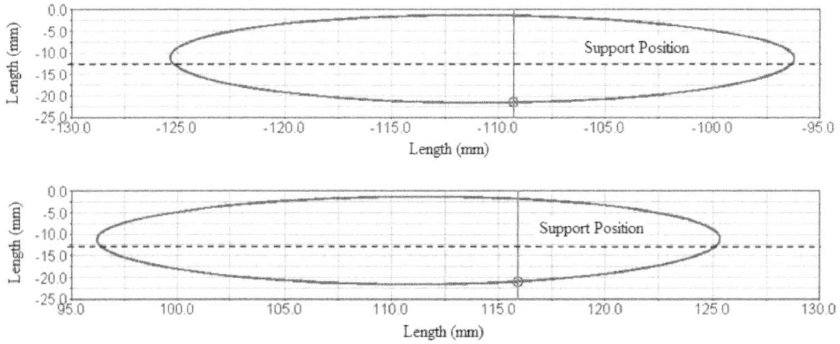

Fig. 7. Trajectories of the foreleg end and hind leg

The support position is shown as the dashed line in Fig.7, the forefoot and rear foot trajectories below the support point can carry out simultaneously. When design the walking mechanism according to the trajectory in Fig.7, the foreleg end and hind leg end, that forefoot and rear foot trajectories are similar to the oval shape. For forefoot and rear foot, the movement time below the support point and above the support point are approximately equal. So feet's duty factors are equal to ensure the inchworm-1ike robot's movement regular and stable, the design can meet the requirements.

When forefoot and rear foot trajectories are above the support point, the maximum height of leg lift is about 14mm, at this time two feet is vacant and the robot is at rest. The bionic inchworm robot's trunk supports the entire body as the support point. When leg ends are below the support point, the robot can walk by its the foreleg and hind leg. Until now the inchworm-like robot complete one gait to move one step forward, it repeats this movement and walk forward further.

4 Kinematic Analysis of the Inchworm-Like Robot

4.1 Moving in the Horizontal Plane

When the inchworm-1ike robot is moving in the horizontal plane, the displacement, velocity and acceleration in the X axis and Y axis are shown in Fig.8 and Fig.9.

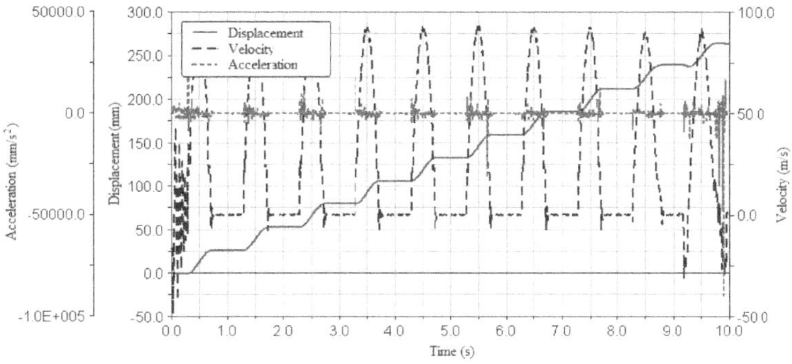

Fig. 8. Displacement, velocity and acceleration in the X axis

From Fig.8, the velocity curve in the X direction is approximate parabola. The average speed is about 50mm / s. The displacement is a ladder-shaped curve, which indicates the robot moves forward in part of time and remains still in another part of time. During the static time, the robot's trunk supports itself on the ground while robot's legs are lifting, which is similar to the movement principle of inchworm. The acceleration has small fluctuations, so the designed walking mechanism can ensure the bionic inchworm robot stable movement.

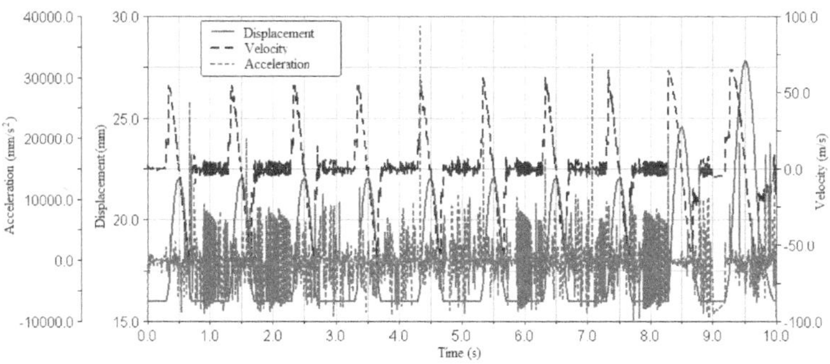

Fig. 9. Displacement, velocity and acceleration in the Y axis

From Fig.9, when the robot is moving, the velocity curve in the Y direction is symmetrical relative to the zero. Because the robot is designed by imitating the inchworm, so the movement in this direction is regularly up and down. The robot achieves the periodic biological rhythms such as the inchworm's arching -dropping - arching – dropping. There are some small speed fluctuations near the center of mass; the displacement curve and the acceleration curve have the similar law of motion. For the bionic inchworm, the amplitude and frequency of fluctuations in the Y direction are higher than that in the X direction; just by changing postures regularly the bionic inchworm robot achieves the stable movement to move forward.

Torques of rotation axis for foreleg and hind leg are measured, the measured curves are shown in Fig.10 and Fig.11. The maximum torque is about 75Nmm, which will make the robot compact and easy to be controlled.

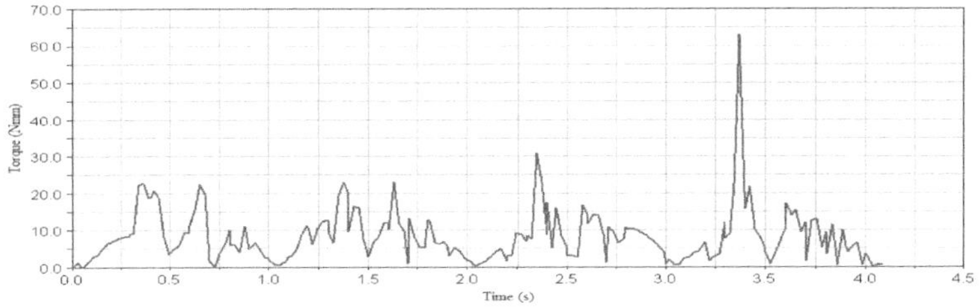

Fig. 10. Torque of rotation axis for foreleg

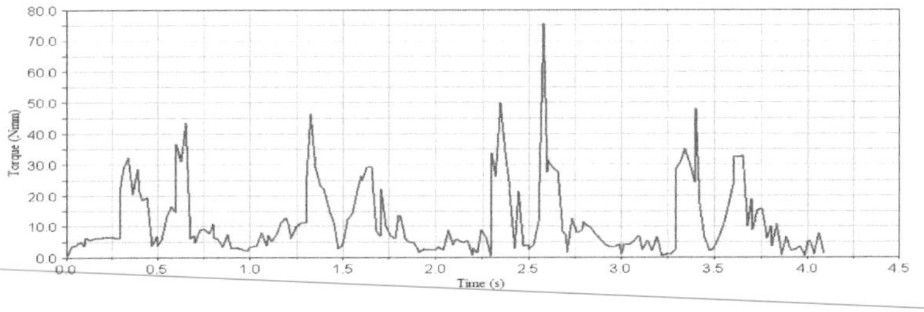

Fig. 11. Torque of rotation axis for hind leg

The walking gaits for the robot moves in the horizontal plane are shown in Fig.12. The designed mechanism makes the robot move like an inchworm, and the movement step of one cycle is 25mm.

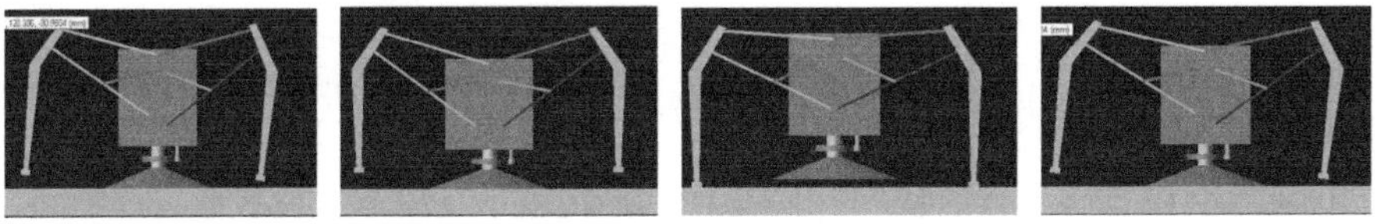

Fig. 12. Walking gaits in the horizontal plane

4.2 Moving on the Slope

The designed robot can walk stably on a 15 ° slope shown in Fig.13. From Fig.14, the required torque for the robot to climb up the slope is about 60Nmm.

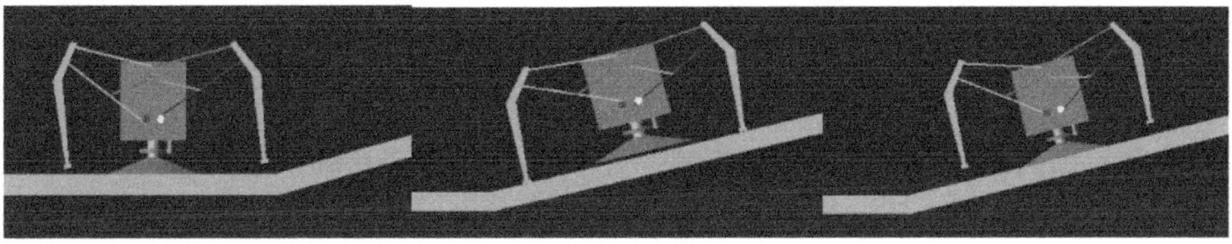

Fig. 13. Walking gaits on the slope

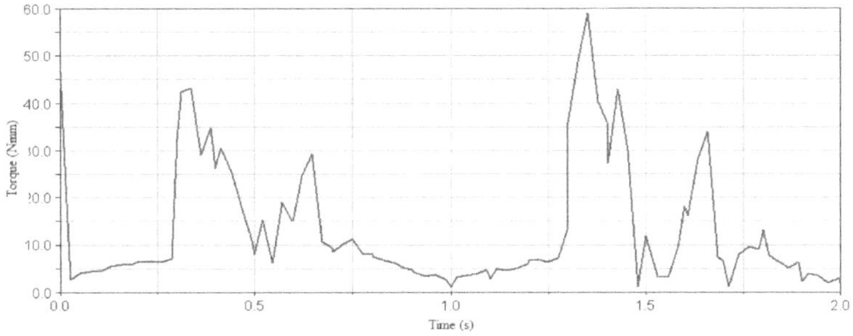

Fig. 14. Torque for climbing up the slope

4.3 Rotating Movement

The designed inchworm-like robot can rotate around itself shown in Fig.15. The robot is ready for the rotation; it is mainly supported by the trunk while its foreleg and hind leg are off the ground. Then the robot is rotating 360° driven by the rotary motor. The rotating movement is smooth and quick. Torque of rotation axis is shown in Fig.16 and the maximum torque is about 40Nmm.

Fig. 15. Rotating movement of the inchworm-like robot

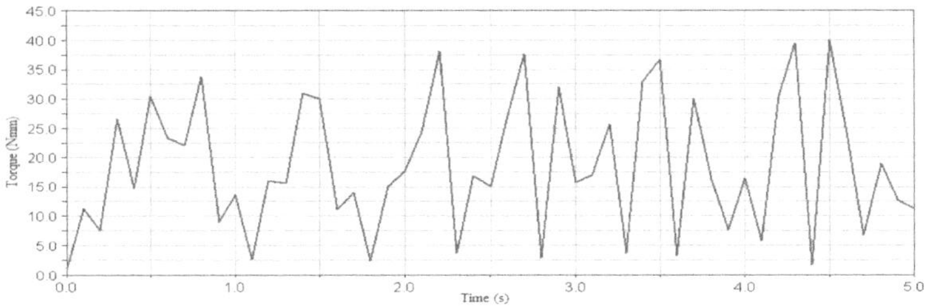

Fig. 16. Torque for rotating

5 Conclusion

In this paper, by studying the inchworm movement principle, the innovative design for inchworm-1ike micro-robot walking mechanism is completed. The robot can achieve the biological rhythmic movement just like inchworm. Some conclusions can be gotten as following by simulation study.

(1) For robot's foreleg end and hind leg end, that forefoot and rear foot, their trajectories are similar to the oval shape, furthermore their movement time below the support point and above the support point are approximately equal, which ensure the same duty factors for the walking mechanism, and the movement of the inchworm-1ike robot is regular and stable, the design can meet the requirements.

(2) The robot can achieve the periodic biological rhythms such as the inchworm's arching -dropping - arching – dropping when it is walking in the horizontal plane. There are some small speed and acceleration fluctuations near the center of mass, just by changing postures regularly the bionic inchworm robot achieves the stable movement to move forward. So the walking mechanism based on the linkage mechanism is effective and reasonable.

(3) The inchworm-1ike micro-robot can move on the slope and rotate successfully and stably.

(4) The maximum torque of rotation axis for inchworm-like robot moving in the horizontal plane is 75Nmm, for climbing up the slope is 60Nmm, and for rotating is 40Nmm, so it is convenient to choose the drive mode. The inchworm-1ike micro-robot is compact, flexible and easy to be controlled.

References

1. Zuo, J., Yan, G.: A Multi-unit Creeping Robot Based on Earthworm Principle. J. Robot 4, 320–324 (2004) (in Chinese)
2. Zhang, X., Zheng, H., Chen, K., Duan, G.: Research on Robotic Bionics. J. Robot 3, 188–192 (2006) (in Chinese)
3. Pennisi, E.: Evolution-Robot Suggests How the First Land Animals Got Walking. J. Science 5817, 1352–1352 (2007)
4. Lin, W., Yan, G., Wang, Z., Jiang, P., Liu, H.: Mechanism Design of an Intestinal Robot Based on Micro-motor. J. Robot 2, 319–323 (2011) (in Chinese)
5. Li, M., Guo, W., Sun, L.: Study of an inchworm-like micro-robot. J. High Technology Letters 21, 165–169 (2006)
6. Qu, D., Chen, L., Li, M., Qiao, S.: Design and Experiment of Micro Positioning Bionic Robot. J. Piezoelectrics and Acoustooptics 5, 557–559 (2005) (in Chinese)
7. Guo, T., Ke, Y., Li, J.: Inchworm -like in-pipe micro-robot actuated by piezoelectric slotted cymbal. J. High Technology Letters 11, 35–38 (2005) (in Chinese)
8. Bayraktaroglu, Z.Y.: Snake-like Locomotion: Experimentations with a Biologically Inspired Wheel-less Snake Robot. J. Mechanism and Machine Theory 44, 591–602 (2009)
9. Ma, G., Yan, G., Wang, K.: Design of a kind of mini creepage robot for the digestive tract of human body. J. Journal of Machine Design 7, 17–19 (2007) (in Chinese)
10. Chi, D., Yan, G.: Motion Characteristics of Earthworm Based Miniature Robotic Component. J. Chinese Journal of Scientific Instrument 5, 475–479 (2002) (in Chinese)

Movement Status Based Vision Filter for RoboCup Small-Size League

Yang Liu and Dong Liu

Vaasa University of Applied Sciences, Wolffintie 30, 65200 Vaasa, Finland
yli@puv.fi

Abstract. Small-size soccer league is a division of the RoboCup (Robot world cup) competitions. Each team uses its own designed hardware and software to compete with othersunder defined rules. There are two kinds of data which the strategy system will receive from the dedicated server, one of them is the referee commands, and the other one is vision data. However, due to the network delay and the vision noise, we have to process the data before we can actually use it. Therefore, a certain mechanism is needed in this case.Instead of using some prevalent and complex algorithms, this paper proposes to solve this problem from simple kinematics and mathematics point of view, which can be implemented effectively by hobbyists and undergraduate students. We divide this problem by the speed status and deal it in three different situations. Testing results show good performance with this algorithm and great potential in filtering vision data thus forecasting actual coordinates of tracking objects.

Keywords: RoboCup, Small-size league (SSL), Vision filter.

1 Introduction

RoboCupis one of the highest level internationalevents to advance robotics and AI research through friendly competitions.RoboCupSmall-size league (SSL) [1] has become one of most challenging leagues in RoboCup competitions from the need of both own designed hardware and software, its highly dynamic nature and team cooperation strategy of multiple robots under unforeseen situations.

Figure 1 illustrates the structure of a typical SSL system used in the match. The vision server (SSL-Vision) and the referee system are set up by the organizing committee so we do not have much to do with them. The strategy receives and processes the data from the aforesaid systems then sends the commands to the robots via wireless communication.

The SSL-Vision software [2], which has been developed by volunteers, analyses the captured live video from the cameras, and calculates the coordinates of the robots and the ball on the field, then encapsulates them via Google's protocol buffers [3] and transmits the package via a UDP multicast address.Unfortunately, there are different kinds of noise existing in nature that influence the data collection. On the other hand, when the data arrive at our strategy system, they are already obsolete because of the delay. Therefore, we need a certain mechanism to filter out the noise, while predicting the current coordinate of the target.

G. Lee (Ed.): Advances in Automation and Robotics, Vol.2, LNEE 123, pp. 79–86.
Springerlink.com

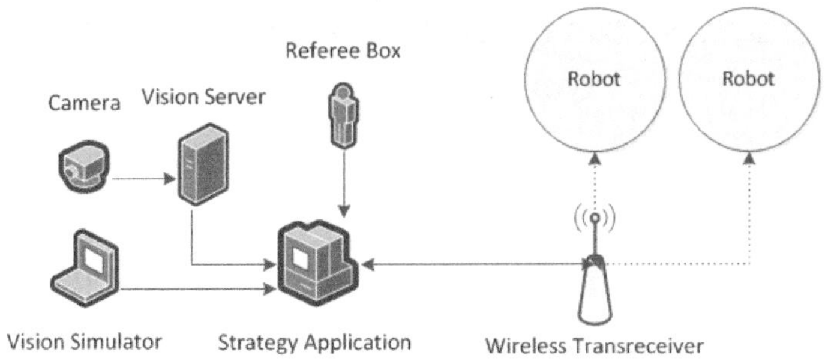

Fig. 1. Typical system architecture of a robot soccer SSL team

2 Visionfilter

2.1 Movement Status Based Filter (MSBF)

Any objects on the field have two states: motionless state and moving state. Thus, the filter is designed to have two different sets of algorithms to deal with these states. Besides, due to the existence of noise and nature resistance, an object might change its state at any time, and its next location can be largely varied, so an intermediate state is designed to deal with such situation, which is named "prediction state".

MSBF is 1-dimensional filter, so it can be applied to any kind of objects in any dimension space if they follow the Newton's laws of motion. For example, assume the filter is f, the coordinate of a robot is (x, y), so we can filter it as (f(x), f(y)); for a ball whose location is (x, y, z), we can filter it as (f(x), f(y), f(z)).

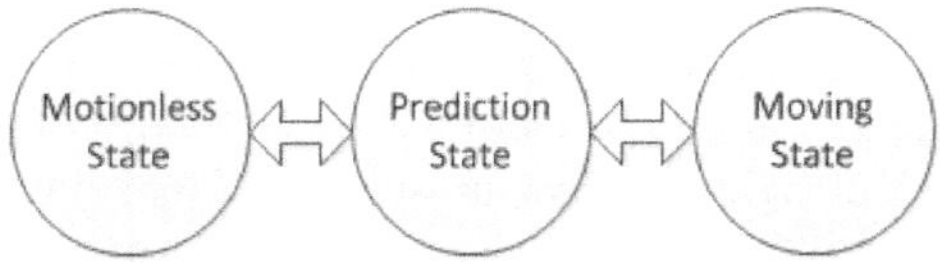

Fig. 2. Overall mechanism of MSBF

2.2 Value Buffers and Control Variables

To store the receiving data, some list containers are initialized inside the constructor function of the filter. There are also two pointers, named "Primary List Pointer" (PLP) and "Secondary List Pointer" (SLP), are used to control the data flow.The container at which the PLP points, is defined as the primary list. The other list will be called the secondary list.

Together with the lists, there are also some other variables which control the mechanisms of the filter:

(1) Minimum Moving Speed (MMS): A threshold value of speed defined as the boundary between motionless state and moving state, which means the object will be regarded as "motionless" if its speed is less than the threshold, even it is still moving.
(2) Delay: The time elapsed when the next piece of data comes. This value can be calculated using the system ticks via "GetTickCount" function.
(3) Speed Limit (SL): This value is calculated using the current speed (v) of the object, the maximum possible acceleration (a) andthe Delay (t)
The formula is as below:

$$SL = v + at \tag{1}$$

(4) State Hold Tolerance (SHT): This value indicates how many continuous values, which might cause a change of the state thatwe shouldignore before changing the state.The bigger default value the SHT is, the better resistance the filter will have against noise. However, its response towards real state changes will be slower.
(5) Status Indicator (SI): Stores the current state.
(6) Buffer Size Limit (BSL): Limits the length of the buffers.
(7) Low Speed Moving Threshold (LSMT): Will be described later.

Furthermore, there are two additional lists used to store speed and acceleration.

2.3 Base Mechanisms and Prediction State

The initial state of the filter is motionless mode. If the primary list is empty, any incoming data will be put into the primary list.If the primary list is not empty, an analyze process will begin. For example, we are currently at motionless state. However when we receive the next value and compare it to the last value of the primary list, we find that the speed is above the MMS. In this case, we will put this value to the secondary list and reduce the SHT;otherwise, we will put the value into the primary list and set the SHT to its original value and clear the secondary list.

When the value of the SHT is less than its default value, the filter will enter a phase called "prediction state". The algorithm in this state is decided by the previous state. However, The SI will not be changed in this state.

When the SHT value is below zero, the state is changed by following steps:

(1) Change the SI;
(2) Clear the primary list and restore SHT to its default value;
(3) Exchange the address stored in the PLP and SLP.

In common situation, the SI will change from either motionless to moving, or from moving to motionless. But there is a special case. In some situation, the speed might exceed the value of SL, which will be considered as noise under normal circumstances. However we will still store them in the secondary list and reduce the SHT. In this case, when the SHT reaches 0, all the procedures above will be executed, but the SI will remain at moving state.After the analyzing process, the filter will calculate the output based on the SI, which will be described in the following chapters.

2.4 Moving State

In the moving state, the Newton's law will be used to predict the output value of the filter:

$$\mathrm{ds} = v_0\Delta t + \frac{1}{2}a(\Delta t)^2 \tag{2}$$

In the formula, "v" is the speed of the object, "a" is the acceleration of the object, "ds" is the correction value which should be applied to the received value, and "□t" here is the delay in seconds.So the output value of the filter should be the last value in the primary list, plus the correction value.

For example, in the moving state the original last value in the primary list is 1, and then a new value 3 comes. Since 3 – 1 = 2 is greater than the MMS and less than the SL, so the speed (v) will be 2; assume the previous speed value is 1 and the delay is 0.1s, so the acceleration will be 2 – 1 = 1.

Thus, the "ds" can be calculates as

$$\mathrm{ds} = 2 * 0.1 + 1 * (0.1)2/2 = 0.2 + 0.005 = 0.205 \tag{3}$$

And the output value is

$$\mathrm{Output} = 3 + \mathrm{ds} = 3 + 0.205 = 3.205 \tag{4}$$

2.5 Motionless State

In the motionless state, the average value of all the elements in the primary list will be used as an output value.

Assume the length of the primary list is N, "a" is the value of the PLP, "k" is the offset of the PLP, and the output value in the motionless state can be represented as:

$$\mathrm{Output} = \sum_{k=0}^{N} a_k \tag{5}$$

2.6 Dynamic Buffer Length Adjusting (DBLA)

As mentioned above, if the speed of the object is less than MMS, it will be considered as motionless even it is moving.Figure 3 is the simulation in the motionless state. The blue line is the raw value we get from the camera, while the red line is the output value of the filter; the time is used as the horizontal ordinate.From Figure 3, we can observe that in the case of low speed movement, the output value is always fall behind. That is because of the old value in the filter, which performed a negative impact on the output.Lessening the MMS seems can be a solution. However this is not possible. Since the existence of the noise, a too small MMS will be easily exceeded by the noise, which makes the rapid status change between the motionless and the movement states. As a result, this will make the output unstable.

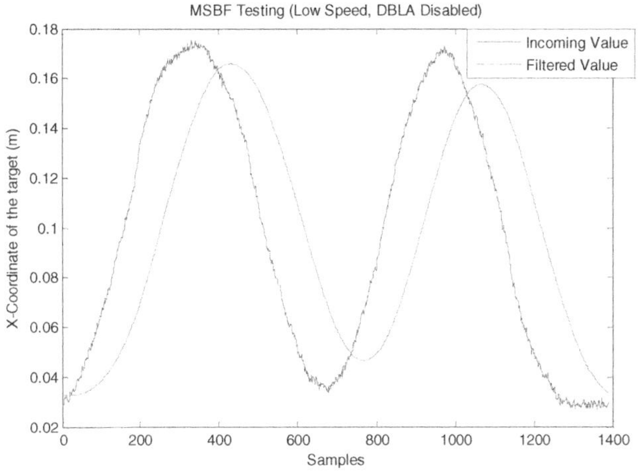

Fig. 3. Low speed object moving test

So a dynamic algorithm has been developed as a countermeasure of this situation. In this algorithm, LSMT is introduced.

First of all, assume that the ball will move slowly for a long time so that the SI will remain the same.When a new value comes, we will compare the value and the last output value of the filter by doing subtraction. If the difference is above 0, a counter will increase by 1; if the difference is below 0, this counter will decrease by 1.And if the value of the counter is bigger than LSMT, the size of the filter will be divided by 2, with first half of the elements abandoned.

2.7 MSBF vs. EKF

In Botnia SSL team [4], an Extended Kalman Filter (EKF) is implemented based on the MATLAB simulation at first when such filtering requirement is raised. However, based on its complexity and flexibility, this algorithm had been abandoned at last.

The reason why we decide to design MSBF to replace EKF is because:

(1) EKF considers every properties of an object into a unique framework. So its portability becomes a big problem when we applying it to different objects.

For instance, each robot have its own rotation value, which should be calculated in the filter, however, the ball's rotation do not need to be considered. On the other hand, the height of the ball should not be processed in the filter but the robots. Therefore, we have to implement different filters to both robots and balls.

(2) The value of the parameters of the EKF, have to be manually adjusted. As for each environment, the best parameters might vary, so the optimizing process will be time-consuming.

(3) The programmer must have a solid mathematics background with a good understanding of the robotics, which is difficult to achieve for beginners.

The design of MSBF solved the issues above since:

(1) It can be applied to any objects by disassembling its location to 2D / 3D space without change.
(2) The filtercan adjust its parameters by itself. Since the filter only take the internal timestamps and the location difference into concern, so it is basically immune to the interference caused by the environment.
(3) Effective but easy to implement. The mechanism can be understood by even high school students, which is useful for popularization ofsuch project.

3 Testing

3.1 Method

In the test, we apply this filter to the x-axis of a ball in the vision simulator and draw the graph on the user interface.In the 2-dimensional graph, the x-coordinates of the ball are presented in the y-axis of the graph and time is the x-axis.The blue line in the graph illustrates the raw value which we get from the vision server / simulator; the red line will indicates the output value from the filter.

3.2 Static Objecttest

First of all, a static ball is used to test the filter. Although a lot of noise has been introduced, the output is steady even the noise is bigger at some point, showing a very good resistance to noise.

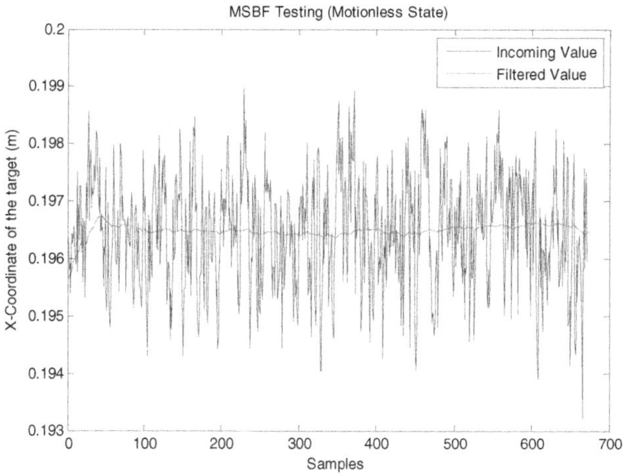

Fig. 4. Motionless state testing

3.3 Low Speed Moving Test

After the implementation of the dynamic buffer, the low speed moving test is done again using the same parameters and a similar moving track. From the result (Figure 5), we can observe that the latency time between the output and the input in this situation have been reduced by about 70%.

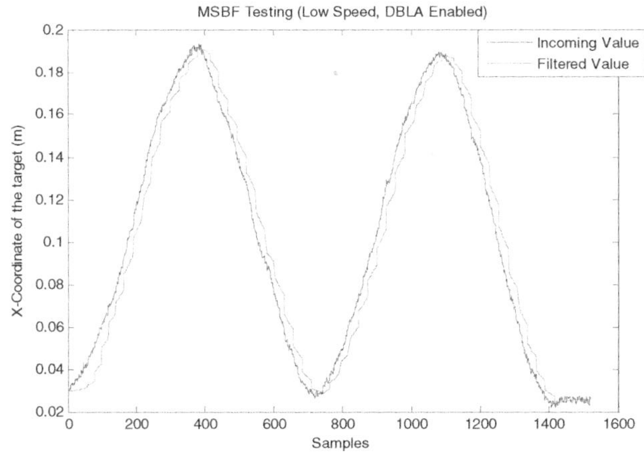

Fig. 5. Low speed moving test with DBLA implemented

3.4 Tests on the Moving Object

Also, a test on the moving object had been performed.In this test, since we need to compare the previous predicted value and the current input value, thus we deleted the first element in the raw value matrix before plotting in the MATLAB, which aligned the N_{th} value in the predict value matrix, with the $(N+1)_{th}$ value in the raw value matrix.The plotting is shown in Figure 6, and we can observe that the filtered value well presented the next incoming value.

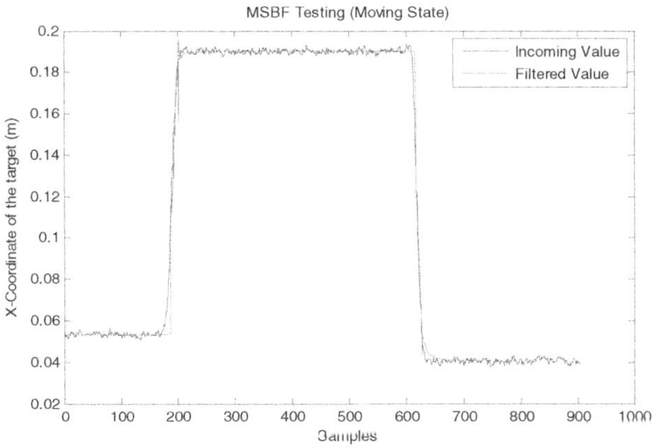

Fig. 6. Moving object test (with noise)

A more detailed test is performed without noise. The application will log the input, output and the error between the current input value and the previous predicted value.Below are some sample values.

```
In = 1.53351 out = 1.52426 error = -0.000588413
In = 1.51447 out = 1.48462 error = -0.00979524
In = 1.50494 out = 1.50711 error = 0.0203153
In = 1.49542 out = 1.48563 error = -0.011693
In = 1.48589 out = 1.47664 error = 0.000261497
In = 1.47637 out = 1.46677 error = -0.000274773
In = 1.46684 out = 1.45649 error = 7.27231e-005
In = 1.44779 out = 1.42013 error = -0.0086988
In = 1.42874 out = 1.4085 error = 0.00861349
In = 1.41922 out = 1.4205 error = 0.0107177
In = 1.40969 out = 1.40019 error = -0.0108112
In = 1.40017 out = 1.39032 error = -2.8072e-005
In = 1.38112 out = 1.35281 error = -0.00920502
In = 1.37159 out = 1.36975 error = 0.0187788
In = 1.36207 out = 1.35357 error = -0.00768522
In = 1.34302 out = 1.31442 error = -0.0105539
```

In the test, the maximum error is about 4cm and the average error is about 1cm.

4 Conclusion and Future Work

From the test, we can see that the MSBF is an effective but easy-to-implement filter, which needs only some basic physics and mathematics knowledge, which is suitable for undergraduate education and hobby robot soccer teams.

Refer to one of its purpose: "all the parameters should be automatically adjusted by the filter algorithm itself", there are plenty of stuff which should be improved in the future work. For example, the MMS and LSMT are using fixed value at the moment, and the LSMT should use some ratio based on the length of the buffer. On the other hand, the algorithm of low speed moving object can be changed to a better one.

Another issue which will probably become the primary future work is the algorithm of moving state which needs to be made better. A third buffer is planned to deal with the unstable output when great changes happen toacceleration, which can be found in Figure 6.

The most important issue is, the algorithm is not yet used in any real competition, thus its full potential have not been reached. We believe that it could perform better if tuned with more experiences.

References

1. Small Size Robot League, `http://small-size.informatik.uni-bremen.de/`
2. SSL-Vision (Carnegie Mellon University), `http://code.google.com/p/ssl-vision/`
3. Protocol Buffers - Google's data interchange format, `http://code.google.com/p/protobuf/`
4. PUV Robotics (Vaasanammattikorkeakoulu), `http://robotics.puv.fi/`

Study of Automatic Separation System of Coal and Gangue by IR Image Recognition Technology

Jie Xu and Fengwen Wang

Automation Engineering Department, Northeastern University at Qinhuangdao, Qinhuangdao, China
xujieandonion@yahoo.cn, wfw0335@126.com

Abstract. Separation of coal and gangue is an important procedure in coal production. Present separating technology has such defects as low efficiency, high pollution, high energy consumption, and high cost. Based on the analysis of grey value of digital images, we prompt a new separation method using image segment. Optimizing ordinary illumination, infrared light is used in lighting part. Algorithm is designed in the basis of histogram character of coal and gangue. Automatic recognition is realized to the segment of coal and gangue, which reach the aim of lower energy consumption and reduce pollutant discharge.

Keywords: coal gangue, automatic separation, grey value, infrared image, energy saving.

1 Introduction

Coal is the main energy source of China at present and in the future, with a yearly output of 30 billion tons, in which contains up to 20% gangue, the byproduct. If the gangue would not be separated in the production procedure, the quality of the coal cannot be guaranteed, and the environment pollution will be hardened and transport capacity will be wasted.

There are two main separation methods, i.e., wet separation and dry separation. Wet separation has the character of complicated procedure, serious water pollution, high production consumption. Dry separation method has such shortcoming as low efficiency, at the same time serious noise and dust pollution. The advanced separation methods include natural γ ray, artificial γ ray, and mechanical vibration. All the mentioned methods have not been widely used in field condition.

A new separation method based on discriminating different grey level between coal and gangue. Infrared image is first obtained by CCD camera. Then the image is analyzed and separated by morphology. The separation method can save energy and reduce pollutants discharge and reduce transportation cost.

2 Grey Level Analysis of Coal and Gangue Image

Generally speaking, the color of coal is deeper than that of gangue, and their constitution has severe differences, which may be seen as the difference in grey level

G. Lee (Ed.): Advances in Automation and Robotics, Vol.2, LNEE 123, pp. 87–92.
Springerlink.com

and textures in image. Under the light, the coal part reflects light and light points can be seen in image whereas the image of gangue is generally medium grey.

Fig.1 is images of coals with different qualities under the infrared light, aligned by carbon content from high to low.

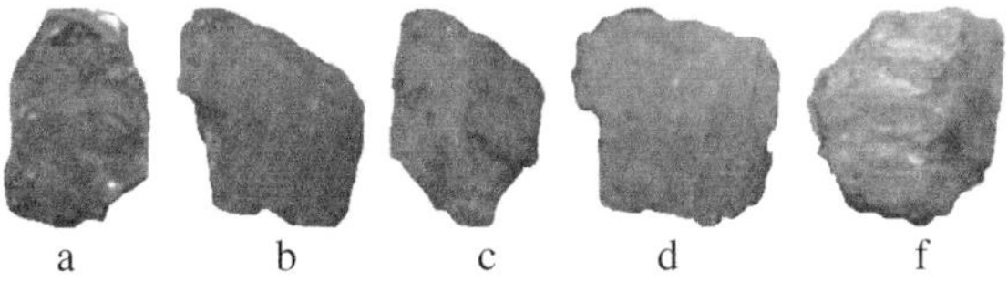

Fig. 1. Images of different quality coal under infrared light illumination

Fig.2 is images of gangue with different qualities under the infrared light, aligned by carbon content from low to high.

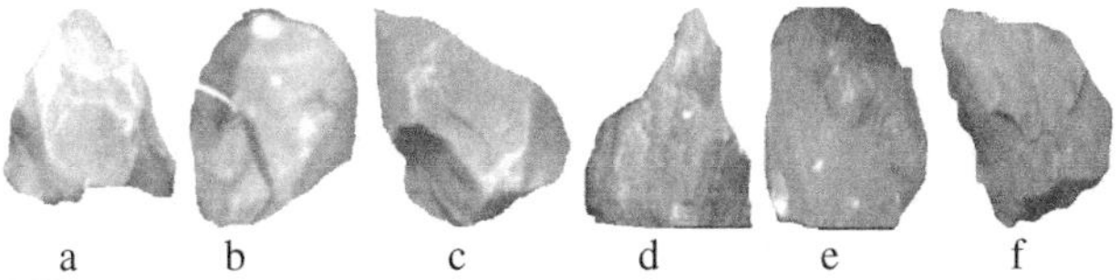

Fig. 2. Images of different quality gangue under infrared light illumination

Fig.3 is the grey level histogram of coal with different qualities processed by MATLAB.

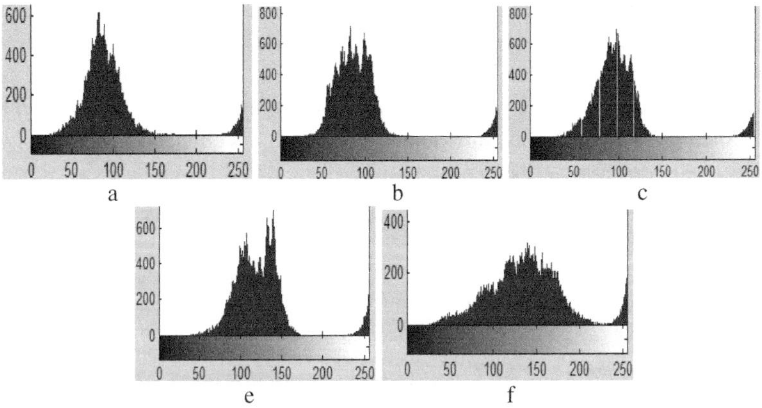

Fig. 3. Grey value histograms of different quality coal

From the above histograms, it can be observed that the grey level of coal mainly centralized from 80 to 130. Coals with different qualities have different grey level distribution and various peak values. The more carbon content it contains, the lower value it has; and vice versa. For example, the histogram of image Fig.1 (a) reaches its peak value at 80, while the peak value of Fig.1 (e) is 130.

Fig 4 is grey level histograms of gangue with different qualities processed by MATLAB language.

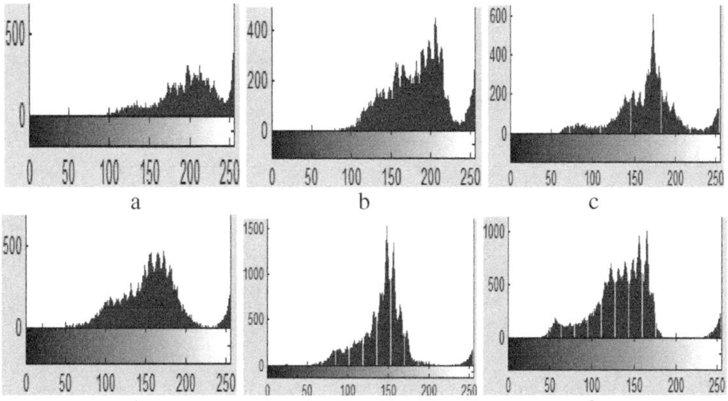

Fig. 4. Grey value histograms of different quality gangue

It can be concluded that grey level of waste grey images are mainly limited between 150 and 200. Different quality gangues have different grey level distribution, especially have different peak values. The more carbon content the gangues have, the low value they have; and vice versa. For example, the histogram of image Fig.4 (f) has the peak value of 150, while the peak value of Fig.4 (f) is 200.

Conclusion can be obtained that there are obvious grey level difference between coal and gangue.

3 Image Recognition Principle of Coal and Gangue

The main differences of digital image of coal and gangue are different probability distribution, especially grey level and peak value. Thus we can separate coal and gangue by above mentioned characters.

Based on this principle, grey levels of coal and gangue samples are stored in the Industrial Personal Computer as standards. Then real values of coal and gangue are compared with the standard values, if the differences are small than presetting threshold, the object can be classified into certain group.

At beginning, normal incandescent lamp is used as light source. In practice, it does not provide satisfied classification results, because incandescent lamp has disadvantages of divergence, light emitting and various light wavelength, etc., which leads up to the problems like the reflecting light from the conveyor was so strong to segment. This makes the images of coal and gangue blurred. Problems such as strong reflecting light, blunt colour contrasting exist, which may influence the efficient recognition effects.

By consulting technical literature, infrared light is used as light source to avoid problems like reflecting light. To decrease influence from nature light, a dark box is set on the conveyor. Adjustable infrared LED array and IR camera are installed in the dark box to acquire infrared image. Furthermore, in order to adapt to different coal and gangue from different region and overcome multiplicity of work field environment, light intensity of IR LED is adjustable. In practical application, best discrimination result can be obtained by adjusting light intensity.

4 Recognition of Coal and Gangue Images

The digital images will be contaminated by noise in the process of recording and transferring. Noise can influence the grey values and grey scale characters. The main rule for discriminating coal and gangue we based is the difference of grey values. In order to get distinct images, sequence processes are adapted which include image smooth, image segment, morphological tab, and character extraction.

4.1 Image Smooth

Gaussian low pass filter is used in image smooth process aim to remove noise. After filtering, high frequency content is eliminated and low frequency content is retained.

4.2 Image Segment

The histogram difference of grey level is the most obvious character. Coal and gangue have different colour in vision, which gives the result of different grey value in image. As a result grey level threshold can be used to segment coal and gangue. Histogram analysis is done on pre-processed image, every grey value are compared to the preset threshold value. The pixels that large than threshold value are set white, and the pixels small than threshold value are set black. After the processing, new binary value images are constructed. OSTU method is selected as the segment algorithm. The main character of OSTU method is automatic threshold value can be obtained using maximum variance between classes.

4.3 Morphological Filter

The original images must have noise. So the binary image have rough boundary. Object areas have some noise hole after threshold process. Morphological algorithm can eliminate the noise induce small area, such as corrosion and dilation. Sometimes several pair (corrosion and dilation) operations should be used to get better result.

Based on the experimental results, two pair morphological operations are done on the threshold images. The morphological operations have the following effects: isolated points and burrs are eliminated, holes in object surface are filled, and general position and shape of object remained.

4.4 Image Labeling and Character Extraction

We do image labelling using function bwlabel in MATLAB in this formulation: [L, num]=bwlabel (BW, 8). The separated area of coal and gangue are transformed to region. Then we use STATS=region props(L, properties) statement to calculate individual area and centroid of coal and gangue. The areas of object are the pixels number the object possesses.

When area are determined, if the area are smaller than specific value, we make conclusion that there are gangue in coal but the percentage is a low value or wrong judgment because of reflected light. The coal is passed. Have the center of gangue been calculated, the distance between coal and gangue is computed.

Fig.5 are typical images of original image, Gaussian smoothed image, threshold segmented image, and morphological filtered image from (a) to (d).

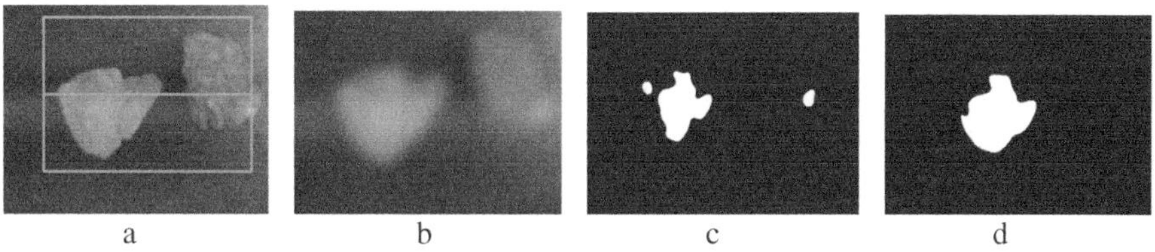

Fig. 5. Infrared images processing result of coal and gangue

5 Conclusions

In this paper, recognition method based on different grey value of coal and gangue is presented. Coal and gangue have different grey value histogram, which provide recognition basis of them. In order to level up recognition success ratio, infrared light is optimized as light source. A lot of experiments have been done to test the validation of mentioned recognition method and statistic data are obtained. In the in situ measure condition recognition success rate can reach 95%, and in line measure condition right recognition ratio can reach 90%. The right recognition ratio is a rather high level.

The colour difference between coal and gangue is just a phenomenon; the essential distinction between them is specific gravity. In order to further increase the segment, X-RAY imaging could be considered.

References

1. Pei, X., Zhang, R., Du, G., Tan, Y.: A synthetical study on application technology of gangue. Coal Mine Security 9, 99–101 (2008)
2. Lan, T., Zheng, H.: Research on separation technology of coal and gangue based on texture features fusion. Journal of Longyan University 26(6), 56–59 (2008)
3. Heng, D., Wang, F., Miao, L.: Recognition of coal and gangue based on coal transport volume control system. Opencast Mining Technology (5), 47–49 (2008)

4. Hobson, D.M., Carter, R.M., Yan, Y., Lv, Z.: Differentiation between Coal and Stone through Image Analysis of Texture Features. Imaging Systems and Techniques, IST 2007 (2007)
5. Li, K., Zheng, C., Zhong, Z., Cheng, J.: Study on dynamical recognition model of coal and gange using dual energy. Coal Preparation Technology (1), 9–12 (2001)
6. Wang, X., Wu, C.: System analysis and design based on MATLAB-digital image processing. Xidian university press, Xi'an (2009)
7. Teague, M.R.: Image analysis via the general theory of moment. Journal of Optical Society of America 70(8), 920–930 (1980)
8. Jiang, X.: Digital image pattern recognition engineering software design. China waterpower press, Beijing (2008)
9. Zhou, M., Weng, L., Feng, G.: The Research of a New Computer Control System of Filtrating of Coal and Rock (1), 45–46 (2002)

Research and Applications on Virtual-Hand Technology

XiaoGang Wang, YueWei Bai, and Kai Liu

Mechanics & Electronic Engineering Faculty, Shanghai Second Polytechnic University
Shanghai, China
victor.hust@gmail.com

Abstract. Virtual-hand operation techniques which combined virtual reality with kinematics will be used broadly in the future. In this paper, the collision detection theory in virtual scene about virtual-hand, and the rules of virtual-hand stably grasping a virtual object based on those theories are discussed. At last, a series of virtual operations of virtual-hand grasping virtual objects are achieved, using data gloves and motion tracker as input devices and based on Open Inventor.

Keywords: virtual-hand, collision detection, grasping rules, oriented bounding box (OBB).

1 Introduction

With the swift development of computer technology especially the development of virtual reality, the virtual reality (VR) technology is widely used in many fields, e.g. virtual assembly, virtual maintenance, virtual manufacturing and virtual reality teaching. As hands play an important role in everyday life, researchers pay more and more attention to virtual-hand operating technology and implementation methods for virtual-hand are presented. There are two key technologies in implementation of virtual-hand. One is collision detection and the other one is grasping rules. But as the existing methods for these two key factors are either too simple or too complicated, in some cases, the methods may cause the results unacceptable and impractical. Therefore methods with effective and accurate, and easy to use for collision detection and grasping rules are sorely needed.

A new method for collision detection based on OBB is proposed. With this method, collision detection of virtual-hand with other virtual objects in virtual environment under the platform of Open Inventor is achieved fast and accurate. The rational rules for grasping not only get over the disadvantages of virtual contact force model not being able to well match the real-time requirements, but also overcome the previous grasping rules are too simple and often grasper the wrong objects or miss the true objects.

G. Lee (Ed.): Advances in Automation and Robotics, Vol.2, LNEE 123, pp. 93–98.
Springerlink.com

2 Collision Detection and Grasping Rules

2.1 Collision Detection Methods Based on OBB

In the virtual hand operation, the virtual hand and virtual objects, virtual objects and virtual environments and other virtual objects, contact and collision may often occurs. The virtual hand operation system based on 3D model must be able to detect such interaction between virtual objects and collision, otherwise the unreal situation in real world of one object passing through the other object or overlapping may be often found. The collision detection has been widely studied and two collision detection algorithms based on geometric model can be divided into two categories: space decomposition and hierarchical bounding box method. The Bounding Box Structure represents the space occupied by a box. The bounding box class is axis aligned. Each face of the bounding box is perpendicular to the x-axis, the y-axis, or the z-axis.

There are several benefits of using the bounding box for collision detection. The bounding box class fits rectangular shapes aligned with the axis very well. Compared to the bounding sphere class, the bounding box class provides a much tighter fit for non-rotated rectangular objects. Because the bounding box class is axis aligned, you can make certain assumptions that result in collision checks between bounding boxes being quicker than a bounding box that can be rotated. Rotating a bounding box causes it to no longer be axis aligned. Because of this, if you rotate a model being bounded, you will need to recreate the bounding box. Doing so can be slow, since all the points in an object are iterated through to get the bounding box. If the model has not changed orientation, you can translate the bounding box instead of recreating it.

2.2 Grasping Rules for Virtual-Hand

Rules of virtual hand grasping operations in the virtual system cannot be too simple or too cumbersome. If they are too simple, error rate of wrong grasping operation will increase, and the hand posture of the virtual hand in the virtual environment is visually not appealing. On the other hand, if the rules is too cumbersome, the time-consuming calculation is too large and results of lagging is serious. This paper presets a set of virtual hand grasping rules based on modified point-contact normal vectors technology. The rules are described as follows.

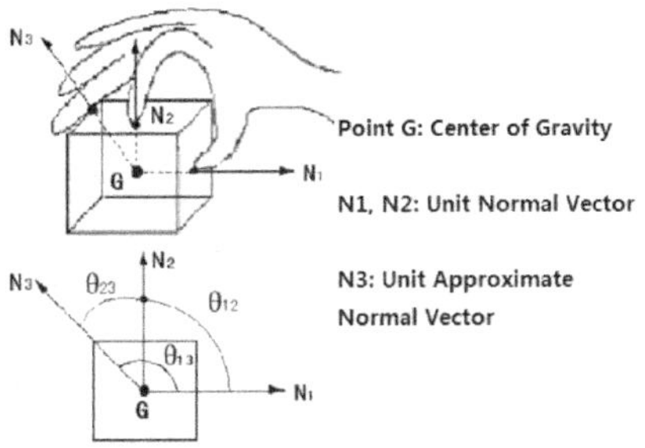

Fig. 1. Criterion of steadily grasping

Rule 1: must have three or more fingers (fingertip or palm) in contact with the object, including at least three contact points are not at the same straight line;

Rule 2: angles of the normal vector of any two contact surfaces (flat or curved) should have at least one or more values that are bigger than the critical angle of a predefined value(here we set this angle is 90 degree).

If the condition satisfies two conditions of rule 1 and 2, the objects could be considered as grasped. As shown in Fig.1, there are three finger tips in contact with the object, and the three contact points are not at the same straight line (satisfying rule 1), meanwhile θ13 of the angles of θ_{12}, θ_{13} and θ_{23} formed by normal vectors N_1, N_2 and N_3 is greater than 90 degree which matches the 2nd rule. So the object in the figure is considered as grasped. These rules have the characteristics of accurately grasping, adapting to a wide range conditions, fast speed determination and stability.

3 Implementation of Virtual-Hand Operation Technology

By solving the two key problems of collision detection and rules for Steady grasping objects for virtual-hand, operators can complete some virtual operations to exchange information of real conditions with virtual environment. The operation process is shown as figure 2 and includes steps as follows.

Step 1: a controller issues the proper motion commands according to the operational task needed to operate.

Step 2: Position trackers estimates the position of the target object (one or two hands in reality), and the data glove gets the primitive data of different finger joints. By using the RS-232 COM interface, these data will be fast transmitted into the virtual operating environment.

Step 3: When getting the primitive data, computer calculates the moving or grasping dynamic parameters. This will update the position or hand shapes of the virtual hand in the virtual environment.

Step 4: collision detection between the virtual hand and the virtual targets. If there is no collision, the procedure comes into the feedback step. Otherwise, the procedure will check the hand and targets obey the rule of Stable grasping conditions.

Step 5: If the virtual hand and virtual objects Satisfy the rules of Stable grasping conditions, then: affixing the coordinate of virtual objects with the virtual hand that means the objects have been grasped by the virtual hand; b. the objects will rotate or translate with the virtual hand; c. results of virtual operations will be processing by the computer in order to give the operator either force/haptic feedback or stereo vision feedback.

Step 6:If the virtual hand and virtual objects do not satisfy the rules of Stable grasping conditions, then: a. if the coordinate is fixed, releasing the fixed coordinate between virtual objects with the virtual hand, that means the objects have been Releasing by the virtual hand; b. If not, the operations of virtual hand will give the operator either force/haptic feedback or stereo vision feedback while the virtual objects keep their own condition.

Step 7: When the operators get feedback from the virtual system, they can change the situation and posture of their hands according to the needs of tasks.

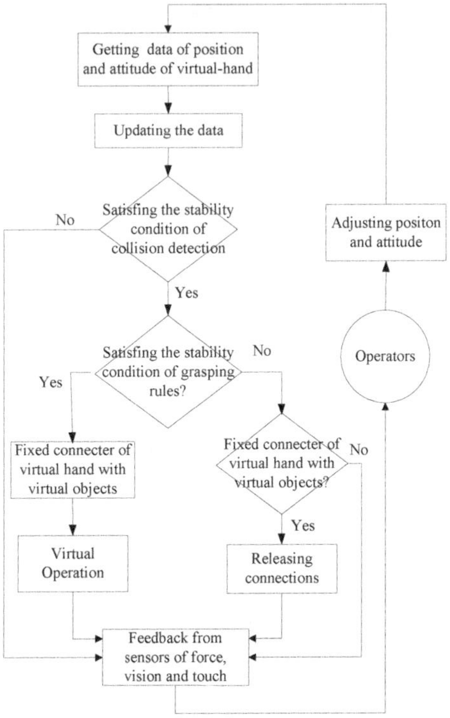

Fig. 2. Process of human-computer interface based on virtual hand

Step 8: When trackers get the newer information, system will update the position or hand posture of the virtual hand in the virtual environment and system will start the new turn.

Step 9: This loop executes continuously until the operational tasks complete.

4 Prototype System and Applications

4.1 Structure of Prototype System

The prototype system is consisted of two parts: hardware system and software system.

The hardware system mainly includes: A graphics workstation based visual system with Intel Pentium processor, 3D Labs Graphics Accelerator, 2D display part, and 3D shutter glasses, E-2 Emitter, 5DT DataGlove and FOB Trackers. The data glove is used for capturing the user's finger configuration to drive the skeleton motion. FOB tracking system is used to track the user's hand movements. The shutter glasses and emitters are used to obtain stereoscopic view. Our experiments show that the virtual hand can perform virtual grasping operations while maintaining its visual realism as well as motion realism. The software system mainly includes NT systems, Visual C++ development system, Open Inventor visual system and 3D modeling system.

4.2 Experiments and Applications

The basic framework of the virtual environment has been implemented and tested using a series of assemblies. The current system runs on NT platforms. All the program development has been written using C++ language. Figures 3 and 4 illustrates the User-OSD Interface and the virtual grasping operations which have been successfully assembled using the constraint-based virtual environment. Fig. 3 shows the virtual environment which consists of the virtual hand, one desk model and two different shapes cubes on the desk. Fig 4 represents a simple Operating process in which situation the virtual hand grasps the small blue cube, moves to the big red one and lay it on the big one. The case study shows the virtual operations such as grasping, moving, and releasing.

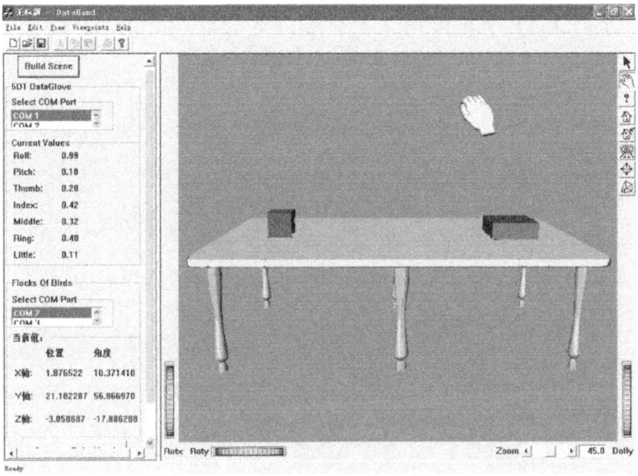

Fig. 3. The User-OSD Interface of the virtual system

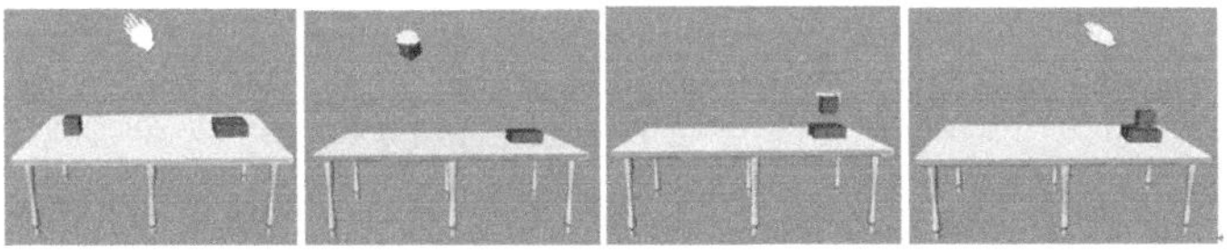

Fig. 4. The virtual grasping operations

Acknowledgments. The authors would like to extend thanks to our project partners for helpful comments. This work is done within the Leading Academic Discipline Project of Shanghai Municipal Education Commission (No.J51802), Project of Shanghai Second Polytechnic University (No. QD209005) and Innovation Program of Shanghai Municipal Education Commission (egd. 10020). The content of this publication is the sole responsibility of the authors.

References

1. Wang, C.-Q., Zhao, Z.-M., Chen, Y.-Q.: Research on 3D Visualization Simulation For Robot Manipulators. Robot 23, 577–588 (2001)
2. Ismaila, N., Bakarb, N.A., Juri, A.H.: Recognition of cylindrical and conical features using edge boundary classification. International Journal of Machine Tools and Manufacture 45, 649–655 (2005)
3. Li, Q., Huang, D.-S., Wang, W.-G.: Modeling and Driving of Virtual Hand Model Based on Data Glove. Computer and Modernization 4, 57–60 (2010)
4. Skubic, M., Bailey, C., Chronis, G.: A sketch interface formobile robots. In: Proceedings of the 2003 IEEE International Conference on Systems, Man and Cybernetics, pp. 919–924. IEEE, Piscataway (2003)
5. Xiong, Y.-J., Li, S.-Q., Wang, W.-T.: Operating Technology of Virtual Robot Based on Data Glove Drive. Mechanical Science and Technology 23, 1433–1436 (2004)
6. Li, K.-B., Wang, S.-Q., Shan, J.-H.: Research and Implementation of Virtual-hand Operating Technology. Computer Simulation 24, 227–229 (2007)

A Method to Align POS Data and Linear Push-Broom Imaging Data

Xiangang Meng[1], Aiwu Zhang[1,*], Shaoxing Hu[2], Weidong Sun[3], and Jilian Yang[3]

[1] Laboratory of 3D Information Acquisition and Application (MOST), Capital Normal University, Beijing 100048, China
[2] School of Mechanical Engineering and Automation, Beihang University, Beijing 100191, China
[3] Department of Electronic Engineering, Tsinghua University, Beijing 100081, China
zhangaw163@163.com

Abstract. In the field of aerophotogrammetry and aerial remote sensing, the linear push-broom imager is increasingly applied for getting the imaging information. What the linear push-broom imager captures is a sequence of line images, therefore, it is essential for imaging to collect the POS data of each line image synchronously. In other words, we need a good method to align the POS data and the line images which are captured at the same time. Using GPS time to align data is a traditional method, but sometimes, GPS does not work, so we present a new method to align the POS data and the line images using computer's high resolution performance counter. What's more, we have tested this method proposed in this paper by ground-based experiments and flight experiments, the test results show this method is robust.

Keywords: time synchronization, POS, linear push-broom imager.

1 Introduction

With the high speed development of scientific technology, in the field of aerophotogrammetry and aerial remote sensing, POS and other sensors (including planar-array digital camera, linear-array CCD camera, two-dimensional laser etc.) are often mounted on the same platform to get quantitative information of ground objects. And what the linear push-broom imager captures is a sequence of line images; hence, it is essential for imaging to collect the POS data of each line image. However the POS data and the line images are from different sensors, how to align POS data and line images is a challenge. Usually, GPS time is used to align data from different sensors, the PPS pulse signal from GPS triggers other sensors and the corresponding time mark will be stamped on the output data from other sensors [1-3]. For example, Regel LMS-Q560 laser scanner has designed internally a 1-MHz counter, the counter re-corrects its value based on the PPS pulse signal and RS232 serial time tag every second. When the LMS-Q560 laser scanner collects one scanline, the data recorder

* Corresponding author.

G. Lee (Ed.): Advances in Automation and Robotics, Vol.2, LNEE 123, pp. 99–106.
Springerlink.com

will record the value of the counter at the acquisition moment of this scanline[4]. While, the method above based GPS time has some fatal weaknesses as follow:

1) GPS can not receive signal blocked by high buildings, high mountains, and so on [5].
2) GPS outputs pulse with frequency of 1PPS. Obviously, the pulse signal in such low frequency is unable to meet the precise data synchronization between GPS and other sensors outputting signal high frequently.
3) The PPS signal from GPS needs to be supported by the hardware interface of its trigger target.

On account of these above defects in the time sync control on the hardware, a method of time sync using computer's high resolution performance counter (below abbreviated as HPC)is presented , which is a method based on computer software.

2 Time Sync Method

The software sync is to record the value of HPC that can be used to match the data of POS and the data of the linear push-broom imager when the computer software collects the data of POS and the data of linear push-broom imager. And both POS and linear push-broom imaging sensor are not necessary to output the pulse signal to third-party equipment.

2.1 Algorithm of the Software Sync

HPC works with the 8253 and 8254-type programmable interval timer chip. It counts number in binary or BDC (Binary Decimal Counter). The counter generates n pulses per second, and the number of the counter will subtract 1 for each pulse. To obtain more information of the counter, Windows operating system offers two API functions: QueryPerformanceFrequency and QueryPerformanceCounter [6]. According to different CPU frequency, the frequency of the counter is variable. The QueryPerformanceFrequency function retrieves the frequency of the HPC. For example, if the type of CPU is P4, 2.8GHz, the frequency of the HPC is 2793050000Hz, if the type of CPU is P4, 3GHz, the frequency of the HPC is 2992530000Hz. QueryPerformanceCounter can get the current value of the counter. If we use CPU P4 3GHz, the interval of the counter is 0.00033 microsecond.

2.2 Implement

As to the above method of time sync, during the development of the collection software, each sensor (including POS and linear push-broom imager) will be designed two counters. One is for recording sampling serial number of the sensor's data, called count1; the other is for saving the value of HPC, called count2, its value dynamically retrieves every time. When the collection software receive one sampling data, the number of count1 will be added 1 immediately, and at same time the collection software call QueryPerformanceCounter to get the current value of count2, the value of count2 corresponds to the number of count1 one to one. Finally, the value of count1 and count2 are saved in pair into a sync file.

Next is to align POS data and line images. First, the values of count2 in the sync files of POS data and line images will be extracted respectively. Second, we calculate the corresponding POS data of each line image by interpolation based on the values of count2 [7]. The pipe of the software sync is shown in fig.1.

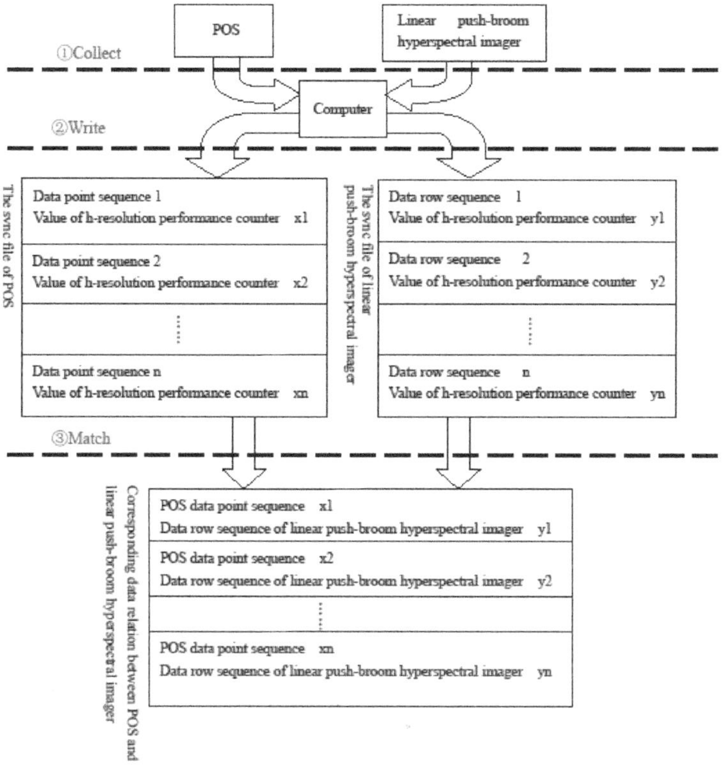

Fig. 1. The pipe of the software sync

2.3 Comparing with the Hardware Sync

Comparing with the hardware sync, the sync on the software using HPC possesses the following advantages:

1) More extensive applicability. The software sync can be applied into the data synchronization between any sensors.

2) Raise the frequency of data sync between POS and other sensors . The software sync using HPC improves the interpolation precision of POS to achieve the data match between POS and other sensors much better.

3) Without hardware interface. As long as the data can be collected by the computer, the sync between the sensors can be implemented.

3 Experiments

3.1 Ground Experiments

Before the flight experiments, we do many experiments on the ground to test the sampling uniformity of the sensor according to HPC value.

1) Test the sampling uniformity of linear hyperspectral sensor. First, hyperspectral sensor is placed individually on a uniform-speed rotating turntable, stop as n line images have been collected; Second, according to the generated sync file of hyperspectral sensor, extract the HPC value of each line image, $x_i (i = 1, 2, ..., n)$. Assume that the average value and standard deviation of the sampling interval between adjacent line images are expressed by m and s, the formulas are as follows:

$$m = \frac{1000}{f} \frac{1}{n-1} \sum_{i=1}^{n-1} (x_{i+1} - x_i) \tag{1}$$

$$s = \sqrt{\frac{1}{n-2} \sum_{i=1}^{n-1} (\frac{1000}{f} (x_{i+1} - x_i) - m)^2} \tag{2}$$

- The raw collected spectral image is smooth without jumping (as Fig. 2 shows).

Fig. 2. Liner push-broom hyperspectral image collected on the ground

■ Standard deviation can be used as a measurement of divergent level of a group of data. Smaller standard deviation indicates that the data are closer to the average value, that is, these data scatter in smaller range. When standard deviation is small enough, these data can be regarded as uniform. The Standard deviation of the experimental data, s, calculated by above formulas, is less than 1.0, which indicates that the interval of data collection is uniform.(as table 1 shows, t is the sampling interval between HPC value of x_{i+1} and x_i, in unit of millisecond).

2) Test on the uniformity of the sampling interval of POS data. First, POS antenna is placed outside with good signal , stop as n items of POS data have been collected; Second, according to the generated sync file of POS, extract the HPC value of each sample, $x_i (i = 1, 2, ..., n)$. Finally, according to the formula (1) and (2), the average value m and standard deviation s can be calculated.

Table 1. Evaluation of uniformity on sampling interval of hyperspectral imager

x_i	x_{i+1}	t
2137896717	2137964714	29.010370
2137964714	2138035000	29.986953
2138035000	2138107701	30.023644
2138107701	2138178073	30.081241
2138178073	2138245943	31.017294
2138245943	2138316358	28.956186
2138316358	2138386865	30.041990
2138386865	2138459475	30.978469
2138459475	2138529805	30.005726
2138529805	2138599847	29.882853
2138599847	2138670193	30.012552
2138670193	2138740498	29.995059
2138740498	2138808580	29.046635
2138808580	2138881148	30.960550
2138881148	2138949326	29.087592
2138949326	2139019649	30.002739
2139019649	2139090037	30.030471
2139090037	2139162711	31.005774
2139162711	2139233006	29.990793
	m =30.0066	
	S = 0.6565	

Table 2. Evaluation of uniformity on sampling interval of POS

x_i	x_{i+1}	t
12218819162	12218840273	9.006837
12218840273	12218863663	9.979154
12218863663	12218889433	10.994562
12218889433	12218915236	11.008641
12218915236	12218936391	9.025610
12218936391	12218959785	9.980861
12218959785	12218983310	10.036751
12218983310	12219006682	9.971475
12219006682	12219030085	9.984701
12219030085	12219053427	9.958675
12219053427	12219074581	9.025183
12219074581	12219097971	9.979154
12219097971	12219123713	10.982616
12219123713	12219144887	9.033716
12219144887	12219168303	9.990247
12219168303	12219194024	10.973657
12219194024	12219215223	9.044382
12219215223	12219240976	10.987309
12219240976	12219262217	9.062301
	m = 9.9487	
	S = 0.7624	

The standard deviation, s, calculated by large numbers of experimental data, is less than 1.0 ,which indicates that the interval of POS data collection is uniform. .(as table 2 shows, t is the sampling interval between HPC value of x_{i+1} and x_i, in unit of millisecond).

3.2 Flight Experiments

The experimental area is at Three Gorges , and the light and small-type hyperspectral imaging system independently integrated by Capital Normal University based on the unmaned airship as the data acquisition equipment.(as Fig.3 shows) .This system integrates hyperspectral sensor, digital camera and POS together. It is able to acquire three kinds of data streams: h-resolution CCD image, hyperspectral imaging data and navigation data. Besides the system is mainly used for low-altitude remote sensing monitoring. The hyperspectral camera is a kind of linear push-broom imaging spectrometer, which is mainly made up of lens, spectrometer and camera. The spectrum covers 400-1000nm; the space resolution is in 1600 pixels. The imaging system provides each pixel with spectrum information of the overall continuous 840-wave-band.

The main parameter of the light and small-type hyperspectral imaging system is given as Fig. 4. With the purpose of getting the direct geographical locating data of each scanning line of hyperspectral image, the hyperspectral camera and POS should make time synchronization for they are two mutually independent sensors. During the experiment, the time sync between POS and the hyperspectral camera is implemented by the computer HPC. Recording the value of HPC and the corresponding data sequence and writing those to sensors' own sync file when the hyperspectral camera and POS collect data. Due to the climate condition of the experimental area is typically windy, foggy and rainy, the sun does not rise until 10 am and the fog is pretty heavy , whick make the original image seriously deform. The deformation lowers the identifiability, so geometrical correction must be carried out for the subsequent data process. In the process of geometric correction, the first step is to extract values of HPC in the corresponding sync files of POS and the hyperspectral camera. The second step is to apply different algorithms to these values and match each scanning line of hyperspectral image with its corresponding POS locating data at the same acquisition time. Finally, after a series of coordinate transformation and some other process, the final orthophoto map forms.

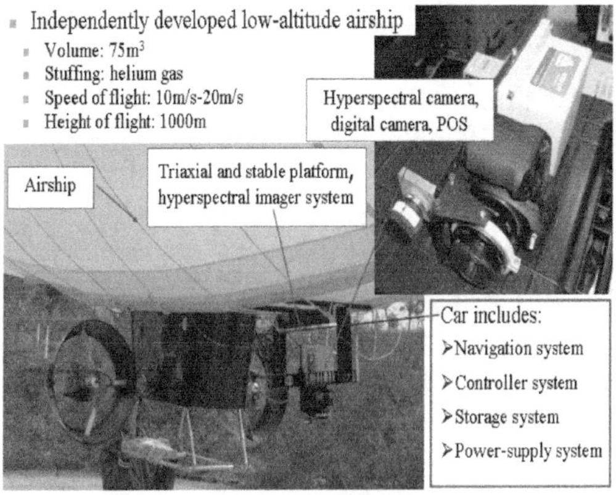

Fig. 3. Light and small–type hyperspectral imaging system based on unmaned airship

Optical properties :
Spectral range : 400-1000nm
Spectral resolution : 2.8nm
Lens parameters :
FOV: 27.2°
IFOV: 0.017°
Track width : 0.52×elevation
Ground resolution @ elevation 1000m: 0.33m
Electronic properties :
bands: 840
Sampling rate : 33fps
Spatial resolution : 1600pixels

Fig. 4. The main parameters of hyperspectral camera

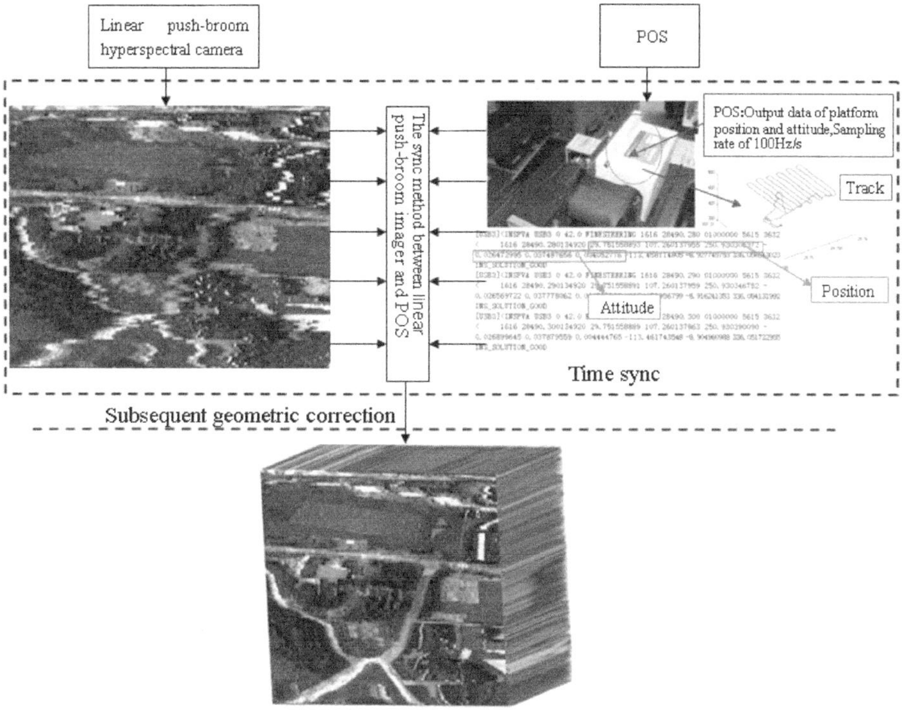

Fig. 5. Aligning the linear push-broom hyperspectral data and the POS locating data

The upper left picture is the original data acquired by the linear push-broom hyperspectral camera. The upper right picture is the original locating data collected by POS. They are originally not related. But the time sync method will align and match the POS locating data and the linear push-broom hyperspectral data. Aligning and matching these two kinds of data one to one is essential for the subsequent geometric correction as Fig.5 shows.

The experimental result showed that the proposed time synchronization method has extensive applicability, and can be used in the subsequent geometry correction, etc [8-9].

4 Conclusions

Based on the time synchronization method of the computer HPC, with the microsecond-level timing accuracy, ensure the reliability of match between POS and linear push-broom imager. Meanwhile, this method embraces more extensive applicability in theory. It can not only be applied to the time sync between POS and linear push-broom imager, but also meet other sensors' requirement of time sync control.

According to this above experiment, the final corrected image met the requirement of scientific research, as the acquired sync file, the POS locating data and the hyperspectral imaging data go through a series of processes. Of course, it is just preliminary, the stability and reliability of the time synchronization put forward by the paper need more studies in the future. Yet further measure and calculation is needed to eliminate the time delay caused by the data transmission between sensors and computer.

References

1. Liu, S.X., Lin, Y., Huang, C.H.: The Effect of 1PPS Time Synchronization upon The Multibeam Echosounding. Hydrographic Surveying And Charting. J. 31(2), 31–33 (2011)
2. Xie, H.: Time Synchronization to Local Network Base on GPS single and NTP Protocol. J. Industrial Control Computer 23(6), 75–76 (2010)
3. Zhu, Z.Q., Wu, Y.H., Yang, Y.X.: Modulari Practice of Time Synchronization in Integrated GPS/ INS Systems. J. Geomatics and Information Science of Wuhan University 35(7), 830–836 (2010)
4. Riegl, C.: Airborne Laser Scanner LMS-Q560 General Description and Data Interfaces. Z.88-94
5. Wu, J.: GPS/INS Supplemental Aerial Photography Measuring Principle and Application Research. Photogrammetry and Remote Sensing, D. Zhengzhou (2006)
6. About Timers, `http://msdn.microsoft.com/en-us/library/ms644900.aspx`
7. Li, X.Y.: IMU/DGPS-Supported Photogrammetry-Theory,Approaches and Practice. Photogrammetry and Remote Sensing, D.Zhengzhou (2005)
8. Norbert, H., Dieter, F., Dirk, S.: On The Performance of Digital Airborne Pushbroom Cameras for Photogrammeric Data Processing – A Case Study. J. Iaprs 33 (2000)
9. Pece, V.G., Paul, E.G.: The Design and The Development of a Hyperspectral and Multispectral Airborne Mapping System. J. ISPRS Journal of Photogrammetry and Remote Sensing 64, 184–192 (2009)

Evaluation Methods of Architecture for Intelligent Robotic System

Wei Xie[1], Jiachen Ma[2], Qi Zhang[1], and Mingli Yang[2]

[1] School of Astronautics, Harbin Insitute of Technology, 150001, Harbin, China
[2] Department of Information Science and Engineering, Harbin Insitute of Technology, 264209, Weihai, China
hitxiewei@163.com

Abstract. The architecture is the backbone of Intelligent Robotic System (IRS). Aimed at the current difficulties for designing architecture, on the basis of previous studies, the evaluation methodologies based on comprehensive integration is provided and then the systematical evaluation criterions including nine quality attributes (Modularity, Niche Targetability, Autonomy, Robustness, Portability, Coordinability, Real-time, Security, Cost and resource constraint) are presented in this paper. Finally, a case study has shown these methods combined of qualitative and quantitative are feasible.

Keywords: Evaluation, Architecture, Quality attributes, Quantitative, IRS.

1 Introduction

The first goal of designing Intelligent Robotic System (IRS) is to consider the architecture, because it can provide hardware and software foundation for realizing intelligent behaviors. Therefore the architecture is a technology for solving complexity of system, which ensures reasonable coordination, openness, and scalability of robotic system. These complex systems are made up of many subsystems or modules, and the goals are so diverse and dynamic that it is not easy to analyze and evaluate. As the complexity and intelligence of IRS applications increase, evaluating performance becomes more difficult [1].

Many researches have thought that the metasynthesis method of combination of qualitative and quantitative is the only effective way for analyzing complex information system [2-3]. The architecture of IRS is similar to "body" of living organisms, so the scientific metasynthesis method for building robotic system is still unclear. Currently, the methods of designing architecture are mainly based on "repeated optimization" and "trial and error" [4], which lack of systematic analysis methodology, and there are not unified modeling theory and tools so far. The design methods based on experience have not ensured the reasonability and correctness, let alone optimization if there are not suitable evaluation mechanisms. Plus to robotic interdisciplinary nature and diversity of application areas, the methods of quantitative analysis and design robot offline are very difficult to seek.

G. Lee (Ed.): Advances in Automation and Robotics, Vol.2, LNEE 123, pp. 107–114.
Springerlink.com

Some progress had been made towards seeking scientific evaluation methods. Luca [5] proposed a simulation tools for the design and evaluation of robot architecture. Akin [6] suggested some desirable qualitative attributes for analyzing behavior–based robotic architecture including support for parallelism, hardware targetability, Niche target ability, support for modularity, robustness, timeliness in development, run-time flexibility and performance effectives. Orebäck.A [7] made a comparative study of some successful software robotic architecture such as Sapphire, Team Bots and BERRA, and then provided some software evaluation norms. Murphy[8] concluded four qualitative evaluation criteria adapted form Akin, which are support for modularity, niche targetability, ease of portability to other domains and robustness. Manuel Kolp[9]evaluated five styles (control loop, layered architecture, task trees, structure-in-5 and joint-venture), and he used four agent software quality attributes including condensability, predictability, failability-tolerance and adaptability to evaluate the developing these architectures for mobile robot control. The result had shown that the Structure-in-5 and Joint-venture based on organization theory had obvious advantages for build robot system. Sigal Berman[1] proposed a methodology for detailed evaluation of AGVS, and he provided some qualitative and quantitative analysis method to model, experiment and compare different AGVS control. The qualitative system analysis methods of architecture were an extension of Akin's theory. D.C.Herath presented a new evaluation methods based on dual task paradigm for thinking heard framework. And he pointed that developing a comprehensive evaluation metric consisting of a series of individual test is the future trends [10].

It is can be seen from previous research that most of evaluation methods emphasized on qualitative analysis, and metrics are not uniform. Our proposal of evaluation tries to merge the merits of ways mentioned above while minimizing their shortcomings, and we expect to explore a new assessment method to guide the design.

2 Evaluation Methodologies

The same as human's organizational structure, the architecture of IRS is complex. In this complex system, there are not only include some quantitative objects but also include some qualitative be difficult to quantify. So the method of comprehensive integration is expected to analyze this complex system. With the development and application of intelligent robot, the design of architecture is becoming more and more important for solving the problems of composition, internet and interaction between components. How to select suitable structure for designing robotic system is problem plagued the people because of the diversity of robotic applications. Most scholars believe that the architecture should be open and scalability, and these evaluation criteria are mainly come from computer's software engineering and control engineering. Most qualitative evaluation criteria are too diversified and one-side, especially the control performance, to reflect overall characteristics of IRS.

Now there are there kinds of architecture including deliberative, reactive and hybrid in designing architecture of mainstream. Comparing to the computer, robot is a combination of software and hardware tightly coupled. As constituent parts of the organism, according to the functions, a robot can be divided into three very general categories (Sense, Plan and Act) which have become accepted primitives in robotics

[8]. Corresponding, the three primitives are always the main line to describe IRS. According to the relationship between them, the structure is classified three paradigms such as deliberative, reactive and hybrid. Whatever in the centralized and decentralized system, the design system may be considered the extension of these three paradigms. With the improving of hybrid paradigm, the non-layer-type structure based on organizational theory and multi-agent system (MAS) are used to design robotic system successfully. Therefore, these three primitives are the main elements of architecture. And also they are the basic starting point of analyzing it.

On the basis of basic architecture based on S,P and A, there are many reference architecture to build IRS. Overall there are three characteristics during architecture development, which are Three-dimensional distribution pattern, Variety of modeling methods and Intelligentization. The hybrid architecture is a comprehensive space structure made up of intelligence, behaviors, information and control, so the distributed and centralized architecture co-exist, and it not only include generalized model system based on information processing but also include narrow model system based control engineering. Now the research topics in robotics have developed form classical motion control for industrial robots to modern intelligent control techniques and social learning paradigms. With the complexity increasing, all kinds of algorithms are fused in the design of functional analysis for build architecture. The hardware and software of system must be considered together. Another main feature of IRS is itelligentization. There are three methods used to study this problem from different sides, which are structuralism represented by the neural network, functionalism represented by expert system and behaviorism represented by sensor-action system. In hybrid system, the implementations of machine intelligence are usually integrated, so appropriate architecture is expected to synthesize those methods.

From the above analysis we can see that it is an indeed challenging work for evaluating this complex system. We must seek for new evaluation ways form the three aspects under metasynthesis [2]. It is fortunate that the computer is the main driving force for constructing robot, and the methods of system analysis and software engineering for it can be extended to the fields of robotics. Therefore the evaluation methodologies for IRS can be summarized as following for a guiding principle.

1) Multi-level,multi-angle integrated
2) Combination of qualitative and quantitative
3) Full evaluation and trade-offs
4) Comprehensive integration of online and offline.

3 Evaluation Criterions

According to the evaluation methodologies above, we select nine quality attributes which reflect the system of quality indicators to evaluate the reasonability. Evaluation requires a set of criteria, therefore how to select the reasonable criteria is still controversy. The evaluation methods will not only ensure design standardization, but also improve development efficiency. The metrics about quality attribute in this paper follow the principles of testability, independence, completeness and consistency. The detailed descriptions about nine quality attributes are shown in Table 1.

Table 1. Evaluation criterions

Nine quality attributes	Qualitative	Quantitative
(1) Modularity	Function independence Structure independence Openness Standard of interface	Degree of coherence(R):[11] $R = A/(A+V)$ Degree of coupling(S) [11] $S = (B-V_1)/B$ $A(B)$: the average distance of all modular samples inside(between) class $V(V_1)$:the MSE of distance of all modular samples inside (between) class
(2) Niche targetability	Function targeted	The proportion of special module $\tau = Ns/(Ns+Ng)$ $Ns(Ng)$ the numbers of special (general) modules
(3) Autonomy	Rank division Ⅰ only include A Ⅱ S,P,A(simple algorithm) Ⅲ S,P,A(complex algorithm)	The complexity of planning algorithms $O(x)$ Time for completing objectives and tasks T Learning ability LA
(4) Robustness	Freedom from jamming Redundancy Fault tolerance	Degree of influence DI $DI = m_a / m_s$ $m_a(m_s)$ Average(Standard) deviation
(5) Portability	Universality of modules	Related assessment techniques of software engineering[12]
(6)Coordinability	Standardization of interface Amount of shared information	Petri net model Information entropy[13]
(7)Real-time	Reflective including $S \to A$ and $S \to P \to A$ The response of emergency	The time for performing the most complex algorithm $R_{t\max}$ The fastest response time $R_{t\min}$
(8) Security	ISOstandard [14] ISO12100,ISO10218,ISO1412	None
(9) Cost and resource constraint	Low cost Endurance	Maximum load capacity P_A Longest continuous run time T_c Maximum operating speed V_m

3.1 Modularity

Modularity is the prior task for designing architecture. The modularity determines the openness and reconfigurable, and the rational modular design will help to improve the efficiency of the system. Now the design of the modules has been started from a base of knowledge or using mathematical methods [15]. Whatever methods are used, the general principles must be considered such as independent, complete relatively, achieve rapid connection and independence in the kinematics and dynamics [16]. By the analysis of existing literatures, we find that the main bases of modular classification are the organization theory based on distributed problem solving and the methodology “division and interaction”. The modular division is related to intelligent forms of organization. The three primitives(S, P and A) have provided the basic classification method form robotic functions. On the basis of qualitative analysis, The strategy “tight cohesion and loose coupling” is used to design guidelines for IRS, so we consider degree of coherence inside the module and degree of coupling among modules as the measure for analyzing openness and use cluster analysis to evaluate this metric.

3.2 Niche Targetability

It is important to solving this problem that how well it works for the intended application. As for this interdisciplinary complex system designs, the methods of functional analysis and functional decomposition constitute the main frame of the design. The more intended application, the more specificity, those are odds at with the modularity, so we must make trade-offs in the design. The functional modules of robot are made up of special and general modules. If the numbers of special modules are majority, they have shown more functional targetability. Therefore, we consider the proportion of special module as this metric.

3.3 Autonomy

Autonomy is the main characteristics of IRS, and it also is named “adaptivity”. In the absence of intervention, robot can generate intelligent behaviors purposefully through adaptivity. Although the autonomy mainly is reflected during the actual operation, the organizational structure and planning algorithm determine the intelligent behaviors of robot in the future. Therefore it is necessary to evaluate and analyze in the early design. According to the organizational level and three primitives, autonomy can be measured by three ranks and three quantitative indicators to assess predictably.

3.4 Robustness

Where is the system vulnerable, duce that vulnerability? As for a robotic system, it works mostly in the unknown environment. So the vulnerability is important design specifications. The relevant indicators of control theory are applied to evaluate reactive and some relatively simple deliberative paradigm. But for the whole architecture, if a single function module fails or meets instantaneous interference, the robot still maintains certain functions and certain performance indicators within the range of uncertainty. Therefore we introduce a measure of randomness variability to quantify the robustness.

3.5 Portability

Portability relates to the openness, standardization and other important characteristics. From the robot engineering point of view, the platform based on computer and embedded systems have created good conditions for realizing portability. Although the diversity and complexity of robotic system make some difficult for reusing code and building unified architecture, the openness and standardization of computer have provided possibility for standardization of robot system integration. The software models such as Robotics studio, DSS and Agent-oriented have brought new hope for integration of intelligent robot system.

3.6 Coordinability

If we consider the robotic system from the organization theory of complex system, the building mechanism of architecture is described as how to coordinate the various modules under different levels. Based on three primitives such as S,P and A, whatever the centralized and decentralized structure are applied, the coordination including cooperation and competition is the key technique for solving complex problems dynamically. Every module shares resources through communication or blackboard structure. Therefore the amount of information communication and algorithm of planning and coordination determine the reactive ability and bandwidth. The Standardization of communication interface, the amount of shared information, Petri net model and information entropy are used to describe coordinability.

3.7 Real-Time

As for the robotics system, the real-time is very important indicator. The robotics system must react to environment appropriately and quickly. And the robot also completes the tasks quickly according to intelligent strategy of planning. It guarantees the safety itself in the case of an emergency or dangerous.

3.8 Security

The general principles of robot design should follow Asimov's (1940) "three principles of robot", which is ethics program for the guidelines of designing robot. In addition, it must comply with the relevant ISO standards [14].

3.9 Cost and Resource Constraint

The specific application of constraints including the robot loads, limited space, carrying a finite energy and achieving cost must be considered. From the perspective of synthetic biology, it is impossible to create a perfect environment to meet a variety of unknown robot. Therefore, we must consider the architecture under the limited resources and cost.

4 Case Study

The architecture is the backbone of complete robotic systems. The right choice of architecture can help us to design robotic system efficiently, conversely, the wrong choice can make it difficult to implement. Therefore, the evaluation methods become so important for designing intelligent robotic system. The right architecture should facilitate the development process by providing beneficial constraints on design and implementation of desired application, without being overly restrictive [17]. And the nine quality attributes have reflected the full range of research content, they have provided some guidelines for designing robotic system. As for a robot, different applications need different architecture to fit these requirements.

The evaluation of architecture is systems engineering. The robotic systems differ from other software applications in many ways. If we have right evaluation methods for architecture, the development and application should become easy relatively. Let us take the example of intelligent handing robot (also named automated guided vehicle) to illustrate the feasibility of these assessment methods.

According to AGV system needs analysis and evaluation methods above, we selected several design schemes. The AGV prototype was shown as Fig.1 and some major technical indicators (due to limited space) were presented in Table 2.

Table 2. Major technical indicators

Attributes	Indicators	Attributes	Indicators	Attributes	indicators
(1)	++ R =0.35 S = 0.86	(4)	++	(7)	++ $R_{t\min}$ =10ms
(2)	+- τ =0.4	(5)	+-	(8)	++
(3)	II	(6)	++	(9)	+- P_A = 100Kg T_c =8h V_m =2.5m/s

Qualitative analysis[9]: ++ sufficient/positive; +- partial/ positive.

Fig. 1. AGV prototype

5 Conclusion

The architecture is the backbone of robotic system. Table 2 has shown the main features of AGVs, these evaluation indictors can help us to design robotic system scientifically from qualitative to quantitative analysis. And the experiment results also have demonstrated its feasibility. In this paper, we endeavor to seek for scientific evaluation methods to design intelligent system in this paper. In fact, it is difficult to ensure the reasonableness of the architecture according to designer's experience. Therefore, these appropriate assessment methods have provided guidelines and insights for designing intelligent robotic system. And the number of principles or evaluation criteria including qualitative and quantitative have provided the basic research for designing robot system off-line or developing analysis of robotics.

References

1. Berman, S., et al.: Evaluation of automatic guided vehicle system. Robotics and Computer-Integrated Manufacturing 25, 522–528 (2009)
2. Yavuz, H.: An Integrated Approach the Conceptual Design and Development of An Intelligent Autonomous Mobile Robot. Robotics and Autonomous Systems 55, 498–512 (2007)
3. Longbin, C., Ruwei, D.: Fundamentals, Concepts, Analysis and Design and Implementation. Post and Telecom Press, China (2008)
4. Nehmzow, U.: Scientific Methods In Mobile Robotics-quantitative Analysis of Agent Behaviour. Spring, London (2006)
5. Luca: Hybrid Modeling and Simulation for The Design of an Advanced Industrial Robot Controller. IEEE Robotics &Automation Magazine 6, 45–51 (1997)
6. Akin, R.C.: Behaviour-based Robotics. MIT press, USA (1998)
7. Orebäck, A., Christensen, H.I.: Evaluation of architecture for mobile robotics. Autonomous Robots 14, 33–49 (2003)
8. Murphy, R.R.: Introduction to AI Robotics. Massachusetts Institute of Technology, Second printing (2002)
9. Kolp, M., Giorgini, P., Mylopoulos, J.: Muti-agent Architecture as Organizational Structures. Auton Agent Multi-Agent Sys. 13(3), 3–25 (2006)
10. Herath, D.C., et al.: Thinking head: Towards human centred robotics. In: 11th International Conference on Control Automation Robotics & Vision (ICARCV), pp. 2042–2047. IEEE Press, Singapore (2010)
11. He, J., Feng, Y., et al.: Clustering Analysis of Robot Controller Openness Based on System on Chip. Journal of Huazhong University of Science and Technology S1, 16–18 (2008)
12. Amoretti, M., Reggiani, M.: Architectural paradigms for Robotics Applications. Advanced Engineering Informatics 24(1), 4–13 (2010)
13. Kong, L.: Evaluation Model Based on Information Entropy and Clustering Analysis. Jisuanji Yu Xiandahua (china) 3, 63–64 (2010)
14. Yasunori, T., Yu, O., Amir, A.F., et al.: A System Design Concept Based on omni-directional mobility, Safety and Modularity for an Autonomous Mobile Soccer Robot. Journal of Bionic Engineering suppl., 121–129 (2008)
15. Pozna, C.: Modular Robots Design Concepts and Research Directions. In: Proceeding of 5th International Symposium on Intelligent System and Informatics, pp. 113–118 (2007)
16. Liu, X., Zhang, M., Liu, W.: Design method to modular robot system. In: International Conference on Mechanisms and Robots, pp. 521–528 (2009)
17. Coste-Maniere, E., Simmons, R.: Architecture, The Backbone of Robotic Systems. In: Proceeding of IEEE International Conference on Robotics & Automation, San Francisco, pp. 67–72 (2000)

Research and Implementation on Resource Saving-Based Server Cluster Scheduling System

Yang Cao, Jienan Cao, and Zeyu Xiong

School of Computer Science, National University of Defense Technology, Changsha 410073, China
James6116@163.com

Abstract. For the traditional server cluster system, even when the work load of the system becomes lighter, the scheduling algorithm will load evenly to all servers, which will cause the server is lightly loaded or idling. It will waste power resources and reduce the effective life of the server. This paper proposed and implemented a resource saving-based server cluster scheduling system (RSSCSS). The system according to the severity of the current work load, enable or disable part of server systems dynamically, with the minimal service resources to meet the user's service requests , so to save the energy consumption of the servers, network equipments and other supporting facilities such as air conditions, lighting and extend the effective life of those equipments.

Keywords: resource saving, server cluster, load balance, scheduling system, magic packet.

1 Introduction

In order to save the costs of server's hardware ,many major Web sites, searching sites and game sites are using the clustering system with N PC servers and the load balancing scheduling algorithm (such as round robin) to meet with the large-scale concurrent access requests and provide reliable network information services[7,8,9]. However, the frequency of user access to networking service is variable in 7 days of a week and 24 hours of the day. So the load of the server is unbalance. For example, In 8:00~9:00 and 19:00~20:00 periods, there may be many people visiting the news sites, while the other time such as after 22: 00, there is a very little access requests even no network traffic. If we still running N Web servers, and distribute the small amount of the load to all servers, it's apparently not necessary (in extreme cases only a single server is sufficient) and not only be a waste of power energy but also shorten the effective life of the electronic devices and does not compatible with the ideas of low-carbon economy and environment-friendly society.

According to IDC data shows that, China's total spending in electric power and cooling for the servers reached 19 billion U.S. dollars in 2007, only Beijing and Guangdong reached 3.2 billion and 3.6 billion U.S. dollars [1]. It is the indisputable facts that the server's power consumption resulted in shortage of power in these areas. In this era of information, IT has become a vital driving force of enterprises, however, the huge amount of costs of IT is a heavy burden of enterprises. It is a new major

G. Lee (Ed.): Advances in Automation and Robotics, Vol.2, LNEE 123, pp. 115–120.
Springerlink.com

problem that how to reduce energy consumption and operating costs of ISP. and the online version of the publication. Every detail, including the order of the names of the authors, should be checked before the paper is sent to the Volume Editors.

2 The Idea of Resource Saving-Based Server Cluster Scheduling System

As we said before, the frequency of access to the service resources is variable with the time, thus the workload of a server is changeful. The idea of RSSCSS is based on the process capacity of the server system and the current performance quotations about the number of requests for service, the network traffics, the server's CPU utilization and memory utilization in a giving period. Under the control of the scheduling system, start or stop part of the server system dynamically (extremely, start N servers or stop N-1 servers), in order to have minim service resources to meet with the user's requests, so to reduce energy consumption and carbon emissions, to extend equipment effectively life, in the end, to cut down the costs of service operating.

Take a portal Web site which has 250 PC servers as an example. In general, a work group server's power is 300W ~ 500W, a departmental server's power is 350W ~ 650W, and a dedicated Web server's power is 650W ~ 850W [2]. If the average energy consumption of a server is 600W, then the total energy consumption of a month is 30 (days) * 24 (hours) * 0.6 (KW) * 250 = 108000 KWH. If we adopt the RSSCSS (a conservative estimate, assuming save 50% running time a day), the total energy consumption will be 108000 KW * 50% = 54000 KWH. Experts estimated that burning one ton of standard coal can generate about 3000 degrees electricity, and produce about 3 tons of carbon dioxide[4], so the portal Web site can save 18 tons of standard coal and decrease 54 tons carbon dioxide emissions per month. If the commercial electricity is ￥1.00 per degree, the operating costs of the company will save ￥54,000 a month, which does not include the benefits of saving of energy consumption of other devices , the benefits of extension of the effective life of the devices and the social benefits of reduce to generate carbon emissions.

3 Server Work Load Metrics

A server's comprehensive process capacity can be measured by the number of times which the server responds to the user access requests in a time unit. TPC (Transaction Processing Performance Council) defined a set of server's performance benchmark such as TPC-C,TPC-H,TPC-E,TPC-W[10], etc. We can use those benchmarks to measure a server's work load. For example, for a Web server, we can count the numbers of HTTP connection in the unit time and take it as the workload of the server. If the current load of the server does not exceed or far below its capacity, the server is worked in the normal state, otherwise, it is in the overloaded or low loaded state.

The server's workloads can also be measured by the server's CPU utilization, RAM utilization, RAM swap utilization, HTTP/SMTP/POP3 service connections or the average response time of the service, in/out network traffic and other metrics, as well as the growth trends of these metrics.

4 Server Work Load Access Method

The work load metrics of a server (include the CPU utilization, RAM utilization, network traffic, network connections, etc) can be fetched according to the SNMP protocol. In Linux operating system, the value of each metrics corresponds to the OID is shown in Tab.1 [5]:

Table 1. The object identity of load metrics

Metrics	OID
CPU utilization	1 minute Load:.1.3.6.1.4.1.2021.10.1.3.1 5 minute Load: .1.3.6.1.4.1.2021.10.1.3.2 15 minute Load: .1.3.6.1.4.1.2021.10.1.3.3
RAM utilization	Total Swap Size: .1.3.6.1.4.1.2021.4.3.0 Available Swap Space: .1.3.6.1.4.1.2021.4.4.0 Total RAM in machine: .1.3.6.1.4.1.2021.4.5.0 Total RAM used:.1.3.6.1.4.1.2021.4.6.0 Total RAM Free: .1.3.6.1.4.1.2021.4.11.0
Network traffics	ifInOctets :1.3.6.1.2.1.2.2.1.10 ifInUcastPkts: 1.3.6.1.2.1.2.2.1.11 ifInNUcastPkts: 1.3.6.1.2.1.2.2.1.12 ifInDiscards :1.3.6.1.2.1.2.2.1.13 ifInErrors :1.3.6.1.2.1.2.2.1.14 ifOutOctets: 1.3.6.1.2.1.2.2.1.16 ifOutUcastPkts: 1.3.6.1.2.1.2.2.1.17 ifOutNUcastPkts: 1.3.6.1.2.1.2.2.1.18 ifOutDiscards: 1.3.6.1.2.1.2.2.1.19 ifOutErrors :1.3.6.1.2.1.2.2.1.20
TCP connections	tcpCurrEstab 1.3.6.1.2.1.6.9 tcpInSegs 1.3.6.1.2.1.6.10 tcpOutSegs 1.3.6.1.2.1.6.11

If the server does not run the SNMP protocol, we can still do it by a private agent program. In the Linux/Unix operating system, the CPU utilization is stored in file /proc/stat and the RAM utilization is stored in file /proc/meminfo[6]. In the Windows operating system, it is also available through the following functions:

```
//get the CPU number and other basic information:
NtQuerySystemInformation(GET_SYS_BASIC_INFO,&SysBaseInfo,
sizeof (SysBaseInfo), NULL);
//get running time:
NtQuerySystemInformation  (GET_SYS_TIME_INFO,&SysTimeInfo,
sizeof (SysTimeInfo), NULL);
//get the CPU idle time:
NtQuerySystemInformation  (GET_SYS_PERF_INFO,&SysPerfInfo,
sizeof (SysPerfInfo), NULL);
//Get storage unit status:
GlobalMemoryStatus (& MemStat);
```

5 Resource Saving-Based Scheduling Algorithm

In order to prevent the server from being scheduled frequently around the scheduling critical point (which is also called scheduling jitter), we designed a low baseline (or threshold) and a high baseline for all the load metrics. The scheduling system gets all of the current load metrics periodically. If the current load metrics is limited in the region which is defined by the low baseline and high baseline, then the server clustering is running in normal state and does not need scheduling. When the current load metrics is lower than the low baseline and the current load is not greater than it of the last time(the trend of the load is declining or unchanged), a appropriate number of (e.g. n = 1,2) servers (including the service process) should be stop; Conversely, When the current load metrics is higher than the high baseline and the current load metrics is not less than it of the last time (the trend of the load is upward or unchanged), then a part of servers (including the service process) are supposed to be started. The mentioned trend of the load changing can prevent the server from being over scheduled. For example, when we start a new server, the average load of all servers will certainly fall down. It will take some time before the average load dropped below the high baseline, so we don't need to worry that a certain server will be started again immediately.

Resource saving-based scheduling algorithm can be described as follow:

(1) Set the capacity value of the server and the high baseline , low baseline of the CPU utilization, RAM utilization, RAM swap utilization, network traffic;

(2) Start work load sampling timer, the interval of the timer is M minutes;

(3) Get the initial values of the CPU utilization, RAM utilization, RAM swap utilization, network traffic and save to the old variables;

```
while (true)
{
    (4) If the sampling timer expired, then:
    {
        (4.1) Get the current values of the CPU, RAM, RAM swap utilization and
        the network traffic and the numbers of the requests or connections be
        finished;
        (4.2)If ((curNumRequest >= Capacity * 80%) ||
          ((curCpuRate>=highCpuBaseline && curCpuRate – oldCpuRate>=0)
        && (curRamRate >= highRamBaseline && curRamRate –
        oldRamRate>=0) && ((curRamSwapRate >= highRamSwapBaseline &&
        curRamSwapRate –oldRamSwapRate >=0) | | (curNicTraffic >=
        highNicTrafficBaseline && curNicTraffic – oldNicTraffic >=0 ))))
              Then start n servers through the control system;
        (4.3) else if ((curNumRequest <= Capacity * 10%) ||
          ((curCpuRate<=lowCpuBaseline &&curCpuRate – oldCpuRate<=0)
        &&(curRamRate <= lowRamBaseline && curRamRate –oldRamRate<=0)
        && ((curRamSwapRate <= lowRamSwapBaseline && curRamSwapRate–
        oldRamSwapRate <=0) | | (curNicTraffic <= lowNicTrafficBaseline &&
        curNicTraffic – oldNicTraffic <=0 ))))
```

Then stop n servers through the control system;
(4.4) Save the current load values to the old load variables;
}
}

After we adopt the RSSCSS's scheduling algorithm, the server cluster system can meet the quality of service with load balancing and high-reliability by the minimal server resources. So as to achieve energy saving and extend device's effective life, finally to reduce operating costs.

6 The Server Control Method

In the above scheduling algorithm, the server control (start or stop) is the key technology, especially when the server is shutdown (while the operating system and IP network protocol is not running), how to start the server remotely is the difficult problem for the scheduling system. There are two alternative technologies:

(1) DPC

DPC (Direct Platform Control Console), which is developed by Intel Corporation, is a server management technology for Intel architecture. The managed servers must connect to a terminal server (or serial interface server) by its serial interface. The manager can access the performance parameters of the managed server and start or stop the server without the support of operating system and network protocols through the terminal server. However, when the number of managed servers is more than 50, we will need more terminal servers, Under this circumstance, the connections between the application servers and terminal servers would become complex. Furthermore, the serial bus communication distance between the terminal server and application servers must be limited in 25 meters.

(2) Wake up on LAN

The motherboard and Ethernet card of the modern PC server supports remote wake-up function. To set up this feature, first of all, we need configure the CMOS settings of the server in the "Power Management Setup" item and set the "Wake Up On LAN" entry to "Enable", and then connect the server to the local area network and let it be powered (the operating system and IP protocol need not running). If we want to start up a special server (assuming the MAC address of the server's network card is 112233445566), the scheduling server can build and send a special broadcast frame which called "magic packets" to the network. The format of the magic packets is shown as below (102 bytes):

FF FF FF FF FF FF 11 22 33 44 55 66 11 22 33 44 55 66 11 22 33 44 55 66
11 22 33 44 55 66 11 22 33 44 55 66 11 22 33 44 55 66 11 22 33 44 55 66
11 22 33 44 55 66 11 22 33 44 55 66 11 22 33 44 55 66 11 22 33 44 55 66
11 22 33 44 55 66 11 22 33 44 55 66 11 22 33 44 55 66 11 22 33 44 55 66
11 22 33 44 55 66

Finally, all of the server's network card will receive and decode the broadcast frame at data link layer, Only when the MAC address in the magic packet is equal to its own MAC address will the motherboard boot up the server and startup its operating system as well as the application services.

If the scheduling server wants to shutdown a managed server at the work state (the operating system and network is running), it can be done with socket communication program (is called agent) running in the managed servers. The scheduling server can send a shutdown command to the agent, and then the agent will call a special function, such as ExitWindowsEx(EWX_POWEROFF, EWX_FORCE) in Windows and the SYSTEM ("/sbin/shutdown-h now") in Linux to shutdown the specified server.

Acknowledgments. The study is supported by "College students' innovative experimental projects" of NUDT in 2010.

References

1. IDC: China's total energy expenditure of the server (2007), http://www.enet.com.cn/article/2008/0128/A20080128136541.shtml (September 2011)
2. The server cluster power management , http://www.cei.gov.cn/index/serve/showdoc.asp?Color=Twelve&blockcode=wnitdt&filename=201006031210 (July 2011)
3. Zhong, J., Zhu, L., Cao, B.: The energy consumption of IT equipments in data center, http://dcteam.blog.51cto.com/691409/228936 (September 2011)
4. How to calculate carbon dioxide emissions, http://www.power120.com/news/ShowInfo.asp?InfoID=330 (July 2010)
5. sun9527, Commonly used SNMP OID in Linux, http://blog.chinaunix.net/space.php?uid=305417&do=blog&cuid=1136607 (September 2011)
6. Nineday, The calculation method of CPU utilization and memory utilization in Linux, http://blog.csdn.net/nineday/article/details/1
7. Zhang, W.: Linux Virtual Server Cluster for IP Load Balancing, http://yp.oss.org.cn/software/show_resource.php?resource_id=225
8. Pan, L., Wang, L.: Virtual Server Based on Cluster Technique. Computer Engineering And Applications 39(14) (2003)
9. Li, W., Guo, S.: An Adaptive Load Balancing Algorithm for Service Composition. Journal of Software 5 (2006)
10. http://www.tpc.org/default.asp

PID Control Based on Generic Alogorithm Froction Compensation in Sero System

Dingdu Wen

College of Electrical and Information Engineering, Hunan University of Technology, Zhuzhou, 412000, China
wdd2008@163.com

Abstract. In servo system, because of the existence of nonlinear friction link, it is very difficult to parameter identification, and the system dynamic and static performance are greatly effected. In order to improve the performance of the control system, the appropriate control method must be adopted to eliminate the influence of friction. Aiming at these problems, PID control based on genetic algorithm friction compensation in servo system is proposed in this paper. The simulation results show that this method is effective.

Keywords: genetic algorithm, friction compensation, parameter identification, servo system.

1 Introduction

In precision and super low-speed mechanical servo system, because of the existence of nonlinear friction link, the system dynamic and static characteristics are greatly affected. The main performance: crawling phenomena appears when speed is low; larger static error or limit cycle oscillation appears when systen is the steady-state[1]. Therefore, to improve the performance of the control system, appropriate control method must be adopted to eliminate the influence of friction torque, such as high power gain PD control methods, adaptive friction compensation method, the compensation method based on interference observer, and the compensation method based on friction model, etc[2-7]. Based on the model of the friction compensation technology, the key is the friction model which can reflect friction nonlinear characteristics. As friction itself is a complex nonlinear phenomenon, and It relates with many other factors, such as load, movement speed, environmental temperature, lubrication condition, and exercise time, etc [8], so it is difficult to obtain a accurate frictional model parameter by classic system identification method.

So far, domestic and foreign scholars has built more than 30 kinds of friction model [9,10]. Overall, classic friction model can be classified into two categories: static friction model and dynamic friction model. Static friction model mainly describes system friction phenomenon in partial lubrication and complete lubrication stage. Although it cannot well describe the friction phenomenon of static friction stage(namely near zero velocity), the static friction model is simple, contains few parameters, and easy to get more precise friction model parameters, so it is often used to the friction compensation of system. The dynamic friction is the change of the

G. Lee (Ed.): Advances in Automation and Robotics, Vol.2, LNEE 123, pp. 121–128.
Springerlink.com

friction force as a response of speed, and it uses differential equation to describe the dynamic characteristics, so it can better describe friction phenomenon of system in all stages stages. Especially in recent years, the friction model is proposed, such as Lucre friction model and the improved form, GMS friction model,etc. It can be quite comprehensively described to the static and dynamic characteristics of friction, and it summarizes most of friction phenomena which can observed in the experiments, including stribeck effect, Dahl effect, presliding displacement, frictional lag, varying away force, etc. But these models have such shortcomings as more parameters, nonlinear parameters and the impassable identification state variables, etc. So it is difficult to get precise friction model.

In 1962, Generic Algorithn (GA) developed by Holland professor, is a parameter searching and optimization technique based on emulating the evolutionary process of the nature. In identification process, genetic algorithm is adopted as optimization tools, and directly replace identifying vector parameter as individual, thus the friction model parameters of system is obtained. The genetic algorithm is used to system identification with wide adaptability, strong robustness, computational stability, identification precision, and genetic algorithm with global optimization ability, need not assume the search space is differentiable or continuous.Based on "the coulomb + viscous friction" model for parameter identification object, the friction model parameter identification method based on genetic algorithm is proposed in this paper.

2 PID Control Based on Genetic Algorithm Friction Model Parameter Identification

In the initial stage of evolution, becuase the selection pressure is bigger, convergence speed quickly evolves to later period, group fitness difference decreases, and convergence rate slows, even appears precocious phenomena. Crossover rate and mutation rate are two important parameters of the genetic algorithm, and its numerical choice affects the optimal searching performance of genetic algorithm in a large extent. The genetic algorithm is studied from cybernetics theory in this paper, and PID control method based on the frictional model parameter identification of the genetic algorithm is proposed, and it is applied to the service system, and good results are obtained.

2.1 Introduction of the Frictional Model

Because the static friction model is simple, contains few parameters, and easy to get more precise friction model parameters, so it is often used to the friction compensation of system. the dynamic model has such shortcomings as more parameters, nonlinear parameters and the impassable identification state variables, etc, so it is difficult to get precise friction model. Therefore "the coulomb + viscous friction" model is selected in this papers, and its mathematical model is described as:

$$F(t) = F_c \operatorname{sgn}(\dot{\theta}(t)) + F_v\dot{\theta} = \operatorname{sgn}(\dot{\theta}(t))kx_1 + kx_2\dot{\theta}(t) \quad (1)$$

where F_c is Coulomb friction coefficient, F_v is viscous friction coefficient, and kx_1 and kx_3 are on-identify parameters.

2.2 Genetic Algorithm Design

System identification with genetic algorithm have two kinds of solution. One is to determine the model structure based on prior knowledge, then the model parameter is optimized by genetic algorithm. Another is to optimize several structure of empirical estimation based on genetic algorithm, and the parameter is estimated at the same time. the first solution is adopted in this paper. Genetic algorithm identification principle diagram is shown in figure 1.

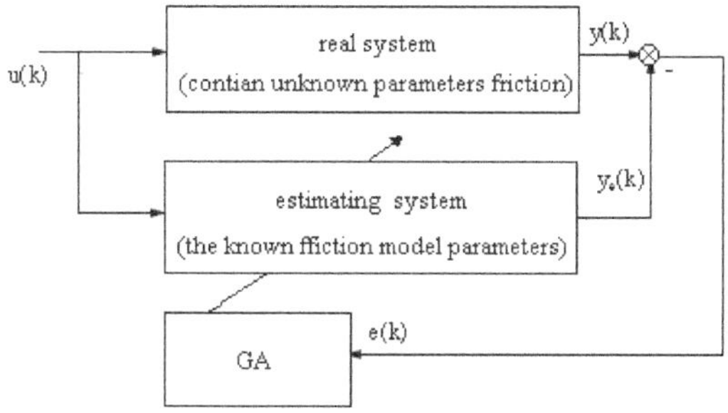

Fig. 1. Genetic algorithm identification principle diagram

where u(k) is system input, y(k) is actual observation value of system output, $y_e(k)$ is estimation model calculation value, e(k) is system output error, and $e(k)=y(k)-y_e(k)$.

2.3 Creating Initial Population

At present, the coding method is mainly binary code and real number code. Binary code cannot reflect the problem of structure, and individual length is big, and it occupies more memory, so it is hard to improve precision. Real number code applies to range larger numerical value, and it is easy to larger space search, and also it is easy to design special algorithm to handle more complicated variable constraint condition. Real number code is used and initial population which randomly generates is 50 in this paper.

2.4 The Objective Function Selection

Identification purpose is to obtain sample parameter of the lowest for the objective function J. Namely, the optimal sample.

In the process of control system parameter optimization or model identification, in general, we take the error function as the objective function, such as error time integral function, etc. In order to obtain satisfactory dynamic characteristics, and prevent excessive control energy, the square of controller output is added to the objective function. The optimal index is described as follows:

$$J = \int_0^\infty (w_1 |e(t)| + w_2 u^2(t))dt + w_3 t_u \tag{2}$$

where e(t) is system error, u(t) is controller output, t_u is rise-time constant, w_1, w_2 and w_3 are weight.

In order to avoid overshoot, the penalty function is adopted. Namely once control system produces overshoot, we take the overshoot as an optimal index. At this time, the optimal index function is described as follows:

$$\text{If } ey(t) < 0, \quad J = \int_0^{\infty} (w_1|e(t)| + w_2 u^2(t) + w_4|ey(t)|)dt + w_3 t_u \tag{3}$$

where w4 are weight, and $w_4 >> w_1$, ey(t)=y(t)-y(t-1).

2.5 Fitness Function Selection

In this paper, random sampling selection method saving optimal individual was used during the selection operation. The initial population is 50. The objective function is such as (3). The fitness function is described as follows:

$$F(k) = \sum_{i=0}^{m} C_{\max} - (e(k-1))^2 \tag{4}$$

where $C_{max}=\max\{e(i)^2\}$, it is used to ensure that the fitness function is nonnegative. m is Sampling step before the 10th sampling.

2.6 Crossover and Mutation

In genetic algorithms, the probabilities of crossover and mutation are the key affecting genetic algorithm behavior and performance, and it directly affects convergence. Crossover probability is bigger, speed of the new individual creation is faster. However, crossover probability is too bigger, genetic mode is more likely destroyed, thus high fitness individual will soon be destroyed. But if crossover probability is too small, it makes the search process slow, even stagnates. For mutation probability, if it is too small, it is difficult to produce new individual structure. Contrarily contrariwise whereas, genetic algorithm becomes pure random search algorithm. For the different optimization problem, this is a tedious work which need repeate test to determine crossover probability and mutation probability, and it is difficult to find the best value to each question [11].

Adaptive genetic algorithm is to let the probabilities of crossover and mutation change with fitness change. When each individual fitness of population tends to be consistent or local optimal, it makes probabilities of crossover and mutation increase. Contrarily contrariwise whereas, it makes probabilities of crossover and mutation decrease. At the same time, corresponding to the lower probabilities of crossover and mutation, the individual which the fitness value is higher than the average fitness value, is protected into the next generation. If not, it is obsoleted. Therefore, adaptive genetic algorithm can provide the best solution for probabilities of crossover and mutation. Meanwhile, it is able to keep the diversity of the population, and it also ensures the convergence of genetic algorithm. Adaptive mutation probability is adopted in this paper.Namely, the fitness is larger, the the mutation probability is smaller. The crossove probabilities (P_c) is 0.9, the crossove probabilities P_m is described as follows:

$$P_m = 0.1 - [1:1:Size] \times 0.01 / Size \tag{5}$$

where Size is sample number.

2.7 Control Algorithm Design

The single PID (proportion-integra-differential) or PD (proportion-differential) controller has many advantages, such as easy adjustment, stable, etc. However, because of the influence of nonlinear friction, it is hard to meet the precision requirement in accuracy tracking control. The PD controller is easy to produce larger steady-state error, and the integral role of PID would lead to the emergence of limit cycles. Therefore, additional friction must compensate. The PD controller is adopted in this paper, and adopting aforementioned friction model compensates friction force, In order to trace expected input signal.

3 Simulation Analysis

The transfer function of a servo system is described as follows:

$$G(s) = \frac{400}{s^2 + 50s} \tag{6}$$

Supposed added controlled system disturbance is a equivalent friction. The friction disturbance of system is simulated by artificial disturbance. The control system diagram is shown in Fig. 2. An input signal of is given, and through simple PID control system, the friction model parameter of system is identified by genetic algorithm. Friction model is described by type (1). In order to avoid the excessive parameter selection range, when the Genetic Algorithm is used, an identification range is chosen according to experience, then parameter is identified in the range. So the blindness of initial optimization is greatly decreased, and calculation is saving. Where sampling time is 1ms, evolution generation is 50 ,100. w_1=0.999, w_2=0.001, w_3=2, w_4=100.

3.1 Simulation without Friction Compensation

Making kx = [0, 0], compensation friction F(t)= 0, System input is $r(t)=0.5\sin(10\pi t)$. Its response is shown in figure 3.

From figure 3, we can conclude that when friction is not compensated, input and output it not only has enormous error between input and output, but also input appears flat-roofed phenomenon. Therefore control effect is poorer.

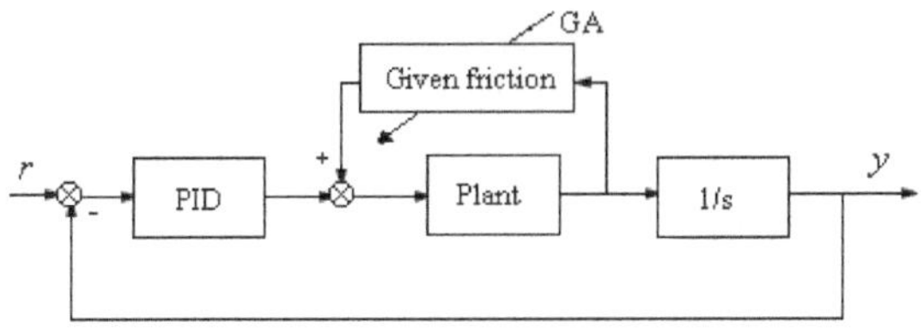

Fig. 2. The control system diagram

3.2 Simulation of Friction Compensation with Genetic Algorithm

Making kx_1=[0.3,1.5], and kx_2=[0.3,1.5], System input is r(t)=0.5sin(10 π t). The optimal sample and the optimal objective function is obtained through the simulation. The system input and output are shown in Fig.4 (a), (b), and the optimization process of objective function J is shown in Fig.5 (a), (b). (Where a is evolution generation as 50, b is evolution generation as 100).

When evolution generation is 50, we obtain: optimal sample Best_s =[0.3649, 1.0729], the optimal objective function values Best_J=25.5887.

When evolution generation is 100, we obtain: optimal sample Best_s =[0.3125, 1.4775], the optimal objective function values Best_J=25.5468.

4 Conclusions and Future Works

Due to widespread friction phenomenon in servo system, PID control based on generic alogorithm friction compensation is put forward in this paper. Simulation of simple friction model parameter identification and servo system friction compensation is achieved. The simulation results show that the precision is higher, and it has certain practical value.

For future works, it will be interesting to design more complex genetic identification algorithm to realize dynamic friction model parameters identification,thus making identification more accurate and rapid. And how will the genetic identification algorithm apply to practical engineering to improve the control precision of servo system.

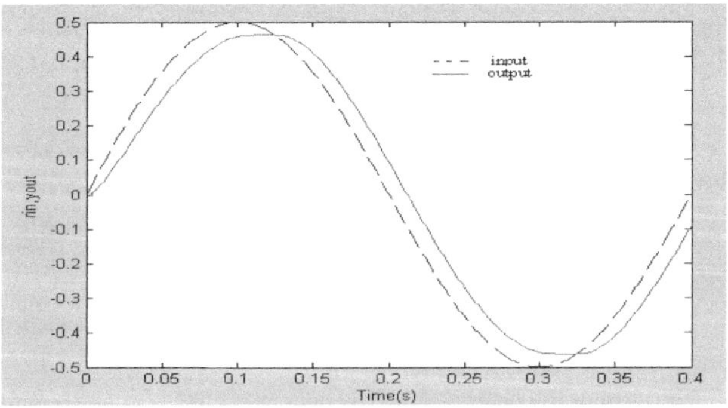

Fig. 3. The response curve without friction compensation

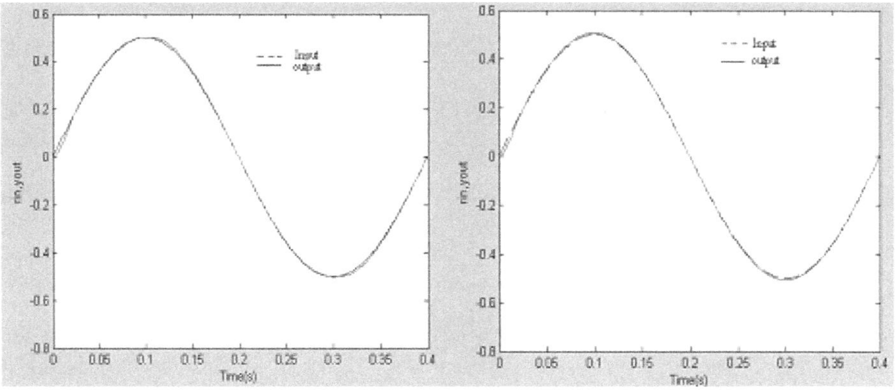

(a) (b)

Fig. 4. The response curve with friction compensation

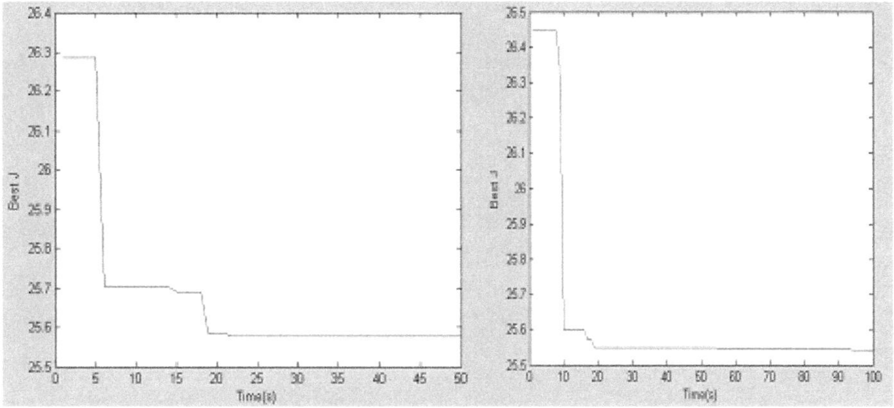

(a) (b)

Fig. 5. The optimization process of objective function J

References

1. Armstrong, B., Dupont, P., Canudas, C.: A survey of models, analysis tools and compensation methods for the control of machines with friction. Automatica 30, 1083–1138 (1994)
2. Rizos, D.D., Fassois, S.D.: Slip Model Based Identification and Control of Systems with Friction. In: 44th IEEE Conference on Decision and Control, pp. 4578–4583. IEEE Press, New York (2005)
3. Makkar, C., Dixon, W.E., Sawyer, W.G., Hu, G.: A New Continuously Differentiable Friction Model for control systems design. In: The International Conference on Advanced Intelligent Mechatronics Monterey, pp. 24–28. IEEE Press, New York (2005)

4. Abender, F., Lampaert, V., Swevers, J.: The Generalized Maxwell-Slip Model: A Novel Model for Friction Simulation and Compensation. IEEE Transactions on Automatic Control 50, 1883–1887 (2005)
5. Ashwani, K.R., Jinhyoung, O., Dennis, S.B.: On the lugre model and friction induced hysteresis. In: The American Control Conference, pp. 3247–3252. IEEE Press, New York (2006)
6. Farid, A.B., Vincent, L., Jan, S.: The generaized maxwell-slip model: a novel model for friction simulation and compensation. IEEE Transactions on Automatic Control 50, 1883–1887 (2005)
7. Liu, H., Gao, W.: Study of the Feedforward Friction Compensation in Servo System. Science Technology and Engineering 7, 614–617 (2007)
8. Oh, J., Ashwani, K.: Duhem Models for Hysteresis in Sliding and Presliding Friction. In: 44th IEEE Conference on Decision and Control, pp. 12–15. IEEE Press, New York (2005)
9. Paprk, E.C., Lim, H., Choi, C.H.: Position control of X-Y table at velocity reversal using presliding friction characteristics. IEEE Transactions on Control Systems Technology 11, 24–30 (2003)
10. Liu, Q., Er, L., Liu, J.: Overview of Characteristics, Modeling and Compensation of Nonlinear Friction in Servo Systems. Systems Engineerinn end Electronics 25, 47–52 (2003)
11. Wen, D., Zeng, H., He, L.: PID paramenter optimization of the Resistance-heated Furnace Temperature Control system based on genetic algorithm. Electrical Automation 30, 6–8 (2005)

The Analytical Research on Construction Project Cost Estimation Model Which Is Based on Artificial Neural Network

Junming Hong

Hangzhou Polytechnic College
1357827972@qq.com

Abstract. The cost of construction projects is more and more complicated, and facing cost reform trend to a certain extent. Construction project's cost estimation model which is based on artificial neural network, can, to some extent, make the auxiliary construction project cost better estimate. It's a big help in related engineering bid.This article makes a analytical research on it.Meanwhile it also discusses the model and application process.

Keywords: Artificial neural network, construction project, cost of construction.

1 Introduction

Construction project cost often use process similar project data, and use certain mathematical model to make cost calculation. But, to a certain extent, because of the construction projects are affected by many factors, and there is complex relationship between each other. Thus, related engineering data generally has the characteristics of the fuzziness and the randomness. Therefore, the use of related data model calculation of construction cost can hardly reach people's requirements of precision. But artificial neural network to the complex nonlinear object modeling which is difficult to precisely describe and computational reasoning operation is relatively easier. So, we can use artificial neural network to establish relevant model, through it we can perform some of the estimates in the cost of construction projects.

Artificial neural network is a new information processing system which imitates the structure and function of the brain cells .It also has cranial nerve structure and has the ability to deal with problems such as thinking brain. Besides, models based on artificial neural network are a complicated nonlinear adaptive dynamic network system, which is united extensive together by a large number of simple processing. Now, the model is being more and more widely used in construction engineering field and is used to estimate the cost of construction projects. To a large extent, neural network model has a distributed storage, parallel computing and adaptive learning characteristic. So it doesn't need to build accurate calculation equation but can directly use the sample data to realize nonlinear mapping between input and output layers. So it is suitable to establish the problems which is hard to establish mathematical model but easy to collect learning samples.

G. Lee (Ed.): Advances in Automation and Robotics, Vol.2, LNEE 123, pp. 129–134.
Springerlink.com

2 The Building of Construction Project Cost Model Based on Artificial Neural Network

Artificial neurons are the most basic cell of artificial neural network. Neurons are the highly abstract and simplified structure model in the brain cells. Neurons can handle the corresponding information and cooperate with other neurons. Figure 1 is a typical artificial neural model.

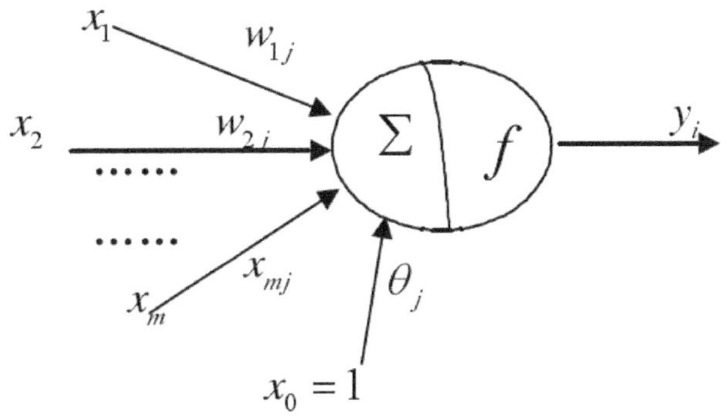

Fig. 1. Typical model of artificial neural typical

Besides, other basic elements of the artificial neural network includes threshold, excitation function, weights between and so on.

Usually, artificial neural network has good approximation ability to nonlinear function. Up to the present day, there are about 40 kinds of them. To a great extent, artificial neural network has become an important technology in nonlinear system's modeling, identification and control. Among them, the most widely applied and be studied by most people, is multilayer neural network BP model. BP neural network is proposed by Rume chart and others in 1985.It are a kind of typical neural network structure which is developed on the basis of the back propagation algorithm. It is also a kind of typical multi-level feedback type network. From the composition structure aspect, the BP neural network is composed by three neurons layers. The lower level is called the input layer, middle layer is called hidden layer, and the top is output layer. Each of the lower layers can realize the connection between neurons. There is no connection between neurons in each layer of the parallel. When making information processing, input information begins to spread from the input layer node and pass the hidden layer of each node one by one. In the end, it make output through output node. Its structure is shown in figure 2.

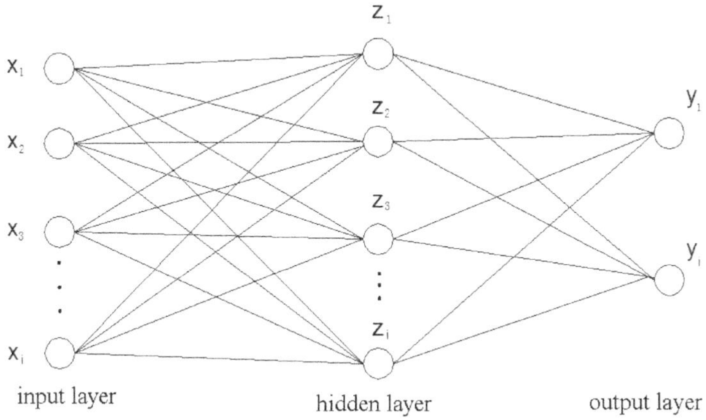

Fig. 2. The typical BP neural network structure model

The steps of building a neural network model for construction project cost forecast by the BP neural network model are as below. First, according to the construction of the project to estimate categories, confirm the main factors which affect the cost as input layer. Generally speaking, factors which influence architecture planning and design includes the cost of design program, artificial cost and the consumption of building material costs. The related research data shows, the main factors which influence these factors are: 1, building structure characteristics; 2, the geological and hydrological conditions of the building; 3, the influence of construction unit; 4, the influence of the construction unit; 5, construction technology and mechanical equipment factors. Among them, Condition 3, 4, 5 are consumption cost factors in the construction phase. It allows no consideration in the design phase, but it can influence the overall cost of the project. From the points in building type, Construction projects types may generally be divided into: office buildings, residential building, and industrial building and so on. There is little difference in the cost of different types in construction projects. For example, to estimate the cost of a residential building by neural network model. The established input layer which is based on neural network's building engineering estimation model can be defined as building construction structure, foundation forms, the whole building area, internal door model form of residential building, type and extent of indoor and outdoor decoration after the completion of the building, the type of ground engineering and Windows and doors engineering. The structure of the output layer's output is the architectural engineering project cost.

The learning algorithm of BP neural network is promotion and development through simple learning laws. Using iterative method, it first presents a set of the initial value, and then output a sample to calculate output value. Next we calculate the difference between the output value and expectative value. Then reduce the difference by modification weights of network. Then repeating a line of iterative process until the difference between output value and expectative value is less than preservative value. The progress in building a neural network model which is forecasted by the construction project cost is shown below:

Set the training sample is $(x^k, y^k), k = 1,2,3,\cdots$, and set m is learning samples in the sample number. Set the input mode vector is $x^k = (x_1^k, x_2^k, \cdots x_n^k)$, the expectative output value of the model's first k iteration is y_k.The specific training process of the model is shown as below:

At first, according to the input conditions, we calculate mold input conditions in the hidden layer of the model.It are shown as below:

$$a_j^k = \sum_{i=1}^{n} \omega_{ij} \times x_i^k,\ j = 1,2,3\cdots p$$

According to the above formula, we calculate the corresponding hidden output conditions by hidden input conditions we have got. In the process of computation, we can use stimulated function to deal with the input X. Excitation function is usually single polarity sigmoid function. Using the function, we get the hidden output conditions. It's shown as below:

$$f(x) = \frac{1}{1+e^{-x}} = \frac{1}{1+\exp(-x)}, and\ b_j^k = f(a_j^k)$$

After the two above steps, we can get the output of the hidden layer. Put the output of the hidden layer to the input conditions of output layer. Using the connection weight ω_j from the hidden layer to the output layer. We can calculate the output condition c^k of neural units in the hidden layer is:

$$c^k = \sum_{j=1}^{p} \omega_j \times b_j^k$$

According to the above calculating results, we can get the output in the corresponding output layer unit, it is:

$$y_k = f(c^k)$$

After calculating the output of output layer unit, compare output results with prediction results. Then we calculate the difference between model's output results and the prediction results. According to this difference, make amend in network weights of BP neural network. From mathematical sense, BP neural network is a mapping which uses map neural samples to realize the goal from N d space vector to Space subset. For example, suppose the difference of squares of the output units is:

$$E^k = \frac{1}{2}(y_k - \hat{y}_k)$$

In it, $\hat{y}_k$ are the actual output results after unit iteration?

So, the total error of each sample can be shown:

$$E = \sum_{k=1}^{m} E^k$$

3 Example Analysis

Generally speaking, the estimates of construction project cost are affected by many factors. All the factors are very complicated. But, experienced engineers, according to the characteristics of the engineering, can estimate the cost of corresponding engineering on the basis of the previous engineering data. Thus they can provide certain basis for the corresponding engineering bid. And, generally speaking, more historical data, more accurate estimate. The artificial neural network which imitates the human brain's way of thinking is good at handling this data.

According to the principle of engineering cost estimation and the relevant historical data, and using the BP neural network model, we can estimate the cost analysis of some high-rise residential houses which is invested by a tourism development company. The residential building has 24 floors altogether, including 1floor underground and 23 floors ground. Its construction area is about 21200 square meters. Its structure is shear wall structure and its building foundation is reinforced concrete bored piles. The exterior wall of masonry part is using imported exterior wall emulsion paint. Its interior wall is ordinary white emulsion paint and its door and window uses model steel structure. The ground is simplified made of cement. The relationship of relevant record data can be shown as form 1 below:

Table 1. The table of the similarity of the different forms of the construction

	Project name	Structure form	Foundation forms	Doors and Windows structure	Exterior wall	Interior wall	ground
A	Residential building	The shear wall structure (0.9)	Concrete pile (0.9)	Model steel structure (0.9)	Imported emulsioni paint (0.85)	White paint (0.85)	Ordinary cement (0.85)
B	Residential building	The shear wall structure (0.8)	Concrete pile (1)	Model steel structure (0.85)	The exterior coating (0.85)	White paint (0.85)	Ordinary cement (0.85)
C	Residential building	Box shear structure (0.8)	Box foundation (0.8)	Aluminum alloy structure (0.8)	Exterior wall emulsioni paint (0.85)	white paint (0.85)	Ordinary cement (0.85)
D	Residential building	Shear wall (0.9)	Concrete pile (0.86)	Aluminum alloy structure (0.8)	Exterior wall emulsioni paint (1)	White paint (1)	Ordinary cement (1)

According to the different similarity, we choose the three engineering above to estimate the cost of the residential building. First, work out the size of the similarity in cost between the above project and the will —construct—project. Then use the engineering cost estimate model based on network model which is introducing above to estimate the revised unilaterally cost is 1453. The estimated results are estimation that is on the basis of not too accurate some detail information's of the engineering. Its characteristics are fuzzy, but the accuracy of results can reach 95%.So, it's relatively satisfactory.

4 Epilogue

With the development of the technology in building industry, the engineering cost of construction engineering is more and more complicated. And it is also facing trend in cost reform. Compare to other evaluation model, the construction project cost estimation model which is based on artificial neural network adapts to construction project cost estimation better. The accuracy of estimation can meet the expected requirements to a certain extent. Besides, to make estimation by neural network model has many advantages, such as convenient modeling and quick calculation. So it will have a good prospect in construction project cost estimation.

References

1. Ke, H., Xia, L.: The discussion on index method of construction and installation's cost. Journal of Tianjin Institute of Technology (2003)
2. Yuan, Z.: Artificial neural network and its application. Tsinghua university press, Beijing
3. Tang, X., Lu, H.: Project cost estimation based on artificial neural network. Journal of Changsha Communications University (March 1999)
4. Chen, Q., Zhai, M.: The research on construction project cost estimation model based on BP neural network. Optimization of Capital Construction (June 2002)

The Design of Stereoscopic Warehouse Stacker' Motion and Control System

Jingtao Geng, Yongxia Lu, Huijing Yang, and Dong Chen

Hebei Chemical & Pharmaceutical College
1357827972@qq.com

Abstract. With the continuous improvement of industrial automation, automated warehouse has been widely used in various industries to improve the production efficiency and Logistics speed to a great extent. As an essential equipment of automated warehouse, stacker' working efficiency and stable performance will affect the usability of the warehouse directly. This essay studies the tunnel type stacking in machine' speed control system of an automated warehouse, and analysis its motion curves. Then we combine it with the stacker's composite material structure to guarantee better motion control accuracy and motion reliability.

Keywords: stereoscopic warehouse, stacker, motion control.

1 Introduction

Modern automated warehouse is a modern warehouse which uses high-dimensional shelf to store materials, and to control management by computers and making access role by automatic stacking trucks. In recent years, stereoscopic warehouses has highly improved related enterprises 'production and logistics speed. Stacker is an important and indispensable equipment in a stereoscopic warehouse, its working efficiency and stable function affects stereoscopic warehouse'efficiency and economic benefits fundamentally. As a whole, stacker's evaluation performance includes : lifting speed, running speed, goods forking speed, and stability and address recognition accuracy. Among all the evaluation performances, the stacker's speed is the essential factor, because it can control can deeply affect the stability, the vibration situation, and address recognition accuracy of stacker's operation. So, making In-depth research in stacker's motion situation and speed control system is extremely important, it can help us to make sackers with high—efficiency, more stable and larger motion range.

2 Motion Control of the Stacker

2.1 Abstract

An stacker's motion control technology mainly reflects in its motion speed control, nowadays there are several methods of stacker's motion speed control, as been showed below :

G. Lee (Ed.): Advances in Automation and Robotics, Vol.2, LNEE 123, pp. 135–139.
Springerlink.com

1) Multi-range speed control mode. This control mode is built in a certain usage foundation. When it's working, we must consider the stacker's actual control need and it also can use sub-file operation mode according to relevant experience. To a large extent, this mode needs coordination exercise control and it also requires dealing with Inverter's performance.

2) Using inverter to complete the closed-loop speed stacker and stacker control displacement's open-loop control, usually, the Stacker's ideal motion curve can be concluded to Standard acceleration and deceleration curves. After one motion's experimental curve being confirmed, it can be stored in a stored and we can use PLC to control the inverter to change the motion curve, thus we achieve the purpose of accurate speed and stop.

3) There are some problems in the turning point in speed and acceleration switch of stacker movement functions. To solve this problem, we can use the fuzzy control technology to make reasonable improvement.

A mechanical processing factory's warehouse is mainly used for mould、the storage of semi-finished products and finished products processing. It has the function of automatic management store items. The deposit product for stacker deposit requirement is taken must be fast, stable and reliable. The control system adopts PLC as the master components; it has a greater flexibility and high reliability. Warehouse stacker uses common stacker machine, the weight is of 0.25-1.5 tons, and its operation speed is from twenty five to one hundred and eighty meters per minute. It mainly composed of the body, the level of operating agencies, vertical hoisting mechanism, telescopic fork loading station and institutions and several other components. The main structure of stacker machine is as shown in figure 1 below:

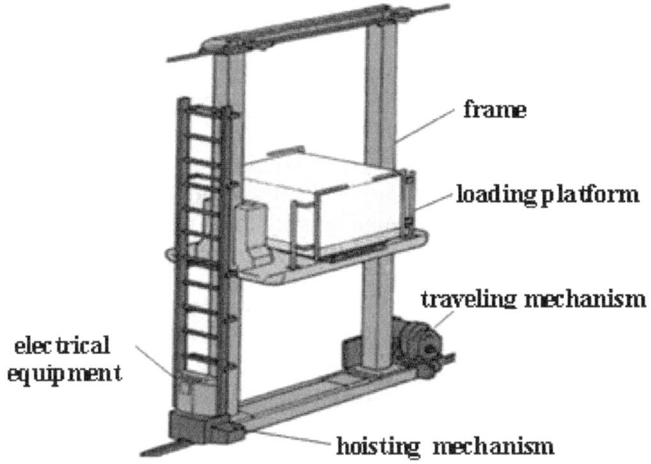

Fig. 1. Stacker machine's main structure

2.2 The Set of the Velocity of the Stacker

From the main structure of stacker machine ,we can also know, tuckers' major sports includes three basic action, these are : level of movement mechanism、 the vertical lifting mechanism's up and down movement、 adjustable action in goods fork telescopic machine. So, stackers' final task efficiency, its movement and performance is also decided by these three movements of the basic control degree. When stacking machine is moving, to ensure the movement is smooth and reliable, we must ensure that the stop registration of movement's accuracy and its movements flatness by some certain control strategy. The method of motion speed is to adjust the operating curve, and then stored it in control memory. When PLC is controlling the motion, through changing the curve which is stored in the inverter we can achieve the purpose of accurate speed and stop.

Level speed setting mechanism is settled for four gears, these are crawling, low speed, middle and high speed. When PLC is controlling the motion, it receives the destination address from a controller, then it Compare the current address to the destination address, and calculate the distance from current address distance to destination address. Make a good's distance as a unit distance. When the distance is larger than six units distance, it use the motion mode of curve a, and use proper acceleration speed when starts and stops, to ensure the precision of the prospective stop. It also use maximum speed to operate in the middle location. Thus it can improve the efficiency. When the distance is less than six units distance yet greater than two units distance, it use the motion mode of curve b. And when the distance is less than two units distance, it use the motion mode of curve c.

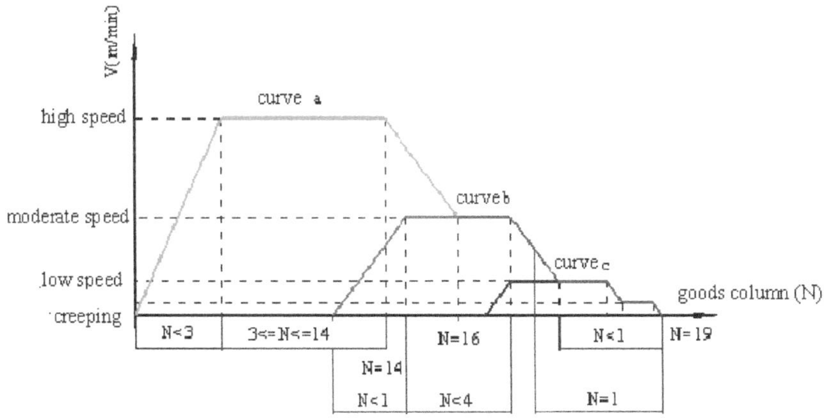

Fig. 2. The graph of level mechanism's speed

Similarly, the vertical lifting operation of lifting mechanism sets three different speeds. When the operation is normal, it adopts high speed to improve efficiency. When it approaches the target, it down to low speed in order to raise the stop precision. Vertical lifting mechanism's speed curve is as shown in Fig. 3:

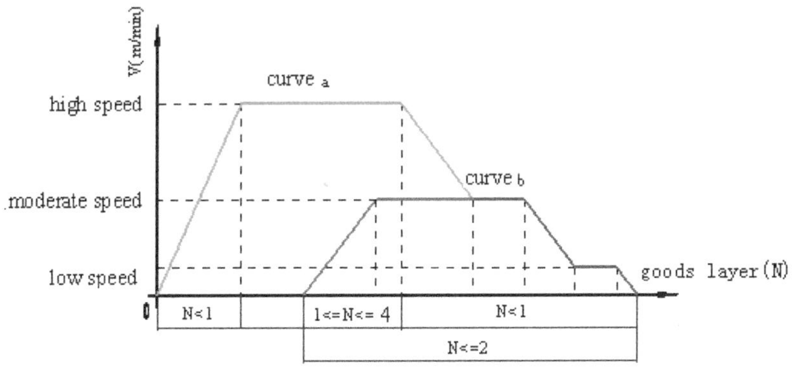

Fig. 3. The graph of Vertical lifting mechanism

The role of the fork telescopic machine goods is through the telescopic movement to put corresponding goods to goods box. The speed of expansion movement can not be too fast, so it can reduce the impact of the equipment action. In this system, adjustable speed of goods fork is been settled to two gears, normal expansion speed is 10 m/min speed, and when is close to the goods the speed is 4m/min.

2.3 Positioning Control of Stacker

Besides controlling the motion speed, stackers' positioning control directly affects the stop precision. So, the position control of stacker system is an important indicator of stacker motion control system.

Stackers positioning control includes the above three positioning accuracy of motion organizations, among them, the level address system is the foundation of stackers motion position. Its function deeply affects the stackers' working efficiency and mechanical properties. In the motion aspect, the stackers' access to goods is very frequent, so it requires high efficiency and positioning correct in motion. From stacker mechanical structure, the stacker which used this system can move heavier goods. So it requires when a stacker is in acceleration and deceleration, column of the deformation must be in the allowable range.

As for stackers' automatic control system, automatic stop bit is the most fundamental question. Now the widely used reorganization address location technology, it can be divided into photoelectric sensor address technology, rotary encoder address technology and laser ranging address technology in general. This system uses the photoelectric sensor obtain address, uses the recognition of address and machine of address to compose the stacker's automatic device of address. It also uses relative obtain address, that means uses photoelectric switch when passing the corresponding goods position, through the photoelectric sensor to produce electrical impulses, according to different moving direction, to address code add one or minus one operation, use PLC count to record the current address,. In the car body, we put two position— different photoelectric switches. When PLC captures the first switch's moves, it makes car body to slow down; when it captures the second switch moves, it makes the car body stop.

The motion control method vertical movement mechanism, goods fork telescopic machine and level control method is similar. The location of the device level control is as shown in Fig.4:

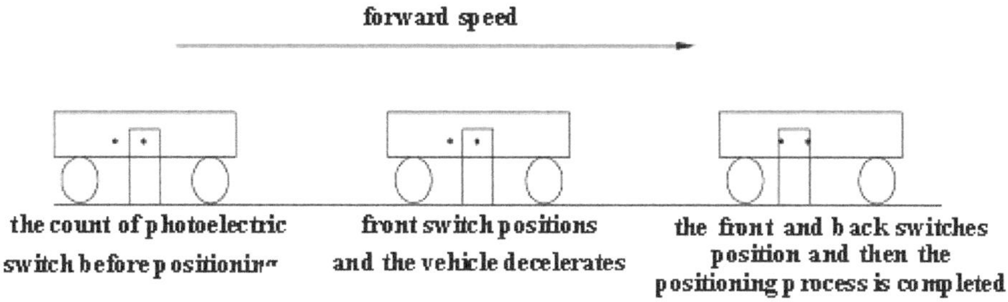

Fig. 4. Level control positioning scheme

As the scheme shows, when it making horizon movement, PLC figures out the distance from current position to the target position. In place of the other position before the orientation patch is used to count, after getting to the destination address, the first action photoelectric switches, car body slows down, through reasonable control of haste and slow acceleration, to ensure when the latter photoelectric switch moves, then the car body just stop. Thus, it can guarantee the car body operates in reasonable speed and acceleration. Slow speed operation time is short. This improves the operation condition of the motor to greater extent. It also has much advantage in reliability and stability of the system.

3 Conclusion

These above design process ensure the stacker machine movement's performance in movement of speed and positioning accuracy. After the motion control method being used in the actual operation of the factory warehouse, it get a good effect. But as for the influence on the performance of the movement in actual movement speed and acceleration, we still need making deeper analysis to a large extent.

References

1. Luo, Z.: The research on controling and designment of automated warehouse stacker, Master's degree thesis. Gansu University of Industry, Lanzhou (2005)
2. JB/T 9018-1999, The design specification of rail tunnel type top shelf warehouse
3. Wang, G.: The designment of automated warehouse stacker's control system. Lifting the Transport Machinery (October 2008)
4. Hu, C., Lu, J., Mei, W.: The designment of parcel's automated warehouse and the designment of management system (January 2007), http://mw.newmaker.com/

Exploring the History of New Energy Vehicles

Shuying Li

Shenyang Polytechnic College
1357827972@qq.com

Abstract. Now days, new energy vehicles is more and more popular in our life. It's very important for us to explore its origin, development, present situation and its prospect. But now we are not sure whether the new energy vehicles have a good development prospect, whether it has difficulty in technology, whether the new energy vehicles can become the main stream of the car industry. All of these are not sure. Through the analysis of the facts, this essay will clear up the development of new energy vehicles, so we can have a good understanding in the prospect of new energy vehicles.

Keywords: New energy, Car, Technology, Development prospects.

1 Traced Back, Exploring the New Energy Vehicle's Development Way

In September 2010, green *car* magazine is published formally, it's the only professional media in new energy vehicles field. In the end of 2010, exploring five hotspots in car market, new energy is on the list, and the most popular one is limit new energy vehicles. On December 1, 2010, the State Council development research center and the China international trade promotion association formally jointly organize*2010 China international new energy vehicle development peak BBS*, is held in Beijing conference center. On September 6 to 10, 2011, *the sixth China intelligence annual meeting and China international traffic intelligent transportation exhibition* and *The 7th international energy saving and new energy vehicles BBS and innovative development exhibition* will be held in Beijing. This activity's theme is *intelligent transportation and new energy vehicle-low carbon green travel* .In recent years, from a lot of subsidies on new energy vehicles to new energy vehicles promotion plan, it includes the forthcoming *Energy saving and new energy vehicles industry planning (2011—2020)* .Our country's policies in the state to new energy vehicles support is unprecedented.

In fact, automobile's new energy research and development is not a new thing. China's automobile new energy industry has come into being in the beginning of the 21st century. In 2001, the research in new energy vehicles was brought into "863" science subject, which is made during "Ten Five-Year Plan". Our country also made a strategy that is from gasoline car to Hydrogen-powered vehicles. Since "Eleven Five-Year Plan", our country has put forward "energy saving and new energy vehicles" strategy, and the government attaches great importance to the development of new energy vehicle and industrialization.

G. Lee (Ed.): Advances in Automation and Robotics, Vol.2, LNEE 123, pp. 141–146.
Springerlink.com

In 2008, new energy vehicles have been in full bloom in our country. 2008 years is China's "new energy vehicles to start the reign of an emperor". On July 1, 2009, *management rules of producing access to new energy vehicles* are coming into effect officially.

2 Analyze of the Present Situation, and Overlook the Prospect of the Development in New Energy Vehicles

All the facts shows that cars with new energy technology, has broad development prospects, but whether things will be so ideal? This needs us to analyze it according to the facts.

To consider present situation and prospects of car's new energy technology development, the first problem we must make clear is : what is the's new energy technology ? Then the problem is emerging.

2.1 What Is the's New Energy Technology Vehicles ?

New energy technology is also can be called unconventional energy, is an energy form other than the traditional energy. Especially it's refer to those energy which are just in development and utilization, or are in actively research, and needs to promote, such as wind energy, geothermal energy, solar energy, oes, biomass energy ,nuclear fusion energy and so on.

2.2 What Is the New Energy Vehicle?

New energy vehicle production enterprise and products access management rules says:" new energy technology car is the vehicle which fuel is unconventional power sources (or use a conventional car fuel、 new on-board power device), combine the dynamic control of vehicle with driving technology. Thus we can get the vehicles with advanced technology principle— new technology, new structure. New energy vehicles include HEV, BEV, FCEV, hydrogen driving—vehicle, other new energy (efficient accumulator, dimethyl ether) vehicles, and so on.

The new energy technology of vehicles is the problem to research in the new energy vehicle's source of power. Electric vehicles are widely considered to be the main force of the new energy vehicles in the future. The difficulty in technology is: battery capacity, engine system and electronic control system. Now, Tsinghua University, Shanghai JiaoTong University and Tongji University, all these university with vehicles departments is the pioneer in the technology.

In 2009, the latest direction in automobile new energy technology research field is "thermal magnetic oscillation power technology", which can be used in portable power equipment field such as the vehicles. When it's working, it may become a hydrogen fuel battery alternative, so it's also called" physical fuel battery". Now it's in early development research phase.

Then, what's development prospects of automobiles with new energy technologies will be? Whether it can be the master of the car sales market ?

In 2010, director of the Development Research Center of the State Council Zhang Yutai said in New energy vehicles BBS" facing the global energy shortage and the pressure of climate warming, the traditional industry will meet serious challenges. All countries in the world seek countermeasures actively, and consider the development of new energy vehicles as an important strategic in improving the car industry's competitiveness and keeping the economy in sustainable development."He pointed that new energy vehicle development needs five innovation, these are: the innovative environment of the system, the innovative technology path, the innovative division of value chain, innovative cooperation mechanism, innovative business model.

The state pies attention to the development of new energy vehicles and give a lot of policy support. Then what's the market reaction?

In Beijing, new energy vehicles are very popular; because Beijing has considered developing new energy vehicles is the most important factor in realizing Beijing's great-leap-forward development in new energy vehicles. By the end of 2012, Beijing plans to promote 30000 Private new energy vehicles. It also plans to build relevant new energy vehicles' filling stations and charging pile for new energy vehicles. Besides, it plans to build management information platform for new energy vehicles demonstration running.

On November 25, 2010, the *Energy conservation and environmental protection industry development planning* is approved by the State Council which is put forward by the National Development and Reform Commission and Ministry of Environmental Protection. It includes specific finance, taxation and financial policy support. Wu Xiaoqing, vice minister of Environmental protection, points that during" Twelve Five-Year Plan", China's environmental protection investment will be more than double than in the past five years. This means it will reach more than 3 trillion yuan. Among them, a big part of it will be filled into automobile new energy research and extension. "This will provide power and guarantee to continuous development of new energy car technologies" says Cao He, National securities auto analyst. He thought policy support usually is only a figure, but it will drive local governments, businesses, and capital market to make more investments. Under such a great plan, the next five to ten years, the investment in new energy vehicle industry will reach several hundred billion yuan.

Although the government's support strength is very large, enterprise development level is advancing, there are progressive difficulty in Beijing's new energy vehicles' popularization and demonstration. Beijing's production capacity of the progress may have good reserves. Located at Beijing dazing district, Beijing new energy vehicles industry base will become the research and development and the production base of electric cars, hybrid cars. Next year it will realize the goal of single class the annual production of 20000 cars. By 2015,it will form 150000 new energy automobiles and will form 150000 sets of driving system and 200000 sets of controller. But independent research and development has problems in input-output ratio. If we can't balance this ratio, independent research and development will fail. The electric car technology is also independent research and development, and not yet accepted the test of the market. If we only pursue great-leap-forward development, we are easy to make mistakes.

Besides, strictly from concept, we will find that new energy vehicles are still very perplexed. There is not a unified concept in this field. Although the ideas and classifications made by technical experts, and government management department has been very clear. But to market, consumers and ordinary people who are focus on the vehicles industry will be confused. They don't have fresh intuitive impression in it.

In fact, in the consumer's point of view, the first new energy vehicles are hybrids vehicles. These vehicles have been relatively mature and perfect in technical terms in nowadays market. From the beginning of the Toyota pries to noted Buick JunYue's hybrid Power World Expo Taxi now everywhere we can see in the street, although hybrid cars is common, but because of the higher price reason, it is still difficult to special more consumer choice. But at least they can make intuitive feel of real existence to ordinary customers.

The problem is, the pace of development in domestic hybrid vehicles is very slow. Except JunYue hybrid vehicles attracts people's attention because of World Expo. Foison S30 hybrid Power vehicles and Volunteer xiang chang's medium hybrid Power vehicles are still can't be seen in the market; even they are in the list of the independent brand new products which is made by 2010 the first letter work new car directory. This made the public even if is interested in real earnest might not be concerned about details of hybrid electric vehicles, let alone to buy. At an advantageous position, hybrid power vehicles is even so, actual sales of other more fresh new energy vehicles are undoubtedly more difficult.

From the above, we can see, new energy vehicles have good prospect of development. And all kinds of resources are also very useful. But the technical breakthroughs need time, and self-identity of consumer needs to be more appropriate propaganda and lead. In a word, *A bright future, but the road turns and twists.*

3 More and More New Energy Vehicles Are Emerging and More Companies Come into the New Energy Vehicles Industry

Major automotive giants are stepping into the new energy vehicles' development and research. This will undoubtedly bring challenges and opportunities to new energy vehicles. On October 14 to 16 ,2010, BMW Group make a vivid show in the vehicles' new energy research and development and the promotion has been achieved to the public *the fourth china (hangzhou) international clean energy and environmental protection industry exhibition.* It also showed the new technology and new products that will change our lives in the near future, including BMW efficient hybrid and BMW 7 series, hybrid X6, electric drive car MINI E and BMW Active. In line with "in China, originated from China, for China" objective, they develop ECHO electric car project, it left deep impression on the relevant government departments, scientific research institutions, industry experts and the general public. And it also brought enlightenment to the development of China's new energy vehicles.

In 2010 at the Hanover commercial auto exhibition, the most attractive thing is so many energy vehicle technology and products. Low carbon economy has come today, through two years' the global financial crisis, the global commercial vehicle industry agreed to coincide the ground that in order to get out of the economic crisis problems and to adapt to the needs of the development of a low carbon economy. So they must develop new energy commercial vehicles. These are breach and the development direction in the future. At this auto exhibition, we can see that enterprises of global mainstream commercial vehicle have done many things in new energy vehicles field. In this field they expect a great development. The automobile enterprise is the main force of the new energy vehicles, hybrid power trucks such as sought DAF,LF; MITSUBISHI FUSO Canter Eco Hybrid, Volvo, in fairs diesel and methane hybrid power technology for the first time, Germany's PLANTOS pure electric light trucks, all of them are the examples. New energy vehicles is not only large enterprise's patents : Allison, supported by the government, it will speed up the development of the hybrid technology. The auto show not only shows a variety of exciting new technology in the new energy vehicles. It also attracted the attention of the enterprise in spare parts. New energy vehicles are just like suddenly appeared, but have a bright future.

These new technologies, new energy vehicles development not only provides the reference for the development of Chinese cars, but also brings great competition. But only China's new energy vehicles industry pass through such a balance, can it be stronger.

4 Look for the Best Way Out

In 2011, at Shanghai international auto showcasing Che, the Daimler group President and Byd group President Wang Chuanfu, agreed the plan in a new generation of pure cooperative production electric cars to 2013. Although in the past byd has always thought it as secrets for business and insist on an electric vehicle and the battery technology blockade. But Benz still adopted byd's technology and make it as the foundation of the cooperation for both sides. This is the beginning of the competition in new energy vehicles field.

Because traditional automobile manufacturers don't grasp core technology, they only can be the system integration. This will be more prevalent in new energy vehicles the era. What we can foresee is, most of the traditional car maker will lose the core competitive power because they don't master the battery, the electronic control and motor the core of the system technology. They may be controlled by battery makers or components groups.

New energy vehicles has become clear to us .The last we know is battery, motor and electronic control system these three key technology. According to the requirements of *The Ten Years Planning,* by 2015, power battery, motor, electronic key parts of core technology will realize democracy. By 2020, energy saving and new energy vehicles and key components technology will reach the international advanced level.

In the core technology of new energy vehicles, what's the situation of Chinese auto company? The personage inside course of study thinks that in the past the ashkenazi enterprises had a overwhelmingly advantage in the traditional motor technology, for electric car motor they don't have absolute advantage. They are also in the early stages of development. But in the electronic control system aspect, Chinese enterprises have fell behind, because car electronic technology has been a big weakness of the auto industry of China.

In the technology of electric car with motor, enterprises at home and abroad are in the same basic starting line. But in the industrial manufacturing and research on basic materials, foreign counties are more advanced. *In engineering application, domestic enterprise still has certain advantages.* A controller of Universal group said.

In addition to the rich ore, lithium mineral resources and rare earth resources, new energy vehicles of motor still needs a kind of rare earth resources, called a nd. This is still very rich in China. This provide a condition for China's development motor system. *But only rich in resources without technology patent, this makes these resources can't sell a good price.* Said by a car company people.

Our country has developed ac induction motor, the permanent magnet synchronous motor switched reluctance motor independently. These core indicators have reached the same power levels of international advanced level. But in product integration, reliability and system application technology, there still exists a relatively large gap between them and the advanced international level. One personage inside course of study says. Lots of much domestic enterprise has made investment to motor system, rare earth resources, and power battery investment. But the difference in the scale of investment is very large; the enterprise strength is also quite different. If cannot grasp core technology, China's new energy vehicles were doomed to failure.

For this, the experts points, the development of new energy vehicles, not only facing the technology and price limits, but also could trigger new problems, for example, power generation could bring more carbon emissions, the battery can also bring more pollution. The personage inside course of study says when the energy price is increased to a certain degree, the advantage of hybrid electric vehicles will be revealed. In the long run, hybrid cars are more likely to achieve industrialization. As a concern about industry, new energy vehicles have a wonderful prospect. But from the reality, hybrid is the most realistic choice of new energy vehicles. And its rapid development will also make a more solid platform for the future of new energy vehicles popularities.

New energy vehicles is not only be environmental protection, but also must adapt to economic development needs and most important is to be consistent with the consumer demand. We might as well think of bold ideas, but make proce cautiously and gradually achieve new energy lift.

References

1. Qi, S.: Like night of spring breeze comes suddenly _ click on the 2010 IAA new energy commercial vehicle. auto salon (December 2010)
2. Xiang, Y.: The present and future of automobile new energy technology showed by BMW Group. special purpose vehicle (November 2010)

Establishment of the Personal Credit System in the Network Environment

Sai Luo

International School, Beijing University of Posts and Telecommunication, Beijing, P.R. China
jp093129@bupt.edu.cn

Abstract. Personal credit system means some intermediary bodies assemble personal credit, reputation and morality together which scattering in different finical institutions and other social organizations. They will process and storage them and form a relevant real personal credit file database. Through e-network transmit, it can provide service to kinds of departments for knowing the personal reputation, asset, tax-payment, morality and so on. The personal credit system construction for our own country is still in a primary level. Therefore we are urging to establish a complete personal credit system and win a positive position for the future competition. This paper will analysis and make a comparison for each country's personal credit system. According to our national conditions, I try to explore a suitable way to establish personal credit system.

Keywords: Personal Credit System, Network Environment.

1 Introduction

We have formed a primary socialist market economy, and the legal situation and competitive mechanism which are the same with the market economy. However, we don't form the corresponding credit system yet. Discredit has become one of the main factories for hindering the healthy development of our economy.

1.1 Credit System and E-Commerce

E-commerce is an activity that trade in the network situation. In recently years, e-commerce is widely used and penetrated in walks of life. As we know, favorable social credit system is a fundamental for the existence and development of e-commerce. In a good credit and legal situation, e-commerce can definitely affect more. Therefore, in the time of prosperous development of economic globalization, network trade and e-commerce, it is urgent to set up the credit system.

1.2 Credit System Construction Situation for Our Country

In recently years, we have started to consider how to establish the credit system. After making efforts, we get some achievements. But, there is still some distance between

G. Lee (Ed.): Advances in Automation and Robotics, Vol.2, LNEE 123, pp. 147–151.
Springerlink.com

the fast development of market economy and a complete credit platform we needed. The credit systems are different in each province and no unified criterion which leave a hidden trouble for the popularization of the whole country.

2 International Experience for Reference to Personal Credit System

Compared to our nation, the personal credit of Europe, America and Japan are relatively advanced. For 150 years development, they have built up perfect personal credit system and mature operation system that guarantee and promote the development of personal credit economy. The typical social credit system mode can be divided into 3 types: market-orientation mode, government-lead mode and member system mode.

2.1 Market-Orientation Mode

First is the market-orientation mode, take America as a typical example, therefore it calls American mode as well. Making profit is the characteristic of this credit system mode. Collect and process personal or enterprise credit information in order to provide independent third-party service for the credit information user. Credit enterprise establishes various credit databases; then they enhance the volume of the database through business acquisition; finally it will come into being few large commercial credit enterprises.

2.2 Government-Lead Mode

Second is government-lead mode. This mode is adopted by many Europe counties such as France, Germanys, Spanish, Italy and Belgium so that it is called Europe mode. This mode is a non-profit organization formed by social credit management system which takes central credit register system as main body. In addition, the mode which takes central credit register system as main body combining social credit system of private credit organization is called nation credit database.

2.3 Membership System Mode

Third is membership system mode, this mode can take Japan as representative so that it calls Japanese mode. This mode takes association as main carrier and sets up the credit information center which can provide the associate members personal and enterprise credit information exchange platform. As a result, it can make the collection and usage of the information come true via internal credit information share system.

Owing to the final purpose of the credit organization is to serve the society. Therefore, the three modes are almost interlinked. No matter market-orientation mode or government-lead mode， the credit organization will be supervised by government strictly. Each nation's government manages credit industry by way of making laws and regulations which includes the content， organization management， evaluation program， evaluation standard， credit report， evaluation legal responsibility and usage responsibility. Making law for supervising the credit industry can not only promote industry standardization development , but also filter into people's heart so that it can set the view of credibility for the whole society.

3 Establishment for China Personal Credit System

3.1 The Existing Problem of Constructing the Personal Credit

3.1.1 Lag of Legal System

So far, we haven't any law or regulations about personal credit. The lack of legal situation will limit the healthy development of personal credit industry and establishment of personal credit system.

3.1.2 Personal Credit Level Evaluation

Currently, the grade work is focus on collection and classification of personal credit data. What's more, it sets up grade mode and national personal credit evaluation regulations and method which is based on it. If there is no reasonable and scientific evaluation approach or criterion, it will lose the fundament of establishment of the credit system.

3.1.3 Personal Credit System Internationalization

With the development of economic globalization and network technology, e-commerce breaks the space limit of individual and enterprises by unique mode which make the business globalized. At the same time, the management mode of credit organization and business activity mechanism should be identical, and the regulations need to keep race with the international credit industry's , avoiding the division of the credit system by individual behavior in order to keep a fast and efficient flow of the information. Therefore the development and establishment of our credit system should integrate with international society and fit it.

3.2 Advantage for Establishing Personal Credit System in China

3.2.1 Advantage of Network

With the development of computer, network and information technology, various credit information can be embodied by e-data.The classification of credit can realize computer softwarization and class automation. The information that scattered in each department could be collected easily, and the database for individual credit system can be built in a short time. At the same time, each department and individuals can check the data that collected through internet. consequently, it can establish the social environment for credit need as soon as possible.

3.2.2 Up-Rising Advantage

Although our credit system starts lately, we can use foreign experience for reference. Because our credit report starts very late, we can chose some innovation type of foreign countries when facing the tide of information technology. When providing credit report we can still create our own production. In this way, we save the process to establish the credit system.

3.3 Analysis and Summary of Personal Credit System Construction in China

As the market economy developing, we certainly will set up personal credit system which fits our own characteristics. We can use the foreign experience for reference as well as considering our own state conditions to create a system mode.

3.3.1 Strengthen Legal Credit Construction

Complete law and regulations can promote the development of personal credit system. Legal system determines the credit organization's characteristics and decides the structure of personal credit system. Legal system is a reflection of the development of social economy. It can not only reflect the developing degree of economy, but also act on economy development. Through the constraint of the law, it can limit the range and usage of the credit. In addition , it can affect the flow and application of the information. As a result, it determines the status of the personal credit organization in social economy. According to our own needs and draw the experience from foreign countries, we need to make the relevant credit system regulations in order to restrict the discredit behavior.

3.3.2 Set Pp Credit Database

Personal credit database is a base for establish the personal credit system. On account of the disperse personal credit information of our own country, government have to combine the credit data in different department as a public resource in social credit system. Using the administrative, legal and economic means as well as advanced information processing technology and network to set up a perfect credit system and database throughout the country. Personal credit data includes personal conditions, job-hunting status and personal public record.

3.3.3 Establish the Unified Credit File and Evaluation Criterion

Personal credit evaluation is a process that intermediate organization uses certain math mode and evaluation criterion to make a data analysis and calculate the credit risk factory , credit point , which quantizes the personal credit and provides a scientific and direct personal credit report.

3.3.4 Establish Discredit Punishment Mechanism and Credibility Reward Mechanism

To be strict with the credit supervision and discredit punishment mechanism, we need to give a reward to the credibility and punish the incredibility, which is a precondition to develop healthily credit system. Having administrative and judiciary punishment or economic sanction to the incredibility, plus revealing the ones who have a bad credit record or lack of record will take a negative effect on his existence and development.

3.3.5 Speed Up Electronization Establishment

Encouraging credit consumption will not only change the way of consumption, but also speed up eletronization establishment and realize information share to the whole society. Moreover, it refers to the backup of electric equipment that we need to manufacture the unified software and make the unified operation program. On the basis of capital input by relevant departments for networking, the comprehensive information can be transmit through modern communication which could realize the sharing of information in order that it can work effectively and serve better for the local finical industry.

4 Conclusion

Because our credit system is still in a primary state and has some problem in management and realization, we are facing a lot of challenges in the future. Therefore, in the process of development, we need to use scientific method to plan and explore, taking full use of other's experience and make our credit system more scientific and systematism. Besides, we need to strength the government supervision and management. The establishment of credit system is a necessary option for our economic development, however the foreign experience have already shown the way. With the establishment of our credit system and the form of the credit environment, the economy of our nation will have a faster development and rising e-commerce will receive a brand-new stage as well.

References

1. Manganiello, A.:The Debt-Free Millionaire: Winning Strategies to Creating Great Credit and Retiring Rich (2009)
2. Rich, J.R.: The Complete Book of Dirty Little Secrets: Money-Saving Strategies the Credit Bureaus Won't Tell You (2009)
3. Jank, W.:Statistical Methods in e-Commerce Research (2008)
4. Cross, F.B.: The Legal Environment of Business: Text and Cases – Ethical, Regulatory, Global, and E-Commerce Issues (2008)

Design and Implementation PLC Real-Time Control Based on NI-VISA

Qingsheng Feng[1] and Hong Li[2]

[1] School of Electrical & Information, Dalian Jiaotong University
[2] Software Institute, Dalian Jiaotong University
{e_fengxd,e_lihong}@163.com

Abstract. PLC is an industry control computer which applies microprocessor to its core 材PLC has been applied to the distributed monitor system which is responsible for high speed data collect of the spot, logic operation, count, PID regulate etc. The computer has been used to manage the monitor system. The HostLink communication protocol of OMRON CP series PLC is introduced in this paper. This article introduces the serial communication control VISA provided by the LabVIEW, the command of the programmable operation of PLC, the link method between PC and PLC, the communication parameter's setup of PLC.The article also introduces how to use the RS-232 serial communication port to realize the communication between PC and PLC. The method for realizing the communication between PLC and the host computer is presented by means of LabVIEW. The communication procedure and program is also introduced in detail in the paper. This method ensures that the communication is real time and accurate. Through reading and writing of the host computer from or to the PLC, automatic collection of field data is realized, and the device can be monitored and controlled effectively. The serial communication software is proved to be reliable and stable through the successful application in actual project. This article discusses the communication method between PC and PLC and the structure of s oft and hardware, applying the visual program language. The method can make the best of resources and characters of the system and improve communication efficiency.

Keywords: PLC, VISA, Serial communication, Host Link.

1 Introduction

The Programmable Logic Controller(PLC) has many advantages such as good reliability、strong anti-interference ability、extend easily、simple operation and easy to program. So it has been used in every field of the industrial control.

But the interactive performance of PLC is far less than the industrial control computer. So a combination of the two can greatly improve the performance of the system in which PLC is the lower computer to complete the data acquisition task and PC is the host computer to display and control the real-time data. In the system PLC is

G. Lee (Ed.): Advances in Automation and Robotics, Vol.2, LNEE 123, pp. 153–158.
Springerlink.com

for the direct control of the spot equipments and the computer is for PLC programming, monitoring and management. And using a human friendly visual monitoring simulation display system, the field staff can timely get the scene information to control and operate the system well.

2 Paper Preparation

In order to ensure the upper and lower computer can exchange the data correctly, we must develop the communication protocol. The HostLink protocol is Omron 's proprietary protocol which uses the computer RS-232C bus and connects with OMRON CP1E for information exchange. The communication uses the master slave polling mode which is the host computer always has an initial transmission priority and all communications are initiated by it. The HOST LINK communication unit receives the command and sends to PLC to execute and then returns the executing result to the host computer. The unit of the data exchange in the upper and lower computer is the frame. When the host computer sends data to the lower computer or reads data from the lower computer ,the data are sent from master in the form of the command frame then the station responses in the form of response frame.

3 Realization of PC and PLC Real-Time Communication

3.1 Serial Port Initialization

According to the communication protocol, communication port has been set: COM4, baud rate:9600bps, 1 start bit, 7 bits of data, 2 stop bits, parity.

The process of PC reads the data from PLC CIO

Command frame format:

Table 1. Command frame format

@	Node Number	RR	Initial Channel Number	Channel Number to Read	FCS	* Enter
	$\times10^1\times10^0$		$\times10^3\times10^2\times10^1\times10^0$	$\times10^3\times10^2\times10^1\times10^0$		

The identification code RR can continuous read the data in the CIO channel started from the specified channel. PLC CIO is a dynamic data storage area, and the channel number is CIO 0 to CIO 289.

Response frame format:

Table 2. Response frame format

@	Node Number $\times 10^1 \times 10^0$	WR	End Code $\times 16^1 \times 16^0$	Initial Channel Data $\times 16^3 \times 16^2 \times 16^1 \times 16^0$	Second Channel Data $\times 16^3 \times 16^2 \times 16^1 \times 16^0$	…	FCS	* Enter

If the command frame is" @ 00RR0200000452", it represents the host computer continuous read the data from 4 channels which are CIO 200 to CIO 203 and 52 is FCS [10] calculated by the program. The data in the CIO 200 channel are put to the response frame in eighth bit to twelfth bit in the form of hexadecimal number, the data in the CIO 201 channel are thirteenth to sixteenth, the data in the CIO 202 channel are seventeenth to twentieth and the data in the CIO 203 channel is the twenty-first bit to 24 bit, the data for each channel are four bits. The data of PLC can be read only in the "channel" form, not in the "bit" form.

The programming steps of the system using LabVIEW are: ①firstly, we must initialize the serial port node. We should set the serial number, baud rate, data bits, stop bits and parity bits. ②Then write the command frame through PC serial 232C to PLC. ③process the response frame automatically returned from PLC serial port, and determine whether PLC normally execute the command frame of PC. If the result is normal, the data in PLC specified channel which extracted from the response frame is sent to PC and shown in the form of decimal. If not, alarm.

PLC executes the command frame and transmits the response frame as the result back to PC serial, using" VISA Write.vi " and" VISA Bytes at Serial Port.vi" ,PC can read the response frame from the serial and display the response frame in" VISA Read.vi ".

After reading the response frame, PC uses "String Subset.vi" to extract the "end code" from the sixth、seventh bit of the response frame and judge whether the result is equal to" 00". If equal, PC should extract the eighth to eleventh bit of the response frame respectively from the CIO channel. Then it uses "Hexadecimal String to Number.vi " to convert them into a decimal number respectively. If the "end code " is not equal to" 00", then the light" error" in the front panel is on, and it shows that the communication command can not be executed normally.

3.2 The Process of Writing the Data from PC to PLC CIO

In" RUN" state the data cannot be written to PLC, so writing the data from PC to PLC is more complex than reading the data. There are three step: ①the on-line operation mode of PLC is set to" MONITOR" or" PROGRAM" status; ②write the data to PLC; ③then the on-line operation mode of PLC is set back to the original " RUN" status. We can directly set the status in the CX-programmer software.

Command frame format and Response frame format Table 3 and Table 4:

Table 3. Command frame format

@	Node Number $\times10^1\times10^0$	WR	Initial Channel Number $\times10^3\times10^2\times10^1\times10^0$	Initial Channel Number $\times16^3\times16^2\times16^1\times16^0$	…	FCS	* Enter

Table 4. Response frame format

@	Node Number $\times10^1\times10^0$	WR	End Code $\times16^1\times16^0$	Initial Channel Data $\times16^3\times16^2\times16^1\times16^0$	Initial Channel Number $\times16^3\times16^2\times16^1\times16^0$	…	FCS	* Enter

If the command frame is 00WR0200000500600072B *, it means that PC will write a hexadecimal number" 5" into the 200 channel of PLC ICO, " 6" into the 201 channel, and" 7" into the 202 channel. The FCS checking subroutines can check the number of the command strings only up to 13, so once it can only write the data into one channel, as shown in Fig.1.

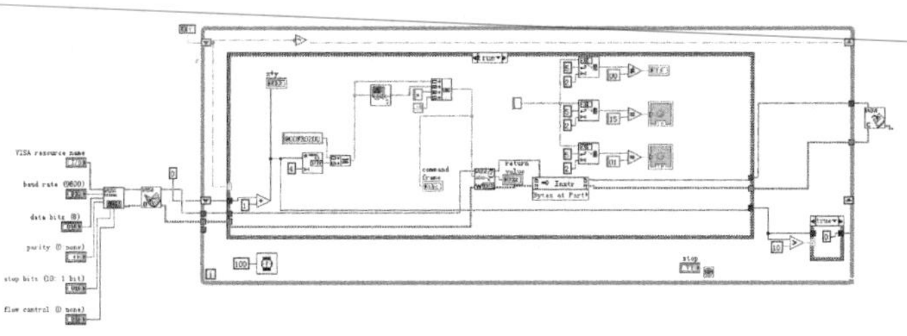

Fig. 1. Writing the data to the memory

3.3 FCS Checking Program

FCS checking code is a portion of the command frame sent by PC. We can calculate FCS checking code of the command frame in such program shown in Fig. 2.

FCS checking implements all the characters from the start symbol" @" to the last one of RS-232C protocol by using XOR operation. Using "String" input each character of the command frame before FCS checking code, converts them into the array of ASCII code and extracts each element of the array respectively. Then the first bit of the array will XOR the second one ,the result we get will XOR the third one. We can operate other bits by analogy. The final operational result should be converted into 2 hexadecimal characters then output and display in " hex integer string".

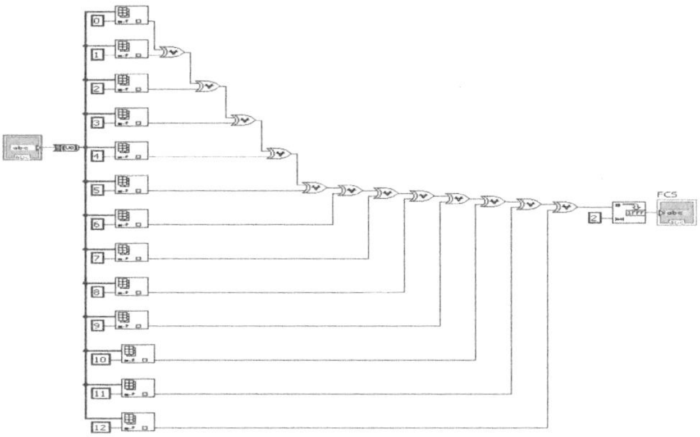

Fig. 2. Checking program chart

The FCS checking program can operate a13 bits string and can meet the need of the read / write operation of PLC with a single channel. If the multiple continuous PLC channels are need for writing the data, the corresponding numbers of array elements are inverted to implement XOR operation.

4 Run Results of Program

4.1 Operation Interface

Based on the LabVIEW, the program can realize the communication between PC and PLC.The data can be written to CIO and read from CIO .The operation interface and program diagram is shown in Fig. 3.

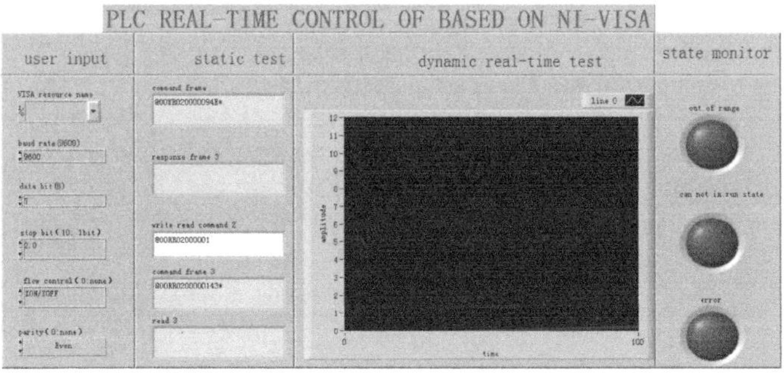

Fig. 3. Operation interface

4.2 Run Results

The host computer program can automatically write the data to PLC CIO and can also automatically read the data from CIO. The result can be seen in Fig.3. From the oscillogram in the figure we can also clearly observe the real-time change of the data in CIO. The change curve of the waveform is shown in Fig. 4.

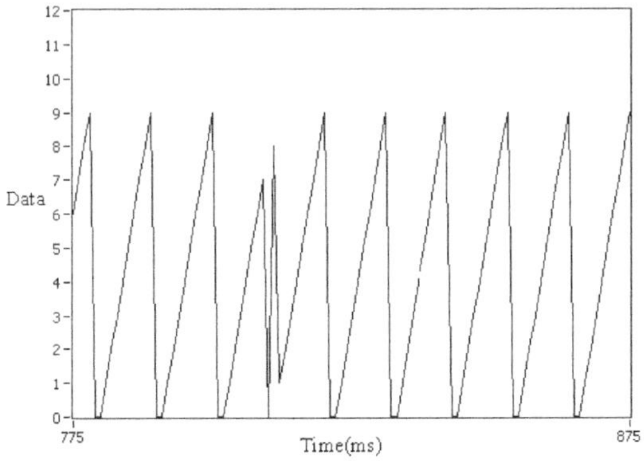

Fig. 4. The run results of the program

5 Summary

With the stability of PLC and the simplicity of LabVIEW, the program can realize the real-time communication of PC and PLC. The system can perform such task as real-time monitoring for some certain parameter and has good practicality.

References

1. Pennec, E.L., Mallat, S.: Sparse Geometric Image Representations with Bandelets. IEEE Transactions On Image Procession 14(4), 423–438 (2005)
2. G Programming Reference Manual, National Instruments Corporation (1998)
3. Huifang, K., Benbing, J.: AMT data acquisition and analyzing system based on LabVIEW. Chinese Joumal of Scientific Instrument 27(6), 442–443 (2006)
4. Lu, H.-Y., Lu, Q.-M., Niu, Q.-Z.: Virtual Logic Analyzer Based on DAQ and LabVIEW. Instrument Technique and Sensor 12, 14–16 (2007)
5. NI-DAQmx Help. USA. National Instruments, (September 2005)

Caculation of Resistance and Inductance of Laser Plasma

Cai Feng Zhao

Electrical Engineering Department, Harbin Institute of Technology,
China, 150006

Abstract. The time-invariant resistance and inductance of laser plasma are related to the nanosecond currents in laser excitation in initial discharge phase, and can be obtained through measurements of the laser chamber voltage and current using experienced formula and voltage equation. It is found that the plasma resistance and inductance decrease rapidly with time, and is also proved that the time-invariant resistance is dependent on the total charge in the discharge area, the time-invariant inductance is dependent on the electric field in the laser chamber.

Keywords: laser plasma, curve proximity, Matlab.

1 Introduction

Various theory of the development of spark discharges and the modeling of the phenomena have been extensively studied, the inductance and resistance of different arc discharges have been investigated[1-3]. Generally, a spark gap switch and laser chamber in a discharge can be replaced by a nonlinear inductor in series with a resistor. H.AKIYAMA characterize these discharge by four phases: initial phase, quasi-stationary phase, transient phase, and relaxation phase. The initial phase appear between the initial time and the time of the current maximum when the pulse current is in the rise edge and the variations of the inductance and resistance of the plasma are the important parameters affecting discharge speed.

1.1 Laser Discharge Circuit

A Blumlein circuit for exciting S_2 laser is shown in Fig. 1. In this figure, K_1 is the spark gap switch, C_S is the transmission capacitor, C_P is the storage capacitor, K_2 is a laser

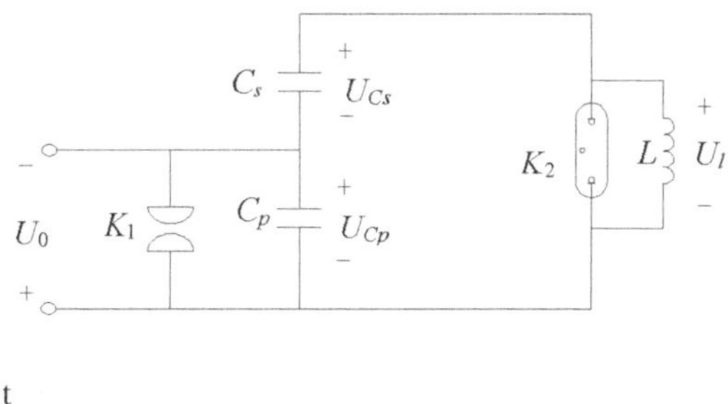

Fig. 1. Blumlein Circuit

G. Lee (Ed.): Advances in Automation and Robotics, Vol.2, LNEE 123, pp. 159–164.
Springerlink.com

chamber, L is a inductor with larger value. A fast oscilloscope Tektron (300M bandwidth) is used to measuring voltage. For no wire in the main circuit, the pulse current can't be measured through amperemeter, instead, we can calculate the value of a pulse current through a titanium film in series with the circuit. The size of the Titanium film is 30cm in length, 1cm in width, 30 μm in thick, For its resistivity is $\rho = 42.0\times10^{-2}\,\Omega\,\text{mm}^{-2}/\text{m}$, then the resistance of the titanium film is

$$R = \rho\frac{l}{s} = 42.0\times10^{-2}\times\frac{0.3}{10\times30\times10^{-3}} = 0.42\Omega$$

1.2 Calculation of the Resistance and Inductance of Laser Plasma at Initial Phase and the Result Analysis

The variation of the resistance and inductance of laser plasma at initial stage can be measured through the breakdown voltage and current waveforms of the laser chamber. The plasma resistance at initial stage can be given as[4]:

$$R = \frac{\sqrt{P}d}{\sqrt{2a\int_0^t i^2 dt}} \tag{1}$$

In the equation, P is the number of atmospheres, d is the electrode distance which is in centimeter. The value of a depends on the kind of gas and is in atmosphere.cm^2/s.volt^2.

In the experiment, 16.0Kpa neon is pumped into a laser chamber, then a is 30 in the equation (1), and the measured electrode spacing is 0.78cm, the breakdown voltage and current waveforms of the laser chamber are shown in Fig. 2a)、b).

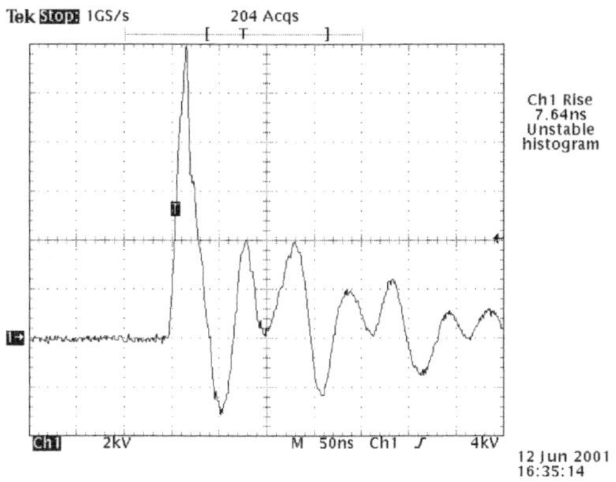

a) current across the laser chamber

Fig. 2. Discharge waveforms of the laser chamber

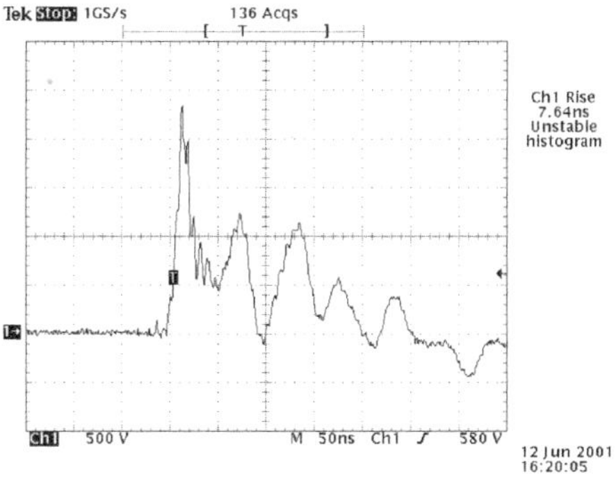

b) current in the circuit

Fig. 2. *(continued)*

From Fig. 2b) the current function at initial phase can be similarly expressed as:

$$i = 2738.1 + 2738.1\sin(0.2094\times10^{9}t - \frac{\pi}{2}),\ 0 \leqslant t \leqslant 15\text{ns} \quad (2)$$

Substituting eq.(2) into eq.(1) , we use matlab to perform curve proximity every 1ns, then get the variation of laser plasma resistance versus current and time respectively. Fig.3 and Fig.4 shows that the resistance decrease rapidly with time and approaches 0.1 Ω .

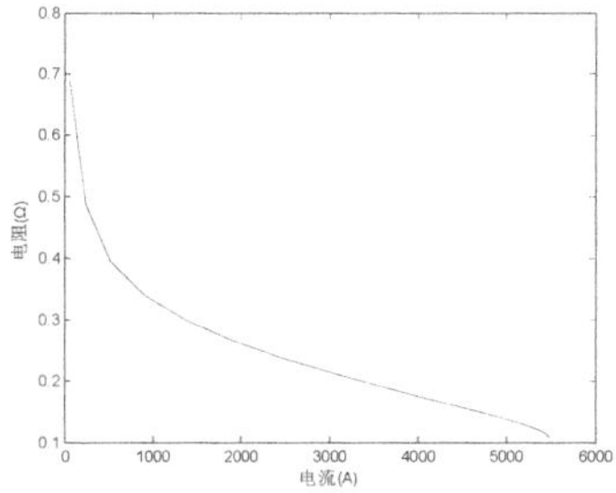

Fig. 3. The laser plasma reisistance versus current at initial phase

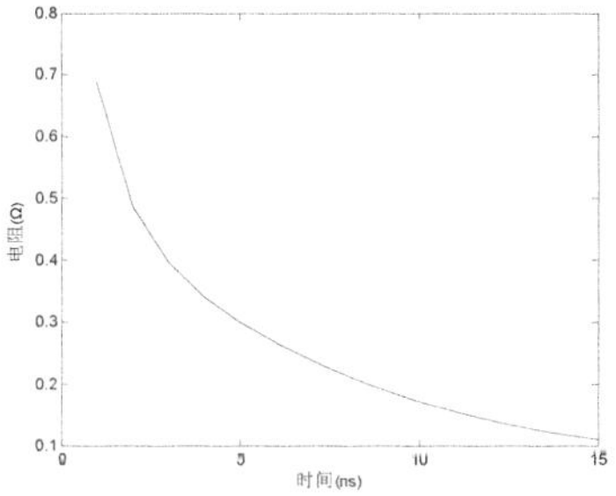

Fig. 4. The laser plasma reisistance versus time at initial phase

From Fig. 2a) the breakdown voltage at initial time can be similarly expressed as :

$$u = 4500 + 7500\sin(0.0898\times10^{9}t + \frac{\pi}{2}), \qquad 0 \leqslant t \leqslant 15\text{ns} \tag{3}$$

The voltage equation can be written as:

$$u = \frac{d(Li)}{dt} + Ri = L\frac{di}{dt} + i\frac{dL}{dt} + Ri \tag{4}$$

In the equation, L is the equivalent inductance of laser plasma. From the above relationship, we proceed curve proximity every 1ns, the equation can be written as :

$$(u - Ri)\Delta t = L\Delta i + i\Delta L \tag{5}$$

let $\Delta i = i_{n+1} - i_n, \quad \Delta L = L_{n+1} - L_n$

Then Eq.5 can be written as

$$(u_n - R_n i_n)\Delta t = L_n(i_{n+1} - i_n) + i_n(L_{n+1} - L_n) \tag{6}$$

when $t_0 = 0$, $n = 0$; when $t_1 = 1\,\text{ns}$, $n = 1$, we can calculate the inductance at initial time

$$L_0 = \frac{u_0}{i_1} = \frac{12000}{59.8} = 200.67\,\text{nH}$$

let $\Delta i = i_n - i_{n-1}, \quad \Delta L = L_n - L_{n-1}$

Then Eq.5 can be written as

$$(u_n - R_n i_n)\Delta t = L_n(i_n - i_{n-1}) + i_n(L_n - L_{n-1}) \tag{7}$$

$$L_n = \frac{(u_n - R_n i_n)\Delta t + i_n L_{n-1}}{2i_n - i_{n-1}} \tag{8}$$

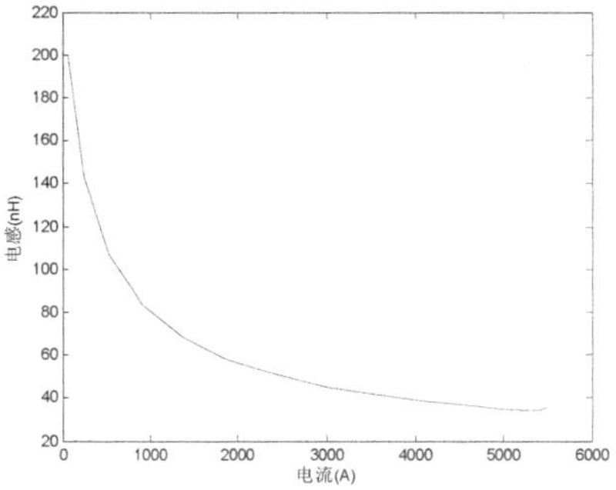

Fig. 5. The laser plasma inductance versus current at initial phase

We use MATLAB to calculate the value of L every 1ns, then get the variation of laser plasma inductance versus current and time respectively. Fig.5 and Fig.6 shows that the inductance decrease rapidly with time and approaches 29nH.

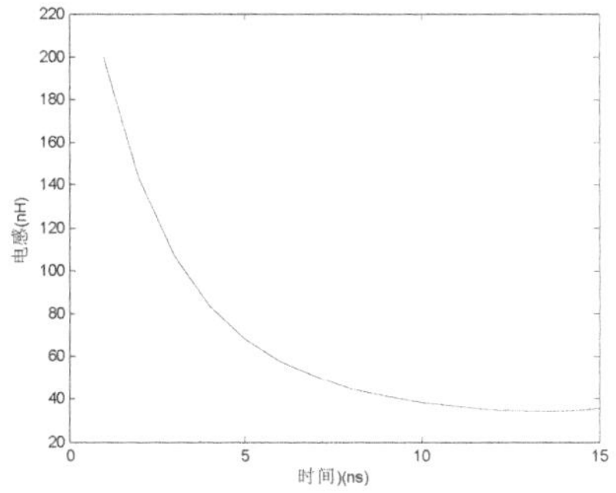

Fig.. 6. The laser plasma inductance versus time at initial phase

The above results shows that the laser plasma resistance and inductance decrease rapidly with time at the laser discharge initial phase, we know that the modeling of the laser plasma resistance leads to the used formula.

$$R = \frac{d}{n_e e \mu S} \tag{9}$$

Where d is electrode spacing, n_e is electron number density, μ is electron mobility, S is cross area of discharge plasma. At initial time the number of charges increase rapidly, the electron number density increase, the conductance of the discharge plasma decrease, so the plasma resistance decrease quickly. The decrease of plasma inductance is related with drift velocity of electric charge. In P. Persephonis's study the greater the drift velocity of the electric charge, the smaller the radius of the discharge channel, or vice versa. The drift velocity also varies with the electric field in the laser chamber, and is proportional with the value of E/P. From Fig. 3a), we know that the breakdown voltage of the laser chamber is very high at initial time, so the value of E/P is very high, then the breakdown voltage decrease, so does E/P. Hence we can conclude that the drift velocity of the electric charge will decrease correspondingly, so will the plasma inductance. It is reasonable to think that the result presented in this work is right.

2 Conclusion

By measuring the waveforms of voltage and current of laser chamber and using experienced formula and voltage equation, this paper provide a method of calculating plasma resistance and inductance versus current and time at initial phase and deeply studying laser discharge characteristics. It is proved to be a more easy way of calculating nonlinear resistance and inductance at discharge initial phase , and a reasonable explaination has been given.

References

1. Engel, T.G., Donaldson, A.L., Kristiansen, M.: The pulsed discharge arc resistance and its functional behavior. IEEE Trans. Plasma Sci. 17, 323–329 (1989)
2. Persephonis, P., Vlachos, K., Georgiades, C., Parthenios, J.: The inductance of the discharge in a spark gap. J. Appl. Phys. 71, 4755–4762 (1992)
3. Persephonis, P., Giannetas, V., Georgiades, C., Parthenios, J., Ioannou, A.: The inductance and resistance of the laser discharge in a pulsed gas laser. IEEE J.Quantum Electronics 31, 573–581 (1995)
4. Rompe, R., Weizel, W.: Ueber das Toeplersche Funkengesetz. Z. Phys. 122, 636–639 (1944)
5. Persephonis, P., Giannetas, V., Georgiades, C., Parthenios, J., Ioannou, A.: The Influence of the External Circuit on Arc-Discharge of a Spark-Gap:Its Application to a Pulsed Gas Laser. IEEE J.Quantum Electronics 31, 567–571 (1995)

WLR-Normal Cryptogroups and LR-Regular Semigroups

Aifa Wang[1], Lili Wang[1], and Yinbin Li[2]

[1] School of Mathematics and Statistics, Chongqing University of Technology
Chongqing 400054, P.R. China
[2] Chongqing Gangcheng Experimental School
Chongqing 400080, P.R. China
{wangaf1980,wllaf}@163.com

Abstract. In this paper, we define the WLR-normal band, WLR-normal cryptogroup and LR-regular semigroup. We give some equivalent conditions for a completely regular semigroup to be a WLR-normal cryptogroup and give the characterization of LR-regular semigroups.

Keywords: Left(right) group, Completely regular semigroup, Strong semilattice.

MSC: (2000) 20M10

1 Introduction

A band B is called a WLR-band [6] if for any $e \in B$, $D_e = R_e$ or $D_e = L_e$. A WLR-band B is called a WLR-normal band, if it is normal. A completely regular semigroup S is a cryptogroup if $\mathcal{H}$ is a congruence. A cryptogroup S is a $\mathcal{C}$-cryptogroup if $S/\mathcal{H}$ belongs to a class of $\mathcal{C}$-bands. An orthocryptogroup S is called a $\mathcal{C}$-orthocryptogroup if $S/\mathcal{H}$ is a $\mathcal{C}$-band.

In [2], Clifford has shown that a completely regular semigroup S is a normal cryptogroup if and only if it is a strong semilattice of completely simple semigroups. The classes of WLR-normal cryptogroups (bands) are important subclasses of normal cryptogroups (bands). In section 3, we give some equivalent conditions for a completely regular semigroup to be a WLR-normal cryptogroup.

In [4], the author has show a π-regular semigroup with a left zero band is a nil-extension of a left group. In section 4, we generalize this result on the class of LR-regular semigroups. The result on the π-regular semigroup can be as a corollary of our result.

The readers are referred to [2] and [6] for notations and definitions not mentioned here.

2 Basic Definitions Preliminaries

In this section, we first state some known results from the literature which will be used in the sequel. We need the following notations:

G. Lee (Ed.): Advances in Automation and Robotics, Vol.2, LNEE 123, pp. 165–171.
springerlink.com

$\mathcal{LG}$	left groups
$\mathcal{RG}$	right groups
$\mathcal{LZ}$	left zero bands
$\mathcal{RZ}$	right zero bands
$\mathcal{S}_{lrg}$	$\mathcal{LG} \cup \mathcal{RG}$
$\mathcal{S}_{lr}$	$\mathcal{LZ} \cup \mathcal{RZ}$

Lemma 2.1. [6, *Lemma* 2.1] *Let B be a band. Then the following conditions are equivalent:*

(1) For any $e \in B$, $D_e = R_e$ or $D_e = L_e$;
(2) For any $e, f \in B$, $ef\mathcal{R}fe$ or $ef\mathcal{L}fe$;
(3) For any $e, f \in B$, $efe = ef$ or $efe = fe$.

A band B is called a WLR-band [6] if for any $e \in B$, $D_e = R_e$ or $D_e = L_e$; and if B is also regular, it is called a WLR-regular band. A band B is called an LR-regular band [7], if for any $e \in B$ at least one of the following conditions holds in B: (i) $efe = ef$ for all $f \in B$; (ii) $efe = fe$ for all $f \in B$.

Definition 2.2. *A WLR-band B is called a WLR-normal band, if B is also normal.*

It is obvious that the class of WLR-normal bands is contained in the class of WLR-regular bands. Next, we cite an example which is a WLR-regular band but not a WLR-normal band.

Example 2.3. *[7] Let $B = \{e, f_1, f_2, g_1, g_2, 0\}$ with the following multiplication table*

	e	f_1	f_2	g_1	g_2	0
e	e	f_1	f_1	g_1	g_2	0
f_1	f_1	f_1	f_1	0	0	0
f_2	f_2	f_2	f_2	0	0	0
g_1	g_1	0	0	g_1	g_2	0
g_2	g_2	0	0	g_1	g_2	0
0	0	0	0	0	0	0

We can easily cheak B is a WLR-regular band. But it is not a WLR-normal band. This is because that $eg_1g_2e = g_1g_2 \neq g_2g_1 = eg_2g_1e$.

Example 2.4. *[7] Let $Y = \{\alpha, \beta, \gamma\}$ be a semilattice with $\gamma = \alpha\beta = \beta\alpha$ such that $\gamma < \alpha$ and $\gamma < \beta$. Let $L_\alpha = \{e_1, e_2\}$ and $L_\gamma = \{g_1, g_2\}$ be two left zero bands, $L_\beta = \{f_1, f_2\}$ a right zero band. Define strong semilettice structure homomorphisms as following:*

$$\Phi_{\alpha,\gamma} : L_\alpha \to L_\gamma, e_i \mapsto g_1; \qquad \Phi_{\beta,\gamma} : L_\beta \to L_\gamma, f_i \mapsto g_2,$$

where $i = 1, 2$. Then we obtain a normal band $B = [Y; L_\delta, \Phi_{\delta,\varepsilon}]$ of rectangular L_δ's and it satisfies the condition (1) of Lemma 2.1, so B is a normal band. But since

$$f_1 f_2 f_1 = f_2 f_1 = f_1 \neq f_2 = f_1 f_2,$$

$$f_1 e_1 f_1 = g_2 g_1 g_2 = g_2 g_1 = g_2 \neq g_1 = g_1 g_2 = e_1 f_1;$$

so B is a WLR-normal band but not an LR-regular band.

Let S be a semigroup and denotes its set of idempotents by E_S. An orthogroup S is called a WLR-(WLR-normal, LR-normal) orthogroup, if E_S is a WLR-(WLR-normal, LR-normal) band. A completely regular semigroup S is called a cryptogroup if $\mathcal{H}$ is a congruence on S; further, if S is also orthodox, S is called an orthocryptogroup. If $S/\mathcal{H}$ is a $\mathcal{C}$-band, S is called a $\mathcal{C}$-cryptogroup; an orthocryptogroup S is called a $\mathcal{C}$-orthocryptogroup if $S/\mathcal{H}$ is a $\mathcal{C}$-band(equivalently E_S is a $\mathcal{C}$-band).

Lemma 2.5. *$S \in \mathcal{S}_{lrg}$ if and only if S is a regular semigroup and $a\mathcal{L}b$ or $a\mathcal{R}b$ for any $a, b \in S$.*

Proof. Necessity. It is obvious.

Sufficiency. Let $e, f \in E(S)$ and $e \leq f$. We have $e\mathcal{L}f$ or $e\mathcal{R}f$. If $e\mathcal{L}f$, then $fe = f$. Again $ef = fe = e$, then $e = f$. Similarly, if $e\mathcal{R}f$, then $e = f$. Thus S contains a primitive idempotent. And S is simple is clear. By [2, *Theorem III*.1.5], we have S is completely simple. Thus $S \cong \mathscr{M}(I, G, \Lambda, P)$ for suitable I, G, Λ and P. For $(i, g, \lambda), (j, h, \mu) \in S$, we have $(i, g, \lambda)\mathcal{L}(j, h, \mu)$ or $(i, g, \lambda)\mathcal{R}(j, h, \mu)$, thus $\lambda = \mu$ or $i = j$. That is, $|I| = 1$ or $|\Lambda| = 1$. By [2, *Theorem III*.5.9] and its dual, $S \in \mathcal{S}_{lrg}$.

Corollary 2.6. *$B \in \mathcal{S}_{lr}$ if and only if $ef = e$ or $ef = f$ for any $e, f \in B$.*

Lemma 2.7. [6, *Theorem* 3.3] *Let S be a completely regular semigroup. Then the following conditions are equivalent:*

(1) S is a WLR-orthogroup;
(2) $D_a = R_a$ or $D_a = L_a$ for any $a \in S$;
(3) $ab\mathcal{L}ba$ or $ab\mathcal{R}ba$ for any $a, b \in S$.

Corollary 2.8. *Let S be a completely simple semigroup. Then S is a WLR-orthogroup if and only if $a\mathcal{L}b$ or $a\mathcal{R}b$ for any $a, b \in S$.*

Proof. Since S is completely simple, $\mathcal{D} = S \times S$ by [2, *Theorem II*.4.5]. It follows that S is a WLR-orthogroup if and only if $a\mathcal{L}b$ or $a\mathcal{R}b$ for any $a, b \in S$, by Lemma 2.7.

Lemma 2.9. *Let S be a completely regular semigroup and S_α be a completely simple subsemigroup of S. Then the following conditions are equivalent:*

(1) $D_a = R_a$ or $D_a = L_a$ for any $a \in S$;
(2) $D_a = R_a$ or $D_a = L_a$ for any $a \in S_\alpha$.

Proof. (1) implies (2). It is obvious.

(2) implies (1). For any $x \in S$ and $a \in S_\alpha$, if $x\mathcal{D}a$, then $x \in S_\alpha$ since S_α is a completely simple semigroup. Thus $x\mathcal{R}a$ or $x\mathcal{L}a$. It follows that for any $a \in S$, $D_a = R_a$ or $D_a = L_a$.

Lemma 2.10. [2, *Theorem IV.*1.6] *The following conditions on completely regular semigroup S are equivalent:*

(1) S is a normal cryptogroup;
(2) S is locally a Clifford semigroup;
(4) S statisfies $\mathcal{D}$-majorization (for idempotents);
(5) S is a strong semilattice of completely simple semigroups.

Lemma 2.11. [2, *Theorem IV.*2.9] *The following conditions on a band S are equivalent:*

(1) S is a normal band;
(2) S is locally a semilattice;
(4) S statisfies $\mathcal{D}$-majorization;
(5) S is a strong semilattice of rectangular bands.

3 WLR-Normal Cryptogroups

Theorem 3.1. *Let S be a completely regular semigroup. Then the following conditions are equivalent:*

(1) S is a WLR-normal cryptogroup;
(2) S is a WLR-normal orthocryptogroup;
(3) S is locally a Clifford semigroup and E_S is a WLR-normal band;
(4) S statisfies $\mathcal{D}$-majorization (for idempotents) and E_S is a WLR-normal band;
(5) S is a strong semilattice of S_α, where S_α satisfying $a\mathcal{L}b$ or $a\mathcal{R}b$ for any $a, b \in S_\alpha$.

Proof. (1) implies (2) By hypothesis, $S/\mathcal{H}$ satisfies the identity $efe = ef$ or $efe = fe$ for any $e, f \in S/\mathcal{H}$, whence every $\mathcal{D}$-class of $S/\mathcal{H}$ is left zero semigroup or a right zero semigroup by Lemma 2.1. Hence every component of S must be a left group or a right group and therefor orthodox. By [2, *Theorem II*.5.3], S is orthodox.

(2) implies (1) It is trial.

(2) implies (3) and (3) implies (4) It is easy by Lemma 2.10.

(4) implies (2) If S statisfies $\mathcal{D}$-majorization (for idempotents), then S is a strong semilattice of completely simple semigroup and hence S is a normal

cryptogroup by Lemma 2.10. Since E_S is a WLR-normal band, we have S is WLR-normal orthocryptogroup.

(1) implies (5) If S is a WLR-normal cryptogroup, then, by Lemma 2.10, S is a strong semilattice of completely simple semigroup S_α. Again S is a WLR-normal cryptogroup, $D_a = R_a$ or $D_a = L_a$, for all $a \in S$. Hence by [2, *Theorem II*.4.5] and Lemma 2.5, for any $a, b \in S_\alpha$, $a\mathcal{L}b$ or $a\mathcal{R}b$.

(5) implies (2) By Lemma 2.5, $S_\alpha \in \mathcal{S}_{lrg}$. If S is a strong semilattice of S_α, then by Lemma 2.10, S is a normal cryptogroup. It is obvious that $D_a = R_a$ or $D_a = L_a$ for all $a \in S$. It follows that S is a WLR-normal orthocryptogroup by Lemma 2.7.

Corollary 3.2. *The following conditions on a band S are equivalent:*

(1) S is a WLR-normal band;
(2) S is locally a WLR-semilattice and S is a WLR-band;
(3) S statisfies $\mathcal{D}$-majorization and S is a WLR-band;
(4) S is a strong semilattice of S_α, where S_α satisfying $ab = cd$ implies $a = c$, or $b = d$ for any $a, b, c, d \in S_\alpha$.

Proof. (1) implies (2) If S is a normal band, then S is locally a semilattice by Lemma 2.11.

(2) implies (3) If S is locally a WLR-semilattice, then S statisfies $\mathcal{D}$-majorization by Lemma 2.11.

(3) implies (4) If S statisfies $\mathcal{D}$-majorization, then S is a strong semilattice of rectangular band S_α by Lemma 2.11. Since S is a WLR-band, for any $a \in S$, $D_a = L_a$ or $D_a = R_a$ and hence $S_\alpha \in \mathcal{S}_{lr}$ since S_α is a completely simple semigroup.

(4) implies (1) By Lemma 4.5, $S_\alpha \in \mathcal{S}_{lr}$, then S_α is a rectangular band. It follows that S is a normal band by Lemma 2.11. Again $S_\alpha \in \mathcal{S}_{lr} \subseteq \mathcal{S}_{lrg}$, by Corollary 2.8, S_α is a WLR-orthogroup. Thus S is a WLR-normal band.

4 *LR*-Regular Semigroups

Let I be an ideal of S. We say S is a nil-extension of I, if for any $a \in S$, there exists $n \in N$ such that $a^n \in I$. In this section, the characterization of LR-regular semigroups will be given.

Lemma 4.1. *[4] A simple semigroup S is completely simple if and only if S is completely π-regular.*

Lemma 4.2. *[4] S is completely π-regular if and only if for every $a \in S$, there exists $m \in Z^+$, such that $a^m \in a^m S a^{m+1}$.*

Lemma 4.3. *[4]The union K of all the minimal bi-ideals of a semigroup S is the kernel of S.*

Definition 4.4. *A π-regular semigroup S is called LR-regular if $E_S \in \mathcal{S}_{lr}$.*

Lemma 4.5. *$S \in \mathcal{S}_{lr}$ if and only if $ab = cd$ implies $a = c$, or $b = d$ for any $a, b, c, d \in S$.*

Proof. Necessity. It is obvious.

Sufficiency. Since $a \cdot a^2 = a^2 \cdot a$, then $a^2 = a$. Thus S is a band. It follows that $aba \cdot aba = a \cdot ba$, we have $aba = a$ or $aba = ba$. If $a \cdot aba = ba$, then $a = b$ or $aba = a$. Thus S is a rectangular band. Let $S = I \times \Lambda$. If $|I| \neq 1$ and $|\Lambda| \neq 1$. Let $i, j \in I, \lambda, \mu \in \Lambda, i \neq j, \lambda \neq \mu$, then $(i, \lambda)(j, \mu) = (i, \mu)(i, \mu)$. It follows that $(i, \lambda) = (i, \mu)$ or $(j, \mu) = (i, \mu)$ a contradiction. Hence $|I| = 1$ or $|\Lambda| = 1$. Now we have $S \in \mathcal{S}_{lr}$.

The notion of an LR-regular semigroup is a generalization of the notion of an π-regular semigroup with a left zero band. By the next theorem a characterization of LR-regular semigroup will be given.

Theorem 4.6. *On a semigroup, the following condition are equivalent:*

(1) S is a LR-regular semigroup;

(2) S is a nil-extension of I, where I is a regular semigroup and for any $a, b \in I$, $a\mathcal{L}b$ or $a\mathcal{R}b$;

(3) there is an idempotent $e \in S$ and S is left archimedean or right archimedean;

(4) for every $a, b \in S$, there exists $m \in Z^+$, such that $a^m \in a^m S a^m b$, or $a^m \in b a^m S a^m$.

Proof. (1) implies (4) By Lemma 4.1, S contains the simple Kernel K. If $E(S) \in \mathcal{LZ}$, then K is a left group. For $a, b \in S$, there exist $m, n \in Z^+$ such that $a^m, b^n \in K$, so $a^m = xb^{n+1}, b^n = ya^m$ for some $x, y \in K$. Since $a^m \in G_e$, for some $e \in E(S)$, we have $a^m = a^m(a^m)^{-1}xya^m b \in a^m S a^m b$. Similarly, if $E(S) \in \mathcal{RZ}$, we can prove that $a^m \in ba^m S a^m$.

(4) implies (2) If for every $a, b \in S$, there exists $m \in Z^+$, such that $a^m \in a^m S a^m b$, then for $a = b$, there exists $m \in Z^+$ such that $a^m \in a^m S a^{m+1}$. Thus by Lemma 4.2, S is completely π-regular. $K = \cup_{e \in E(S)} G_e$ is the kernel of S by Lemma 4.3, which is a left group, Since for $e, f \in E(S), f = fxfe$ for some $x \in S$, so $fe = (fxfe)e = f$, i.e. $E(S)$ is a left zero band. Hence S is a nil-extension of a left group. If for every $a, b \in S$, there exists $m \in Z^+$, such that $a^m \in ba^m S a^m$, then for $a = b$, $a^m \in a^{m+1} S a^m$. Thus by Exercises 1, S is completely π-regular. $K = \cup_{e \in E(S)} G_e$ is the kernel of S by Lemma 4.3, which is a right group, Since for $e, f \in E(S), f = efxf$ for some $x \in S$, so $ef = e(efxf) = f$, i.e. $E(S)$ is a right zero band. Hence S is a nil-extension of a right group. By Lemma 2.5, (2) holds.

(2) implies (3) It is clear that S contains an idempotent. If for any $a, b \in I$, $a\mathcal{L}b$, then there exist $m, n \in Z^+$ and $x \in I$, such that $a^m = xb^n$, for every $a, b \in S$, and so S is left archimedean. Similarly, if for any $a, b \in I$, $a\mathcal{R}b$, then S is right archimedean.

(3) implies (1) Let $e \in E(S)$. If S be left archimedean, then the intersection K of all the ideals of S is nonempty, since $e \in K$. Furthermore, for every $a \in S$ there exist $m \in Z^+$ and $x \in S$ such that $a^m = xe$, so $a^m \in K$. For $e, f \in E(S), f = fe$, so $fe = (xe)e = xe = f$, i.e. $E(S)$ is a left zero band. Now,

K is a simple semigroup with a primitive idempotent, so by Lemma 4.2, S is a π-regular semigroup with $E(S) \in \mathcal{LZ}$. Similarly, if S be right archimedean, we can prove that S is a π-regular semigroup with $E(S) \in \mathcal{RZ}$. Thus $E(S) \in \mathcal{S}_{lr}$.

A characterization of π-regular semigroup with a left zero band can be as a corollary.

Corollary 4.7. *[4] On a semigroup, the following condition are equivalent:*

(1) S is a π-regular semigroup and $E(S)$ is a left zero band;
(2) S is a nil-extension of a left group;
(3) there is an idempotent $e \in S$ and S is left archimedean;
(4) for every $a, b \in S$, there exists $m \in Z^+$, such that $a^m \in a^m S a^m b$.

Acknowledgements. The authors would like to thank the referee for his or her useful comments relating to this paper. The research was supported by Scientific Research Foundation of Chongqing University of technology(2010ZD18). Corresponding author: Lili Wang.

References

1. Petrich, M.: The structure of completely semigroups. Trans. Am. Math. Soc. 189, 211–236 (1974)
2. Petrich, M., Reilly, N.: Completely regular semigroups. Wiley, New York (1999)
3. Clifford, A.H.: Bands of semigroups. Proc. Amer. Math. 42, 499–504 (1941)
4. Stojan, B.: Semigroups with a system of subsemigroups. Trans. Am. Math. Soc. 189, 211–236 (1974)
5. Kong, X.Z.: Completely regular semigroups with generalized strong semilattice decompositions. Algebra Colloquium 12(2), 269–280 (2005)
6. Zhang, J.G., Song, G.T., Liu, G.X.: WLR-regular orthgroups. Semigroup Forum 77, 463–473 (2008)
7. Guo, Y.Q., Shum, K.P., Sen, M.K.: LR-normal orthgroups. Chin. Sci. Bull. 35(12), 1384–1396 (2005)
8. Du, A.H., Liu, Y.: Weak LR-bands. Shandong Sci. 18(2), 1–5 (2005)

The Inspecting and Controlling System for the Testing Platform of CO_2 Air Conditioning Based on LabVIEW

Bingqiang He[1], Huijuan Li[2], and Chunling Liao[1]

[1] Guangdong Jidian Polytechnic, Guangzhou, Guangdong, 510515
[2] Guangzhou Academy of Special Equipment Inspection & Testing, Guangzhou, Guangdong, 510640

Abstract. From the functional requirements of the inspecting and controlling system for the testing platform of CO_2 air conditioning, the inspecting and controlling system based on LabVIEW has been designed in this paper. The inspecting and controlling system based on the virtual instrument is composed by the hardware and the computer equipment software system. The hardware includes the data acquisition card, the mass flow sensors, the pressure and temperature sensors, etc. And the computer equipment software system is self-developed, which has taken the LabVIEW software as the platform and that has the virtual panel and testing functions. The inspecting and controlling system can collect, process and store the data of experimental platform as well as control the parameters of air-conditioning testing platform.

Keywords: CO_2, Air conditioning, Inspecting, Control, LabVIEW.

1 Introduction

The carbon dioxide refrigerant naturally exists in the nature and the ozone depletion potential (ODP) is zero. The global warming potential is small (GWP=1), which is just $\frac{1}{2000} \sim \frac{1}{1000}$ of the global warming potential of HFC as well as the mixed refrigerant. At the same time, as a refrigerant, the carbon dioxide has a lot of advantages. For example, it is environmentally friendly, safe and cheap. The refrigerating capacity in unit volume is large and it doesn't have to be recycled. Therefore, the people have conducted a large number of applied researches on CO_2 air-conditioning system[1~8].

According to the characteristics of CO_2 air-conditioning system, the inspecting and controlling system for the testing platform of CO_2 air conditioning based on LabVIEW has been developed in this paper from the study of the performance requirements of CO_2 air-conditioning system and the gas cooler of micro-channel with parallel flow (GCMCPF).

The LabVIEW based on the graph development, the debugging and the running programs is the integrated virtual instrument development software which is introduced by American National Instrument Enterprise. The designed virtual instrument can be separated from the LabVIEW development environment, which has the same operability as the actual instruments[9].

G. Lee (Ed.): Advances in Automation and Robotics, Vol.2, LNEE 123, pp. 173–179.
Springerlink.com

2 The Testing Platform of CO_2 Air Conditioning

The testing platform of CO_2 air conditioning is composed by the trans-critical CO_2 recycling system and the inspecting and controlling system, the structure of which can be shown in Figure 1. The working principle of this system is: the CO_2 fluid with low temperature and low pressure will be compressed into the supercritical CO_2 gas with high temperature and high pressure through the CO_2 compressor. Then the gas will be cooled down by wind in GCMCPF. After the cooling, the gas with high pressure will be sent to the regenerator through the drying filter and the temperature will be continuously decreased. The temperature and the pressure of the gas will be reduced by the throttling valve. Then the two mixtures after throttling will be evaporated and vaporized in evaporator. After the evaporation, the working fluid will be firstly through the gas-liquid separator. Then it will be heated in regenerator and finally it will be sent to the suction port of CO_2 compressor for re-compression. After that, the continuous working process has been formed.

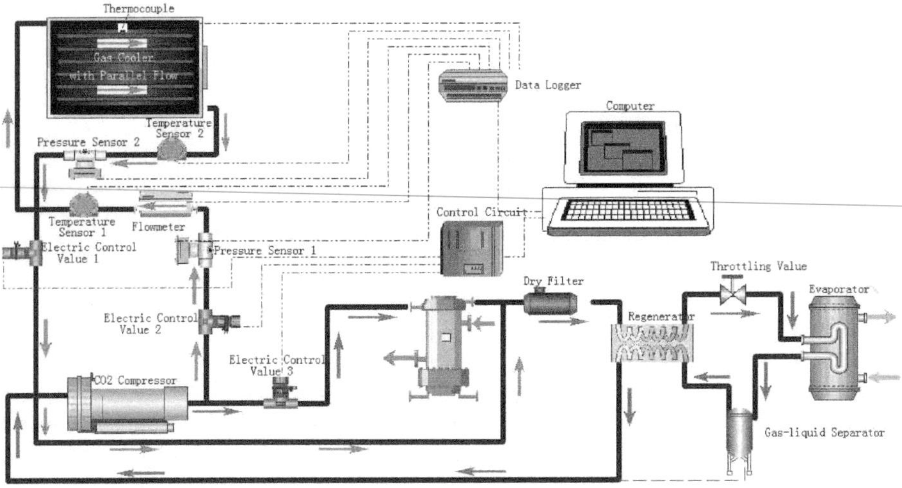

Fig. 1. The system principle of testing platform

3 Inspecting and Controlling System as well as the Design Program

In order to meet the study of CO_2 air-conditioning system and its main components, this inspecting and controlling system has employed the virtual instrument system which is easy to be maintained, expanded and upgraded. The virtual instrument (VI) refers to a computer equipment system which is designed and defined by the users and that has the virtual panel and testing functions on the hardware platform in which the general computer is regarded as the core. Its testing function can be achieved by the software.

The users can simply click on the virtual panel and then they can operate the "instrument" which takes the computer as the platform[9].

The design of the inspecting and controlling system for the testing platform of CO_2 air conditioning based on virtual instrument can be shown in Figure 2.

From Figure 2 we can find that the inspecting and controlling system of this testing platform is mainly composed by the sensor, the data acquisition card and the virtual instrument. The sensors are used to detect the real-time temperature, the pressure, the flow and the wind speed of this system. The detection and the analysis of system have taken the computer as the platform. When the system works, the computer will collect the system temperature, the pressure and the mass flow of CO_2 fluid through the data acquisition card. After the processing and the analysis, the results will be displayed on the monitor of computer.

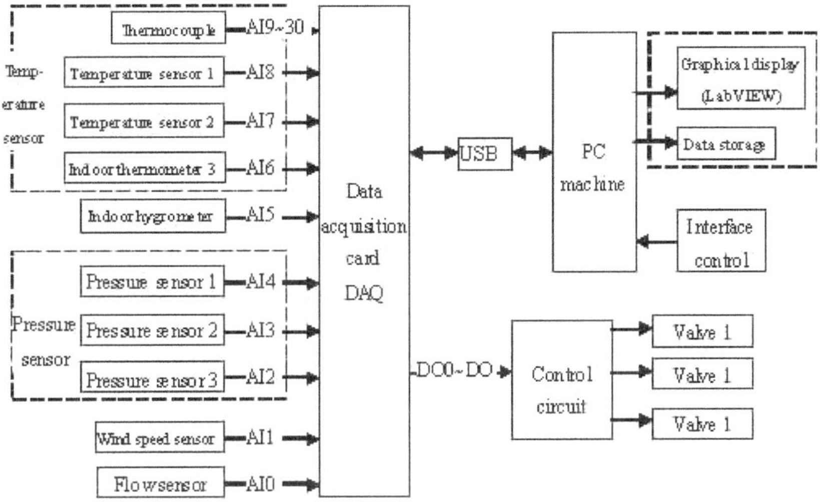

Fig. 2. The program of inspecting and controlling system

4 The Construction of Inspecting and Controlling System

In order to establish an inspecting and controlling system which meets the needs, it is required to reasonably select the data acquisition and the control products and take full account of the field signal conditioning and transmission as well as the development of computer inspecting and controlling system.

4.1 The Selection of the Main Hardware of System

4.1.1 The Data Acquisition Card

Due to the powerful inspecting and controlling functions, the LabVIEW can be combined with the data acquisition card which supports the LabVIEW language platform. On the testing platform of CO_2 air conditioning, the collected signals should be the analogue values, so the acquisition board should have the A/D function. The USB2080 data acquisition card which is introduced by Beijing Art Technology Development Limited Company can meet the requirements of data acquisition on this platform.

2. Sensor

On this platform, the F100S high-quality flowmeter which is introduced by Emerson Micro Motion Enterprise has been employed to measure the CO_2 flow when CO_2 runs through the GCMCPF. It has been equipped with the 1700 transmitters.

The pressure of experimental system working is 7~9.5Mpa. The U5171-100BG pressure sensor which is introduced by Beijing York Instrument Enterprise has been adopted, whose measurement range is 0~10Mpa.

In order to obtain the inlet pressure and the outlet pressure as well as the inlet temperature and the outlet temperature of GCMCPF when CO_2 flows through the GCMCPF, it is required to install the appropriate pressure sensor and temperature sensor at the exist and the entrance of GCMCPF. By comparison, the ETL-1T-375(M) temperature-pressure integration sensor which is introduced by Beijing York Instrument Enterprise has been adopted.

In order to measure the temperature in GCMCPF surface and the inlet and outlet pipe walls, it is required to arrange the processed copper-copper nickel (constantan) thermocouple in GCMCPF surface and the inlet and outlet pipe walls. In order to get the wind speed in GCMCPF surface and accurately master the temperature and humidity changes of experimental environment, it is necessary to employ the EE65 wind speed sensor and the EE21 high-precision temperature and humidity sensor which are introduced by Austria E+E Enterprise.

4.2 The Development of System Software

The software of inspecting and controlling system is developed by the LabVIEW language and the software is mainly composed by the drives of Art USB2080 data acquisition card, the voltage acquisition and the display routines. The drive of data acquisition card is developed by the drive of acquisition card so as to generate the dynamic link library which is controlled by the main program of system that is written by LabVIEW. Then the data acquisition function will be achieved.

4.2.1 The Inspecting Function

The data such as the temperature, the pressure, the flow and some others have been collected by the data acquisition card and the sensors. And they will be transmitted to the computer through USB bus. The input signals of data acquisition card are all the signals with 0~10V voltage and the collected analog signals must be put into the signal conditioning circuit through the program. After being amplified, they will be stored in memory through the data acquisition circuit. The whole process is controlled by the LabVIEW software and the collected data will be analyzed and processed by Excel and Matlab software. The process of data collection of Art USB2080 data acquisition card has been shown in Figure 3. In this system, the continuous data acquisition method has been employed.

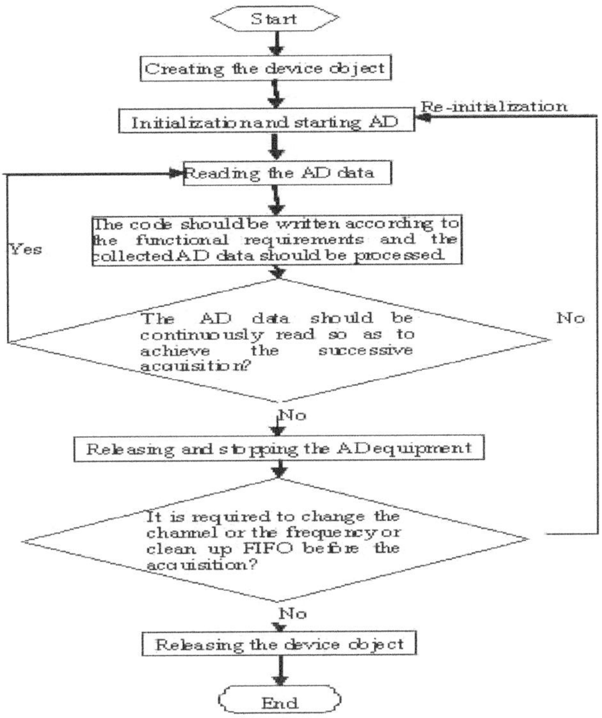

Fig. 3. The process of data acquisition

4.2.2 The Controlling Function

The control module which is composed by the data acquisition card and three electric control valves will control the CO_2 flow when CO_2 runs through the GCMCPF. The process that the output of digital switching value of data acquisition card has controlled the electrical control valve can be shown in Figure 4.

4.2.3 The Data Processing

The file performance function of LabVIEW is a group of powerful document processing tools with strong flexibility. They can read and write the data, move and rename the files and catalogues. The measurement functional program has collected the data in the corresponding channels of data acquisition card. And the Write LabVIEW Measurement File Express VI has been employed to achieve the input of files. Based on the setting path and the file names, the collected data including the temperature, the pressure and the flow has been stored as the spreadsheet data files according to the one-dimensional array.

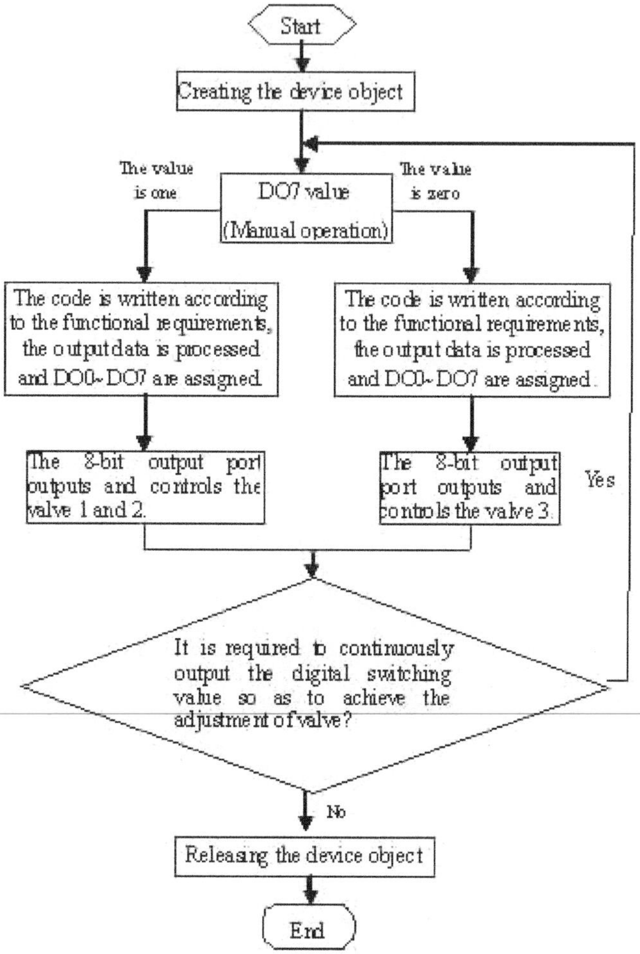

Fig. 4. The control process of electric control valve

5 Conclusion

From the functional requirements of the inspecting and controlling system for the testing platform of CO_2 air conditioning, the inspecting and controlling system based on LabVIEW has been designed in this paper according to the discussion of the implementation of functional modules. The system software is developed by the LabVIEW language, which is composed by the drives of Art USB2080 data acquisition card, the voltage acquisition and the display routines. The drive of data acquisition card is developed by the drive of acquisition card so as to generate the dynamic link library which is controlled by the main program of system that is written by LabVIEW. Then the data acquisition function will be achieved.

In the running process of this system, the data acquisition value can be real-timely displayed and the experimental data can be also saved. Then it is required to employ the data processing software such as Excel and Matlab to analyze the data and control the valves. The designed inspecting and controlling system has the good interactive interface and the strong functional extension. At the same time, the cost of this system is low and it is easy to be maintained and upgraded.

References

1. Lorentzen, G.: Revival of carbon dioxide as a refrigerant. International Journal of Refrigeration 17(5), 292–301 (1994)
2. Pettersen, J., Hafner, A., Skaugen, G.: Development of compact heat exchangers for CO2 air-conditioning systems. International Journal of Refrigeration 21(3), 180–193 (1998)
3. Mehendale, S.S., Jacobi, A.M., Shah, R.K.: Fluid flow and heat transfer at micro and meso-scales with application to heat exchanger design. Applied Mechanics Review 53(7), 175–193 (2000)
4. Lorentzen, G., Pettersen, J.: A new efficient and environmentally benign system for car air-conditioning. International Journal of Refrigeration 16(1), 4–21 (1993)
5. Barreau, M., Blanc, J.: Present European environmentally-friendly CFC &HCFC substitutes for refrigeration and air conditioning applications. In: Twentieth International Congress of Refrigeration, Sydney, Australia (1999)
6. Brown, J.S., Yana-Motta, S.F., Domanski, P.A.: Comparitive analysis of an automotive air conditioning systems operating with CO2 and R134a. International Journal of Refrigeration 25, 19–32 (2002)
7. Pitla, S.S., Robinson, D.M., Groll, E.A.: Heat Transfer from Supercritical Carbon Dioxide in Tube Flow: A Critical Review. HVAC&R Research 4(3), 281–30 (1998)
8. Pettersen, J., Hafner, A., Skaugen, G.: Development of compact heat exchangers for CO2 air conditioning systems. International Journal of Refrigeration 21(3), 180–193 (1998)
9. LabVIEW Measurements Manual[M/OL] (NationalInstruments) (April 2003), `http://www.ni.com/China`

An Efficient Path Planning Method Based on State Automata Model for Mobile Robots

Jingdong Yang[1], Jinghui Yang[2], and Weiguang Wang

[1] Dept. of School of Optical-Electrical and Computer Engineering, Shanghai University Science and Technology, Shanghai 200093, China
eerfriend@163.com
[2] School of Business Management, Shanghai Second Polytechnic University, Shanghai 201209, China
yangjinghui99@yahoo.com.cn

Abstract. In the item of autonomous mobile robots of Robocup game, the tasks of attack and defense are results of some series state or action transition, including division of roles, finding the ball, keeping the ball, etc. Therefore, the key is how to control and describe the state or action of mobile robots in the competition effectively. Finite Automata model is adopted when designing the decision-making system for mobile robots. In the first place, the robot and environmental information all around are all abstracted into some special symbols and states. Meanwhile, the transition functions of the robots state are set using the perception system of vision and transformed to the inputting symbols of Finite Automata. Finite Automata enters into the next state or remains its original state, according to the preset state transition functions. In the end, it will transmit the order to kinematic mechanism. The experiment proved that the method was able to link and smooth all kinds of action or state of mobile robots efficiently; kinematic mechanism of mobile robots was provided with higher real-time and accuracy.

Keywords: Automata, Finite automata, States transformation functions, Mobile robots.

1 Introduction

The theory of finite automaton (FA) is an important branch of research in computer science with wide applications in fundamentals of compiling, genetic algorithm, pattern identification, cryptology, artificial intelligence, etc. Almost all the finite state systems can be represented by using FA. FA offers a computation model to recognize language and control process, etc. The key technique is to control robots to finish transformation among some main actions in real-time and accurately, such as finding the ball, taking the ball, defense, shooting, and so on. That process needs unified description, modeling and analysis, algorithm implementation.

The paper used FA model to design the decision-making system. This method shows that the decision-making system can control attack or attack and defense for

G. Lee (Ed.): Advances in Automation and Robotics, Vol.2, LNEE 123, pp. 181–188.
Springerlink.com

mobile robots in real time. The idea of system modeling and sequencing can improve efficiency and immediacy of system as well.

2 Definite Finite Automata (DFA)

The description of FA is given as the followings.

As a formalization model, FA can be applied to many fields of science. And DFA is used in this paper. A DFA is defined as an 5-tuple: A =(Q, $\sum$, δ, q_0, F)

① Q- a finite state sets
②$\sum$- a finite input symbol set
③ δ- State transformation function, including two variables, state and input symbols, and return a state. δ is denoted by the sign, arrow or arc.
④ q_0- an initial state, a state in Q
⑤ set F- a final state or accepted state, a son set in Q

3 FA Model of Forward

During the Robocup Game, the forward player need attack or defend, while the opposite player offensing. The significance of the forward player decides the complexity of decision-making system. Therefore the design of Q, $\sum$, δ decides the complexity of FA, too.

According to the rules of Robocup game, FA of the forward player includes 8 states: finding ball, approaching ball, obstacle avoidance, taking ball, shooting, defense, colliding, and goal.

(1) Finding ball: it is denoted as q_0, the initial state of whole FA. Enabled state while the whole system starts or the ball is not in the vision.

(2) Approaching ball: it is denoted as q_1, Enabled state, while the forward player finds the ball and the ball is farther from it. The robot remains the state, when it approaches the ball, and there are no obstacles in front.

(3) Taking ball: it is denoted as q_2, enabled state, while the ball appears in the view and the distance between the ball and the robot satisfies some constraints, the robot controls the ball and moves to the opposite goal.

(4) Shooting- q_3, the conditions satisfy the shooting constraints.

(5) Obstacle avoidance- q_4, enabled avoiding, when the robot meets the opposite player or its own player in the process of approaching to the ball, or taking the ball.

(6) Defense- q_5, enabled this state, the robot moves to its own goal, and defend when the ball is not in the view of robot and the current time satisfies some conditions.

(7) Colliding- q_6, enabled this state, when the robot defends or attacks, and collides with others. The state can transform from the defense, because the collision lead to the goal in the defense. But in the attacking, the collision will be neglected.

(8) Goal- q_7, enabled state, when the forward player shoots and goals. The state is the final state of FA. The robot will reset, return to the initial situation, and wait for the next beginning in this state.

The above state sets make up the finite state sets of FA, Q, $Q=\{q_0, q_1, q_2, q_3, q_4, q_5, q_6, q_7\}$, in which, q_0 is the initial state and q7 is the final state.

When symbol set, $\sum$,is designed, process the data from the vision system. Then combine or abstract them as an input of FA. For example: the symbol set, from taking ball to shooting, is composed of the information from the ball, the goal and opposite goal-keeper. There are 14 input symbol sets together.

(1) w_1 : The ball is found, but is not under control (The ball is far away from the robot)

(2) w_2 : The ball disappears from the view of robot.

(3) w_3 : The ball is under control(It is in front of the robot, and the distance is short.)

(4) w_4 : The shooting conditions are satisfied.

(5) w_5 : The ball has entered the opposite goal

(6) w_6 : The ball is not under control with obstacles between the robot and the ball.

(7) w_7 : The ball is not under control with no obstacles between the robot and the ball.

(8) W_8 : The ball is under control, but there are some obstacles when the robot takes the ball, and moves to the opposite goal.

(9) w_9 : The ball is under control, and there are no obstacles when the robot takes the ball, and moves to the opposite goal.

(10) w_{10} : The ball is not in the view, and the robot cannot find the ball after one rotation. The robot is not in the defense zone.

(11) w_{11} : The ball is not in the view, and the robot collides with other robots.

(12) w_{12} : The ball is not in the view, and the robot does not collide with other robots.

(13) w_{13} : The ball is in the view, and the robot does not collide with other robots.

(14) w_{14} : The ball is not in the view, and the robot in the defense zone.

The transformation conditions, w_1, w_2, ... w_{14}, define the input symbol set, $\sum=\{w_1, w_2, \ldots w_{14}\}$,which will lead to state transformation, according to the following state transform functions:

$\delta(q_0, w_1)=q_1$; $\delta(q_0, w_9)=q_2$; $\delta(q_0, w_{10})=q_5$;
$\delta(q_1, w_2)=q_0$; $\delta(q_0, w_3)=q_2$; $\delta(q_0, w_6)=q_4$;
$\delta(q_2, w_1)=q_1$; $\delta(q_2, w_8)=q_4$; $\delta(q_2, w_9)=q_0$;
$\delta(q_2, w_4)=q_3$; $\delta(q_3, w_5)=q_7$; $\delta(q_3, w_2)=q_0$;
$\delta(q_3, w_1)=q_1$; $\delta(q_3, w_3)=q_2$; $\delta(q_4, w_7)=q_2$;
$\delta(q_4, w_9)=q_2$; $\delta(q_4, w_4)=q_3$; $\delta(q_5, w_1)=q_1$;
$\delta(q_5,w_{11})=q_6$; $\delta(q_5,w_{14})=q_0$; $\delta(q_6,w_{12})=q_5$; $\delta(q_6,w_{13})=q_1$。

The above functions constitute the state transformation functions of FA, δ,

$\delta=\{\delta(q_0, w_1)=q_1, \delta(q_0, w_9)=q_2, \delta(q_0, w_{10})=q_5, \ldots\delta(q_6, w_{12})=q_5, \delta(q_6, w_{13})=q_1\}$.

Because there is only one input state for DFA, it is feasible for state transformation between any two states. So if the robot is in a certain state, which satisfies the above condition of state transformation function, it will enter the next state; if the state condition does not satisfy the function condition, the robot is still in the current state, and runs the corresponding command.

According to the above state, symbol set, state transformation function, initial state and final state, a DFA is able to be composed as the followings:

A={{q_0, q_1, q_2, q_3, q_4, q_5, q_6, q_7}, {w_1, w_2, ··· w_{15}}, {$\delta(q_0, w_1) = q_1$, $\delta(q_0, w_9) = q_2$, $\delta(q_0, w_{10}) = q_5$, ···$\delta(q_6, w_{12}) = q_5$, $\delta(q_6, w_{13}) = q_1$}, q_0, q_7}。

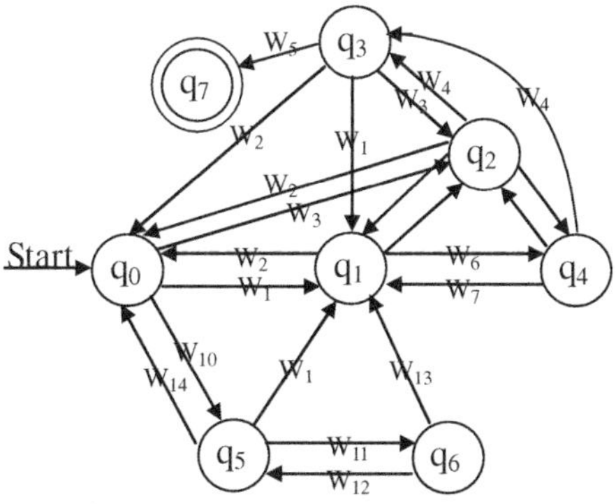

Fig. 1. The state transition graph of the forward's DFA

The language of DFA, A=(Q, Σ, δ, q_0, q_7), for forward player is L(A) = {ω | $\delta(q_0, \omega) = q_7$}, Namely, language A is one of the string set, w, which transforms from the initial state, q_0 to the final state, q_7. $L(A)$ is a regular language. The state of DFA for forward player includes the series of decision-making and actions, corresponding with the current state. When the forward player enters the state, the state transformation will occur, namely, the corresponding actions are completed. For example, the received input condition is $\omega = w_1 w_3 w_4 w_5$, the state will change q_0->q_1->q_2->q_3->q_7, namely, the executive actions: Finding-ball->approaching-ball->taking-ball->shootting->goal.During the state transformation, the robot completes commission and its corresponding tasks.

4 Algorithm Analysis of Game Action for Mobile Robot

The algorithm of shooting is not an independent algorithm, but the series of state transformation, which is led by the sentences, received by the corresponding DFA. And it is the groups of actions, corresponding to every state, and is the result from the combination with vision system, making system, and bellow control system. Several main state process is given as the followings.

(1) Finding Ball Algorithm

The state of finding ball is the initial state of DFA, which is the basis of attack or defense. The vision system adopts objective recognition based on colors, which is more influenced by natural brightness. So often it may not recognize objective. The full dimension system is much more easily influenced than monocular system, so the robot is frequently in the state of finding ball. The finding ball algorithm is able to be designed, with the reliable monocular system and the flexible transformation method between attack and defense.

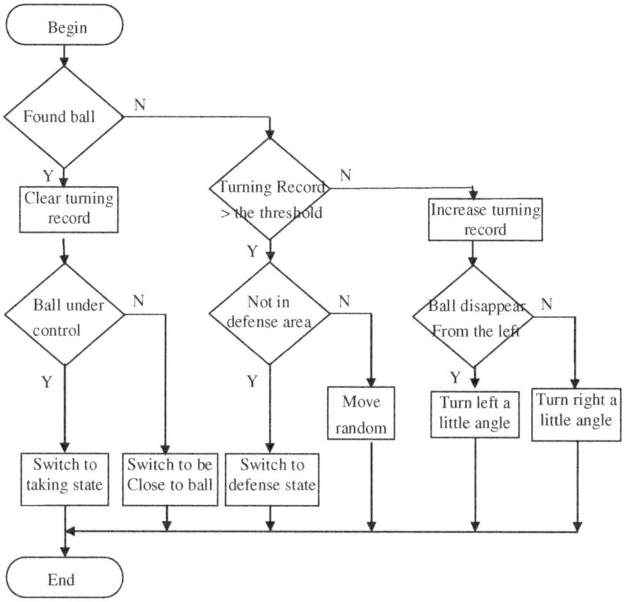

Fig. 2. The flow chart of searching ball

When the ball disappears from the view, the robot turns around on site to find ball for the full dimension vision; if there are no shelter, the monocular system can find ball while turning around on site. If the vision system can not find the ball in one rotation, it is more possible that the ball is sheltered, or the ball is taken to attack by the opposite player. So the robot must go back. If the robot is in the field of defense, the robot moves randomly in order to improve the view of vision; If the ball is found, the distance between robot and the ball is measured, then enter the state of taking ball or approaching ball.

(2) Approaching ball algorithm

The algorithm of approaching ball is executed when the robot approaches the ball. When the robot finds the ball, it will move to the ball, and control the ball. That is the effective method to wield the initiative in the fields. The algorithm will be executed repeatedly, while the robot approaches the ball.

The curvature of left curve line or right curve line is determined by the position between the ball and the robot, the ratio is decided by the distance between the robot and the ball. So the larger angle between the ball and the robot results in the larger curvature, farther the distance between the ball and the robot, the higher rate. The algorithm is executed repeatedly, while the robot approaching the ball. So the robot completes these actions smoothly for every rotation.

(3) Taking ball algorithm

The taking ball algorithm is executed when the ball is in the control, and is taken by the robot. If the transformation condition is satisfied during taking ball, the robot will change the state from the current state, taking ball, to the other state, and executes corresponding actions. Otherwise, the robot adjusts its pose to face the ball in the view of 180 degree, and takes the ball in some curve. The robot need adjust the distance and angle among itself, the ball, and the goal, until satisfies the shooting constraints.

In the process of taking ball, the robot moves without the ball in a short time. Because the algorithm of taking ball is executed, the ball is under control, the robot moves forward snakelike. The method of taking ball makes the ball close to the ball-taking machine. Even when the speed is lower, the ball is not able to break away from the taking-ball machine, and is still under control.

5 The Experiment Results and Analysis

(1) Shooting experiment

The shooting experiment process is the following descriptions. The ball is about 2.5m in front of the goal, and the robot is near the center line of the field. After the programming of forward player is activated, the ball is found by the vision of the robot. Then it approaches the ball, take the ball, and then shoot . the process is given as the following Figure 3.

The vision window is intercepted about every 10 seconds, including the forward vision and full dimension. As the figure 7 are shown, those pictures are put in order of time.

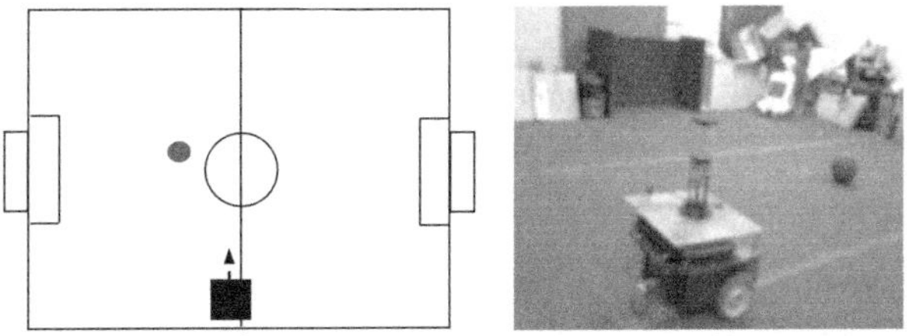

a) b)

Fig. 3. Shooting experiment

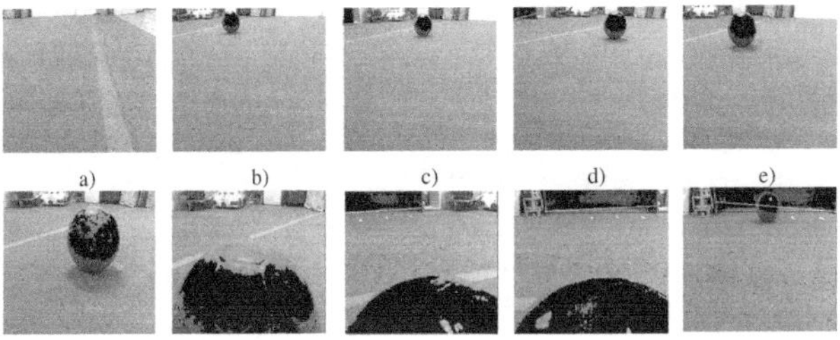

Fig. 4. The forward vision

In the figure 4 Figure a) shows the robot is looking for the ball in monocular system. So the robot approaches the ball. The process is shown as Fig. b),Fig.c),Fig.d),Fig.e),Fig.f)., The ball is still in the view of monocular system when approaching ball. Thus the ball is sure to be in front of the robot. The different picture position in the left or right is because that the robot approaches the ball snakelike. The snakelike movement benefit for the approaching the ball dynamically. For the figure g), the distance between the robot and the ball is shorter, and the robot can shift to the state of taking ball, such as fig. h) and fig. i). The task is to adjust the distance among the robot, the ball, and the goal. From Fig. h) and Fig. i) in Figure 4, the ball is found in the right of the robot, because the robot turns to the left goal, taking the ball. When the distance and angle satisfy the constraints of shooting, the robot enters the state of shooting, and kicks the ball to the goal.

The monocular system is able to recognize the ball clearly in Fig.4. From comparison of both images, the monocular system can recognize the near objective more efficiently than the full dimension system. From lots of shooting experiments: Most shooting experiments are successful, and are able to satisfy the Robosot game of 3 vs 3. However, it is easy to take the ball out of fields for the robot while the ball is near the borderline. This is because the forward robot can not localize its own pose in the global scope. It is also the key research in the future.

(2) The Errors Analysis of State Modeling

The experiment environment is shown as Fig.5. The contrast data of pose and angle errors after the autonomous navigation for 20 min are given shown as Fig.9, in which, the red line denotes the pose errors (SMF), and the blue one denotes the angle errors(SMB) after state modeling.

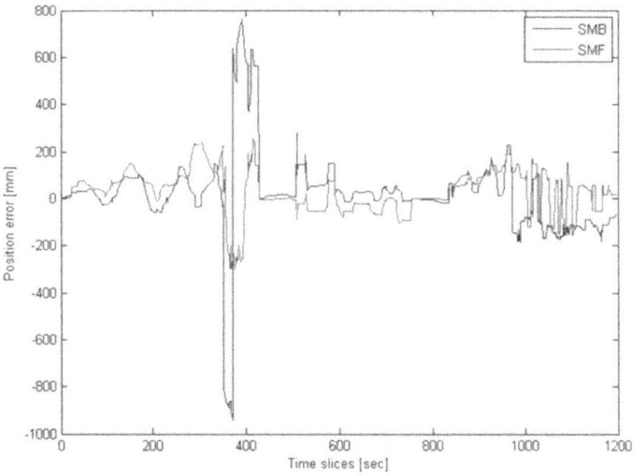

Fig. 5. The pose errors before and after state modeling

It is shown that the pose errors is still in the field of 300mm. When the robot moves at the time of 350s, it meets obstacles or opposite players, so it enters the state of obstacle avoidance. From the contrast data, the robot can not find the optimal route for obstacle avoidance, and some shocking happened before state modeling. And the pose error is increasing by a large margin, and the error drifting of odometer happened. Finally, Odometric errors are amended and global localization is updated by the vision or infrared sensors, until the localization errors become stable. After state modeling, the robot finds the most efficient route by the state transform instruct, avoids the obstacles accurately. The whole path is smooth and safe. So It is shown that the most efficient solution at the current time can be obtained under the state model, and shocking phenomena, result from the optimal solution, are reduced. The real-time and reliability is improved after state modeling.

Acknowledgments. The authors would like to thank national natural science foundation of china, under the grant No. 60874002, for funding the research. And this project is also supported by Innovation Program of Shanghai Municipal Education Commission (No.10YZ207) and Innovation Program of Department of School of Optical-Electrical and Computer Engineering, Shanghai University Science and Technology (No.GDCX-Y-107).

References

1. Moravec, H.: In: Rover visual obstacle avoidance. In: International Joint Conference on Artificial Intelligence, pp. 785–790. Carnegie Mellon University (1981)
2. Harris, C., Stephens, M.: A combined corner and edge detector. In: Fourth Alvey Vision Conference, pp. 147–151. University of Manchester, Manchester (1988)
3. Crowley, J.L., Parker, A.C.: A representation for shape based on peaks and ridges in the difference of low-pass transform. IEEE Transaction on Pattern Analysis and Machine Intelligence 6(2), 156–170 (1984)
4. Lowe, D.G.: Districtive Image Features from Scale-Invariant Keypoints. International Journal of Computer Vision 60(2), 91–110 (2004)
5. Mutch, J., Lowe, D.G.: Object class recognition and localization using sparse features with limited receptive fields. International Journal of Computer Vision 1, 45–57 (2008)
6. Lowe, D.G.: Object recognition from local scale-invariant features. In: International Conference on Computer Vision, Corfu, Greece, pp. 1150–1157 (1999)
7. Arya, S., Mount, D.M., Netanyahu, N.S., et al.: An Optimal Algorithm for Approximate Nearest Neighbor Searching in Fixed Dimensions. In: the Proceedings of the Fifth Annual ACM-SIAM Symposium on Discrete Algorithms,University of Maryland, USA, pp. 573–582 (1994)
8. Tao, Y., Zhang, J., Papadias, D., Mamoulis, N.: An efficient cost model for optimization of nearest neighbor search in low and medium dimensional spaces. IEEE Transactions on Knowledge and Data Engineering 10(16), 1169–1184 (2004)

Parallelization of the Training for Face Detection with Transactional Memory

Kun Zeng

Key Laboratory and technology for National Defence of Parallel and Distributed Processing
School of Computer
National Univiersity of Defense Technology
ChangSha, Hunan, China
kunzeng.nudt@gmail.com

Abstract. The development of multi-core processor makes the parallelization of traditional sequential algorithms increasingly important. Meanwhile, transactional memory serves a good parallel programming model. This paper takes the advantage of software transactional memory to parallelize the Multi-Exit Asymmetric Adaboost algorithm for face detection. The parallel version is evaluated on three different implementations of software transactional memory. The experiment results show that the transactional memory based parallelization outperforms the traditional lock based approach. A speedup of nearly seven is achieved on a eight-core machine on an eight-core system.

Keywords: Transactional Memory, Face Detection, Machine Learning, Optimistic Algorithm, Pessimistic Algorithm.

1 Introduction

With the development of multi-core processors, the parallelization of traditional sequential algorithm is becoming increasingly important. The Adaboost algorithm is a machine learning algorithm widely used in pattern preconization. It is meaningful to take the advantage of multi-core processors to speedup the Adaboost training.

Transactional memory is a new parallel programming scheme which greatly simplifies the concurrency control of parallel programs [8]. It serves an easier parallel programming model compared to the traditional lock based approach. Currently the research works about transactional memory is still in the laboratory phase. In most of the researches, only small kernel benchmarks are used to evaluate the effectiveness of the implementations of transactional memory.

This paper takes the Multi-Exit Asymmetric Adaboost algorithm [10] for face detection as the target algorithm, makes use of the software transactional memory programming model, proposes a parallelized Multi-Exit Asymmetric Adaboost algorithm. On the one hand we try to find out a proper way to parallelize the Multi-Exit Asymmetric Adaboost algorithm, on the other hand, we evaluate the implementations of software transactional memory through a practical workload. This paper evaluates the parallelized Multi-Exit Asymmetric Adaboost algorithm on three different software transactional memory implementations: the pessimistic software transactional memory, the

G. Lee (Ed.): Advances in Automation and Robotics, Vol.2, LNEE 123, pp. 189–196.
springerlink.com

write-through optimistic software transactional memory and the writ-buffered software transactional memory. The experiment results show that transactional memory based parallelization outperforms the traditional lock based approach. Among the three implementations of software transactional memory, the write-buffered approach performs the best.

2 The Multi-exit Asymmetric Adaboost Algorithm

Adaboost is an aggressive machine learning algorithm for object detection. It selects a number of weak classifiers from a large number of candidates to form a strong classifier. The Adaboost algorithm for face detection trains a strong classifier to classify face pictures from non-face pictures.

Algorithm 1. The Multi-Exit Asymmetric Adaboost Algorithm

Require:
a generator G that produces training examples
maximum false acceptance rate α_0
maximum false refuse rate β_0
a stopping condition D

$M_0 = m = 0$
$M = \emptyset$
$\lambda = \alpha_0/\beta_0$
Generate a trainig set $Z = \{(x_n, y_n)\}_{n=1}^{N_1+N_2}$ of N_1 positive examples and N_2 Negative examples
repeat
  Initialize weight:$\omega_n = e^{-(S_m(x_n)-0.5log\lambda)}$
  repeat
    $m = m + 1$
    Normalize weights ω_n
    Select the best feature $f_m(x) : \{x_i\} \to \{-1, +1\}$ with respect to the weighted error: $e_m = \min_i e_i = \sum_{k=1}^{N_1+N_2} \omega_k \times 1_{[y_k \neq f_m(x_k)]}$
    Compute the corresponding coefficient:
      $c_m = 0.5 \times \log(1 - e_m/e_m)$
    Update weights for each examples:
      $\omega_n = \omega_n \times e^{-y_n c_m f_m(x_n)}$
    Estimate the training error rates:
      $\alpha_m = 1/N_2 \sum_{i:y_i=-1} 1_{[H_m(x_n)\geq 0]}$
      $\alpha_m = 1/N_1 \sum_{i:y_i=1} 1_{[H_m(x_n)<0]}$
  until $D or \alpha_m \leq \alpha_0 and \beta_m \leq \beta_0$
  if not D **then**
    $M_0 = m$
    $M = M \cup \{f_m\}$
    Re-generate $Z = \{(x_n, y_n)\}_{n=1}^{N_1+N_2}$ of N_1 positive examples and N_2 Negative examples such that $H_m(x_n) \geq 0 \forall m \in M, n \in \{1, ..., N_1 + N_2\}$
  end if
until D
return the strong classifier $\sum_{f_i \in M} f_i \times c_i$

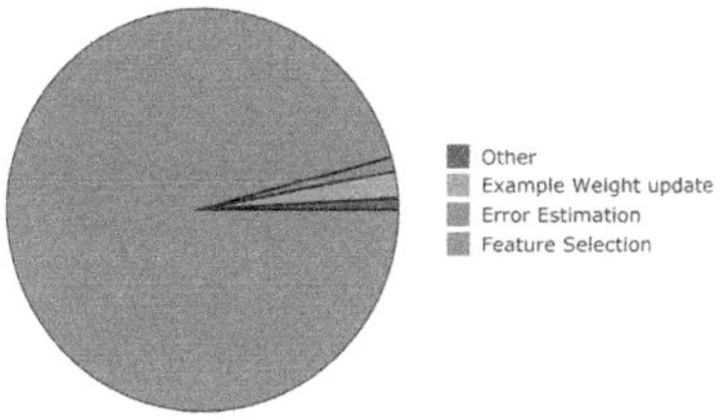

Fig. 1. The Execution Time Break Down

There are a number of available algorithms, such as Viola-Jones [11], Boosting chain [13], Nested cascade [12], etc. The target algorithm of this paper is the Multi-Exit Asymmetric Adaboost algorithm proposed by Pham [10]. The pseudo code of the algorithm is shown in Algorithm 1

There is a main loop in the Multi-Exit Asymmetric Adaboost Algorithm. The main loop consists of three steps: the selection of the best weak classifier, error rate estimation and the update of the example weights. The selection of the best weak classifier needs to scan the whole feature set. For each feature three sub-steps are taken: firstly, the whole example set is scanned to get their vote values; secondly, the example set is sorted according to the vote values; thirdly, the sorted example set is caned to get the optimal threshold for the feature. The error rate estimation is done by a loop with a counter. The update of the example weights contains a write operation for each example.

The core data structure of the algorithm is the feature set and the example set. During each iteration, the whole feature set is tested to get the feature with the lowest error rate. Then the optimal feature is added to the strong classifier. The example weights are updated according to the error rate of the optimal feature.

3 Parallelization of the Algorithm Based on Transactional Memory

there are strong dependences between adjacent iterations, as each iteration needs the example weights generated by the last iteration. So it is not possible to parallelize the outer-most loop. The only choice is to parallelize the inner loops for each sub-step. We analyze the time break down of the training process, which is shown in Figure 1. We can see that the selection of the optimal feature takes more than 95 percents of the time. So we put our most energy in parallelizing the selection of the optimal feature.

The selection of the optimal feature needs three operations to be done for each feature. Firstly, the feature is used to evaluate every example. Secondly, the examples are sorted according to their vote values. Thirdly, the sorted examples are scanned to get the optimal threshold. The first operation is easy to parallelize because the calculation of the vote values are independent from each other and no shared data structure is needed. The second operation is completed through the quick sort algorithm. It is not easy to parallelize the quick sort algorithm for it contains a lot of recursive calls, which incurs data dependency. The third operation is to scan the sorted examples in order. The adjacent iterations incur complex data dependencies. So we will details the parallelization of the latter two operations in the following sections.

The parallelization of the quick sort algorithm has been studied in detail [4]. We borrow the idea proposed by Edahiro [4], which start a new thread when a recursive call is made. The difference is that we take the advantage of transactional memory and enclose the recursive call with a new transaction.

The parallelized quick sort algorithm is shown in Algorithm 2. The phase "#Start_Transaction_Thread" denote that the following procedure call will start a new thread to execute the transaction. Such phases just tell the runtime environment that a new thread can be started. But when and how the thread is started is totally managed by the runtime environment. This will make sure the number of active threads corresponds to the number of cores available.

Algorithm 2. The Parallelized Quick Sort Algorithm

Require:

a vector of training examples starting index of the vector q ending index of the vector r

Procedure QuickSortExamples(v, q, r)

```
if r − q = 1 then
    return
end if
mid=v[r].weight
i=q-1
j=r
repeat
    repeat
        i++
    until v[i].weight¡mid
    repeat
        j–
    until v[j].weight¿mid
    if i¿j then
        break
    end if
    swap(v[i],v[j])
until true
#Start_Transaction_New_Thread
QuickSortExamples(v,q,i-1)
#Start_Transaction_New_Thread
QuickSortExamples(v,i+1,r)
```

As is shown in Algorithm 3, the parallelized threshold computing take the advantage of transactional memory to ensure the consistency of the shared data $thresh$ and $error\_min$. The phase "parallel_for" indicates that the different iterations of the loop can be executed in different threads. The phases #Begin_Transaction and #End_Transaction indicate the beginning and the ending of the transaction.

Algorithm 3. The Parallelized Algorithm for Threshold computing

Require:

```
a sorted vector of the training examples v of size N
a weak classifier f
thresh = v[0].value
error_min = ∞
parallel_for( i=0; i< N-1; i++ )
{
   tmp_thresh = v[i].value
   error = compute_error(tmp_thresh)
   #Begin_Transaction
   if( error < error_min ) {
      error_min = error
      thresh = v[i].value
   }
   #End_Transaction
}
return thresh
```

4 Experiment Environment

The software transactional memory platform employed in our work is the RSTM software transactional memory platform. RSTM serves a flexible C++ library which supports the semantics of transactional memory. It supports varied implementation algorithm for software transactional memory. In our experiment, three kinds of implementation algorithms are tested, namely the pessimistic algorithm the write-through optimistic algorithm and the write-buffered optimistic algorithm.

The pessimistic algorithm is a conservative implementation of software transactional memory. During the concurrency execution of transactions, no operations which may lead to conflicts are allowed. More specifically, if a shared data item has been written by a certain transaction , no other access to the data item from other transactions are allowed; if a shared data item is read by a certain transaction, no other write operations to the data item from other transactions are allowed. Pessimistic algorithm only allows shared read. No other forms of concurrent access are allowed.

The optimistic algorithms are more aggressive implementations of software transactional memory. Compared to the pessimistic algorithm, optimistic algorithms allow operations which may lead to inter-transaction conflicts. They detect the conflict at the time when the transactions commit. The transactions with conflicts are aborted and restarted. More specifically, if a share data item is written by a certain transaction, read accesses from other transactions can still complete; if a shared data item is read by a certain transaction, write access from other transactions can still complete. The runtime environment has to maintain the new and old value of the same share data item, as the modification of the transactions may be aborted. According to the way to maintain the two values, software transactional memory systems can be classified into two classes: the write-through algorithm and the write-buffered algorithm. The write-through algorithm writes the new value in place and maintains the old value in the log, while the write-buffered algorithm maintains the new value in a thread specific buffer and keep the old value unmodified.

Our experiment evaluate the parallelized Multi-Exit Asymmetric Adaboost algorithm with above three different software transactional memory algorithms to find out the impact of the performance by the different implementations.

The maximum example set we used on the real hardware contains 100,000 22×22 pictures. And the maximum example set we used on the simulated platform contains 5000 22×22 pictures. The feature set we used contains 50952 different features.

5 Experiment Results

Two set of experiments are carried out. Firstly, we fix the size of the dataset and vary the number of threads started. Secondly, we fix the number of threads and vary the dataset size.

Figure 2 shows the speedup over sequential algorithm achieved by the parallel Multi-Exit Asymmetric Adaboost algorithm. We can see that pessimistic software transactional memory shows little performance benefit over the lock based approach, while the optimistic software transactional memory outperforms the lock based approach significantly. Because the runtime overhead of software transactional memory is bigger than the lock based approach, the three software transactional memory approaches achieve a speedup less than 1 when the number of threads is 1. The advantage of optimistic software transactional memory approaches gets more obvious as the increase of the threads started.

Generally speaking, parallelized Multi-Exit Asymmetric Adaboost algorithm does not achieve linear speedup. A speedup of less than 7 is achieved when the number of threads is 8. It is because that the quick sort algorithm, which takes a lot of computation time is not easy to parallelize. The recursive calls incurs data dependences, which makes

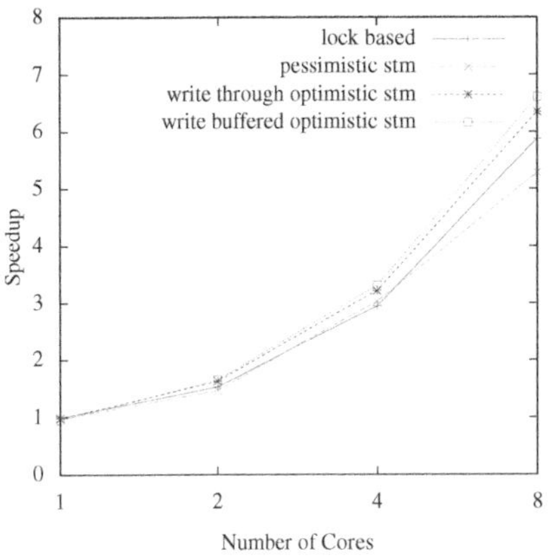

Fig. 2. Speedup on the Real Hardware

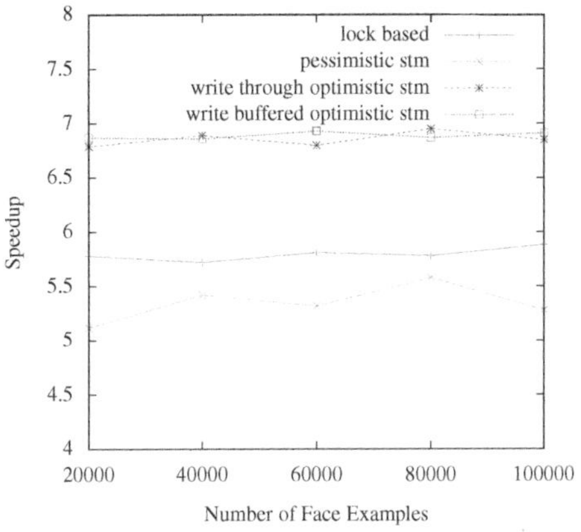

Fig. 3. Speedup vs. Dataset Size on the Real Hardware

the parallelization incomplete. What is more, the threads are dynamically started, which incurs more runtime scheduling overhead.

Figure 3 shows the speedup achieved by the parallelized Multi-Exit Asymmetric Adaboost algorithm for different dataset sizes. The speedup achieved does not vary much with the dataset size. Generally speaking, the speedup achieved increases with the increase of the dataset size. It is because that the increase of the dataset increases the ratio of the parallelized execution. It is shown that when the size of the dataset reaches 20000, the performance of the parallelized algorithm can be fully utilized. Further increase of the dataset does little good to the speedup achieved.

6 Related Work

The Adaboost algorithm is widely used in pattern recognition and a number of researchers has studied its parallelization. Chen uses OpenMP to parallel the Adaboost algorithm [1]; Merler runs the Adaboost algorithm on a cluster [9]; there are also tries to use FPGA to speedup the algorithm [2]. This paper differs from above work in that we use software transactional memory to ease the parallelization of the algorithm. We not only parallelized the algorithm, but also compare different implementation strategies of software transactional memory.

Transactional memory is firstly proposed by Herly in 1993 [7], it is used to ease the concurrency control of share resources in parallel programming. Recently, transactional memory draws a lot of attentions and multiple implementations are proposed, including hardware bsed, such as TCC [6], software based such as RSTM [5] and hybrid implementations, such as HybridTM [3]. This work is based on software transactional memory, which requires no extra hardware supports. The software transactional memory system used in this work is RSTM [5]. It supports multiple implementation algorithms, which makes it a good platform for our work.

7 Conclusion and Future Work

This paper employs software transactional memory to parallelize a popular machine learning algorithm. On the one hand we try to find out a proper way to parallelize the Multi-Exit Asymmetric Adaboost algorithm, on the other hand, we evaluate the implementations of software transactional memory through a practice work load.

This work directly applies software transactional memory to the sequential version of the Multi-Exit Adaboost algorithm. No platform specific optimization is done to the algorithm. Limited by the performance of the parallelized quick sort, we do not get linear speedup. Further researches on how to parallelize a practice algorithm on a transactional memory platform is still needed.

References

[1] Chen, Y.K., Li, W., Tong, X.: Parallelization of adaboost algorithm on multi-core processors. In: IEEE Workshop on Signal Processing Systems, SiPS: Design and Implementation. pp. 275 – 280. Washington, DC, United states (2008),

http://dx.doi.org/10.1109/SIPS.2008.4671775, adaBoost; Adaboost algorithms; Computation powers; Core processors; Load imbalances; Parallel processing;Parallelization; Pattern classification; Person detections

[2] Cho, J., Mirzaei, S., Oberg, J., Kastner, R.: Fpga-based face detection system using haar classifiers. In: Proceeding of the ACM/SIGDA International Symposium on Field Programmable Gate Arrays (FPGA 2009), pp. 103–112. ACM, New York (2009), http://doi.acm.org/10.1145/1508128.1508144

[3] Damron, P., Fedorova, A., Lev, Y., Luchangco, V., Moir, M., Nussbaum, D.: Hybrid transactional memory. SIGOPS Oper. Syst. Rev. 40, 336–346 (2006), http://doi.acm.org/10.1145/1168917.1168900

[4] Edahiro, M.: Parallelizing fundamental algorithms such as sorting on multi-core processors for eda acceleration. In: Proceedings of the 2009 Asia and South Pacific Design Automation Conference (ASP-DAC 2009), pp. 230–233. IEEE Press, Piscataway (2009), http://dl.acm.org/citation.cfm?id=1509633.1509698

[5] Felber, P., Fetzer, C., Riegel, T.: Dynamic performance tuning of word-based software transactional memory. In: Proceedings of the 13th ACM SIGPLAN Symposium on Principles and Practice of Parallel Programming (PPoPP 2008), pp. 237–246. ACM, New York (2008), http://doi.acm.org/10.1145/1345206.1345241

[6] Hammond, L., Wong, V., Chen, M., Carlstrom, B.D., Davis, J.D., Hertzberg, B., Prabhu, M.K., Wijaya, H., Kozyrakis, C., Olukotun, K.: Transactional memory coherence and consistency. SIGARCH Comput. Archit. News 32, 102–113 (2004), http://doi.acm.org/http://doi.acm.org/10.1145/1028176.1006711

[7] Herlihy, M., Moss, J.E.B.: Transactional memory: architectural support for lock-free data structures. SIGARCH Comput. Archit. News 21, 289–300 (1993), http://doi.acm.org/10.1145/173682.165164

[8] Larus, J., Rajwar, R.: Transactional Memory (Synthesis Lectures on Computer Architecture). Morgan Claypool Publishers (2007)

[9] Merler, S., Caprile, B., Furlanello, C.: Parallelizing adaboost by weights dynamics. Comput. Stat. Data Anal. 51, 2487–2498 (2007), http://dl.acm.org/citation.cfm?id=1221579.1221709

[10] Pham, M.T., Hoang, V.D.D., Cham, T.J.: Detection with multi-exit asymmetric boosting. In: 26th IEEE Conference on Computer Vision and Pattern Recognition, CVPR, Anchorage, AK, United states (2008), http://dx.doi.org/10.1109/CVPR.2008.4587738; asymmetric boosting, Boosted classifiers, Conventional methods, Operating points, Significant reduction, Training time, Weak classifiers

[11] Viola, P., Jones, M.: Rapid object detection using a boosted cascade of simple features. In: Proceedings of the IEEE Computer Society Conference on Computer Vision and Pattern Recognition, Kauai, HI, United states, vol. 1, pp. I511–I518 (2001); object detection

[12] Wu, B., Ai, H., Huang, C., Lao, S.: Fast rotation invariant multi-view face detection based on real adaboost. In: Proceedings of the Sixth IEEE International Conference on Automatic Face and Gesture Recognition (FGR 2004), pp. 79–84. IEEE Computer Society, Washington, D C, USA (2004), http://dl.acm.org/citation.cfm?id=1949767.1949784

[13] Xiao, R., Zhu, L., Zhang, H.J.: Boosting chain learning for object detection. In: Proceedings of the Ninth IEEE International Conference on Computer Vision - (ICCV 2003), pp. 709–715. IEEE Computer Society, Washington, D C, USA (2003), http://dl.acm.org/citation.cfm?id=946247.946591

Research and Application on Seismic Image Enhancement Based on Wavelet Transformation

Ma Chang-Qing

Information Center of Daqing Oilfield Company NO.9 Production Plant,
Daqing 163853, China

Abstract. In this paper, digital image enhancement method is studied and the theory of choosing hard-threshold and soft-threshold in wavelet transform is analyzed, and according to the trait of seismic data, the wavelet threshold is improved. Through the verification of the seismic images in Daqing oil field the new threshold function can effectively remove Gaussian white noise and impulse noise, and the image enhancement effect is remarkable.

Keywords: wavelet transform, image enhancement, seismic image, wavelet threshold.

1 Introduction

In the course of geological exploration and development, seismic image is an effective way to obtain geological information. However, due to the role of defects or other factors of seismic data acquisition instruments itself seismic images obtained inevitably contain noise or blur, which greatly reduce the efficiency of seismic interpretation, so seismic image edge enhancement is becoming increasingly important. In recent years, wavelet transform has been widely used in the field of image enhancement, high frequency enhancement methods, unsharp masking method [1] [2] and other methods and so on, which have become a better application. In this paper, a common traditional wavelet threshold algorithm [3] is improved, combined with characteristics of seismic images and using local threshold to construct wavelet threshold function. It is used to construct a seismic image edge enhancement system. The results show that, after using this method, the more obscure, details of poor contrast has been enhanced in the original seismic images. The results are more satisfactory compared with traditional methods.

2 General Image Enhancement

2.1 Mean Filter and Median Filter

General image enhancements are mainly through the mean filter and median filter [3] methods to smooth and denoise seismic images. Both of these two methods are all belong to template-based image enhancement method, usually choosing a square template of which the side length number is odd. Besides the two methods both have a

G. Lee (Ed.): Advances in Automation and Robotics, Vol.2, LNEE 123, pp. 197–202.
Springerlink.com

strong ability to suppress noise, however to some extent, it will lead to image blur, so the interpretation effect will be affected.

2.2 Histogram Equalization

The basic idea of Histogram equalization [4] is to transform the original image histogram to the uniform distribution form, which can increase the dynamic range of pixel gray value, thereby to achieve the desired effect of enhancing the overall image contrast. Histogram equalization is the main method of image enhancement at present, but there are some drawbacks of its own, they are:

(1)The gray-scale reduced after transform, some details are disappeared;

(2)Some images, of which the histogram has peaks, the contrast is unnaturally enhanced after treatment.

2.3 Image Enhancement Based on Fuzzy Set

Fuzzy Image Enhancement [5] is an image enhancement method combined with fuzzy theory. The basic idea is to transform the image to the fuzzy domain from airspace by membership function, and through this we can obtain fuzzy feature plane, in which images are enhanced. At last, transform them back to airspace and get final enhanced images.

3 Seismic Image Enhancement Based on Wavelet Transform

3.1 The Basic Idea of Wavelet Transform

Assume $f(x,y)$ is image of space $L^2(R)$.First, we do wavelet forward transform to get several wavelet coefficients. Then determine the wavelet thresholding according to the image model. We use this threshold to denoise and filter wavelet coefficients. The wavelet coefficients retained after filtering contain most of the image's real effective information. Then determine the gain factor by using human-computer interaction way, and enlarge wavelet coefficients through this factor. Finally, transform the wavelet coefficients after amplification back by inverse wavelet transform, Therefore, we can gain seismic image after enhancement.

From the basic idea of wavelet transform we know that choosing wavelet thresholding λ plays a very important role in the final image enhancement. If the value is too small, then there will be a large part of the noise which can not be removed. If the value is too large, there will be more non-noise wavelet coefficients are set to 0, which greatly affects the authenticity of the image, eventually causes the image not only has not been enhanced but blurred. This article will analyze and research the hard threshold and soft threshold function, then final lydetermine a suitable wavelet threshold function for seismic images.

3.2 Hard and Soft Threshold Function and Their Shortcomings

Threshold function is important to the reconstructed signal's continuity and accuracy, and has a great influence on wavelet-based image enhancement. At present, the choice of threshold mainly contains hard threshold and soft threshold[6] [7].The expression of two threshold functions are as follows:

Hard threshold function:

$$\overline{W}_{j,k} = \begin{cases} W_{j,k} & |W_{j,k}| > \lambda \\ 0 & |W_{j,k}| \le \lambda \end{cases} \quad (1)$$

$w_{j,k}$ is the wavelet coefficient after wavelet forward transform.

Disadvantages: Hard threshold function in the wavelet domain is not continuous, there exists discontinuous point between λ and $-\lambda$.In practice we often have to do derivation on the threshold function. The existence of discontinuous point makes derivation can not be done. This makes the method have some limitations. Meanwhile only the wavelet coefficients which are less than the threshold value are processed, the greater not disposed. so It is not match with the actual case that, the value of the wavelet coefficients is greater than the threshold, and there also exists noise.

Soft threshold function:

$$\overline{W}_{j,k} = \begin{cases} \mathrm{sgn}(W_{j,k})(|W_{j,k}| - \lambda) & |W_{j,k}| > \lambda \\ 0 & |W_{j,k}| \le \lambda \end{cases} \quad (2)$$

sgn (n) is the sign function, it is:

$$\mathrm{sgn}(n) = \begin{cases} 1 & n > 0 \\ -1 & n < 0 \end{cases} \quad (3)$$

Disadvantages: Soft threshold function is continuous in the wavelet domain, but its derivative is not continuous, because it is very difficult when doing the Higher Order Derivative, and that the threshold function does a constant compression to wavelet coefficients, which is not match with the case that the noise component increases with the wavelet coefficients' decreasing.

3.3 Improved Threshold and the Threshold Function

Traditional global threshold in λ is determined from the function proposed by Donoho formula: $\lambda = \sigma\sqrt{2\ln N}$ [8] (σ is the noise standard deviation, N is the signal of the size or length)is clearly unreasonable that wavelet coefficients use the same threshold for different scales. With the increase of decomposition scale, wavelet coefficients are gradually reduced. In response to this feature, a new threshold is given in this article, that is: $\lambda_j = \sigma\sqrt{2\ln N}/\ln(j+1)$, $j(j \ge 1)$ is decomposition scale. Do derivation on λ_j, and the result is: $\lambda_j' = -\sigma\sqrt{2\ln N}/(j+1)*\ln^2(j+1) < 0$,which shows that λ_j is reducing with the increase of decomposition scale j ,and this is consistent with the case that wavelet coefficients decreases with the increasing of decomposition scale.

Since we can only compare $|w_{j,k}|$ with λ_j to estimate a point's wavelet coefficients of an image, therefore, the noise element was saved of which the wavelet coefficients are greater than λ_j.In order to improve it, we can let

$$\delta_{j,k} = \sum_{m=-2^j M}^{2^j M} W_{j\,k-m} \quad (4)$$

Of which, M is the length of high-pass filter, for the wavelet Hear, take M = 1.So the improved hard threshold function becomes:

$$\hat{W}_{j,k} = \begin{cases} W_{j,k} & |W_{j,k}| \geq \lambda_j 且 |\delta_{j,k}| \geq \lambda_j \\ 0 & |W_{j,k}| < \lambda_j 或 |\delta_{j,k}| < \lambda_j \end{cases} \tag{5}$$

The improved soft threshold function becomes:

$$\hat{W}_{j,k} = \begin{cases} \operatorname{sgn}(W_{j,k})(|W_{j,k}| - \lambda) & |W_{j,k}| \geq \lambda 且 |\delta_{j,k}| \geq \lambda \\ 0 & |W_{j,k}| < \lambda 或 |\delta_{j,k}| < \lambda \end{cases} \tag{6}$$

The two improved threshold function can not only effectively remove Gaussian white noise but also remove the impulse noise at the same time. Gaussian white noise and impulse noise are precisely the two most common noises a seismic image has. Thus the improved wavelet threshold and threshold function both played a better denoising effect on seismic images.

3.4 Enhancement of Wavelet Coefficients

After processed by threshold and threshold function above, wavelet coefficients preserved contain most of the information of the image. However, if this time we directly do inverse wavelet transform on the wavelet coefficients after processed, then what we got is only the image after denoising, while can not be enhanced. Based on the actual needs of seismic exploration, we did inverse wavelet transform after enhancing the wavelet coefficients. Thus the wavelet coefficients were effectively enlarged (enhanced), which is helpful to seismic interpretation. After repeatedly comparing among numbers of enhancement methods, in this paper the method that a seismic image is enhanced by single threshold value enhancement is proposed and the algorithm are mainly through the following formula:

$$W_{out} = \begin{cases} W_{in} + \lambda(G-1), & W_{in} > \lambda \\ W_{in} - \lambda(G-1), & W_{in} < -\lambda \\ G \cdot W_{in}, & W_{in} < \lambda \end{cases} \tag{7}$$

λ is threshold for the threshold value, G is the gain factor, W_{in} is the wavelet coefficients before transform and W_{out} the after.In specific test,the values of λ and G are given by Way of human-computer interaction.The algorithm is combined with wavelet threshold functions selected in this paperand amplified most of the denoised wavelet coefficients ,which contain most of the image information.The experiment proved that the enhance effect of the algorithm is significantly obvious and plays a very good enhancement effect on the seismic images.

3.5 Simulation Results and Analysis

In this paper, we used MATLAB to do simulation experiment for the seismic images of pu and shj and many other areas of Daqing. n order to overcome the shortcoming that traditional seismic image enhancement algorithm doesn't do the work of denoising and filtering before being dealt with, we first denoise and filter the seismic images, then apply the enhancement algorithm to the seismic images. Due to limited space, we take only shi area as an example, to show the enhancement effect comparison chart after various enhancement algorithms.

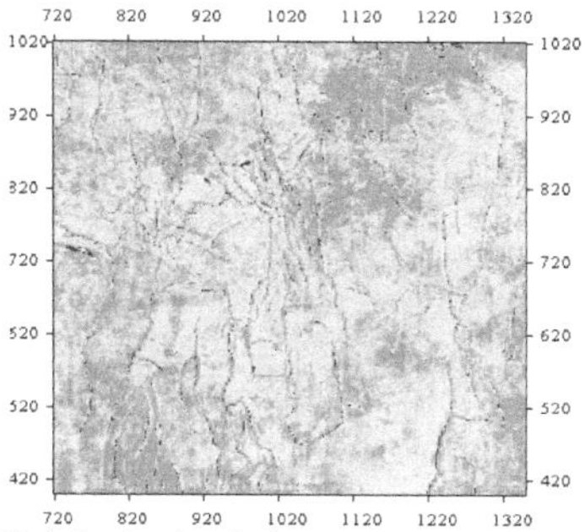

Fig. 1. Original seismic image

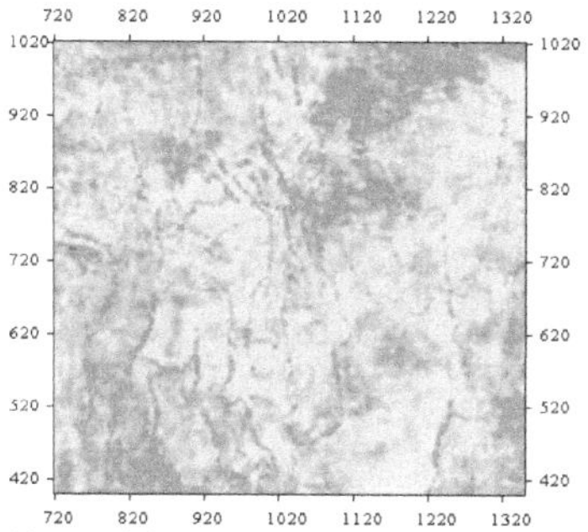

Fig. 2. After average filter(5*5)

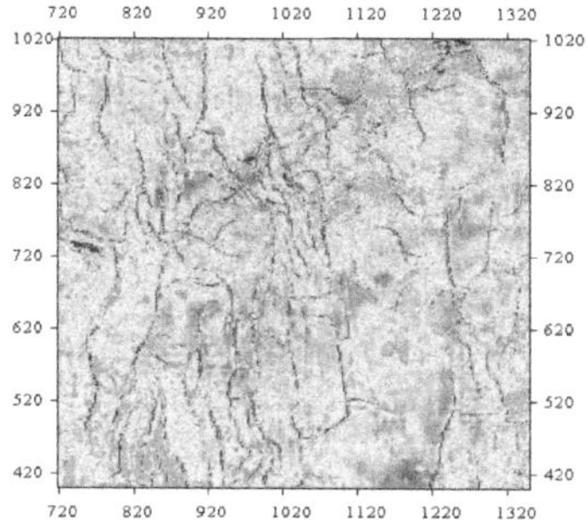

Fig. 3. After histogram equalization

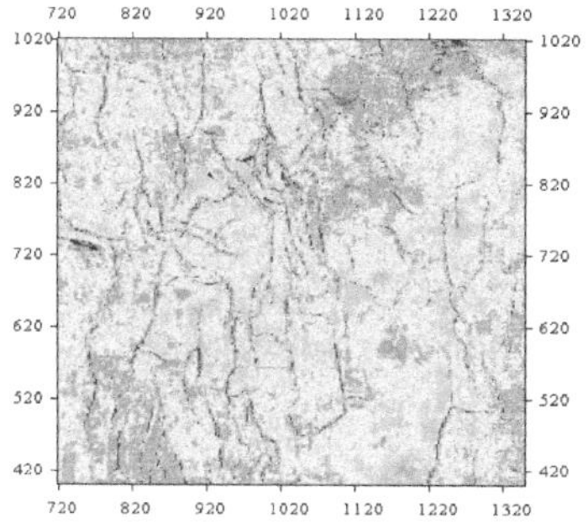

Fig 4. After improved enhanced algorithm

Figure 1 shows the original seismic image which contains noise and the color is monotonous. Figure 2 is the image after being filtered by 5*5 template. Figure 3 shows the image processed by histogram equalization from which we can see that the image contains noise and the color is complex. Figure 4 is processed by wavelet transform and image noise is obviously reduced the contrast is enhanced, the edge becomes more clear. Obviously, the image enhancement algorithm based on wavelet transform can make a better enhancement effect.

4 Conclusions

Based on analyzing and studying the traditional image enhancement methods, we did research on wavelet-based image enhancement method, and improved the common threshold, the determined by traditional wavelet transform according to the seismic image characteristics. To a large extent, the improved algorithm overcomes the shortcomings of traditional image enhancement algorithm, but has the advantages of simplicity and fast. This article applied the enhancement algorithm after being improved to many blocks in Daqing and did a large number of tests. Experiments show that the seismic image contrast markedly improved after this algorithm, edge details of low gray values maintain well, which greatly improves the efficiency of seismic interpretation.

References

1. Li, Z.: Image Enhancement Method Based on Wavelet Transform. Mechanical Engineering & Automation 2, 16–17 (2009)
2. Xu, K., Song, S.: Research on the Application of Wavelet Analysis in Signal De-noising. Ship Electronic Engineering 3, 185–187 (2010)
3. Ting, C., Guo, C., Huang, W.: Wavelet threshold denoising method. Software Guide 2, 166–167 (2010)
4. Song-Tao, L., Gang, W.: A Novel Method for Image Enhancement Based on Generalized Histogram. Optic and Control 3, 12–14 (2010)
5. Tang, S., Nie, M., He, L.: Applying Improved Fuzzy Enhancement Algorithm to Seismic Image. Science Technology and Engineering 23, 7040–7044 (2009)
6. Berkner, K., Wells, R.O.: Smoothness estimates for soft-threshold denoising via translation-invariant wavelet transforms. Applied and Computational Harmonic Analysis 12(1), 1–24 (2002)
7. Claypoole, R.L., Davis, G.M., Sweldens, W.: Non linear wavelet transform for image coding via lifting. IEEE Trans on Image processing 13(12), 1449–1459 (2003)
8. Qu, W.W., Gao, F.: Study on Wavelet Threshold Denoising Algorithm Based on Estimation of Noise Variance. Mechanical Engineering 2, 28–31 (2010)

Structural Optimization Method of Key Part of High Speed Machining Center

Liu Yu, Liu Chunshi, Mao Lieqian, and Zhang Yimin

Northeastern University
Box 319, Northeastern University, Wenhua Road, Heping District,
Shenyang, Liaoning, 110819, China
Liuyu_nx@163.com

Abstract. Aimed at the dynamical design of the column for a high speed machining center, an optimization design is carried on under the restrain of mass and frequency. Dynamical stiffness is the optimization target. Using the Ansys Workbench, the optimal thickness of ribs in column is solved, and the increment of dynamical stiffness is also given. Optimal model is analysised to verify the optimization design. It is a method to dynamical design typical parts in machining tools.

Keywords: High Speed Machining Center, Column, Optimization Design, Ansys Workbench.

High speed manufacturing technology is one of the key development areas of our country in recent years. And it has a great role in the manufacturing of aerospace structures, auto parts, moulds. High speed machining center is the major equipment. In the design of high speed machining center, dynamic characteristic is more important than ever. It has a new target. Firstly, the machine must have a high vibration resistance. Since the periodical unbalance force because of high speed is great, the machining center should have the ability to resist the force. Secondly, the mass addition because of movable parts need extra power of driving motor. Hence, the study of light mass designing is necessary.

In view of above, many factors need to be considered during designing of a high speed machining center. Optimization design is more important. Paper [1, 2] study the optimization issue in designing of machine tools. Paper [3, 4] study the optimization method in the designing of automobile, crane. Paper[5] adopt the wavelet theory and neural net to optimize the structure. In this paper, the column of a high speed machining center is the background, the optimization method is studied using the Ansys workbench.

1 Optimization of Column

The stand column studied is as figure1. It is a foundry part, the outside layer is 20mm in thickness. There is a long hole in the x axle direction. Inside of column, there is a rib plate in the XZ plane, 3 rib plates in the YZ plane. The optimization target of column is maximum dynamic stiffness.

G. Lee (Ed.): Advances in Automation and Robotics, Vol.2, LNEE 123, pp. 203–208.
Springerlink.com

Design parameters: $X1$, $X2$, $X3$, $X4$
Initial value: $X1=X2=X3=X4=20$mm
Constrain conditions :

$$\omega_a \approx \omega$$

$$m_a \approx m$$

$$15 \le X_1 \le 40$$

$$15 \le X_2 = X_3 = X_4 \le 40$$

Where: ω_a ——optimized frequency.

ω ——before frequency

m_a ——optimized mass

m ——before mass

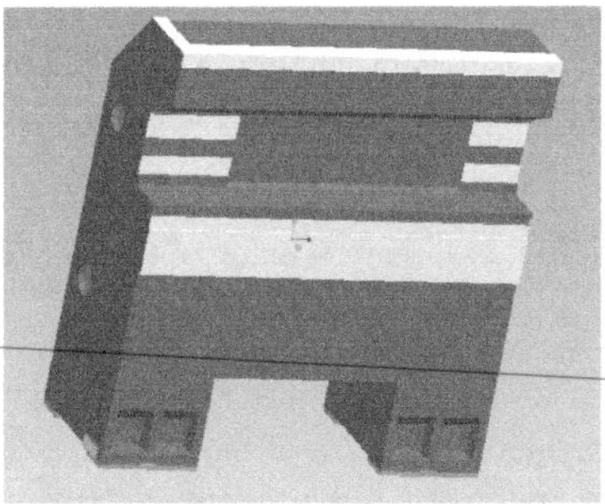

Fig. 1. 3D model of column

2 Solving of Optimization Issue

2.1 Obtaining of Optimization Target

The optimization target is dynamic stiffness has maximum value. The response analysis is carried in this paper on the platform of ANSYS WORKBENDCH, the amplitude of vibration in specified direction can be obtained. From the definition of stiffness in a linear system, we can know:

$$k = X / F$$

Where : X——amplitude of response ;
F——amplitude of stimulation force.

2.2 Parameters Solving

In this paper the analysis model named GDO in the Design exploration is used, the scope of parameters are: rib plate thickness in XZ plane is $15 \le X_1 \le 40$, rib plate thickness in XY plane is $15 \leq X_2 = X_3 = X_4 \leq 40$. The checking points is set by Ansys workbench, a group of parameters are get. Every group of parameters is calculated and analyzed. the varies of parameters and mass and varies of response amplitude are got.

2.3 Setting up of Optimization Parameters

In this paper, the parameters involved are X1,X2,X3,X4, where X2, X3, X4 have same characteristic. According to the fact, set the X1 and X2 to X4 are input parameters, set the mass, natural frequency, maximum amplitude in X, Y, Z direction are output parameter.

3 Result and Analysis

To do the response analysis of the column of the vertical machining center and take the maximum amplitude of vibration as the index of dynamic stiffness. Import the entity model of column to the Ansys Workbench, apply the load at slide block in the X, Y, Z direction. Set the frequency scope is 0 to 600Hz, for the maximum spindle speed is 30000 rpm.That is the maximum frequency is 500Hz. It is reasonable to set the maximum analysis frequency is 600Hz. The vibration checking point is surface of guideway.

The results are as following:

(1) The amplitude of vibration in x direction firstly increase and then decrease as the thickness of rib plate in XY plane is increasing. The maximum point is at 25mm.

(2) The amplitude of vibration in y direction decreases as the thickness of rib plate in XY plane is increasing, as figure 3.

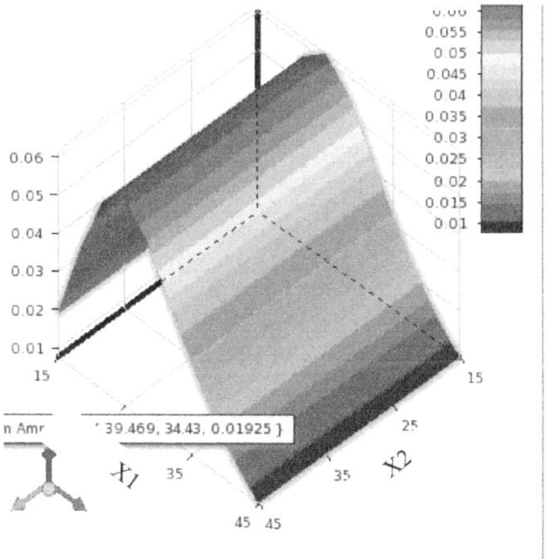

Fig. 2. Varies of maximum of response in x direction

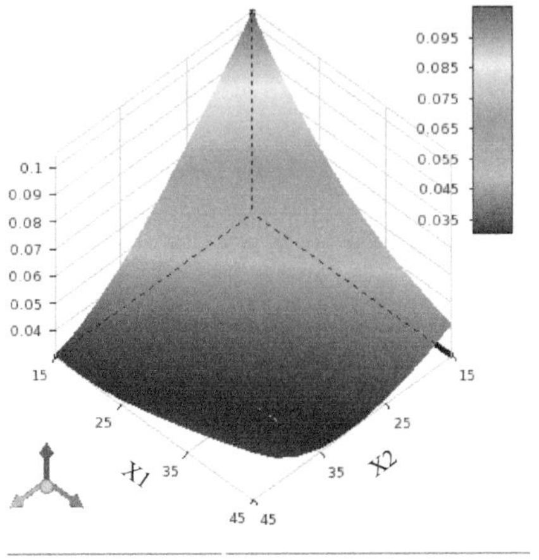

Fig. 3. The maximum amplitude of response in y direction

(3) the optimization results

Comparing of the optimization group and initial group, the first group is excellent for the obvious effect. The third group results in the decreasing of dynamic stiffness. The second group has a moderate optimization effect and the increase of mass is acceptable. According to the mass control, one can choose the first group as final group.

Table 1. Optimization Group of Parameters

	X1	X2~X4	X amplitute of response	Y amplitute of response	natural frequency	mass
initial	20	20	0.034	0.01935	138	1625.8
1	29.55	16.56	0.0285	0.00602	140.58	1676.5
2	24.75	15.62	0.0226	-0.00043	142.41	1634.1
3	26.25	34.369	-0.3947	0.05293	138.26	1759.8

(4) After optimization X1=29.55mm, X2 = X3 = X4 =16.56mm

Now, the mass is 1676.52Kg, the maximum deform in x direction is 0.028mm,the maximum deform in y direction is 0.056mm, and the first natural frequency is 140.58Hz.

Table 2. Dynamic Characteristic After Optimization

	before	after	increasing
mass /kg	1625.8	1676.52	3.1%
X maximum amplitude	0.034	0.028	17.6
Y maximum amplitude	0.072	0.056	22.2%
First order natural frequency	143.46	140.58	2%

In the condition that mass of column and natural frequency vary not beyone 3.5%, the dynamic stiffness increases over 17%. The maximum increasing 22.2%.that is to say the optimization is acceptable.

4 Validation of the Thickness of Rib Plate

After the optimization, in order to validate the dynamic characteristics meet the demand. It is necessary to modify the structure and take a response analysis. Ansys Workbench has the function to modify the structure automatively, one can get the best model to validate the optimization. Figure 4 show the vary curve of maximum amplitude of response.

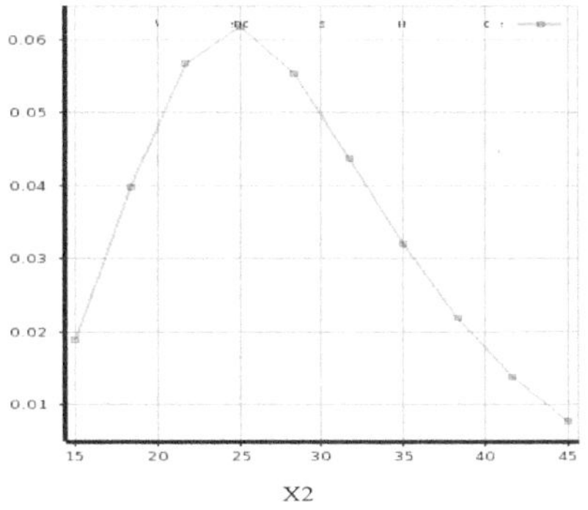

Fig. 4. Iinfluence of X2 to maximum amplitude in X

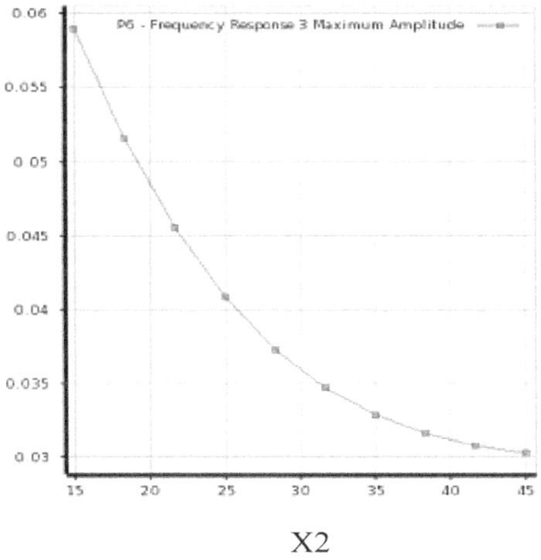

Fig. 5. Influcnce of X2 to maximum amplitude in Y

5 Conclusion

In the paper a method to optimization the column of high speed machining center is proposed. And based on the Ansys workbench, the optimization is carried. after the analysis ,we can get:

(1) It is a effect method to choose the dynamic stiffness to optimization in dynamic design.
(2) Using the ansys workbench to validate the algorithm, it is prove the method is correct.
(3) Though the modification of parameter of structure to validate the response amplitude, one can see the method is correct.

Acknowledgements. We would like to express our express our appreciation to program for Changjiang Scholars and Innovative Research Team in University (No.IRT0816), Key National Science &Technology Special Project on "High-Grade CNC Machine Tools and Basic Manufacturing Equipment" (2010ZX04014-014), Chinese National Natural Science foundation (51135003, 51105067, 50875039) and Key Projects in the National Science &Technology Pillar Program in the Eleventh Five-year Plan Period(2009BAG12A02-A07-2).

References

1. Feng, C.Z., Li, C.H.: High Speed Bridge Machine Tools Structural Design and Optimization. J. Digital Manufacturing Industry 18, 25–30 (2011)
2. Weule, H., Fleischer, J., Neithardt, W.: Structural Optimization of Machine Tools Including the Static and Dynamic Workspace Behavior. The 36th CIRP-International Seminar on Manufacturing Systems 12, 56–60 (2003)
3. Feng, H.: Anasys Based a Car Skeleton Structure of Ansys Finite Element Analysis And Optimization. J. Huazhong University Of Science And Technology Graduate Thesis (2007) (in Chinese)
4. Sun, C.: Based on the Ansya Crane Structure Optimization of the Principle and Method of Study. J. Graduate Thesis of Wuhan University of Thechnology (2010) (in Chinese)
5. Gholizadeh, S., Salajegheh, E., Torkzadeh, P.: Structural optimization with frequency constraints by genetic algorithm using wavelet radial basis function neural network. Journal of Sound and Vibration. 212, 316–331 (2008)

Development of the Analysis Software of Reliability Based on the Matlab

Liu Yu, Liu Chunshi, Li Zhichao, and Zhang Yimin

Northeastern University
Box 319, Northeastern University, Wenhua Road, Heping District,
Shenyang, Liaoning, 110819, China
Liuyu_nx@163.com

Abstract. In this paper, the analysis of reliability software was developed using the MATLAB language. This software has the functions as following: creating sample data with the specify Probability distribution types and parameters, test sample data analysis, solving the reliability problem with 2 to 4 stochastic variable. Reliability function is input by user which enhances the applicability. The method of Monte Carlo is presented to verify the result. The development of this software is an application of computer to the engineering.

Keywords: FOSM, MATLAB, reliability, Monte Carlo.

1 Introduction

Mechanical Reliability design is called mechanical probability design. Since 1940s the reliability theory is proposed, the technology increasingly cause the attention of people[1-3]. Different from tradition design, the main characteristic is quantification the quality of parts. It is more accuracy than ever. So far, the method of reliability design is widely used in the industry of automobile, aerospace etc. important parts and other parts[4,5]. The reliability calculation is the major content of reliability design. In practice, the common used methods include second order moments methods, response surface, Monte Carlo method and NNT based reliability analysis method.[6-9]

Reliability calculation involved the differential of function, lots of matrix operation etc. So based the computer software, reliability analysis will improve the analysis efficiency and precision. Zhang developed a linear cutting machining using the MATLAB GUI.

2 General Design of Moment Method Reliability Software

Considering of the main contents of mechanical reliability analysis, it includes below models: data distribution fitting and statistic parameters calculation. Monte Carlo reliability validation, probability distribution drawing.

G. Lee (Ed.): Advances in Automation and Robotics, Vol.2, LNEE 123, pp. 209–214.
Springerlink.com

The software structure is shown in figure 1.

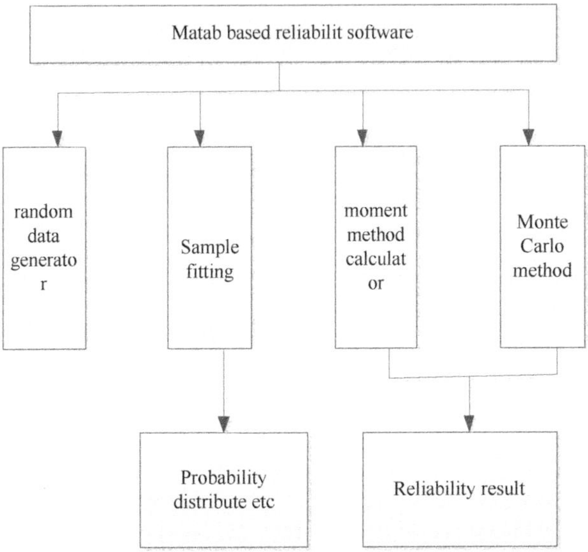

Fig. 1. Software structure of reliability analysis

(1) Fitting of data sample distribution

Data fitting can get the distribution of test data, and calculate the parameters. Distribution would be norm, extreme distribution, exponential distribution, Belta distribution, lognormal distribution, Poisson distribution, uniform distribution, weibull distribution etc. The type of distribution including first and second order moments, probability density function, calculative distribution function display in a graphic form.

(2) Generation of random sample

The software can generate sample data according to the mean, variance and coefficient of variation set by user. All the data is saved in a file with *.xls postfix.

(3) Reliability calculating using Moments method

Method of second order moments is used to calculate the reliability. The function model is input by user. Two, three, or four parameter are acceptable in the function model. The first and second order moments are input by user which can be get by fitting the data.

(4) Monte Carlo method of reliability calculate

Monte Carlo method is adopted to calculate the reliability for the validation with that calculated by moment method.

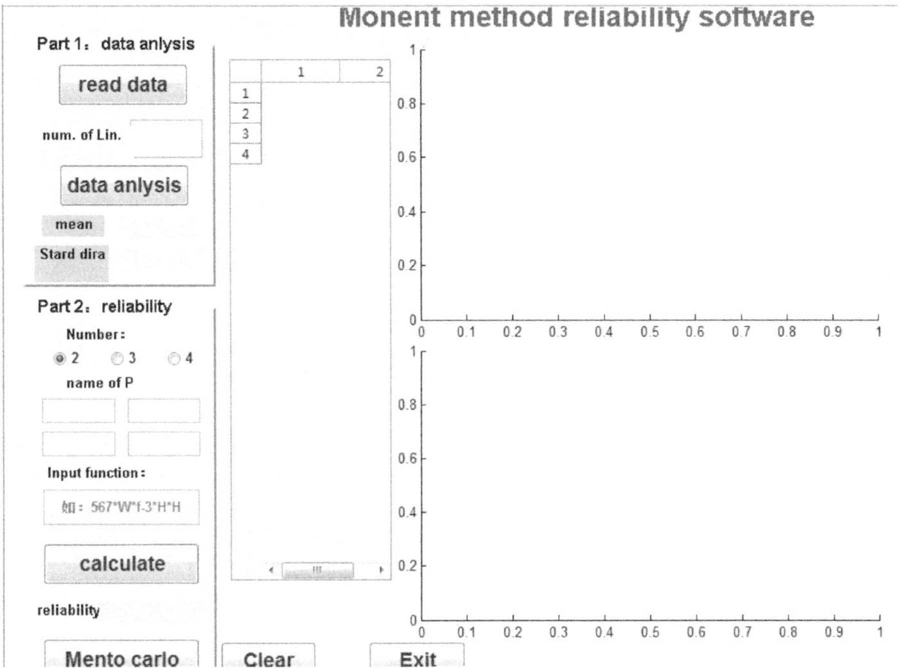

Fig. 2. Interface of the software

3 Key Issues on the Software

The software is developed on the MATLAB platform, the interface is shown in the figure 2. MATLAB has the GUI model. It has two steps to complete a GUI task. One is layout of components, the other is program. GUIDE can generate all the two files: one is Fig file which includes figure windows and all the sub-components (user ActiveX and coordinate axis). The other one is M file.

3.1 Sample Data Generation

Sample data is generated by random number generator. For norm distribution the function is

$$X=normrnd(Mu,Sigma,m,n)$$

the role is to generate a norm random number group with the mean Mu and standard diviation is Sigman. The number of lines is m, number of comlums is n. For the lognorm distribution the function is

$$X=lognrnd(Mu,Sigma,m,n)$$

Different from the norm distribution, Mu and Sigma are not mean and standard diviation but parameter. So Let M S are mean and standard diviation, then

Mu=log(M)-S^2/2; Sigma=sqrt(log(1+(S/M)^2)).

For the extreme distribution the generator function is

X=-evrnd(Mu,Sigma,m,n)

Similiar with the norm distribution ,parameter is calcuted as

Mu=-psi(1)*Sigma-M;Sigma=sqrt(6)*S/pi

3.2 Sample Data Save

Uiputfile command is used to save sample data, the instruction is:

[FileName, PathName, FilterIndex] = uiputfile(FilterSpec, DialogTitle, DefaultName)

The rols is display a dialogue, choose a file. The dialogue display all the file ad directory. If you cancel the dialogue, the return is zero. If a file is choosen, the file name will be return. "uiputfile" is the file name and Default Name is path plus filename. Display is only effect to the proper file with correct postfix.

The data is saved with *.xls files, using the command XLSWRITE.

4 Reliability Analysis Software

Calculating reliability part acquire the user input, and determine whether enough information, if enough to points much operations of the situation discussion.

4.1 Two Independent Variables

Function has two variable calculation is divided into 6 kinds of cases, respectively for two variables are normal distribution a variable is normal distribution another variable to logarithmic normal distribution a variable is normal distribution another variable to obey the extreme I type distribution are two variables to logarithmic normal distribution a variable to logarithmic normal distribution.

When all of the random variable is normal distribution, by average a first order second moment method (center points), when the normal distribution is used when existing JC method algorithm to solve.

4.2 Multi-variables

When there are three random variables, divided into 10 kinds of distribution types of combination, the four variables the situation more complex, there are 15 kinds of distribution types of combination of calculation may.

4.3 Partial Derivation

In reliability calculation, the need to solve function of random variables, the partial derivative MATLAB "Diff" function realize this function. because the random parameter is set by user, the parameter is defined SStr1=sym(Str1); SStr2=sym(Str2), str1 is the notation of first parameter input by user, str2 is the noatation of second parameter input by user. The function is expressed in subs function, the form is

g=subs(Expression,{SStr1,SStr2},{x(1),x(2)})

the meaning of the function is repleacing SStr1 and SStr2 in expression with x(1), x(2) Then calculate the partial derivation, the expression is:

Diff1=diff(Expression,SStr1); Diff2=diff(Expression,SStr2)
Using subs function again to get the partial derivation
gX=[subs(Diff1,{SStr1,SStr2},{x(1),x(2)});
subs(Diff2,{SStr1,SStr2},{x(1),x(2)})]

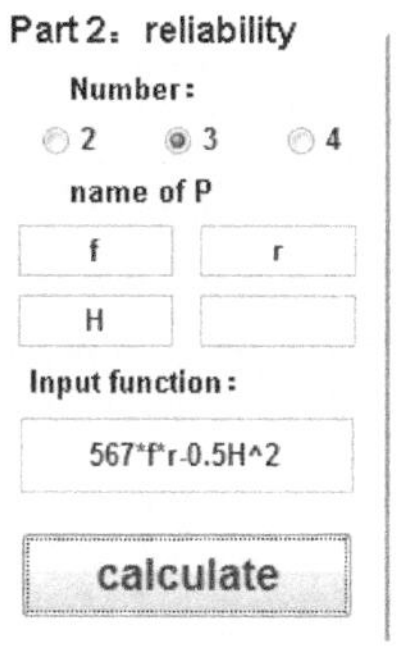

Fig. 3. Reliability calculate

5 Application Example

The function is 567fr-0.5H^2=0, f belongs to norm distribution, $\mu_f = 0.6$, $V_f = 0.131$.r belongs to norm distribution $\mu_r = 2.18$, $V_r = 0.03$ .H belongs to lognormal distribution, $\mu_H = 32.8$, $V_H = 0.03$. Set the number of parameters is 3 and the value of mean and variant coefficient. The reliability is 0.97527, JC method is adopted. The reliability value is 0.97452 by the Mento carlo algorithm.

6 Conclusion

(1) Development a reliability analysis software for moment method in the MATLAB GUI platform, the software can solve the issues with 2 to 4 parameters.
(2) The result is validated with Monte Carlo method.
(3) The software has a function to generate the random parameters with common distribution. It can fit the distribution, get the type, parameter, probability density function curve, calculative distribution probability.

Acknowledge. We would like to express our express our appreciation to program for Changjiang Scholars and Innovative Research Team in University (No.IRT0816), Key National Science &Technology Special Project on "High-Grade CNC Machine Tools and Basic Manufacturing Equipment" (2010ZX04014-014), Chinese National Natural Science foundation (51135003, 51105067, 50875039) and Key Projects in the National Science &Technology Pillar Program in the Eleventh Five-year Plan Period(2009BAG12A02-A07-2).

References

1. Sun, Z.L., Yan, Y.T.: Three Dimensional Modeling of Mechanical Parts and Integration of Reliability Software. J. China Mechanical Engineer. 17, 376–379 (2006)
2. Zhang, Y.M., Liu, Q.L.: Reliability Based Design of Automobile Components. Proceedings of the Institution of Mechanical Engineers Part D Journal of Automobile Engineering 216, 455–471 (2002)
3. Zhang, Y.M., Liu, Q.L., Wen, B.C.: Quasi Failure Analysis on Resonant Demolition of Random Structural Systems. American Institute of Aeronautics and Astronautics Journal 40, 585–590 (2002)
4. Zhang, Y.M., He, X.D., Liu, Q.L., Wen, B.C., Zheng, J.X.: Reliability Sensitivity of Automobile Component Switch Arbitrary Distribution Parameters. Proceedings of the Institution of Mechanical Engineers Part D Journal of Automobile Engineering 219, 165–182 (2005)
5. Zhang, M.: Analysis of Structure Reliability——Method and Program, Beijing (2009)
6. Zhang, Y.M.: Connotation and Development of Mechanical Reliability-based Design. Journal of Mechanical Engineering 46, 168–188 (2010)
7. Yang, Z., Zhang, Y.M.: Reliability-based Sensitivity Design of Automobile Familiar Springs with Non-normal Random Parameters. J. Vehicle & Power Technology 112, 38–44 (2008)
8. He, X.D., Zhang, Y.M., Liu, Q.L.: Reliability-Based Robust Optimization Design for Banjo Flange. Journal of Mechanical Strength 26, 666–669 (2004)
9. Zhang, Y.M., He, X.D., Liu, Q.L., Wen, B.C.: Reliability based optimization and Robust Design of Coil Tube Spring With Non-Normal Distribution Parameters. Proceedings of the Institution of Mechanical Engineers Part C, Journal of Mechanical Engineering Science 219, 567–576 (2005)

Study of B-Spline Interpolation, Correction and Inverse Algorithm

Liu Chunshi, Liu Yu, and Zhang Yimin

Northeastern University
Box 319, Northeastern University, Wenhua Road,
Heping District, Shenyang, Liaoning, 110819, China
Liu_chunshi@163.com

Abstract. This paper aims to make a detail analysis of B-spline interpolation and inverse algorithm that are widely used in CAD/CAM. It points out the nature of two definitions of B-spline curve as well as five correction methods of one B-spline interpolation. In this paper, it explores B-spline inverse algorithm and gives the scope of the algorithm. In addition, it verifies this algorithm by writing a program using VB 6.0.

Keywords: B-Spline Interpolation, B-Spline Inverse Algorithm, B-Spline Correction.

1 Introduction

The direct interpolation of free-form curve is a major issue in CNC machine tool research. B-spline curve is featured with intuition, convex and can be partial modified, which is widely adopted in CAD, CAM and numeric control system. For instance, FUNAC 16i numeric control system supports non-uniform rational B-spline interpolation function, while Delta-Tau Company's PMAC motion control card supports both uniform and non-uniform B-spline curve interpolation function. In engineering application, there are two common types of issues: one is known B-spline control point to get B-spline curve such as numeric control machine tool interpolation issue; the other is known series of data points and solving B-spline curve via known data points such as robot path issue.

2 Several Issue about B-Spline

2.1 B-Spline Curve's Mathematical Expression

In 1971, De Boor and Cox gave standard algorithm of B-spline calculation independently.

$$V(t)=\sum_{i=0}^{n}P_iN_{i,k}(t)$$

G. Lee (Ed.): Advances in Automation and Robotics, Vol.2, LNEE 123, pp. 215–221.
Springerlink.com

In which, $P_i(i=0,1,\cdots n)$ is the vertex of control polygon, $N_{i,k}(t)(i=0,1,\cdots,n)$ is order K (k-1 degree) B-spline primary function.

$$N_{i,1}(t)=\begin{cases}1, & t_i \le t \le t_{i+1}\\ 0, & \text{ot her}\end{cases}$$

$$N_{i,k}(t)=\frac{t-t_i}{t_{i+k-1}-t_i}N_{i,k-1}(t)+\frac{t_{i+k}-t}{t_{i+k}-t_{i+1}}N_{i+1,k-1}(t)$$

In which $\mathbf{T}=[t_0,t_1,\cdots t_{n+k}]$ is the node vector.

B-spline De Boor-Cox formula shows, B-spline is determined by control point P, order K and node vector T.

2.2 Classification of B-Spline

According to different distribution of node vector T's each node, B-spline can be uniform periodic B-spline, open and non-uniform B-spline and B-spline curve.

In which uniform periodic B-spline refers to the distance is an arbitrary constant value between node vectors. Through derivation, it is known that this kind of B-spline can be displayed by a simpler matrix form so as to be widely applied. For example, multi-axis motion controller PMAC designs a uniform periodic B-spline interpolation algorithm which can reduce the amount of real-time computing.

Three uniform periodic B-spline matrix form is shown as:

$$V_{i,3}(t)=\frac{1}{6}\begin{bmatrix}t^3 & t^2 & t & 1\end{bmatrix}\begin{bmatrix}1 & 3 & -3 & 1\\ 3 & -6 & 3 & 0\\ -3 & 0 & 3 & 0\\ 1 & 4 & 1 & 0\end{bmatrix}\begin{bmatrix}P_i\\ P_{i+1}\\ P_{i+2}\\ P_{i+3}\end{bmatrix} \quad 0\le t\le 1 \quad i=0,1,\cdots N-3 \tag{1}$$

Uniform periodic B-spline can be defined only by control point vector P and order K. Each section of cubic B-spline is decided by four control point vectors. Therefore, n control points can decide three B-spline of n-3section. By using of VB programming language, according to equation (1), written uniform periodic B-spline curve interpolation results shown in Figure 1. There are six control points in Figure 1, we can get three B-spline respectively from six control points with different colors.

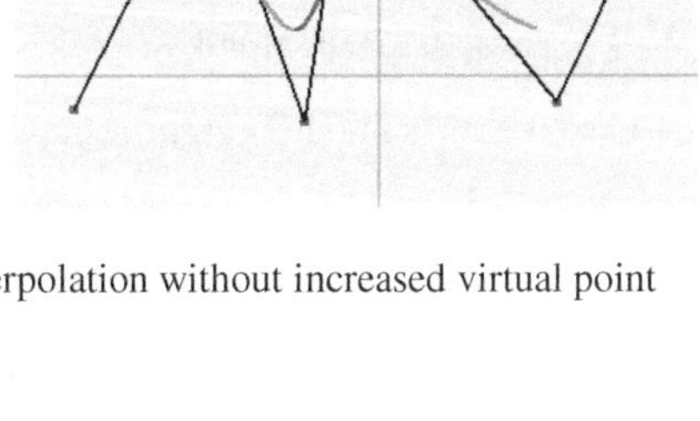

Fig. 1. B-spline curve interpolation without increased virtual point

Non-uniform B-spline vectors distribute freely, and each primary function must be calculated separately. However, non-uniform B-spline curve can interpolate, delete and edit nodes freely, which facilitates the control of the local shape of the curve.

2.3 Interpolation Method to Increase the Virtual Point of B-Spline

Figure 1 shows that B-spline curve that defined by four control points didn' t pass the first and the last control point, which is not suitable for many applications. Because it requires to start with the starting point in robot trace planning or CNC system interpolation task and reach the end point determined by control polygon.

We add control points, make B-spline pass through the first and last control point, and be tangent with the first and last control polygon.

$$P_{-1} = 2P_0 - P_1$$
$$P_{n+1} = 2P_n - P_{n-1}$$

The coordinate of the virtual point added is

$$P_{-1} = 2P_0 - P_1$$
$$P_{n+1} = 2P_n - P_{n-1}$$

After adding 2 virtual points, B-spline increase 2 sections, and pass through the start point and ending point. An shown in figure 2, it can be proven that the new added section is tangent with the control polygon $\overline{P_0P_1}$, $\overline{P_{n-1}P_n}$.

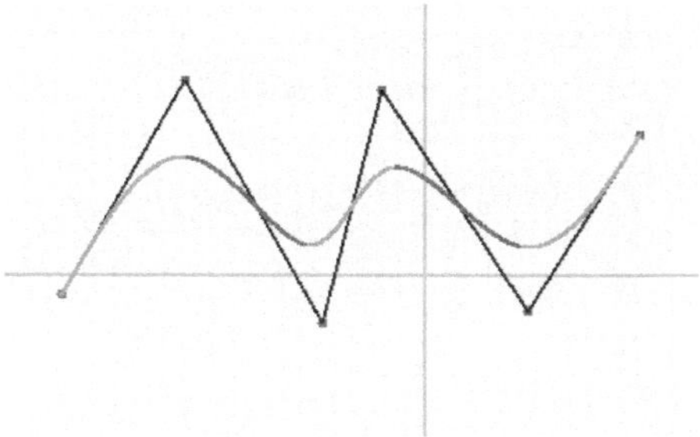

Fig. 2. B-spline after adding the virtual control point

2.4 Five Points Correction for B-Spine

For the trajectory planning of robot based on B-spline, as shown in figure 2, the distance is much large from the spline and control polygon. Although, the B-spline is nth derivable, the distance is too much.

We can adopt five points correction to solve the problem. The idea is to calculate a group of new control points to instead the old group ones. Except for the first and last points, other points can be calculate using expression (2).

$$P_n' = \frac{-P_{n-1} + 8P_n - P_{n+1}}{6} \tag{2}$$

As shown in figure 3, the B-spline between red and green curve, the blue one is new spline after correction, we can see that, the new spline is more near the control points.

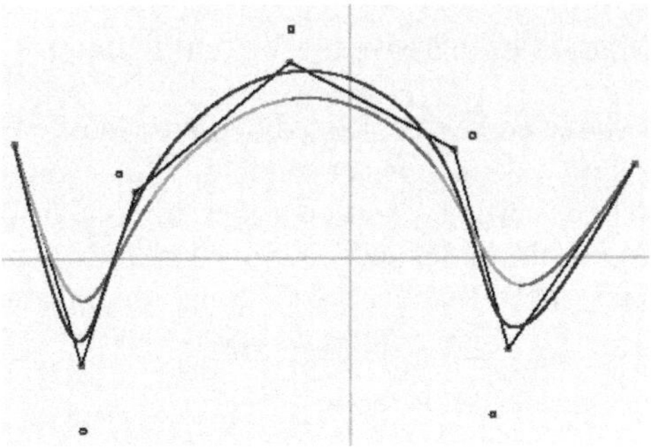

Fig. 3. B-spline with 5points correction

2.5 Inverse Calculation of B Spline

In the application in CAD/CAM or trajectory planning for robot, there is another issue. The condition given is not control point but type point. From expression (1), the control points must be exact parameter before the B spline interpolation, but the type point. We must get the control points from given type points.

When the type points are given, to solve the control points, it is the inverse calculation issue. N-1 type points define n-2 sections B-spline, it is needed to calculate n+1 control points, we can use expression (2) to solving B spline. For any $i = 1, 2, \cdots n-2$, when t=0, from(2), we can get

$$V_{i,n}(0) = \frac{1}{6}(P_i + 4P_{i+1} + P_{i+2})$$

That is

$$P_1 + 4P_2 + P_3 = 6V_1$$

$$P_2 + 4P_3 + P_4 = 6V_2$$

$$\vdots$$

$$P_{n-1} + 4P_n + P_{n+1} = 6V_{n-1}$$

In matrix form

$$\begin{bmatrix} 1 & 4 & 1 & 0 & \cdots & 0 & 0 & 0 \\ 0 & 1 & 4 & 1 & \cdots & 0 & 0 & 0 \\ \vdots & \vdots & \vdots & \vdots & \ddots & \vdots & \vdots & \vdots \\ 0 & 0 & 0 & 0 & 0 & 1 & 4 & 1 \end{bmatrix}_{(n-1)\times(n+1)} \bullet \begin{bmatrix} P_1 \\ P_2 \\ P_3 \\ \vdots \\ P_n \\ P_{n+1} \end{bmatrix}_{(n+1)\times 1} = 6\begin{bmatrix} V_1 \\ V_2 \\ \vdots \\ V_{n-1} \end{bmatrix} \quad (3)$$

There is n-1equations to solve n+1 unknown parameters, we need to add two boundary conditions. Often, there are 3 kind of boundary conditions, one is free boundary conditions. That is to let

$$\begin{matrix} P_1 = P_2 \\ P_n = P_{n+1} \end{matrix},$$

Substitute into (3)

$$\begin{bmatrix} 5 & 1 & 0 & \cdots & 0 & 0 & 0 \\ 1 & 4 & 1 & \cdots & 0 & 0 & 0 \\ \vdots & \vdots & \vdots & \ddots & \vdots & \vdots & \vdots \\ 0 & 0 & 0 & \cdots & 1 & 4 & 1 \\ 0 & 0 & 0 & \cdots & 0 & 1 & 5 \end{bmatrix}_{(n-1)\times(n-1)} \bullet \begin{bmatrix} P_2 \\ P_3 \\ \vdots \\ P_{n-1} \\ P_n \end{bmatrix}_{(n-1)\times 1} = 6 \begin{bmatrix} V_1 \\ V_2 \\ \vdots \\ V_{n-1} \end{bmatrix} \tag{4}$$

From equation group (4), we can get n-1control points, according to the boundary condition, we can get P_1 and P_{n+1}.

As shown in figure 4, it is an example of inverse calculation for B-spline, the type points are V_1 [-301.95, -29.791], V_2 [-232.21, 92.59], V_3 [-154.11, 90.98], V_4 [-180.61, -44.28], V_5 [-84.38, -50.73], V_6 [0.69, 63.61], V_7 [120.64, 70.05], V_8 [170.85, -57.16]￿ Control points after inverse calculation are P_1 [-313.97, 60.67], P_2 [-313.97, 60.67], P_3 [-241.87, 124.63], P_4 [117.75, 123.88], P_5 [-225.94, -79.23], P_6 [-66.32, -77.51], P_7 [16.04, 90.57], P_8 [143.66, 103.35], P_9 [176.29, -89.27], P_{10} [176.29, -89.27], The blue points if figure 4 are the control points. We know that, through adding boundary conditions we can get a B-spline though the type point.

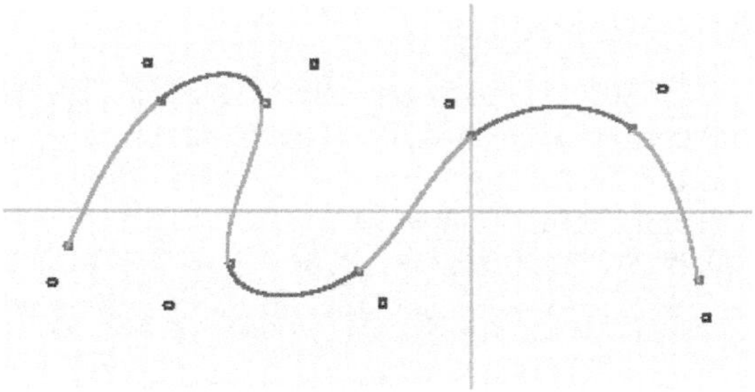

Fig. 4. Inverse calculation of B-spline

It should be known, the inverse calculate above is aimed to the uniform B spline, the B-spline calculated is uniform B-spline.

3 Relationship between B Spline and NURBS

NURBS is Non-uniform rational B-spline, the expression is

$$P(t)=\frac{\sum_{i=0}^{n}\omega_i P_i N_{i,k}(t)}{\sum_{i=0}^{n}\omega_i N_{i,k}(t)}, i=0,1,\cdots n \tag{5}$$

Where P_i is control points, ω_i is weight factor of P_i; t is parameter. Often, it is turn into [0, 1]; $N_{i,k}(t)$ is k_{th} B spline basic function.

From the properties, we can know

$$\sum_{i=0}^{n} N_{i,k}(t)=1 \tag{6}$$

From expression (6), we can know that when all the weight is equal, the NURBS is simplified to uniform B spline.

4 Conclusion

The B spline interpolator and five points correction are studied in this paper. On the platform of VB, the program are written. It is proven that five points correction has an advantage that much near the control points that traditional one. At last, the relationship between B spline and NURBS is studies. NURBS is developed from B-spline for the latter cannot express 2 order curve. B spline has some advantages same with NURBS, and it is easy to calculate, so it is valuable to study.

Acknowledgement. We would like to express our express our appreciation to program for Changjiang Scholars and Innovative Research Team in University (No.IRT0816), Key National Science &Technology Special Project on "High-Grade CNC Machine Tools and Basic Manufacturing Equipment" (2010ZX04014-014), Chinese National Natural Science foundation (51135003, 51105067, 50875039) and Key Projects in the National Science &Technology Pillar Program in the Eleventh Five-year Plan Period(2009BAG12A02-A07-2).

References

1. Tang, Z.S.: Procederal Elements for Computer Graphics, Beijin (1995) (in Chinese)
2. Luo: Procederal Elements for Computer Graphics, Beijin (2003) (in Chinese)
3. Bedi, S., Ali, I., Quan, N.: Advanced Techniques for CNC Machines. Journal of Engineering for Industry 1, 329–336 (2009)
4. Shpitalni, M., Koren, Y., Lo, C.C.: Realtime Curve Interpolators. J. Computer Aided Design 26, 832–838 (1994)

5. Syh, S., Pau, L.: Adaptive-Feedrate Interpolation for Parametric Curves With a Confined Chord Error. J. Computer-Aided Design 34, 229–2372 (2002)
6. Sung-Ho, N., Min, Y.Y.: A Study on a Generalized Parametric Interpolator With Real-Time Jerk-Limited Acceleration. J. Computer-Aided Design 36, 27–36 (2004)
7. Tsai, M.C., Cheng, C.W.: A Real-time Predictor- Corrector interpolator for CNC Machining. Journal of Manufacturing Science and Engineering 125, 449–460 (2003)
8. Marchenko, T., Tae, J.K.: NURBS Interpolator For Constant Material Removal Rate In Open Nc Machine Tools. International Journal of Machine Tools and Manufacture 44, 237–245 (2004)

Content-Related Features for Video Quality Assessment Based on Bit Streams

Caihong Wang, Xiuhua Jiang, and Yuxia Wang

Information Engineering School, Communication University of China,
Beijing 100024, China
{rainbow,jiangxiuhua,yuxiaw}@cuc.edu.cn

Abstract. Quality assessment plays an important role in video business. The bit-stream-layer quality assessment model becomes popular in recent years because of its low-complexity advantage. This paper defines and analyzes some content-related bit-stream features derived from encoded MPEG-2 HD videos. We divide the features into three categories: spatial features, temporal features and compression settings. 9 HD videos with different types of content were used in our experiment. They were encoded at 6 kinds of bit rates. Subjective test and feature extraction were performed on all the bit-rate samples. Then we analyzed all the data by correlation matrix. The results show some compression setting features have very high linear correlation (up to 89.89%) with subjective perceived scores. At the same time, some spatial and temporal features are graphed as functions of bit rate to show their ability of depicting spatial-temporal characteristics of videos. In future, these features will be used for building a bit-stream-layer quality assessment model.

Keywords: Content-related Features, HD, MPEG-2, bit-stream-layer, video quality assessment.

1 Introduction

Along with the promotion of network bandwidth and performance ability of computer, video services have become more and more popular. Video streams possibly undergo various impairments before of display in user end, typically caused by video compression and network transmission. Video quality assessment and monitoring is beneficial to find and solve problems as soon as possibly. The emerging quality assessment method is to analyze compressed video streams in transmitting node or user terminal [1]. Compared with the methods performed on pixels, the way based on compressed streams can eliminate most of decoding process, so it has the advantage of low computing complexity and is suitable for real-time network video quality monitoring. These models which analyze compressed video streams can be further divided into parametric packet-layer model and parametric bit-stream-layer model [2]. Packet-layer model uses packet-header information to evaluate video quality, while bit-stream-layer model uses both packet-header information and media-related payload information.

G. Lee (Ed.): Advances in Automation and Robotics, Vol.2, LNEE 123, pp. 223–230.
Springerlink.com

The subjective quality of the video depends on its content [3]. Bit-stream-layer model, which considers video content information, will be more consistent with subjective perceived quality than packet-layer model [4]. When researching on compression impairments, bit rate is usually used as the only parameter in packet-layer model. However, the videos with the same bit rate maybe have large perception quality differences. It is mainly dues to different video contents. The encoder will go through different pressure when encoding videos with different spatial and temporal complexity. For CBR (Constant Bit Rate) mode, after encoded, video will impair seriously, if the content of video is complex. So we should reasonably define and select features from compressed streams, in order to build an effective parametric bit-stream-layer quality evaluation model. These features should be able to depict spatial-temporal characteristics of video well.

Many parameters can be extracted from bit streams complied with MPEG-2 or H.264 syntactic structure, such as bit rate, QP (quantization parameter), DCT coefficient, MV (motion vector) and macro-block type, etc. Then a lot of statistical methods can be applied to each parameter. There are various features in current bit-stream-layer video quality assessment metrics. [5] looks at MV, calculating mean size of all the MVs, standard deviation of MV sizes, histograms of MV directions, variance of MV directions, proportion of horizontal movement and proportion of dominant MV direction. But MV can only reflect motion characteristics of video. The metric hasn't considered video spatial texture features. [4] calculates the maximum QP in each frame as the only bit-stream feature, while in [6], mean value of QP for each frame is used as one of features. For DCT coefficients, average and standard deviation of 15 AC transform coefficients are defined as features [3], while in our previous research [7], AC energy of AC transform coefficients is introduced as our metric's feature. So from the above, we can find that there aren't recognized and uniform features for bit-stream-lay models.

In this paper, features are defined based on compression parameters extracted from MPEG-2 HD video streams. Statistics features recommended by ITU-R BT.1210-3 [8] are introduced as a part of our features. In the recommendation, these features are intended to choose videos used in subjective experiment. Considering these features can well reflect the spatial and temporal characteristics of videos, they are innovatively used for our objective assessment. A statistic of CBP (Coded Block Pattern) is also included in our features set. Other researches usually do a statistic of skipped macro-block in macro-block layer [9], the skipped block statistic show higher linear correlation with subjective perception, which can be found in Section 4. We think statistics based on block are more elaborative than based on macro-block.

2 Influence of Video Content on Perceived Quality

We can observe that video content clearly has an influence on the perceived quality from our subjective video quality test. Nine HD videos are used in the experiment. Each one has more than 10s duration. They are encoded at bit rates of 9, 12, 15, 18, 21 and 25 Mbps by a MPEG-2 video encoder. The 9 video contents used for the test are representative of various TV programs. They differ in their amount of details and complexity of structures and movements. The screenshots of 9 videos are shown in

Fig.1. Among them, the video *A: dolphin* has fast panning along with dolphin swimming. There is a scene cut in *C: market*. *E: festival* has slow panning. *H: branch* presents slow zoom. The video *D: harbor* has both slow panning and slow zoom. The other videos have relatively fixed lens.

Fig.2 shows subjective results of the 9 videos as a function of the bit-rate. Subjective evaluation scores are from subjective experiments with DSCQM according to the recommendation ITU-R BT.500-11 [10]. The uncompressed original video is used as a hidden reference in the test. The statistical analysis is performed on all the D-values between original and processed video to obtain an average value for each video. From Fig.2, we can observe that two videos, *D: harbor* and *H: branch*, have the minimum D-value in most of bit rates, in other word, the best subjective perceived quality. And three videos, *E: festival*, *F: fire* and *G: leaves*, have the worst quality in most cases. There is smaller difference of perceived quality at higher bit rate for 8 videos (except for *G: leaves*). Content influence is more important at the lower bit-rate. Especially for 9 Mbps, video content leads to a variation of up to 13 scores. Thus, it is obvious that content-related features need to be introduced into video quality model to adjust the predicted video quality.

Fig. 1. Screenshots of 9 videos used in our paper

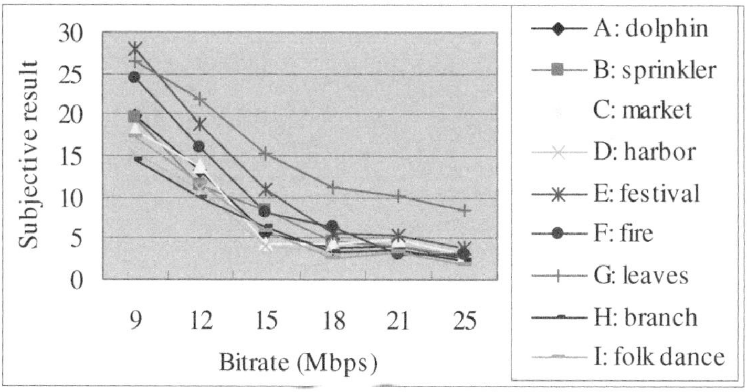

Fig. 2. Subjective result as a function of bit-rate for 9 video contents

3 Content Features from Bit-Streams

3.1 Spatial Features

DCT converts image signal from spatial domain to frequency domain. Spatial complex video usually has more high frequency AC components in I-frame. Though quantization process loses some AC components, we can still try analyzing DCT coefficients to acquire the descriptor of spatial complexity. Two statistics features are defined in [8] based on DCT coefficients, which are AC energy and spectral entropy expressed by equation (1) and (2) (3). C(m,n) denotes the DCT coefficients of a 8*8 block. AC Energy represents the degree of picture activity. Spectral entropy represents the degree of randomness of the DCT coefficients.

$$ac_k = \sum_{m=0}^{7}\sum_{n=0}^{7} C(m,n)^2 - C(0,0)^2 \tag{1}$$

$$se_k = -\sum_{m=0}^{7}\sum_{n=0}^{7} \frac{|C(m,n)|}{A} \log_2 \left[\frac{C(m,n)}{A}\right] \tag{2}$$

$$A = \sum_{m=0}^{7}\sum_{n=0}^{7} |C(m,n)| \tag{3}$$

In the paper, we calculate the average value of AC energy and spectral entropy for all the blocks averaged over each I-frame and over the whole video. We label the two features as *AC* and *SE* for short.

3.2 Temporal Features

Motion vector can reflect the degree of motion of the picture, i.e. temporal complexity. We integrate horizontal and vertical component of motion vector into one scalar $\sqrt{MVx^2 + MVy^2}$. Two kinds of statistics are applied in the scalar in the paper, which are mean magnitude averaged over a frame (labeled as *MVa*) and standard deviation within the frame (labeled as *MVs*). *MVa* represents the degree of motions as a whole, while *MVs* represents the degree of non-uniform motions. In the paper, we calculate *MVa* and *MVs* for all the macro-blocks with non-zero motion vector over each P-frame, then average over all the P-frames in the whole video.

For each P-frame, we also analyze motion-compensated frame-difference signal (prediction error). We calculate the power of the prediction error, ep_k, in DCT domain, expressed by equation (4) for each macro-block. This is equal to calculate prediction error power in spatial domain according to the law of conservation of energy. But the calculation in DCT domain is low-complexity because of avoiding inverse DCT process. Then we calculate the average value of ep_k for all the macro-blocks averaged over each P-frame and over the whole video as the final feature *EP*.

$$ep_k = \frac{1}{4*8*8} \sum_{block=0}^{3}\sum_{m=0}^{7}\sum_{n=0}^{7} C(m,n)^2 \tag{4}$$

There are many kinds of macro-block types in P-frame and B-frame. Macro-block will take intra coded mode if content change in adjacent frames is so large that macro-block can not search out the matching block. So a lot of intra coded macro-blocks in a P-frame mean high temporal complexity. If macro-block's prediction with the default motion vector is good and all the DCT coefficients are quantized to zero, the macro-block will become "skipped", because nothing need to be transmitted. In the paper, the percentage of intra coded macro-block (*MBintra*) and skipped macro-block (*MBskipped*) in each P-frame is calculated. Then they are averaged over the whole video as the final features.

All the temporal features are only calculated for P-frame. On the one hand, this simplifies calculation complexity (because B-frame doesn't need to be decoded). On the other hand, One-way predicted P-frame can well reflect the variations between two frames.

3.3 Compression Settings

The parameter bit-rate (*BR*) is usually looked as a benchmark of video quality. We can abstract it from sequence layer of MPEG-2 bit-streams.

Quantization is the source of video impairments in compression process. Rate control mechanism adjusts quantization parameter to keep the constant bit rate regardless of high or low temporal/spatial complexity. Higher complexity is usually corresponding to the larger quantization step. The parameter indicating quantization degree can be extracted from video streams. We calculate mean value and standard deviation of quantiser-scale factors over each frame over the whole video, labeled as *QPa* and *QPs*.

A statistic of skipped block in a frame can be carried out when getting CBP parameter from streams. A skipped block means that all the DCT components in the block are zero. While a macro-block becomes "skipped" needing the macro-block' motion vector is the default, in addition to all the coefficients of each block in the macro-block are zero. The skipped block statistics can more subtly reflect compression degree than skipped macro-block statistics. We calculate the proportion of skipped luminance block to all the luminance blocks for each P-frame, then average over the whole video as the final feature *Blum*. And the feature *Bchr* is a similar statistic for chroma blocks. Such statistics can't be found in other algorithms, but they show good experimental result.

4 Adequacy of Content Descriptors

After subjective experiment, we can obtain subjective perceived scores (*SCOREsub*) of all the videos. By partial decoding for video streams, we can obtain all the feature values. The Person Correlation Coefficients between different features (including subjective scores) are analyzed.

4.1 Compression Settings

From Table 1, it can be observed that compression setting parameters have very high linear correlation with subjective perception. Among them, *Blum* has the highest correlation up to 89.89%. Carefully observing the five compression setting parameters, we can find that the correlation between *QPs* and *QPa* is 94.62% and the correlation between *Blum* and *BR* is 92.61%. So we can remove *QPs* and *BR* which have lower correlation with subjective scores than *QPa* and *Blum* from our features set. Fig.3 show two features *QPa* and *Blum* as functions of bit-rate. Compared Fig.3 with Fig.2, it can be found that the position of videos has rough similarity. *G: leaves* and *F: fire* are in the upper position of the figures. *D: harbor* and *C: market* are in the lower position.

Table 1. Correlation matrix of compression settings

Compression Settings	*BR*	*QPa*	*QPs*	*Blum*	*Bchr*	*SCOREsub*
BR	1	-0.6519	-0.6419	-0.9261	-0.535	0.8091
QPa	-0.6519	1	0.9462	0.7561	0.1566	-0.8485
QPs	-0.6419	0.9462	1	0.7191	0.1895	-0.8154
Blum	-0.9261	0.7561	0.7191	1	0.3638	-0.8989
Bchr	-0.535	0.1566	0.1895	0.3638	1	-0.2453
SCOREsub	0.8091	-0.8485	-0.8154	-0.8989	-0.2453	1

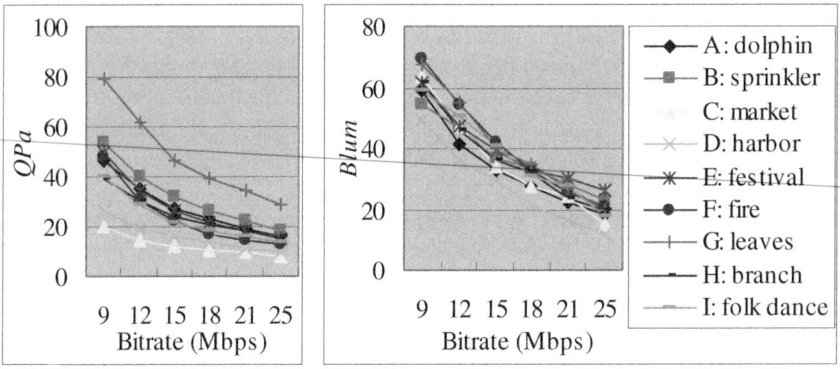

Fig. 3. *QPa* and *Blum* as functions of bit-rate for 9 video contents

4.2 Spatial Features

From Tab.2, we can find that *SE* has higher correlation with subjective scores than *AC*. The correlation between *SE* and *AC* is 82.15%. So we can select *SE* as the ultimate feature. Fig.4 shows *SE* is as the function of bit-rate. The videos *H: branch* with rich details, *A: dolphin* and *D: harbor* with complex backgrounds, have the larger *SE*. While the video *F: fire* with the simple content has the lowest *SE*.

Table 2. Correlation matrix of spatial features

Features	*AC*	*SE*	*SCOREsub*
AC	1	0.8215	0.1878
SE	0.8215	1	0.5571
SCOREsub	0.1878	0.5571	1

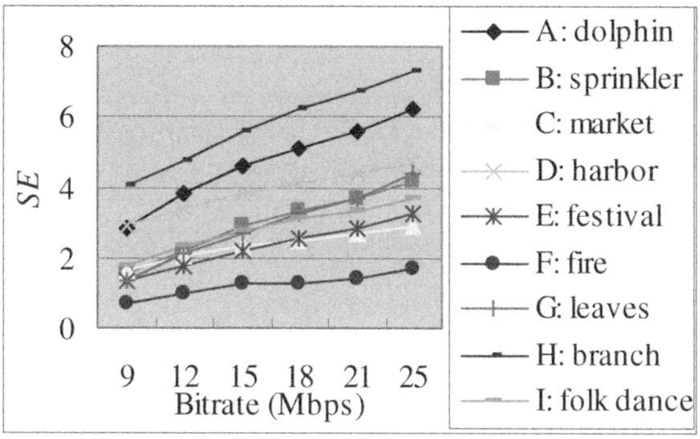

Fig. 4. *SE* as a function of bit-rate for 9 video contents

4.3 Temporal Features

Most of temporal features do not vary in function of the bit-rate, so the correlations with subjective scores are low. It can be observed that the correlation between *EP* and *MBintra* is 90.71% from Tab.3, so we can remove *EP* from our temporal features set.

The features *MVa* and *MVs* as the functions of bit-rate are shown in Fig.5. *H: branch* and *D: harbor* with relatively static contents and slow lens movement, have the smallest values, in other word, the lowest temporal complexity. It can be understood that when many small objects move chaotically in a coding scene, the standard deviations of motion vectors are high, so *F: fire* with random moved flame has the largest *MVs*. While *A: dolphin* with fast panning has the largest *MVa*.

Table 3. Correlation matrix of temporal features

Features	*MVa*	*MVs*	*EP*	*MBintra*	*MBskipped*	*SCOREsub*
MVa	1	0.7087	-0.0201	0.0311	-0.1483	-0.058
MVs	0.7087	1	0.4497	0.462	-0.3044	-0.0632
EP	-0.0201	0.4497	1	0.9071	-0.387	-0.2961
MBintra	0.0311	0.462	0.9071	1	-0.5491	-0.4545
MBskipped	-0.1483	-0.3044	-0.387	-0.5491	1	0.3037
SCOREsub	-0.058	-0.0632	-0.2961	-0.4545	0.3037	1

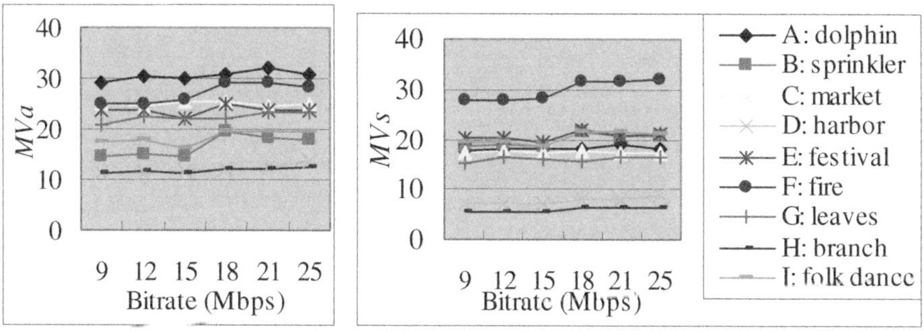

Fig. 5. *MVa* and *MVs* as functions of bit-rate for 9 video contents

5 Conclusion

In this paper, twelve spatial and temporal features are defined based on MPEG-2 HD video streams. Combining with our subjective perceived result, we can find these features highlight a strong adequation with perceived quality. Our experimental results show that a statistic of proportion of skipped block (*Blum*) has nearly 90% correlation with subjective quality. So it is easy to achieve higher correlation when integrating with other spatial-temporal descriptors. In future work, we will combine these features to build a perceived video quality model.

References

1. Winkler, S., Mohandas, P.: The Evolution of Video Quality Measurement: From PSNR to Hybrid Metrics. IEEE Trans. on Broadcasting 54, 660–668 (2008)
2. Takahashi, A., Hands, D., Barriac, V.: Standardization activities in the ITU for a QoE assessment of IPTV. IEEE Communications Magazine 46(2), 78–84 (2008)
3. Garcia, M.N., Schleicher, R., Raake, A.: Towards a content-based parametric video quality model for IPTV. In: Proceedings of the 3rd International Workshop on Perceptual Quality of Systems, PQS 2010 (2010)
4. Watababe, K., Yamagishi, K., Okamoto, J., Takahashi, A.: Proposal of New QoE Assessment Approach for Quality Management of IPTV Services. In: IEEE ICIP, pp. 2060–2063 (October 2008)
5. Ries, M., Nemethova, O., Rupp, M.: Video quality estimation for mobile H.264/AVC video straming. Journal of Communications 3(1), 41–50 (2008)
6. Rossholm, A., Lovstrom, B.: A new low complex reference free video quality predictor. In: IEEE Workshop on Multimedia Signal Processing, pp. 769–772. IEEE (2008)
7. Wang, C., Jiang, X., Wang, Y., Meng, F.: Quality Assessment for MPEG-2 Video Streams. In: DEIT, pp. 493–496 (2011)
8. Rec. ITU-R BT. 1210-3, Test materials to be used in subjective assessment (2004)
9. Garcia, M.N., Raake, A., List, P.: Towards content-related features for parametric video quality prediction of IPTV services. In: Proc. of ICASSP (2008)
10. REC. ITU-R BT.500-11, Methodology for the subjective assessment of the quality of television pictures (2002)

Research of Non-equilibrium Urban Traffic Control System

Shulin He

Liaoning Police Academy, Dalian 116036, China
heping2000@163.com

Abstract. The problem of traffic control is studied according to the system self-organization theory, and from the non-equilibrium traffic system point of view. The rule of the traffic system from existence to evolution is analyzed using the theories. Based on this study, the concept of non-equilibrium traffic control system is introduced, and resulted in the establishment of a fundamental frame work for the non-equilibrium traffic system. Moreover, a practicable methodology is presented.

Keywords: traffic control, non-equilibrium system, existence and evolution, management Policy, cooperated control.

1 Introduction

It is a common knowledge that traffic is a very sensitive and complicated social phenomenon and has significant impacts on the economic development of a society. Governments and their police forces have to handle the problem seriously, and to find out efficient measures to control and crack down on the traffic activities. However, owing to the uncertainties and the complexities of the traffic system, it is difficult to study the problem. The traditional ways of studying are relied on empirical models and qualitative analysis. Effective traffic control is based on the research of traffic statistics, traffic predictions and hidden cases. Traffic statistics is a research means which can provide instructive information to the traffic control practice. For example, it can give the indications on the scales of the inhabitant involvement in traffic and provide the basic data for tracing the causes of the traffic jams. The influences of the traffic economical factors (e.g., industrialization, motorization, urbanization, inflation and unemployment) on the traffics can be observed by comparing the traffic statistical data with the population data. According to traffic statistics, it is possible to judge in what population and which region the rate of traffic jams and the rate of traffic accidents are higher than the average rate, and predict the future trends. Therefore, traffic statistics and predictions provide the necessary information for the establishment of the traffic control policies and regulations. However, the present methodology of traffic statistics and predictions do has its limitations, and is far from perfection. Much latent information on traffic is impossible to collect due to the lack of unified procedures in the analysis of the traffic system.

G. Lee (Ed.): Advances in Automation and Robotics, Vol.2, LNEE 123, pp. 231–237.
Springerlink.com

Our study shows that the mechanisms of the traffic system activities should be analyzed by the system self-organization theory in order to gain the information on the traffic system uncertainty, and to get the processing method. In this way a purpose-oriented traffic control is achievable. The system theory consisted of Systematology, Cybernetics, and Informatics, has become the theoretical foundation for the research methods to many problems since 1970's. Recently, new branches of the system theory have appeared, such as Dissipation Structure Theory, Synergetic, Mutation Theory and Super circulation Theory. These theories can be used out only to reveal the rules of the system's evolution dynamically, but also emphasize on the relationship between systems and their environments. Harken's Synergetic theory is a new theory, and it describes how a system can reach the highly Synergetic states. The theory stresses the point of the correlations between the systems under study. Based on this point the self-organized behavior and the self-adjusted mechanism of a system are formed, as well as the thought of system control for the cooperative management and control. From the view point of system and environment, Prigogine analyzed the rules of the evolution of a system from ordered states to disordered states or vice versa. He pointed out that a dissipation structure is formed in the exchange process of information, energy and substance between a system and its environment. If the control system for the social traffics forms a dissipation structure, it can efficiently control a system changing its state from a disordered state to an ordered one. Tom's mutation theory pointed out that system can be grouped into seven kinds, and he studied the mutation of a system in different environments. The author of this article studied the system non-optimization theory in depth. A practicable decision-making system was resulted from the comprehensive analysis of the controlled optimization process for a social system according to its optimal and non-optimal characters. This article introduces a novel thought into the analytical process for traffic management and control, and resulting in a non-equilibrium traffic control frame.

2 The Mechanisms of the Traffic Management and Control

Traffic management and control is an extremely complicated issue, and has the features of random mutation according to its occurrence, development and the trend of its evolution. From the viewpoint of government management, traffic safety and traffic prevention is the aim of traffic control. According to the existing concept on traffic control, the control process is determined by three factors:

(1) Determine the possible space and time of traffic jams and accidents;
(2) Select some states from the possible space as targets;
(3) Create the necessary conditions to make the traffic control reach the preset aim;

It is known that the term "possible space" is the assembly of all the possibilities faced in the development process of an object. The possible space of traffic jams and accidents is determined by the conditions leading to traffic control. Generally speaking, the possible factors related to traffic accidents involve the victims and the runner, the motives and the opportunities, and the means and the scenes for traffic system. These factors have their characteristic possible spaces, differing from each other in the amounts and the forms, and may interchange from one to another. When a possible space of traffic jams is developing into a certain state, it may turn into a new possible space. The occurrence of several possible spaces in the development of the traffic system makes the traffic process appearing in different stages. In other words, the target of traffic control changes as the possible traffic space varies.

For a complicated traffic control process, the possible traffic space may have many states. Moreover, these states may have different ways of appearances, and the conditions influencing the form of the traffic system are sophisticated and complicated as well. Accordingly, the control-selecting process is complex. Conditions and factors to be controlled vary as the stages of the traffic change. Furthermore, the concurrence between the conditions and the factors, as well as the interactions between the developing states, has to be taken into account. As an example, the urban environment, space and the social economy, together with the formation of the traffic conditions, directly determine the ways of traffic control. It accords the laws of the development of a society to use cybernetics to process the problem of the urban traffic. Cybernetics adopts the view points of system theory and kinetics, structure and function, action and target, as well as information and cybernetics.

In general, however, it has not been practiced, up till now, to use the methods of Cybernetics for a systematic analysis of the urban traffic problems for a society as a whole, and to give a quantitative description. At present, human society has entered the information era. The so-called information highway has extended itself into every field in the society. Briefly, the information society is characterized by the exchange, processing, and storage of the information. The Social traffic systems are complicatedly structured, possess various features and interrelated to a large extent. Therefore, it is necessary to treat the traffic control problem with Informatics and Cybernetics. To sum up, the mechanism of the traffic control is to use Informatics and Cybernetics in the possible traffic spaces in order to achieve the purpose of the traffic management and control.

3 Non-equilibrium Traffic Control System

Using non-equilibrium system theory to study the problems of traffic control is relatively new. Traffic is a common social phenomenon. Traffic management and control originate as traffics occur. The extent of the urban stability is dependent on what level the traffic is under control. Reducing the urban traffics to the lowest extent means that the society advancing.

Philosophically, equilibrium exists temporarily and relatively. Non-equilibrium exists commonly and absolutely. The equilibrium achieved in the traffic control system refers to the unified behavior of the control activity under the predictive traffic model. This is a static point of view for the traffic and their control. Equilibrium can be understood as a process of adapting with the traffic and adjusting traffic control, from the kinetic point of view. This reflects the characters of the non-equilibrium process. Since the urban traffic appear to be kinetic in nature as the society is developing, traffic control system should be a non-equilibrium process. Both equilibrium state and steady state do not change with time. But the two states are different in definition. Equilibrium state refers to traffic and control, and steady state is the symbol of the urban stability. Ordered structure is a kind of steady state. However, not all the steady states are in equilibrium states. For instance, ordered structure in a non-equilibrium state is a kind of steady state, but not in an equilibrium state.

The phenomena of order and disorder are a contradiction commonly existed in human society, and are interchangeable. A disordered social state can be converted into an ordered and stable one if the rules of the urban traffic are understood.

3.1 Understanding the Rules of the Urban Traffic

The urban traffic can be treated as a complicated system. There are rules to follow as the system changes and develops. The traffic control will be effective once these rules are known. The rules of the urban traffic play a dominate role in the existence and development of the traffic. These rules are a type of relationship correlating to the occurrence of the traffic in a society. Inevitability and contingency determine the development of all systems. The two factors coexist in the urban traffic, including the causality relationship in the occurrence of traffic management. Inevitable traffic are a phenomenon resulted from the non-coordination and inadaptability factors in the field of economy, policy, management and education. This type of traffic dominates the urban traffic system. Contingent traffic are caused by some specific reasons. These traffic problems may or not occur, may occur in this way or the other way. In short, the inevitable traffic can be realized through a large number of the contingent traffic reflect the inevitable relationship in the urban traffic. In the modern societies, the rules in the traffic activities can be found out by studying a large number of the random and contingent cases. The role of traffic statistics is to describe the properties of the urban traffic in macroscopic scale based on the statistical analysis of the traffic on a microscopic scale.

However, the role of the traffic statistics should not be exaggerated. It has to be pointed out that the objectives processed by the statistics should be independent incidences and large numbers and have a random feature. Strictly speaking, the individual cases in a macroscopic traffic study do not fully fulfill the above mentioned assumptions.

The statistics used before are not accurate and complete enough in describing the urban traffic problem as a whole. Therefore, the needs are rising for a better theory to describe the urban traffic thoroughly, and to take both microscopic and macroscopic viewpoints into consideration. From the view point of non-equilibrium system theory, the combination of statistics with kinetics should be useful in solving the problem mentioned before. The cooperation theory applies statistics for both microscopic and macroscopic investigations for the objectives under study. The need for a large number of objectives is no long a necessity. It is statistically meaningful on one hand, and it emphasizes the interactions in the urban traffic on the other. Synergetic can be used to reveal the rules of the evolution of the traffic problems from the point of time, space and forms. Therefore, the urban traffic problems can be described in a stricter and complete ways according to this theory, and the effective traffic control can be found with the help of this comprehensive statistical method.

3.2 The Characteristics of the Urban Traffics from the Existence to Its Evolution

The evolution of the urban traffic problems follows the sequence of quantitative change to qualitative change to qualitative change to qualitative change, etc. The evolution has a waving type of pattern. The mutation theory points out that the conversion of one state to another can be completed in two ways, mutation and gradation. These different ways of evolution generate different impacts on the social stability.

The evolution of the urban traffic problems is partly dependent on the changes of the traffic environment, and partly dependent on the control means and conditions of the government. According to the analysis of the mutation model, it is clear that the traffic state model for studying the evolution pattern can be established by finding out the state variables representing the quantitative change, and the condition variables (i.e. controlling variables) which cause the change of the state of the traffic-affected state in the social development.

For the analysis of the local traffic problem, the same principles apply. There are a number of factors which affect the traffic variations. The difficulty is to find out the dominant variables. All the traffic variables can be divided into two groups: direct variables (directly affect the occurrence of the traffic). Once the relationship between the direct variables and the state variables is known, the mathematics model can be obtained, and the quantitative analysis can be carried out for the rules of the changes of the traffic state.

4 The Management Policy and Cooperated Control on the Urban Traffic

Traffic management and control has the nature of socialites and is a kind of government administrations. Its execution is dependent on the coordination from different departments and divisions. Based on system analysis, and in combination with the principle of Synergetic, the main frame for the cooperated control on the urban traffics is formed. The following factors should be taken into consideration.

4.1 Non-equilibrium Factors

It is known that an urban traffic system is a non-equilibrium open system. According to Synergetic, an equilibrium system hardly has any advance to make. In an opened urban traffic management system, exchanges of substance, energy, and information between system and environment are constantly occurring. Hence, the system will deviate from equilibrium. Each of the functional divisions in the system is not equally distributed, but showing a non-equilibrium and a non-linear type of interaction. This type of behavior is called non-equilibrium behavior.

Non-equilibrium traffic management system can gain information from outside, and lowering the confusing states within the system. Thus, opening and non-equilibrium are the two prerequisites for the development of a system. Since the relationship between the traffic management system has a non-linear character, the cooperation function can be generated, and an ordered, function-enlarged self-organized structure can be formed in the system.

4.2 The Factors Involved in Information

Gaining the information and processing it are the preconditions for traffic control. The key point for a security protection system is the establishment of the traffic information system. The decision-maker's task is to receive all the sufficient and necessary information on the urban traffic, make a proper judgment and issue the necessary traffic control information. The effective traffic control is dependent on the reliable information on traffic management. The information provides the platform for the prediction of the future trends in the urban traffics. The blindness in security protection can be overcome if the information can be gained in time, and used defectively for both original information and the high-quality derived information. The determination of the reliability of the information also is an important factor.

The decision-making on traffic management needs a feed-back adjustment. It has to be predetermined of the "post-effect" of the decision. One of the effects is delaying, and the other is enlarging. Delaying in making the decision will make the measures of traffic control failed. To prevent its occurrence, the importance of the management on the traffic information has to be stressed. The effects of enlarging result from the cooperation between different quarters of the society. In this case, an ordered structure is formed, leading to the enlargement of the functions of the traffic control system.

4.3 The Points and Factors on Driving Force

The traffic control and traffic safety protection are kinetic in nature. It is important to know the kinetic characters in the urban traffic. Through kinetic analysis, the future trends on traffic can be predicted, and the control regulation can be optimized. The number of the random traffic cases will be under control with the study on the wave rule. The motion of the traffic system can be regarded as a traffic space from the Cybernetics point of view. Accordingly, studying all the factors involved in the traffic space, e.g., timing of the cases, short-term waves and long-term waves in the local traffic will enhance the efficiency of traffic control.

5 Conclusions

To sum up, the adoption of non-equilibrium system theory into the studies on the problems of the urban traffic is a kind of experimenting, and may result in useful discussions and argumentations in the field. However, this new thought and analytical methodology can be used to solve some of the problems in traffic management, for example, traffic statistics, correlated analysis, the kinetic and evolution characteristics of the urban traffics. The concept of traffic control can be formed by introducing non-optimum into optimization. In addition, the thought of self-organization in non-equilibrium state, is an effective methodology for the traffic management system. Traffic control and safety protection will fit in with the needs of the society's development by using this methodology.

References

1. Bursik Jr., R.J.: Urban dynamics and ecological studies of delinquency. Social, Forces 63 (1984)
2. Bursik Jr., R.J.: Community change and patterns of delinquency. American Journal of Sociology 88, 24–42 (1982)
3. Derrida, J.: Margins of philosophy, Bass A (Trans.). University of Chicago Press, Chicago (1982)
4. National Center for Biotechnology Information (1985), `http://www.ncbi.nlm.nih.gov`
5. Ritchie, S.G.: A knowledge-based decision support architecture for advanced traffic management. Transportation Research Part A: General 24, 27 (1990)
6. Zhang, H., Ritchie, S.G.: Real-Time Decision Support System for Freeway Management and Control. Journal of Computing in Civil Engineering 8, 35–51 (1994)
7. Peled, A., Hakkert, A.S.: A PC-oriented GIS application for road safety analysis and management. Traffic Engineering and Control 34, 355–361 (1993)
8. Sabel, C.E., Gatrell, A.C., et al.: Modelling exposure opportunities: estimating relative risk for Motor Neurone Disease in Finland. Social Science & Medicine 50, 1121–1137 (2000)
9. Steenberghen, T., Dufays, T., Thomas, I., Flahaut, B.: Intra-urban location and clustering of road accidents using GIS: a Belgian example. IJGIS 18, 169–181 (2004)

On the Offense and Defense Game in the Network Honeypot

Hongxia Li, Xiaoqiong Yang, and Lianhua Qu

Chongqing Technology and Business University, Nan' an District, Chongqing, 400067, China
lihongxia@ctbu.edu.cn

Abstract. By using the method of game theory, this paper has analyzed the information asymmetry game of both the offensive and defensive parts in the network honeypot through the complete but imperfect information dynamic game model. From the perspective of game theory, it has conducted the analysis of feasibility and necessity on the intrusion detection system of introducing the honeypot technology and has described the general intrusion and the intrusion detection behavior. Through the analysis on such characteristics as information security asymmetry, incomplete information, rationality and repeated game, and the analysis on game model elements and general model, we have constructed the offense and defense game model in the network honeypot and game process model in the intrusion detection. Meanwhile we have conducted a thorough study and analysis on the game process and then put forward the effective measures of improving the network system security and the attackers' attacking difficulty.

Keywords: network honeypot, offensive and defensive game, model, intrusion detection, dynamic game.

1 Introduction

From the early nineties when the concept honeypot was put forward to 1998 or so, honeypot was just confined to a kind of thought, usually used by the network management staff to achieve the purpose of tracking by cheating the hackers. The honeypots in this phase were actually some hosts and systems attacked by the hackers. Since 1998, the technology of honeypot has gradually attracted the attention of a group of security researchers and developed some relevant tools especially for cheating the hackers. The honeypot in this phase can be called virtual honeypot. That is to say, the developed honeypot tools can simulate virtual operating system and network services and respond to the hackers' attacks, thus cheating the hackers. After 2000, the security researchers are more inclined to use the real hosts, operating system and application programs to set up the honeypot. Different from the virtual honeypot tools in which exist such problems as the low level of interaction and likeliness to be identified by the hackers, the honeypot established by the real host has more powerful tools of data capture, data analysis and data control and thus the researchers can more easily track the hackers who have intruded the honeypot and then amylase their attack

G. Lee (Ed.): Advances in Automation and Robotics, Vol.2, LNEE 123, pp. 239–246.
Springerlink.com

behaviors. During the process of struggling against network attacks, network honeypot technology is reaching maturity and is playing an increasingly important role.

2 Network Honeypot

Network honeypot is a network resource especially designed to be scanned, attacked and invaded. Network security staff deploy the network honeypot in the network server room, simulating a variety of operating systems or system vulnerabilities, and confusing and inducing the attacker to attack the network in order to achieve the targets of transferring the network attack and then to protect the real protective measures commonly used in network servers. By studying the honeypot attack logs we can restore the entire process of the attackers' attacking and master the attacking steps, methods and tools, thus facilitating the security defense against other potential attacking targets [1],[2],[3],[4],[5].

To exert its roles, network honeypot must possess such three basic conditions as honeypot bait, abilities to detect and analyze and the means of recording. First, it should be ensured that the honeypot can lure the attackers to attack on their own initiatives, that is, that the network honeypot should have loopholes or flaws that attract attackers to attack, or that the honeypot should have bait. Then, when the attackers attack the honeypot, the latter should be able to identify the attacking behaviors and interact with the attackers to induce a certain in-depth attack, which means that the honeypot should have the abilities to detect and the analyze the attacking behaviors. Besides, in the process of tracking the attackers, the honeypot generates large amounts of data. For these data, the honeypot should have a complete set of storage preservation mechanism, which means the honeypot needs to have a sound means of recording the attacking behaviors.

Honeypot technology can allow an attacker unknowingly to be in a trap set by the honeypot designers, and its behavior in the honeypot can be captured and analyzed by the honeypot designers. On the other hand, an attacker can also conduct an analysis of the honeypot through anti-honeypot technology and find the evidence that honeypot is not a real system. After the attacker realizes that he is in the honeypot system, he will try every means to break the shackles of honeypot system and in turn gain the control of the operating system of the honeypot, even attacking other real servers in the network using the system of the honeypot as a springboard.

As the offensive and defensive parts in the network honeypot, there exists serious information asymmetry in the battle of network security, which is manifested in the following aspects. First, for the attackers, there exists asymmetry in grasping the attacking targets' information. Before attacking the targets and to prepare for the attacking, the network attackers will try to understand such information of the target server as its company's address, operation range, office phone or even the server administrators' telephone and mailbox, whereas, the defender will try to hide the information and use sniffing tools to detect vulnerabilities but cannot determine (at least at the beginning of the attack) whether the attacking target is the physical host or target honeypot. Second, the attacker and the defender fail to understand each other's technology means. Actually, before attack occurs, the defender knows nothing about the attacker's information. Only in the deployment of honeypot system and when honeypot logs have not been destroyed by the attacker can the defender learn about the attacker's attacking tools and measures through the honeypot logs. Meanwhile, the attacker can identify the fingerprint information of the honeypot system only through the analysis of the feedback information of the attacking targets and then decide whether the attacking target is the the physical host or target honeypot [5] [8].

3 Basic Offensive and Defensive Techniques

In most cases, one side of the two cannot have a complete understanding of the other, in other words, they have no idea of the measures and countermeasures their opponents will take. Therefore, network honeypot has provided a way for both sides to attack and defend, ultimately learning the attacking methods and purposes.

3.1 Internet Fraud

Internet fraud means that an intruder believes there exist valuable security vulnerabilities which can be used in the network systems and meanwhile set up some sensitive information which can be stolen in order to lead the admission to these information resources. Network fraud technology can rapidly detect the intruders' attacks and record and analyze their methods and intentions. Therefore, when coming across the similar attacks in the future, they can take effective defensive measures in advance. Besides, the workload of the invaders and invasion difficulty and uncertainty have been increased. Such technology as IP fraud, analog service port, analog system vulnerabilities, organize information fraud and network services are commonly used in the network fraud technology[5][9].

3.2 Data Capture

Data capture is to monitor and record all behaviors of the intruders attacking the honeypot host. Through the analysis of these records, we can acquire such information as the intruders' attacking tools, tactics and motives. The difficulty lies in collecting real data as much as possible while not being detected by the intruders. In general, the capture structure adopts different levels to capture data because such hierarchical structure can combine the steps of the intruders' activities for easy analysis of the data. In addition, the captured data cannot remain in the local honeypot host because the honeypot features will make the intruders easy access to the data kept in the honeypot, resulting in the disclosure of the identity of the honeypot. Therefore, the captured data should be kept in a separate security system[5][9].

3.3 Data Control

Data control is the norm to restrict the intruders' behaviors. What needs to be ensured is that, once the system is attacked by an intruder, honeypots within the honeypot should not do harm to any other system outside the honeypot. The key to data control is to control data flow without being doubted by the intruder. On the other hand, the honeypot system must provide enough freedom for the intruder to carry out attacks.

The more the intruders' activities are allowed, the more they are understood. But it is known that the higher the degree of the invaders' freedom is, the greater the risk will be. Therefore, each honeypot system should strike a balance [5] [11].

3.4 Data Analysis

Data analysis is to analyze from the captured data the intruders' intention, the specific attacks and attacking tools, including network behavior analysis, network data traffic analysis and network attacking characteristics analysis. Currently, data analysis is still a problem in the honeypot system [5].

4 Offense and Defense Game Model

In the network honeypot, because of the information asymmetry and the different operation modes between the offensive and defensive sides in the honeypot, we can make use of the game tree method to analyze the process. In constituting a game process, the players, strategies and benefits are the most basic elements, and the players, actions and results are generally known as the rules of the game.

4.1 Brief Introduction to the Model

Offense and defense game model in the network honeypot is shown in Figure 1, in which the number 1 is the defender, and the number 2 is the attacker. Among the benefit array, the form is the defender's gains and the latter is the attacker's. The game process is divided into two stages. At the first stage, the defender will decide whether to create the honeypot or not since building the honeypot will spend a certain cost (q).

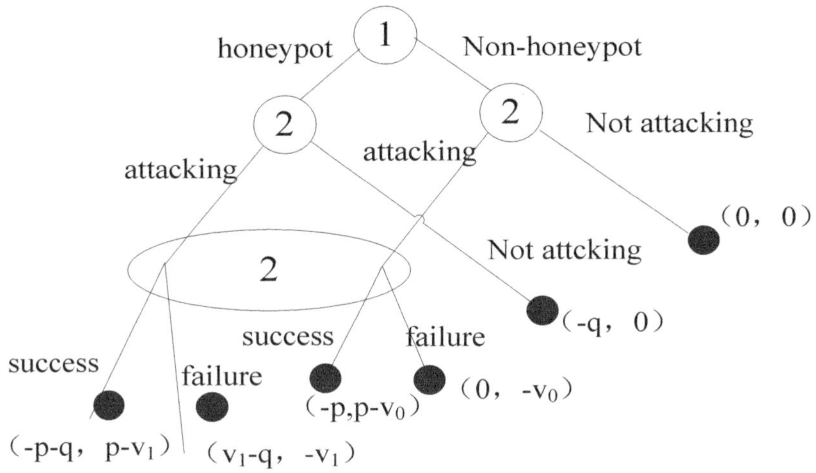

Fig. 1. Offense and defense game model in the network honeypot

At the second stage, the attacker will decide whether to attack or not. If the attacker decides not to attack, the attacker's gains will be zero and the defender's will be the cost of expenditure in the previous stage. If the attacker decides to attack, when the attacking target is not the honeypot system, the attacker's gains will be p and the defender's will be –p after the successful attack, while the attacker's gains will be -v0 and the defender's will be 0 after the failure in attacking; when the attack target is the honeypot system, the attacker's gains will be p-v1 (v1 is the attacking cost of the attacker) and the defender's will be -p-q after the successful attack, while the attacker's gains will be -v1 and the defender's will be v1-q after the failure in attacking.

4.2 Model Hypothesis

First, each variable in the model is positive.Second, the attacking cost spent on the honeypot system (v1) should be higher than that spent on the non-honeypot system (v0), namely, v1>v0. Otherwise, the case that the attackers give up the attacks when encountering the honeypot system will not occur.

4.3 Model Analysis

Before the attack, the attacker should conduct an analysis on the attacking target because different judgments lead to different gains. Therefore, the above-mentioned is a complete but imperfect information dynamic game model. In the following, we are to analyze it in detail.

In Figure 1, the information set of the defensive side contains two decision nodes, indicating that in the decision-making of this stage the defender is unaware of the first stage of the attacker. That is to say, when the attacker chooses attacking at the second stage, the defender has no idea where the attacker comes and then cannot make accurate decisions according to the two different paths. Though the defender has only two possibilities of success and failure, the result has four possibilities (c.f. Figure 1). When the attacker chooses attacking, the defender may choose the honeypot or the non=honeypot. Therefore, it is advisable that the defender should know which game path the attacker comes from, namely, honeypot or non-honeypot. Of course, the defender should not judge at random, because if random judgments do not conform to reality, there will be mistakes in decision-making.

Therefore, we can judge the probability of attack and defense of the offensive and defensive parts in the network honeypot. P (g|a) is used to indicate the conditional probability of a honeypot while P (b|a) indicates the conditional probability of a non-honeypot when the attacker chooses to attack. The two kinds of conditional probability are the judgments the attacker should conduct.

Before making reasonable judgments, the defender should first know whether or not the first phase is honeypot no matter whether the attacker attacks. We use P(g) to indicate the probability of a honeypot and P (b) indicates the probability of a non-honeypot. Obviously, the two kinds of probability can only be inferred from the experience or the general cases. Besides, the attacker should also know the probability of choosing to attack when the defender is honeypot and the probability of choosing to attack when the defender is non-honeypot. We can use P (a|g) and (a|b) respectively to indicate the two kinds of probability.

According to Bayes Rule, we have the following formula:

$$P(g|a) = [P(g) * P(a|g)] / P(a)$$
$$= [P(g) * P(a|g)] / [P(g) * P(a|g) + P(b) * P(a|b)]$$

Based on the formula (P (b|a) =1-P (g|a)),it is easy to get P (b|a).

From the defender's gains in the game tree, it can be seen that because choosing attack is the best strategy, the attacker will choose to attack the network in the honeypot system, namely, P (a|g) =1. Similarly, the attacker will attack the network in the non-honeypot system, namely, P (a|b) =1. Thus, the above formula can evolve into the following ones:

$$P(g|a) = [P(g) * P(a|g)] / [P(g) * P(a|g) + P(b) * P(a|b)]$$
$$= P(g) / [P(g) + P(b)] = P(g)$$

Likewise,

$$P(b|a) = P(b).$$

The defender is aware that the attacker will choose attacking at the second stage no matter whether it is honeypot or not, but because the information is not completely symmetrical, the attacker cannot identify the existence of the honeypot. Thus, the attacker can only make judgments on P (g) and P (b) under the general situations, and on such basis, choose the strategy which can maximize its own expectation payment. Suppose the attacker generally makes such judgments based on the experience as P(g)=P (g|a) =P*, the attacker's expectation payment of successful attacking in its own decision-making information sets will be:

$$E(R) = P^{**}(p-v1) + (1-p^*)*(p-v0)$$
$$= [(p-v1)-(p-v0)]p^* + (p-v0)$$

$$\text{If } E(R) = [(p-v1)-(p-v0)]p^* + (p-v0) > 0,$$

That is, if the attacker's expectation payment is positive, the attacker will succeed in attacking, for the expectation payment of failure is negative.

$$\text{If } E(R) = [(p-v1)-(p-v0)]p^* + (p-v0) < 0,$$

The attacker will fail. In other words,

$$\text{If } p^* > -(p-v0)/[(p-v1)-(p-v0)],$$

The attacker will succeed in attacking; whereas, if $p^* < -(p-v0)/[(p-v1)-(p-v0)]$, the attacker will fail in attacking.

5 Intrusion-Detecting Game Model

The intruders often cannot accurately estimate its purpose, time, manner and behavior to be taken, but the real invasion generally has the same steps, namely, detecting, entering the system, uploading shell or opening ports, getting the file, clearing the traces, maintaining the post-visit and so on. A successful invasion can not only make the intruder obtain unauthorized access to information, but also make the system shield other intruders and provide the secrete channel for the invaders with the initial success sometime in the future. During the process of implementing the intrusion, the intrusion detection system will send quick alarms and take measures to deal with the apparently simple attacking behaviors in accordance with the presetting, which we call pre-response [13].

On encountering the complex invasion, the simple pre-response is far from perfect and what needs is to enter the game decision-making process designed by game and to make the decision response after the trade-off thinking according to a complex set of strategies.

Invasion process will get more benefits in the shortest possible time. Honey-IDS needs to consider whether the current gains are adequate for its maintenance costs and quickly makes pre-response and decision-making response. Good results of the game should be that the attacker wins the game advantages and feels satisfied with its gains and then leaves with the information obtained, and meanwhile the honeypot succeeds in recording the invasion but without being detected by the attacker.

The whole invasion process can be expressed by using the following model.

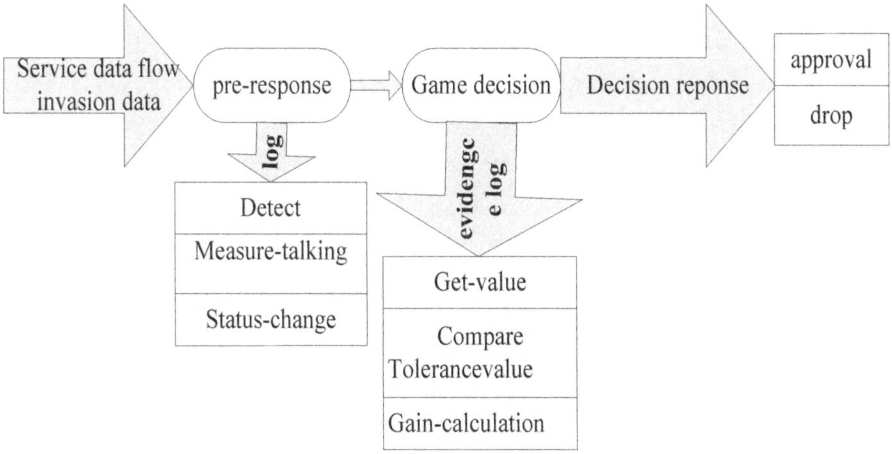

Fig. 2. Intrusion-detecting game model in the network honeypot

Acknowledgments. The research described in this paper was substantially supported by a project (70901080) from the National Natural Science Foundation of China. The authors would like to thank anonymous referees for their helpful comments and valuable suggestions which improved the content and composition substantially.

References

1. Stoll, C.: The cuckoo's egg: tracking a spy through the maze of computer espionage. The Bodley Head Ltd. (1990)
2. Artail, H., Safa, H., Sraj, M., et al.: A hybrid honeypot framework for improving intrusion detection systems inprotecting organizational networks. Computers and Security 25(4), 274–288 (2006)
3. Spitzner, L.: Honeypots Tracking Hackcr. AddisonWesley, America (2003)
4. Anagnostakis, K., Sidiroglou, S., Akritidis, P., et al.: Detecting targeted attacks using shadow honeypots. USENIX Security (2005)
5. Yin, C., Li, M., Ma, J., et al.: Honeypot and scan detection in intrusion detection system. In: Canadian Conference on Electrical and Computer Engineering (2004)
6. Kim, M., Kim, M., Lee, H.K.: Design of active honeypot system. In: Kumar, V., Gavrilova, M.L., Tan, C.J.K., L'Ecuyer, P. (eds.) ICCSA 2003. LNCS, vol. 2668, Springer, Heidelberg (2003)
7. Thakar, U.: Honey analyzer-analysis and extraction of intrusion detection patterns & signatures using honeypot. In: Proceedings of the Second International Conference on Innovations in Information Technology (2005)
8. Syverson, P.E.: A different look at secure distrubuted Computation. In: Proceedings of the 1997 IEEE Computer Security Foundations Workshop, pp. 109–115. IEEE Computer Society, Washington (1997)
9. Lye, K.-W., Jeannette, M.W.: Game strategies in network security. International Journal of Information Securiy

Virtual Tourism Interactive Product Design for Senior Citizens

Zhongyan Hu, Zaihui Cao, and Jinfa Shi

Department of Art and Design, Zhengzhou Instiute of Aeronautical Industry Management
450015 Zhengzhou, China
huzohngyan@zzia.edu.cn, czhhn@126.com

Abstract. Considering the expectation of enjoying a tour and many movement function limitations of senior citizens, the author analyzed the advantages and restrictions of the virtual tourism net work for them. The design methods, principles, theories and impression drawings, which meet the physical and mental demands of senior citizens, has been fully detailed.

Keywords: senior citizen, virtual tourism, Interactive product, physical, mental.

1 Introduction

Older the rapid population growth has become the important features of the Chinese population growth, the older product design will become aged population social services in the key and difficult. Tourism, as a city the old physical fitness, increase knowledge, edify sentiment, looking for life new hopes to a project, no doubt for rich old man late years provides a choice. But the physical, psychological and objective conditions, and many other factors, many old people will travel intend to deep heart, even from his own life in the dictionary to delete. If not, the old people have no tourism and the desire of appeal, but did not factor is too much, excessive demand.

2 Main Factors of Restricting Urban Elderly Travel

According to the 2008 national health service investigation analysis report provides information on urban proportion of chronic diseases in older people with age increasing, prevalence of chronic diseases and the 2003 national health service a further upward trend compared to survey data (or even higher in big cities), as shown in table 1.Health is one of the important factors affecting senior tourism[1].

Table 1. Elderly population age don't chronic disease prevalence (%)

age	2008 year	2003 year
60-69	46. 8	44.5
70-79	58. 8	55.4
80+	60.9	52.2

G. Lee (Ed.): Advances in Automation and Robotics, Vol.2, LNEE 123, pp. 247–253.
Springerlink.com

Psychological factors.Older growth, old in intelligence, memory is somewhat recession, this apparent defects, take older persons in his familiar environment measures to prompt such as write notes, sticking note, with hard everyday life, improved life quality.Once out of their familiar environment, ability to adapt to the environment for the elderly will be decreased significantly, defects of cognitive decline will be exposed.So in the old heart and unwilling to leave their familiar environment, disrupts the original pace of life.Psychological characteristics of the elderly this special, as well as most elderly people despite the willingness to travel, but does not actually put into action the main reason.

Objective factor. Currently transport, safety and accommodation and restaurants to other factors, mostly tourist scenic spot planning and environmental infrastructure construction, almost considered elderly tourists--this group special needs, let alone specially designed specifically for elderly tourists [2] times.Therefore truly meet the elderly mental and physiological needs very few tourist products of, mostly middle-aged and young tourism products or its subsidiary products as the old lack of attraction.Caused mainly due to lack of tourism products in the elderly: high cost of travel services, tourism operations is difficult, such as the high cost of travel insurance, must be equipped with the Mission of care doctors and equipment.Risky business, with low fees, profit less.So many in the tourism market, there is no travel agency specializing in senior tourism service [3].These objective factors, to some extent restricted the travel of older people wish.

3 Advantages and Limitations of Virtual Tour of the Elderly

Under the influence of these objective factors, senior tourism will increase with age, fading demand.Established realistic landscape on the basis of virtual tourism, with the help of virtual technology, network technology, simulation of three dimensional simulation of tourism scene, let tourists get there and travel fulfillment means of travel, realizing the city elderly will no doubt be a travel dreams the best alternative.Especially virtual tourism, high security, low cost, no boundaries and autonomy, accessibility to the superiority of the Super, interactivity, and more aware of the feature [4][5] determines that it will become one of the senior tourism development.To sum up, for the aged group, there are several advantages of virtual tour:

(1) Relaxation of travel requirements of the health status of the elderly;

(2) Reduced travel costs, affordable price more in line with the consumer psychology of the elderly;

(3) Eliminates old psychological burden of outdoor activities, avoid travel risks;

(4) Widening the connotation and extension of activity in the elderly to meet the spiritual pursuit of a higher level in elderly.

In summary, virtual tour to meet the travel needs of the elderly, and to overcome the current lack of tourist market of various, realization of tourist wishes of elderly people is undoubtedly the best choice. But according to China Internet network information center released in 2011 27th report on the State of China's Internet development statistics show that only uses the network of the elderly aged 60 and over the country's Internet population 1.9%[6], actual participation in the online virtual tour of the old people are negligible.The main reason is, is unfamiliar to most elderly people on the computer, network, virtual tour does not understand, re-learn has a certain difficulty, that is, at present this network travel for the elderly is not very suitable.Second long onset of senile chronic diseases induced by computer will likely not suitable for the physical needs of the elderly.Therefore to change the current virtual tour of how older persons familiar with the popular fitness equipment with a combination of network virtual tour scene and fitness equipment control tourism scene change, is the key to virtual tour interactive product design.

4 Virtual Tour Interactive Product Design for the Old Ideas and Design Principles

4.1 Interactive Product Design Ideas

Virtual tour interactive product design the overall idea is to use an existing virtual travel simulation results for the three-dimensional travel scene, focusing on research for the aged group of virtual scene control mode.Virtual tour in the 1990 of the 20th century research to date, attracted many Chinese and foreign scholars to study and exploration of virtual tourism, are now available at the website virtual tour systems, virtual tour systems, virtual reality simulation and achieved a number of research results, particularly in terms of three-dimensional virtual scene simulation, both technically and in the application, and have a lot of new breakthroughs.Borrowing these existing research results, restrictive control part of elderly participating in virtual tourism to improve, designed for older physiological and psychological requirements inclusive fitness and traveling as one of interactive product, involving the elderly healthy, while in virtual tour, enjoy modern convenience of technology.

4.2 Interactive Product Design for the Old Principle

Interactive product design for the old principle: meet the functional characteristics of product design, use, features, security features and aesthetic characteristics.Function is the product's survival is all about.Most thrifty practice your life in the elderly home, seeking, and Chase vanity elements are relatively small.Main purpose of the functional requirement is to purchase products.They hoped the product features or make up for the inadequacy of the physical, or psychological on the lack of, help them solve their practical difficulties in everyday life.Characteristics of product design for the old function is particularly important.Also consider specific physiological and psychological characteristics of the elderly, interface simple and clear, easy to learn and use, safe and reliable, avoid complex operations, increase the memory of the elderly afford (memory decline in the elderly often have intermittent easy to forget the physiological phenomena, too many steps will add a lot of inconvenience, and caused to them as a psychological pressure).Common operations as far as possible without changing the lifestyle inherent in the elderly, reducing the use of the product need to learn new content (physiological degradation of organs, the elderly difficult to absorb new knowledge).Design many familiar elements into their generally recognized helps stimulate empathy projection of aesthetic interest, won the elderly on product acceptance [7] times.

5 Virtual Tour Interactive Product Design for the old Principle and Effect Chart

5.1 Design Principle

Fitness equipment and other electronic devices in interaction design has been relatively successful examples.Such as DDR and TV connection, while dancing and electronic display control device, and so on.

Study on how it works, is currently electronic device manually control into the foot motion sensing controls.Follow this approach, by studying the relative motion of a computer mouse and face works, will face fitness equipment sports surface as the mouse feet movement instead of hand move mouse action, borrowing an existing mouse works, through the electro-optical device, snap fitness equipment face motor speed and direction of movement, achieve fitness equipment and computer virtual scene interactive campaigns.

5.2 Effect Drawing of Human-Computer Interaction Interface

In accordance with these design ideas, design two old virtual tour interactive product, as shown in Figure 1, Figure 3.Figure 2 is another shape design of Figure 1. Figure 1 and Figure 32 product to control the same way, when using Start button on the handle switch, by means of a batch level, automatically bring up virtual travel procedures.People moving on the walking belt, walking belt driven transmission to simulate mouse moves back and forth relative motion, virtual scenes before and after image transformation.Placement of support rod inserted in handle fixed bushing, simulating bikers works transform direction of travel, with connecting rod driving mechanism, by control devices in the lower face free rotation, implementing virtual scenes about the direction of image transformation.These activities are housed in the base through the optical signal is passed to the optical mouse chip and attached to the computer through a standard USB interface, interactive virtual scenes and sports.This mode of activity, increases both the aged sport fun, meet the tourism aspiration of older persons.

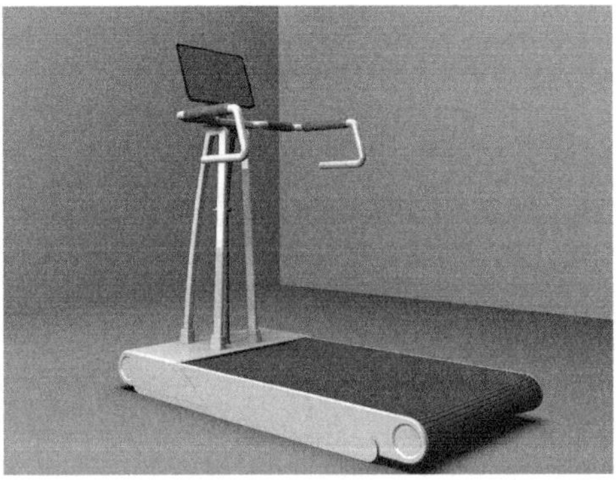

Fig. 1. Schematic Diagram of an Interactive Product

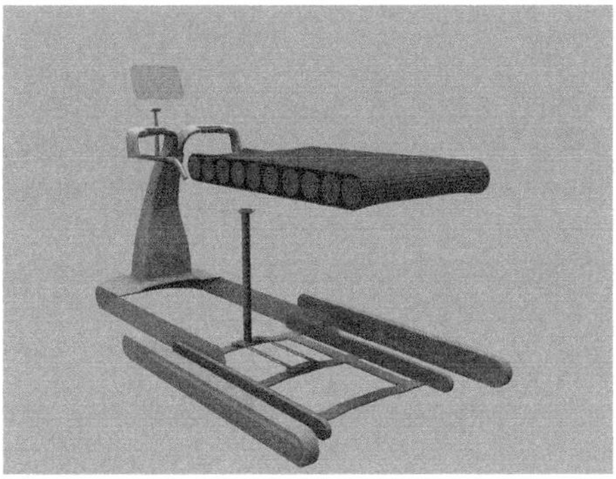

Fig. 2. Internal Drive Rod Shape Shape Effect Diagram

These two models of contour design, surface appearance of the figure 1 more fashionable, but from a production perspective, irregular surface processing needs to die, the production process more complicated.Figure 2 comparison of Rod-like shape simplicity, technology is relatively simple, the user can according to their preferences, the freedom to choose.The features of both products without distinction on the principle and methods of operation.

Meet elderly sports audience, plan II from perspectives of more human ageing physiological and psychological requirements, provides modular design.The model can be used alone or combination of two groups for use by the elderly husband and wife jointly, can also make multiple sets of combinations and space size as needed.Product design drew on old exercise bike design concept in interactive product has been added to seats, action-so that's not convenient or breakdown of the old man, sitting on a seat, for exercise (such as the old man of slight side hemiplegia, can make use of the seat, and limited rehabilitation).The design of the seat can be folded, choose whether to use their own health status of the elderly.

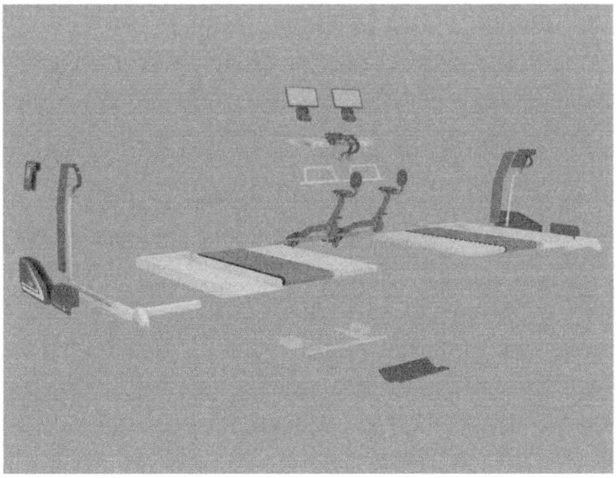

Fig. 3. Scheme ii rendering

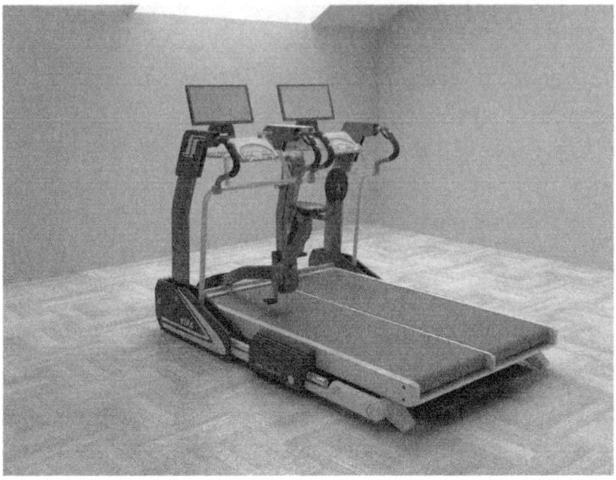

Fig. 4. Scheme ii explosion figure

The above two paragraphs of the scheme are common interface, considering the use of the old habits, and the secondary interface to choose. The first level for the purpose of tourist province (or countries), the second choice for tourism destination that choice. Whatever level 1 or secondary interface choice, operation mode are the same. If choose target in this page appears directly, according to determine the key, such as not to flip can form looking for.

6 Conclusion

Older virtual tour interaction products, considering the old this special group of physical and psychological needs, encourage the old people in the exercise as much as possible. The product is simple to use and control mode and experience of life most older people is consistent, easy for operation, is out of the movement of the simple old analog. Compared with old people go out, movement of sports limited more small. No matter, Yin qing chun xiaqiu winter changes in temperature, the old man could use this product to exercise, and interest sex is stronger, in athletic process not only enjoy the natural beauty, and the growth of knowledge, edify sentiment, have more spirit to meet.

Virtual tour eased travel to the old man of the conditions of the body, eliminate the requirement senior tourism encountered in the process of all kinds of inconvenience, from psychological completely removed the old people all kinds of participation in tourism concerns. By leveraging virtual tourist interactive products, old people in the family or community for online virtual tour, do not need to leave the old man familiar with living environment, and don't affect older people in the original life habit, and on the basis of a new offer leisure entertainment. This way the old people further enrich the life of old age in China, to deal with the arrival of aging peak various kinds of social contradictions, no doubt, provides a new solution.

Acknowledgement. This paper supported by the department of education of HeNan province natural science research program(No.2011A630062),the department of education of HeNan province humanities and social science research project(No/2011-ZX-233),Henan government decision-making tender (No.2011A630062).

References

1. In 2008 The fourth national health services survey analysis [EB / OL] (2008), http://www.moh.gov.cn/publicfiles/business/htmlfiles/mohwsbwstjxxzx/s8560/201009/49165.htm (September 21, 2010)12:02:20
2. Liu, L.: On the old problems in the tourism market and develop countermeasures. Business Economics 09(334), 106–108 (2009)
3. Wang, L.: Old city of Chengdu tourism market research and development. Sichuan Normal University, a master's degree thesis (2007)
4. Li, W.: Analysis of virtual tourism. Scientific and Technical Information 20, 544 (2009)
5. Li, L., Yang, P., et al.: "Virtual tourism" new theory of right and wrong. DC Economic 02, 142–143 (2009)
6. 27th China Internet Development Statistics Report. [EB / OL], http://www.chinanews.com/it/z/cnnic27/
7. Yan, H., Cao, Y.: Old psychological factors in product design. Packaging Engineering 05, 133–135, 143 (2008)

Staged Soft-Sensor Modeling for Batch Fermentation Process

Qiangda Yang

School of Materials & Metallurgy, Northeastern University 345#,
110819 Shenyang, China
daqiang9666@163.com

Abstract. Crucial biochemical parameters are hard to be measured on-line in real-time in batch fermentation process, which brings difficulties to control and optimization of this process. To solve this problem, soft-sensor technique is applied to implement the on-line estimation of crucial biochemical parameters, and a staged modeling method is presented. Firstly, the modeling data are classified according to their stages by using fuzzy c-means clustering algorithm. Then, a model for the on-line identification of fermentation stages and some local soft-sensor models corresponding to each stage are developed by using neural network. Lastly, simulation is performed based on the production data from Nosiheptide batch fermentation process. The results show the effectiveness of the presented soft-sensor modeling method.

Keywords: fermentation, soft-sensor, staged modeling, fuzzy c-means clustering, neural network.

1 Introduction

Fermentation is a common mode of production in modern process industry, which has been widely used in the production of medicine, food, chemical products, and so on [1]. During batch fermentation process, the real-time acquisition of some crucial biochemical parameters such as biomass and substrate concentration is of great significance for process optimization and control. However, until now there are no practical on-line sensors that can be used to measure them directly. In the industrial production, crucial biochemical parameters are mostly measured through off-line analysis [2]. Off-line analysis has large time-delay, which brings many difficulties for the real-time monitoring of crucial biochemical parameters and the direct quality control of this process. Therefore, for batch fermentation process, one of the key problems to be solved is how to fulfill the on-line measurements of crucial biochemical parameters under existing technical conditions.

Soft-sensor is a technique that can estimate the hard-to-measure variables using easily available process variables, which provides an effective way for solving the above problems. Many soft-sensor modeling methods have been proposed over the last 30 years for fermentation process, and one of the typical modeling methods is developing a global soft-sensor model using ANN [3-5]. However, in batch fermentation process, there are apparent stage characteristics and the process

G. Lee (Ed.): Advances in Automation and Robotics, Vol.2, LNEE 123, pp. 255–264.
Springerlink.com

expresses different characteristics in different stages [6,7], so attempting to model the process using a global method would often lead to complicated network structure with limited estimation accuracy.

In this paper, a staged modeling method is presented to develop the soft-sensor model of crucial biochemical parameters for batch fermentation process, furthermore the on-line estimation of these parameters are realized. Firstly, based on the off-line assay values of crucial biochemical parameters which can reflect the stage characteristics of batch fermentation process, the modeling data are classified according to their stages by using fuzzy c-means clustering algorithm. Then, a model for the on-line identification of fermentation stages and some local soft-sensor models corresponding to each stage are developed by using neural network. Lastly, simulation is performed based on the production data from Nosiheptide batch fermentation process, and the results show the effectiveness of the presented soft-sensor modeling method.

2 Soft-Sensor of Crucial Biochemical Parameters Based on Staged Modeling Method for Batch Fermentation Process

For practical fermentation production industry, the batch fermentation process generally experiences three stages including lag stage, exponential stage and stationary stage [8]. The influence factors upon cells are different in different stages, which results in the batch fermentation system exhibiting different nonlinear characteristics in different stages. So developing a global soft-sensor model for batch fermentation process would often lead to complicated network structure with limited estimation accuracy. To simplify the model structure and improve the model estimation accuracy, staged modeling strategy is adopted in this paper.

The structure of the staged modeling method based soft-sensor model is shown in Fig. 1. $\boldsymbol{x}$ are the measured values of secondary variables. $\hat{\boldsymbol{y}}$ are the estimation values of crucial biochemical parameters by using the staged modeling method based soft-sensor model (It should be pointed out that the crucial biochemical parameters include biomass and substrate concentration, but may also include other biochemical parameters such as production concentration). $\hat{\boldsymbol{y}}_1$, $\hat{\boldsymbol{y}}_2$ and $\hat{\boldsymbol{y}}_3$ are the estimation values of crucial biochemical parameters by using local models corresponding to lag stage, exponential stage and stationary stage respectively. $\hat{\boldsymbol{y}}_{cs} = [\hat{\rho}_c, \hat{\rho}_s]$ are the estimation values of biomass and substrate concentration by using the staged modeling method based soft-sensor model. $\hat{\boldsymbol{\mu}} = [\hat{\mu}_1, \hat{\mu}_2, \hat{\mu}_3]$ are the estimation values of the membership degrees of fermentation process to lag stage, exponential stage and stationary stage respectively by using the model for the on-line identification of fermentation stages. k, k-1, k-2 and k-3 are time k, k-1, k-2 and k-3 respectively. When modeling off line, firstly, based on the off-line assay values of biomass and substrate concentration, the modeling data are classified into three groups according to their stages by using fuzzy c-means clustering algorithm, and then, a model for the on-line identification of fermentation stages and three local soft-sensor models are developed by using neural network. When estimating crucial biochemical parameters on line, firstly, the fermentation stages are identified on line, and then, according to the results of stage identification, the estimation values of crucial biochemical parameters are obtained.

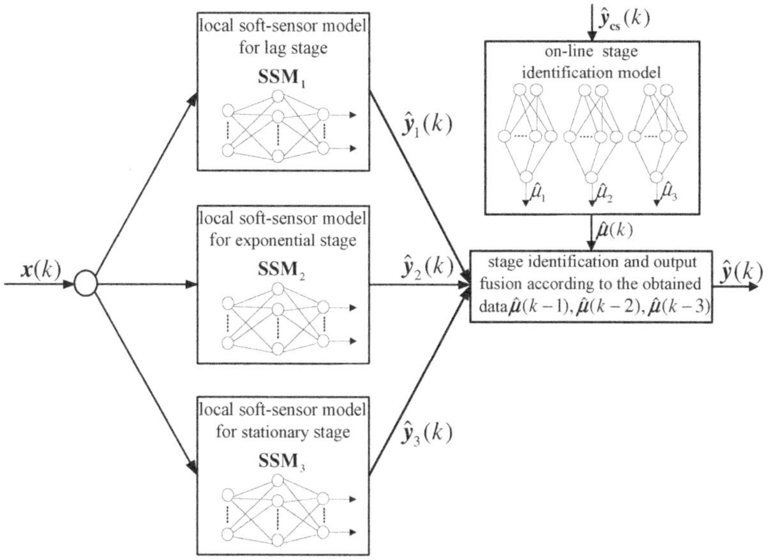

Fig. 1. Staged modeling method based soft-sensor model for batch fermentation process

2.1 Development of Staged Modeling Method Based Soft-Sensor Model

The development of staged modeling method based soft-sensor model contains two steps. The first step is classifying the modeling data into three groups according to their stages. The second step is developing three local soft-sensor models for crucial biochemical parameters.

In addition, this paper uses neural network to identify the fermentation stages on line, so an on-line stage identification model based on neural network also needs to be developed.

2.1.1 Classification of Modeling Data

To Classify the Modeling Data According to their stages, it first needs a basis of stage judgment. Essentially, the stage characteristics shown in fermentation process are physiologic accommodations of cells to the changes of their growing environment. In different fermentation stages, the biochemical shape, physiological characteristics and cell components change obviously. Therefore, the most direct method is to identify the fermentation stages according to the analysis of biochemical shape, physiological characteristics and cell components [9]. This method has the advantages of definite physiological meaning, high accuracy and so on, but it needs complicated operation and relevant professional knowledge. In addition to the above method, another adopted method for stage identification is to use expert experience and fuzzy sets [10], [11]. This expert experience and fuzzy sets based stage identification method needs relevant expert knowledge, but for most of batch fermentation processes, the expert knowledge of stage characteristics has very big limitations. Therefore, it's hard to classify the modeling data of batch fermentation by using the two above-mentioned methods.

As pointed out by [12], there are certain biochemical variables which are indicative of the stage characteristics in batch fermentation process. As mentioned above, the batch fermentation process experiences three stages. The first stage is lag stage, during which the biomass is relatively low and the substrate concentration is high. The second stage is exponential stage, during which the biomass is increasing whereas the substrate concentration is decreasing. The third stage is stationary stage, during which the biomass is relatively high and the substrate is almost consumed up. Hence, the biomass and substrate concentration could be used to classify the modeling data of batch fermentation process according to their stages. In this paper, fuzzy c-means clustering algorithm is used to classify the modeling data of batch fermentation process according to their stages, and the concrete steps are as follows.

① Let $\boldsymbol{p}_{mn}(m=1,2,\cdots,B; n=1,2,\cdots,b_m)$ represent the n-th data pairs of the m-th batch of modeling data, then it can be written as:

$$\boldsymbol{p}_{mn}=[\boldsymbol{x}_{mn} \mid \boldsymbol{y}_{mn}]=[\boldsymbol{x}_{mn} \mid \boldsymbol{y}_{\mathrm{cs}\,mn} \quad \boldsymbol{y}_{\mathrm{o}\,mn}] . \tag{1}$$

Where $\boldsymbol{x}$ are the secondary variables, $\boldsymbol{y}$ the crucial biochemical parameters, $\boldsymbol{y}_{\mathrm{cs}}=[\rho_c,\rho_s]$ the biomass and substrate concentration, $\boldsymbol{y}_{\mathrm{o}}$ the other crucial biochemical parameters, B the batch of modeling data, b_m the number of data pairs of the m-th batch of modeling data.

② Let $\mathbf{Y}_{\mathrm{cs}\,m}(m=1,2,\cdots,B)$ represent the matrix constructed by biomass and substrate concentration of the m-th batch of modeling data, then it can be written as:

$$\mathbf{Y}_{\mathrm{cs}\,m}=\begin{bmatrix} \boldsymbol{y}_{\mathrm{cs}\,m1} \\ \boldsymbol{y}_{\mathrm{cs}\,m2} \\ \vdots \\ \boldsymbol{y}_{\mathrm{cs}\,mb_m} \end{bmatrix}=\begin{bmatrix} \rho_{c\ m1} & \rho_{\mathrm{s}\ m1} \\ \rho_{c\ m2} & \rho_{\mathrm{s}\ m2} \\ \vdots & \vdots \\ \rho_{c\ mb_m} & \rho_{\mathrm{s}\ mb_m} \end{bmatrix} . \tag{2}$$

③ Cluster the normalized $\mathbf{Y}_{\mathrm{cs}\,m}(m=1,2,\cdots,B)$ by using fuzzy c-means clustering algorithm with the cluster number being three respectively, and gain the membership degrees of each point of $\boldsymbol{y}_{\mathrm{cs}\,m1}, \boldsymbol{y}_{\mathrm{cs}\,m2},\cdots,\boldsymbol{y}_{\mathrm{cs}\,mb_m}$ to the three stages, meanwhile used as membership degrees of each modeling data pairs to which each point of $\boldsymbol{y}_{\mathrm{cs}\,m1}, \boldsymbol{y}_{\mathrm{cs}\,m2},\cdots,\boldsymbol{y}_{\mathrm{cs}\,mb_m}$ belongs to the three stages.

④ According to the maximum membership degree principle, classify $\boldsymbol{p}_{mn}(m=1,2,\cdots,B; n=1,2,\cdots,b_m)$ into three groups corresponding to three stages.

2.1.2 Development of On-Line Stage Identification Model

Based on the above discussion, it can be seen that there is no suitable on-line stage identification method for most of batch fermentation processes. For that, an on-line stage identification method based on cluster results and neural network is presented in this paper.

According to the clustering results, it can be seen that each set of modeling data pairs corresponds to three membership degrees. So there is the following function relationship:

$$\begin{cases} \mu_1 = F_1(\rho_c, \rho_s) \\ \mu_2 = F_2(\rho_c, \rho_s) \\ \mu_3 = F_3(\rho_c, \rho_s) \end{cases} . \tag{3}$$

Where ρ_c and ρ_s are the biomass and substrate concentration respectively. μ_1, μ_2 and μ_3 are the membership degrees of modeling data pairs to lag stage, exponential stage and stationary stage respectively, which are obtained by clustering.

Neural network is used to fit the functional relationships F_1, F_2 and F_3 in this paper. The inputs of neural network are normalized ρ_c and ρ_s, and the outputs of neural network are μ_1, μ_2 and μ_3. After the train of neural network is completed, the on-line stage identification can be implemented by using the neural network, namely the on-line stage identification model.

2.1.3 Development of Local Soft-Sensor Models

After the modeling data are classified into three groups, three local soft-sensor models are developed by using neural network. When the constructions of all the local soft-sensor models are completed, a distributed estimation network is determined as shown in Fig. 1. Local model SSM_1, SSM_2 and SSM_3 correspond to lag stage, exponential stage and stationary stage respectively.

2.2 On-Line Estimation of Crucial Biochemical Parameters

When estimating crucial biochemical parameters on line by using staged modeling method based soft-sensor model, firstly, the on-line stage identification is carried out, and then according to the identification results, the outputs of local soft-sensor models are integrated on the basis of certain data fusion methods, furthermore, the on-line estimation values of crucial biochemical parameters are obtained. The concrete steps are as follows.

① Firstly, the crucial biochemical parameters are estimated by using the local model SSM_1, that is to say, the outputs of SSM_1 are chosen as the estimation values of crucial biochemical parameters. And then the normalized estimation values of biomass and substrate concentration are inputted into the on-line stage identification model, furthermore three membership degrees μ_1, μ_2 and μ_3 are obtained. When all of $|\mu_1(i_1) - \mu_2(i_1)|, |\mu_1(i_1+1) - \mu_2(i_1+1)|$ and $|\mu_1(i_1+2) - \mu_2(i_1+2)|$ are less than or equal to σ_1 which is subject to $\sigma_1 \geq 0$, the batch fermentation process is thought to step into the transition stage between lag stage and exponential stage from time $i_1 + 3$.

② During the transition stage between lag stage and exponential stage, the estimation precisions of the local models SSM_1 and SSM_2 are thought to be the same. Therefore, from time $i_1 + 3$, the crucial biochemical parameters are estimated by using the local models SSM_1 and SSM_2. Assuming that the estimation values of SSM_1 and SSM_2 are $\hat{y}_1$ and $\hat{y}_2$ respectively, from time $i_1 + 3$, the estimation values of the crucial biochemical parameters are obtained by Formula 4. And then the normalized estimation values of biomass and substrate concentration are inputted into the on-line stage identification model, furthermore three membership degrees μ_1, μ_2 and μ_3 are obtained. When all of $\mu_2(i_2) - \mu_1(i_2)$, $\mu_2(i_2+1) - \mu_1(i_2+1)$ and $\mu_2(i_2+2) - \mu_1(i_2+2)$ are more than σ_1, the batch fermentation process is thought to step into the exponential stage from time $i_2 + 3$.

$$\hat{y} = \frac{\hat{y}_1 + \hat{y}_2}{2} . \tag{4}$$

③ From time $i_2 + 3$, the crucial biochemical parameters are estimated by using the local model SSM_2. And then the normalized estimation values of biomass and substrate concentration are inputted into the on-line stage identification model, furthermore three membership degrees μ_1, μ_2 and μ_3 are obtained. When all of $|\mu_3(i_3) - \mu_2(i_3)|$, $|\mu_3(i_3+1) - \mu_2(i_3+1)|$ and $|\mu_3(i_3+2) - \mu_2(i_3+2)|$ are less than or equal to σ_2 which is subject to $\sigma_2 \geq 0$, the batch fermentation process is thought to step into the transition stage between exponential stage and stationary stage from time $i_3 + 3$.

④ During the transition stage between exponential stage and stationary stage, the estimation precisions of the local models SSM_2 and SSM_3 are thought to be the same. Therefore, from time $i_3 + 3$, the crucial biochemical parameters are estimated by using the local models SSM_2 and SSM_3. Assuming that the estimation values of SSM_2 and SSM_3 are $\hat{y}_2$ and $\hat{y}_3$ respectively, from time $i_3 + 3$, the estimation values of crucial biochemical parameters are obtained by Formula 5. And then the normalized estimation values of biomass and substrate concentration are inputted into the on-line stage identification model, furthermore three membership degrees μ_1, μ_2 and μ_3 are obtained. When all of $\mu_3(i_4) - \mu_2(i_4)$, $\mu_3(i_4+1) - \mu_2(i_4+1)$ and $\mu_3(i_4+2) - \mu_2(i_4+2)$ are more than σ_2, the batch fermentation process is thought to step into the stationary stage from time $i_4 + 3$.

$$\hat{y} = \frac{\hat{y}_2 + \hat{y}_3}{2} . \tag{5}$$

⑤ From time $i_4 + 3$ until the end of the batch fermentation, the crucial biochemical parameters are estimated by using the local model SSM_3.

3 Results and Discussion

In this paper, simulation study is carried out based on the Nosiheptide batch fermentation process. Biomass, substrate concentration and production concentration are three crucial biochemical parameters in Nosiheptide batch fermentation process, but there are no practical on-line sensors that can be used to measure them directly, so only off-line analysis method could be adopted. Off-line analysis has large time-delay, which brings many difficulties for the timely operation and the direct quality control. Therefore, it is of important practical significance for the implementation of the on-line estimation of biomass, substrate concentration and production concentration by soft-sensor technique for Nosiheptide batch fermentation process.

3.1 Secondary Variables and Modeling Data

Based on deep analysis of the mechanism of Nosiheptide batch fermentation process, the dissolved oxygen concentration, one ordered derivative of dissolved oxygen concentration, CO_2 evolution rate, pH value, temperature, speed of agitator and flow rate are selected as the secondary variables in this paper.

The simulation date are collected from the Nosiheptide batch fermentation process in a fermentation laboratory of Northeastern University. The production cycle is about 96 hours, and the sampling cycle is about 3 hours. Biomass, substrate concentration and production concentration are obtained by off-line analysis, and the other needed valuables are all obtained through DCS system. In this paper, 6 batches of data are used to develop the soft-sensor model, and 1 batch of data are used to verify the effectiveness of the developed soft-sensor model.

3.2 Development of Soft-Sensor Model

Based on the above data, an on-line stage identification model and three local soft-sensor models of crucial biochemical parameters are developed by using neural network in this paper. As one kind of artificial neural network, RBF neural network has the nature of global approximation and the best approximation performance [13], so RBF neural network is used to develop the above on-line stage identification model and local soft-sensor models in this paper. The number of the hidden layer node and the center of each hidden layer node are determined by the improved k-means clustering algorithm [14], and the spread of each hidden layer node is determined by neighbor rules [13]. The link weights from hidden layer to output layer are determined by the least-square algorithm. The concrete steps are as follows.

① Make corresponding pretreatment for the modeling data. The pretreatment mainly includes error treatment and data transform. The error treatment includes elimination of rude error and data smooth. The data transform means the normalization of each variable.

② Classify the modeling data into three groups by using the method presented in Section 2.1.1.

③ Develop the on-line stage identification model by using RBF neural network. The structure of the on-line stage identification model is $2 \times 21 \times 3$.

④ Develop three local soft-sensor models of crucial biochemical parameters using RBF neural network. The inputs of all the neural networks are normalized secondary variables selected in Section 3.1, and the outputs of all the neural networks are normalized biomass, substrate concentration and production concentration. It must be noted that, when using the developed soft-sensor model to estimate crucial biochemical parameters on line, reverse normalization is needed, and the estimation values of crucial biochemical parameters given in the following sections are the results of reverse normalization. The structure of the three local soft-sensor models are $7\times12\times3$, $7\times10\times3$ and $7\times13\times3$ respectively.

3.3 Performance Analysis of Soft-Sensor Model

Based on practical situation of Nosiheptide batch fermentation process, σ_1 and σ_2 are chosen to be equal to 0.13 and 0.16 respectively. After the staged soft-sensor model is developed, the normalized secondary variables of verification data are inputted into this model, and then the estimation results of biomass, substrate concentration and production concentration are obtained, as shown in Fig. 2 (*solid line*).

From the curves of assay values of crucial biochemical parameters and estimation values of the staged soft-sensor model as shown in Fig. 2, the staged soft-sensor model developed in this paper can estimate the crucial biochemical parameters of Nosiheptide batch fermentation process with high accuracy.

In addition, in order to reflect the advantage of the staged soft-sensor modeling method, estimation results are compared between the staged soft-sensor model and the global soft-sensor model in this paper. Based on the same training method, the same modeling data and the same verification data, the global soft-sensor model is developed, and the structure of the global soft-sensor model is $7\times25\times3$.

The estimation results of biomass, substrate concentration and production concentration based on the global soft-sensor model are shown in Fig. 2 (*dotted line*). The average relative generalization errors of the staged soft-sensor model and the global soft-sensor model are shown in Table 1.

Table 1. Error comparison using two soft-sensor models (%)

Model	Biomass	Substrate concentration	Production concentration
Staged model	3.96	3.75	4.22
Global model	6.19	5.92	6.87

From Fig. 2, the outputs of the two soft-sensor models can all reflect objectively the trends of biomass, substrate concentration and production concentration in Nosiheptide batch fermentation process. From Table 1, each estimation error of the three crucial biochemical parameters of the proposed staged soft-sensor is less than the global soft-sensor model, which clearly demonstrates that the proposed staged soft-sensor model is superior to the global soft-sensor model.

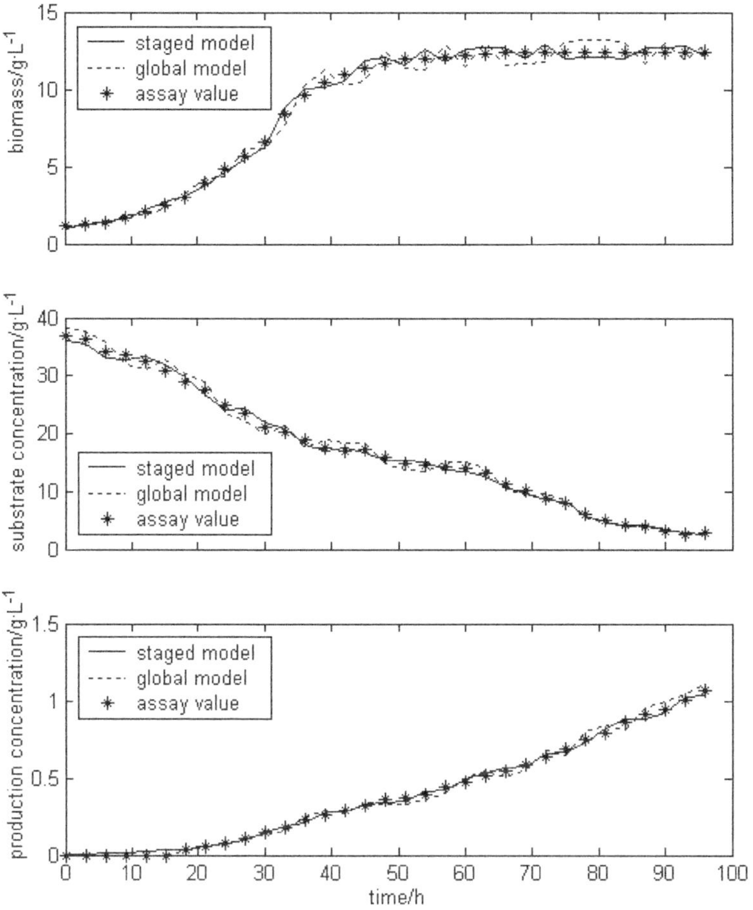

Fig. 2. Estimation results of biomass, substrate concentration and production concentration

4 Conclusion

Combined with stage characteristics of batch fermentation process, a staged modeling method is presented in this paper. The modeling data of batch fermentation process are classified into three groups according to their stages by using fuzzy c-means clustering algorithm with biomass and substrate concentration being used as input, and then a model for the on-line identification of fermentation stages and three local soft-sensor models are developed by using neural network. The simulation results based on Nosiheptide batch fermentation process show that the estimation performance of soft-sensor model based on staged modeling method is good, and compared with soft-sensor model based on global modeling method, the estimation accuracy of the staged soft-sensor model is improved obviously and the staged soft-sensor model is more consistent with stage characteristics of batch fermentation process.

Acknowledgments. This work is supported by "the Fundamental Research Funds for the Central Universities" (No. N090302006).

References

1. Montague, G., Morris, J.: Neural-Network Contributions in Biotechnology. Trends in Biotechnology 12, 312–324 (1994)
2. Schügerl, K.: Progress in Monitoring, Modeling and Control of Bioprocesses During the Last 20 Years. Journal of Biotechnology 85, 149–173 (2001)
3. Shene, C., Diez, C., Bravo, S.: Neural Networks for the Prediction of the State of Zymomonas Mobilis CP4 Batch Fermentations. Computers and Chemical Engineering 23, 1097–1108 (1999)
4. Tholuder, A., Ramirez, W.F.: Neural Network Modeling and Optimization of Foreign Protein Production by Induced Bacterial. American Institute of Chemical Engineering Journal 45, 1660–1670 (1999)
5. Adilson, J., Rubens, M.: Soft Sensors Development for On-Line Bioreactor State Estimation. Computers and Chemical Engineering 24, 1099–1103 (2000)
6. Foss, B.A., Johansen, T.A., Sorensen, A.V.: Nonlinear Predictive Control Using Local Models Applied to a Batch Fermentation Process. Control Engineering Practice 3, 389–396 (1995)
7. Karim, M.N., Yoshida, T., Rivera, S.L., Saucedo, V.M., Eikens, B., Oh, G.S.: Global and Local Neural Network Models in Biotechnology: Application to Different Cultivation Processes. Journal of Fermentation and Bioengineering 83, 1–11 (1997)
8. Qi, Y.Z., Wang, S.X.: Biochemical Reaction Kinetics and Reactor (I). Chemical Industry Press, Beijing (1996)
9. Doan, X.T., Srinivasan, R., Bapat, P.M., Wangikar, P.P.: Detection of Phase Shifts in Batch Fermentation via Statistical Analysis of the Online Measurements: A Case Study with Rifamycin B Fermentation. Journal of Biotechnology 132, 156–166 (2007)
10. Muthuswamy, K., Srinivasan, R.: Phase-Based Supervisory Control for Fermentation Process Development. Journal of Process Control 13, 367–382 (2003)
11. Chen, Y.Q., Wang, S.Q., Chen, Q.: Multivariate Statistical Analysis for Streptomycin Fermentation. Chinese Journal of Biotechnology 15, 369–374 (1999)
12. Konstantinov, K., Yoshida, T.: Physiological State Control of Fermentation Processes. Biotechnology and Bioengineering 33, 1145–1156 (1989)
13. Moody, J., Darken, C.J.: Fast Learning in Networks of Locally-Tuned Processing Units. Neural Computation 1, 281–294 (1989)
14. Li, C.F., Wang, G.Z.: A Clustering Method and Its Application to the Modeling of RBF Networks. In: 6th International Conference for Young Computer Scientist: in Computer Science and Technology in New Century, pp. 173–177. Science Press, Beijing (2001)

High Accuracy Dynamic Adjusting Voltage Source for Cell Impedance Tomography Application

Nan Li, Hui Xu, Yang Yang, Jietao Diao, Hongshan Nie, Hongqi Yu, and Haijun Liu

ESSS Center, School of Electronic Science and Engineering
National University of Defense Technology, 410073, Changsha, China
linan@nudt-esss.com

Abstract. This paper presented a high accuracy voltage source with current limiting function. A high-speed (250 MHz), high-resolution (16 bit) dual DAC is used for generating stimulate voltage by DDS technique. The amplitude of source could be adjusted by wide bandwidth, low noise variable gain amplifiers from -11 to +51 dB. The 4-Opamp architecture is applied to realize a high gain, high-accuracy variable gain instrument amplifier circuit for tiny current sensing. By using analog comparator to convert over-current time into length of digital pulse, the FPGA could monitor the amplitude of current in real-time. Evaluation results show that system has good SFDR performance and fast response speed. With the advantages of accuracy and bandwidth, it is suitable for cell impedance tomography application.

Keywords: Cell, Voltage Source, Impedance, Tomography.

1 Introduction

The trend in modern biological and medical research is towards the micro-scale: smaller samples, faster readout, and greater detail at the level of cells and molecules. The response of cells to a chemical or biological agent in terms of their impedance changes in real-time is a useful mechanism that can be utilized for a wide variety of biomedical applications [1]. Measurement of the electrical impedance characterize of cell provides a effective means to examine biological events, and enables detection of physiological changes in cell-cell, cell matrix interactions caused by the effect of viral and bacterial infections, environmental parameters, toxicity and the effect of pharmaceutical compounds [2].

Electrical Impedance Tomography (EIT) is a medical imaging technique which displays the spatial distribution of the complex impedance inside a tissue or body [3]. It has long been used in medical diagnosis applications such as breast cancer, epilepsy, pulmonary edema and intracerebral hemorrhage [4]. Recently, EIT technique was introduced to cell-level for imaging the kinetics in the cell growth, migration, and morphology changes, .etc [5].

In cell impedance measurements, when large electric current pulses are applied to cells, thermal energy may be generated at the electrode site causing undesired heating of cells. In order to keep cells survive during the measurements, the stimulated current

G. Lee (Ed.): Advances in Automation and Robotics, Vol.2, LNEE 123, pp. 265–272.
Springerlink.com

need be limited in 100 nA level [1]. The traditional method of current limiting is connecting a large resistor in series with cell-electrodes configurations. Although it's easy-to-implement and cost-efficient, the series resistor and the parasitic capacitor in circuits will be low-pass filter which limits the system bandwidth. In order to solve this problem, we designed a high accuracy dynamic adjusting voltage source. With the current monitor function, it can limit the stimulate current in a desired level for protecting cell viability without losing bandwidth performance.

2 Hardware Design

The voltage source circuits mainly consists of a field program gate array (FPGA) device, a dual digital-to-analog convertor (DAC), buffers (BUF), reconstruction filters (FLT), variable gain amplifiers (VGA), instrument amplifier circuit (INA), analog comparator (CMP) and voltage reference (REF). A brief hardware functional diagram of circuits is shown in Fig.1.

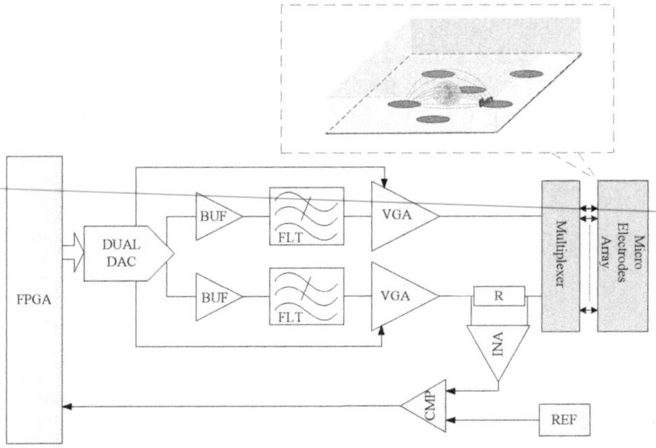

Fig. 1. Hardware architecture of dynamic adjusting voltage source

The dual DAC is used to generate a pair of DDS-based differential sine-wave excitation signals. The amplitude of excitation signals could be dynamically adjusted via variable gain amplifiers. Instrument amplifier circuit measures the differential voltage across the current sensing resistor, this voltage will compare with a user-defined threshold voltage which is set by voltage reference. FPGA could adjust the excitation voltage amplitude according to the pulse length from output of comparator.

2.1 Digital-to-Analog Convertor

In our system, we use AD9747 as signal generator which could be controlled by FPGA. The AD9747 is a high dynamic range, dual digital-to-analog converter with 16-bit resolution and sample rate of up to 250 MSPS. With its low noise and intermodulation distortion (IMD) performance, it is suitable for high quality signal synthesis application. It also integrates dual low-speed auxiliary DACs which could be used to control variable gain amplifiers. The function diagram of AD9747 is shown in Fig.2.

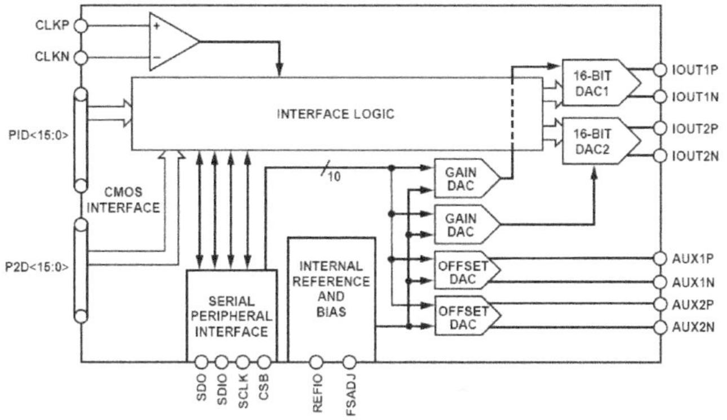

Fig. 2. Function diagram of AD9747

2.2 Reconstruction Filter

In typical DDS applications, a low-pass filter is utilized to suppress the effects of the image responses in the output spectrum. In order to keep the cutoff requirements on the low-pass filter reasonable, it is an accepted rule to limit the f_{OUT} bandwidth to approximately 40% of the f_{CLOCK} frequency. In our application, a passive 6th order elliptical filter is designed, which offers the steepest pass band to stop band transition among other filter types. The pass band of the elliptical filter is 5 MHz and the stop band is 7 MHz with 60 dB attenuation. The circuit of this filter is shown in Fig.3.

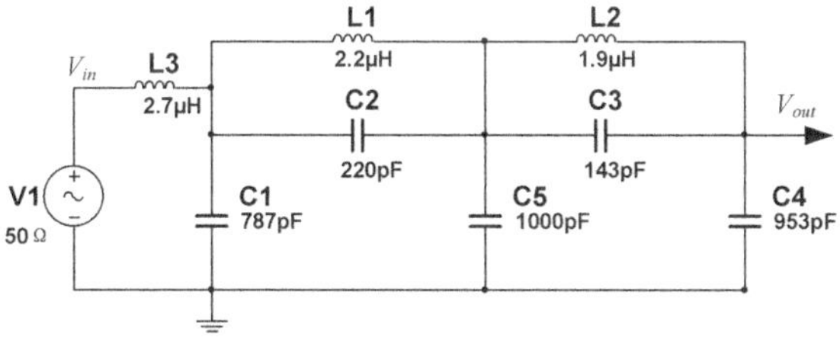

Fig. 3. Schematic of 6th order elliptical filter

2.3 Variable Gain Amplifier

For adjusting gain of voltage amplitude, we applied variable gain amplifiers in our circuits. The AD603 is a low noise (1.3nV√Hz noise spectral density), voltage-controlled amplifier which could provides accurate, variable gains of −11 dB to +31 dB with a bandwidth of 90 MHz or +9 dB to +51 dB with a bandwidth of 9 MHz.

The decibel gain is linear in dB, accurately calibrated, and stable over temperature and supply. The gain scaling is 25 mV/dB. It only requires a gain control voltage of 1 V to span the central 40 dB of the gain range. The gain control response time is less than 1 μs for a 40 dB change.

2.4 Current Sensing Circuit

In practice, due to the parasitic capacitor of circuits, the actual injecting current into cells is different from the measured current. By simulation results of a simple circuit model (Fig.4), we could see this effect intuitively. Although this error could be calibrated by software algorithms, we still need an adjustable current sensing circuit to deal with this wide dynamic range changes (from 100nA to 10uA).

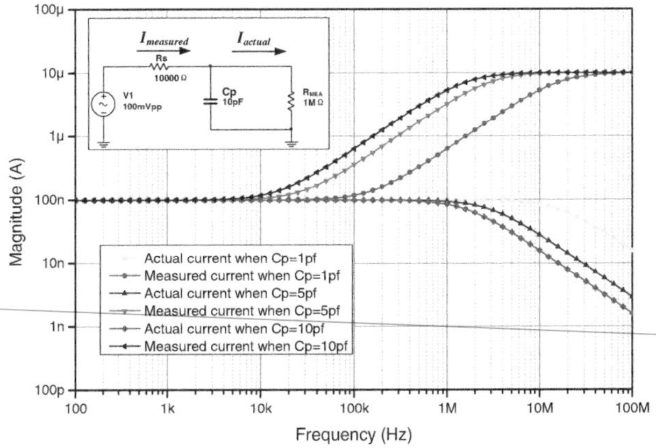

Fig. 4. The influence caused by parasitic capacitor

We applied 4-Opamp architecture to realize a high-speed, high-accuracy variable gain instrument amplifier for this tiny current sensing. This architecture provides matched, high impedance inputs so that the circuits will have a minimal effect on current path. As the Fig.5 shows, high input impedance operational amplifiers A1 and A2 constituent the input stage. A fully-differential programmable gain amplifier (PGA) is gain stage which provides wide gain adjusting range for adopting different current levels. The output stage consists of a high common rejection ratio (CMRR) differential amplifier. If the value of R5 = R6, R1 = R3, and R2 = R4, G_{A3} is the gain of PGA, then the total output voltage could be calculated by Eq.1:

$$V_{out} = (V_{in2} - V_{in1})(1 + 2\frac{R_5}{R_G})G_{A3}(\frac{R_2}{R_1}) \tag{1}$$

Digital-programmable gain amplifier AD8369 is used in this circuit. The AD8369 is a high performance PGA which provides digitally controlled variable gain from -5 dB to +40 dB in 3 dB Steps. It has excellent distortion performance: the two-tone,

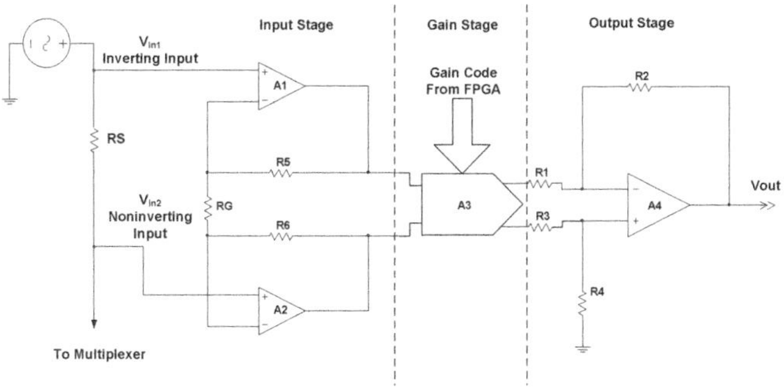

Fig. 5. Current sensing circuit

third-order inter-modulation distortion is -69 dBc at 70 MHz. Even at maximum gain, AD8369 has a nominal noise figure of 7 dB. Digital control of the AD8369 is achieved using serial interface and be driven with standard CMOS logic levels by FPGA. The overall bandwidth performance of whole current sensing circuit is shown in Fig.6.

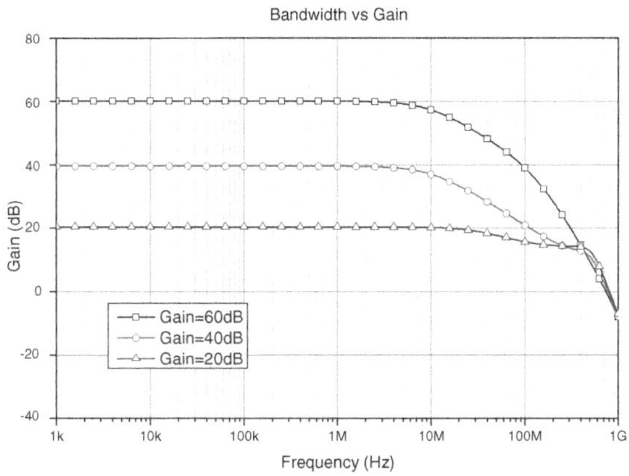

Fig. 6. Current sensing circuit bandwidth. Even at 60dB gain, the bandwidth is beyond 5 MHz.

2.5 Comparator

Analog comparator is used to compare the sensing voltage and threshold voltage for converting over-current time of excitation signal to logic signal. In our application, the frequency of excitation signal is in 10 kHz to 5 MHz, so we need a high-speed comparator to meet this bandwidth requirement. The AD8611 is a high-speed comparator with latch function and complementary output. Fast 4 ns propagation delays for rising and falling signals are closely matched and tracked over temperature, which makes that a good choice for our application.

3 Software Design

Digital dynamic current adjusting method is an easy-to-implement, high-speed, cost-efficient way to reduce the over current to safety range. In aspect of hardware design, different from traditionally ADC method, it only needs an analog comparator to monitor the amplitude of high speed signal. The FPGA could calculate the amplitude of current amplitude by internal digital signal processing resources at high speed (>100 MHz). The brief principle of this monitor is shown in Fig.7.

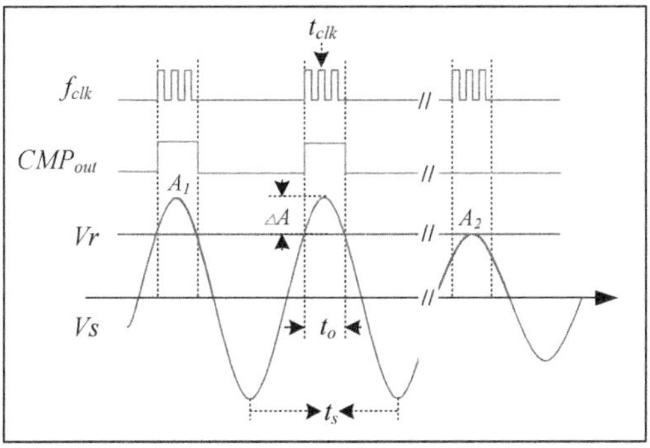

Fig. 7. Principle of digital dynamic current adjusting

As Fig.7 shows, the high-speed comparator will compare the output voltage amplitude between instrument amplifier (V_s) and the voltage reference (V_r). When V_s higher than V_r, which means that injecting current is over safety limit, the comparator will output logic signal '1' to FPGA. When V_s fall below V_r, the comparator output will turn to logic '0'. Consequently, the pulse length of logic '1' approximately equals to the time of the over-current state (t_o). This pulse length could be count by internal clock in FPGA. Assuming the clock frequency of FPGA is f_{clk}, the length of t_o can be calculated following Eq.2:

$$t_o \approx n \cdot \frac{1}{f_{clk}} = n \cdot t_{clk} \tag{2}$$

In Eq.2, the variable n is count number of counter in FPGA and t_{clk} is the period of FPGA's counting clock. Therefore, the amplitude of original over-current voltage (A_1) can be calculated by Eq.3:

$$A_1 = \frac{V_s}{\sin(2\pi f_s t)} = \frac{V_r}{\sin(2\pi f_s(\frac{t_s}{4} - \frac{t_o}{2}))} = \frac{V_r}{\cos(\pi f_s t_o)} \tag{3}$$

Once the adjusting gain value is calculated, the FPGA will send the control words to auxiliary DAC to change the gain of VGA. And the injecting current will be reduced along with the output voltage.

4 System Evaluation

4.1 Output SFDR Performance

Spurious-Free Dynamic Range (SFDR) is a critical parameter of a DDS-based voltage source. In order to test the SFDR performance, we acquired a sine-wave signal from the voltage source and use fast Fourier transform (FFT) algorithm to plot frequency spectrum on computer. The test results are shown in Fig.8. When we generated a 1 MHz stimulate signal at 250MHz clock speed, the 2nd order harmonic is about -69.6 dBm, 3rd order harmonic is about -65.1 dBm (SFDR) and the base-line noise is approximately under 75dB.

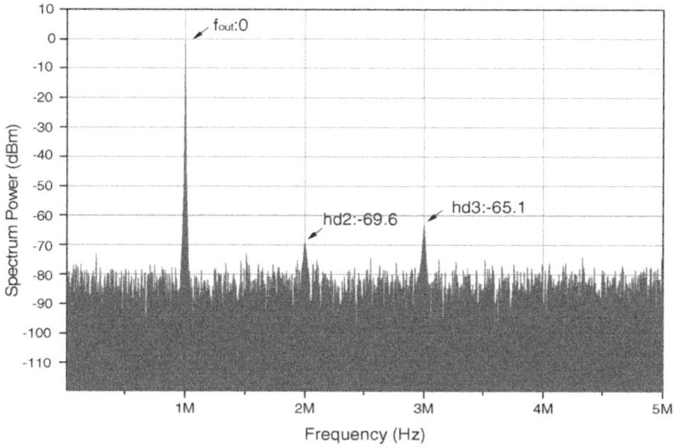

Fig. 8. SFDR Performance of voltage source

4.2 Current Adjusting Response Time

We tested current adjusting function of our design. The result is shown in Fig.9. The output voltage is 5 MHz sine wave which amplitude is suddenly changed inside FPGA to verify the response of current adjusting function. The red line is measured voltage on the output of a VGA. Graph shows that the response time is about 1.2us.

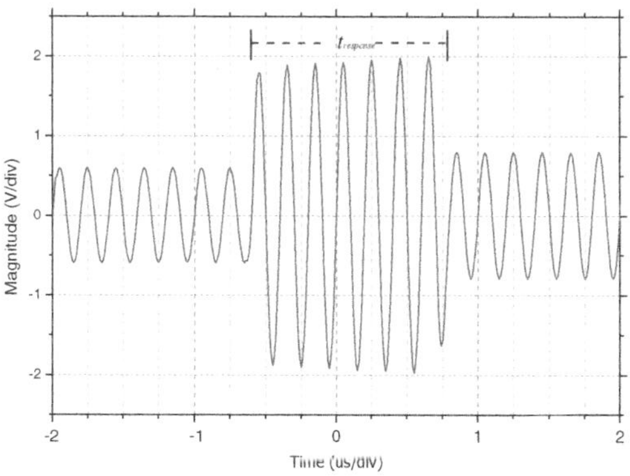

Fig. 9. Over-current adjusting response time

5 Conclusion

In order to restrict the current density magnitude to a 100nA level that does not damage cell structure or function in cell impedance tomography application, a high accuracy dynamic adjusting voltage source has been developed. It mainly consists of 16bit high-resolution DAC, low-noise variable gain amplifier and high dynamic range instrument amplifier for limiting over-current in wide frequency bandwidth with fast response. The circuit is suitable for protecting cell viability without losing detectable bandwidth. Our next step is integrating this voltage source into our cell impedance tomography system.

References

1. Zhang, M.Q., Asphahani, F., Wang, K., Thein, M., Veiseh, O., Yung, S., Xu, J.A.: Single-cell bioelectrical impedance platform for monitoring cellular response to drug treatment. Phys. Biol. 8(1) (February 2011)
2. Bhansali, S., Rahman, A.R.A., Lo, C.M.: A micro-electrode array biosensor for impedance spectroscopy of human umbilical vein endothelial cells. Sensor Actuat B-Chem. 118(1-2), 115–120 (2006)
3. Brown, B.H., Barber, D.C., Wang, W., Lu, L.Q., Leathard, A.D., Smallwood, R.H., Hampshire, A.R., Mackay, R., Hatzigalanis, K.: Multifrequency Imaging and Modeling of Respiratory-Related Electrical-Impedance Changes. Physiol. Meas. 15, A1-A12 (1994)
4. Hinz, J., Hahn, G., Quintel, M.: Electrical impedance tomography - Ready for routine clinical use for mechanically ventilated patients? Anaesthesist. 57(1), 61–69 (2008)
5. Sun, T., Tsuda, S., Zauner, K.P., Morgan, H.: On-chip electrical impedance tomography for imaging biological cells. Biosens Bioelectron. 25(5), 1109–1115 (2010)

Study on a New Approach of Face Detection under Video Environment

Qing Lin[1,2], Xiang Zhao[1], Yi-jia Xu[3], and Min Wu[4]

[1] College of Computer Science & Communication Engineering, Jiangsu University, Zhenjiang, China
[2] NanJing University of Science and Technology, NanJing, China
[3] Dct Project Department,Shanghai Automobile Gear Works, Shanghai, China
[4] Personnel Department, Jiangsu University, Zhenjiang, China
zxyangyu@163.com, milam0479@126.com, walkingmu@gmail.com, milam@ujs.edu.cn

Abstract. Face detection in the image is an important research branch of face recognition. For the purpose of detecting the faces in images efficiently, a new method of face detection, which is based on AdaBoost merging with the eye contour feature by combining statistics-based with feature-based approach, is proposed. Experimental results show that the algorithm can gain higher face detection efficiency.

Keywords: face detection, AdaBoost, Laplacian operator, sharpening, closing operation.

1 Introduction

Face detection [1] is a process that determines whether or not there exists face and the face location, size, posture in the input image. With the development of Pattern Recognition and Computer Vision technology, face detection has become an important research focus. The methods can be classified into the following categories [2]:

Approach based on the feature: The advantage of such method is its fast speed in face detection, and the disadvantage is the unstable detection results under complex situation, so the biggest difficulty is to find a relatively stable characteristic.

Approach based on the template matching: Template itself is a rough model, so the detection accuracy of these methods is not high and the amount of calculation in such methods is great.

Approach based on the statistical: The advantages of such method are as following: (1) depending on none of the prior knowledge and face model parameters, they can avoid errors causing by inaccurate or incomplete knowledge; (2) they can expand the range of detection mode and improve the detection robustness of the system by increasing the learning examples. The disadvantage of the method is that diverse "non-face" samples which are obtained by collecting the misclassified samples from the original classifier and adding them into the "non-face" sample to train iteratively, occupy a large space, require more iteration, and make convergence very difficult.

G. Lee (Ed.): Advances in Automation and Robotics, Vol.2, LNEE 123, pp. 273–280.
Springerlink.com

The face detection methods under video environments need to adapt to complex environments to meet the real-time and high accuracy. Any kind of face detection approaches mentioned above is hard to meet the challenge completely. In this paper, according to advantages and disadvantages of all the approaches mentioned above, it proposes a method based on AdaBoost merging with the eye contour feature by combining statistics-based with feature-based approach. Main steps are as follows: firstly, face detection with AdaBoost is initially used for video images. Secondly, local image sharpening is made to the preliminary detection results, and the human eyes will appear if they exist in the image. An approach by combining the gray-scale distribution property with the Laplacian operator is given out in the operation of the local sharpening. After the image binarization and morphological operations, the similarity of contour area and the position of the contour are used to determine whether or not there are human eyes. Once the human eyes exist, the detected image can be considered as a face image. The experiment results show the method proposed in this paper has a good performance.

2 Adaboost Algorithm

In 2001, Viola and Jones [3] proposed a face detection method based on AdaBoost algorithm and established the first true real-time face detection system that is an effective and fundamental solution to the problems in human face detection speed. The basic idea of AdaBoost algorithm is that a large number of classifications and capacity of the weak classifier are stacked up to generate a strong classifier classification capability by a certain way. Algorithm reference [4][5][6] is described as follows:

Given a series of training samples as $(x_1, y_1),...,(x_n, y_n)$, where $y_i = \{0,1\}$ $(i = 1,2..., n)$ is corresponding to the false and true samples, n is the total number of training samples. The initial distribution of the sample is $w_{1,i} = \frac{1}{2m} \cdot \frac{1}{2l}$, then cycle the following steps for $t = 1,2,3,...,T$.

1) Normalize the weight factor, the equation is obtained as $w_{t,i} = \frac{w_{t,j}}{\sum_{j=1}^{n} w_{t,j}}$, and w_t constitutes a sample distribution.

2) The eigenvalue f_j of each input characteristic has a simple binary classifier. The J-characteristic of the weak classifier is composed by a threshold θ_j, a eigenvalue f_j and a direction of bias directed inequality p_j (only two cases ± 1) . The formula as follows:

$$h_j = \begin{cases} 0, & p_j f_j < p_j \theta_j \\ 1, & \text{otherwise} \end{cases} . \tag{1}$$

3) In a sample distribution w_t, the error of each weak classifier is described as $\varepsilon_j = \sum_{i=1}^{n} w_i \left| h_j(x_i) - y_i \right|$.

4) Selecting $h_i(x) = h_k(x)$, where $k = \arg\min_j(\varepsilon_j)$, $\alpha_i = \frac{1}{2}\ln(\frac{1-\varepsilon_k}{\varepsilon_k})$.

5) Updating the weights, $w_{i+1,j} = w_{i,j}\beta_i^{1-e_i}$, where $\beta_i = \varepsilon_i/(1-\varepsilon_i)$. If the sample i was correctly classified, then $e_i = 0$, otherwise $e_i = 1$.

6) After training for T times, a strong classifier which consists of T weak classifier in parallel is formed.

$$h(x) = \begin{cases} 1, & \sum_{i=1}^{T} \alpha_i h_i(x) \geq 0.5 \sum_{i=1}^{T} \alpha_i \\ 0, & \text{ohterwise} \end{cases}, \tag{2}$$

α_i in equation (2) is described as $\alpha_i = \log(1/\beta_i)$.

Classical AdaBoost algorithm is based on Haar features and it is one of the algorithms with the highest correct rates. Figure1 shows the results of the face detection of two frames in a video by the algorithm. Human faces in two images have been seen from the test results, but some non-faces are mistaken for human face, which is due to the shortcomings of the statistics-based approach itself. The next operation will combine feature-based approach to cut non-face region in detection results based on AdaBoost algorithm.

Fig. 1. Results of Adaboost Algorithm

3 Eye Contour Positioning

Since the human eye in the face has the obvious individuality and distribution characteristics, the characteristics of the human eye are used in this paper. The existence of the human eye is determined by the similarity of the eye contour areas and their location in the image. The process of face detection is completed by cutting non-face region, which will be described in subsequent sections.

3.1 Image Preprocessing

Image definition in a video is usually not enough, so the image is enhanced basically by using equalization. In order to get better effect in locating the eyes, the image noise need to be got rid. Gaussian filtering method is chosen in this paper.

3.2 Local Sharpening

Sharpening is designed to enhance the definition of eyes area as much as possible in the paper, but the experiment shows that a negative impact is occurred on the region of eyes in the binary image after sharpening, thus, a combination of the gray-scale distribution property with the Laplacian operator is proposed to get the local sharpening effect.

3.3 Laplacian Operator

Laplacian, with the property of rotation-invariant, is the simplest isotropic derivative operator. It [7] use highlights gray-level discontinuities in an image and deemphasizes regions with slowly varying gray levels. This will tend to produce images that have grayish edge lines and other features can be recovered while still preserving the sharpening effect of the Laplacian operation simply by adding the original and Laplacian images. For image functions $f(i, j)$, the discrete function of the Laplacian is defined as:

$$\nabla^2 f(i,j)=f(i+1,j)+f(i-1,j)+f(i,j+1)+f(i,j-1)-4f(i,j)\,. \tag{3}$$

Equation (3) can be implemented by using the mask shown in Figure 2(a), which gives an isotropic result for rotations in increments of 90°. When the diagonal directions are incorporated in the definition of the digital Laplacian, (3) changes as follow:

$$\begin{aligned}\nabla^2 f(i,j)=&f(i+1,j-1)+f(i+1,j)+f(i+1,j+1)+f(i-1,j+1)+\\&f(i-1,j)+f(i-1,j-1)+f(i,j+1)+f(i,j-1)-8f(i,j)\end{aligned}\,. \tag{4}$$

Equation (4) can be implemented by using the mask shown in Figure 2 (b), which gives an isotropic result for rotations in increment of 45°. The other two masks shown in Figure 2(c) and (d) are frequently used in practice.

$$\begin{bmatrix} 0 & 1 & 0 \\ 1 & -4 & 1 \\ 0 & 1 & 0 \end{bmatrix} \begin{bmatrix} 1 & 1 & 1 \\ 1 & -8 & 1 \\ 1 & 1 & 1 \end{bmatrix} \begin{bmatrix} 0 & -1 & 0 \\ -1 & 4 & -1 \\ 0 & -1 & 0 \end{bmatrix} \begin{bmatrix} -1 & -1 & -1 \\ -1 & 8 & -1 \\ -1 & -1 & -1 \end{bmatrix}$$

(a) 4 neighborhood sharpening mask (c) 4 neighborhood sharpening mask

(b) 8 neighborhood sharpening mask (d) 8neighborhood sharpening mask

Fig. 2. Laplacian mask

3.4 Gray-Scale Distribution Property

The core area of the face, which has its unique gray-scale distribution, includes glasses, nose and mouth. Certain rules can be established on the basis of the gray-scale distribution property and then the detecting to the human face in the image is transformed into detecting to the image blocks which meet those rules very well.

3.5 Algorithm Description

Eyes gray value is lower than the average gray value in the image. In the local sharpening operation, the gray average is calculated by a mask with the size of 3×3. If the gray average of the eye area is not lower than that of the whole image, there is nothing to happen; otherwise, the region will be sharpened by Laplacian operator in Figure 2(d). Figure 3 shows that the eye contour in the image after local sharpening is more obvious than the one obtained by sharpening to the whole image.

(a) result of overall sharpening (b)result of local sharpening

Fig. 3. Results of sharpening

3.6 Binarization

Image binarization is intended to highlight the area of the eyes, and the key of the binarization processing is to choose an appropriate threshold. The principle of threshold selection is to choose the gray value which is close to the eye's gray value and a variable threshold is used in this paper. The gray values of the image are arranged in a descending sequence as $\{T_{-N}, T_{-N+1}, T_{-N+2}, \ldots, T_0, T_1, T_2, \ldots T_N\}$, where T_{-N} and T_N are the minimal and maximal gray values, respectively. $T_0 = \sum_{i=-N}^{N} T_i$ is defined as the gray value of the image. The threshold is set as $T_0 \times a$ where a $(0<a<1)$.

3.7 Closing Operation

As the gray values in certain parts of the eyebrow, nose and mouth are close to that of the eyes, the binary face image may contain the eyes as well as the residual images of the eyebrow, nose and mouth. The morphological closing operation is used in the paper to eliminate such noise impact on the follow-up eye detecting. The residual images of the eyebrow, nose and mouth disappear after the dilating operation, but the region of eyes becomes smaller. And then the erode operation for expanding the region of eyes is selected to detect the eyes in a convenient way.

The dilate operation works in a rectangular structure with the size of 2×2 and the erode operation runs in a cross structure with its size of 2×2, and other kinds of structures make little influence in the paper. The results of the close operation can be seen in Table 1, which shows that the most obvious eye contour is just detected out by the close operation.

Table 1. Results of closing operation

	Binarization	Dilate Operation	Erode Operation
a=0.3			
a=0.6			
a=0.9			

3.8 Locating the Eyes

The detected results contain more than one contour, so assuming that the contour sequence is as $\{c_0, c_1, c_2, ..., c_N\}$, the sequence construction and the area of each contour can be acquired through the algorithm provided by OPENCV. These areas constitute a sequence as $\{area_0, area_1, area_2, ... area_N\}$.

According to the structural characteristic of the eye, that is, the areas of two eyes are similar and two array elements are compared successively from the largest to the smallest contour area .When the area ratio is within a certain threshold, two eyes can be detected out except that several groups of the contours are approximately equal in size.

Operations provided by OPENCV can not only detect out the contour area, but also obtain the coordinates of any point on the contour. If the coordinates of any two points on the contours are (x_i, y_i) and (x_j, y_j), where $x_i \in (0, \frac{w}{2}), y_i \in (\frac{2}{5}h, \frac{3}{5}h)$, $x_j \in (\frac{w}{2}, w), y_j \in (\frac{2}{5}h, \frac{3}{5}h)$. w and h are the width and height of the region of interest in the image respectively and they are the relative coordinates in practice. The areas of the two contours are approximately equal and thus the two contours can be determined as the human eyes.

4 Experiment Results

UF-5307 surveillance network camera with the resolution of 640×480 is chosen in recording the tested video. Some experiment results obtained by such algorithm have been mentioned in the above sections and certain results are shown in Figure 4 which can provide better detecting results compared with Figure1.

Fig. 4. Experiment results

5 Conclusion

The method based on AdaBoost merging with the eye contour feature has obtained some good results, but it also has certain problems. The number of the contour in the video image increases after the binarization and the area decreasing makes a great impact on the positioning of the human eyes contour under the condition of strong lights. Thus, the way to reduce the negative impact on the positioning of the human eyes contour when the light is strong will be the next focus of the study. In addition, adaptively adjusting the threshold on the basis of the changes in the environment of video image and obtaining an ideal binary image will also be one of the priorities of the follow-up.

Acknowledgments. This work is partially supported by National Natural Science Foundation of China Under Grant No.61170126 and Jiangsu Provincial Natural Science Foundation (BK2009199 and BK2010339).

References

1. Wei, C., Sun, T.: Face Detection Based on Half Face-template. In: ICEMI, pp. 54–58 (2009)
2. Tian, Y., Yu, F.-Q.: A Survey on Human Face Detection. Computer Security, 76–79 (May 2009)
3. Viola, P., Jones, M.: Rapid object detection using a boosted cascade of simple features. In: Proc IEEE Conference on Computer Vision and Pattern Recognition, Kauai, Hawaii, USA, pp. 73–113 (2001)
4. Ruan, J., Yin, J.: Multi-pose face detection based on facial features and AdaBoost algorithm. Journal of Computer Applications 30(4), 967–970 (2010)
5. Guo, L., Wang, Q.: Research of Face Detection Based on Adaboost Algorithm and OpenCV Implementation. Journal of HARBIN University of Science and Technology 14(5), 123–126 (2009)
6. Qian, Z., Xu, D.: Fast Adaboost Training Algorithm. Computer Engineering 35(20) (October 2009)
7. Rafael, C., Gonzalez, Richard, E., Woods: Digital Image Processing, 2nd edn. (2002)

A Novel Automated Experimental Approach for the Measurement of On-Chip Speed Variations through Dynamic Partial Reconfiguration

Hasan Baig[1], Jeong-Gun Lee[2], and Jeong-A Lee[1,*]

[1] Chosun University
[2] Hallym University, South Korea
hasan.baig@chosun.kr, jalee@chosun.ac.kr
jeonggun.lee@hallym.ac.kr

Abstract. In this research we have developed a complete automated experimental setup for the measurement of on-chip delay variations through dynamic partial reconfiguration. The experiment is performed on two different Virtex-5 devices on the basis of which intra-die and inter-die speed comparisons have been made. We developed an on-chip sensor map consisting of 60 sensors out of which only alternating sensors remains active at a time. This is done to avoid affects of inter-sensor heat dissipation. Once finished collecting the data, on-chip processor communicates with the sensor hardware to extract and encode the data, and send it to the online Graphical User Interface implemented on the LabVIEW platform. Afterwards, the processor dynamically loads another bitstream from the Compact Flash Memory Card to activate another alternate group of sensors to perform remaining analysis. While using the same sensor topology we found the intra-die speed variations of up to 6-10%. However, inter-die speed comparison result depicts that one chip is 2-10% faster than another one.

Keywords: Automation, FPGA, Dynamic Partial Reconfiguration, delay measurement, on-line sensing, reliability, ring oscillator.

1 Introduction and Related Work

Field Programmable Gate Arrays (FPGAs) based devices are now an alternative to design high-performance digital systems. Like any other semiconductor device, FPGAs based design's speeds are also degraded – it depends on many factors that could affect the speed of a digital circuit implemented on FPGA. These factors includes environmental/on-chip temperature, operating voltage and also due to process variations. Kenneth *et al.* [1] developed a system consisting of MicroBlaze processor which was used to supervise the complete hardware. The FPGA was statically reconfigured each time to place the sensor at different locations. Their approach was very time consuming and frantic.

* Corresponding author. This work was supported in part by the National Research Foundation of Korea (NRF) grant funded by the Korean government (MEST) (No. 2010-0009303).

G. Lee (Ed.): Advances in Automation and Robotics, Vol.2, LNEE 123, pp. 281–290.
Springerlink.com

The best approach to solve this problem is to use partially reconfigurable design. Partial reconfiguration is the practice of reprogramming only a portion of an FPGA. More precisely, dynamic partial reconfiguration denotes the ability to reprogram a portion of a circuit while it is operating. Majority of the FPGA systems have proposed hundreds of applications of dynamic partial reconfiguration. This research work is one of the best applications of dynamic partial reconfiguration in which speed variations throughout the chip can be observed effectively in just two runs.

2 Proposed Work

The use of ring oscillator is a standard technique for measuring variation in semiconductor fabrication processes, because of its easy on-chip implementation and high sensitivity to process parameters [2],[3],[4]. Li *et al.* reported the use an array of ring oscillators in an FPGA to measure variation in the gate length critical dimension [5]. The results obtained were then used to improve the fabrication process to reduce variations. However, useful information was not been disclosed and the experimental techniques were also not clearly expressed. Katsuki et al. presented at-speed testing model in a custom 90nm LUT array for measuring within-die variations [6]. Much of the literature on testing delays in FPGAs focus on revealing delay faults [7], [8].

In this research, we followed the traditional approach of measuring the delay variations using Ring Oscillator (RO) as a probe. The experimental system comprises of on-chip hardware including MicroBlaze Processor Unit, custom designed sensor interfaced with MicroBlaze through PLB, embedded software to communicate with sensor and online software designed on LabVIEW platform for real time data analysis through UART communication protocols. The experiments are performed on 65nm Virtex-5 (XC5VLX110T) FPGA devices at XUPV5-LX110T platform.

3 Experimental Setup

The automated experimental system presented here is divided into two categories; on-chip hardware and software, and online LabVIEW Graphical User Interface (GUI). We will first discuss the on-chip implemented setup and then will discuss the GUI.

3.1 On-Chip Implemented Hardware

The top level diagram of on-chip system implementation is shown in Fig. 1, which clearly depicts two different sensor maps shown in red and green. Eight extra sensors are incorporated in static module (central MicroBlaze area) to cover the centre part of the die. These eight sensors, inside MicroBlaze boundary (Fig. 1 (a)), are not partially reconfigurable and will remain on the chip throughout the experiment. When green sensors are activated, red sensors are disabled (reconfigured as a black box) or vice versa. Fig. 1 (b) elaborates the clear picture of how neighboring sensor heat dissipation doesn't affect each other. The map is designed in such a manner that neighboring sensor will always be disabled while taking the observations from one set of sensor arrays. In this case (Fig. 1(b)), red sensors are loaded as a black box.

The entire implemented system block diagram is shown in Fig. 2 which also depicts the components of designed sensor. For clarity only one sensor is shown which consists of a Ring Oscillator (RO), a counter and a timer.

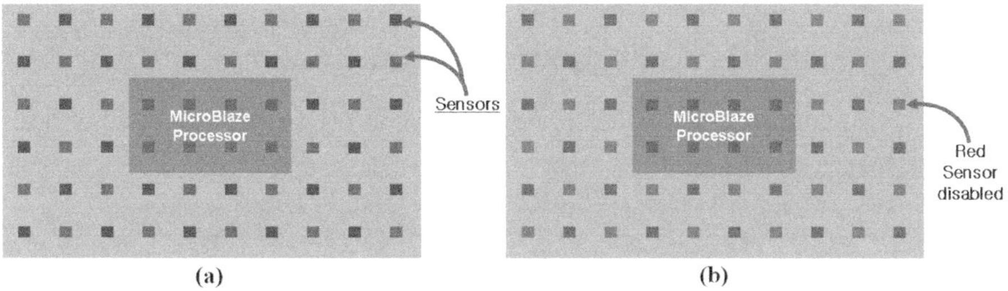

(a) Two sensor arrays. (b) Green sensors are active and red map is disabled.

Fig. 1. Implemented on-chip system

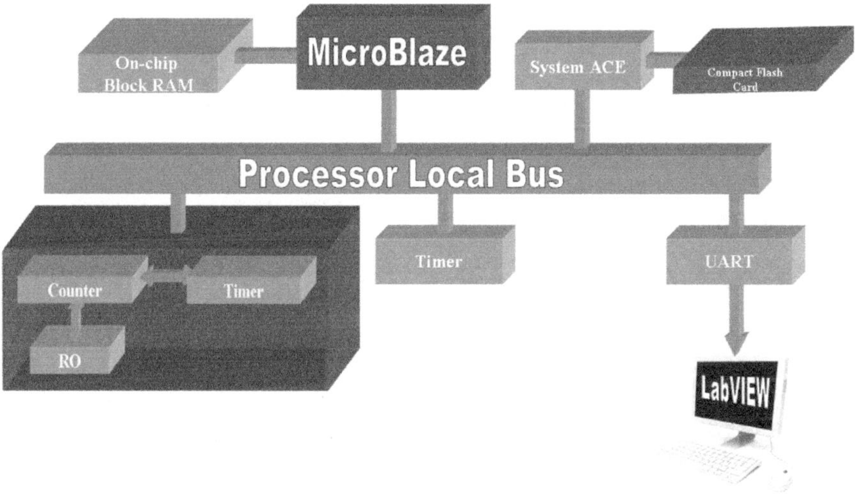

Fig. 2. Complete Implemented System: Designed sensor along with its components

Virtex-5 devices comprises of 160x54 CLBs. Each sensor is accommodated in 5x4 CLBs. Ring oscillator used in this research is based on 3 inverters, each being implemented in one-input LUT and a controlling AND gate implemented in 2-input LUT (depicted in Fig. 2). Comparatively, ring oscillators for thermal sensing usually use higher number of stages including 48 [9] and 75 [10]. The reason is to minimize self-heating since the slow and sparse design has lower heat density than a fast and dense one. Since, we are interested in finding out the maximum efficiency of FPGA - this is why we utilized possible least number of inverters in RO design.

3.2 On-Line Graphical User Interface (GUI)

GUI is developed on LabVIEW platform to automate the whole experimentation. GUI developed here for this project is a powerful and unique in a way that a user can see which sensor map (partial bitstream) is running on board graphically. The screenshot of GUI is shown in Fig. 3. It consists of three different tabs (Fig. 3) which are;

- Process monitor and control
- Statistical Analysis
- Graphical Comparison

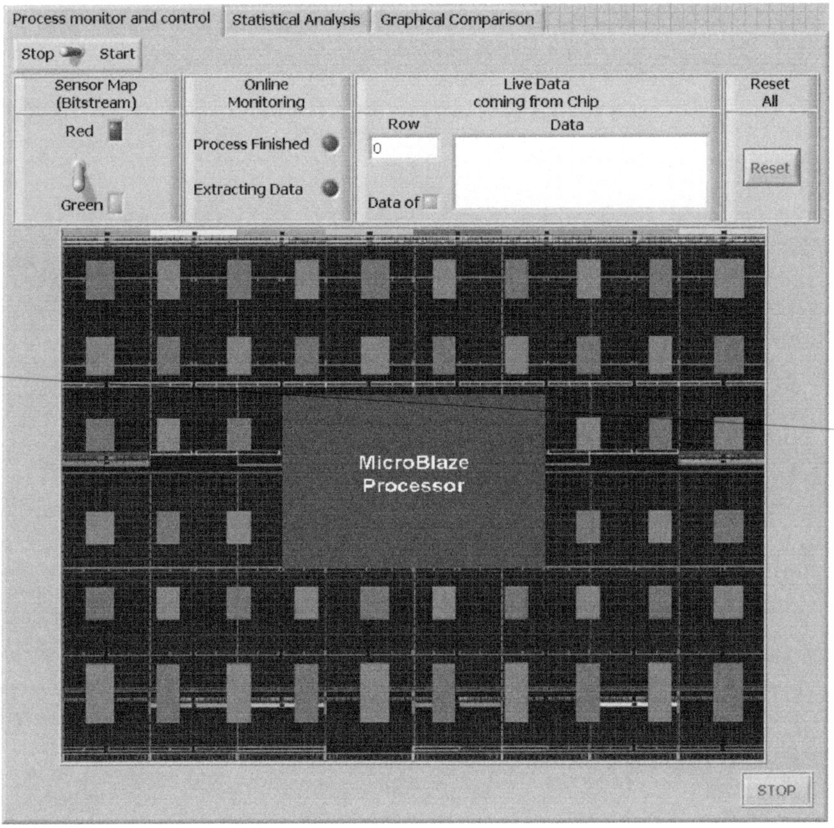

Fig. 3. Graphical User Interface of implemented system

Process monitor and control tab consists of two different jpeg images of the chip. Purpose of these images is to indicate which sensor map (or a bitstream) is currently running on chip. Fig. 3 indicates the partial bitstream, which is currently loaded and running on chip, is the partial bitstream of green sensors array. MicroBlaze loads this bitmap (default configuration) when the process first starts. Soon after the process finishes (Section 4) an indicating signal is generated from MicroBlaze processor that can be noticed on GUI's "Process Finished" LED indicator. After that GUI acknowledge MicroBlaze and sends a request to begin start sending data row-by-row. The live incoming data appears in a window illustrated in first tab of GUI and store it in a temporary buffer. Bitstream of red sensors now can be loaded manually with the help of a toggle switch (Fig. 3). MicroBlaze acknowledges GUI when bitstream is successfully loaded on chip. This signal in turn fed to GUI which changes the image showing the array of red sensors as an active one.

4 ASMD of Embedded Software

The brief Algorithmic State Machine Description (ASMD) of on-chip embedded software is pictured in Fig. 4. Each on-chip sensor iterates for the 5 times and stores its data in a separate register. All sensors run in parallel, for a second, in each iteration and stores observed data in a soft register (that are accessible to both hardware and embedded software) whose generalized form is shown in Equation 1.

$$\mathrm{SoftregXY}_i = \mathrm{CountXY} \quad (1)$$

Where,
X is the number of row.
Y is the number of sensor in row X.
i is the number of iteration.

For example, $\mathrm{Softreg23}_3$ denotes the register holding 3rd iteration data of the sensor # 3 placed in a second row of active sensor map. If we combine both the maps, this sensor will be considered to be placed at (X,Y) = (2,6), see Fig. 3. GUI takes care of this convention while generating the table of Statistical analysis. The reason for assigning this convention is to use same set of registers for both partial bitstreams. Five observations are taken consecutively each after a delay of 1 sec to figure out the affects of speed variations due to self heating. MicroBlaze first encodes the data in a frame and then sends to GUI row-by-row. Data packet used in this experiment is mentioned below;

Table 1. The standard data packet used in this experiment

Map	Row	Sensor	Data1	Data2	Data3	Data4	Data5	Sensor	...	EOP

In the packet frame, depicted in Table 1, *Map* corresponds to the partial bitstream. Red denotes the partial bitstream of sensors map depicted in Red in Fig. 1. and vice versa for the green. *Row* indicates the row number of which the data is incorporated in a current frame. *Sensor* field points to the number of sensor the next five fields of iterated data belongs to. Since each row has 5 sensors, thus the frame contains 33 comma-separated fields in total.

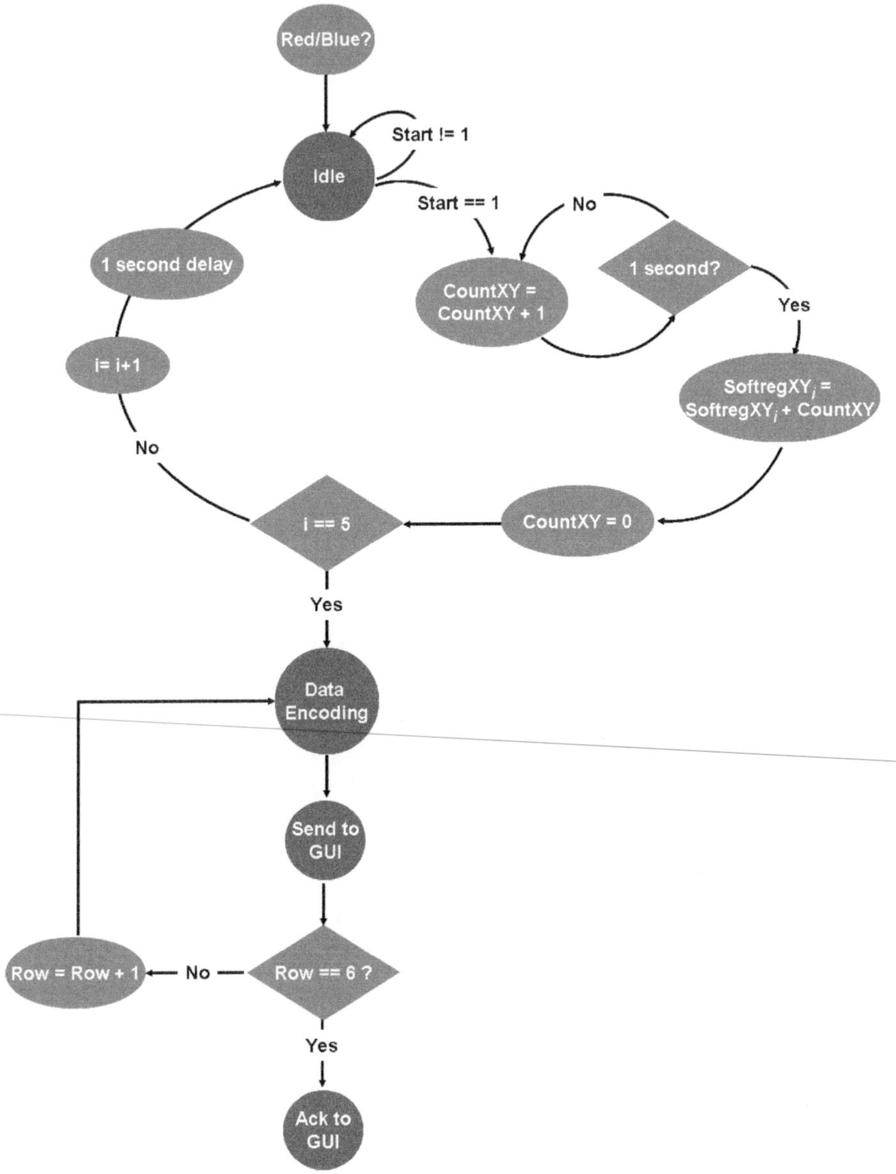

Fig. 4. ASMD of on-chip embedded software

5 Experimental Observations

We have compared the speed variations of a die in two different ways. We made a separate comparison of inter-die and intra-die speed variations that we will discuss alongside. All the readings have been taken at ambient temperature. The internal die temperature and voltage, reported by System Monitor (ChipScope Pro) was 66.5 ^{0}C and 0.998V respectively.

We kept the topology of RO constant throughout the die using RLOC attribute. Also the hierarchy of logic gates implemented in LUTs was kept constant (top-to-down) using KEEP_HIERARCHY attribute, as shown in Fig. 5. The average expected RO frequency turned out to be 470 MHz according to the "Post-Place and Route" net delays estimated by XILINX ISE tool. This auto routing, shown in Fig. 5, though not much, but somehow contributes a part in intra-die speed variation.

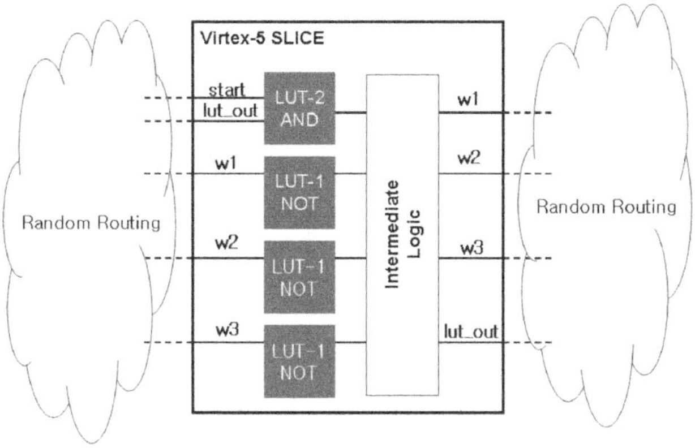

Fig. 5. Virtex-5 Slice – depicting maintained top-to-down hierarchy of RO-logical units and its surrounding random routing

5.1 Statistical Analysis and Comparison

Five readings were taken from each sensor (Fig. 4) after a delay of one second. GUI algorithm performs the statistical analysis, according to equation 2 and 3, to find out the mean and deviation of all five observations taken at each location.

$$\mu = \sum_{i=0}^{4} \frac{x_i}{5} \tag{2}$$

$$\sigma = \sqrt{\sum_{i=0}^{4} \frac{(x_i - \mu)^2}{5}} \tag{3}$$

It then generates a table of observation exactly in a same manner as the sensors are distributed on each die (See Fig. 1 (a)), and reflects the results on the second tab of GUI, *Statistical Analysis* as shown in Fig. 6.

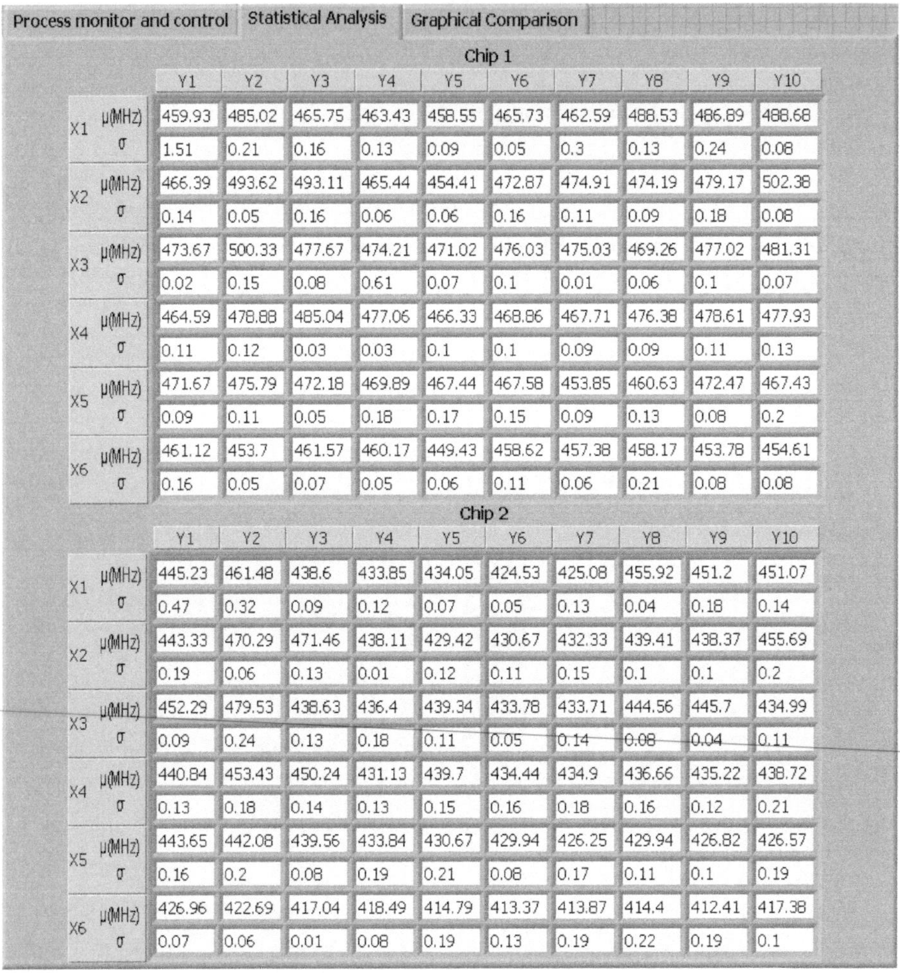

Process monitor and control | Statistical Analysis | Graphical Comparison

Chip 1

		Y1	Y2	Y3	Y4	Y5	Y6	Y7	Y8	Y9	Y10
X1	μ(MHz)	459.93	485.02	465.75	463.43	458.55	465.73	462.59	488.53	486.89	488.68
	σ	1.51	0.21	0.16	0.13	0.09	0.05	0.3	0.13	0.24	0.08
X2	μ(MHz)	466.39	493.62	493.11	465.44	454.41	472.87	474.91	474.19	479.17	502.38
	σ	0.14	0.05	0.16	0.06	0.06	0.16	0.11	0.09	0.18	0.08
X3	μ(MHz)	473.67	500.33	477.67	474.21	471.02	476.03	475.03	469.26	477.02	481.31
	σ	0.02	0.15	0.08	0.61	0.07	0.1	0.01	0.06	0.1	0.07
X4	μ(MHz)	464.59	478.88	485.04	477.06	466.33	468.86	467.71	476.38	478.61	477.93
	σ	0.11	0.12	0.03	0.03	0.1	0.1	0.09	0.09	0.11	0.13
X5	μ(MHz)	471.67	475.79	472.18	469.89	467.44	467.58	453.85	460.63	472.47	467.43
	σ	0.09	0.11	0.05	0.18	0.17	0.15	0.09	0.13	0.08	0.2
X6	μ(MHz)	461.12	453.7	461.57	460.17	449.43	458.62	457.38	458.17	453.78	454.61
	σ	0.16	0.05	0.07	0.05	0.06	0.11	0.06	0.21	0.08	0.08

Chip 2

		Y1	Y2	Y3	Y4	Y5	Y6	Y7	Y8	Y9	Y10
X1	μ(MHz)	445.23	461.48	438.6	433.85	434.05	424.53	425.08	455.92	451.2	451.07
	σ	0.47	0.32	0.09	0.12	0.07	0.05	0.13	0.04	0.18	0.14
X2	μ(MHz)	443.33	470.29	471.46	438.11	429.42	430.67	432.33	439.41	438.37	455.69
	σ	0.19	0.06	0.13	0.01	0.12	0.11	0.15	0.1	0.1	0.2
X3	μ(MHz)	452.29	479.53	438.63	436.4	439.34	433.78	433.71	444.56	445.7	434.99
	σ	0.09	0.24	0.13	0.18	0.11	0.05	0.14	0.08	0.04	0.11
X4	μ(MHz)	440.84	453.43	450.24	431.13	439.7	434.44	434.9	436.66	435.22	438.72
	σ	0.13	0.18	0.14	0.13	0.15	0.16	0.18	0.16	0.12	0.21
X5	μ(MHz)	443.65	442.08	439.56	433.84	430.67	429.94	426.25	429.94	426.82	426.57
	σ	0.16	0.2	0.08	0.19	0.21	0.08	0.17	0.11	0.1	0.19
X6	μ(MHz)	426.96	422.69	417.04	418.49	414.79	413.37	413.87	414.4	412.41	417.38
	σ	0.07	0.06	0.01	0.08	0.19	0.13	0.19	0.22	0.19	0.1

Fig. 6. Frequency statistics chart generated by GUI for both Chip 1 and Chip 2

It can easily be observed that Chip 1 is faster than Chip 2 which reflects non-uniformity in fabrication process and thus results in inter-die speed variations. The speed variations between neighboring sensing block varies from 6-10% that could be the combine effect of process variations and auto (random) routing performed by the tool. However, variations in successive readings (represented as standard deviation) can not be the affect of auto-routing algorithm because once the bitstream is generated; the only source of speed variations is either self-heating of sensor that degrades the speed or due to process variations from which die has already been affected during fabrication process. There is no special pattern found in consecutive readings from each location.

GUI interface is also capable to represent the extracted data in the form of 3D Histograms and 2D surface plots as shown in Fig. 7.

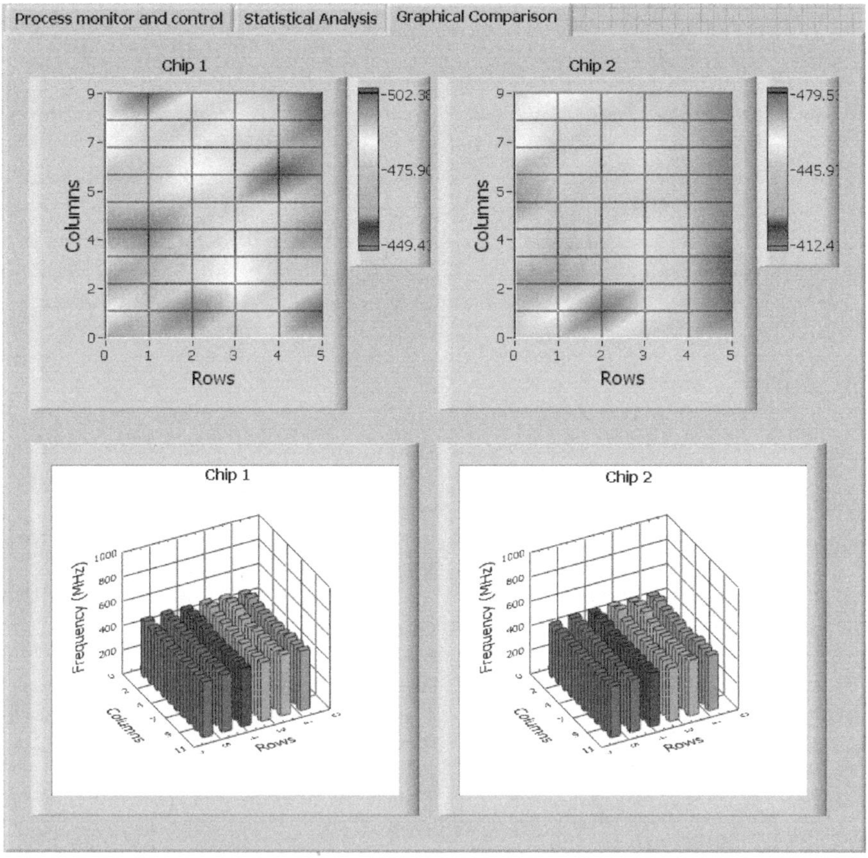

Fig. 7. Graphical plotting performed by GUI

The surface plots, pictured in Fig. 7. shows that a frequency of Chip 1 varies between 449MHz and 502MHz whereas on the other hand speed of Chip 2 varies between 412MHz and 479MHz. More precisely, Chip 1 is 3-10% faster than Chip 2.

6 Conclusions and Future Enhancements

In this research we presented a fully automated system to measure on-chip speed variations. We developed two sensor maps which run after each other through dynamic partial reconfiguration. All the sensors in each of the map run in parallel and sense the speed of FPGA die at the specific locations they are placed on. Graphical User Interface developed for this experimentation efficiently monitors the on-chip process. The designed GUI is smart enough to decode the incoming data from each sensor map and tabularized the results of each sensor according to its physical location on die.

Experimental observations points to the fact that speed varies differently in each die as an inter-die speed variation of 3-10% has been noticed. We also observed intra-die speed variation of about 6-10% which is the combine effect of self heating and process variations. Auto-routing also play an important role in the speed variations between neighboring sensing block. Longer the wire between neighboring components a tool makes, longer will be the routing delay evidently.

This research work can be further enhanced at both chip and host level. Sensors can be improved to have uniform routing at each location. This can be done through hard macros. Static sensors placed within MicroBlaze boundary can be designed partially reconfigurable that could be a challenging task. Alongside, GUI can be further improved to graphically indicate on-chip hotspots.

References

1. Zick, K.M., Hayes, J.P.: On-Line Sensing for Healthier FPGA Systems. In: ACM/SIGDA International Symposium on Field Programmable Gate Arrays, FPGA 2010 (February 2010)
2. Bhushan, M., Gattiker, A., Ketchen, M.B., Das, K.K.: Ring oscillators for CMOS process tuning and variability control. IEEE Trans. Semicond. Manuf. 19(1), 10–18 (2006)
3. Panganiban, J.: A Ring Oscillator Based Variation Test Chip. In: M.Eng. thesis, Dept. Elect. Eng. and Comput. Sci. MIT, Cambridge (2002)
4. Ketchen, M., Bhushan, M.: Product-representative "at speed" test structures for CMOS characterization. IBM J. Res. and Dev. 50(4/5), 451–468 (2006)
5. Li, X.-Y., Wang, F., La, T., Ling, Z.-M.: FPGA as process monitor–an effective method to characterize poly gate CD variation and its impact on product performance and yield. IEEE Trans. Semiconduct. Manufact. 17(3), 267–272 (2004)
6. Katsuki, K., Kotani, M., Kobayashi, K., Onodera, H.: Measurement results of within-die variations on a 90nm LUT array for speed and yield enhancement of reconfigurable devices. In: Proc. Asia and South Pacific Design Automation Conference (2006)
7. Abramovici, M., Stroud, C.E.: BIST-based delay-fault testing in FPGAs. Journal of Electronic Testing: Theory and Applications 19(5), 549–558 (2003)
8. Girard, P., Héron, O., Pravossoudovitch, S., Renovell, M.: High quality TPG for delay faults in look-up tables of FPGAs. In: Proc. IEEE International Workshop on Electronic Design, Test and Applications (2004)
9. Jones, P.H., et al.: Adaptive thermoregulation for applications on reconfigurable devices. In: Int'l Conf. on Field-Programmable Logic and Applications, pp. 246–253 (August 2007)
10. Chen, P., Shie, M., Zheng, Z.-Y., Zheng, Z.-F., Chu, C.: A fully digital time-domain smart temperature sensor realized with 140 FPGA logic elements. IEEE Trans. Circuits and Systems I 54(12), 2661–2668 (2007)

Low-Dimensional Local Self-Similarity Descriptor for Image Matching

Jingneng Liu and Guihua Zeng

State Key Laboratory of Advanced Optical Communication Systems and Networks, Department of Electronic Engineering, Shanghai Jiaotong University, Shanghai 200240, China
bzljn@163.com, ghzeng@sjtu.edu.cn

Abstract. The advanced descriptor, i.e., Local Self-Similarities (LSS), is successfully adopted for object classification and scene recognition. However, it is only invariance against small geometric and photometric transformations. In this paper, two low-dimensional descriptors for extracting distinctive invariant features from interest regions are presented, i.e., PCA and Local Self-Similarities feature based descriptors, namely PCA-LSS and PLSS. They are achieved by applying PCA on LSS feature field and the improved LSS descriptors of normalized patches, respectively. The performance of these proposed PCA-LSS and PLSS descriptors to image matching is studied through extensive experiments on the INRIA Oxford Affine dataset. Empirical results indicate that the proposed PLSS achieves more transformation invariance, significantly outperforms the original LSS, and also outperforms the SIFT.

Keywords: Local Self-Similarity, SIFT, PCA, Region description, Image matching.

1 Introduction

A critical aspect of computer vision research involves the detection and description of interest points. Nowadays, local image feature detection and description, which are distinctive and yet invariant to many kinds of geometric and photometric transformations, have been gaining more and more attention because of their promising performance. There are two different ways of utilizing local features in applications [1], i.e., the traditional utilization, involved the feature detection, feature description, and feature matching, and the bag-of-features [2], included the feature detection, feature description, feature clustering, and frequency histogram construction for image representation. Today, they are the preferred strategy for solving a wide variety of problems, for such tasks as image retrieval [29], wide baseline matching [30], object recognition [6], texture recognition [31], object category classification [13] and robot localization [32].

Recently for image and video matching, Eli Shechtman and Michal Irani [14] explored Local Self-Similarity descriptor based on Log-Polar location grid, namely LSS (LP). This descriptor has been quickly adopted in the object detection and classification community as yet another local descriptor in bag-of-visual-words frameworks [16, 17, 19, 20] or in nearest-neighbor classifiers [15]. Junejo et al. also

G. Lee (Ed.): Advances in Automation and Robotics, Vol.2, LNEE 123, pp. 291–300.
Springerlink.com

performed human action recognition in video by using temporal Self-Similarities [18] extended from LSS (LP). Thomas Deselaers and Vittorio Ferrari proposed Global and Efficient Self-Similarity (GSS) [21] for object classification and detection. However, there are still some limitations in the original LSS method, such that this descriptor is only invariance against small local affine and non-rigid deformations, and insensitive to small translations. To address these problems, in this paper, we propose two new image descriptors based on LSS texture features and Principle Component Analysis (PCA), namely PCA-LSS and PLSS, which are more suitable for different computer vision tasks.

In this paper, we propose two new and low-dimensional interest region descriptors, denoted as PCA-LSS and PLSS descriptors, which combine the good properties of the PCA-SIFT, PSIFT and LSS (LP). They are achieved by adopting the PCA-SIFT and PSIFT algorithm and using the novel LSS feature instead of original gradient feature, i.e., applying PCA on LSS feature field and the improved LSS descriptor of normalized patches, respectively. To the best of our knowledge, we are the first to improve the original LSS (LP) descriptor and explore it for image matching. The performance of these proposed PCA-LSS and PLSS descriptors to image matching is studied through extensive experiments on the INRIA Oxford Affine dataset. Empirical results only measuring by Euclidean distance indicate that the proposed low-dimensional PLSS descriptor can yield more stable and robust results, significantly outperform the original LSS (LP) descriptor, and also outperform the SIFT in these experimental evaluations with various geometric and photometric transformations.

1.1 Related Works

A local feature consists of a feature detector and a feature descriptor. Most of the existing detectors can be categorized into three types [3]: i) corner detectors, e.g. Harris, SUSAN detector, ii) blob detectors, e.g. Hessian, Hessian-Affine detector and iii) region detectors, e.g. MSER detector. Existing detector performance evaluation [4] has offer more information on interest region detection.

Given invariant region detectors, the remaining question is to describe interest regions around the feature points. Many of the proposed descriptors are distribution-based, i.e., they use histograms to represent different characteristics of appearance or shape. Lowe [6] developed a Scale-Invariant Feature Transform (SIFT) descriptor based on the gradient distribution in the detected regions. The most similar and an extension to the SIFT descriptor is the GLOH descriptor [5], which replaces the Cartesian location grid used by the SIFT with a log-polar one, and applies PCA to reduce the dimension of the descriptor. Similar to GLOH reduced dimension by PCA is the PCA-SIFT [8]. The up to date and very attractive descriptor is center-symmetric local binary pattern (CS-LBP) [7], which performs favorably compared to the SIFT. Belongie et al. [9] proposed the shape context descriptor, which is scale and rotation invariant. The SURF descriptor [10] builds on the strengths of the leading existing detectors and descriptors. In Ref. [11], geodesic sampling is used to get neighborhood samples for interest points and then a geodesic-intensity histogram (GIH) is used as a deformation invariant local descriptor.

There are several recent comparative studies on performance evaluation of local region descriptors [5,12]. Almost all the experimental comparison results revealed that the best discriminative descriptors are distribution-based descriptors such as GLOH, SIFT and CS-LBP. Although good performances are achieved by SIFT and other distribution-based descriptors (e.g. GLOH and CS-LBP), their dimensions are high (usually 128) which make them not comfortably practical for limited-memory systems and large-scale computation systems. Thus, lower-dimension descriptors are commonly a better option. Zhao et.al [26, 27] proposed another PCA-based version of SIFT, named PSIFT (36 dimension), which pre-computes the eigenspace based on SIFT descriptors extracted from a training set. The authors claimed that PSIFT can achieve almost the same performance as SIFT. Similar to PSIFT, Yang Duanduan et.al [28] proposed two more low-dimensional SIFTs (namely LPP-SIFT and SPCA-SIFT) by incorporating manifold subspace and sparse eigenspace learning techniques (Locality Preserving Projection and Sparse PCA are used as the exemplary implementations). Their results show that LPP-SIFT and PSIFT generally outperform PCA-SIFT and SPCA-SIFT, and achieve almost the same performance as SIFT, while their dimensions are significantly lower. Hongping Cai et.al [23] proposed a strategy to not only reduce the feature dimensionality but also yield better results than the original SIFT and PCA-SIFT in various application scenarios by learning Linear Discriminant Projections (LDP). However, experiments show that no general discriminant projections would improve matching performance on any type of data. Recently Eva Hörster and Rainer Lienhart etc. [13] have studied the influence of different types of local feature descriptors in the context of scene recognition based on pLSA image models. They argued that the commonly used SIFT descriptor is outperformed by the two other feature descriptors: the geometric blur feature and the LSS (LP) feature. Although a thorough comparison of local region descriptors in the context of matching and recognizing the same object or scene is presented elsewhere, a detailed evaluation about the advanced local region descriptor of LSS (LP) descriptor in this context (image matching) is still missing. This leads to necessary evaluations to investigate the performance of the LSS for image matching.

2 SIFT and Local Self-similarity Descriptors

2.1 SIFT and Low-Dimensional SIFTs

SIFT: SIFT [6] descriptors are computed for normalized image regions. A SIFT descriptor is a 3D histogram of gradient with location and orientation, where location is quantized into a 4x4 Cartesian location grid and the gradient angle is quantized into eight orientations. The resulting descriptor is of dimension 128. Each orientation plane represents the gradient magnitude corresponding to a given orientation. To obtain illumination invariance, the descriptor is normalized by the square root of the sum of squared components.

PCA-SIFT: The PCA-SIFT [8] descriptor adopts PCA to linearly project high-dimensional feature field of gradient magnitude onto a low-dimensional feature space. For application, this projection (encoded by the patch eigenspace) can be pre-computed once and stored. The dimensionality of the feature vector is significantly smaller than the standard SIFT feature vector, and it can be used with the same matching algorithms. In PCA-SIFT, the input vector is created by concatenation of the horizontal and vertical gradient maps for the 41x41 patch centered to the keypoint, which has 2x39x39=3042 elements. According to PCA-SIFT, fewer components requires less storage and will result to a faster matching, they choose the dimensionality of the feature space, n = 20 or 36, which results to significant space benefits.

PSIFT: The PSIFT [26, 27] descriptor utilizes PCA as well, but in a different method. The eigenspace of PSIFT is pre-computed offline based on SIFT descriptors extracted from a training set. Empirically, PSIFT uses the eigenvectors of the first 36 largest eigenvalues of SIFT descriptors for the linear projection. Thus, PSIFT tries to capture the major variances of SIFT descriptors. The PSIFT has similar performance as SIFT, and better performance than PCA-SIFT.

2.2 Local Self-similarity Descriptor

LSS (LP): The idea of this LSS method is associated with that measuring similarity between visual entities from images can be based on matching the internal layout of Local Self-Similarities. The LSS (LP) [14] descriptor showed in Fig. 1, captures the internal geometric layout of Local Self-Similarities (LSS) and can be compared across images which appear substantially different at pixel level. To derive the LSS feature d_q associated with an image pixel q, the surrounding image patch (typically patch size: PxP=5x5) is compared with a larger surrounding image region centered at q (typically radius 20, region size: NxN=41x41), using simple sum of square differences (SSD) between patch colors (They used CIE L*a*b* space). The resulting distance surface $SSD_q(x, y)$ is normalized and transformed into a "correlation surface" $S_q(x, y)$:

$$S_q(x,y) = \exp\left(-\frac{SSD_q(x,y)}{\max(\mathrm{var}_{noise}, \mathrm{var}_{auto}(q))}\right) \tag{1}$$

where var_{noise} is a constant that corresponds to acceptable photometric variations (in color, illumination or due to noise), and $\mathrm{var}_{auto}(q)$ takes into account the patch contrast and its pattern structure, such that sharp edges are more tolerable to pattern variations than smooth patches. In implementation $\mathrm{var}_{auto}(q)$ is the maximal variance of the difference of all patches within a very small neighborhood of q (of radius 1) relative to the patch centered at q.

The correlation surface $S_q(x, y)$ is then transformed into log-polar coordinates centered at q, and partitioned into 80 bins (20 angles, 4 radial intervals). Invariance against small local affine and non-rigid deformations is achieved by the log-polar representation. The maximal correlation value in each bin is selected. The maximal values in those bins form the 80 entries of local "self-similarity descriptor" vector d_q associated with the pixel q. By choosing the maximal correlation value in each bin, the descriptor becomes insensitive to small translations. Finally, this descriptor vector is normalized by linearly stretching its values to the range [0…1], in order to be invariant to the differences in pattern and color distribution of different patches and their surrounding image regions.

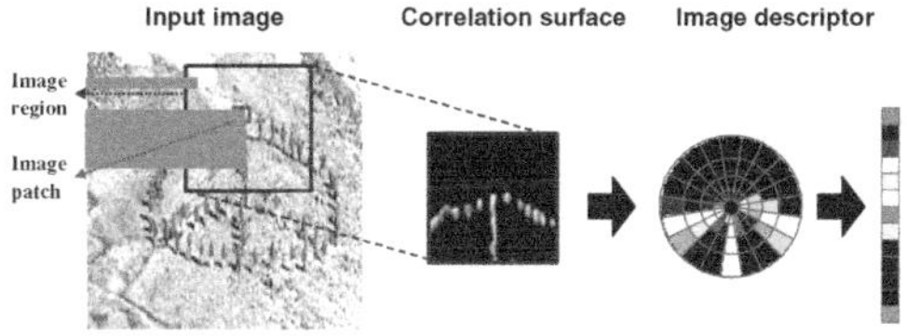

Fig. 1. Extracting the local "self-similarity" descriptor

2.3 Proposed Local Self-similarity Descriptor

In this sub-section, we presents two new low-dimensional LSS descriptors, i.e., PCA-LSS and PLSS, which compute the subspace of LSS feature field and LSS descriptor similarly to PCA-SIFT and PSIFT, respectively.

PCA-LSS: Similar to PCA-SIFT, the proposed PCA-LSS descriptor also directly performs PCA on the feature image of interest region, and replaces the gradient magnitude feature used in PCA-SIFT with the novel LSS feature, which has 1x41x41=1681 elements. Obviously PCA-LSS vector is significantly smaller (usually 36 dimensions) than the standard LSS (LP) descriptor (usually 80 dimensions), and can be used with the same matching algorithms.

PLSS: The proposed PLSS descriptor utilizes PCA as well, but unlike PCA-LSS. The eigenspace of PLSS is pre-computed offline based on LSS descriptors extracted from a training image set. Empirically, the top 36 eigenvectors of LSS descriptors for linear projection are retained, which correspond to the first 36 eigenvalues sorted in descending order. Thus, these 36 eigenvectors basically capture the major variances of LSS descriptor. Consequently, the PLSS is also a descriptor in 36 dimensional space similar to PCA-LSS.

In the proposed low-dimensional PLSS descriptor, the basic high-dimensional descriptor used for PCA projection is not the original LSS (LP), but our improved Orientated Local Self-Similarity descriptor based on Cartesian location grid, namely OLSS (C).

OLSS (C): The improved and oriented OLSS (C) descriptor is the modified version of the well-known LSS (LP). In order to get more robust translations invariance, the OLSS (C) descriptor uses distribution-based histogram representation in each cell, rather than choosing the maximal correlation value in each bucket in the Log-Polar location grid in the natural LSS (LP) descriptor. These steps are depicted in Fig. 2. Empirically, the OLSS (C) descriptor is computed for normalized image region. It uses a SIFT-like Cartesian location grid as pooling strategy and replaces SIFT's gradient feature with LSS based feature. A OLSS (C) descriptor is also a 3D histogram of LSS feature with location and direction, where location is quantized into a 4x4 Cartesian location grid and the direction is quantized into 8 bins based on gradient orientation. The resulting descriptor is of dimension 128. Specially, we propose a new binning strategy to create feature histogram, where the angle of gradient orientation is calculated by the diagonal pixel intensity from a more larger image patch, here 5 × 5, which is different from the standard SIFT. And then the descriptor is normalized to unit length to obtain illumination invariance.

(a) (b) (c)

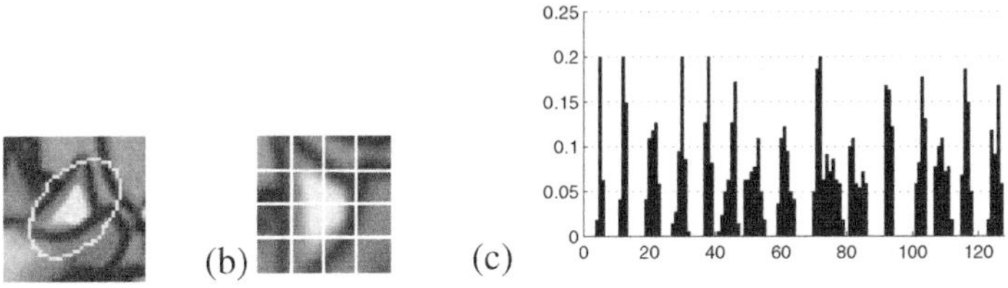

Fig. 2. The OLSS (C) descriptor. (a) An elliptical image region detected by Hessian-Affine detector. (b) The region with Cartesian location grid after affine normalization. (c) The resulting OLSS (C) descriptor computed for the normalized region.

3 Experimental Evaluation

3.1 Image Matching Protocol

Experimental evaluation for the proposed PCA-LSS and PLSS descriptors is based on the well-known protocol, i.e., feature descriptor matching protocol, which is freely available on the Internet. This protocol is a image matching protocol that is designed for matching interest regions between a pair of images [22]. For region detection and computing SIFT descriptors we use the software routines provided by the matching protocol [22]. The interest region detection methods used in the experiments is the Hessian-Affine (HesAff) detector [24]. This evaluation protocol takes recall versus 1-precision graphs as the evaluation criterion based on the number of correct and false matches between a pair of images [25]. Recall and 1-precision can be determined as the followings:

$$recall = \frac{\#correct\ matches}{\#correspondences},\quad 1 - precision = \frac{\#\,false\ matches}{\#all\ matches} \tag{2}$$

where the #correspondences stands for the ground truth number of matching regions between the pair of images. The curves are obtained by varying the distance threshold, and a good descriptor would produce a small decay in this curve, reflecting the maintenance of a high precision while matching more image regions.

3.2 Matching Results

This section presents the results of matching experiments on several descriptors for Oxford Affine dataset, i.e., the SIFT, PCA-SIFT, PSIFT, LSS (LP), PCA-LSS and PLSS, where the dimensionality is 128, 36, 36, 80, 36 and 36, respectively, and validates the performance of the proposed descriptor for HesAff regions. Note that the distance matrix for all these matching experiments is the simple Euclidean distance.

For the five different changes in imaging conditions, i.e., image blur, viewpoint changes, zoom and rotation, illumination, and JPEG compression, we firstly use bikes and trees datasets to evaluate the matching performance of these descriptors involving image blur change. Fig. 3 (a) and (b) show the results of the pair of bikes and trees images. These results show that the proposed low-dimensional PLSS achieves the best performance, and clearly outperforms the classical SIFT and PSIFT. The PCA-LSS and the original LSS (LP) perform not so stably, but still better than the PCA-SIFT. The PSIFT is slightly outperformed by the classical SIFT. Note that the PCA-SIFT achieves the worst performance. We hypothesize that these results can be attributed to several factors. Firstly, the PLSS descriptor based on OLSS (C) is essentially more invariant with respect to image blur transformations by distribution-based histogram representation and binning based on gradient orientation. Secondly, the PCA-SIFT is known to be notoriously sensitive to registration error [8].

Then we use graffiti and wall datasets to evaluate matching performance of these descriptors. The graffiti and wall datasets involve viewpoint change which is more challenging to the robustness of descriptor than other image transformations. Fig. 3 (c) and (d) plot the results of the graffiti and wall datasets. Although the perspective transformation with the high degree of projective warp is a challenging task to feature descriptors, these results demonstrate that the proposed low-dimensional PLSS also achieves more affine invariance, yields more stable and robust results than order descriptors, and noticeably outperforms the original LSS (LP) descriptor. However, the proposed PCA-LSS performs worst in the wall test case.

We also use bark and boat datasets to evaluate the matching performance of these descriptors on zoom + rotation, leuven dataset on illumination, and ubc dataset on JPEG compression. Fig. 3 (e), (f), (g) and (h) show these experimental results. The experimental results on the performance comparison of the proposed descriptors for zoom + rotation are similar to those for viewpoint change. In the test cases with illumination changes, the orientation-based descriptors, i.e., the proposed PLSS, performs best, followed by the classical SIFT, and then PSIFT. However, the PCA-LSS, LSS (LP) and the PCA-SIFT are much more sensitive to illumination changes. Essentially, the OLSS (C) and SIFT accumulate the histogram of the LSS feature and the gradient magnitudes to eight bins according to the pixel's orientation, respectively. Thus, the PLSS derived from the oriented OLSS (C) achieves more discriminative performance to the illumination change than the non-oriented descriptors, i.e., the PCA-LSS, LSS (LP) and the PCA-SIFT. For JPEG compression, the PLSS still performs better than the SIFT, and the LSS (LP) dramatically obtains the best test results. However, the difference is very small and the improvement magnitude only about 0.01 in AUC.

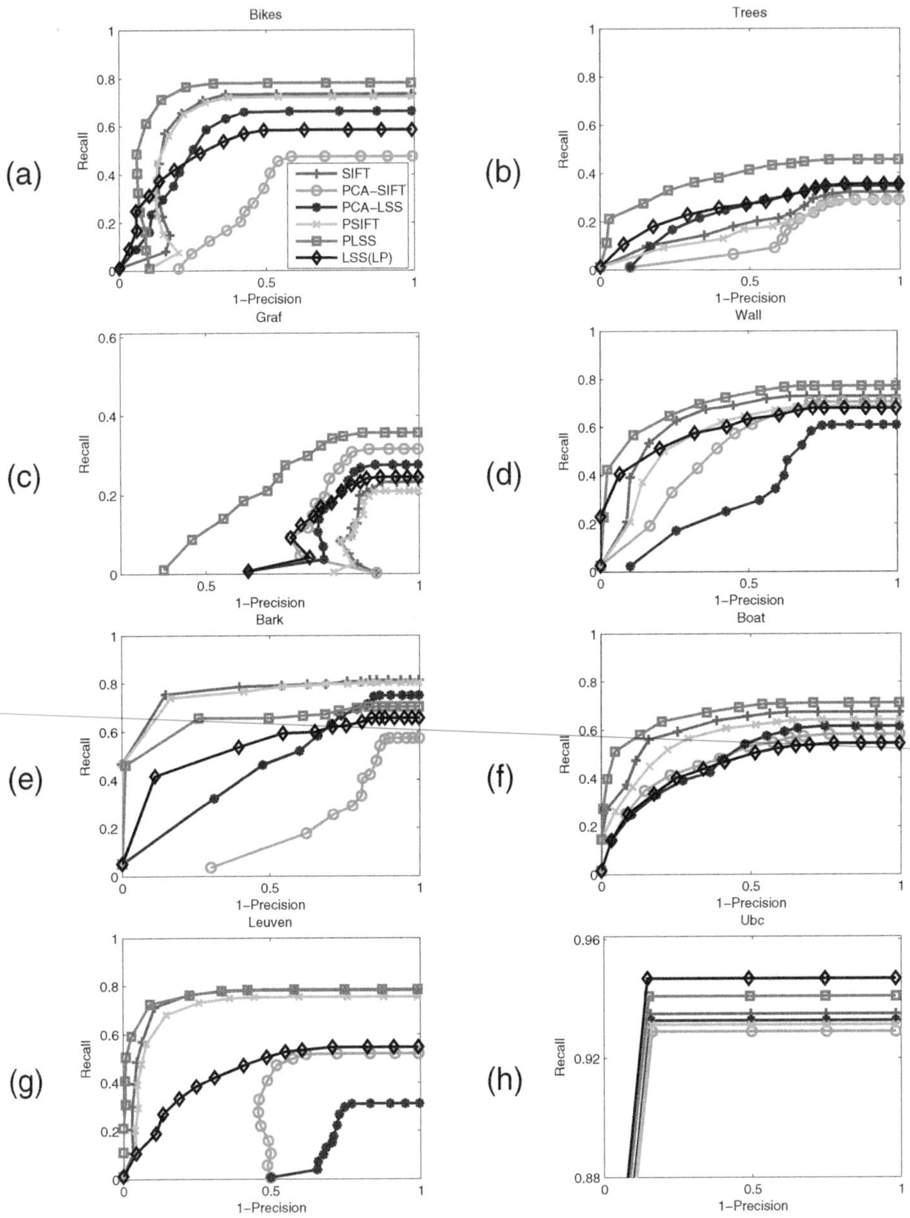

Fig. 3. Performance evaluation of several descriptors (SIFT, PCA-SIFT, PSIFT, LSS (LP), PCA-LSS and PLSS) on INRIA Oxford Affine dataset, and resulting in Recall and 1-Precision. The pair image is the 1st and 4th.

4 Conclusions

In this paper, we present two new low-dimensional approaches for describing image interest regions, i.e., PCA-LSS and PLSS descriptors, which are achieved by applying PCA on LSS feature field and the improved OLSS (C) descriptors of normalized patches, respectively. Empirical results indicate that the proposed PLSS descriptor obtains more transformation invariance than the SIFT. Thus, in addition to the low computational cost, the superior performance makes the PLSS as a substitute for the classical SIFT.

Acknowledgments. This work was supported by the National Natural Science Foundation of China (Grant No. 60773085, No. 60970109), and NSFC-KOSEF International Collaborative Research Funds (Grant No. 60811140346, and No. F01-2008-000-10021-0).

References

[1] Li, J., Allinson, N.M.: A comprehensive review of current local features for computer vision. Neurocomputing 71(10-12), 1771–1787 (2008)

[2] Sivic, J., Zisserman, A.: Video google: A text retrieval approach to object matching in videos. In: ICCV (2003)

[3] Tuytelaars, T., Mikolajczyk, K.: Local Invariant Feature Detectors: A Survey. Foundations and Trends in Computer Graphics and Vision 3(3), 177–280 (2008)

[4] Mikolajczyk, K., Tuytelaars, T., Schmid, C., Zisserman, A., Matas, J., Schaffalitzky, F., Kadir, T., Van Gool, L.: A Comparison of Affine Region Detectors. Int'l J. Computer Vision 65(1/2), 43–72 (2005)

[5] Mikolajczyk, K., Schmid, C.: A Performance Evaluation of Local Descriptors. IEEE Trans. Pattern Analysis and Machine Intelligence 27(10), 1,615-1,630 (2005)

[6] Lowe, D.: Distinctive Image Features from Scale-Invariant Keypoints. Int'l J. Computer Vision 2(60), 91–110 (2004)

[7] Heikkilä, M., Pietikäinen, M., Schmid, C.: Description of interest regions with local binary patterns. Pattern Recogn. 42(3), 425–436 (2009)

[8] Ke, Y., Sukthankar, R.: PCA-SIFT: a more distinctive representation for local image descriptors. In: IEEE Conference on Computer Vision and Pattern Recognition, vol. 2, pp. 506–513 (2004)

[9] Belongie, S., Malik, J., Puzicha, J.: Shape matching and object recognition using shape contexts. IEEE Trans. Pattern Anal. Mach. Intell. 24(4), 509–522 (2002)

[10] Bay, H., Tuytelaars, T., Gool, L.V.: SURF: speeded up robust features. In: European Conference on Computer Vision, vol. 1, pp. 404–417 (2006)

[11] Ling, H., Jacobs, D.W.: Deformation invariant image matching. In: 10th IEEE International Conference on Computer Vision, vol. 2, pp. 1466–1473 (2005)

[12] Moreels, P., Perona, P.: Evaluation of features detectors and descriptors based on 3D objects. In: 10th IEEE International Conference on Computer Vision, vol. 1, pp. 800–807 (2005)

[13] Hörster, E., Greif, T., Lienhart, R., Slaney, M.: Comparing Local Feature Descriptors in pLSA-Based Image Models. In: 30th Annual Symposium of the German Association for Pattern Recognition (DAGM) 2008, Munich, Germany (June 2008)

[14] Shechtman, E., Irani, M.: Matching Local Self-Similarities across images and videos. In: Proc. CVPR (2007)
[15] Boiman, O., Shechtman, E., Irani, M.: In defense of nearest-neighbor based image classification. In: CVPR (2008)
[16] Chatfield, K., Philbin, J., Zisserman, A.: Efficient retrieval of deformable shape classes using Local Self-Similarities. In: NORDIA Workshop at ICCV 2009 (2009)
[17] Hörster, E., Lienhart, R.: Deep networks for image retrieval on large-scale databases. In: ACM Multimedia (2008)
[18] Junejo, I.N., Dexter, E., Laptev, I., Pérez, P.: Cross-View Action Recognition from Temporal Self-Similarities. In: Forsyth, D., Torr, P., Zisserman, A. (eds.) ECCV 2008, Part II. LNCS, vol. 5303, pp. 293–306. Springer, Heidelberg (2008)
[19] Lampert, C.H., Nickisch, H., Harmeling, S.: Learning to detect unseen object classes by between-class attribute transfer. In: CVPR (2009)
[20] Vedaldi, A., Gulshan, V., Varma, M., Zisserman, A.: Multiple kernels for object detection. In: ICCV (2009)
[21] Deselaers, T., Ferrari, V.: Global and Efficient Self-Similarity for Object Classification and Detection. In: CVPR 2010 (2010)
[22] Mikolajczyk, K.: Affine Covariant Features. Visual Geometry Group, University of Oxford (2004), `http://www.robots.ox.ac.uk/~vgg/research/affine/`
[23] Cai, H., Mikolajczyk, K., Matas, J.: Learning Linear Discriminant Projections for Dimensionality Reduction of Image Descriptors. IEEE Trans. Pattern Anal. Mach. Intell. 33(2), 338–352 (2011)
[24] Mikolajczyk, K., Schmid, C.: Schmid, Scale & affine invariant interest point detectors, Int. J. Comput. Vision 60(1), 63–86 (2004)
[25] Mikolajczyk, K., Schmid, C.: An Affine Invariant Interest Point Detector. In: Heyden, A., Sparr, G., Nielsen, M., Johansen, P. (eds.) ECCV 2002. LNCS, vol. 2350, pp. 128–142. Springer, Heidelberg (2002)
[26] Zhao, W.-L., Ngo, C.-W.: Scale-rotation invariant pattern entropy for keypoint-based near-duplicate detection. IEEE Trans. on Image Processing 18(2), 412–423 (2009)
[27] Zhao, W.-L., Tan, S., Ngo, C.-W.: Large-scale Near-duplicate Web Video Search: Challenge and Opportunity. In: Int'l Conference on Multimedia & Expo (ICME), Workshop on Internet Multimedia Search and Mining, Cancun, Mexico (July 2009) (Oral)
[28] Yang, D., Sluzek, A.: Performance evaluation of low-dimensional sifts. In: ICIP 2010, pp. 2729–2732 (2010)
[29] Mikolajczyk, K., Schmid, C.: Indexing based on scale invariant interest points. In: 8th IEEE International Conference on Computer Vision, vol. 1, pp. 525–531 (2001)
[30] Tuytelaars, T., Gool, L.V.: Matching widely separated views based on affine invariant regions. Int. J. Comput. Vision 59(1), 61–85 (2004)
[31] Lazebnik, S., Schmid, C., Ponce, J.: A sparse texture representation using local affine regions. IEEE Trans. Pattern Anal. Mach. Intell. 27(8), 1265–1278 (2005)
[32] Se, S., Lowe, D., Little, J.: Global localization using distinctive visual features. In: IEEE/RSJ International Conference on Intelligent Robots and System, vol. 1, pp. 226–231 (2002)

Fast Algorithm for Lunar Craters Simulation

Li Jian-Sheng[1,3], Liu Wei-min[2], Lan Chao-zhen[1], Mao Xiao-yan[3,4], and Lu Zhan-wei[1]

[1] Institute of Surveying and Mapping, Information Engineering University, Zhengzhou 450052, China
[2] 62217 Troops, Geermu 816000, China
[3] Beijing Institute of Control Engineering, Bei Jing,100190, China
[4] National Laboratory of Space Intelligent Control, Bei Jing, 100190, China

Abstract. In order to solve the lack of high resolution imagery and DEM in the accurate simulation of lunar terrain, a fast algorithm for simulating the lunar terrain is presented in this paper. The algorithm fits simple shapes of lunar craters based on rotational equation, and simulates the true distributing of craters using random function. The experimental results demonstrate that the proposed algorithm is not only effective in the simulating of craters which accord with the true lunar terrain well in shape and distributing but also actualizes the secondary impact of craters.

Keywords: simulation, crater, rotational equation, random function, lunar.

1 Introduction

The moon is abundant in resources, so the lunar exploration is strategic. The space powers have developed ambitious lunar exploration programs. The United States has issued a project of "Return to the moon", planning to re-send aerospace members to the moon within 20 years and establish a lunar base. Recently, Russia, European Union, Japan and India have also launched new lunar satellites. China proposed a grand blueprint of "three-step exploration of the moon - around, down, back to the moon" in the early 21st century [1]. In 2007 China has successfully launched the "Chang-e One" orbiting satellite, initially completed the first lunar probe mission. During the second phase of "Chang-e" probe, China has launched a self-developed lunar rover landing on the moon, taken science search instrument to carry on tour search and sent probing data to the Earth[2].

The development and design of lunar rover needs a large number of experiments and validation .It's no doubt that it's a fast, cost-cutting means to adopt the computer simulation technology .When we conduct performance test of the lunar rover with the computer, which needs to imitate an fine and nearly actual lunar surface geography to meet the demand of kinematics and dynamics simulation experiment of the lunar rover circulating on uneven surface [3]. For the lack of stitching detection means, China is still unable to obtain high-resolution lunar image and DEM data. So it's a good method by using computer simulation techniques to simulate high-resolution topographic data to imitate an environment which is similar to the lunar surface.

G. Lee (Ed.): Advances in Automation and Robotics, Vol.2, LNEE 123, pp. 301–306.
Springerlink.com

As surface is covered with various sizes of meteorite collision pits, thus it attaches great importance to building a model of crater to reflect the real lunar surface.

During the "Apollo" program, the United States Aeronautics and Space Administration (NASA) and related scholars have published large numbers of literature [4-7] on the lunar environmental surfaces, the literature provide a theoretical and technical basis for the lunar environment simulation. Since China has proposed lunar exploration program, many domestic colleges and research institutions began related studies. In Shanghai Jiaotong University, Yang Yanchun, Bao Jinsong, proposed a method of constructing a virtual lunar surface [3] [8], which means simulating the crater through plane rotating method, but the specific algorithm process was not mentioned.

According to some scholars' statistical and computing experience, this thesis uses the method of plane rotating to simulate the lunar crater' shape, uses random function to simulate distribution of lunar craters, and achieves a secondary impact crater effect.

2 Shape and Distribution of the Lunar Impact Craters

According to the data [9] [10], the topography of the lunar surface mainly includes: crater, sea, land and mountains on the lunar surface, the lunar surface radiation pattern, Valley (lunar gap), and so on. The Ring Mountain is the lunar impact crater, which takes on a bowl-shaped concave structure, among which there are more than 33000 whose diameter is greater than 1 km. It is one of the obvious features of the lunar surface .There are two main reasons for the Crater formation: one is the result of meteorite impacts, which accounts for the majority; the other is volcanic activity. Figure 1 shows the surface of the crater taken by the lunar probe satellite in the United States.

Lunar surface is dotted with a great deal of craters. Crater can be roughly divided into two categories: simple crater and complex craters. Simple crater forms a circular bowl-shaped and edge bumps projection. Generally speaking, the craters with diameter less than 15 kilometer on the lunar are regarded as simple. We need to construct small-region topography on the lunar surface based on the simulation demand, so we only simulate simple craters in this thesis.

Simple craters profile approximation diagram is shown in Figure 2, in which: Hr is the height of the crater edge, H is the the depth of crater, and D is the diameter of crater. Through a large number of measurement and calculations of the lunar craters, Pike (1974) deduced the following empirical formulas:

$$\begin{aligned} H &= 0.196D^{1.01} \quad if \quad D < 11km \\ Hr &= 0.036D^{1.014} \quad if \quad D < 21km \end{aligned} \tag{1}$$

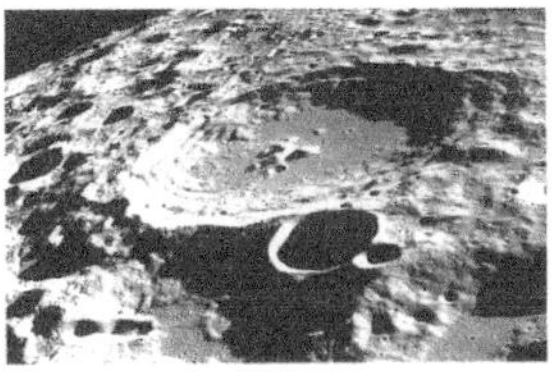

Fig .1. Surface of the lunar crater

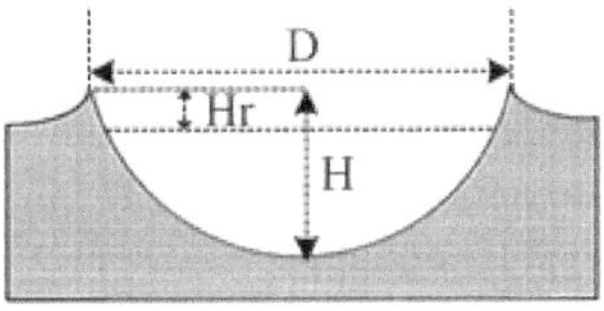

Fig. 2. Simple approximate profile of the crater [4]

In order to make the lunar crater simulation more realistic, it's required to consider the probability of the crater's distribution on the lunar surface .According to the literature [5], the distribution of craters on the lunar surface abide by the following empirical formula:

$$N=cD^{-b} \tag{2}$$

Among them: N is the number of the lunar craters which diameter is less than D, c and b are both constants, b constantly values 1.8.

3 Modeling of the Simple Craters

Due to the regular shape of the simple craters, we can generate craters by using the method of plane rotating. Assuming the crater's center coordinate is (x_1, y_1), we adopt the following formula to calculate the elevation deviation of the point (x, y):

$$\Delta h = k \cdot [(d \;/ r^2 - \frac{1}{4}) \cdot r] \cdot (1 - d / r\;) \tag{3}$$

And $d = \sqrt{(x-x_1)^2 + (y-y_1)^2}$, stands for the distance from the point (x, y) to the crater center, r is the radius, K is a control parameter for controlling the depth of the pit.

Based on the above elevation deviation formula, we draw a curve as shown in Figure 3

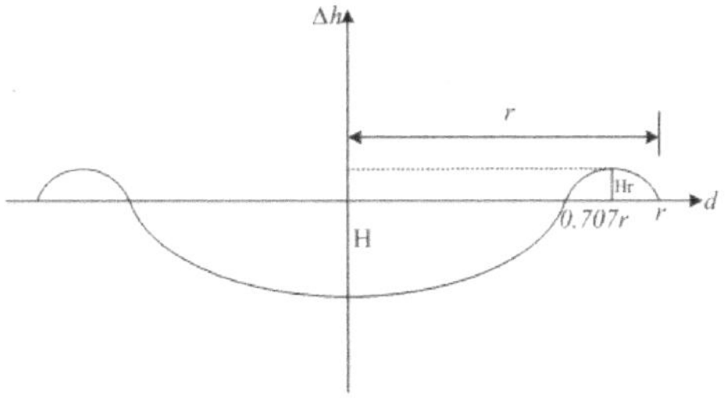

Fig. 3. Curve shape

Through analyzing we come to:

When $d=0$, $\Delta h = r/4$, means the depth of the crater center accords with $H = r/4$; when $d = \sqrt{2} \cdot r/2$ or $d = r$,the depth is 0, forms the edge of crater bumps .The shape of the crater drawn through equation (3) accords with the real shapes of the simple crater as well as Pike's empirical formula, and the crater depth can be modified according to the actual situation.

4 Random Function in Simulating Lunar Craters Distribution and Algorithm Steps

After a single simple crater model has been built, we used the random function to randomly distribute the lunar craters on the lunar surface. In order to meet the real lunar crater distribution, this article used a number of cycle parameters to control the number of craters, as needed to generate different density crater distribution. Furthermore, in order to avoid the crater randomly generating multiple stacking, we should use different seeds for each crater.

The algorithm steps of lunar crater simulation is shown in Figure 4, to facilitate the programming calculation, the algorithm firstly normalized the foundation topography that would join the craters, making the horizontal and vertical coordinates between 0 and 1; then according to the size of the topography to generate specific number of craters and their random distribution; Finally, according to the crater rotation equation traverse the entire topography, calculate elevation deviation within the craters, then integrate the calculated deviation and original topography, thus generated a new topography with craters.

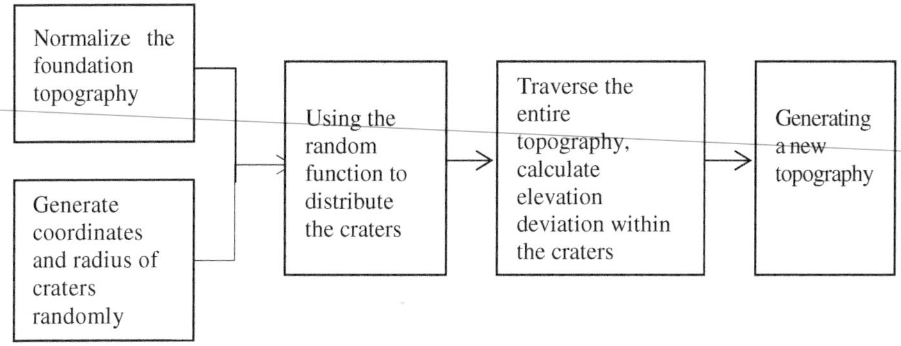

Fig. 4. Algorithm steps

5 Experimental Results and Analysis

In this paper, the simulation platform is the Intel Xeon (TM) processor, frequency 3.0G, Display Card for NV Quadro 1G, FX 560, memory 128M.

Figure 5 is the experimental results by using the given algorithm to simulate the lunar craters, figure 5-a shows a dense distribution effect, figure 5-b shows the crater volume reduction effect, figure 5-c shows the secondary impact effect, but this phenomenon should not be too much, for probability of meteorites impacting on the same place twice is lower.

In order to confirm that the algorithm is simple and fast, this thesis used the 100 * 100, 200 * 200, 300 * 300, 500 * 500, 1000 * 1000 five kinds of mesh to test ,generated moderate density of the lunar craters, then compareed their desired generation time, CPU occupancy rate and frame rate when roaming.

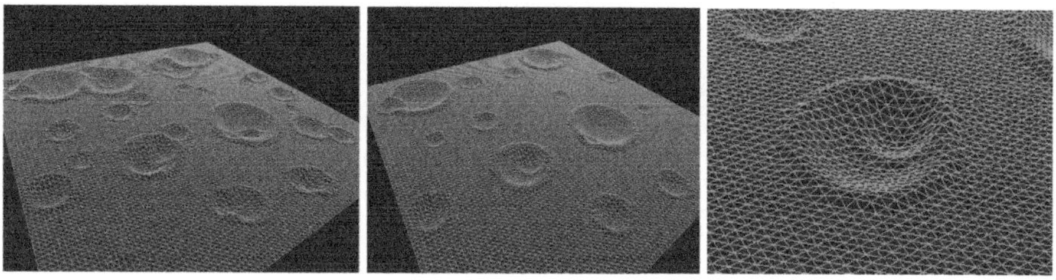

(a) A dense distribution effect (b)The crater volume reduction effect (c)Secondary impact crater effect

Fig. 5. Simulation effect of the craters

From table 1 we conclude that, with the increase in the number of grid, craters generated time increased, but which was acceptable. Because the hole has been generated before rendering, CPU occupancy rate was low when roaming and the frame rate was not affected greatly.

Table 1. Craters generation time and CPU occupancy rate of data mesh Simulating

Mesh size	Time needed (ms)	frame rate (fps)	CPU **occupancy rate**
100*100	47	4023	2%
200*200	203	1320	2%
300*300	438	423	3%
400*400	765	312	4%
500*500	1203	210	5%

Figure 6-a is the real moon local low resolution image derived from the "Appollo" spacecraft, figure 6-b is the rebuild lunar crater morphology based on the low resolution DEM through using the algorithm in this thesis ,and Figure 7 is the simulation effect of close observation of lunar crater. Compared with the original image, they better reflects the pit details, that's to say they also correspondingly increase image resolution .It is proved that this method can be applied in real lunar environment high accuracy simulation.

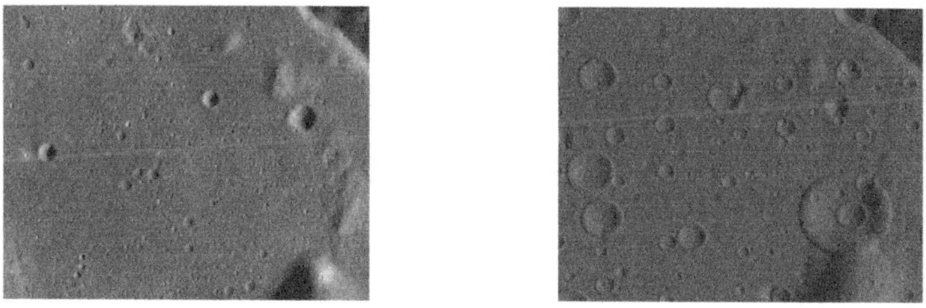

(a)Real images of the lunar surface (b)Simulating result

Fig. 6. Comparison between real and Simulating result

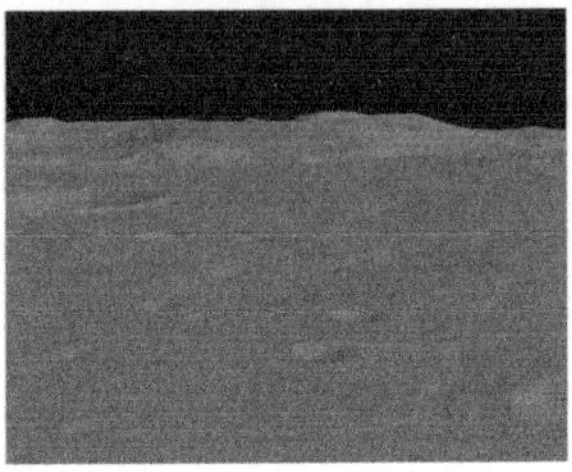

Fig. 7. Simulating effect of the lunar surface through the given method

6 Summary

This thesis presents the algorithm of the lunar craters' simulation. Experimental results show that the simulation results accord with the real situation of the lunar craters, and depth, size, distribution density of the crater can be adjusted according to the actual situation. It can be easily applied in the realistic virtual lunar surface environment simulation; meanwhile this method can also be applied in other stars meteorite crater simulation, such as the simulation of the surface of Mars meteorite crater. However this method can only simulate simple craters on the moon, for those complex craters, this algorithm is not yet effective.so the next research will be focused on this.

References

1. Moon car demonstration group. Lunar Rover Feasibility Demonstration report (2004)
2. Ziyuan, O.: Chinese lunar exploration project target and development strategy. Advance in Earth Sciences (2004)
3. Yang, Y., Bao, J., Ye, J.: A realistic virtual lunar surface modeling method. Journal of System Simulation 19(11) (2007)
4. Pike, R.J.: Depth / diameter relations of fresh lunar craters: Revision from spacecraft data. Geophysical Research Letters (S0094-8276), 291–294 (1974)
5. Melosh, H.J.: Impact Cratering - A Geologic Process. Oxford University Press (1989)
6. Lunar Surface Models: Space Vehicle Design Criteria (Environment). NASA SP-80231969
7. Cross, C.A., Fisher, D.L.: The Computer Simulation of Lunar Crater. Mon. Not. R. Astro. Soc., 261–272 (1968)
8. Bao, J., Hu, X., Wang, W., Yu, D., Ye, J.: Lunar Rover Virtual Simulation System with Autonomous Navigation. World Academy of Science, Engineering and Technology 53 (2009)
9. Wu, G.: The surface of the moon exploration. Space Exploration 1(12) (2007)
10. Sun, J.: Lunar features [EB / OL]. Place of publication (2006), http://www.Hyedu.net/moden/students/sjy-yzq/websaves/1/4.htm

A Research on the Relationship between Share Price of Chinese Gold Enterprises and Gold Prices

Anyu Cai, Luzhuang Wang*, Dandan Li, Shuyang Wang, and Yi Zhong

Zhejiang University City College, Hangzhou, China
zuccjohn@hotmail.com

Abstract. By qualitative analysis and stepwise regression, authors explored the relationship between share prices of Chinese gold companies and the gold prices both internationally and domestically. The results show that the share prices of gold enterprises are generally positively related to the domestic and international gold prices, with the former price impacting more obviously than the latter one, and the gold price in Euro is more powerful than that in US dollar in term of influencing the share price. The rule above is slightly different for different gold companies.

Keywords: Share price of gold enterprise, International price of gold, Domestic gold price, Empirical Study, Stepwise regression.

1 Introduction

In recent years, the world gold price shows a sharp increase, starting a wave of "gold fever". Many investors put their interest on the gold. Playing a decisive role the gold listed companies are the largest enterprises in Chinese gold industry. However, there are few researches about gold enterprises for the moment. Seeking the relationship between the share prices of a country's gold listed companies and the domestic and international gold prices, investors will analyze the market better and grasp investment opportunities. Gold listed companies can respond quickly to market fluctuations based on the gold price. And it will be helpful for the government to raise relevant policies.

The western countries' researches about gold market have been long-standing, both theoretically and empirically, and most of them are about gold futures contracts for the majority of international gold transactions are futures trading [1].
After the cancellation of "the exclusive operation" management system April 2001 in China, the Shanghai Gold Exchange set up in October 2002, which means Chinese gold market has become comprehensively open. According to preliminary statistics, the annual units output of gold of Shanghai Gold Exchange accounts for 75% of the country, the amount of gold use occupies 80% of the country and the smelting capacity takes up 90%. It shows that China's gold trading has been marketized.

The marketization of gold in China didn't last for long, and the researches on it are quite few. Hu Entong's empirical research in 2005 showed that there was no

* Corresponding author.

G. Lee (Ed.): Advances in Automation and Robotics, Vol.2, LNEE 123, pp. 307–314.
Springerlink.com

significant correlation among the real gold yield and GDP as well as other major macroeconomic variables[2]. He also proved and explained the causes of weak correlation between the benefits of gold and other financial assets, because gold income is lack of correlation with economic activity, while stock and other financial assets are relevant with economic activities [3]. Based on the empirical analysis of the factors affecting gold price, Li Yuhan revealed several gold price-affecting factors and provided advice to investors in gold investment [4].

After the establishment of the Shanghai Stock Exchange, the price of gold in domestic began to converge with the international. One after another, many scholars studied on a unilateral problem of the gold market. Yan Yi (2006) analyzed the trend of international gold price from many aspects such as the supply and demand in gold market, dollar, oil price and so on [5]. In 2008, Sun introduced the idea of establishing inter-bank gold market as a new platform for open-market operations of the central bank, to stabilize the currency and promote economic growth [6].

Due to the late opening of Chinese gold market, the researches till now in gold field were focused on discussing the construction of China gold system. Therefore, it needs much more empirical research and quantitative exploration for the subject.

2 Research Methodology

Sample Selection. To make the study on Chinese gold enterprises more representative, large gold companies in Shanghai and Shenzhen Securities Markets are chosen as research samples. In addition, taken into account empirical research needs a certain amount of data transactions, this research selects the enterprises which listed over five years. That is to say, there are only two companies (ZHONGJIN GOLD and SHANDONG GOLD) in China meet the study criteria.

The full name of Zhongjin Gold is ZhongJin Gold Corp.,LTD. , which was established on June 23, 2000 by the main sponsor, China National Gold Group Corporation (formerly China National Gold Corporation), join with other six co-sponsored enterprises. On 14 August 2003, it organized an Initial Public Offering of RMB common stock with 100 million shares in Shanghai Stock Exchange, becoming "China's first gold stocks." Producing high gold, standard gold, electrolytic silver, copper and sulfuric acid and other products, it is a large company with the ability of gold mining, dressing, smelting, processing and comprehensive capabilities. The stock refers to as "ZhongJin Gold" and stock code is "600489."

Shandong Gold Group was established in 1996. The company was registered in Shandong Province Administration Bureau for Industry and Commerce in 2000, with registered capital of 100 million Yuan. The primary businesses include gold mining, smelting and processing, which is combined with silver and sulfur ore concentrate production. Company has successfully gone public on August 28, 2003. The number of Shandong Gold Shares is 600547.

Data Source. The share price of gold enterprise is assumed to be the closing quotation price of every month of public company. International gold prices come from the World Gold Council which offers average month gold prices since December 1978. Both the performance of US dollar and that of Euro shall be taken into consideration. Domestic gold prices come from the Shanghai Gold Exchange which offers monthly report about Chinese gold marketing quotation from January, 2008 to August, 2011.Additionally, there is a website called KITCO, which owned loads of data about gold price. A part of the data in this paper comes from this website. The entire research range is from August, 2003 to March, 2011.

Model Analysis. With software including Eviews and SPSS, analysis is conducted on the data using the methods of stepwise forward estimation and multiple linear regressions. The relationships between the share price of each two chosen companies and each gold price in US dollar, Euro and RMB are analyzed univariatelly. Simultaneously, multivariated analyses of regression are taken to identify the relationships between the company's share price and each three gold prices. The relationship models about public gold company with domestic and international gold prices have been listed as followed.

$$Y= c1 + a1X1, \quad (1)$$

$$Y= c2 + a2X1+ a3X2+ a4X3 \quad (2)$$

Among them, Y lists the gold company Gold's share price, and X1 lists the China's domestic gold prices, X2 and X3 list the international gold price (Euros or US).

3 The Empirical Results

Single Factor Analysis. Eviews calculates regression analysis about international prices and Shandong Gold monthly closing quotation prices since 2003. The data is then in form of a scatter diagram. A clear positive variation correlation can be found in the scatter diagram.

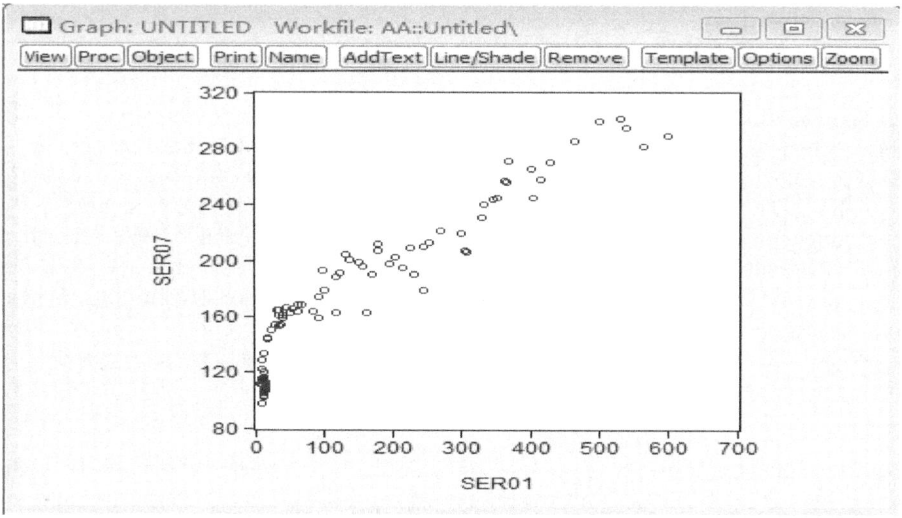

Fig. 1. Domestic gold price scattering with Shandong Gold share price

The analysis of regression suggests that the coefficient of determination, F-value and t-value be significant and the significance level high. R2, the coefficient of determination drawn from this model value is 0.89, indicating a high fitting degree while the DW value, saying 0.29, proves a positive self-correlation problem with residual sequences in the autocorrelation test. Also, the coefficient valuing 2.69 in this model indicates that the share price of Shandong Gold Company has been under a positive influence of the domestic gold price.

The figure 2 lists the relationship between Shandong Gold share price and international gold price (US), an obvious positive correlation can be found from the scatter diagram, although some areas seem a little scattered.

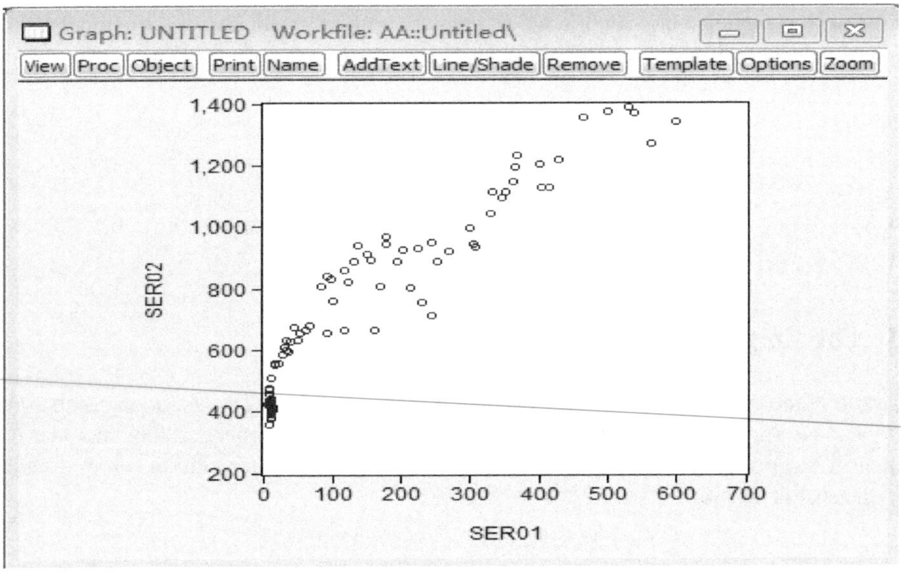

Fig. 2. International gold price (USD) scattering with Shandong Gold share price

The analysis of regression suggests that the coefficient of determination, F-value and t-value be significant and the significance level high. R2, the coefficient of determination drawn from this model value is 0.89 (as same as the former), indicating a high fitting degree while the DW value, saying 0.32, proves a positive self-correlation problem with residual sequences in the autocorrelation test. Also, the coefficient valuing 0.51 in this model indicates that the share price of Shandong Gold Company has been under a positive influence of the international gold price. When international gold price rises, Shandong Gold share price will rise too.

When international gold price is settled in euro, it represents positive correlation between Shandong Gold share price and international price. It is worth mentioning that the empirical research about Zhongjin Gold also obtains similar results.

Multivariate Analysis. In order to comprehensively consider the impacts of both domestic and international gold prices upon share prices of gold enterprises, authors adopted the stepwise regression to analyze the relationship between share price of gold enterprises and gold price. The results have been listed in Table 1.

Table 1. The relationship between Shandong Gold share price and gold price

Model Summary					
Model		R	R Square	Adjusted R Square	Std. Error of the Estimate
1		.945a	0.893	0.892	53.33862
a. Predictors: (Constant), GPrmb					

Coefficientsa						
Model		Unstandardized Coefficients		Standardized Coefficients		
		B	Std. Error	Beta	t	Sig.
1	(Constant)	-322.994	18.158		-17.788	0
	GPrmb	2.688	0.099	0.945	27.218	0
a. Dependent Variable: SPsd						

Excluded Variablesb					
Model		Beta In	t	Sig.	Partial Correlation
1	GPusd	.417a	1.363	0.176	0.144
	GPeur	.281a	1.591	0.115	0.167
a. Predictors in the Model: (Constant), GPrmb					
b. Dependent Variable: SPsd					

It can be found in the effects of three different gold prices, the domestic gold price is the most significant one. Relatively speaking, the international gold price can be ignored. Therefore, the relationship of the company's gold shares with domestic and international gold prices can be expressed as:

$$SPsd= -322.994 + 2.688GPrmb \tag{3}$$

Among them, SPsd lists the gold company Shandong Gold's closing quotation share price, and GPrmb lists the China's domestic gold prices.

Similar statistical analysis showed the relationship between Zhongjin Gold share price and domestic/international gold prices can be calculated. The figure has been listed in Table 2.

It can be found that in the effects of three different gold prices, the domestic gold price is the most significant one, relatively speaking, the international gold price (euros) was relatively small, while the international price of gold (U.S. $) can be ignored. Therefore, the share price of the company's relationship with domestic and international gold prices can be expressed as:

$$SPszjh= -41.949 + 1.537GPrmb -0.345GPeur \tag{4}$$

Among them, SPzjh listed the gold company Gold's closing stock price, and GPrmb listed the China's domestic gold prices, Gpeur listed the international gold price (euros).

Thus, while on the single-factor, the gold price and foreign enterprises have a positive correlation between the share price and gold price, but when considering them together, the stock price is affected mainly by the domestic price of gold, and the impact of the price of international gold is very small, although the situation is different in different enterprises.

Table 2. The relationship between Zhongjin Gold share price and gold prices.

Model Summary						
Model		R	R Square	Adjusted R Square	Std. Error of the Estimate	
	1	.569a	0.323	0.316	25.21244	
	2	.730b	0.533	0.522	21.06972	
a. Predictors: (Constant), GPrmb						
b. Predictors: (Constant), GPrmb, Gpeur						
Coefficientsa						
Model		Unstandardized Coefficients		Standardized Coefficients		
		B	Std. Error	Beta	t	Sig.
1	(Constant)	-17.212	8.583		-2.005	0.048
	GPrmb	0.304	0.047	0.569	6.522	0
2	(Constant)	-41.949	8.183		-5.126	0
	GPrmb	1.537	0.2	2.871	7.681	0
	GPeur	-0.345	0.055	-2.347	-6.28	0
a. Dependent Variable: SPzjh						
Excluded Variablesc						
Model						Collinearity Statistics
		Beta In	t	Sig.	Partial Correlation	Tolerance
1	GPusd	-1.122a	-1.462	0.147	-0.154	0.013
	GPeur	-2.347a	-6.28	0	-0.556	0.038
2	GPusd	.578b	0.82	0.415	0.088	0.011
a. Predictors in the Model: (Constant), GPrmb						
b. Predictors in the Model: (Constant), GPrmb, GPeur						
c. Dependent Variable: SPzjh						

Validation. In order to verify the validity of the above regression analysis, authors compared the real share price with the calculated share price based on regression expression, and plotted the results in Figure 3 and Figure 4.

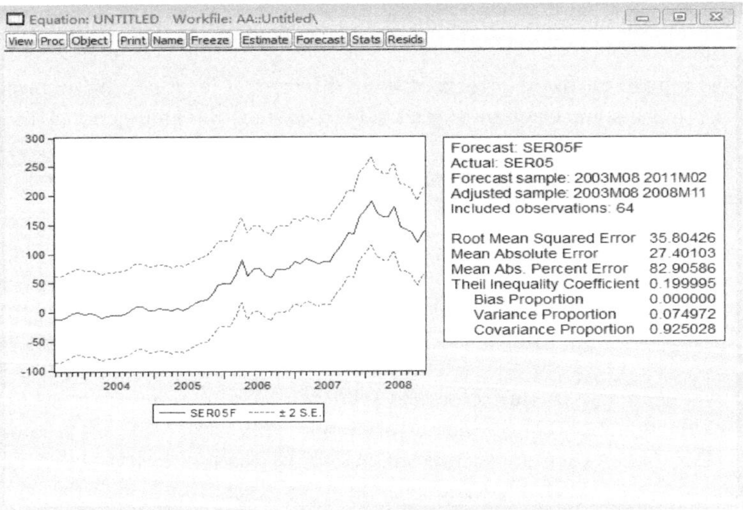

Fig. 3. Shandong Gold 08-11 year forecast

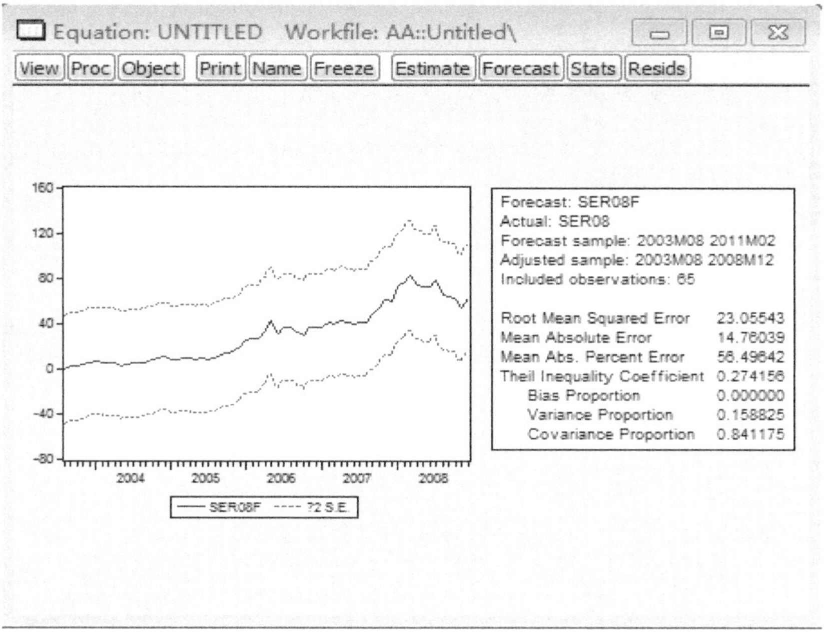

Fig. 4. Zhongjin Gold 08-11 year forecast

From the results in Fig. 3, the Root Mean Squared Error is 35.80, the Mean Absolute Error is 27.40, the Mean Abs. Percent Error is 82.90, Theil Inequality Coefficient is 0.2000, the proportion of variance and covariance rates are 0.0748 and 0.9250. These results suggest that based on the 2003 to 2008 data, it has a greater consistency between the predicted 08 to 11 year results and actual results.

The forecasted results for Zhongjin Gold in Fig. 4 show that the Root Mean Squared Error is 23.06, the Mean Absolute Error is 14.76, the Mean Abs. Percent Error is 56.50, The Inequality Coefficient is 0.2741, the proportion of variance and covariance rates are 0.1588 and 0.8412. These results suggest that based on the 2003 to 2008 data, the expected 08-11 year's results have a greater consistency with the actual results, thus they can provide some reference to investors.

4 Conclusions

The empirical researches in the recent 5 years and earlier, which study the share price of mainland China gold companies listed on main board, show that share price of the company generally positively correlate with domestic and international gold price, and the impact on stock prices of domestic gold price are significantly greater than the international gold price. Relatively, the impact of the international gold price (in Euro) is greater than the international gold price (in U.S. dollars). Rules for different companies are slightly different.

References

1. Zheng, X., Liu, J.: Literature review of China's gold market, market structure and prospects. Reforms and Strategies (26), 179–201 (2010)
2. Hu, E.: The dual property of their gold price determination mechanism. Gold Science and Technology (5), 1–7 (2005)
3. Hu, E.: Unlike other gold assets, an empirical analysis of the Cause. The Financial World (3) (2005)
4. Li, Y.: Empirical analysis of price fluctuations of gold investment and its influencing factors. Economic Research Guide (2) (2010)
5. Yan, Y., Zhou, S., Zhao, Y.: Recent years the international gold price trend analysis and forecasting. Store Modernization (12), 187–188 (2006)
6. Sun, J.: The creation of inter-bank gold market and to achieve the target model approach. Taxation of Financial (5), 93–95 (2008)

Mixed Generalized Hermite-Legendre Spectral Method

Zhang Xiao-yong

Department of Mathematics, Shanghai Maritime University, Pudong Avenue, 1550, Shanghai, 200135, China
zhxiyo@gmail.com

Abstract. In this paper, we propose mixed generalized Hermite-Legendre spectral method for partial differential equations set in a infinite strip. We investigate mixed generalized Hermite-Legendre approximation. We propose the new algorithm for various boundary condition problems.The global convergence of proposed algorithms are proved.

Keywords: Mixed generalized Hermite-Legendre approximation, Spectral method, General boundary condition problem.

1 Introduction

The spectral methods developed rapidly in the past three decades. The main merit of spectral methods is the high accuracy. However, the usual spectral methods are available only for periodic problems and problems on bounded rectangle domains. But many practical problems are set on unbounded domains. For these problems, the simplest way is to confine our calculation to a sufficiently large subdomain with certain artificial boundary condition. However, it causes additional errors. Thus it seems better to solve such a problem directly.

Recently, some authors developed the Hermite spectral method for differential equations defined on the half line and Hermite spectral method for differential equations defined on the whole line. But so far, Most of existing work about unbounded domain concern only one dimensional problems. Unfortunately, many differential equations are set on multi-dimensional unbounded domains. So, it is also interesting to consider spectral method for these problems.

The aim of this paper is to develop the mixed generalized Hermite-Legendre spectral method for differential equations defined in an infinite strip. We investigate various orthogonal projections and derive some mixed generalized Hermite-Legendre approximation results which serve as the mathematical foundation of related spectral methods for unbounded domains. As examples of their important applications, we propose the new numerical algorithms, in which the corresponding discrete linear system for unknown coefficients of numerical solution is symmetric and sparse and so can be resolved very efficiently. We introduce a general framework for error estimates of various boundary condition problems defined in an infinite strip. By using some results on this new mixed generalized Hermite-Legendre approximation, we obtain the error estimates for the proposed spectral scheme. we provide the numerical results, which demonstrate the high accuracy of this method.

G. Lee (Ed.): Advances in Automation and Robotics, Vol.2, LNEE 123, pp. 315–322.
Springerlink.com

2 Generalized Hermite Function Approximation

In this section, we first introduce some notations. Let $R=\{x|-\infty<x<+\infty\}$ and $\chi(x)$ be a certain weight function in the usual sense. For simplicity, we denote $\partial_x^k v(x)=\dfrac{d^k v(x)}{dx^k}$. We define

$$L_\chi^2(R)=\{v \ \ is \ \ measurable \ \ on \ \ R \ \ and \ \ \|v\|_{\chi,R}<\infty\}$$

with the following inner product and norm

$$(u,v)_{\chi,R}=\int_R u(x)v(x)\chi(x)dx, \ \ \|v\|_{\chi,R}=(u,v)_{\chi,R}^{\frac{1}{2}}.$$

For any integer $m\geq 0$, we define the space

$$H_\chi^m(R)=\{v\,|\,\partial_\chi^k v\in L_\chi^2(R), 0\leq k\leq m\},$$

equipped with the following inner product, semi-norm and norm

$$(u,v)_{m,\chi,R}=\sum_{0\leq k\leq m}(\partial_x^k u,\partial_x^k v)_{\chi,R},$$

$$|v|_{\mathrm{m},\chi,\mathrm{R}}=\|\partial_x^m v\|_{\chi,R}, \ \|v\|_{\mathrm{m},\chi,\mathrm{R}}=(v,v)_{m,x,R}^{1/2}.$$

For any $r>0$, the space $\mathrm{H}_\chi^r(R)$ and its norm $\|v\|_{\mathrm{r},\chi,\mathrm{R}}$ are defined by space interpolation as in[1].

Let $\omega_\theta(x)=e^{-\theta x^2}, \theta>0$. The generalized Hermite polynomials of degree l are defined by $\quad H_l^\theta(x)=(-1)^l e^{\theta x^2}\partial_x^l e^{-\theta x^2}, \mathrm{l}=0,\ 1,\ 2,\ \ldots$

Let $\widetilde{\omega}_\theta(x)=e^{-\frac{\theta x^2}{2}}, \theta>0$. The generalized Hermite functions of degree l are defined by $\quad \widetilde{H}_l^\theta(x)=\widetilde{\omega}_\theta(x)H_l^{(\theta)}(x), \mathrm{l}=0,\ 1,\ 2,\ \ldots$

The set $\widetilde{H}_l^\theta(x)$ is the complete $L^2(R)$-orthogonal system, namely,

$$(\widetilde{H}_l^\theta(x),\widetilde{H}_m^\theta(x))_R=\begin{cases}\gamma_l^{(\theta)}=2^l\theta^{l-\frac{1}{2}}l!\sqrt{\pi}, l=m,\\ 0, \qquad\qquad\qquad\qquad l\neq m\end{cases}$$

Now, let M be any positive integer and $P_M(R)$ is the set of all algebraic polynomials of degree at most M ▷we define Hermite function space $Q_{M,\theta}$ as

$$Q_{M,\theta}=\left\{e^{-\frac{\theta x^2}{2}}q(x)\,|\,q(x)\in P_M(R)\right\}.$$

The orthogonal projection $\widetilde{P}_{M,\theta}: \ L^2(R) \to Q_{M,\theta}$ is defined by

$$(\widetilde{P}_{M,\theta}v - v, \phi) = 0, \forall \phi \in Q_{M,\theta}.$$

Now we define $\qquad Av(x) = \partial_x v(x) + \theta x v(x)$.

For technical reasons, we introduce the space $H_A^r(R)$,which are defined by

$$H_A^r(R) = \left\{ v \ \ is \ \ measurable \ \ on \ \ R \ \ and \ \ \|v\|_{r,A} < \infty \right\}.$$

The norm $\|v\|_{r,A} = \|A^r v\|$. For any real $r > 0$, $H_A^r(R)$ and its norm are defined by space interpolation.

Theorem 2.1. For any $v \in H_A^r(R)$ *and* $r \ge \mu \ge 0$,

$$\left\|\widetilde{P}_{M,\theta}v - v\right\|_{\mu} \le c(\theta M)^{\frac{\mu - r}{2}} \|v\|_{r,A}.$$

Now we define the orthogonal projection $\widetilde{P}^1_{M,\theta}: \ H_A^1(R) \to Q_{M,\theta}$ *as*

$$(\partial_x(\widetilde{P}^1_{M,\theta}v - v), \partial_x\phi) + (\widetilde{P}^1_{M,\theta}v - v, \phi) = 0, \ \ \forall \phi \in Q_{M,\theta}.$$

Theorem 2.2. For any $v \in H_A^r(R)$ *and* $r \ge 1$,

$$\left\|\widetilde{P}^1_{M,\theta}v - v\right\|_1 \le (\theta M)^{\frac{1-r}{2}} \|v\|_{r,A}$$

Proof. From the projection theorem and theorem 2.1, it comes

$$\left\|\widetilde{P}^1_{M,\theta}v - v\right\|_1^2 = \inf_{\phi \in Q_{M,\theta}} \|\phi - v\|_1^2 \le \left\|\widetilde{P}_{M,\theta}v - v\right\|_1^2 \le c(\theta M)^{1-r} \|v\|_{r,A}^2.$$

The proof is complete.

Next, we define the orthogonal projection $\widetilde{P}^2_{M,\theta} : H_A^2(R) \longrightarrow Q_{M,\theta}$ by

$$\left(\partial_x^2\left(\widetilde{P}^2_{M,\theta}v - v\right), \partial_x^2\phi\right) + (\widetilde{P}^2_{M,\theta}v - v, \phi) = 0, \forall \phi \in Q_{M,\theta}$$

and $$\|v\|_{0,2}^2 = \left\|\partial_x^2\phi\right\|_{L^2(R)}^2 + \|v\|_{L^2(R)}^2 \ , \forall \phi \in Q_{M,\theta}.$$

Theorem 2.3. For any $v \in H_A^r\left(R\right)$*and* $r \ge 2$,

$$\left\|\widetilde{P}^2_{M,\theta}v - v\right\|_{0,2} \le c\left(\theta M\right)^{\frac{2-r}{2}} \|v\|_{r,A}.$$

Proof. Thanks to the projection theorem and theorem 2.1, one verifies readily that

$$\left\|\widetilde{P}^2_{M,\theta}v - v\right\|_{0,2}^2 = \inf_{\phi \in Q_{M,\theta}} \|\phi - v\|_{0,2}^2 \le \left\|\widetilde{P}_{M,\theta}v - v\right\|_2^2 \le c\left(\theta M\right)^{2-r} \|v\|_{r,A}^2.$$

This ends the proof.

3 Mixed Generalized Hermite-Legendre Approximation

In this section, we investigate some results about the mixed generalized Hermite-Legendre approximation.Let

$$I=\{\zeta \mid -1\le\zeta\le 1\}\, and\Omega = R\times I.$$

Since the measure dΩ is a product $dxd\zeta$, the following identify holds

$$L^2(\Omega)=L^2(I,L^2(R))andL^2(\Omega)=L^2(R,L^2(I)).$$

The following imbedding Lemma is obvious.

Lemma 3.1. For any nonnegative real numbers r and s, the following imbedding hold:

$$H^s(\Omega)\subset H^r(I,H^{s-r}(R))and\quad H^s(\Omega)\subset H^r(R,H^{s-r}(I)).$$

The following Lemma is useful.

Lemma 3.2. For any nonnegative integer m, we have the identity

$$H^m(\Omega)=H^m(I,L^2(R))\cap H^m(R,L^2(I))$$

The proof see (cf. [4]).
There also, a simple interpolation argument allows for deriving the general result: for any nonnegative real numbers s , the following identity holds

$$H^S(\Omega)=H^S(I,L^2(R))\cap H^S(R,L^2(I)).$$

Lemma 3.3. For any $v\in_0 H^1(\Omega)$, we have

$$\|v\|_{L^2(\Omega)}\le c|v|_{1,\Omega.}$$

Lemma 3.4. For any $v\in_0 H^2(\Omega)$, we have

$$\|v\|^2_{2,\Omega}\le c\left(\left\|\frac{\partial^2 v}{\partial x^2}\right\|^2_{L^2(\Omega)}+\left\|\frac{\partial^2 v}{\partial\zeta^2}\right\|^2_{L^2(\Omega)}\right).$$

Next, we recall the some results on Legendre orthogonal approximation. The Legendre polynomial of degree l is defined by

$$L_l(\zeta)=\frac{(-1)^l}{2^l l!}\partial^l_\zeta(1-\zeta^2)^l,\quad l\ge 0.$$

The set of $L_l(\zeta)$ is the complete $L^2(I)$-orthogonal system. namely

$$(L_l(\zeta),L_m(\zeta))_I=\begin{cases}\gamma_l=\left(l+\frac{1}{2}\right)^{-1}, & l=m,\\ 0, & l\ne m\end{cases}$$

Let ${}_0P_N^k(I)=\{\phi|\phi\in P_N(I)and\partial^l_\zeta\phi(+1)=\partial^l_\zeta\phi(-1)=0,0\le l\le k-1\}.$

We define the projection $\pi_N^{k,0} : {}_0H^k(I) \to {}_0P_N^k(I)$ such that

$$\int_{-1}^{1}\left(\partial_\zeta^k v - \partial_\zeta^k \pi_N^{k,0} v\right)\partial_\zeta^k \phi = 0 \quad \forall \phi \in {}_0P_N^k(I).$$

and the projection $\tilde{\pi}_N^k : H^k(I) \longrightarrow P_N(I)$ such that

$$\partial_\zeta^l \tilde{\pi}_N^k v(-1) = \partial_\zeta^l v(-1) \text{ and } \partial_\zeta^l v(1) = \partial_\zeta^l \tilde{\pi}_N^k v(1), 0 \le l \le k-1$$

The following results characterize the properties of $\tilde{\pi}_N^k$ and $\pi_N^{k,0}$.

Theorem 3.1. Let k is a integer. For any real numbers r and $s, 0 \le r \le k \le s$, there exists a positive constant c depending only on s such that, for any function v in $H^s(I)$,thefollowing estimate holds $\left\| v - \tilde{\pi}_N^k v \right\|_{H^r(I)} \le cN^{r-s} \|v\|_{H^s(I)}$.

The proof see (cf. [2]).

Theorem 3.2. Let k is a integer. For any real numbers r and $s, 0 \le r \le k \le s$, there exists a positive constant c depending only on s such that, for any function v in ${}_0H^s(I)$,the following estimate holds $\left\| v - \pi_N^{k,0} v \right\|_{H^r(I)} \le cN^{r-s} \|v\|_{H^s(I)}$.

The proof see (cf. [2]).

Next, we define the space $\tilde{P}_N^{1,0}(I) = \left\{ v \in P_N(I) \middle| \partial_\xi v(-1) = \partial_\xi v(+1) = 0 \right\}$.

and $$H^{1,0}(I) = \left\{ v \in H^1(I) \middle| \partial_\zeta v(-1) = \partial_\zeta v(+1) = 0 \right\}.$$

Let $\tilde{P}_N^{0,1}$ be the orthogonal projector $H^{1,0}(I) \to \tilde{P}_N^{1,0}$ defined by

$$\left(\partial_\zeta\left(\tilde{P}_N^{0,1} v - v\right), \partial_\zeta \phi\right) + \left(\tilde{P}_N^{0,1} v - v, \phi\right) = 0 \quad \forall \phi \in \tilde{P}_N^{1,0}(I).$$

Then, we have the following error estimate.

Theorem 3.3. For any nonnegative real numbers $2 \le s$,there exists a positive constant c depending only on s such that, for any function φ in $H^s(I) \cap H^{1,0}(I)$, the following estimate holds $\left\| \varphi - \tilde{P}_N^{0,1} \varphi \right\|_{H^1(I)} \le cN^{1-s} \|v\|_{H^s(I)}$.

Next, we define $P_{M,N}^1(\Omega) = \left\{ \varphi\phi \middle| \varphi \in Q_{M,\theta}, \phi \in \tilde{P}_N^{1,0}(I) \right\}$.

And $$P_{M,N}^{k,0}(\Omega) = \left\{ \varphi\phi \middle| \varphi \in Q_{M,\theta}, \phi \in {}_0P_N^k(I) \right\}.$$

The orthogonal projector $\tilde{\Pi}_{M,N}^1 v = \tilde{P}_{M,\theta}^1 \tilde{P}_N^{0,1} v$.

The orthogonal projector $\Pi_N^{k,0} : {}_0H^k(\Omega) \to P_{M,N}^{k,0}(\Omega)$ is defined by

$$\left|v-\Pi_N^{k,0} v\right|_{H^k(\Omega)} = \inf_{v_{M,N}\in P_{M,N}^{k,0}(\Omega)} \left|v-v_{M,N}\right|_{H^k(\Omega)}.$$

In order to describe the approximation results, we define the following space:

$$B^{s,s}(\Omega) = L^2(R,H^{1,0}(I)) \cap H^1(R,H^s(I)) \cap H_A^s(R,H^s(I)) \cap H^1(I,H_A^s(R))$$

with the norm $\|v\|_{B^{s,s}(\Omega)}^2 = \|v\|_{H^1(R,H^s(I))}^2 + \|v\|_{H_A^s(R,H^s(I))}^2 + \|v\|_{H^1(I,H_A^s(R))}^2$.

Using this norm, we obtain that

Theorem 3.4. For any nonnegative real number $2 \le s$, there exists a positive constat c depending only on s such that, for any function φ in $B^{s,s}(\Omega)$, the following estimate holds

$$\left\|v-\widetilde{\Pi}_{M,N}^1 v\right\|_{H^1(\Omega)} \le c(M^{\frac{1-S}{2}} + N^{1-S})\|v\|_{B^{S,S}(\Omega)}.$$

Theorem 3.5. Let k be a positive integer, $0<k\le 2$. For any nonnegative real numbers $k \le s$, there exists a positive constant c depending only on s such that, for any function v in $H^s(\Omega)\cap {}_0H^k(\Omega)$, the following estimate holds

$$\left\|v-\Pi_N^{k,0}v\right\|_{H^k(\Omega)} \le c\left(M^{\frac{k-s}{2}} + N^{k-s}\right)\|v\|_{H^S(\Omega)}.$$

4 Applications

In this section, we apply the mixed generalized Hermite-Legendre approximation to various boundary condition problems.

We consider the following problem
$$\begin{cases} -\Delta u = f, & in\Omega \\ u = 0, & on\ \partial\Omega \end{cases} \qquad (1)$$

If the distributionis f given in $H^{-1}(\Omega)$,it admits the following equivalent variational formulation: Find u in ${}_0H^1(\Omega)$ such that

$$a(u,v) = (f,v), \quad \forall v\in {}_0H^1(\Omega).$$

By virtue of Lax-Milgram Theorem, it has a unique solution. where the bilinear form is defined by $a(u,v) = \int_\Omega (gradu)(gradv)d\Omega.$

The spectral scheme for (4.1) is to find $u_{M,N} \in P_{M,N}^{1,0}(\Omega)$

$$a(u_{M,N}, v_{M,N}) = (f, v_{M,N}), \quad \forall v_{M,N} \in P_{M,N}^{1,0}(\Omega).$$

We have the following Theorem

Theorem 4.1. Let us assume the solution u belong to ${}_0H^s(\Omega)$, for any real number $s \geq 1$. Then, the following error estimate holds from Lemma 3.3

$$\left\|u-u_{M,N}\right\|_{H^1(\Omega)} \leq c\left|u-u_{M,N}\right|_{H^1(\Omega)} \leq c(M^{\frac{1-s}{2}}+N^{1-s})\left\|u\right\|_{H^s(\Omega)}.$$

and for any real number $r \geq 1, M \sim N$

$$\left\|u-u_{M,N}\right\|_{L^2(\Omega)} \leq c(M^{-\frac{s}{2}}+N^{-s})\left\|u\right\|_{H^s(\Omega)}.$$

Combining the above two reasults and using an interpolation argument, we derive the general estimate, $0 \leq r \leq 1 \leq s$:

$$\left\|u-u_{M,N}\right\|_{H^r(\Omega)} \leq c(M^{\frac{r-s}{2}}+N^{r-s})\left\|u\right\|_{H^s(\Omega)}.$$

Next, we consider the following fourth-order Dirichlet problem

$$\begin{cases} \Delta^2 u = f, \quad in\Omega \\ u=0, \quad \dfrac{\partial u}{\partial n}=0 \quad on \quad \partial\Omega \end{cases} \tag{2}$$

Let $\qquad a(u,v)=\int_\Omega (\Delta u)(\Delta v)d\Omega.$

When the distribution f is given in $H^{-2}(\Omega)$,a weak formulation of (4.2) is to find u in ${}_0H^2(\Omega)$ such that $a(u,v)=(f,v), \quad \forall v \in {}_0H^2(\Omega).$

By virtue of Lax-Milgram Theorem, it has a unique solution.

The spectral scheme for (4.2) is to find $u_{M,N} \in P^{2,0}_{M,N}(\Omega)$ such that

$$a(u_{M,N},v_{M,N})=(f,v_{M,N}), \quad \forall v_{M,N} \in P^{2,0}_{M,N}(\Omega).$$

Theorem 4.2. Let us assume the solution u belong to $H^s(\Omega)$, for any real number $s \geq 2$, the following error estimate holds from Lemma 3.4

$$\left\|u-u_{M,N}\right\|_{H^2(\Omega)} \leq c\left|u-u_{M,N}\right|_{H^2(\Omega)} \leq c(M^{\frac{2-s}{2}}+N^{2-s})\left\|u\right\|_{H^s(\Omega)}.$$

and $\qquad \left\|u-u_{M,N}\right\|_{L^2(\Omega)} \leq c(M^{-\frac{s}{2}}+N^{-s})\left\|u\right\|_{H^s(\Omega)}.$

The above estimates lead to the following Theorem

Theorem 4.3. Let us assume the solution u belong to $H^s(\Omega)$ for any real number $0 \leq r \leq 2 \leq s$: $\quad \left\|u-u_{M,N}\right\|_{H^r(\Omega)} \leq c(M^{\frac{r-s}{2}}+N^{r-s})\left\|u\right\|_{H^s(\Omega)}.$

Finally, we consider the following Neumann problem

$$\begin{cases} -\Delta w = F, & in\Omega \\ \dfrac{\partial w}{\partial n} = c & on \ \partial\Omega \end{cases} \tag{3}$$

Let $u = w + \frac{c}{2} e^{\zeta(1-\zeta^2)}$, we derive $\begin{cases} -\Delta u = f, & in\Omega \\ \dfrac{\partial u}{\partial n} = 0 & on \ \partial\Omega \end{cases}$ (4)

Next, we define ${}^0H^1(\Omega)$ as

$${}^0H^1(\Omega) = \left\{ v \middle| v \in H^1(\Omega), \partial_\zeta v(x,+1) = \partial_\zeta v(x,-1) = 0 \right\}.$$

We assume that the function f belong to $L^2(\Omega)$. Then, problem admits the equivalent variational formulation: find

$$a(u,v) = (f,v), \quad \forall v \in {}^0H^1(\Omega).$$

where the bilinear form is defined by

$$a(u,v) = \int_\Omega (gradu)(gradv) d\Omega + \mu \int_\Omega uv d\Omega.$$

The spectral scheme for (4.4) is to find $u_{M,N} \in P^1_{M,N}(\Omega)$ such that

$$a(u_{M,N}, v_{M,N}) = (f, v_{M,N}), \quad \forall v_{M,N} \in P^1_{M,N}(\Omega).$$

Theorem 4.4. Let us assume the solution u belong to $H^s(\Omega)$. For any real number $s \geq 1$, we obtain the following error estimate form Theorem 3.4

$$\left\| u - u_{M,N} \right\|_{H^1(\Omega)} \leq c(M^{\frac{1-s}{2}} + N^{1-s}) \left\| u \right\|_{B^{s,s}(\Omega)}.$$

Obviously, $\left\| u - u_{M,N} \right\|_{L^2(\Omega)} \leq \left\| u - u_{M,N} \right\|_{H^1(\Omega)} \leq c(M^{\frac{1-s}{2}} + N^{1-s}) \left\| u \right\|_{B^{s,s}(\Omega)}.$

References

1. Bergh, J.: Lofstrom, Interpolation Spaces, An Introduction. Springer, Berlin (1976)
2. Guo, B.-Y.: Spectral Methods and Their Applications. World Scientific, Singapore (1998)
3. Hardy, G.H., Littlewood, J.E., Plya, G.: Inequalities. Cambridge University Press, Cambridge (1952)
4. Lions and Magenes, Problemes aux Limites non Homogenes et Applications, vol. 1, Dunod, Paris (1968)
5. Zhang, X.-Y., Guo, B.-Y.: Spherical harmonic-generalized Laguerre spectral method for exterior problem. J. Sci. Comp. 27, 523–537 (2006)

Artificial Neural Networks Based on Fractal Growth

Li Zhao*, Weidong Li, Liqing Geng, and Yanzhen Ma

Tianjin Key Laboratory of Informational Sensing & Intelligent Control,
Tianjin University of Technology and Education
300222 Tianjin, China
{jinshihui,dfbblwd}@163.com

Abstract. Artificial Neural Networks(ANN) has been increasingly used by researchers. An ANN model based on fractal growth is introduced in this paper. The design was inspired by the physiology of biological nervous system. Driven by data, fractal structure generates and evolves. Classification interface forms from the fractal structure, which makes the classifier have the capacity on following changes. And a tailor method was designed to change the fractal structure which strengthened the following capacity. The voting framework brings the networks a strong antijamming capability by means of considering several targets at the same time. The design was realized on platform VC/MFC and applied on classification of SEMG. The classification accuracy rate was 91.59%. This model is expected to work well in fields like pattern recognition of weak signal, financial forecast, etc..

Keywords: Fractal artificial neural networks, Fractal growth, Pattern recognition, SEMG, Voting framework.

1 Introduction

Artificial neural networks is an distributing parallel information processing algorithm simulating biological neural networks[1]. Now it has been developed into a widespread application and used in many fields[2] such as pattern recognition, data mining, etc.. Nevertheless, it has drawbacks like poor antijamming ability, complicated design and poor dynamical following ability. In our work, the idea of growth was brought into the design of artificial neural networks which improved these drawbacks to some extent.

The early artificial neural networks had a constant topological structure. But the real problems often change over the time, thus the past design idea was restricted. In addition, it was hard for developers to design the structure for every application. The growing neural networks can adjust its structure to follow through the changes, so it's especially suitable for the occasions where the structure of the problems are not clear and the mathematical models are hard to create. Now there are already some growing neural networks, commonly known as Evolutionary Artificial Neural Networks (EANN)[3,4,5].

* Corresponding author.

G. Lee (Ed.): Advances in Automation and Robotics, Vol.2, LNEE 123, pp. 323–330.
Springerlink.com

Fractal was first put forward by French-American mathematician professor Benoit B.mandelbrot in 1975.The core characteristic of fractal is the self-similarity[6,7].

Fractal is a general growing mechanism. Road network, root system, veins, bronchial, valley, etc., almost any growth can be described by fractal. Growth is the expansion of a individual in a specific space, and fractal structure is useful to growing stability and efficiency.

The neural networks based on fractal growth proposed in this paper learns a lot from brain. The process of growing imitates the brain's studying new knowledge, and the process of pruning imitates the brain's oblivion. The studying with oblivion ensures the brain keep pace with the times and not be jammed by the massive amount of information. Our design which imitates the brain has the analogous advantages.

2 Design of Fractal Artificial Neural Network

Generally, artificial neural networks needs a certain amount of empirical data to train itself. The feature vector including the input value and the class information is abstracted from the empirical data. As a classifier, the fractal artificial neural networks introduced in this paper was designed with a similar idea. When several targets are considered and each target has a set of features, we may abstract a group of values which make up the input feature vectors. Each target has a feature space with a same dimension with the target's feature vector. For example, the feature space with two features is a plane.

Take four classification problem with three targets for example, the framework of the fractal artificial neural networks is shown below (Fig.1) . The networks consists of three parts. Perception layer(feature space) is on the left. For each target, two characteristic values are extracted, so the feature space is a plane. The node *Cij* (Subscript *i* is target number, *j* is class number) in the plane is the initial class center. The fractal structure will grow in the feature space which corresponds to the perception layer of the neural networks.

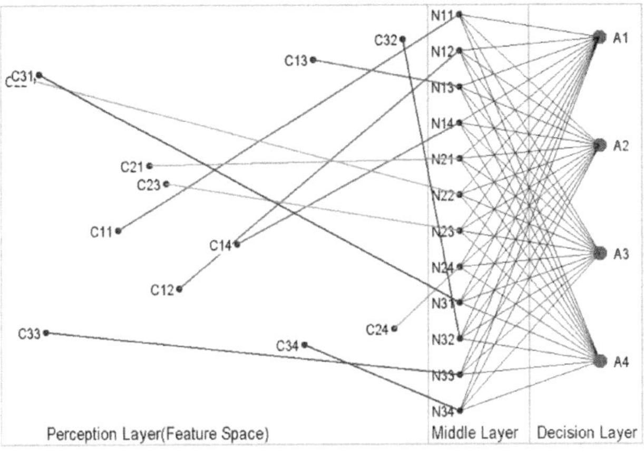

Fig. 1. Fractal artificial neural networks framework

This model uses a "voting framework ", that is, each target can contribute to each class, but for each input, it can only contribute to one class. This contribution is similar to a democratic vote. Voting framework provides a redundant feature, which ensures the validity and stability of the classification and improves the anti-interference ability of the model.

The middle layer is right next to the perception layer. In Fig.1, *Cij* is connected to middle layer node *Nij* one by one with a weight *Wij*. The values of the middle layer node can be calculated by:

$$N_{ij} = f(v) * W_{ij} \tag{1}$$

In the formula above , W_{ij} is a weight representing the stability and validity of the path, $f(v)$ is a function of v , which stands for the cost of a new node's joining up to the fractal structure. In brief , the further the distance, the bigger the cost, and the smaller the $f(v)$.

The decision layer is right after the middle layer. Each node's value of the decision layer is calculated as the weighted sum of the nonzero value in the middle layer, which is to measure the probability of input's belonging to a certain class . Comparing the outputs of the classifier with the feedbacks, we may make changes to the weights.

Fractal growth is reflected in the evolution of the classification structure. In Fig.2, the plane is divided into four areas which means the fractal growth leads to a classifier.

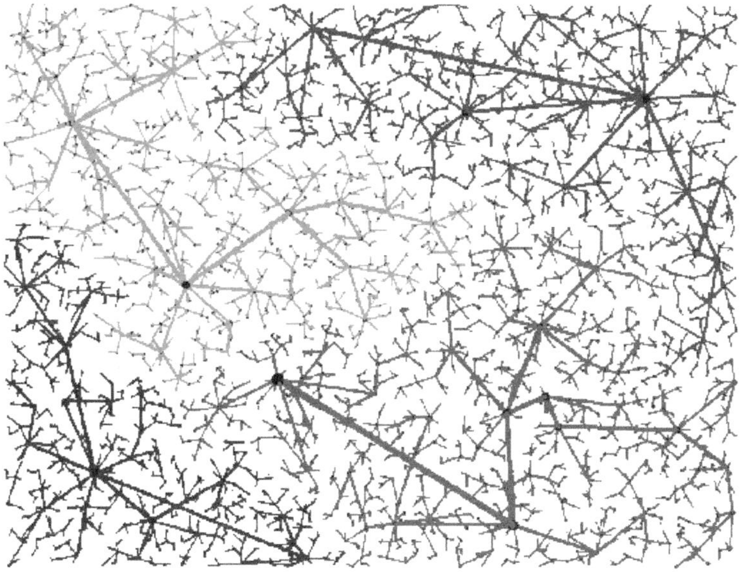

Fig. 2. Fractal growth leads to a division of the plane.(The data generated randomly)

When a new data point was input to the feature space, different operations would be took in different stages:

1. In the stage of initialization, the average value of a certain number of data with the same class information was calculated, and the average value would be set as *Cij*(the center of the class), which was shown in Fig. 1.
2. In the training stage, a new data point was still carrying class information. It searched for the shortest path to connect to its class structure and then become a new branch of the structure.
3. In the testing stage, by comparing the class calculated by the classifier with the feedbacks, we got the classification accuracy rate, then determine whether the model can be put into service;
4. In the using stage, new data point no longer carried class information. This model uses a pruning algorithm to make the networks dynamically follow the changes of the situation.

The division of the feature space was depicted in Fig. 2. In the final model, several targets were considered, and the algorithm weight each target and then make final decision.

3 SEMG Pattern Classification Based on Fractal Artificial Neural Networks

The design was realized on platform VC/MFC, and verified with SEMG(surface electromyogram signal) pattern classification problem.

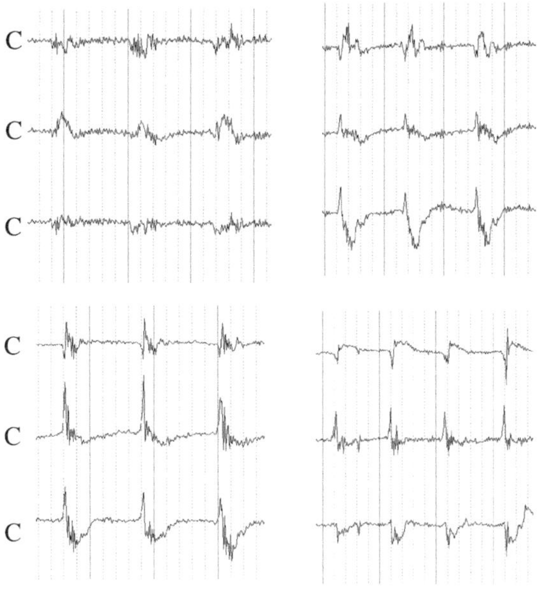

Fig. 3. SEMG accompanying the pressing key action. Top left : little finger ; top right : Ring finger ; down left : index finger ; down right : middle finger.

SEMG is a kind of nerve electrical signal[8], which produces along with the muscle action. Different EMG signal patterns accompany different actions. In this paper, the classification of SEMG signal that produced by actions of pressing key with the all fingers but thumb was completed.

The silver/silver chloride powder electrode were used in the experiment. Because of the good auto-adapted ability of the classifier, accurate location were not required. The data used in this paper was a collection of off-line data. And the SEMG signal waveform of pressing key actions of four fingers are shown in Fig. 3.

The process of using this algorithm for the SEMG pattern classification is expounded as follows:

(1) Action detection

The task is to detect the occurrence of the target action. At present, algorithms such as the short-time Fourier method, the self-organizing artificial neural network method and moving average method are commonly used in action detection. Considering the actual processing effect and the calculation efficiency of embedded system application performance, the moving average method was adopted in our work. When the short-term energy broke through a empirical threshold, an action was supposed to occur, and then the subsequent steps would be carried out.

(2) feature extraction

The purpose of feature extraction is to describe the signal with a group of values which can be representative attribute of it's class, and take the values as the characteristics of this class. In order to facilitate classification, there should be significant differences among the characteristics extraced from the different action types. 100 data points were collected after each action started (sampling rate is 1000Hz, 0.1s of data), take the average absolute value(*MAV*) and the average energy value *E* as the two characteristic values:

$$MAV_k = \frac{1}{N}\sum_{i=1}^{N}|S_k(i)| \tag{2}$$

$$E_k = \sum_{i=1}^{N} S_k^{\,2}(i) \qquad 3)$$

In these formulas, *N* is the data points, *S* is the signal sequence , *k* is channel serial number. *MAV* and *E* have certain correlation, but as a second order parameter, *E* can improve the influence of the large value points. The cooperation between the two parameters can play the role of distinguishing different actions types. Standardization process, that is, the operation of mapping the value of the parameters, *MAV* and *E*, to a specific range, should be performed before inputting them to the neural network , In order to facilitate the calculation and visualization, values of both parameters are mapped to an data region of 0 to 600.

The distribution of feature vectors got from the data which were collected in the experiment was shown in Fig.4. It is obvious that there are some overlaps among the different action types in a single channel. Thus it would be difficult to distinguish these four action types using a single channel accurately with the feature extraction method adopted. However, when multiple channels were considerded together, different actions can be distincted at higher accuracy by adjusting the weights between middle layer and decision layer of the networks.

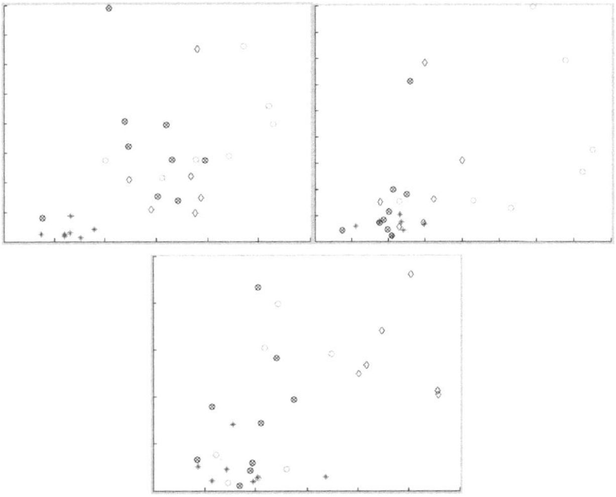

Fig. 4. Distribution of the feature vectors. Each map stands for a channel(target), and each color(point type) stands for one action type(class).

(3) Networks training

Input the feature vectors that obtained from the off-line data to the fractal neural network, then perform the initialization, training, and pruning according to feedback. The distribution of the feature vectors is shown in Fig. 5.

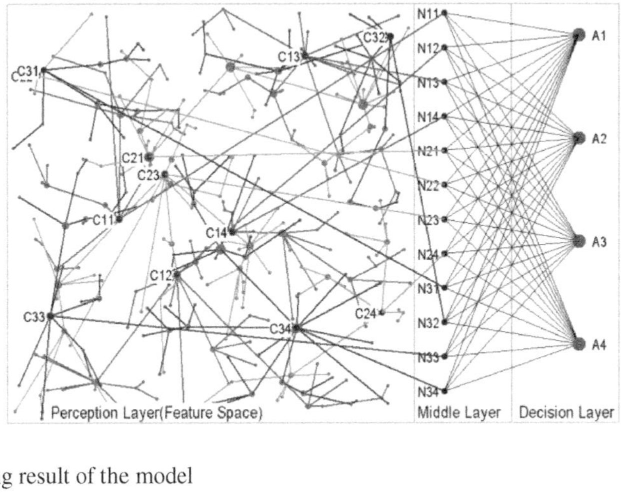

Fig. 5. Training result of the model

(4) Model testing

Take 600 sets of offline data (150 sets for each action type) to test the model. The evolutionary result of the network model is shown in Fig. 6 and the result is shown in Table 1.

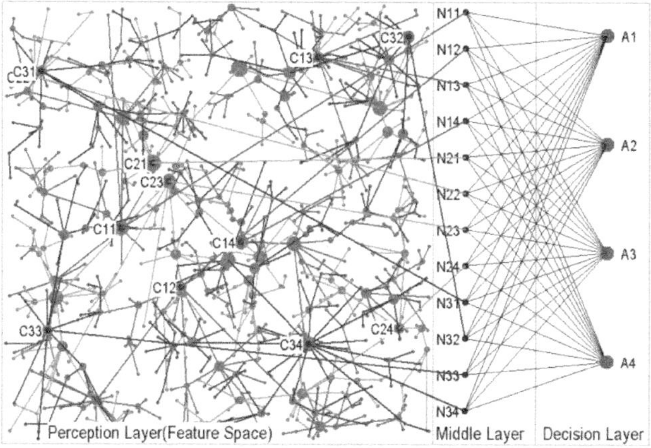

Fig. 6. Evolutionary result after testing

Table 1. The result of the experiment

Action	Index finger	Middle finger	Ring finger	Little finger	total
False alarm rate (%)	4	8	3.33	4.67	5
Missing-report rate (%)	6	2.67	12	10.67	7.83
* accuracy rate (%)	94.07	94.03	93.70	84.25	91.59

***Note**: The final classification accuracy rate considered the false alarm action, but not considered the missing-report action.

For the SEMG that with low amplitude and high noise, a classification accuracy rate of 91.59% is a good validation for the feasibility and efficiency of the model. In Table.1, the little finger action type has a relatively high missing-report rate. By adjusting the electrode position, a higher accuracy rate can be expected. The two characteristics used in this experiment have a certain correlation, by adopting other characteristics such as power spectrum ratio, Mean Power Frequency (MPF) and Median Frequency (MF), a improvement is expected in further research.

4 Conclusion

The essence of pattern classification is to achieve the division of a feature space through constructing classification interface. In this paper, the division of the feature space is realized through fractal growth driven by data (Fig.2), and combined with the advantages of artificial neural networks, a fractal neural networks was designed and realized. It has the following advantages:

1) The classification interface based on fractal growth can be any shape without the limitation of connected set or convex set. So the model is especially suitable for linear inseparable problem.

2) The voting framework provides a redundant feature which enhance the validity and stability of the classification.

3) The model has strong self-adaptive ability and simple parameters, so it's convenient to use.

4) By using the pruning algorithm, the model can follow the change of the condition in time, while in the other hand, keep a proper scale that ensures the speed of operation.

Based on these advantages, this model has a wide prospect in many fields, especially suitable for pattern recognition problems in complex dynamical systems, such as weak signal pattern recognition, stock forecasting and economic forecasting, etc..

Acknowledgment. This work was supported by the National Natural Science Foundation of China (Grant No. 61178081), also supported by the 2011 issues of the National Science Education "Tenth Five-Year Plan" (BFA110049)and Tianjin Municipal Science and Technology Plan Project Application(09ZCKFGX02100). And thanks for Tianjin Key Laboratory of Informational Sensing & Intelligent.

References

1. Qiu, D.: A genetic neural network model based on a trend examination method and engineering application 40(3) (2010)
2. Sta, N., Davoodi, M.: Designing a multivariate-multistage quality control system using artificial neural networks. International Journal of Production Research 47(1), 251–271 (2009)
3. Yao, X.: A review of evolutionary artificial neural networks. International Journal of Intelligent Systems 8(4), 539–567 (1993)
4. Brown, A.D., Card, H.C.: Evolutionary artificial neural networks[J], Electrical and Computer Engineering. In: IEEE 1997 Canadian Conference on Electrical and Computer Engineering, vol. 1, pp. 313–317 (1997)
5. Rivero, D., Dorado, J., Rabuñal, J., Pazos, A.: Generation and simplification of Artificial Neural Networks by means of Genetic Programming. Neurocomputing 73(16-18), 3200–3223 (2010)
6. West, B.J.: Fractal physiology and the fractional calculus: a perspective. Frontiers in Physiology 1, 1–17 (2010)
7. Circum, S.: Mechanical study on the formation of floc fractal structures. Electronic Publishing (2007)
8. Dong-yue, X.U., Ke-yong, L.: The Analysis Method and Application of Electromyographic Signal. Biomedicine 10(13), 2593–2596 (2010)

Phase Transitions of Traffic Flows Based on Traffic Flow Simulation in the Merging Section on Freeways

Xingju Wang* and Xiaoming Xi

School of Traffic and Transportation, Shijiazhuang Tiedao University, Shijiazhuang 050043, Hebei, China
wangxingju@stdu.edu.cn
xxm19871031@sina.com

Abstract. In order to alleviate traffic congestions and the influence on environmental impact, developing a model to evaluate under various geometric and traffic conditions is desirable, which is the objective of this study. This study develops a novel traffic simulation model consisting of car following model and lane change model based on desired speed on freeways. Phase transitions of traffic flows are reproduced by using this simulation model.

Keywords: following model, lane change model, traffic flow simulation, traffic congestions, phase transitions.

1 Introduction

Increasing dependence on car-based travel has led to the daily occurrence of recurrent and nonrecurrent freeway congestions not only in the China but also around the world. Congestions on freeways form when the demand exceeds capacity. Recurrent congestion reduces substantially the available infrastructure capacity at rush hour, i.e., at the time this capacity is most urgently needed, causing delays, increasing environmental pollution, and reducing traffic safety. Similar nonrecurrent congestion caused by traffic incidents such as crashes, disabled vehicles, effects are observed in the frequent case of adverse weather conditions, work zones, special events and other temporary disruptions to the freeway transportation system. Therefore, a dynamic traffic flow simulation model on freeways which is applied to emulate state of traffic flow is needed.

Corresponding to the physical parts of a highway network, the objects of simulation can be classified into three types: a single highway road section, a single merging section, and the whole network. According to these objects, there are three classes of simulation models. The traffic simulation models can also be divided into a macro model and a micro model, according to the method for describing the traffic flow. A macro model treats vehicle flows as continuous physical variables, while a micro model simulates the behavior of individual vehicles and the interaction among adjacent vehicles.

As a major assessment measure, the traffic flow simulation has been widely used to conduct a sensitivity analysis in merging section on freeways. To evaluate travel time,

* No.17 North 2nd-Ring East Road, P.O. BOX 050043, Shijiazhuang, Hebei, China.

G. Lee (Ed.): Advances in Automation and Robotics, Vol.2, LNEE 123, pp. 331–338.
Springerlink.com

speed and density under various geometric and traffic conditions, a dynamic traffic flow simulation is developed. Various parameters output, including travel time, traffic flow rate, mean speed and density, can be applied to explore an optimal profile of ramp length for minimum congestion subject to inflow of a merging section on freeways. This paper proposes a micro model for describing the traffic flow in simulation model consists of Intelligent Driver Model [1] and lane change model presented by this study. Congestion on freeways generally exhibits distinct patterns according to temporal variations in traffic flow. Phase Transitions of traffic flows are reproduced, and sensitivity analysis of acceleration lane length is conducted. It is shown that when traffic flow rate of mainline was less, the effectiveness of acceleration was good, and vice versa phase. Transitions are also reproduced [2], [3].

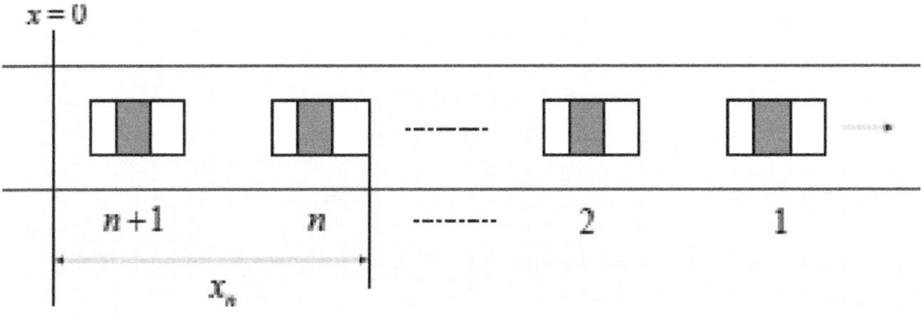

Fig. 1. Car following behavior

2 Theories

2.1 Car Following Model

In a micro-simulation model, a modeled fundamental behavior is the "car following" which adjusts the driver's characteristics: the distance between two adjacent cars, the relative speed, etc.

In 1953, Pipes proposed the following basic differential equation model for car following behavior [4]:

$$\ddot{x}_{n+1}(t) = a[\dot{x}_n(t) - \dot{x}_{n+1}(t)] \tag{1}$$

where $\ddot{x}$, $\dot{x}$ and x denote the acceleration, speed and distance from the reference point of vehicle n respectively; and a is a constant. In the model, the acceleration of the vehicle which follows a leading vehicle is proportional to the speed difference between the vehicles. It is assumed that the delay of time the vehicle responds to the speed difference is so small that it can be neglected. To remove this drawback, Chandler introduced a reactive delay time T [5].

Based on the rationale that the acceleration of the following car is also influenced by its speed and the distance between the vehicles, Gazis, Herman and Rothery proposed the type of car following model [6]:

$$\ddot{x}_{n+1}(t+T)=\frac{a[\dot{x}_{n+1}(t+T)]^m[\dot{x}_n(t)-\dot{x}_{n+1}(t)]}{[x_n(t)-x_{n+1}(t)]^l} \quad (2)$$

Newell proposed the following model in which the acceleration is propositional to an exponential function of the distance between the vehicles, based on the measurement result [7]:

$$\ddot{x}_{n+1}(t+T)=a_1[\dot{x}_n(t)-\dot{x}_{n+1}(t)]\times\ell^{-(a_2/[x_n(t)-x_{n+1}(t)-a_3])} \quad (3)$$

Although the above modifications have improved the reality of car following model, they have two drawbacks, as following below.

(1) When the proceeding vehicle does not exist (equivalently, the distance to the preceding car approaches infinity, the acceleration is 0, independent of the speed and the speed difference) this implies that a car will maintain an initial speed if there is not a preceding car.
(2) When the speed difference is 0, the acceleration is 0, independent of the distance between two cars at all. This implies the unrealistic phenomenon that the following car will not apply the brake even when the distance to the preceding car approaches 0, and will not accelerate even if the distance is very long.

To solve the above-mentioned problems, Treiber and Helbing introduced the Intelligent Driver Model, in which a desired speed and a shortest distance between cars are introduced. The IDM is given as:

$$\dot{v}_n=a[1-(\frac{v_n}{v_0})^\delta-(\frac{s^*(v_n,\Delta v_n)}{s_n})^2] \quad (4)$$

$$s^*(v,\Delta v)=s_0+\max(Tv+\frac{v\Delta v}{2\sqrt{ab}},0) \quad (5)$$

$$s_n(t)=[x_{n-1}-x_n-l] \quad (6)$$

$$\Delta v_n(t)=[v_n(t)-v_{n-1}(t)] \quad (7)$$

where x =distance; n =the n th car; v =velocity; l =length of car; s_0 =desired minimum gap; a =maximum acceleration; s^* =effective gap; b =comfortable deceleration $(a \le b)$; δ =parameter; T=time gap; and v_0 =desired velocity.

2.2 Lane Change Model

2.2.1 Physical Condition

To simulate driver's behavior in the merging section on freeways and the merging behavior in the weave section, etc. the lane change model is needed [8]. We propose anew lane change model in which driver's behavior is described by judgment functions. We focus on a vehicle approaching to a confluence point and describe its behavior with several variables: the relative speed between the vehicle and vehicles in current lane, the locations of both the main line cars and the on-ramp cars, Driver's judgment functions for changing his lane and driver's desired speed. The driver's judgment function for the free merging is different from the judgment function for the forced merging. A free merging implies that a car on ramp can merge into the main line without influences and vehicles on the main line are not interfered. The forced merging occurs when cars on the main line decelerate or accelerate by inflows from the ramp and the drivers are compelled to change their behaviors. As in Figure 2, the lane change model with driver's judgment function is expressed as:

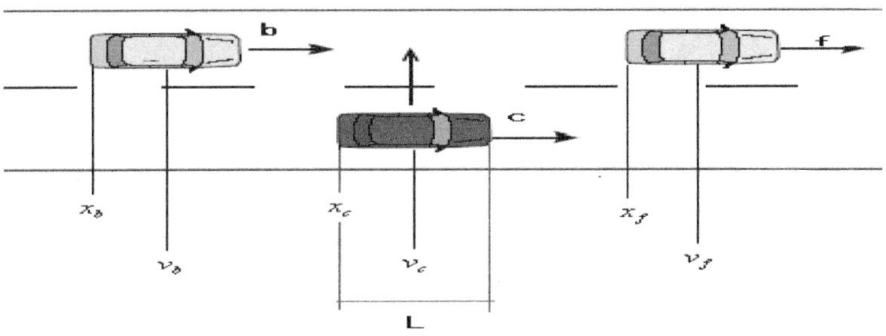

Fig. 2. Lane change behavior

$$h = x_f - x_c - L + (v_f - v_c)t + (-A + B)t^2 / 2 + \delta \frac{(v_{0f} - v_f)}{v_{0f}} S + \zeta \frac{(v_{0c} - v_c)}{v_{0c}} S \geq S \tag{8}$$

$$g = x_c - x_b - L + (v_c - v_b)t + (A - B)t^2 / 2 - \theta \frac{(v_{b0} - v_f)}{v_{b0}} S - \xi \frac{(v_{c0} - v_c)}{v_{c0}} S \geq S \tag{9}$$

$$0 \leq A \leq e \tag{10}$$

$$0 \leq B \leq d \tag{11}$$

where h, g= judgment function; x = distance from reference point; v= speed; L=length of a vehicle; t=the judgment time; v_0=desired speed, subject to normal distribution; $\delta, \zeta, \theta, \xi$ ($\delta, \zeta, \theta, \xi \in [0,1]$) = adjustment coefficient; A= rapid acceleration with upper bound e; and B= rapid deceleration with upper bound d. Note that parameters A and B are associated with vehicle c's judgment functions for lane change and decide the free merging or the forced merging. Since vehicle c judges to accelerate or decelerate to merge into the main line, two events are mutually exclusive.

The function h judges whether vehicle c accelerates or decelerates to merge according to given space and speed conditions between vehicles f and c. Similarly, the function g is applied to judge in the relationship between vehicles c and b.

If both A and B take 0, the distance between two vehicles f and b is large enough for vehicle c to be accommodated to move into the main line, then the free merging occurs (no acceleration or deceleration behavior is required for vehicle c). Conversely, in the case of the forced merging, we need to examine whether the solution of Inequality equations (8) to (11) exists. If A and B are mutually exclusive, then the following two conditions (I) and (II) are got:

(I) When a rapid brake event B does not exist, then B=0 and only an event A could happen;

(II) When a rapid acceleration event A does not exist, then A=0 and only an event B is approved. The lane changing behavior of vehicle c could happen when (I) or (II) has a solution.

2.2.2 Psychological Constraints

Psychological constraints describe driver's motivations on lane change [9],[10].

(1) If the present car has not reached the desired speed;

(2) If the predicted speed of lane change is greater than that of no change, or gain speed advantage;

The following inequality systems show the above problems,

$$V_c < V_0 \tag{12}$$

$$V_1 < V_0 \tag{13}$$

$$V_2 < V_0 \tag{14}$$

$$V_1 < V_2 \tag{15}$$

where V_c =present speed ; V_0=desired speed; V_1 and V_2= predicted speed of lane change and no lane change respectively. The predicted speed equation is shown as:

$$V_1 = V_c + a_1 t \tag{16}$$

$$V_2 = V_c + a_2 t \tag{17}$$

where t= unit time; a_1 and a_2 =predicted acceleration of lane change and no lane change; respectively. a_1 and a_2 are given from the inequalities(4),(5),(6)and(7). Then the inequality (15) can replace by

$$a_1 < a_2 \tag{18}$$

If the inequalities (12), (13), (14) and (18) have a solution, the driver has maneuvers of changing the current lane to the target lane. Conversely, the driver does not conduct the lane changing maneuvers.

Lane change behaviors can be characterized as a sequence of three stages:

(1) The ability of lane change (Physical condition);
(2) The motivation of lane change (Psychological constraints); and
(3) The execution of lane change.

When constraint models of psychological condition and physical condition are both satisfied, the driver conducts the above-mentioned three stages. Otherwise, the driver continues the car following behavior without lane change behaviors.

2.3 Traffic Flow Simulation Model

We develop a traffic flow simulation model consisting of car following model, called IDM, and lane change model. The basic concept of car-following theories is the relationship between Stimuli and Response. In the classic car-following theory, the stimuli is represented by the relative speed of following and leading vehicle, and the response is represented by the acceleration (or deceleration) rate of the following vehicle[11]. The car following model describes following behaviors that drivers follow each other in the traffic stream on only one lane. To reproduce the traffic flow in two or more lanes, lane change model which explores lane change behaviors is needed. By using the car following model and lane change model, we express dynamic and complex traffic behaviors in two or more lanes[12], [13].

3 Data and Measures

To conduct sensitivity analysis of acceleration lane length generating time varying input flows and demonstrate features of traffic flows, numerical traffic flow simulations in merging section on freeways are given in this study by java. In our simulation, we set main line length on freeways to 1000m, ramp length to 200m, and length in merging sections of mainline and ramp to100m. Three combinations of traffic flows on main line and ramp are listed as shown in Table 1.

Table 1. Trafficflow of phase transition

Case	Group numbe r	Traffic flow rate on ramp(pcu/h)	Traffic flow rate of lane 1 on mainline(pcu/h)	Traffic flow rate of lane 2 on mainline(pcu/h)
	1	900	1500	1300
A	2	800	1200	1200
	3	700	1500	1500
	1	1100	1300	1300
B	2	1000	1500	1500
	3	1150	1800	1400

4 Results and Discussions

Case A in table 1 presents free merging behaviors in merging section on freeways, and Case B in Table 1 explores the localized cluster phenomenon of traffic congestions.

Moreover, Case B in Table 1 shows the homogeneous congestion. In addition, the simulations reproduce the phase transitions of traffic flows called the Synchronized flow-wide moving jam (S-J) when the parameters settings of 1150 pcu /h on ramp ,1800 pcu /h on lane 1 of main line and 1400 pcu /h on lane 2 of main line. Curves of phase transitions are shown in Figure 3.The left of phase transitions point is the synchronized flow, and the right is wide moving jam. In the initial stage of simulations, congestions firstly occurs on main lane 1 on freeways, and as time goes on, lane 2 goes into congestion state that density and speed are homogeneous and synchronized.

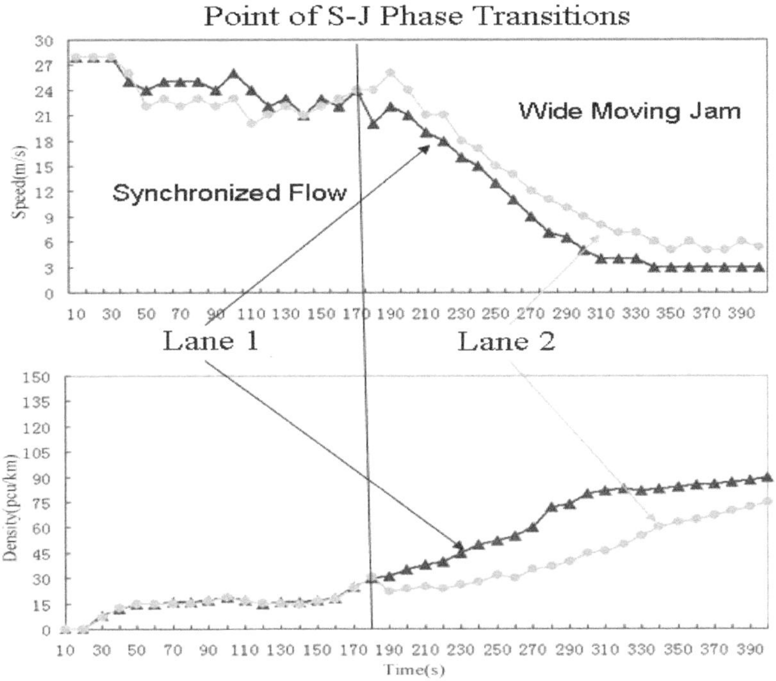

Fig. 3. S-J phase transitions

5 Conclusions

A traffic flow simulation model is developed in this study, which provides a useful tool to simulate car movements based on driver's behaviors, calculate the average speeds and estimate the density. Moreover, according to user specified time interval, this model is applied to conduct a sensitivity analysis of acceleration lane to demonstrate effectiveness of acceleration lane extension for alleviating traffic congestions. In addition, the phase transitions of traffic flows called the Synchronized flow-wide moving jam is reproduced by using the traffic simulation model.

As discussed in simulations, the proposed traffic flow simulation model has demonstrated itself a potential approach to explore and quantify the benefit of optimal ramp metering on various acceleration lane alignments and input flows in the main line and the ramp. The results indicated that the effectiveness of acceleration lane length was good when traffic flow rate of mainline was less, and vice versa. In addition, phase Transitions of traffic flows are reproduced by using this simulation model when the parameters settings of 1150 pcu /h on ramp ,1800 pcu /h on lane 1 of main line and 1400 pcu /h on lane 2 of main line. In the initial stage of simulations, congestions firstly occurs on main lane 1 on freeways, and as time goes on, lane 2 goes into congestion state that density and speed are homogeneous and synchronized.

Acknowledgement. This research is founded by the National Natural Science Foundation of China (Grant No. 51008201). And this research is also sponsored by the Scientific Research Foundation for the Returned Overseas Chinese Scholars, State Education Ministry of China. Moreover, this research is also the key project supported by the Scientific Research Foundation, Education Department of Hebei Province of China(Grant No. GD2010235).

References

1. Treiber, M., Hennecke, A., Helbing, D.: Congest traffic state in empirical observations and microscopic simulation. Physical Review E 62, 1805–1824 (2000)
2. Kerner, B.S.: Synchronized flow as a new traffic phase and related problems for traffic flow modeling. Mathematical and Computer Modeling 35, 481–508 (2002)
3. Kerner, B.S.: The Physics of Traffic: Empirical Freeway Pattern Features, Engineering Applications, and Theory. Springer, Germany (2004)
4. Pipes, L.A.: An Operational Analysis of Traffic of Traffic Dynamics. Journal of Applied Physics 24(3), 274–287 (1953)
5. Chandler, R.E., Herman, R., Montroll, E.: Traffic Dynamics: Studies in Car Following. Oper. Res. 6(2), 165–184 (1958)
6. Gazis, D.C., Herman, R., Rothery, R.W.: Nonlinear Follow-the-Leader Models of Traffic Flow. Oper. Res. 9(4), 545–567 (1961)
7. Newell, G.F.: Nonlinear Effects In the Dynamic of Car Follow. Oper. Res. 9(2), 209–229 (1961)
8. Peter, H.: Modeling vehicle interactions in microscopic simulation of merging and weaving. Transportation Research Part C: Emerging Technologies 13(1), 37–62 (2005)
9. Heydecker, B.G., Addison, J.D.: Analysis and modelling of traffic flow under variable speed limits. Transportation Research Part C 19(2), 206–217 (2011)
10. Treiber, M., Kesting, A.: Evidence of Convective Instability in Congested Traffic Flow: A Systematic Empirical and Theoretical Investigation. Procedia-Social and Behavioral Sciences 17, 683–701 (2011)
11. Castelli, L., Pesenti, R., Ranieri, A.: The design of a market mechanism to allocate air traffic flow management slots. Transportation Research Part C 19(5), 931–943 (2011)
12. Sumalee, A., Zhong, R.X., Pan, T.L., Szeto, W.Y.: Stochastic cell transmission model(SCTM): A stochastic dynamic traffic model for traffic state surveillance and assignment. Transportation Research Part B 45(3), 507–533 (2011)
13. Soriguera, F., Robusté, F.: Estimation of traffic stream space mean speed from time aggregations of double loop detector data Original Research Article. Transportation Research Part C 19(1), 115–129 (2011)

Multi-Agent Systems Simulation Base on HLA Framework

Xuehui Wang and Lei Zhang

School of Computer. 17,
National University of Defense Technology
Changsha, Hunan province, P.R. China
findyanzi@126.com

Abstract. Recently there has been considerable interest in Multi-agent systems(MAS). However, it is very difficult to accurately design and implement simulation of MAS, in respect that the systems are often extremely complex and based on autonomous software and hardware components, termed as agents, which cooperate within an environment to perform some task. Obviously interoperability is one of the key roles to play in the investigation and development of Multi-agent systems. This paper begins with introduction to High Level Architecture; following, provides an emphatic introduction to agent architecture. Then we outline an approach how to make the MAS simulation accord with the HLA framework. Especially show how the agents simulated on different machines can become a legal Agent Federate joining the flexibly MAS simulation federation. Also emphatically describe the simulation cycle of the MAS system.

Keywords: Multi-agent systems, high level architecture, interoperability, Agent Federate, simulation cycle.

1 Introduction

As we known, an agent in general can be viewed as a self-contained, concurrently executing thread of control, and it encapsulates some state and communicates with environment and other agents via some sort of mechanism. Multi-agent systems are often extremely complex and difficult to exactly verify their attributes. As a result, simulation is an important tool and plays a key role in the research of agent-based systems, even MAS, it allows the designer to learn more about the behavior of a system or to investigate the implications of the agent architectures, also the researcher can probe the relationships between the agent architectures, the environments and the behaviors. The application of simulation is more effective to partially control experimental conditions and facilitates the replication of results.

Agent-based simulation modeling, generally referred to multi-agent system modeling, rapidly activate in a wide range of areas, including the business modeling, computer games, control of mobile robots and the military simulations. Furthermore, the MAS offer much more great advantages when independently developed components must operate among heterogeneous environments. However the relative new trend to use of intelligent

G. Lee (Ed.): Advances in Automation and Robotics, Vol.2, LNEE 123, pp. 339–346.
Springerlink.com

agents has increased dramatically during the last decade of years. Such systems based on the clever software and/or autonomous hardware components which are termed as the agent. And then it cooperates within environment to perform some task. It is the urgent requirement for agents to communicate and cooperate in MAS; accordingly the agents cooperate with one another to achieve a specific mission. There is an obvious requirement to find an efficient ways and means of the information exchange between agent and environment.

Commonly used simulation techniques often do not address the special concerns of the artificial intelligence. In particular, existing simulation methods used in artificial intelligence research are often non–reused. In fact, multi-agent systems typically are the parallel and distributed event simulation. Therefore many MAS simulation are able to take advantage of existing work and production in parallel and distributed simulation for design and realization distributed, efficient, and repeatable MAS simulations.

Within the modeling and simulation domain, the High Level Architecture (HLA) is established to facilitate the interoperability between all types of models and simulations, e.g. behavior modeled as intelligent agents. The Runtime Infrastructure (RTI) is the basal software component of HLA; the RTI makes it possible for the different simulations to interoperate. And also it provides commonly required services to each simulation for coordination of operations and data exchange during a HLA federation execution. These services are specified in the HLA Interface Specification.

2 HLA Framework and RTI

The High Level Architecture (HLA) developed by the Department of Defense in the United States first describes to provide a concrete approach to integrate, or federate, separate, autonomous simulators into a single, distributed simulation system. It aims to promote the reuse and interoperation of the various simulations. The intent of the HLA is to provide a structure that supports reuse of different simulation systems, ultimately reducing the required cost and time to create a synthetic environment for a new purpose.

Figure 1 illustrates the HLA framework on the whole. The HLA consists of three parts: the HLA rules, the Object Model Templates (OMT) and the interface. The standard of HLA does not prescribe a specific implementation, nor does it request to use the any particular set of software or programming language. It assumes that new technological improvement become available, and different implementations be possible within the framework. A federation of HLA consists of several interacting simulations, termed as federate. It may be computer simulation, a manned simulator, a supporting sensor, or a live interface. All objects represent the internal states of the federate. The HLA requires every federate to allow their objects to interact with else objects in other federates in virtue of exchange of data.

Data exchange and a variety of other services are realized by Runtime Infrastructure (RTI). The RTI is, in effect, a distributed operating system for the federation. The RTI provides a general set of the services that support the simulations in carrying out these federate-to-federate interactions and federation management support functions. All interactions among federates entirely go through the RTI.

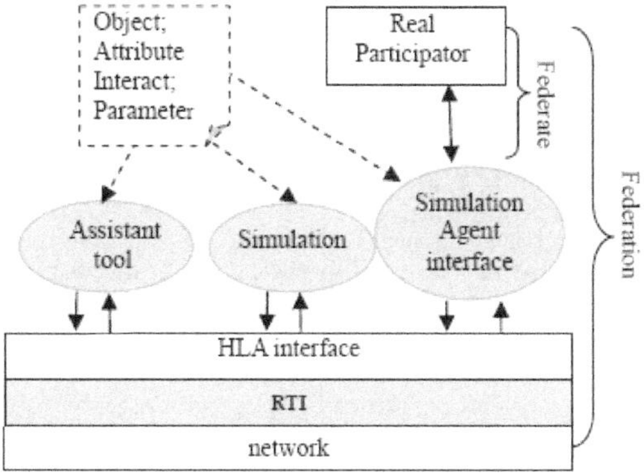

Fig. 1. HLA Framework

3 Multi-Agent System Initialization

In order to achieve flexibility and adaptability, most federates are more or less configurable to be able to participate in different federations and scenarios. Behavior, characteristics and responsibilities of a federate can be modified to suit a specific federation execution by providing initialization information.

Multi-agent system initialization is composed of two parts. On one hand, Agent Federate initialization is the necessary activities performed to make the federation ready for scenario; On the other hand, scenario initialization is account for exchange of data to ensure that MAS Federation is ready to run the scenario.

Taking account of Agent Federate come from partially autonomic, self-contained agent, the initialization of agent not just include initial values of public Federation Object Model(FOM) attributes and specifying internal objects, but also contain configuration the internal KB with initial values. In addition, the information about responsibility for register public objects as well as update object attributes still must be provided.

In Agent Federate a Simulation Object Model (SOM) is associated with description of the agent's sensory system, namely input and output. The SOM is based on the type description of information that the agent can reason about. Together with the specified SOM, it is enough to extract which object instances are internal to the Agent Federate however which instances are external public to the federation.

And then we have selected offline methods and design patterns for providing information and performing initialization for Agent Federates. So-called offline is that the initialization information is furnished with external files offering the Agent Federate to read during its initialization. The files are generated offline through separate equipment or tools with a universal or special format. Offline pattern gives the federate developer the flexibility or selectivity to provide different scenario information without modifying the federate itself. Initial values of both private and public objects and their attributes are specified when agents modeling the behavior and relationships. Therefore Agent Federate can produce the necessary initialization information.

4 Brief Introduction of Agent Group

Capability of single agent is very limited, so that need cooperate with each other in order to fulfill a complicated mission. Therefore, the thought state of the appointed Agent Group is one of the key problems to be settled in MAS research. Mainly include unite faith and unite intention, etc.

(1) Union faith: They commonly are denoted with mode logic. The unite faith of multi-agent is not simple addition one by one among the faith of single agent, but is recursive definition, which format express as follows:

Agent x toward Agent y together faith:
$M, \omega(BMB \cdot xyp) \equiv \forall n \forall (\omega, \omega') \in B[x, y, n] M, \omega' p$
B is faith arithmetic operators,
$B[x, y, n]$ is recursively defined as $B[x, y, 1] = B_l$
$B[x, y, n+1] = B_l \bullet B[y, x, n]$
Agent x toward Agent Group τ together faith:
$(BMB \cdot x\tau p) \equiv \forall z \tau(z) \supset (BMB \cdot xzp)$
Agent Group τ toward Agent x together faith:
$(BMB \cdot \tau xp) \equiv \forall z \tau(z) \supset (BMB \cdot zxp)$
The unite faith of Agent Group:
$(MB\ \tau_1 \tau_2 p) \equiv (BMB\ \tau_1 \tau_2 p) \cap (BMP\ \tau_2 \tau_1 p)$

(2) Union intention: Among a multi-agent system , in the event that every agent undertakes to fulfill a collective action, it generally indicates that unite intention of come into being. The Joint Persistent Goal JPG is the unite intention of whole member. Here also being conclusion:

(JPG x y p)→ (PGOAL x p) ∧(PGOAL, y p)

The meaning is that if agent X and agent Y both have sustained unite intention P, we can educe that agent X and/or agent Y severally has the sustained intention P; However, contrariwise, agent X and/or agent Y severally has the sustained intention P not always have sustained unite intention P.

5 Simulation of MAS Accord with HLA

Various approaches to parallel simulation have been developed in the field. Decentralized, event driven distributed simulation is particularly suitable for modeling systems with inherent asynchronous parallelism, such as multi-agent systems.

5.1 Time Management

Time management usually assumes the simulation consists of a collection series of logical processes (LPs) that communicate by exchanging time-stamp messages or events. Each simulation Agent Federate can be viewed as a single LP. The goal of the synchronization mechanism is to ensure that each LP deals with events in time-stamp order. This requirement is referred to as the Local Causality Constraint (LCC).

In traditional Parallel and discrete event simulation, the set of state variables of a simulation model is decomposed into subsets. Each of these is assigned to a logical process that manages the corresponding sub state. Namely this approach seeks to divide the simulation model into a network of concurrent LPs. In other words, parallel simulation consists of a number of LPs, typically each one running on a separate processor. The simulation progress is ensured by the processes scheduling new events to be executed in the future, and execute these events in the time-stamp order. A process can schedule an event for itself, or remotely, for another process. In the latter case, a message is sent via the network to the remote process.

Each process maintains a separate local simulation time clock, called the Local Virtual Time(LVT). This value represents the process's local view of the global simulated time and denotes how far in simulated time the corresponding process has progressed.

Synchronous approaches utilize global synchronization schemes to force the LPs to advance together in lock step and thus to keep them tightly synchronized. In contrast, LPs in asynchronous simulation advance at completely different rates, simultaneously processing events which occur at different simulated times. This approach has greater potential for speedup, but additional synchronization mechanisms are required to ensure that the LPs adhere to the LCC and process messages in increasing timestamp order. Asynchronous simulation approaches are further categorized as conservative and optimistic. Briefly, conservative algorithms take precautions to avoid the possibility of processing events out of time stamp order, i.e., the execution mechanism avoids synchronization errors. On the other hand, optimistic algorithms use a detection and recovery approach. Events are allowed to be processed out of time stamp order, however, a separate mechanism is provided to recover from such errors.

We look each agent as a single Agent Logical Process (Agent-LP). Similarly, we assume that objects and processes within the agents' environment are modeled as one or more Exterior Logical Process (Exterior-LP). And we adopt an optimistic mechanism of events execution in MAS federation. Thus Agent-LP does not idly wait for potential events from other processes, such as Exterior-LP and other Agent-LP. But assumes that the current set of scheduled events is accurate and executes them.

5.2 Time Management

In a conventional event-driven distributed simulation, the processes interact with each other in a small scope and in a little number. The layout of the simulation is determined by the decomposed processes of the simulated system, and mostly is static. However, the particular characteristic of multi-agent systems is often difficult to determine which decomposing approach is more appropriate. Consequentially, a simulation of MAS usually requires a very large set of shared variables which could be accessed or updated by the agents, such as whether they were at the right position in the right time etc. Which variables the agents can access/update in fact, the variables depend on both the state of the agents and the state of the environment. The completely resulting communication of the shared state variables is extremely costly and results in the loss of the advantages of distributed simulation.

As we discussed above, each process deal with a portion of the state space of the system. State changes are modeled as time stamped events in the simulation. Here two types of events are distinguished, namely internal events and external events. The former has a causal impact only to the state variables of one LP, and the latter may also have an impact on the states of other LPs, namely more than one LPs, which are typically modeled as time stamped messages exchanged between the LPs involved. If a state variable is read or updated by external events that are generated by more than one logical process, then the variable is said to be shared.

Agent-LP interacts with Exterior-LP and other Agent-LP via external events. The purpose of this interaction is to exchange information regarding the values of the shared state variables, which define the agent's environment and interfaces between the Exterior-LP.

The method to avoiding broadcast communication is termed Data Distribution Management (DDM), which mainly has been explained in the context of real time large scale simulations in HLA. This technique utilizes filtering mechanisms to provide each process with the subset of information which is relevant to it (e.g., its location or other needed specific attributes). Our approach currently avails of the DDM services provided by the RTI.

5.3 The Simulation Cycle of MAS

The Figure 2 illustrates the MAS Federation Execution step. Now we provide an emphatic introduction to this execution cycle and work flow.

Step1: Agent Execution runs on RTI/Agent server, in order to establish Federation MAS execution. At first, Agent Federate Execution (AFE) is set up by a certain Agent Federate, which is contact with Agent Execution using prior promissory communication channels. Secondly, AFE registers itself on Agent Execution, to make other Agent Federate known who is manager, moreover apply to join or exit federation execution. Simultaneously this Agent Federate becomes a manager in the local AFE during the whole system execution.

Step2: Simulation system circulates itself to wait for a period of time, which gives the opportunity for other Agent Federate to join the federation. At the same time the manager tracks record concerned information of Agent Federate, such as entity type, entity number, and entity response to environment, etc. All of these are basic information for the manager to control the process.

Step3: It will publish or subscribe the information of the Agent Federate. MAS directly employ the public and subscribed characteristic of HLA. This approach helps to filter redundant information by utilizing the announcement of publish and subscribe.

Hereto, it is ready for MAS simulation to implement. Thus every Agent Federate begins to run respective program, and exchange information through agent software. Moreover under RTI/Agent uniform control, Agent Federate advances its time according to others.

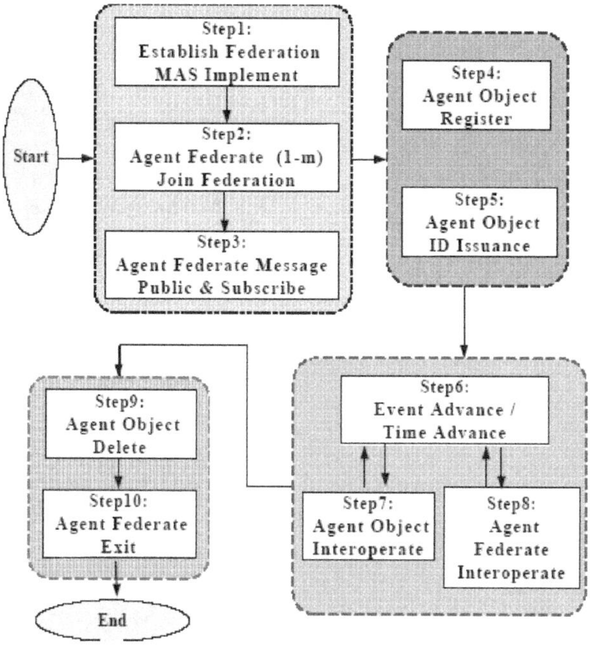

Fig 2. MAS Simulation Federation Work Flow

Step4: Total Agent Object (AO) register. A group of agent constitutes Agent Federate, who deals with Agent Object according to the generated object. When an AO comes into being during Agent Federate Execution (AFE), AFE register this AO to Agent software, and then get the only ID of this AO from RTI/Agent.

Step5: Publish the ID of the AO. Agent software informs other Agent Federate by sending messages, including the original state and data of AO.

Step6, step7 and step8: Event and time advance. Agent Federate Execution can be regarded as the process that Agent Federate circularly deals with itself interior entity object in MAS. Entities receive and dispose exterior event, including entity attribute value update and interaction information. All these missions are realized through the common time management server interface.

Step9: When an AO is not any longer being, it will be eliminated from simulation system, in order to release the system resource possessed. Similarly Agent Federate also need notify Agent software at the same time, afterward Agent software notifies to other Agent Federates by sending messages to eliminate the interior representation of this object.

Step10: The Agent Federate who manages the Federation MAS is responsible to termination the MAS simulation.

6 Summary

In this paper, we present a novel approach to simulate the multi-agent systems. To sum up, this approach makes partly use of ways and means in parallel and distributed simulation, at the same time adopts interoperability idea which embody in High Level Architecture framework. For one thing, we design and construct the agent architecture and explain the function of its subassembly. For another thing, we give some fundamental conceptions of Agent Group which are available during the whole MAS simulation. Meanwhile we show how the agents can become a legal Agent Federate joining the flexibly MAS simulation federation. Further more this paper emphatically expatiates on the simulation cycle of the MAS system.

However the further work is required to characterize the performance of the MAS simulation based on HLA with different kinds of agents and environments. Another area for the future work is inter-operation, using HLA to integrate MAS simulation with other simulators. This would allow investigating the different agent architectures and environments by the different simulators in a straightforward way. Besides the key issue is one of specific interfaces for the sensor and the data. Currently we are in the process of developing a set of inter-operability guidelines and data structure for MAS simulations.

Acknowledgments. This research is supported by the National Natural Science Foundation of China under grant numbers 60803100 and 61103190.

References

1. Lees, M., Logan, B.: Simulating Agent-Based Systems with HLA: The Case of SIM AGENT—Part II. Euro Simulation Interoperability Workshop, 03E–SIW–076 (2003)
2. HLA IEEE 1516, `http://www.ieee.org`
3. Logan, B., et al.: Distributed Simulation of Agent-Based Systems Case for Support
4. Löfstrand, B., Byström, J., et al.: HLA Design Patterns for Agent Based Federates. In: Euro Simulation Interoperability Workshop, 03E–SIW–076 (2003)
5. Namatame, A., Morita, H., Matsuyama, K.: Agent-Based Modeling for the Study of Diffusion Dynamics. In: Proceedings of the 2009 Spring Simulation Multiconference, San Diego,USA, pp. 1–14 (2009)
6. Craenen, B.G.W., Theodoropoulos, G.K.: Interfacing Multi-Agent Models to Distributed Simulation Platforms: The Case of Pdes-Mas. In: Proceedings of the 2010 Winter Simulation Conference, pp. 587–594 (2010)
7. Iskra, K.: Parallel Discrete Event Simulation Issues with Wide Area Distribution. In: ASCI course a9, March 7 (2003)

A Novel Feature Selection and Extraction Method for Sequence Images of Lip-Reading*

Chuanzhen Rong and Zhenjun Yue

Institute of Sciences, PLA Univ. Of Sci. & Tech. Nanjing, China
rcz@foxmail.com, yuezhenjun2002@yahoo.com.cn

Abstract. Lip-reading and lip identification are the hot spots in the technology of human-machine interaction. Researchers have been committed to work out the key issue that how to extract the most effective features for lip reading. This paper selects feature points and feature parameters for lip-reading based on FDP and FAP defined by MPEG-4. This paper uses an improved snake model on the location and extraction of the feature points. The algorithm has better robustness and has improved the accuracy of contour extraction. Then the SFFS method and correlation analysis are exploited to obtain the quicker optimal sub-aggregate and eliminate the redundancy features. Experimental results of the real data show that the features extracted from the sequence image have a high recognition rate in lip-readings recognition system.

Keywords: lip-reading, feature selection, feature extraction, snake, SFFS.

1 Introduction

Lip-reading and lip-identification are the hot issues of human-computer interaction. Lip-movement technique is used to help pepole who are weak in hearing to learn and understand the expression of normal persons, it can also be used to obtain information on specific occasions (such as intelligence, etc.). Research on lip-movement includes basic lip movements, visual features of lip-movement, feature extraction, description and representation, as well as the lip movement recognition and understanding, and the comprehensive understanding of information from multiple channels. Feature extraction plays a core role and has a significant impact on recognition rate in

Methods of Chiou[1] and Yao Hongxun[2] which adopt color information is the better solution to the separate issue. U.S. AT & T Labs Potamianos use features based on contour and features based on pixel-based image transformation[3](such as discrete wavelet transform, discrete cosine transform, Walsh transform, KL transform-PCA), and combine HMM with GMM. He created a new idea of feature extraction, and had a good result in image sequence processing. In order to explore more representative and effective lip features, studying robust feature extraction method is very meaningful. This paper selects lip feature points and lip feature vectors based on the FDPs(face defined parameters) and FAPs(facial animation parameters) defined by

* This work is supported by pre-study fund of PLA Univ. Of Sci. & Tech(No.20110504).

G. Lee (Ed.): Advances in Automation and Robotics, Vol.2, LNEE 123, pp. 347–353.
Springerlink.com

lip-movement. So how to extract the most effective features for lip-reading is a key issue for the researchers.

MPEG-4[4][5][6]. Use an improved snake model[7]to locate and extract the lip feature points. This method has high precision and better robustness. At the same time, use the SFFS (Sequential Floating Forward Search)[8]combined correlation analysis to search the optimal feature subset. Reduced the feature dimension and system complexity. Experimental results show that the selected feature subset in this paper has a high recognition rate of the lip language recognition system.

2 Selection, Location and Extraction Lip Feature Points

This paper selects lip feature points based on FDPs(as Figure 1 shows) defined by MPEG-4. According to the characteristics of lip-reading, just select feature points from the Region of Interest. Select 21 feature points(including 8 inner lip feature points,10 outer lip feature points, the bottom points of the chin, the left corner points of the chin, right corner points of the chin). These points are obtained through a large number of experimental analysis, can truly reflect the changes in facial expression, with high robustness and high flexibility.

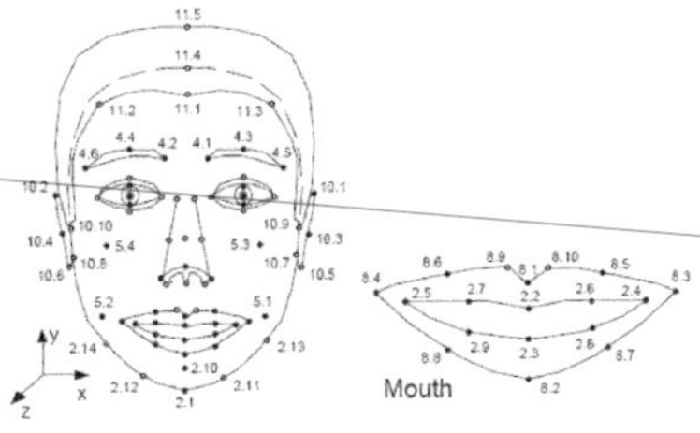

Fig. 1. FDPs

In the location and extraction of feature points, this paper use a new active shape model . In this algorithm, the energy function written as:

$$E=\sum_{i=0}^{n-1}\left[\alpha(i)E_{\text{int}1}(i)+\beta(i)E_{\text{int}2}(i)+\gamma(i)E_{image}(i)+\sigma(i)E_{hard}(i)\right] \quad (1)$$

Where n is the number of control points on the contour. $E_{\text{int}1}$ is the first order continuity constraint; $E_{\text{int}2}$ is the second-order continuity constraint; E_{image} is gradient amplitude, control some image features. E_{hard} is the energy generated by the hard constraint, here use the radio of color components R and G to describe it. Parameters $\alpha,\beta,\gamma,\sigma$ used to balance the relative interaction of the above four parameters. Defined as: $\alpha=1$, β is 0 or 1(depending on whether the point is a corner point), $\gamma>1.0$. This option makes E_{image} have a great impact on deciding the movement of a point on the contour than $E_{\text{int}1}$. σ depends on the characteristics of specific problem and the imposing force, if impose constraint fore on some points, σ take positive values, otherwise 0. This algorithm has strong adaptability, quicker convergence, and can also change the number of control points adaptively to make the Snake's control points better described the shape and ensure a stable track. The result as flows:

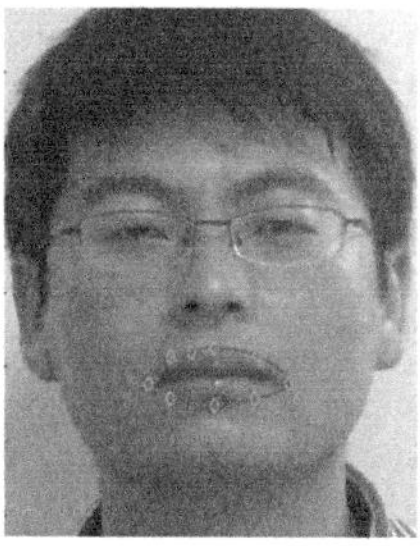

Fig. 2. Initial state

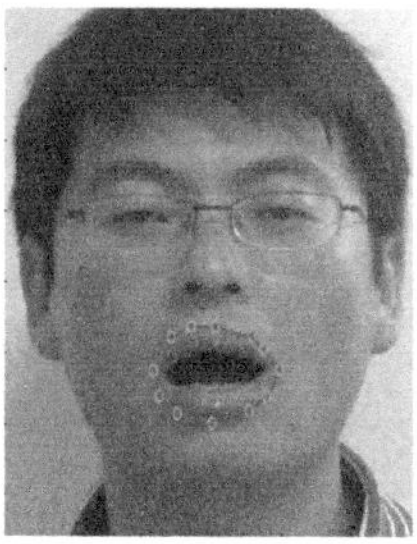

Fig. 3.. The state of 'a'

3 Selection and Obtaining of Lip Feature Parameters

Considering the versatility and flexibility of the description, we use the FAPs (Facial Animation Parameter) defined by MPEG-4 as the feature parameters for lip-reading. Considering that the position of the moth can be affected when we pronounce, we choose 24 FAPs, as Table 1 shows:

Table 1. Feature parameters for lip-reading

FAP#	name	FAP#	name
3	open_jaw	16	push_b_lip
4	lower_t_midlip	17	push_t_lip
5	raise_b_midlip	51	lower_t_midlip_o
6	stretch_l_cornerlip	52	raise_b_midlip_o
7	stretch_r_cornerlip	53	stretch_l_cornerlip_o
8	lower_t_lip_lm	54	stretch_r_cornerlip_o
9	lower_t_lip_rm	55	lower_t_lip_lm_o
10	raise_b_lip_lm	56	lower_t_lip_rm_o
11	raise_b_lip_rm	57	raise_b_lip_lm_o
12	raise_l_cornerlip	58	raise_b_lip_rm_o
13	raise_r_cornerlip	59	raise_l_cornerlip_o
14	thrust_jaw	60	raise_r_cornerlip_o

In these parameters, FAP3 and FAP14 define the displacement of the jaw; FAP4, FAP5 and FAP8~FAP13 define the vertical displacement of the 8 points of the inner-lip; FAP6, FAP7 and FAP53, FAP54 define the horizontal displacement of the inner lip corner and the outer lip corner respectively; FAP51, FAP52 and

FAP55~FAP60 define the vertical displacement of the 8 points of the outer-lip; FAP16 and FAP17 define the depth displacement of the top middle lip and the bottom middle lip.

Method on acquisition of the FAPs for the lip-shape set is: when the people pronounce, first we use a camera to take the frontal image of the given lip-shape, then specify the given feature points of the model(as figure 4 and figure 5 show), in the end we can get the values of the 24 FAPs by calculating the displacement between the lip-shape and silent model[9]. As different people has different mouth shape, so it is not representative just collected the FAPs from specifical person. In this paper, we take images of different people, and calculate multi-group FAPs and use the average as the final FAPs. The formula of calculating the FAPs as follows:

$$fap'(n,i) = \frac{\sum_{j=0}^{m-1}(fap_j(n,i))}{m}, n = 1,2,\ldots,22, i = 1,2,\ldots,24 \quad (2)$$

Among them, $fap(n,i)$ represents the ith FAP of the nth lip-shape for the people, m represents the number of the people. So we can get the FAPs for the basic lip-shape. Table 2 shows the FAPs of the lip-shape 'a':

Fig. 4. The natural image

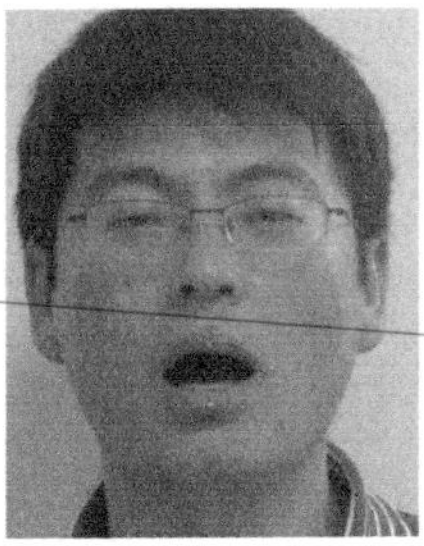

Fig. 5. The 'a' image

Table 2. FAPs For Lip-shape 'a'

FAP#	value	FAP#	value	FAP#	value	FAP#	value
3	520	9	-100	16	-30	55	-50
4	-105	10	-610	17	0	56	-50
5	-700	11	-610	51	-100	57	-510
6	-30	12	-150	52	-510	58	-510
7	-30	13	-150	53	-45	59	-560
8	-100	14	0	54	-45	60	-260

We can get the FAPs for rest lip-movement using the same method. In the end, we get the feature parameters for lip-reading.

4 Reduction of the Feature Vectors

There is redundancy in the feature vectors obtained by the above method. We need to reduce the relevant vectors to further reduce the feature dimensions, and thus can select a small number of effective features to constitute an optimal feature subset, and also can reduce the complexity of the system. Sun and other methods use SFFS to search feature subset[10], and use within-class scatter matrix S_W and between-class scatter matrix S_B to define the cost function:

$$J = T_r(S_W^{-1} S_B)$$

Where Tr (•) represents the trace of the matrix, S_W and S_B are defined as follows:

$$S_W = \sum_{c=1}^{L} \sum_{n=1}^{N_c} (y_c^n - m_c)(y_c^n - m_c)^T$$

$$S_B = \sum_{c=1}^{L} N_c (m_c - m)(m_c - m)^T$$

y_c^n is the sequential feature vectors of the samples for all types, n=1,2,..., N_c, c=1,2,...L, and $N_{total}=N_1+N_2$; $m_c = \frac{1}{N_c}\sum_{n=1}^{N_c} y_c^n$ is the mean vector of various types of samples. The SFFS can avoid nesting problems, but can not remove the correlation of the selected feature subsets. In this paper, use SFFS to search the feature subsets, and at the same time use correlation analysis to remove the redundancy of the feature subsets, the algorithm process shows in Figure 6 [11]:

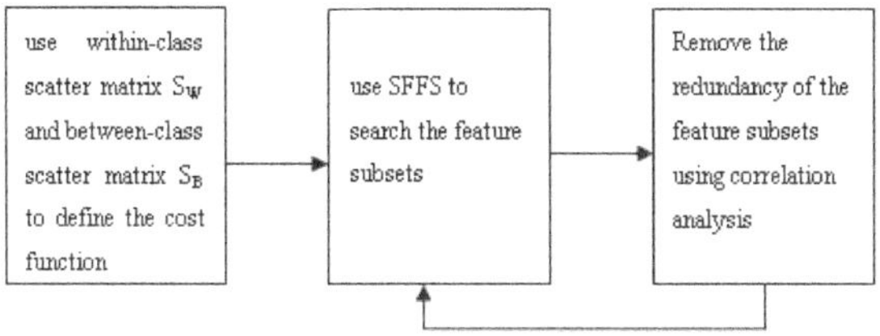

Fig. 6. The flow of SFFS algorithm

Correlation analysis: use $\wedge_{M-1}^{Old}$ as the current feature subset (where $\wedge_M$ is the feature set, M is the number of features), calculate the cost function $J(\wedge_{M-1}^{Old})$ of $\wedge_{M-1}^{Old}$;(2)Select a feature f from the feature set by sequence, and calculate correlation coefficients between f and the various features of $\wedge_{M-1}^{Old}$, if the coefficients are all less than the given threshold, indicates that f is not associated with $\wedge_{M-1}^{Old}$, so add f to $\wedge_{M-1}^{Old}$ get the new feature set $\wedge_M^{New}$ and calculate the cost function $J(\wedge_M^{New})$, otherwise delete f from the feature set; (3) if $J(\wedge_M^{New}) < J(\wedge_{M-1}^{Old})$, indicate that f reduce the identification ability of $\wedge_{M-1}^{Old}$, delete f from the feature set and delete $\wedge_M^{New}$, go step2, search feature subset based on $\wedge_{M-1}^{Old}$;(4) If $J(\wedge_{M}^{New}) > J(\wedge_{M-1}^{Old})$, indicate that f enhance the identification ability of $\wedge_{M-1}^{Old}$, go to step2, search feature subset based on $\wedge_{M}^{New}$ and delete $\wedge_{M-1}^{Old}$; (5) until the search the last feature, get the optimal feature subset. Search feature subset using correlation analysis and SFFS, its purpose is to select the feature subset for global optimality. The Following experimental data verify the validity of the methods presented.

5 Experimental Results and Analysis

We can get a feature subset which contains only 16 feature vectors using correlation analysis and SFFS search algorithms, as shown in table 3:

Table 3. Feature subset for lip-reading

FAP#	name	FAP#	name
3	open_jaw	16	push_b_lip
4	lower_t_midlip	17	push_t_lip
5	raise_b_midlip	51	lower_t_midlip_o
6	stretch_l_cornerlip	52	raise_b_midlip_o
8	lower_t_lip_lm	53	stretch_l_cornerlip_o
10	raise_b_lip_lm	55	lower_t_lip_lm_o
12	raise_l_cornerlip	57	raise_b_lip_lm_o
14	thrust_jaw	59	raise_l_cornerlip_o

To verify the effectiveness of the algorithm, this paper builds a lip language recognition system. Compared the recognition results of the two feature vectors. The image database used in the experiment is obtained by a digital camera in a simple background and moderate light. We only collect the mouth region, the resolution of the image is 240*160. In the experiment, collect 10 peoples' images who read {a,o,e,u} and for every single vowel each read 6 times. We totally get 240 moth shape images.

Each pronunciation images are divided into two groups, one group is used for training and another is used for recognition. The recognition rate is 81.7% using talbe1 as the feature vectors, and 80.5% using talbe3 as the feature vectors. Use the method presented in this paper, we can get a feature subset, which has fewer vectors and lower redundancy.

6 Conclusions

This paper focuses on fearture selection and extraction method for sequence images of Lip-reading. First, select lip feature points according to FDPs defined by MPEG-4, and select FAPs as the feature parameters for lip-reading. The SFFS method and correlation analysis are exploited to fast search the optimal sub-aggregate and eliminate the redundancy features. Experimental results of the real data show that the features extracted from the sequence image have a high recognition rate in lip-readings recognition system.

References

1. Chiou, G.I., Hwang, J.N.: Lipreading by using snakes,principal component analysis and hidden Markov models to recognize color motion video. IEEE Trans. on Image Processing 6(8), 1192–1195 (1997)
2. Yao, H.-X., Liu, M.-B., Gao, W.: Method of Face Locating and Tracking Based on Chromatic Coordinates Transformation of color Images. Chinese J. Computers 23(2), 158–165 (2000) (in Chinese)
3. Potamianos, G., Graf, H.P., Cosatto, E.: An Image Transform Approach for HMM Based Aotumatic Lipreading. In: Proceeding of the International Conference on Image Processing, Chicagao, vol. III, pp. 173–177 (1998)
4. ISO/IEC 14496-1 Systems (2001)
5. ISO/IEC 14496-2 Visual (2001)
6. ISO/IEC 14496-2 Audio (2001)
7. Yang, Y., Zhang, T.-W.: A New Active Contour Tracking Algorithm. Chinese J. Computers 21, 297–302 (1998) (in Chinese)
8. Sun, Y., Li, J.: Landmine Detection Using Forward-Looking Ground Penetrating Radar. In: Proc. of SPIE, vol. 5794, pp. 1089–1097 (2005)
9. Sun, X., An, D., Liu, X.: Arbitray Lip Deformation Driven by Bopomofo Text. Journal of Computer-Aided Desin & Computer Graphics 20(12), 1063–1068 (2008) (in Chinese)
10. Sun, Y., Li, J.: Adaptive Learning Approach to Landmine Detection. IEEE Transactions on Aerospace and Electronic Systems 41(3), 973–985 (2005)
11. Yang, Y.-G., Zhou, Z.-M., Song, Q.: A nove feature evaluation and selection method for sequence images of VFGPIR. Signal Processing 25(10), 1574–1581 (2009) (in Chinese)

Pattern-Discovery-Based Response Time Prediction

Lingshuang Shao[1,*], Yi Guo[1], Xiangping Chen[2], and Yanxiang He[1]

[1] School of Computer, Wuhan University, China
[2] Institute of Advanced Technology, Sun Yat-sen University, China
shaolsh@gmail.com, 376878279@qq.com,
lowlowcxp@gmail.com, yxhe@whu.edu.cn

Abstract. Response time is one of the most important metrics quality of service and response time prediction has always been the a hot topic. Based on random process, there are many common methods for predicting response. However, we find that there are also some features of response time data of service running. According to these features, we summarize the the patterns of response time and propose a novel algorithm for predicting response time. To evaluate our algorithm, we compare PIPA with other algorithms. Experimental results show that PIPA can efficiently improve the accuracy of the prediction.

Keywords: QoS, Response Time, QoS Prediction.

1 Introduction

In many situations, such as service composition and selection, it is necessary to make QoS predictions dynamically[1,2]. QoS prediction is important because of the instability of QoS, which is caused of the environment (such as network load), the repair or upgrade of the service itself, or the changes of resource allocation or load on the platform of the service deployment. In this situation, it is impossible to capture the characteristics and the short-term trends which reflect the dynamic changes of QoS simply rely on the history of the static-measurement method. QoS prediction aims at exploring the characteristics and short-term trends of these dynamic changes and supporting the selection and assembly of dynamic services.

However, previous research related to service selection and composition has not yet taken these characteristics into consideration. Most of the studies have directly used the method that measured the attributes of the QoS for the prediction of QoS.

In this paper, we proposes a method for predicting response time, which is one of the important characteristics of quality of web service. Based on the random process, there are some algorithms of prediction which are commonly used, such as Moving Average (MA), Exponentially Weighted Moving Average (EWMA), Self-Adaptive Exponentially Weighted Moving Average (SAEWMA), and so on.

And then, based on the specific pattern of response time, we propose a novel algorithm of the prediction, which is called Pattern-Discovery-based Response Time Prediction Algorithm (PIPA). There are two main characteristics of this algorithm: (1)

[*] Corresponding author.

G. Lee (Ed.): Advances in Automation and Robotics, Vol.2, LNEE 123, pp. 355–362.

processing the outliers from the response time of services to prevent them from leading to greater errors, (2) adding a recognizing process of the specific pattern to the prediction of response time in order to improve the accuracy of the prediction after the patterns were determined.

The rest of the paper is organized as follows. Section 2 describes the characteristics of the Web Services response time. Section 3 introduces the algorithm of prediction based on the specific pattern of response time. Section 4 evaluates our algorithm. Section 5 gives an overview of the related work. Section 6 concludes of this paper.

2 The Characteristic of Web Services Response Time

By closely examining the operating records, we disclosed the repeated characteristic on the data of the response time. The figures describing the various characteristics in the following subsections are based on the actual data sections derived from transferring the Google Web Search.

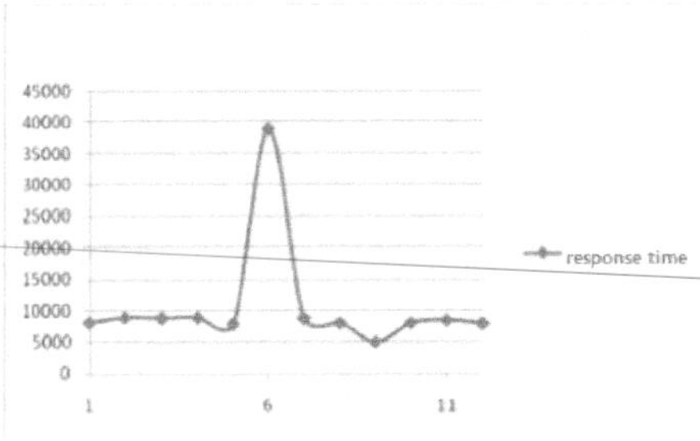

Fig. 1. Sample of the outlier

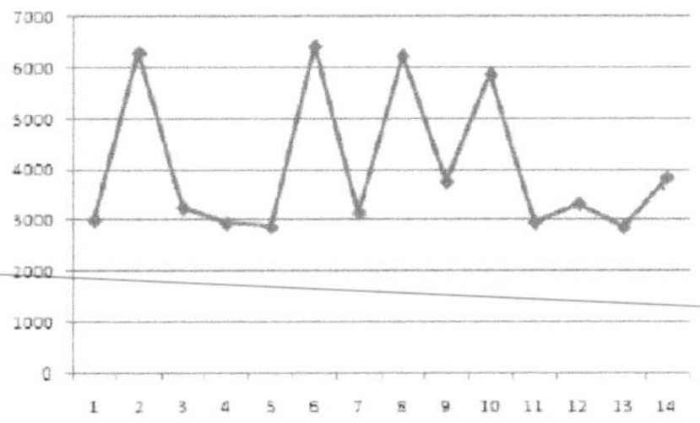

Fig. 2. Sample of the bump pattern

2.1 Outliers

Due to the unstable characteristic of the service, when transferring the service, it may result in the phenomenon that the single response time is overlong. Figure 1 is an illustration of the outlier.

Figure 1 involves 12 transfers. The response time of the sixth transfer is far longer than that of the other transfers before and after it. Therefore, the time data derived from the sixth transfer can be regarded as the outlier. Since the outliers seriously impact the accuracy of response time prediction, the appropriate solution is necessitated.

2.2 Statistical Characteristics of Service Response Time

Pattern 1: Stable pattern

When the data is in the stable pattern, its value is comparatively stable and its fluctuation range is generally small.

Pattern 2: Bump pattern

When the data is in the bump pattern, as illustrated by figure 2, the general fluctuation range of the values is comparatively great and there is no obvious increasing or decreasing trend.

Pattern 3: Gradual increase/ Gradual decrease pattern

When the data is in the gradual increase/gradual decrease pattern, as illustrated by figure 3 and 4, the fluctuation range of the values is comparatively large as well as the values in the latter half of time are manifestly larger or smaller than that in the former half of time.

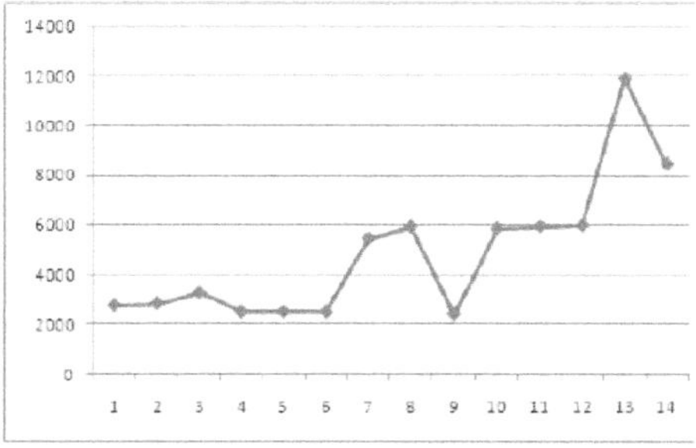

Fig. 3. Sample of the gradual increase pattern

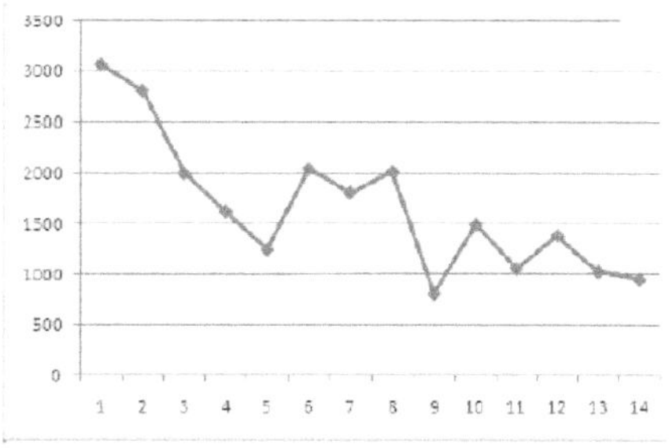

Fig. 4. Sample of the gradual decrease pattern

3 Pattern-Discovery-Based Response Time Prediction Algorithm (PIPA)

Based on the former discovery, this paper proposes a Pattern-Discovery-based Response Time Prediction Algorithm (PIPA). The main advantages of the algorithm are: 1. it deals with the outliers resulting from the service response time, which avoids the considerable prediction errors; 2. the identification for the specific pattern is added to the response time prediction, which improves the prediction accuracy after the patterns are determined.

3.1 Identification and Treatment for Outliers

In order to identify the outliers, PIPA sets a threshold of t%: Among all of the response time, the biggest transfer records of t% are considered as the outliers. Once the data is regarded as the outlier, PIPA would adjust the value of the outlier to 'a' and the value range for 'a' is the maximum after removing the outliers.

3.2 Identification and Treatment for the Specific Pattern

3.2.1 Stable Pattern

The identification for the stable pattern is determined by the variance of the data. When the variance of the data of recent k transfers is lower or equal to 'r', PIPA considers the time data being in the stable pattern. When the data is found to be in the stable pattern, PIPA predicts the response time as the formula,

$$F_k = (t_{k-1} + t_{k-2} + t_{k-3} + \cdots + t_{k-n}) = n \quad (1)$$

as well as the mean value of recent k transfers.

3.2.2 Gradual Increase/Gradual Decrease Pattern

There are two requirements for the judgment of the gradual increase/gradual decrease pattern: requirement one, the variance of the response time of the recent k transfers>r; requirement two, among recent k transfers, the values in the latter half of time are evidently bigger or smaller comparing to the values in the former half of time, which is demonstrated as below (take the gradual increase pattern as an example):

Setting the sequence of the response time of the recent k transfers as t, $t = \{t_1, t_2, \cdots, t_k\}(k > 0, k \in 2N)$, *ti* is the response time of one time. Dividing t into two sub-sequences respectively as $t` = \{t_1, t_2, \cdots, t_{\frac{k}{2}}\}$ and $t`` = \{t_{\frac{k}{2}+1}, t_{\frac{k}{2}+2}, \cdots, t_k\}$.

Setting the mean of the sequence t as *t (as the formula 1)*. When the number of t` less than the element of t is greater than i, and the number of t`` less than the element of t is greater than i, the judgment is true. In the experiment, it is usually setting the value of i between 50% to 80% of t` element numbers. Following this requirement, it guarantees the values of the t` sequence are mostly lower than the mean, while the values of the t`` are mostly higher than the mean. As a result, they collectively assure the establishment of the requirement two.

When the data sequence t is identified as the gradual increase pattern, the prediction result of the next transfer is the maximum of t, as shown by the formula 2, which p is the prediction result. Similarly, when the data sequence t is identified as the gradual decrease pattern, the prediction result of the next transfer is the minimum of t,

$$p = MAX\{t_1, t_2, \cdots, t_k\} \quad (2)$$

3.2.3 Bump Pattern

When the data is neither in the stable pattern nor the gradual increase/gradual decrease pattern, the data is attributed to the bump pattern, and then PIPA predicts the response time as the formula 1, which is the mean value of recent k transfers.

4 Experiment and Results Analysis

The experiment objectives in this part are 4 services arranged on the internet, which are issued by Google, Yahoo and Amazon. The client program of the transfer services are performed by both JDK 1.5 and Axis 1.4. Before starting the client program, transferring the client program of these services does not generates the historical record. The client program automatically stops after it finishes 10000 transfers for each service.

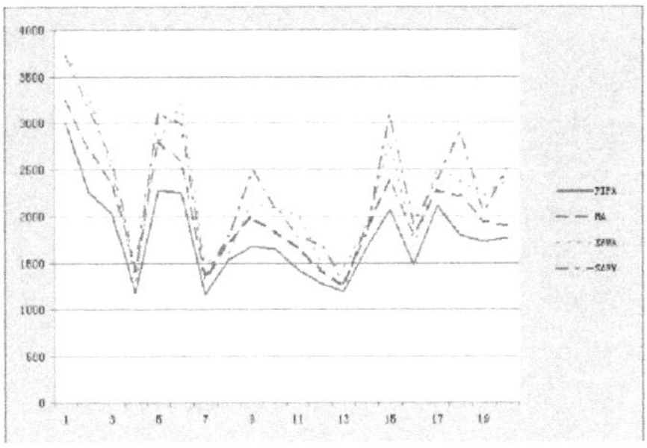

Fig. 5. The comparison between the errors of the response time prediction: service 1

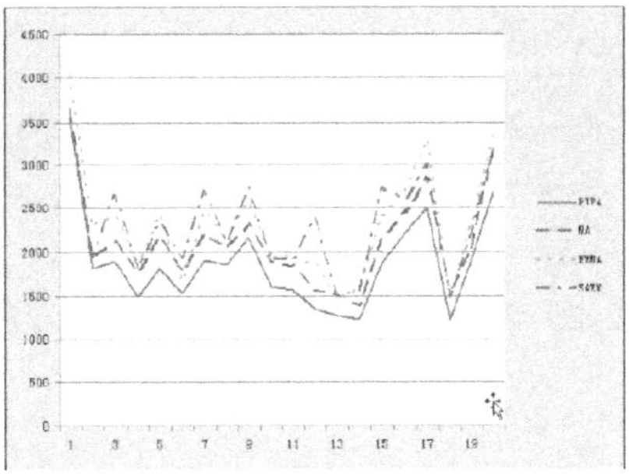

Fig. 6. The comparison between the errors of the response time prediction: service 2

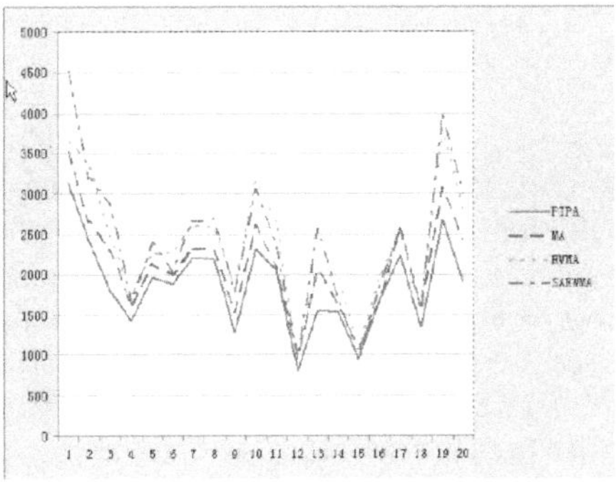

Fig. 7. The comparison between the errors of the response time prediction: service 3

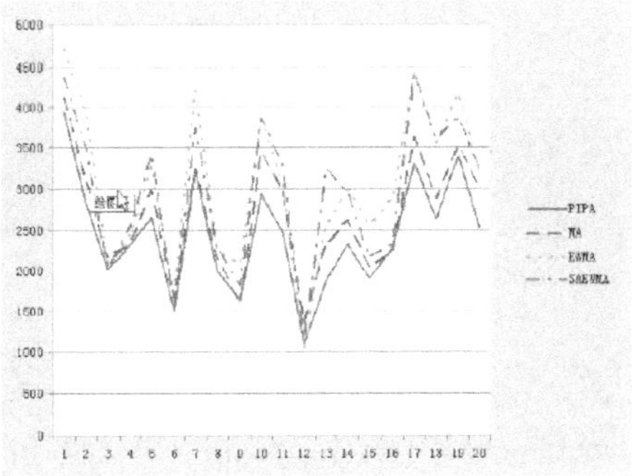

Fig. 8. The comparison between the errors of the response time prediction: service 4

The experiment compares the PIPA proposed in this paper to the prediction accuracy of the three algorithms, MA, EWMA and SAEWMA. Taking the transfer k as an example, the predictive value is pk□ the actual transfer value is rk□ then the error value is ek.

Figure 5, 6, 7 and 8 illustrate the comparison between the prediction errors of the above algorithms. The comparison objectives cover MA algorithm, EWMA algorithm, SAEWMA algorithm and PIPA proposed in this paper. As the figures shown, axis y represents the mean error of 50 continuous transfers (the unit is millisecond); axis x stands for the transfer of the service, each unit represents 50 transfers.

As the figures shown, under the most of circumstances, the response time prediction of PIPA is the most accurate. From the figure 5 and 6, under the conditions that the internet is fine and the response time is comparatively stable, it is manifest that PIPA algorithm is evidently advantageous in comparison to the other three algorithms; as can be seen from the figure 7 and 8, if the internet condition is poor and the data of response time fluctuate greatly when transferring the services, the advantage of the PIPA algorithm is relatively small.

It can be found from the figures that, the MA algorithm and PIPA algorithm are almost equal in terms of error. The reason for this is that once the PIPA determines whether the data is in the bump pattern or stable pattern, the predictive equation for the response time of the next transfer is same as that used by MA algorithm. However, under most of circumstances, the data of response time generally results from the stable pattern and bump pattern. Consequently, the predictive results of the MA algorithm and the PIPA algorithm are almost same in most of cases. Also, since the essence of the above algorithms follows the MA algorithm, the trends of the figures are close in most of cases.

5 Related Work

In web service domain, QoS prediction is always a hot topic. The QoS prediction methods could be divided into four groups.

1) (Machine) Learning-based

These approaches find out the feature in two types of information: invoking records from consumer side and server run-time information, the latter is optional. Invoking records from consumer side provides historical QoS experiences. From experiences we can automatically learn the rules about the tendency of QoS changing.

If the server run time information is also available (I think this is possible when the service selection happens inside enterprises so the server can provide such information), we can setup the relationship between server status and QoS experiences, this relationship is also useful when predicting QoS.

2) Collaboration-based

As we have stated in related work, in researches in recommendation systems, it is proved that setting up relationship among consumers helps to recommend items to them. The process of setting up relationship for consumers can be divided into two categories: automatically and manually. Collaborative filtering is a technique of automatically setting up relationship. Researches use manually setting up relationship to improve predicting QoS include IC-Service[22].

The relationship between consumers include: consumers similarity, trust relationship, belief network and preferences grouping. All these forms may help improve making prediction.

3) Knowledge Reasoning-based

Knowledge reasoning-based prediction uses semantic web and ontologies to setup the relationships between consumers' information and probable QoS experiences. Consumers' information includes consumers' profile (such as name, age, education, etc.) and consumers' preferences. Consumers' preferences are domain oriented knowledge and have very close relationship with the application. For example, when consumers query for a service of booking restaurant, preferences may include "the time of providing Chinese food" or "not many people when rush time"[8][9][10]. In mobile service domain, some researchers describe consumers preferences as "location" and "time"[11].

These researches are also generalized as context-aware computing[12][15]. Semantic and ontologies are the most widely used techniques in these researches.

PIPA is a novel statistics-based method and it is based on the characteristics of service running data. We summarize the pattern of service response time and proposes the PIPA based on these patterns, which means PIPA is a domain-specific method for web services response time prediction.

6 Conclusion

The prediction of QoS is a important technology which support the selection and combination of QoS. However, the most existing technology of prediction which directly used the method of quality measurement does not take into account the characteristics to the dynamic change of QoS. Focused on the important quality attribute, response time, and based on the actual operation data of services, we found the implied model that reflected the data of response time and propose a new algorithm of prediction. Finally, to evaluate our algorithm, we compared it with other algorithms in an experiment, the results of which show that this new algorithm can efficiently improve the accuracy to the prediction of the response time.

Acknowledgement. The work described in this paper was fully supported by National Nature Science Foundation of China(SN: 61003072) and Guangdong Natural Science Foundation (SN:S2011040005180).

References

1. Erradi, A., Maheshwari, P.: A Broker-based Approach for Improving WS Reliability. In: ICWS 2005 (2005)
2. Menasce, D.A.: QoS Issues In Web Services. IEEE Internet Computing (2002)
3. Keidl, M., Kemper, A.: Context-Aware Adaptable Web Service. In: www 2004 (2004)
4. Sahai, A.: Automated SLA Monitoring for Web Services. In: IEEE/IFIP DSOM (2002)
5. Nahrstedt, K., Xu, D., Wichadakul, D., Li, B.: QoS-Aware Middleware for Ubiquitous and Heterogeneous Environments. IEEE Comm. Magazine 39(11), 2–10 (2001)
6. M Keidl, A Kemper. Towards context-aware adaptable web services. WWW 2003 (2003)
7. Birukou, A., Blanzieri, E., D'Andrea, V., Giorgini, P., Kokash, N., Modena, A.: IC-Service: A Service-Oriented Approach to the Development of Recommendation Systems
8. Balke, W.-T., Wagner, M.: Through Different Eyes – Assessing Multiple Conceptual Views for Querying Web Services. In: WWW 2004 (2004)
9. Balke, W.-T., Wagner, M.: Cooperative Discovery for User- centered Web Service Provisioning. In: ICWS 2003 (2003)
10. Balke, W. T.: Towards Personalized Selection of Web Services. In: WWW 2003 (2003)
11. Maamar, Z., Bataineh, E., AlKhatib, G.: On Coordinating Personalized Web Services Using CooPS. In: SAINT 2005 (2005)
12. Barkhuus, L., Dey, A.K.: Is Context-Aware Computing Taking Control away from the User? Three Levels of Interactivity Examined. In: Dey, A.K., Schmidt, A., McCarthy, J.F. (eds.) UbiComp 2003. LNCS, vol. 2864, pp. 149–156. Springer, Heidelberg (2003)
13. Zeng, L., Benatallah, B., Dumas, M., Kalagnanam, J., Sheng, Q.Z.: Quality driven web services composition. In: www 2003 (2003)

14. Menascé, D.A., et al.: Preserving QoS of E-commerce Sites Through Self-Tuning:A Performance Model Approach. In: EC 2001 (2001)
15. Kemper, A., Keidl, M.: Towards Context-Aware Adaptable Web Services. In: www 2004 (2004)
16. Tsai, W.T.: UServices-Oriented Dynamic Reconfiguration Framework for Dependable Distributed Computing. Process Monitoring & Management of Web service Composition
17. http://schist.und.nodak.edu:8899/SiteScope/docs/WebServiceMonitor.htm
18. Sahai, A.: Automated SLA Monitoring for Web Services. HP Tech. Report
19. Verheecke, B., Cibrán, M.A., Vanderperren, W., Suvee, D.: AOP for Dynamic Configuration and Management of Web Services. International Journal of Web Services Research (2004)
20. Barbon, F., Traverso, P., Pistore, M., Trainotti, M.: Run-Time Monitoring of Instances and Classes of Web Service Compositions. In: ICWS 2006, pp. 63–71 (2006)
21. Menascé, D.A., Barbará, D., Dodge, R.: Preserving QoS of e-commerce sites through self-tuning: a performance model approach. In: Proceedings of the 3rd ACM Conference on Electronic Commerce (2001)
22. Birukou, A., Blanzieri, E., D'Andrea, V., Giorgini, P., Kokash, N., Modena, A.: IC-Service: A Service-Oriented Approach to the Development of Recommendation Systems. In: The ACM Symposium on Applied Computing, Special Track on Web Technologies (WT), Seoul, Korea (March 2007) (full paper, to appear)

A Method to Check the Structure of Sequential Business Process Models

Jian Deng[1], Zhi Chen[2], and Jiazhi Zeng[1]

[1] School of Computer Science & Engineering, University of Electronic Science & Technology of China, Chengdu 610054, China
[2] National Key Laboratory of Science and Technology on Communications, University of Electronic Science & Technology of China, Chengdu 610054, China
jockeydj@gmail.com,
{chenzhi,jzzeng}@uestc.edu.cn

Abstract. To satisfy the requirement of user customized business processes, a special kind of computation independent model named well-structured sequential business process model was researched. The method to check the structure of sequential business process models was also proposed. The well-structured sequential business process model was designed in business process modeling notation 1.1 version. It was converted to flowchart model in Windows Workflow Foundation 4.0 version directly, and the tool for model checking was implemented on Windows platform. Experimental results showed that this method supported model converting, and improved the quality of business process models.

Keywords: model-driven software development, business process model, business process modeling notation, windows workflow foundation, flowchart model, model checking.

1 Introduction

Complex and changing business requirements have made software development facing enormous difficulties, so when the Object Management Group proposed using model-driven architecture (MDA) [1] to solve this problem, the model-driven software development has become the research focus in software engineering [2] [3] [4]. The key technologies of MDA [5] involved with Computation Independent Model (CIM), Platform Independent Model (PIM), Platform Specific Model (PSM), and automatic model conversion.

In the domain of business process modeling, the model can be illustrated by Business Process Modeling Notation (BPMN) [6]. In order to facilitate the use of the same model with different modeling tools, Workflow Management Coalition proposed using XML Process Definition Language (XPDL) [7] as the file format for BPMN models.

The BPMN model is a computation independent model. Business Process Execute Language (BPEL) [8] model is a platform independent model. Microsoft's Windows

G. Lee (Ed.): Advances in Automation and Robotics, Vol.2, LNEE 123, pp. 363–371.
Springerlink.com

Workflow Foundation (WWF) [9] is designed specifically for workflow modeling on Windows platform.

The approach to convert BPMN models into BPEL models has been proposed during the development of BPMN standards. Weqqing Lin [10] analyzed the workflow patterns supported by XPDL 2.0. Zapletal Marco [11] researched the control-flow expression in Windows Workflow (WF) 3.5 version, and Lin Miao [12] developed a method to transform the XPDL model to WF 3.5.

However, due to the limitations of WF 3.5, such as the deficiency of modeling multiple output branches and the lack of support for the graphical layout information of sequential workflow models, reference [12] did not discuss the multi-output gateway. The gateway types were limited to AND-gateway and XOR-gateway. The resulted WF models were significantly different in graphical layout comparing to the original XPDL models. WF 4 supports modeling business processes in flowcharts. It has many advantages, such as the ability to specify a following task to any task, direct support for gateways with more than two branches, the ability to specify a node's position and the connection lines between nodes.

To our knowledge, there is no published conversion from the BPMN models to Windows Workflow flowcharts. In order to converting a BPMN model to a flowchart in WF 4 directly, there are certain requirements for the structure of the business process model. Although references [13] and [14] reported that Petri nets, automata and process algebras are three methods to carry out formal checking, and references [15] [16] and [17] have researched these methods in detail, WF 4 has even more stringent requirements to the structure of business models than the requirements of the above-cited formal checking methods.

To this end, this paper presents a method to check the structure of the BPMN models in preparation to conversion into flowchart models in WF 4.

2 Sequential Business Process Model

BPMN defines four categories of modeling elements, including the process objects, connection objects, swimming pools, and artifacts. Each category was divided into several subclass elements.

The process objects include Event (*E*), Task (*T*), Subprocess (*S*), and Gateway (*G*). The connection objects include sequence flow, message flow, and association link. The swimming pools include Pool (*O*) and Lane. The artifacts include data objects (*D*), groups, and text annotations. To facilitate the analysis, the definitions in reference [17] can be extended as the followings.

Definition 1. *A process in a business process model is an eight-tuple* $P = (E, T, S, G, O, D, C, \varphi)$, *where* φ *is a function mapping from subprocess to process.*

The gateways in Definition 1 include XOR-gateway, AND-gateway, OR- gateway, and Complex-gateway. Each type of gateway further includes types of Split and Merge, so $\forall g \in G$, Type (g) $\in$ {*XOR_Split, XOR_Merge, AND_Split, AND_Merge, OR_Split, OR_Merge, Complex_Split, Complex_Merge*}.

Definition 2. *If a business process model does not include the message flow and association link but only includes the sequence flow connection, then it is called a Sequential Business Process Model (SBPM).*

When analyzing and converting process models, the pool, message flow, and association link can be omitted.

Definition 3. *In a BPMN model, let s be the split type of gateway, m be the merge type of gateway, it is called that the type of s is matched with type of m when one of the following conditions is true.*

(1) Type(s) =$XOR_Split \wedge$Type (m) =XOR_Merge
(2) Type(s) =$AND_Split \wedge$Type (m) =AND_Merge
(3) Type(s) =$OR_Split \wedge$Type (m) =OR_Merge
(4) Type(s) =$Complex_Split \wedge$Type(m) =$Complex\_Merge$

Definition 4. *Let there be n paths from split gateway s to merge gateway m in a BPMN model, select any two paths of the n paths, let* $Px=sc_1...m$, $Py=sc_2...m$, *if* $Px \cap Py=\{s, m\}$ *when* $c_1 \neq c_2$, *that is, every path from s to m is an internal disjoint path, then it is called that s, m is a pair of reducible gateways, denoted as* <*s, m*>. *It is also said that each path of s to m is an internal reducible path.*

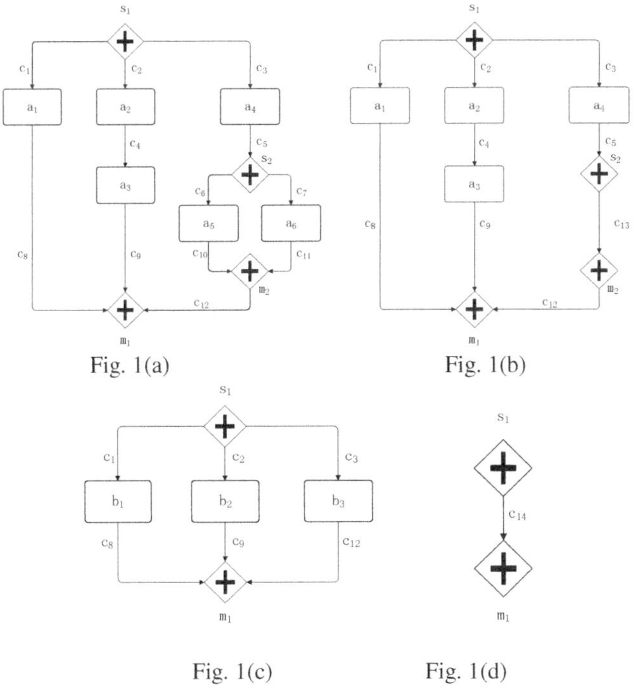

Fig. 1. The reduction of two pairs of reducible gateways

In Figure 1, Figure 1(a) includes two pairs of reducible gateways $<s_1,m_1>$ and $<s_2,m_2>$, after reducing twice in Figure 1 (b) and Figure 1 (d), $<s_1,m_1>=(\{s_1, m_1\},\{c_{14}\})$, $<s_2,m_2>=(\{s_2, m_2\}, \{c_{13}\})$. Figure 1 (c) shows the reduced path information after the first reduction. The information will be used in model transforming, where $b_1=(\{a_1\},\phi)$, $b_2=(\{a_2, a_3\}, \{c_4\})$, $b_3=(\{a_4, s_2, m_2\}, \{c_5, c_{13}\})$, then there are three reduced paths from s_1 to m_1, respectively, $P_1=s_1c_1b_1c_8m_1$, $P_2=s_1c_2b_2c_9m_1$, $P_3=s_1c_3b_3c_{12}m_1$.

Definition 5. *If a sequential business process model meets the following conditions, then it is called a Well-Structured Sequential Business Process Model (WSSBPM).*

(1) $\forall g\in G$, $(|in(g)|\le 1)\vee(|out(g)|\le 1)$. The gateway cannot be both a split and a merge. The number of input and output cannot be greater than 1 at the same time.

(2)When designing a model, the AND-gateway, OR-gateway, COMPLEX-gateway must be used in pairs, so that the requirements of Definition 3 are met.

(3)Each path from a split gateway must end at the same merge gateway in (2).

(4)Each path from split to merge in (2) must be an internal reducible path as in Definition 4.

(5)There exists only one start node. $|\{x|x\in E\cup T\cup S\cup G, in(x)=0\}|=1$.

(6) $\forall x\in E\cup S$, $|in(x)|\le 1\wedge |out(x)|\le 1$. Event and Subprocess cannot act as gateways.

(7) $\forall x\in T$, $(|in(x)|\le 1 \vee Type(x)= XOR\_Merge)\wedge |out(x)|\le 1$. Task can act as a XOR-merge, but not any other merge gateway or split gateway.

When a XOR-gateway is located in one branch in (2), this branch is converted to a flowchart.

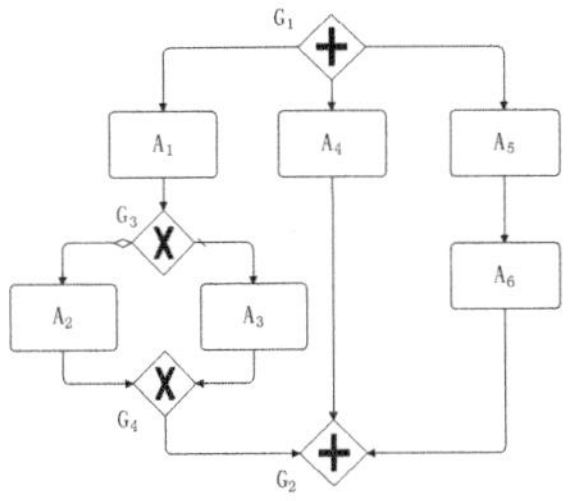

Fig. 2. BPMN model

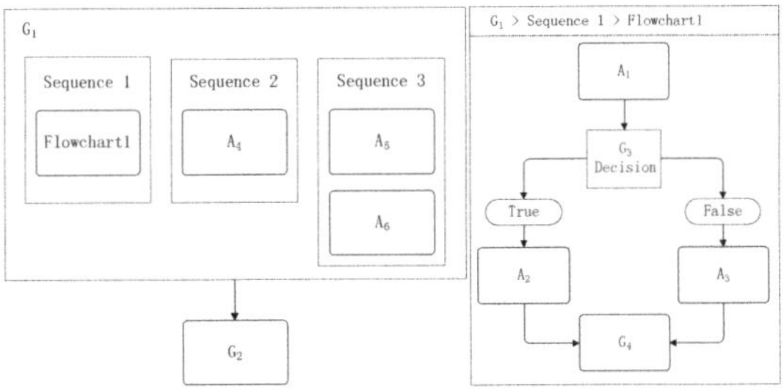

Fig. 3. The flowchart model converted from Fig. 2

In Figure 2, $<G_1, G_2>$ is a pair of reducible gateways, $<G_3, G_4>$ is a pair of reducible gateways located in one branch of $<G_1, G_2>$, and Figure 3 is the converted workflow model.

3 The Method for Checking and Adjusting SBPM

The structure of a sequential business process model can be checked using Definition 5. If it fails the check, the designer should adjust the model. Tables 1 through 6 show the typical cases and the corresponding adjustment methods.

In Table 1, A_1 is a task with multiple-input and multiple-output. It acts as a merge and a split gateway at the same time, so it can be redesigned using a XOR-gateway and a split gateway G_2. The split type of G_2 should be determined according to the actual business requirements.

Table 1. Model with multiple-input multiple-output task

Original Model	Improved Model
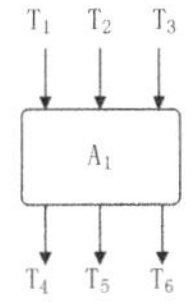	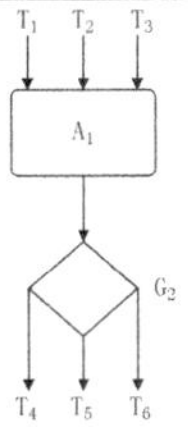

In Table 2, the type of G_3 is *AND_Split* and it does not match the type *XOR_Merge* of G_4. This will lead G_4 to be enabled many times. This situation is usually a mistake of designers not paying enough attention. To fix it, the type of G_4 must match with the type of G_3.

Table 2. Model with mismatched split and merge

Original Model	Improved Model
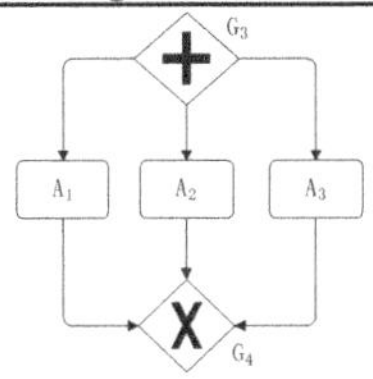	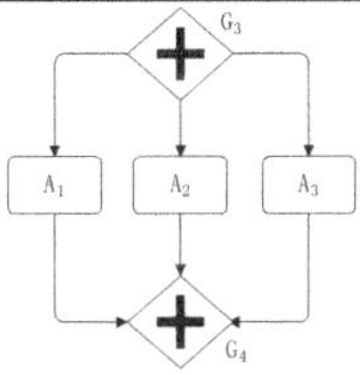

The initial model in Table 3 is logically correct. However since WF 4 has restrictions in the flowchart, one method to overcome the software limitation is to add the gateway G_7. Another method is to combine G_6, G_8 into one gateway G_6.

Table 3. Model with split paths not ending in the same merge

Original Model	Improved Model 1	Improved Model 2
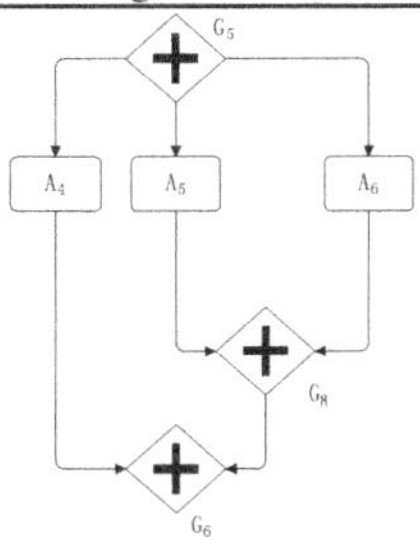	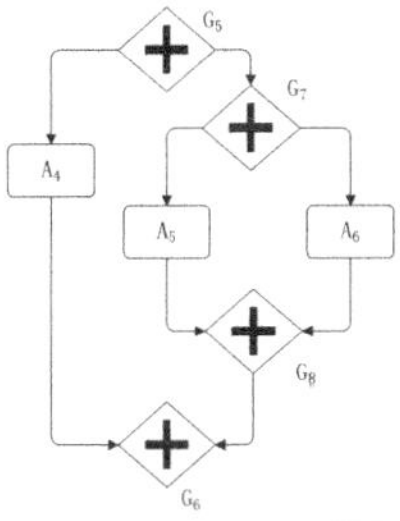	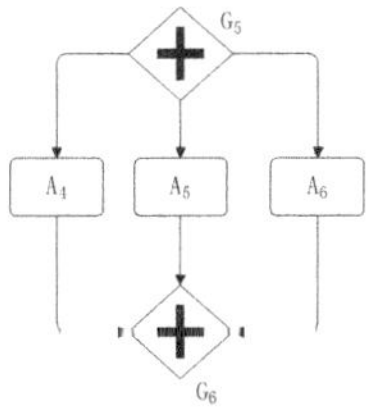

In Table 4, the two branches of G_9 are intersected at G_{12}. WF4 does not allow parallel activities to cross between branches, so the G_{11}, G_{12} should be removed.

Table 4. Model with same node in both branches

Original Model	Improved Model
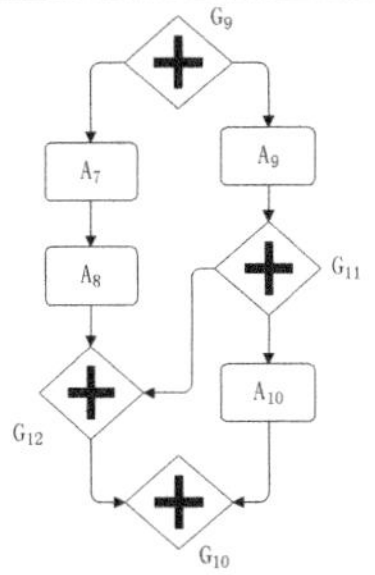	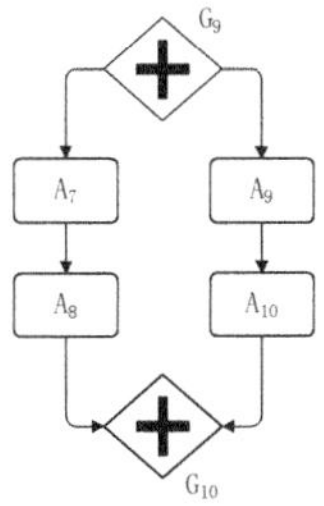

In Table 5, the successor of A_{12} should be G_{16}, otherwise when G_{15} selects the A_{12} branch, G_{14} cannot merge successfully.

Table 5. Model with nested path not ending in the same gateway

Original Model	Improved Model
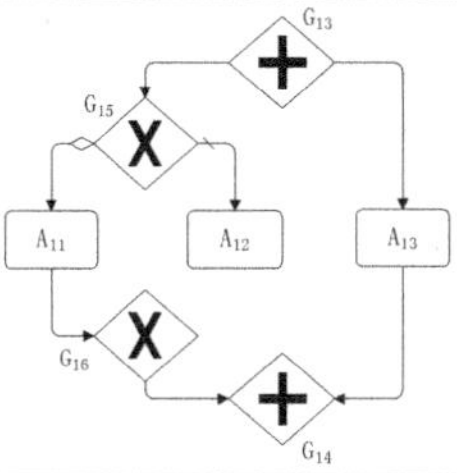	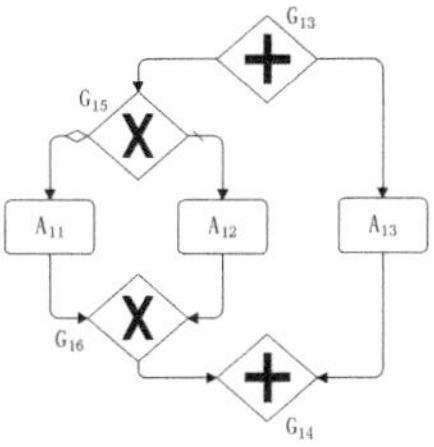

In Table 6, there are two start events e_1 and e_2. It can be redesign using e_0, A_1, A_2 and G_{15}, in which task A_1 corresponds with event e_1, task A_2 corresponds with event e_2, e_0 is the added start event, and G_{15} is gateway for running A_1 and A_2 in parallel.

Table 6. Model with multiple start events

Original Model	Improved Model
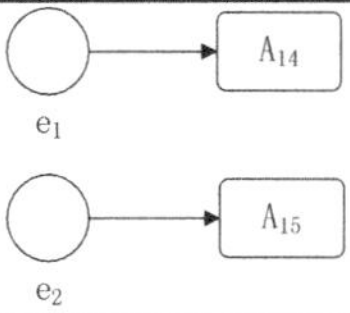	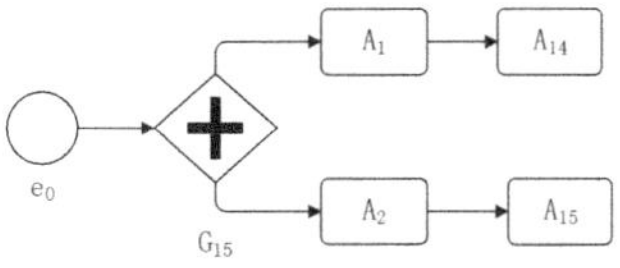

4 Case Study

Figure 4 is the original model taken from reference [12]. By applying our check method, two problems are identified in the model:

(1) The task "Confirm info" acts as an *OR_Split* gateway. When there is no condition specified on its output branches, all branches will be enabled. However, the "Cancel order" and "Receive payment" tasks should not both be enabled in the real world. The model can be adjusted using the method in Table1.

(2) The task "Merge path" acts as a *XOR_Merge* gateway, but its input branches have some branches from G_2. It will be enabled many times. Therefore, the model should be adjusted using the method in Table 2.

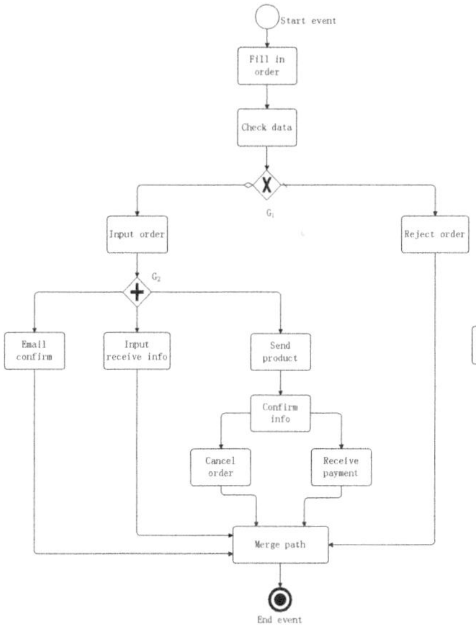

Fig. 4. Original Model

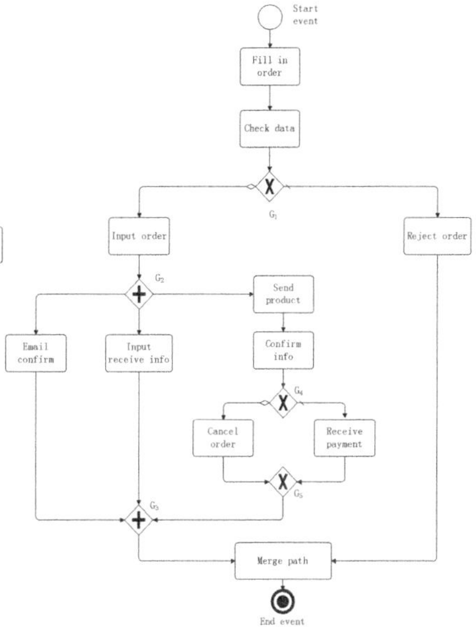

Fig. 5. Improved Model

After these modifications the improved model is shown in Figure 5. It can be converted to a flowchart in WF 4. Figure 6 shows the overall flowchart. The details of G_2 in the Figure 6 can be further expanded into a sequence and a flowchart, which are shown in Figure 7.

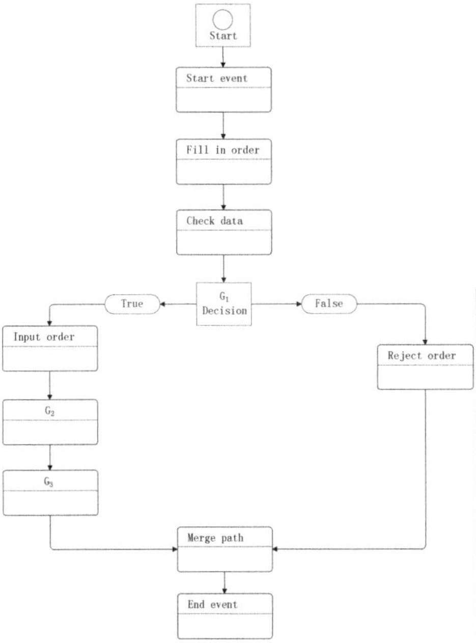

Fig. 6. Flowchart converted from Fig. 5

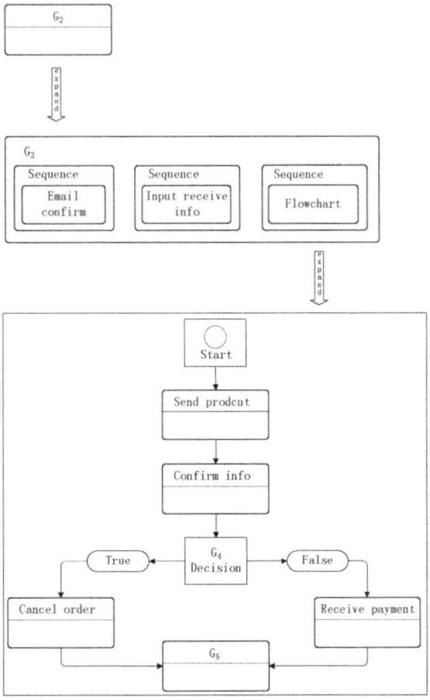

Fig. 7. Details of G_2 in Fig. 6

5 Conclusion

This paper presents a method to check the structure of business process models. The innovation is that it considers the platform programming features of the WF 4 flowchart model, and specifies the structure requirement so that the BPMN models can be directly converted to flowcharts. Experiments show that our method can detect certain errors in business process models, and improve the quality of the business process models. While there are so many modeling features in the BPMN standard, this article only deals with the structural check of sequence flow connections. Future researches can include: (1) to research the checking method for the message flow and association link; (2) to combine the structural check method with Petri nets, automata, process algebra and other methods. Further studies may also include checking models designed by BPMN 2.0.

Acknowledgments. This paper is supported by the project "Enterprise resource cooperation and research of key technologies" (No. 2007AA040801). The authors are extremely grateful to Professor Sun Linfu, Associate Professor Wang Shuying and Liao Weizhi for their help.

References

1. Soley, R.: Model Driven Architecture, http://www.omg.org/cgi-bin/doc?omg/00-11-05.pdf
2. Meservy, T., Fenstermacher, K.: Transforming software development: An MDA road map. Computer 38(9), 52–58 (2005)
3. Braganca, A., Machado, R.J.: A model-driven approach for the derivation of architectural requirements of software product lines. Innovations in Systems and Software Engineering 5(1), 65–78 (2009)
4. Jindan, F., Dechen, Z., Lanshun, N., et al.: Modeling business object platform independent model and its completenes. Computer Integrated Manufacturing Systems 17(6), 1308–1316 (2011)
5. Sendall, S., Kozaczynski, W.: Model transformation: The heart and soul of model-driven software development. IEEE Software 20(5), 42–45 (2003)
6. Object Management Group.DTC/2008-01-17 Business Process Modeling Notation, http://www.omg.org/spec/BPMN/1.1/PDF
7. Workflow Management Coalition. WFMC-TC-1025 Process Definition Interface – XML Process Definition Language, http://www.wfmc.org/xpdl.html
8. Jordan, D., Evdemon, J.: Web Services Business Process Execution Language Version 2.0, http://docs.oasis-open.org/wsbpel/2.0/OS/wsbpel-v2.0-OS.pdf
9. Microsoft. Windows Workflow Foundation, http://www.microsoft.com/visualstudio/en-us/products/2010-editions
10. Weqqing, L., Jian, W.: Analysis of process control pattern of XPDL 2.0. Computer Integrated Manufacturing Systems 13(9), 1839–1846 (2007)
11. Zapletal, M., Aalst, W.M.P., Russell, N., et al.: An Analysis of Windows Workflow's Control-Flow Expressiveness. In: Eshuis, R., Grefen, P., Papadopoulos, G. (eds.) Proceedings of the 7th IEEE European Conference on Web Services, pp. 200–209. IEEE Computer Society, Los Alamitos (2009)
12. Lin, M., Tao, J., Jianmin, W.: Transformation Technology from XPDL Model to WWF Mode. Journal of Computer Research and Development 46(suppl.), 165–171 (2009)
13. Morimoto, S.: A Survey of Formal Verification for Business Process Modeling. In: Bubak, M., van Albada, G.D., Dongarra, J., Sloot, P.M.A., et al. (eds.) ICCS 2008, Part II. LNCS, vol. 5102, pp. 514–522. Springer, Heidelberg (2008)
14. Lohmann, N., Verbeek, E., Dijkman, R.: Petri Net Transformations for Business Processes – A Survey. LNCS Transactions on Petri Nets and Other Models of Concurrency 2, 46–63 (2009)
15. Dijkman, R.M., Dumas, M., Ouyang, C.: Semantics and analysis of business process models in BPMN. Information and Software Technology 50, 1281–1294 (2008)
16. Zha, H., Yang, Y., Wang, J., Wen, L.: Transforming XPDL to Petri Nets. In: ter Hofstede, A.H.M., Benatallah, B., Paik, H.-Y. (eds.) BPM Workshops 2007. LNCS, vol. 4928, pp. 197–207. Springer, Heidelberg (2008)
17. Jian, D., Zhi, C., Jiazhi, Z.: Formal verification of business process models using Petri nets. Computer Integrated Manufacturing Systems 17(5), 1110–1119 (2011)

Short-Term Wind Speed Forecast Based on Best Wavelet Tree Decomposition and Support Vector Machine Regression*

Deliang Zeng, Yu Liu, Jiwei Liu, and Jizhen Liu

North China Electric Power University, School of Control And Computer Engineering
102206 Beijing, China
ncepuly@126.com

Abstract. Wind power industry developed rapidly in recent years. wind power is a type of power with randomness and fluctuation. Accurate wind speed forecasting can reduce the impact of wind power. Paper analyzed the wind speed signal with wavelet packet decomposition method from low-frequency and high-frequency, selected the optimal wavelet tree through the principle of minimum entropy. Short-term wind speed prediction model is built with support vector machine regression. This algorithm has advanced and better accuracy by comparing the results.

Keywords: wind speed forecast, wavelet transform, wavelet packet, support vector machine regression.

1 Introduction

The share of wind power in the grid gradually increase with the rapid development of wind power in recent years. Wind power impact on power grid is growing because of its randomness and volatility characteristics. It is necessary to make accuracy prediction of wind speed for the grid scheduling department and wind farm.

Currently, wind speed forecast more than 3 hours need to consider the nature of atmospheric motion with basic numerical weather prediction; wind speed forecast of 0 to 3 hours sustain by the atmospheric conditions. Modeling methods[1-3] are continuous prediction method, random time series, Kalman filtering and neural networks, support vector machines and other intelligent algorithms. These methods combined with wavelet decomposition and empirical mode decomposition and other signal analysis methods that can better improve the forecasting accuracy.

Wind speed time series fluctuated in time domain, however the wind speed time series is decomposed into different frequency components, The same frequency component has the similar cycle characteristics with a certain regularity. It can increase the accuracy of wind speed prediction to establish prediction model for wind speed at different frequencies. Traditional wavelet analysis method analyzes the low frequency signal only. It ignores the information of the high frequency signal. In this

* The Ministry of science and technology support program(2011BAA04B03).

G. Lee (Ed.): Advances in Automation and Robotics, Vol.2, LNEE 123, pp. 373–379.
Springerlink.com

paper, wavelet packet decomposition is used to analyze wind speed time series from low and high frequency. The optimal wavelet tree is calculated by entropy. Model in every frequency is established by SVR. The short-term wind speed prediction model is established through all of above. It can improve the accuracy by comparison with the previous method.

2 Support Vector Machine Regression Algorithm

SVM is a statistical learning method developed with the structural risk minimization principle developed. It seeks for the optimal of the complexity and learning ability trough limited number of samples. SVM algorithm selects the appropriate kernel function in order to avoid the choice problem of neural network structure, simplifies the nonlinear problem.

Support vector regression solves the nonlinear problems with support vector machine. In SVR, input x is mapped to high dimension feature space F trough the nonlinear mapping Φ(x), A linear model is established to estimated regression function in this feature space.

$$f(x) = w \cdot \Phi(x) + b, \Phi : R^n \to F, w \in F \tag{1}$$

b is the threshold. The sum of risks affecting w in high dimension $\|w\|^2$:

$$R(w) = \frac{1}{2}\|w\|^2 + \sum_{1}^{l} \varepsilon(f(x_i) - y_i) \tag{2}$$

l is the number of samples. $\varepsilon(f(x_i) - y_i)$ is the cost function.:

$$\varepsilon(f(x_i) - y_i) = \begin{cases} 0, & |f(x_i) - y_i| < \varepsilon \\ |f(x_i) - y_i| - \varepsilon, & \text{others} \end{cases} \tag{3}$$

Set up slack variable $\xi_i, \xi_i^* \geq 0$. According to structural risk-based criteria, SVR calculate w and b though minimizing the objective function:

$$R(w, \xi_i, \xi_i^*) = \frac{1}{2}\|w\|^2 + C \sum_{i=1}^{l} (\xi_i + \xi_i^*)$$
$$s.t. \begin{cases} y_i - w \cdot \Phi(x_i) - b \leq \varepsilon + \zeta_i \\ w \cdot \Phi(x_i) + b - y_i \leq \varepsilon + \zeta^* \\ \zeta, \zeta^* > 0 \end{cases} \tag{4}$$

Set up Lagrange $\alpha_i, \alpha_i^*, \lambda_i, \lambda_i^*$, establish Lagrange equation:

$$L(w,b,a,\xi) = \frac{1}{2}\|w\|^2 + C\sum_{i=1}^{l}(\zeta_i + \zeta_i^*) - \sum_{i=1}^{l}\alpha_i(\varepsilon + \zeta_i^* - y_i + w\cdot\Phi(x_i) + b) - \sum_{i=1}^{l}\alpha_i^*(\varepsilon + \zeta_i^* + y_i - w\cdot\Phi(x_i) - b) - \sum_{i=1}^{l}(\lambda_i\cdot\zeta_i + \lambda_i^*\cdot\zeta_i^*) \quad (5)$$

Obtain the minimum of the above equation:

$$\begin{cases} \dfrac{\partial L}{\partial w} = w - \sum_{i=1}^{l}(\alpha_i - \alpha_i^*)\cdot\Phi(x_i) = 0 \\ \dfrac{\partial L}{\partial b} = \sum_{i=1}^{l}(\alpha_i - \alpha_i^*) = 0 \\ \dfrac{\partial L}{\partial \zeta} = C - \alpha_i - \lambda_i = 0 \\ \dfrac{\partial L}{\partial \zeta^*} = C - \alpha_i^* - \lambda_i^* = 0 \end{cases} \quad (6)$$

Then get the dual problem:

$$\min \frac{1}{2}\sum_{i,j=1}^{l}(\alpha_i - \alpha_i^*)(\alpha_j - \alpha_j^*)(\Phi(x_i)\cdot\Phi(x_j)) - \sum_{i=1}^{l}\alpha_i(\varepsilon - y_i) - \sum_{i=1}^{l}\alpha_i^*(\varepsilon + y_i)$$

$$s.t. \begin{cases} \sum_{i=1}^{l}(\alpha_i - \alpha_i^*) = 0 \\ \alpha_i, \alpha_i^* \in [0, C] \end{cases} \quad (7)$$

Solve the quadratic programming problem, get w represented by data points:

$$w = \sum_{i=1}^{l}(\alpha_i - \alpha_i^*)\Phi(x_i) \quad (8)$$

Regression function can be calculated:

$$f(x) = w\cdot\Phi(x) + b = \sum_{i=1}^{l}(\alpha_i - \alpha_i^*)(\Phi(x)\cdot\Phi(x_i)) + b = \sum_{i=1}^{l}(\alpha_i - \alpha_i^*)K(x, x_i) + b \quad (9)$$

$K(x, x_i)$ is called as the kernel function. Different support vector machine can be established by choosing different kernel function.

3 Optimal Wavelet Decomposition Tree

Wavelet packet analysis is a more refined decomposition method compared with the wavelet analysis. It is not only include the low-frequency signal decomposition, also analyze the high-frequency signal. Wavelet packet decomposition algorithm and the reconstruction algorithm is as follows:

Set up $g_j^n(t) \in U_j^n$, $g_j^n(t)$ can be expressed as:

$$g_j^n(t) = \sum_l d_l^{j,n} u_n\left(2^j t - l\right) \tag{10}$$

Wavelet packet decomposition algorithm:

$$\begin{cases} d_l^{j,2n} = \sum_k a_{k-2l} d_k^{j+1,n} \\ d_l^{j,2n+1} = \sum_k b_{k-2l} d_k^{j+1,n} \end{cases} \tag{11}$$

Wavelet packet reconstruction algorithm:

$$d_l^{j+1,n} = \sum_k \left[h_{l-2k} d_k^{j,2n} + g_{l-2k} d_k^{j,2n+1} \right] \tag{12}$$

Entropy can describe the system, the greater the entropy, the more chaotic the system with the less information. Shannon entropy is the most common:

$$E_n = \sum_j^n d_{j,k}^{2n} \log d_{j,k}^{2n} \tag{13}$$

If the sum of entropy of high frequency and low frequency signal is greater than the entropy before decomposition, this decomposition is inappropriate. On the contrary, continue to achieve the optimal wavelet decomposition tree.

4 Simulation

The experiment uses the wind tower speed data in a wind farm, the data is the length of one month, 1000 points. Front nine hundred points is the training sample, the remaining one hundred points is the test samples. The wind speed series is decomposed with wavelet packet method. Decomposition level is 4 as figure 1. Optimal wavelet tree is selected by entropy criterion. Figure 1 shows the gray parts are the optimal decomposition parts. High frequency part(3,7) is no more decomposed.

Every signal from best wavelet tree make the reconstruction process as figure 2.

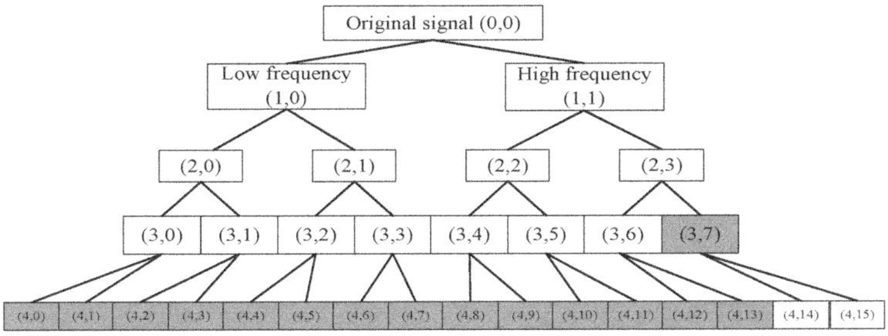

Fig. 1. Four-level decomposition structure of wavelet packet

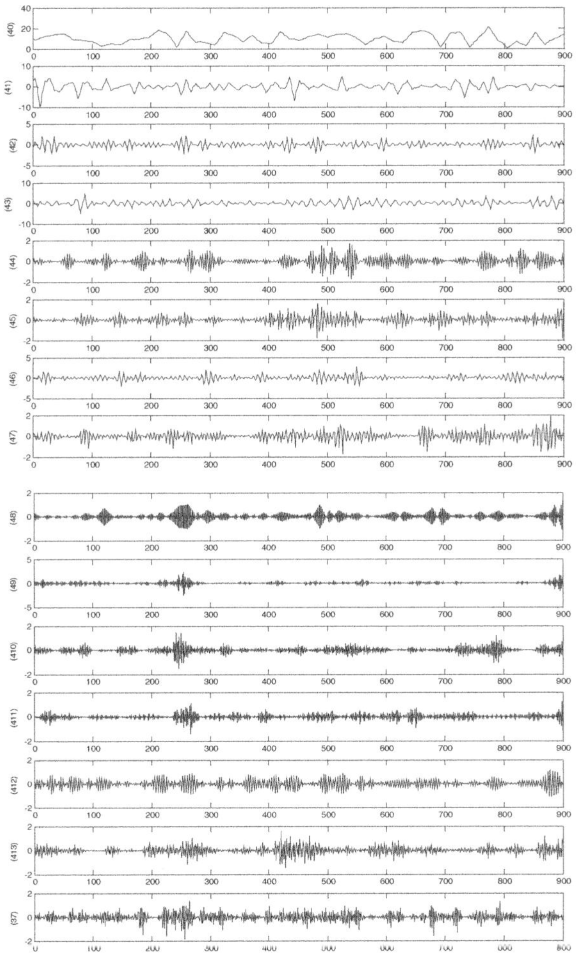

Fig. 2. Decomposed the optimal signals

The fifteen decomposition signals are respectively used to establish the model by support vector machine regression algorithm. The kernel function use RBF kernel function. The parameters of SVR, g and C are calculated by the grid optimization method. The 100 test points were tested by every frequency model. The sum of fifteen outputs is the final wind speed prediction as figure 3. The average absolute error is 0.3235m/S, the average relative error is 5.49%, better than the method of wavelet only analysis of low frequency(The average absolute error is 0.7213m/S, the average relative error is 7.81%). The approach improves the prediction accuracy, shows the performance of advanced.

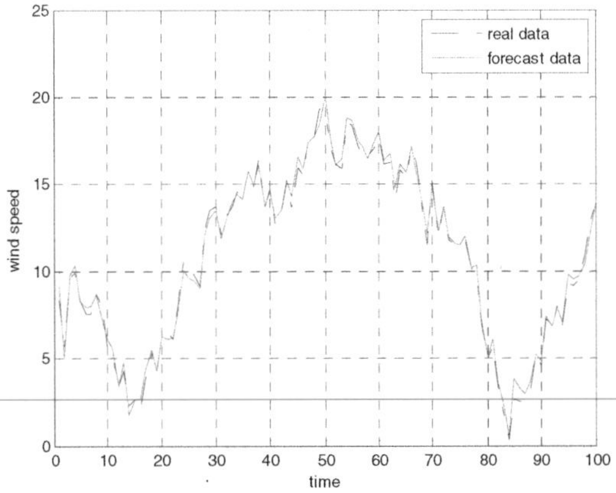

Fig. 3. Result of the short-term wind speed prediction

5 Conclusions

The wind speed time series is analyzed from the frequency domain. In comparison with the traditional algorithms, paper analyzed the high frequency wind signal, used entropy to calculate the best wavelet tree, established the short-term wind speed forecast model with the method support vector machine regression. This method further improves the prediction accuracy.

References

1. Alexiadis, M., Dokopoulos, P., Sahsamanoglou, H., et al.: Short-term Forecasting of Wind Speed and Related Electrical Power. Solar Energy 63(1), 61–68 (1998)
2. Bossanyi, E.A.: A Short-term wind prediction using Kalman. Wind Engineering 9(1), 1–8 (1985)
3. Wang, L.-J., Dong, L., Liao, X.-Z., et al.: Short-term Power Prediction of a Wind Farm Based on Wavelet Analysis. In: Proceedings of the CSEE, vol. 29(28), pp. 30–33 (2009) (in Chinese)

4. Li, Y., Fang, T., Yu, E.: Study of support vector machines for short-term load forecasting. Proceedings of the CSEE 23(6), 55–59 (2003) (in Chinese)
5. Wickerhauser V.I.: Lectures on Wavelet Packet Algorithms (1991)
6. Brown, B.G., Katz, R.W., Murphy, A.H.: Time serial models to simulate and forecast wind speed and wind power. Journal of Climate and Applied Meteorology 23(8), 1184–1195 (1984)

The Realization of Autonomous Roaming Combined with Eagle Eye Navigation in Large Virtual Scene

YiJia Xu, Xiang Liu, and ZiWei Liu

College of Computer Science, Shanghai University of Engineering Science, Shanghai, China
{Walkingmu,liuxiang.sh.cn}@gmail.com, liuzhiwei0326@hotmail.com

Abstract. In a large virtual environment, people always tend to lost themselves because of they are unfamiliar to the VE. Autonomous roaming combined with the eagle eye navigator could assist the positioning in the larger scene. This article sets up a bridge between three-dimensional virtual scene and the two-dimensional plane using the web scripting language - JAVA Script. Thus they are synchronized to the HTML page. It allows the visitor to understand the current position of the large virtual scene and anyone can roam unimpeded in the scene.

Keywords: virtual reality, eagle eye, JAVA Script, VRML.

1 Introduction

In the large virtual environment, there are several roaming ways, such as fixed-point roaming, navigation roaming and independent roaming. Each way has its drawbacks and inconvenience. It is because in fixed-point roaming, the experiencer only has one view at each view point that makes the system lacking continuity and systematic experience. In navigation roaming, the users couldn't choose their direction or location. And in large environment, independent roaming sometimes could make users who are not familiar with the map lost their direction or even lost their way.

In this article, we apply the eagle eye technology to independent roaming, which we called eagle eye independent roaming. It is like presenting a virtual 2 dimensional map in the virtual environment, and on this map, the location of the tourist is marked on this map in order to guide the user so they can easily wandering around even if it's their first time use.

2 General Roaming Way

Roaming in virtual environment could be divided into independent roaming and dependent roaming, categorized by the autonomy in the whole movement.

2.1 Dependent Roaming - Fixed-Point Roaming and Navigation Roaming

Multi-views and jumping mode are used in fixed-point roaming which means there is several duplicate viewpoints. Each viewpoint binds a view, and the best direction and

G. Lee (Ed.): Advances in Automation and Robotics, Vol.2, LNEE 123, pp. 381–386.
Springerlink.com

angle is assigned to it. This roaming way is the fastest and straightforward way; it can help the user get the most important information in the virtual scene in least time.

Navigation roaming is made by guiding animation. Guided by a preset auto direct system, the user can roam in the virtual environment without any operation. For example, the user can follow a car, roaming in the VE following a predefined path.

2.2 Independent Roaming

Comparing to dependent roaming, independent roaming has more freedom, the user would be like personally on the scene. The user can move by their wish in the current VE, if the flying mode is opened, even the whole VE could be overlooked. The least limit for users is the most advantage part in independent roaming; users can walk on every inch of the VE or watch every single detail in the scene.

3 The Principle and Kernel of the Eagle Guiding System

The purpose of using the VRML as the modeling language for Virtual Reality is to build a VRML virtual scene. Although VRML provide a convenient way to design a 3D virtual environment, VRML has its drawbacks too. The VE totally designed by VRML cannot interact with the users immediately. In order to solve this problem, using JavaScript as the bridge, this could allow user interaction between the 3D VE.

Next we will present the kernel. We should first need VRML VE and a 2D map or Java Applet application which could represent the 3D VE. In order to introduce the principle and kernels, here we present the 2D map as an example; figure 1 is the structure of the HTML web page.

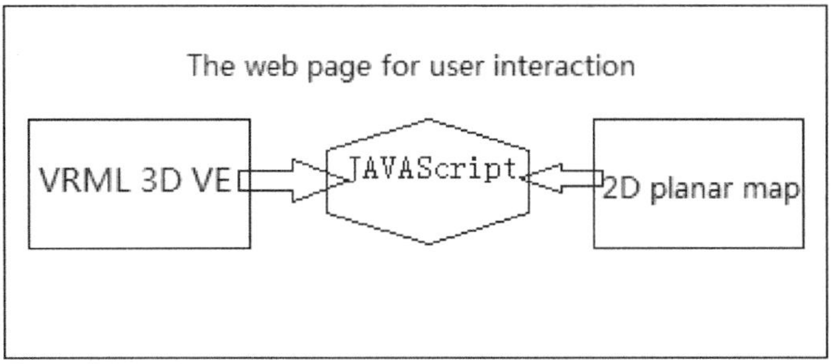

Fig. 1. Structure of Web Page

3.1 Load VRML 3D VE, 2D Map and 2D Substitution into HTML Webpage

Firstly, using the OBJECT tag to insert a 3D VE, Cortona is the plugin for VRML roaming.

```
<OBJECT STYLE="position: absolute; left: 300; top: 0"
  name=Cortona WIDTH="700" HEIGHT="600"
  CLASSID="CLSID:86A88967-7A20-11d2-8EDA-00600818EDB1">
<param name="Scene" value="../vrml/play_free.wrl">
<param name="VRML_DASHBOARD" value="FALSE"></OBJECT>
```

Next, we use IMG tag to insert the 2D map and substitution.

```
<img id="image1" STYLE="position: absolute;left:0;top:
100; z-index: 1" border="0" src="ditu.jpg" width="266"
height="300">
<img id="image2" STYLE="position: absolute; left: 850;
top: 200; z-index: 1" border="0" src="man.GIF"
width="20" height="20">
```

3.2 Transformation between 3D Space and 2D Map

By using coordinate transform twice, the position in a 3D space can be projected onto a 2D map in the webpage.

Firstly, projection matrix will be applied to 3D space transforming to the 2D map.

$$Z_C\begin{bmatrix} x \\ y \\ 1 \end{bmatrix} = \begin{bmatrix} f_x & 0 & 0 & 0 \\ 0 & f_y & 0 & 0 \\ 0 & 0 & 1 & 0 \end{bmatrix}\begin{bmatrix} X_C \\ Y_C \\ Z_C \\ 1 \end{bmatrix} \tag{1}$$

f_x, f_y is the scaling factor in x, y direction.

Then, the offset from the up left corner to the picture control is (u_0, v_0), the size of 2D map is $W_m \times H_m$, and the size of the substitution is $W_p \times H_p$, then the location of the substitution on the top left corner is

$$\begin{bmatrix} x_p \\ y_p \\ 1 \end{bmatrix} = \begin{bmatrix} 1 & 0 & u_0 + \frac{W_m}{2} - \frac{W_p}{2} \\ 0 & 1 & v_0 + \frac{H_m}{2} - \frac{H_p}{2} \\ 0 & 0 & 1 \end{bmatrix}\begin{bmatrix} x \\ y \\ 1 \end{bmatrix} \tag{2}$$

From equation (1), (2), we have

$$Z_C\begin{bmatrix} x_p \\ y_p \\ 1 \end{bmatrix} = \begin{bmatrix} f_x & 0 & u_0 + \frac{W_m}{2} - \frac{W_p}{2} & 0 \\ 0 & f_y & v_0 + \frac{H_m}{2} - \frac{H_p}{2} & 0 \\ 0 & 0 & 1 & 0 \end{bmatrix} \begin{bmatrix} X_C \\ Y_C \\ Z_C \\ 1 \end{bmatrix} \tag{3}$$

The disp1 function implemented the coordinate transform, $f_x = f_y = 0.5$, so from the equation(1) we can know (u_0, v_0) is (0,100), the size the 2D map is 266 * 300, the size of the substitution is 20 * 20. So converting the 3D coordination into 2D points,

```
x= 0.5 * Cortona.Engine.Nodes
     ('PS').Fields('position_changed').x;
y=0.5 * Cortona.Engine.Nodes
     ('PS').Fields('position_changed').y
```

PS is the defined node in VRML; position_change field is the domain of the node. Then in order to make the substitution show on the corresponding location, we assign the 2D point to the substitution of the 2D map.

```
image2.style.left = y + 100 + 300/2 - 20/2;
image2.style.top = x + 0 + 266/2 - 20/2;
```

3.3 Update the User Location in Spare Time in JavaScript

To get the user coordinate in 3D VE by using JAVA Script, and update the coordinate in spare time.

```
// run disp1() each a second, update the position
window.setInterval("disp1()", 1000);
```

This would synchronize the display between the VRML 3D VE and 2D plane.

4 Independent Roaming in Virtual College Scene with Eagle Eye System

Base on the analyses, we present the VE of Shanghai University of Engineering Science, using the eagle eye system to run an independent roaming.

A large scene model cannot be built in one day, for example, when dealing with various buildings in schools and colleges, the buildings should be considered and implemented as an object, and model them individually as Shape, then import into

VRML, and using Translation, Rotation and other geometric transformation to implement set up decorations, lights, docks, sounds, substitutions and collision detection. Arranging the model tree, scale calibration, location calibration and background modification should be adjusted when combing models. The large scene would be presented after finished the above procedure.

In a large VE model, the number of facets on the model would affect the fluency of the independent roaming. In order to reduce the facets without reducing vivid effect; we use images instead of facets. The size of the image should be controlled under 30 million pixels, using duplicated image could reduce hardware requirement, improve the fluency to an acceptable level.

During the implementation of the eagle eye, we embedded Java Script into the HTML, and the setInterval function would automatically be called in order to update the coordinate, which would allow the eagle system get the location of the substitution. To computing the location of the substitution, we use the equation 3.

To sum up, this paper presents a system which has guiding function to independently roaming in the college. Figure 2 and Figure 3 are the scenes when visiting the library and the gymnasium.

Fig. 2. Library

Fig. 3. Gymnasium

5 Conclusion

In large virtual environment, due to the variety and complexity of the models and topography, people who are unfamiliar to the VE usually could only roam in a predefined path, which violates the meaning of virtual reality. After eagle eye is deployed, visitors will effectively locate their position in VE. By using EAI in VRML, combining the independent roaming, available space is maximized for the visitors which improved the regular roaming way.

References

1. Davies, E.R.: Machine Vision: Theory, Algorithms, Practicalities. Posts and Telecom Press, Beijing (2009)
2. VRML committee, `http://www.vrml.org/specifications/vrml97`
3. Cyber technology studio: Programming in VRML and Java. Posts and Telecom press, Beijing (2002)
4. Wei, W., Cheng, Y.: Virtual Campus Roaming System Based on VRML. Journal of Beijing University of industry and commerce 24, 54–57 (2006)
5. Li, X., Li, X.: The research of virtual digital campus modeling based on VRML. Journal of Zhejiang Normal University (nature science) 04, 402–406 (2005)
6. Wan, J.: 3D scene modeling in virtual school modeling base on VRML. Computer Applications and Software 21, 82–84 (2004)
7. Wang, J., Kang, L.: Virtual reality in the application of digital campus. Computer Simulation 06, 79–81 (2003)
8. Ge, Y., Li, W.: YanHong Ge, WenFeng Li: Human interaction with VRML virtual environment based on Java. Computer and Communications 2, 35–37 (2003)
9. Cheng, H., Wang, T.: The design and implementation of Virtual building environment in a real-time roaming system. Journal of Southwest Jiaotong University 36, 53–56 (2001)
10. Song, H.: Huiling Song: Geography Virtual Museum Roaming System Based on VRML. Computer Engineering 32, 239–241 (2006)

A Survey on Multimodal Biometrics

Yu Wang and Zaiwen Liu

School of Computer Science and Information Engineering,
Beijing Technology and Business University, 100048, Beijing, China
wangyu@th.btbu.edu.cn

Abstract. Biometrics has become one of the most promising authentification techniques in last few years. This paper summarizes the multimodal biometrics techniques which identify an individual from others and also describes a lot of multimodal biometrics systems. Many literatures show that multimodal biometrics techniques have performed more reliability and security than monomodal biometrics ones as they take more than one physiological or behavioral characteristics of the person into account to identify and verify that person. Multimodal biometrics have become one of inevitable trends in the future.

Keywords: Biometric, Multimodal, Fusion, Normalization.

1 Introduction

As the network age coming, the chance requiring people to contact face to face is less and less. Stealing identification frequently occurs because of this and has been recently paid more attention to by people in society. Only in America about one million people occupied 5% of the population above eighteen years old were stolen identifications in 2003 [1]. Now authentification has become a hot issue in research and application. Because biometric technologies make full use of physiological or behavioral characteristics of the body or part to authenticate individual from others, they are more safe, secret and convenient than traditional methods. At the same time they aren't easy to be forgotten, faked or stolen, and whenever and wherever they can be used.

On the other hand, monomodal biometric techniques often confronts blank wall for practical application in true life because of unforeseen variability of objective condition. For example, some people can't be collected fingerprint; face image may change with the variability of time and illumination; iris, DNA and fingerprint make users feel uncomfortable even offensive. In addition, in some application field requiring high security, it is difficult for monomodal biometric to obtain prospective result. Because multimodal biometrics are based on diversiform biometric traits and combine data fusion techniques, they not only improve the recognition rate, but also enlarge application range of system, and decrease the risk of system. Therefore multimodal biometrics technologies have been an inevitable trend in the future.

This paper is organized as follows. Section 2 explains the fusion of multimodal biometrics including fusion ways, fusion levels, normalization techniques, fusion

G. Lee (Ed.): Advances in Automation and Robotics, Vol.2, LNEE 123, pp. 387–396.
Springerlink.com

methods and operational modes. Section 3 introduces some multimodal biometrics systems and databases. A conclusion is drawn in section 4.

2 Fusion of Multimodal Biometrics

The fusion of multimodal biometrics includes many aspects such as fusion ways, fusion levels, normalization techniques, fusion methods and operational modes etc. We will introduce these problems one by one.

2.1 Fusion Ways

Generally biometric system consists of sensors, feature extraction, matching score and decision-making modules. The combination of single biometric and the fusion of multimodal biometrics may occur at any stage of this process. So multimodal biometrics systems can be designed by five ways as shown Fig.1 [2, 3]. Some of them may not involve multiple modalities but imply a fusion at some points. They are described as follows.

a) Single biometric and multiple sensors: the same biometric can be obtained using different sensors, and combined to improve the recognition rate.
b) Multiple biometrics: multiple biometrics of the same person can be obtained and fused to improve the recognition rate. This is a real sense of 'fusion'.

c) Single biometric and multiple units: this way uses the same biometric but different unit. For example, we can collect fingerprint of the same person but different finger.

d) Single biometric and multiple representations: the same biometric unit can be obtained several times by a same sensor and represented by multiple ways. Every representation has its classifier, and matching scores generated by classifier are computed and combined to improve the recognition rate.

e) Single biometric and multiple matchers: the same biometric is obtained by a single sensor. Different approaches of feature extraction and matching methods are combined to improve the recognition rate.

2.2 Fusion Levels

Biometric systems have four important components, and information fusion or combination can occur at every stage. This paper uses an example of face and fingerprint to explain the fusion levels as shown Fig.2.

Features extraction: if the extracted biometrical features or traits are independent each other, they can be fused or combined to form a unique feature vector.
Matching score level : matching score which describes the similarities between the acquired biometrics and their templates is put out by every biometric system and combined. This kind of fusion requires normalization of the matching scores in order to assure that matching scores belong to the common domain [3, 4]before they are fused or combined. Generally two steps are performed: statistical estimation of the scores distribution and translation into a common domain.

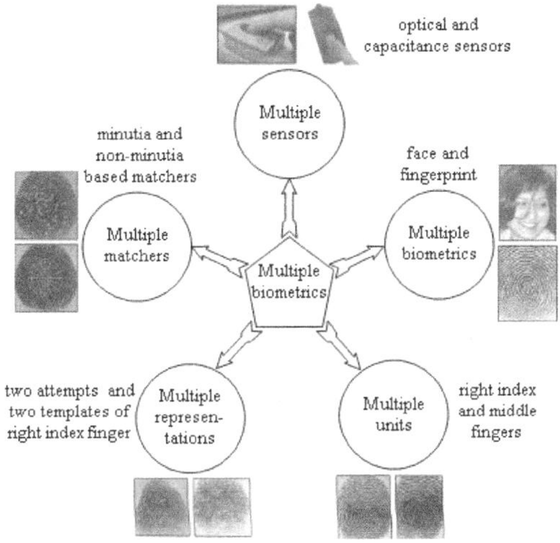

Fig. 1. Fusion ways [2].

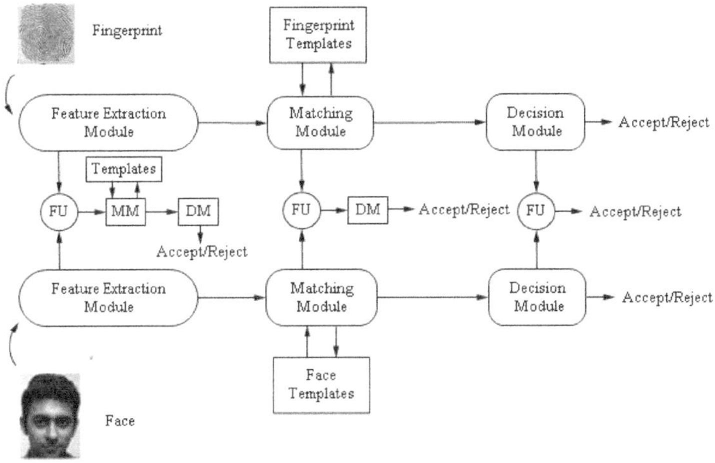

Fig. 2. Possible fusion levels (FU: fusion; MM: matching module; DM: decision module) [4].

Decision level: every decision obtained by every system is combined to complete a final decision.

For a biometric system, the earlier the fusion or combination operates, the more remarkable the effect is. Therefore there will be a better performance if the fusion or combination operates at the features extraction level. However it is very difficult for biometric system to combine at this level [4]. Thus matching score level is a preferred choice. In addition, it is inevitable for a both reliable and robust biometric system to adopt a suitable decision combination approach.

2.3 Normalization Techniques

In a multimodal system, scanning traits, features extraction and classification require various schemes, so the scores must be changed into normalization form before fusion or combination. For example, the range for matching score of face generally is 50 to100, while the range for that of ear often is 5 to10. If all the scores from different source have to be taken into account in a same system, normalization is very important. Different normalization techniques often are used. Some of common normalization methods are described below. We denote the output score by s and normalized score by s'.

a) Min-max : This method normalizes the generated score on the scale of 0-1.

$$s' = {(s - min)}/{(max - min)}, \tag{1}$$

where *min* denotes minimum score range and *max* denotes maximum score range.

b) Z-score : This method normalizes the scores to a distribution with mean of 0 and standard deviation of 1.

$$s' = {(s - \mu)}/{\sigma}, \tag{2}$$

where μ is the mean of scores and σ is the standard deviation of scores.

c) Tanh : This method normalizes scores to the scale with range 0 to 1. This method is considered as one of the robust techniques of normalization.

$$s' = \sqrt{tanh\left(0.01(s - \mu)/\sigma\right) + 1}, \tag{3}$$

where μ and σ is the mean and standard deviation respectively for the generated scores. In addition, there are else other methods to normalize scores such as median adaptation, twoquadrics, logistic and quadric- line-quadric etc.

2.4 Fusion Methods

Fusion techniques are the important component in a multimodal biometric system. The normalized scores are mapped on a single dimensional plane to fuse and make final decision. Decision must be made depending on the defined threshold. If the generated value after fusion is less than the defined threshold, the user will be rejected, or user will be accepted. It is more important and complex for the fusion or combination of matching scores than normalization. In multimodal biometric system, not every biometric feature shares equal measure of accuracy. Some of them are more vulnerable to suffer from the effect of noisy and error than others. Some common fusion methods are introduced below.

In matching score level, some methods such as simple sum rules [3], weighted averaging [2], produce rules, k-NN classifiers, decision trees and Bayesian methods are used to fuse or combine the scores and obtain excellent performance. In addition, Roli et.al. [5] classified decision fusion to two sorts: fixed rules and trained rules. Majority voting and sum rule belong to fixed rules, but weighted averaging and behavior knowledge space are trained rules [3, 6].

2.5 Operational Modes

Furthermore, Multimodal biometrics systems can operate in three different modes from the view of operational aspects [7]: parallel mode, serial mode, and hierarchical mode as shown Fig.3.

Parallel mode: biometric traits can be fused or combined simultaneously in multimodal systems.

Serial mode: biometric traits can be fused or combined one by one in multimodal systems.

Hierarchical mode: This operational mode operates in a hierarchical scheme, like a tree structure, when the number of classifiers is large.

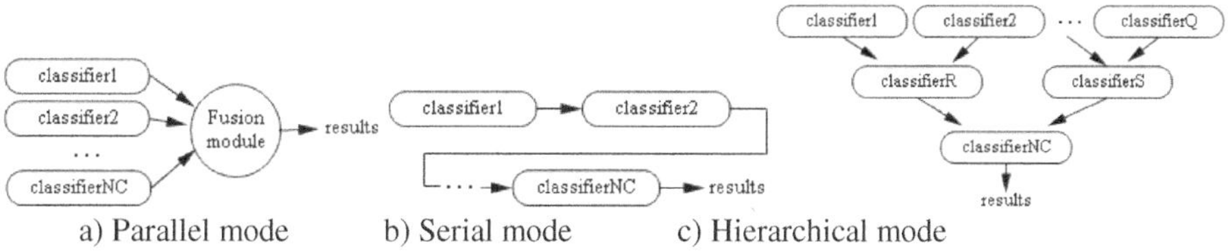

Fig. 3. Operational modes [2, 8]

3 Multimodal Biometrics Systems and Databases

This paper summarizes current multimodal biometrics systems and databases according to the fusion ways, which are listed as follows.

3.1 Single Biometric and Multiple Sensors

In the following literatures, face is used to identify ID. Images obtained by different sensors are fused effectively at matching score level or decision-making level. Experimental results show that using fusion techniques can improve the recognition rate to some extent. Detail contents are described in Table1.

3.2 Multiple Biometrics

The fusion of multiple biometrics is a well-named multimodal biometrics fusion. Table2 summarizes existing multimodal biometrics fusion systems and corresponding databases.

Table 1. Single biometric and multiple sensors

Literature	Database	Image sorts	Fusion levels	Fusion methods	Results
[9]	50 subjects	Gray and range	Matching score	Weighted-sum rule	An improvement of the EER of about 2-5% after fusion comparing to the monomodal approaches.
[10]	278 subjects	Color and range	Matching score	Weighted-sum rule	Recognition rate of 92.8% after fusion, while those of only about 83.1% for 2D and 83.7% for 3D.
[11]	240 subjects	Visible-light and infrared light	Matching score and decision	Unweighted rank based, logarithmically rank transformation, score based	All these combinations methods outperforms the monomodal approaches, and the score based strategy outperforms the other fusion methods.

Table 2. Multiple biometrics

Literature	Database	Modalities	Fusion levels	Fusion methods	Results
[12]	100 subjects	Face, fingerprint, hand geometry	Matching score	Simple-sum-rule, max-rule, min rule	The best monomodal system fingerprint in this study obtained a genuine acceptance rate of 83.6% at a FAR of 0.1%, while the multimodal approach obtained that of 98.6%.
[13]	972 subjects	Face, fingerprint	Matching score	Simple-sum, min/max score, matcher weighting, user weighting	The EER of fingerprint and of face was respectively 2.16% and 3.76%, while the fusion approach obtained an EER of 0.63%.
[14]	50 subjects	Face, fingerprint, signature	Matching score	Simple-sum, SVM	The EER of face, signature and fingerprint was respectively 10%, 4% and 3%. While the sum-rule, the SVM user-independent and the SVM user-dependent fusion approaches obtained EERs of 0.5%, 0.3%, and 0.05%, respectively.
[15]	100 subjects	Palmprint, hand, geometry	Feature extraction, matching score	Max-score	The best monomodal system palmprint in this study obtained a FAR of 4.49% at a FRR of 2.04%, while the multimodal approach obtained a FAR of 0% at a FRR of 1.41%.

Table 2. *(continued)*

[4]	100 subjects	Face, fingerprint, hand, geometry	Matching score	Sum-rule, decision trees, LDA	The best monomodal system fingerprint obtained a FRR of 25% at a FAR of 0.01%, while the combination approach obtained a FRR of 1.78% at a FAR of 0.03%.
[16]	130 subjects	Face, iris	Matching score	Sum rule, neural networks	All the fusion strategies outperform the monomodal approaches.
[17]	236 subjects	Face, fingerprint	Decision	—	At a FAR of 0.01%, the monomodal systems obtained a FRR of 61.2% and 10.6% for face and fingerprint respectively. For the same FRR, the fusion approach obtained a FRR of 6.6%.
[18]	—	Face, speech	Matching score	Weighted geometric average	Voice and face obtained a correct identification rate of 88% and 91% respectively, while bimodal approach obtained that of 98%.
[19]	150 subjects	Face, voice, lip movement	Matching score	—	Experiments demonstrated a decrease below 1% of the FAR.

In addition, there are the BIOMET[20], the BANCA[21] and the MCYT[22] databases for the research about multimodal biometrics fusion. Among these databases, BIOMET includes face, voice, fingerprint, hand geometry and signature five different modalities of 327 people. BANCA contains 208 users and includes two different modalities face and voice acquired in four European languages. MCYT contains 330 users and includes fingerprint and signature two different modalities. Results in table2 show that the effective fusion of multiple biometrics can improve remarkably the performance of the systems.

3.3 Single Biometric and Multiple Matchers, Units or Representations

In following literatures multiple matchers, units or representations of single biometric are combined. Experimental results show that the final recognition rate of the systems can also be improved by these kinds of fusion ways.

Table 3. Single biometric and multiple matchers, units or representations

Literature	Database	Fusion ways	Fusion levels	Results
[23]	100 subjects	Multiple units (two different fingers)	Feature extraction	Fusion approaches obtained about 1.8% of the FAR and 1.9% of the EER.

Table 3. *(continued)*

[24]	206 subjects	Multiple representations (face, PCA, ICA and LDA)	Matching score	Monomodal systems obtained the recognition accuracy of 79.1%, 88.1% and 81.0% for the PCA-based, ICA-Based and LDA-based respectively, while combination approaches obtained that of 90.0% and 90.2% for the sum-rule and the RBF-based respectively.
[25]	167 subjects	Multiple units (four different fingers)	Decision	The combination of all the matchers improved the performance more than 3%, the combination of multiple representations or multiple units of a same modality also improved the performance.
[26]	167 subjects	Multiple matcher (fingerprint, Hough trans-formation, string distance and 2D dynamic based)	Matching score	With multiple matchers, the performance improved significantly.

4 Conclusion

To sum up, multimodal biometrics systems have more advantages than monomadal biometric systems. Existing problems like non-universality, spoofing, inefficiency and inaccuracy in identification can be solved effectively in those systems. In addition, fusion and normalization techniques are helpful to further improve the accuracy and efficiency of multimodal biometrics systems. Although multimodal biometrics systems also exist some disadvantages such as considerable cost and long run time, with the improvement of the technologies in hardware and the great potential of the systems in public areas like internet banking, e-commerce and forensic applications, multimodal biometrics systems must become the mainstream among identification technologies in the future.

Acknowledgments. This work is supported by the National Natural Science Foundation of China under the Grant No. 61171068.

References

1. http://www.people.com.cn
2. Jain, A.K., Ross, A., Prabhakar, S.: An introduction to biometric recognition. IEEE Transactions on Circuits and Systems for Video Technology, Special Issue on Image- and Video-Based Biometrics 14(1), 4–20 (2004)
3. Ross, A., Jain, A.K.: Multimodal biometrics: An overview. In: 12th European Signal Processing Conference, Vienna, pp. 1221–1224 (2004)
4. Ross, A., Jain, A.K.: Information fusion in biometrics. Pattern Recognition Letters 24, 2125–2215 (2003)

5. Roli, F., Kittler, J., Fumera, G., Muntoni, D.: An Experimental Comparison of Classifier Fusion Rules for Multimodal Personal Identity Verification Systems. In: Roli, F., Kittler, J. (eds.) MCS 2002. LNCS, vol. 2364, pp. 325–335. Springer, Heidelberg (2002)
6. Jain, A.K., Ross, A.: Learning user-specific parameters in a multibiometric system. In: Proceedings International Conference on Image Processing (ICIP), vol. 1, pp. 57–60. Rochester, New York (2002)
7. Maltoni, D., Maio, D., Jain, A.K., Prabhakar, S.: Handbook of Fingerprint Recognition. Springer, New York (2003)
8. Dessimoz, D., Richiardi, J.: Multimodal biometrics for identity documents. Research report, PFS 341-08.05. Université de Lausanne (2006)
9. Tsalakanidou, F., Malassiotis, S., Strintzis, M.G.: Integration of 2D and 3D images for enhanced face authentication. In: 6th IEEE International Conference on Automatic Face and Gesture Recognition (FGR 2004), Seoul, pp. 266–271 (2004)
10. Chang, K.I., Bowyer, K.W., Flynn, P.J.: Multi-modal 2D and 3D biometrics for face recognition. In: The IEEE International Workshop on Analysis and Modeling of Faces and Gestures (AMFG 2003), pp. 187–194. IEEE Computer Society, Nice (2003)
11. Chen, X., Flynn, P.J., Bowyer, K.W.: Visible-light and infrared face recognition. In: 2003 Workshop on Multimodal User Authentication, Santa Barbara, pp. 48–55 (2003)
12. Jain, A.K., Nandakumar, K., Ross, A.: Score normalization in multimodal biometric systems. Pattern Recognition 38, 2270–2285 (2005)
13. Snelick, R., Uludag, U., Mink, A., Indovina, M., Anil, J.: Large scale evaluation of multimodal biometric authentication using state-of-the-art systems. IEEE Transactions on Pattern Analysis and Machine Intelligence 27(3), 450–455 (2005)
14. Fierrez-Aguilar, J., Ortega-Garcia, J., Garcia-Romero, D., Gonzalez-Rodriguez, J.: A Comparative Evaluation of Fusion Strategies for Multimodal Biometric Verification. In: Kittler, J., Nixon, M.S. (eds.) AVBPA 2003. LNCS, vol. 2688, pp. 830–837. Springer, Heidelberg (2003)
15. Kumar, A., Wong, D.C., Shen, H.C., Anil, K.J.: Personal Verification Using Palmprint and Hand Geometry Biometric. In: Kittler, J., Nixon, M.S. (eds.) AVBPA 2003. LNCS, vol. 2688, pp. 668–678. Springer, Heidelberg (2003)
16. Bigün, E.S., Bigün, J., Duc, B., Fischer, S.: Expert Conciliation for Multimodal Person Authentication Systems using Bayesian Statistics. In: Bigün, J., Borgefors, G., Chollet, G. (eds.) AVBPA 1997. LNCS, vol. 1206, pp. 291–300. Springer, Heidelberg (1997)
17. Hong, L., Jain, A.K.: Integrating faces and fingerprints for personal identification. IEEE Transactions on Pattern Analysis and Machine Intelligence 20(12), 1295–1307 (1998)
18. Brunelli, R., Falavigna, D.: Person identification using multiple cues. IEEE Transactions on Pattern Analysis and Machine Intelligence (PAMI) 17(10), 955–966 (1995)
19. Frischholz, R.W., Dieckmann, U.: BioID: A multimodal biometric identification system. Computer 33(2), 64–68 (1998)
20. Garcia-Salicetti, S., Beumier, C., Chollet, G., Dorizzi, B., Leroux les Jardins, J., Lunter, J., Yang, N.: Petrovska-Delacrétaz: BIOMET: A Multimodal Person Authentication Database Including Face, Voice, Fingerprint, Hand and Signature Modalities. In: Kittler, J., Nixon, M.S. (eds.) AVBPA 2003. LNCS, vol. 2688, pp. 845–853. Springer, Heidelberg (2003)
21. Bailly-Baillière, E., Bengio, S., Bimbot, F., et al.: The BANCA Database and Evaluation Protocol. In: Kittler, J., Nixon, M.S. (eds.) AVBPA 2003. LNCS, vol. 2688, pp. 625–638. Springer, Heidelberg (2003)

22. Ortega-Garcia, J., Fierrez-Aguilar, J., Simon, D., Gonzalez, J., Faundez-Zanuy, M., Espinosa, V., Satue, A., Hernaez, I., Igarza, J.J., Vivaracho, C., Escudero, D., Moro, Q.I.: MCYT baseline corpus: A bimodal biometric database. In: IEE Proceedings - Vision, Image and Signal Processing, Stevenage, Herts, vol. 150(6), pp. 395–401 (2003)
23. Yanikoglu, B., Kholmatov, A.: Combining multiple biometrics to protect privacy. In: ICPR-BCTP Workshop 2004, Cambridge (2004)
24. Lu, X., Wang, Y., Jain, A.K.: Combining classifiers for face recognition. In: International Conference on Multimedia and Expo (ICME), vol. 3, pp. 13–16. I.E.E.E. Press, Baltimore (2003)
25. Prabhakar, S., Jain, A.K.: Decision-level fusion in fingerprint verification. Pattern Recognition 35(4), 861–874 (2002)
26. Jain, A.K., Prabhakar, S., Chen, S.: Combining multiple matchers for high security fingerprint verification system. Pattern Recognition Letters 20(11-13), 1371–1379 (1999)

Population Bursts Management in Digital City

Xiaojian Wang[1], Siyuan Liu[2], Ce Liu[3], and Yanping Liu[1]

[1] School of Economics and Management
Beijing Jiaotong University, Beijing, China
[2] iLab, Heinz College, Carnegie Mellon University, Pittsburgh, USA
[3] University of Pittsburgh, Pittsburgh, USA

Abstract. In this paper, we study a novel problem, population burst detection, one of the important issues related population monitoring and management within a city. Based on the detected population bursts, we can trace and analyze these bursts, which can help security departments to infer the cause of bursts and be prepared for the future possible bursts. Though it is useful, it is not trivial to detect population bursts, especially under the condition that we can hardly get the population samples within a city from time to time. To address the difficulties of lacking real population data, we take the advantage of communication networks, specifically, mobile phone networks, which offer enormous communication data between peoples. Most importantly, we find the fact that we can use these communication data to infer the population samples. Therefore, we propose an effective and efficient algorithm to detect bursts over call volumes. We verify the performance of our proposed mechanism with an onsite case study and real calling data.

Keywords: population burst, burst detection, clustering.

1 Introduction

The growing urban modernization demands extensive IT services to provide a better living environment for citizens. Our Digital City Project includes pervasive data collection, efficient data transmission, effective information retrieval, and novel knowledge discovery. Based on Digital City, we can efficiently and comprehensively monitor, manage and develop the physical city. This paper concerns with the population data collection and analysis in a city. Population data is clearly dynamic in both temporal and spatial domains. If we know the temporal and spatial population distributions, it will certainly help in many aspects, such as city planning and emergency management.

Note that even for a medium size city, many unexpected events may occur. Most of these events are directly or indirectly related to the population burst. Informally, a population burst is defined as a population distribution situation that is abnormal compared to the normal steady state based on historical data. For example, in Obama's inauguration, more than 2 million populations crowded in Capitol Hill of Washington DC on 20th January, 2009. It was an impressive population burst. Population bursts bring great challenges to city monitoring, emergency management,

G. Lee (Ed.): Advances in Automation and Robotics, Vol.2, LNEE 123, pp. 397–406.
Springerlink.com

and traffic control. It is, therefore, essential to monitor and trace the burst events, investigate the trajectories of bursts, and predict the upcoming bursts. In our work, we mainly focus on detecting the temporal and spatial population burst. The ultimate objective of this research is to provide an in-depth study of the population burst distribution via quantitative research on the real data of a city in China. To achieve this goal, we perform the task in three major steps. We first compute the density of the population. Then we derive the density of bursts. Finally, we construct the burst trajectories. However, it is always a great challenge to shed light to the population burst, by taking into account varieties of movements, extremely distributed situations, and great temporal and spatial diversification.

Due to the lacking of population data samples, it is not possible to detect population bursts directly on population data. Fortunately, in most modern cities, nearly every person brings along one or more mobile phones, especially in China. These phones can be considered as life mates and sensors of a person. Through mobile phones, we can learn the information about locations, temporal and spatial communications of mobile phone users in the daily life. We were able to collect cell phone data from a city in China from 1st January, 2008 to 31st December, 2008. The area of the city is about 8700 km^2 covered by 4120 cell base stations. There are about 3.6 million mobile phone users.

Though we have one year's calling data of the mobile phone networks, it is still not trivial to detect burst on calling volumes due to the following reasons:

1) The population bursts or burst events not only temporally and spatially distribute, but also dynamically and diversely evolve, which causes the calling volume data dynamically changed.
2) We have extremely limited samples and incomplete data with respect to each person. The traditional density-based clustering method does not work due to limited number of samples.
3) The locations of mobile phones are lossy and inaccurate.
4) Frequent and high change rate of the mobile phone usage.

Confronting with the challenges from unique data and application scenario, we take advantage of the following observations from our collected communication data and domain knowledge to detect population burst effectively and efficiently. The first one is that the cell location (the location of the base station in charge of communication in the cell) is accurate. Even though the mobile phone location is extraordinary inaccurate, the cell location is always accurate to the tenth significant digits of latitude and longitude. Moreover, the cell radius is always obtainable. Thus, in this study, we use cells as our observation unit. The cell distribution resolution is good enough for us to monitor and detect population bursts. The second one is that the burst is caused by focal incentives and abnormal accumulation. Especially the more mobile phones online, there are more people in the area. That is, the abnormal mobile phone usage volume will reflect the abnormal population distribution. Fortunately, the volume statistics is accurate and can be collected in real time within cellular networks. The third one is that daily volume spectrums of the city or cells are relative steady. It will offer us a great opportunity to detect and observe the call volume bursts in the city.

Based on the above observations, we first mathematically prove the relation between call volume and population density. Moreover, we propose a novel reflection-based burst detection model. It is able to not only detect fuzzy bursts but also compute the burst strength and lifetime. Then, we quantify and study the temporal and spatial massive burst distribution and evolution.

The rest of the paper is arranged as follows. Section 2 discusses the reflection based burst detection method. In Section 3, we discuss the population burst density issues in more practical environments, and propose the advanced method. We validate the method and model through field studies in real life situations in Section 4. Section 5 gives an overview of the related work. We conclude the remarks and outline the directions for future work in Section 6.

2 Reflection-Based Burst Detection

As we stated earlier, it is not possible to have the population samples all the time within a city. Thus, in order to detect the population burst, we propose a bust-detection method based on the call volume data derived from mobile phone networks. We call this method reflection-based population burst detection method because we want to use call volume change to reflect the population update within a city.

2.1 Relation between Population Density and Call Volume

Definition 1: The current call volume $v^{(t)}(l)$ is defined as the summation of all the personal call volume v_m in cell l at time instance *t*.

$v^{(t)}(l) = \sum_{m \in n} v_m$, where *n* is the number of persons in the cell.

We notate the *hT* time ago call volume in l as $v_{hT}^{(t)}(l)$, where *T* is the time period, and *h* is the number of periods.

Definition 2: Given a cell l, $V^{(t)}(l)$ is the set of all reported history call volume in l corresponding to the time instance *t*, i.e.,

$$V^{(t)}(l) = \left\{ v_{hT}^{(t)}(l), v_{(h-1)T}^{(t)}(l), \ldots, v_{2T}^{(t)}(l), v_1^{(t)}(l) \right\}.$$

The call volume spectrum at all the time instances is $V(l) = \bigcup_{t \in [0,+\infty]} V^{(t)}(l)$.

Assume the number of persons that make phone calls in a time window, denoted as *u*, follows a generative probabilistic model, which is a model based on an unknown probability distribution. With the generative probabilistic model, we can compute the probability of the number of persons that make phone calls in a time window, denoted as $P^{(t)}(u)$.

Lemma 1: $P^{(t)}(u)$ follows the *binomial distribution*.

The number of persons in cell l at t is $n^{(t)}(l)$, hence the population density is $n^{(t)}(l)$, also notated as n, without ambiguity. According to Definition 1, we will get the call volume in the area unit l, $v^{(t)}(l) = n^{(t)}(l)P^{(t)}(u)$.

Lemma 2: $v^{(t)}(l)$ is an increasing function of $n^{(t)}(l)$.

Lemma 3: When $n^{(t)}(l)$ is large enough, $P^{(t)}(u)$ will approximately follows the normal distribution.

2.2 Influence Factors on Relation

Theorem 1: Given n and p, $v^{(t)}(l)$ is the linear function of $n^{(t)}(l)$.

Based on Theorem 1, we can conclude that when there is a burst in the call volume, there is a corresponding burst in the population. Hence, we can quantify and study population bursts by call volumes in the cells.

2.3 Reflection-Based Burst Detection

Based on our burst definition and real applications, burst detection can be divided into two categories. The first one is the fuzzy burst detection, that is, we just try to know whether there exists a predefined burst. The second one is the quantifiable burst detection, that is, we can quantify bursts. When we want to detect the predefined bursts or try to select and study the near-predefined bursts, we use the fuzzy burst. In other words, if we want to detect bursts satisfying certain conditions, we use fuzzy burst detection. When we want to qualify the bursts in the city, we use the burst strength and lifetime. Then we can rank the bursts temporally and spatially.

Definition 3: Let X is the universal set, the function $\mu : X \rightarrow [0,1]$ is a fuzzy set in X, notated as A. μ is the membership function of A [6].

$\mu(x)$ stands for the membership grade of x ($x \in X$) in A. We define fuzzy burst as follows,

$$\mu(x) = \begin{cases} 0, & \gamma_2 < x < \gamma_1 \\ \dfrac{v_{\max} - x}{(v_{\max} - v_{\min})/2}, & \gamma_1 < x \le v_{\max} \\ 1 - \dfrac{x - v_{\min}}{(v_{\max} - v_{\min})/2}, & v_{\min} \le x < \gamma_2 \\ 1, & x \notin (v_{\min}, v_{\max}) \end{cases}.$$

In our study case, x is the call volume. γ_1 and γ_2 are bound thresholds from our settings. In case 1, the call volume is in the normal situation, so there is no burst. $\mu(x)$ is 0. In case 4, x is out of the history call volume spectrum, which means a burst happened, thus $\mu(x)$ is 1. In the rest two cases, the call volume is out of the bound by thresholds. But comparing to the history situation, it is still not that obvious bursty. Hence we introduce the fuzzy concept into the definition. In case 2, even though x is larger than γ_1 , it is still less than $v_{\max}(l)$. We will calculate a value between 0 and 1 to $\mu(x)$. The same situation is as the case 3. Now the problem is how to get γ_1 and γ_2 . The two coefficients should reflect the aggregation degree and distribution of the history data. We can determine them by pre-processing the history data. Based on our domain knowledge, the call volume is relative stable in normal days. Hence we regress the history data in normal days by Support Vector Machine (SVM). From the regression, we get the regressed data as the baseline for history data, and then set γ_1 and γ_2 . There are many other approaches that can be used to pre-process history data, such as neural networks and kernel-based learning methods. We believe that most of them will give us satisfactory results. In this work, we just select SVM because it can already offer good enough results. Furthermore, selecting which learning methods is not the focus of this paper. To quantify bursts, we define the strength and lifetime of the burst as follows.

Definition 4: Given a burst $B^{(t)}(l)$, the strength $S^{(t)}(l)$ is the power degree of the burst.

If $\mu(x)$ in Definition 4 is 1, then

$$S^{(t)}(l)=\begin{cases}1+\dfrac{v_m^{(t)}(l)-v^{(t)}(l)}{v_m^{(t)}(l)}, & v_m^{(t)}(l)\geq v^{(t)}(l)\\[2ex] 1+\dfrac{v^{(t)}(l)-v_m^{(t)}(l)}{v_m^{(t)}(l)}, & v_m^{(t)}(l)\leq v^{(t)}(l)\end{cases}.$$

$v_m^{(t)}(l)$ is the call volume derived from the regression line in cell l and at time t. $v^{(t)}(l)$ is the current call volume in cell l and at time t. The lifetime of the burst is the time interval it exists, which is denoted as I. $I=I^{(0,1]}+I^{[1,2)}$. $I^{(0,1]}$ is the time interval when $\mu(x)\in(0,1]$. $I^{[1,2)}$ is the time interval when $S\in[1,2)$. The key idea of reflection-based burst detection is to utilize call volume in cells to detect population bursts. At first we incrementally process the data stream, if we want to find the predefined or select the near-predefined bursts, we will use Step 1. If we want to qualify the bursts we find, we will use Step 2 and 3 to get the burst strength, lifetime and the burst with the time stamp identifying the time when the burst happens. In our application scenario, the call volume and cell are accurate because of the cellular communication requirements. We maintain the history information for each cell. The time complexity of the algorithm is $O(n^2)$.

3 Advanced Burst Detection

3.1 Cell combination

Call volumes in different cells in the city appear notable non-uniform distribution patterns. 90% cells have relative small call volumes, while no more than 1% cells contribute the most call volumes in the city. The methodology of cell combination is to combine and group cells by certain rules and criteria, and efficiently choose the tradeoff between accuracy and speed. The priority in the rules is defined as the importance grade of a cell in our system. The basic process is including three steps. The first step is collecting the data stream from the cells in the city. The second step is calculating and determining the priorities and weights for different cells. The calculation is based on rules and criteria. The third step is to construct the cell tree. The criteria and rules to construct the tree are listed as follows. Three basic rules to describe the global architecture: The upper level is the subset of lower one; priority criteria could break the first rule; combining all the levels, we get the complete cell set. This design of cell tress offers us an opportunity to make a trade-off between accuracy and efficiency. We take the cell topology as the initial input the algorithm. The cell distribution is divided by Voronoi Diagram [25]. Every node represents the base station. Basically, we first choose candidate cells by heuristic rules, and we get the initial cell tree. When bursts happen, we will move the cells with bursts one level up. If the cell does not have bursts for a given time length, we will move the cell one level down. When the height of the tree is larger than a threshold, it causes higher time cost. We put the cells in the highest level into the lower level. When the breadth of the tree is larger than a threshold, it causes lower accuracy. We put the cells in the lowest level into the upper level. The thresholds in this method are set by setting and experiments. The branch factor will directly impact the accuracy. Then the lower bound is $\frac{1}{\mu}$, and the upper bound is 1. The time complexity of the algorithm is $O(n)$.

3.2 Scan Resolution

In our original burst detection method, we treat every cell equally. We have to scan it every second to detect the bursts. It costs huge time, space and computation resources. However, in the real situation, there is no need to perform in this way. Firstly, the spatial features of the cells make us realize that the scan time interval in the urban area should not be treated equally as mountain areas, the sensitive areas should not be treated equally as normal ones, etc. Secondly the call volumes in cells are non-uniform. Based on the spatial and numerical features of cells, we try to design a scalable scan method to achieve a better performance. The scan resolution consists of two parts. The first one is the cell resolution, which is the cell coverage gravity. It is by default in cellular networks system, so we do not have the ability to improve it. The second one is the time window of the scan resolution. In practice, the time window is essential to the system performance. We design the heuristic procedure as below. At first, we refer to the cell combination to set the initial scan time window for the cell. If the cell in high level, it gets a small scan time window. The cell is scanned extensively. When an event happens, it triggers scan window change. The cell is scanned with a smaller time window. The cells adjacent to this cell are accordingly scanned with a new time window. When the event ends, the cells will be scanned in a new time longer window. The time complexity of the algorithm is $O(n)$. The scalable scan resolution is beneficial to detect event without losing accuracy. In practice, we perform cell combination at first, and then scan resolution method.

4 Empirical Experiment and Evaluation

4.1 Dataset in Real Life

Our experiments are performed on a 2.53GHz Duo PC with 4 GB of main memory. We elaborate our methods and system by the following datasets.

The first one is the call detail record in the city from 1st January, 2008 to 31st December, 2008, as the training dataset.

The second one is the Olympic Torch Relay day dataset, which will help us to study the bursts.

The third one is the empirical controlled experiment with 310 persons in the city we conducted from 28th March, 2009 to 31st March, 2009, including the call detail record and real population situation.

The second and third datasets are test dataset. In the Olympic Torch Relay, based on our domain knowledge, there were 67 observation and surveillance stations, with 960 bursts monitored. In the empirical experiment, we introduce 6 different temporal and spatial events per day. One event with 130 persons involved, three events with 40 persons involved, and two events with 30 persons involved. Totally there are 118 bursts in the experiment. We record and have the complete knowledge of the empirical experiment and the real situation, so we can take it as the ground truth.

4.2 Burst Detection Evaluation

At first we evaluate the cell combination and scan resolution. At last we study the burst detection evaluation.

In Fig. 1 (a) and (b), we evaluate the accuracy and time cost in naive method and advanced method. In (a), x-coordinate is the number of levels, and y-coordinate is the accuracy. In (b), x-coordinate is the number of levels, and y-coordinate is the time cost. We define time cost as the time it takes to scan the entire city. We define accuracy as the call volume scanned in the city. In the advanced method, we want to detect bursts as fast as possible. The results show that when level is 5, it will get to the optimal status. The time cost is the least and the accuracy is more than 90%. For the branch factor in the cell tree, based on our experiment, it should be limited between 5 and 9.

In Fig. 1 (c) and (d), we study the effectiveness of the burst detection algorithms. We evaluate the naive method and the advanced one, with cell combination and scan resolution. We define burst purity as the percentage of bursts we detected that are true comparing to the ground truth. In (c), we process the Olympic Torch Relay data, and the accuracy is more than 90%. The reason for this result is that in the Olympic Torch Relay route, the cells are in the urban area and the resolution is around 100 meter. In (d), we process the empirical experiment data, and the naive method is accurate with more than 90%, while for the advanced method with cell combination and scan resolution, we trade the accuracy with speed and storage, and actually, the accuracy is still acceptable, around 80%. The reason for this result is that some of the controlled population are not in the area with high resolution of the cells.

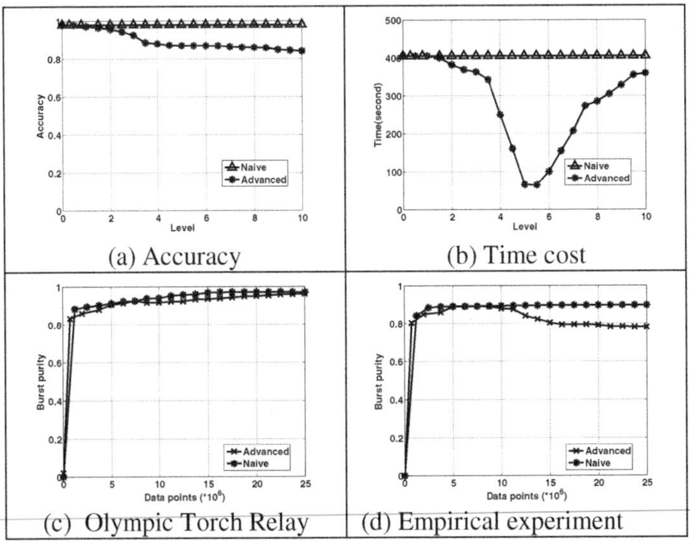

Fig. 1. Burst detection evaluation

5 Related Work

Our work is related to clustering moving objects, which becomes a hot topic recently. In the research work Li et al. [20] discussed the clustering of moving objects and extended the concept of micro-cluster. High quality moving micro-clusters are dynamically maintained which leads to fast and competitive clustering results. Chakrabarti et al. [21] proposed evolutionary clustering which is able to well deal with the mobile clusters in low dynamic environments. Chen et. al. [22] proposed a framework, D-Stream, to efficiently identify outliers of clusters. Extensive recent research efforts are also devoted to the clustering with uncertain data, data streams, uncertain data streams, complex event processing, co-clustering on large datasets, and ranking queries [12, 19-26]. All these existing works are density-based approaches. Especially in their study scenario, they all consider the density or quantity of the objects is enough to cluster. Thus when the density or quantity of the objects is not that good enough for special application scenarios, they will fail. In our application scenario, these methods do not work due to the new arising features, such as extreme less samples and notable data point location errors.

Burst detection has been studied extensively [1-5, 7, 10, 11, 13]. The current works could be categorized as parameter based, and parameter free. In the parameter based literature, the works focus on how to tune a proper threshold, or how to tune a proper time window to detect bursts. Zhu et al. [2] presented a general data structure for detecting interesting aggregates over such elastic windows in near linear time on wavelet. Zhang et al. [7] presented an algorithmic framework for elastic burst detection. Singh et al. [5] propose detecting bursts in streaming, distributed time series data by wavelet. In our application scenario, it's hard to plug it in these methods, due to the unique characteristics of the mobile phone data and the spatial environment with real time.

6 Conclusion and Future Work

In this paper, we study population bursts detection through detecting bursts in mobile phone communication data. In the real life situation and experiment, we extensively evaluated and justified our method. In our work, we mainly focus on the single event burst trajectories.

In the future work, we will try to detect multiple burst event trajectories, not only the correlations but also evolutions. Furthermore, with respect to the privacy concerns in mobile phone communication data, we study the privacy-preserved burst detection over communication data.

References

1. Vlachos, M., Wu, K., Chen, S., Yu, P.S.: Correlating burst events on streaming stock market data. Data Min. Knowl. Discov. 16(1), 109–133 (2008)
2. Zhu, Y., Shasha, D.: Efficient elastic burst detection in data streams. In: Proceedings of the Ninth ACM SIGKDD International Conference on Knowledge Discovery and Data Mining. KDD 2003, pp. 336–345 (2003)
3. Parikh, N., Sundaresan, N.: Scalable and near real-time burst detection from eCommerce queries. In: Proceeding of the 14th ACM SIGKDD International Conference on Knowledge Discovery and Data Mining, pp. 972–980
4. Lee, J., Han, J., Li, X.: Trajectory Outlier Detection: A Partition-and-Detect Framework. In: Proceedings of the 2008 IEEE 24th International Conference on Data Engineering, pp. 140–149 (2008)
5. Singh, L., Sayal, M.: Privately detecting bursts in streaming, distributed time series data. Data Knowl. Eng. 68(6), 509–530 (2009)
6. Montogomery, D.C., Runger, G.C.: Applied Statistics and Probability for Engineers, 2nd edn. John Wiley & Sons, Inc. (1999)
7. Zhang, X., Shasha, D.: Better Burst Detection. In: Proceedings of the 22nd International Conference on Data Engineering
8. Li, T., Li, Q., Zhu, S., Ogihara, M.: A survey on wavelet applications in data mining. SIGKDD Explor. Newsl. 4(2), 49–68 (2002)
9. Seshadri, M., Machiraju, S., Sridharan, A., Bolot, J., Faloutsos, C., Leskove, J.: Mobile call graphs: beyond power-law and lognormal distributions. In: Proceeding of the 14th ACM SIGKDD International Conference on Knowledge Discovery and Data Mining, pp. 596–604
10. Böhm, C., Faloutsos, C., Plant, C.: Outlier-robust clustering using independent components. In: Proceedings of the 2008 ACM SIGMOD International Conference on Management of Data, pp. 185–198 (2008)

11. Zhang, Q., Liu, J., Wang, W.: Approximate Clustering on Distributed Data Streams. In: Proceedings of the 2008 IEEE 24th International Conference on Data Engineering, pp. 1131–1139 (2008)
12. Aggarwal, C.C., Yu, P.S.: A Framework for Clustering Uncertain Data Streams. In: Proceedings of the 2008 IEEE 24th International Conference on Data Engineering, pp. 150–159 (2008)
13. Wang, X., Zhai, C., Hu, X., Sproat, R.: Mining correlated bursty topic patterns from coordinated text streams. In: Proceedings of the 13th ACM SIGKDD International Conference on Knowledge Discovery and Data Mining, pp. 784–793
14. Fung, G.P., Yu, J.X., Yu, P.S., Lu, H.: Parameter free bursty events detection in text streams. In: Proceedings of the 31st International Conference on Very Large Data Bases, pp. 181–192
15. Sherkat, R., Rafiei, D.: On efficiently searching trajectories and archival data for historical similarities. Proc. VLDB Endow 1(1), 896–908
16. Karnstedt, M., Klan, D., Pölitz, C., Sattler, K., Franke, C.: Adaptive burst detection in a stream engine. In: Proceedings of the 2009 ACM Symposium on Applied Computing, pp. 1511–1515 (2009)
17. He, Q., Chang, K., Lim, E.: Analyzing feature trajectories for event detection. In: Proceedings of the 30th Annual international ACM SIGIR Conference on Research and Development in Information Retrieval, pp. 207–214
18. Aggarwal, G., Feder, T., Kenthapadi, K., Khuller, S., Panigrahy, R., Thomas, D., Zhu, A.: Achieving anonymity via clustering. In: Proceedings of the Twenty-Fifth ACM SIGMOD-SIGACT-SIGART Symposium on Principles of Database Systems, pp. 153–162
19. Li, Y., Han, J., Yang, J.: Clustering moving objects. In: Proceedings of the Tenth ACM SIGKDD International Conference on Knowledge Discovery and Data Mining, pp. 617–622
20. Chakrabarti, D., Kumar, R., Tomkins, A.: Evolutionary clustering. In: Proceedings of the 12th ACM SIGKDD International Conference on Knowledge Discovery and Data Mining, pp. 554–560
21. Chen, Y., Tu, L.: Density-based clustering for real-time stream data. In: Proceedings of the 13th ACM SIGKDD International Conference on Knowledge Discovery and Data Mining, pp. 133–142
22. Yoon, H., Shahabi, C.: Robust Time-Referenced Segmentation of Moving Object Trajectories. In: Proceedings of the 2008 Eighth IEEE international Conference on Data Mining, pp. 1121–1126
23. Kao, B., Lee, S.D., Cheung, D.W., Ho, W., Chan, K.F.: Clustering Uncertain Data Using Voronoi Diagrams. In: Proceedings of the 2008 Eighth IEEE International Conference on Data Mining, pp. 333–342 (2008)
24. Jin, C., Yi, K., Chen, L., Yu, J.X., Lin, X.: Sliding-window top-k queries on uncertain streams. Proc. VLDB Endow 1(1), 301–312 (2008)
25. Chen, L., Özsu, M.T., Oria, V.: Robust and fast similarity search for moving object trajectories. In: Proceedings of the 2005 ACM SIGMOD International Conference on Management of Data, pp. 491–502 (2005)
26. Liu, S., Liu, Y., Ni, L., Fan, J., Li, M.: Towards Mobility-based Clustering. In: Proceedings of the 16th ACM SIGKDD International Conference on Knowledge Discovery and Data Mining, pp. 919–928

Population Burst Trajectory Retrieval in Smart City

Wei Zhang[1], Xiaojian Wang[1], Siyuan Liu[2], Ce Liu[3], and Yanping Liu[1]

[1] School of Economics and Management
Beijing Jiaotong University, Beijing, China
[2] iLab, Heinz College, Carnegie Mellon University, Pittsburgh, USA
[3] University of Pittsburgh, Pittsburgh, USA

Abstract. In this paper, we study a novel problem, population burst trajectory retrieval, one of the important issues related population monitoring and management within a city. Based on the detected population bursts, we can trace and analyze these bursts until retrieve the burst trajectories, which can help the city departments to infer the cause of bursts and be prepared for the future possible bursts. Though it is useful, it is not trivial to retrieve population burst trajectories, especially under the condition that we can hardly get the population samples within a city from time to time. To address the difficulties of lacking real population data, we take the advantage of communication networks, specifically, mobile phone networks, which offer enormous communication data between peoples. Most importantly, we find the fact that we can use these communication data to infer the population samples. Therefore, we propose an effective and efficient algorithm to mine burst trajectories with the help of geographical information systems. We verify the performance of our proposed mechanism with an onsite case study and real calling data.

Keywords: population burst trajectory, spatio-temporal analysis, mobile phone.

1 Introduction

Nowadays, even for a medium size city, many unexpected events may occur. Most of these events are directly or indirectly related to the population burst. Informally, a population burst is defined as a population distribution situation that is abnormal compared to the normal steady state based on historical data. For example, in Obama's inauguration, more than 2 million populations crowded in Capitol Hill of Washington DC on 20th January, 2009. It was an impressive population burst. Population bursts bring great challenges to city monitoring, emergency management, and traffic control. It is, therefore, essential to monitor and trace the burst events, investigate the trajectories of bursts, and predict the upcoming bursts. In our work, we mainly focus on retrieving the temporal and spatial population burst trajectory. The ultimate objective of this research is to provide an in-depth study of the population burst trajectory via quantitative research on the real data of a city in China. However, it is always a great challenge to shed light to the population burst, by taking into account varieties of movements, extremely distributed situations, and great temporal and spatial diversification.

G. Lee (Ed.): Advances in Automation and Robotics, Vol.2, LNEE 123, pp. 407–416.
Springerlink.com

Due to the lacking of population data samples, it is not possible to detect population bursts directly on population data. Fortunately, in most modern cities, nearly every person brings along one or more mobile phones, especially in China. These phones can be considered as life mates and sensors of a person. Through mobile phones, we can learn the information about locations, temporal and spatial communications of mobile phone users in the daily life. We were able to collect cell phone data from a city in China from 1st January, 2008 to 31st December, 2008. The area of the city is about 8700 km^2 covered by 4120 cell base stations. There are about 3.6 million mobile phone users. Fig. 1 illustrates a portion of the city and the corresponding cellular network including the geographical information and the cell boundary covered by each base station. The red line is the boundary of cells and the red dot is the base station.

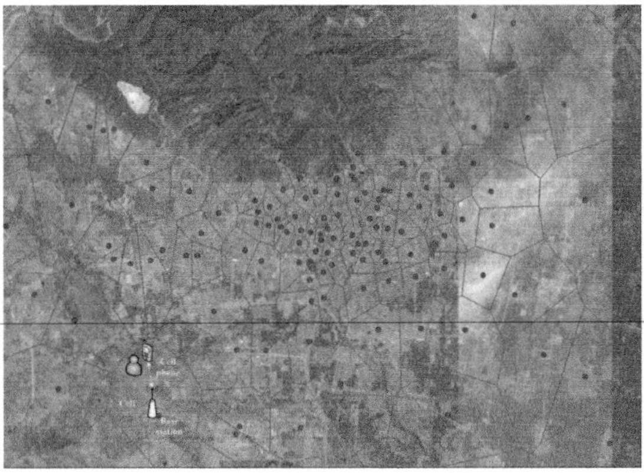

Fig. 1. The cellular network in the city we study

Though we have one year's calling data of the mobile phone networks, it is still not trivial to detect burst on calling volumes due to the following reasons: The population bursts or burst events not only temporally and spatially distribute, but also dynamically and diversely evolve, which causes the calling volume data dynamically changed. We have extremely limited samples and incomplete data with respect to each person. The traditional density-based clustering method does not work due to limited number of samples. The locations of mobile phones are lossy and inaccurate. Frequent and high change rate of the mobile phone usage.

Confronting with the challenges from unique data and application scenario, we take advantage of the following observations from our collected communication data and domain knowledge to detect population burst effectively and efficiently. The first one is that the cell location (the location of the base station in charge of communication in the cell) is accurate. Even though the mobile phone location is extraordinary inaccurate, the cell location is always accurate to the tenth significant digits of latitude and longitude. Moreover, the cell radius is always obtainable. Thus, in this study, we use cells as our observation unit. The cell distribution resolution is good enough for us to monitor and detect population bursts. The second one is that the burst is caused by focal incentives and abnormal accumulation. Especially the more mobile phones online, there are more people in the area. That is, the abnormal mobile phone usage volume will reflect the abnormal population distribution. Fortunately, the volume statistics is accurate and can be collected in real time within cellular networks. The third one is that daily volume spectrums of the city or cells are relative steady. It will offer us a great opportunity to detect and observe the call volume bursts in the city.

Based on the above observations, we define the burst trajectory and design an effective algorithm to mine burst trajectories. Based on the empirical experiments on real life data, the accuracy of our burst trajectory detection method is around 80%, and the efficiency of the approach is more than 3 times better compared to the naïve one.

The rest of the paper is arranged as follows. We provide the preliminary information of our study over real life data in Section 2. In Section 3 we define and investigate burst trajectory. We validate the method and model through field studies in real life situations in Section 4. Section 5 gives an overview of the related work. We conclude the remarks and outline the directions for future work in Section 6.

2 Preliminary

2.1 Raw Dataset Characteristics

The data that we collected from each base station of the cell, called Call Detail Record (CDR) and Cell Detail List (CDL).

Basically, CDR is the dynamic data. It contains two kinds of information. The first one is the information related to mobile phones, such as mobile phone ID, IMEI (International Mobile Equipment Identity), city ID, roam city, roam type, and toll type; the second one is communication related information, such as call type ID, opposite mobile phone ID, start time, call duration, adversary ID, opposite city, opposite roam city, and call moment. CDL is the static data. It is the cell related information, such as cell ID, cell (base station) location, cell direction.

The data are collected from a city in China. The size of the city is around 8700 km^2. Within this city, there are 4120 cells serving nearly 3.6 million mobile phones. The size of raw dataset that we collected from 1st January, 2008 to 31st December, 2008, is around 1 Terabytes (for confidentiality and requirements from data providers, we cannot specify more detailed information in the paper). Several interesting characteristics of the raw data are listed as follows:

The first one is that we have extremely limited samples and incomplete data with respect to each person. In our application scenario, we expect to get the temporal and spatial population density distribution.

The second one is lossy and inaccurate locations of mobile phones (objects). As a result of GPRS (General Packet Radio Service) equipment limitations, the mobile phone location error will be added up to hundreds of meters. Moreover, the lossy and extreme inaccurate locations of objects will make the uncertain data streaming method inapplicable.

The third one is the highly dynamic change rate in the data. In a city scale, the rate of mobile phone usage highly changes, not only temporally but also spatially. It is a spontaneous obstacle for us to determine and detect bursts.

Motivated by these new challenges, we propose reflection- based burst detection approach to settle the temporal and spatial burst detection. Furthermore, we design burst trajectory mining algorithms to study the burst evolution.

2.2 Observations and Design Principles

From our data and domain knowledge, we take advantage of the following observations and opportunities. In our daily life, almost every person possesses a mobile phone, especially in the city. Hence, we safely assume that the number of calls from mobile phones will reflect the population. In the following discussion, if without special notes, the mobile phone is the person. The first one is that even though the mobile phone location is extraordinary inaccurate, while the cell location and radius is accurate enough to sense the situation in the city. The second one is that the burst is caused by focal incentives and abnormal accumulation, especially the more mobile phones online, the more people in the area. That is, the abnormal mobile phone usage volume will reflect the abnormal population distribution. Furthermore, the volume statistics is accurate in cellular networks. The third one is that call volumes in different cells in the city appear notable non-uniform distribution patterns. The fourth one is that daily call volume spectrums of the city or cells are relative stable. It will offer us a great opportunity to detect and observe bursts.

3 Burst Trajectory Mining Model

In this section, we investigate the burst trajectory, which indicates the temporal and spatial evaluation of the burst, and the correlation between events and bursts. Note that limited to the pages, we only introduce the burst trajectory mining model without the details about burst detection which are elaborated in our another work.

3.1 Burst Trajectory Mining

With the progress of a burst event, the burst locations will form a spatial trajectory. We call this spatial trajectory burst trajectory, which is formally defined as follows.

Definition 1: Given an initial burst with strength s_o, time t_o and location l_o, the burst trajectory is defined as the burst sequence and trace with the same initial burst.

$[s_o, t_o, l_o], [s_1, t_1, l_1], \ldots, [s_n, t_n, l_n]$, where $t_o < t_1 < \ldots < t_n$.

The end of the trajectory is end of the burst on strength s_n, time t_n and location l_n.

The burst trajectory mining algorithm is described in Algorithm 1. The main idea is that for every cell we maintain an adjacent matrix. If there is a burst in a cell, we record it and check whether there is burst in the adjacent cells.

The time complexity of the algorithm is $O(n^2)$. In our application scenario, efficiency is strictly required. In Algorithm 1, we construct the trajectory every time instance. When it works for a long period, the efficiency is low. In the following part, we will design the advanced burst trajectory mining algorithm.

Algorithm 1. Burst trajectory mining

```
Input : { the burst location set: L; the adjacent matrix M; the time series: T}.
Output : { the burst trajectory set: J}
1   Begin
2   Step 1. Initialization
3   Maintain the adjacent matrix M for each base station l.
4   Step 2. Burst trajectory heap
5   Algorithm 1;// burst detection
6   if l is a burst location
7     go to l's adjacent nodes in M
8           search the burst in M
9     if there is a burst location l' in the adjacent nodes
10        put l' into the heap: P;// burst trajectory heap
11  Go to 6: iteration
12  Step 3. Burst trajectory
13  P->J with time stamp t.
14  End.
```

3.2 Advanced Burst Trajectory Mining

In the burst trajectory mining algorithm, the time complexity is $O(n^2)$, and the accuracy is low. The reason for that is as follows. Firstly, the method is inefficient because it calculates burst trajectories at every time instance. Secondly, the burst happens not as what we expected to be a sequence. In practice, the bursts are complicated. It is hard to investigate the sequence in time and space. We have to design methods to recover the trajectory by domain knowledge and observations. Observed from our data, we find the following burst trajectory properties.

The burst trajectory is temporally and spatially distributed. We clarify them based on our observation as follows.

- Sequential burst trajectory: in the adjacent burst nodes, sequential burst trajectory is the trajectory that is on the same direction with the event.
- Obsequent burst trajectory: in the adjacent burst nodes, obsequent burst trajectory is the trajectory that in on the reversing direction with the event.
- Diversified burst trajectory: in the adjacent burst nodes, the burst trajectory is diversified.
- Merging burst trajectory: in the adjacent burst nodes, the burst trajectory is merging.
- Preliminary burst trajectory: the same situation as sequential burst trajectory, but the difference is the nodes are far away from each other. In other words, the nodes are not adjacent.
- Couple burst trajectory: the bursts emerge at the same time, and there is a connection between them. In the mobile phone communication, each call is between two mobile phones. The burst may cause the other burst consequently.

In general, all these trajectories have the following common properties.

- Burst strength. The higher strength, the longer burst time and trajectory.
- Burst diversification. The burst diversification always indicates the beginning or end of the trajectory, which is the beginning or end of the event.
- Burst adjacency. The adjacency is two-folds. The first is temporal adjacency, and the second one is spatial adjacency. The better adjacency indicates the better possibility to be the same burst trajectory or event.

In the real life data, we investigate burst trajectories, the correlation between bursts and event trajectories, and the impact of the event on the population and bursts. To reduce the time cost, we use incremental techniques. When new data item arrives, we just process the new data items incrementally. The number of time instances in the new time interval is approximately the constant. On the other hand, we have to enhance the accuracy, so we utilize the burst trajectory properties to refine the detection. The details are in Algorithm 2. The main idea is that we mine the trajectory in the new time instances, and do not calculate in the old time instances, which we have processed. After we get the trajectory, there are noises in it. The noises are from obsequent trajectories, merging trajectories, diversified trajectories, and preliminary trajectories. At last we prune the noises and get the trajectories. The time complexity of the algorithm is $O(n)$.

Algorithm 2. Advanced burst trajectory mining

Input : { *the burst location set: L; the time series processed: T; the new time series to be process: T'* }.
Output : { *the burst trajectory set: J*}

Begin
Step 1. Incremental process
$t_0, t_1, t_3, \ldots, t_k, \ldots, t_n$ in T
$t'_0, t'_1, t'_3, \ldots, t'_k, \ldots, t'_n$ in T'
for a time interval $[t'_k, t'_n]$
if t'_n is out of T
  if t'_k is in T
  Recall Algorithm 4 in time series $[t_n, t'_n]$;
  else
   Recall Algorithm 4 in time series $[t'_k, t'_n]$;
Step 2. Update adjacent matrix
Recall cell combination tree in Algorithm 2;
Update M;
Step 3. Trajectory pruning
Recall burst trajectory mining algorithm;
  Case 1: l in the last step M or not in the M;
   Find obsequent burst trajectory;
  Case 2: more than one bursts in l's M
   Find diversified burst trajectory;
  Case 3: more than one bursts in l's last M

```
      Find merging burst trajectory;
    Case 4: one burst is out of l's M
      Find preliminary burst trajectory;
  Update P, M;
  Step 4. Burst trajectory
  P->J.
  End.
```

4 Empirical Experiment and Evaluation

4.1 Dataset in Real Life

Our experiments are performed on a 2.53GHz Duo PC with 4 GB of main memory. We elaborate our methods and system by the following datasets. The first one is the call detail record in the city from 1st January, 2008 to 31st December, 2008, as the training dataset. The second one is the Olympic Torch Relay day dataset, which will help us to study the bursts. The third one is the empirical controlled experiment with 310 persons in the city we conducted from 28th March, 2009 to 31st March, 2009, including the call detail record and real population situation. The second and third datasets are test dataset. In the Olympic Torch Relay, based on our domain knowledge, there were 67 observation and surveillance stations, with 960 bursts monitored. In the empirical experiment, we introduce 6 different temporal and spatial events per day. One event with 130 persons involved, three events with 40 persons involved, and two events with 30 persons involved. Totally there are 118 bursts in the experiment. We record and have the complete knowledge of the empirical experiment and the real situation, so we can take it as the ground truth.

4.2 Burst Trajectory Evaluation

In this section, we evaluate the burst trajectory detection methods. We evaluate it quantitatively, spatially and temporally. In Fig. 2, we evaluate the burst trajectory detection in with respect to the efficiency and the accuracy, by the Olympic Torch Relay data and the empirical experiment data. In Fig. 3, we report the spatial features and distribution of the Olympic Torch Relay burst trajectories. In Fig. 2, we evaluate the burst trajectory detection method. We define burst trajectory purity as the similarity between our burst trajectory and the event trace. If the Euclidean distance between the two trajectories is under a threshold, we consider it as similar. And it is the pure burst trajectory comparing to the ground truth. In (a) and (b), we evaluate the accuracy and efficiency with progression of data stream in Olympic Torch Relay day. The advanced method (Algorithm 2) is obviously better than the naive method (Algorithm 1) by around 3 times. From the aspect of accuracy, the advanced method is better than the naive one by around 3.5 times. In the practice, every second there are 1.1×10^4 to 5.5×10^4 data points in the stream, while our method can handle around 5×10^4 to 5.6×10^4 data points per second. So the advanced method is good enough to handle the data stream. In (c) and (d), we process the empirical experiment data. The efficiency of the advanced method is 2.5 times better than the naive one. The accuracy of the advanced method is 2.7 times better than the naive one. We report the results by geographic information in the real city situation in Fig. 3. In the figure, we mainly indicate the spatial features of the trajectory and the correlation with the event. To enhance the impact of the results, we visualize the trajectories in different colours in the geographical map. We indicate the trajectories in seven different colours. The white line is the Olympic Torch Relay route in which the torch is relayed on foot. The Yellow line is the Olympic Torch Relay route in which the torch is relayed by vehicles. The blue line is the preliminary burst trajectory. The black line is the sequential burst trajectory. The green line is the obsequent burst trajectory. The purple line is the negative burst trajectory. The orange line is the couple burst trajectory. At the beginning and in end of the event (marked by white boxes in the figure), the trajectories show obvious diversifications. In the yellow line, the torch is relayed by the vehicle, so less population crowed there, and appears negative burst trajectories, as purple line indicated. The burst trajectories are mainly around the event trace as white line indicates. The preliminary burst trajectories appear around the beginning and end of the event. The negative burst trajectories are always before the positive burst trajectory.

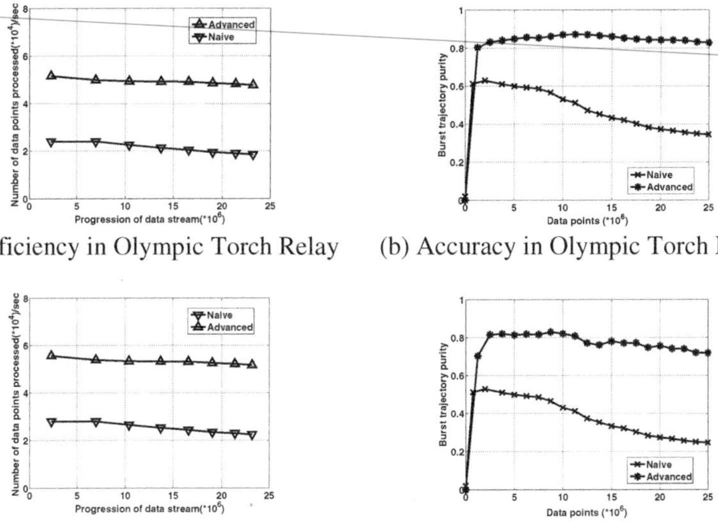

(a) Efficiency in Olympic Torch Relay (b) Accuracy in Olympic Torch Relay

(c) Efficiency in the empirical experiment (d) Accuracy in the empirical experiment

Fig. 2. Burst trajectory detection evaluation

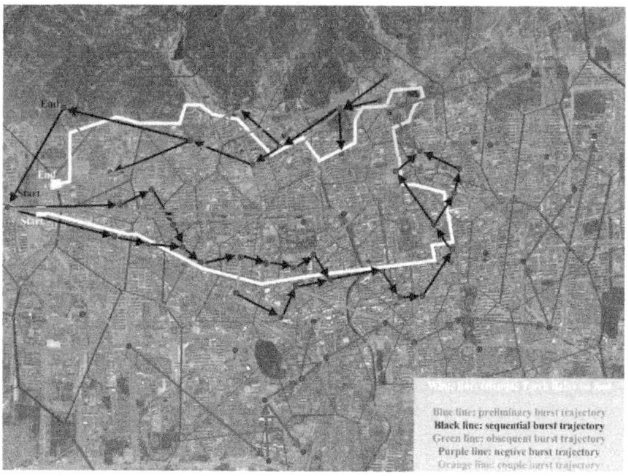

Fig. 3. Burst trajectories in the field

5 Related Work

He et al. [14] considered the problem of analyzing word trajectories in time and frequency domains. And on the moving objects, the current literatures are mainly on moving object traces. The angle of the investigation is from the object itself. Lee et al. [1] proposed the trajectory outlier detection by the partition and detection framework. Their focus is on the outliers in the trajectory cluster and our focus is on the burst trajectory from the event and population evolution. Chen et al. [23] studied the similarity search for moving object trajectories. By the Edit Distance on Real sequence, they tried to efficiently retrieve similarity based moving object trajectories.

On the burst trajectory, to the best our knowledge, we are the first to address and investigate this problem.

6 Conclusion and Future Work

In this paper, we study population burst trajectory retrieval through detecting bursts in mobile phone communication data. Based on the burst detection, we defined the burst trajectory to investigate the temporal and spatial evolution of bursts, and proposed efficient and accurate algorithms to detect burst trajectories. In the real life situation and experiment, we extensively evaluated and justified our method. Eventually, we mapped the burst trajectory to the geographical information map, to study the temporal and spatial properties and features.

In our work, we mainly focus on the single event burst trajectories. In the future work, we will try to detect multiple burst event trajectories, not only the correlations but also evolutions.

References

1. Lee, J., Han, J., Li, X.: Trajectory Outlier Detection: A Partition-and-Detect Framework. In: Proceedings of the 2008 IEEE 24th International Conference on Data Engineering, pp. 140–149 (2008)
2. Singh, L., Sayal, M.: Privately detecting bursts in streaming, distributed time series data. Data Knowl. Eng. 68(6), 509–530 (2009)
3. Montogomery, D.C., Runger, G.C.: Applied Statistics and Probability for Engineers, 2nd edn. John Wiley & Sons, Inc. (1999)
4. Zhang, X., Shasha, D.: Better Burst Detection. In: Proceedings of the 22nd International Conference on Data Engineering
5. Li, T., Li, Q., Zhu, S., Ogihara, M.: A survey on wavelet applications in data mining. SIGKDD Explor. Newsl. 4(2), 49–68 (2002)
6. Seshadri, M., Machiraju, S., Sridharan, A., Bolot, J., Faloutsos, C., Leskove, J.: Mobile call graphs: beyond power-law and lognormal distributions. In: Proceeding of the 14th ACM SIGKDD International Conference on Knowledge Discovery and Data Mining, pp. 596–604
7. Böhm, C., Faloutsos, C., Plant, C.: Outlier-robust clustering using independent components. In: Proceedings of the 2008 ACM SIGMOD International Conference on Management of Data, pp. 185–198 (2008)
8. Zhang, Q., Liu, J., Wang, W.: Approximate Clustering on Distributed Data Streams. In: Proceedings of the 2008 IEEE 24th International Conference on Data Engineering, pp. 1131–1139 (2008)
9. Aggarwal, C.C., Yu, P.S.: A Framework for Clustering Uncertain Data Streams. In: Proceedings of the 2008 IEEE 24th International Conference on Data Engineering, pp. 150–159 (2008)
10. Wang, X., Zhai, C., Hu, X., Sproat, R.: Mining correlated bursty topic patterns from coordinated text streams. In: Proceedings of the 13th ACM SIGKDD International Conference on Knowledge Discovery and Data Mining, pp. 784–793
11. Fung, G.P., Yu, J.X., Yu, P.S., Lu, H.: Parameter free bursty events detection in text streams. In: Proceedings of the 31st International Conference on Very Large Data Bases, pp. 181–192
12. Sherkat, R., Rafiei, D.: On efficiently searching trajectories and archival data for historical similarities. Proc. VLDB Endow 1(1), 896–908
13. Karnstedt, M., Klan, D., Pölitz, C., Sattler, K., Franke, C.: Adaptive burst detection in a stream engine. In: Proceedings of the 2009 ACM Symposium on Applied Computing, pp. 1511–1515 (2009)
14. He, Q., Chang, K., Lim, E.: Analyzing feature trajectories for event detection. In: Proceedings of the 30th Annual International ACM SIGIR Conference on Research and Development in Information Retrieval, pp. 207–214
15. Liu, S., Liu, Y., Ni, L., Fan, J., Li, M.: Towards Mobility-based Clustering. In: Proceedings of the 16th ACM SIGKDD International Conference on Knowledge Discovery and Data Mining, pp. 919–928

Study on Transforming Formulas from Interval Valued Data to Vague Valued Data

Guxin Li[1], Hongxu Wang[2], Weixia Li[3], and Chengyi Zhang[3]

[1] Nanjing Technical College of Special Education, Nanjing, 210038,China
[2] Depart. of Electronic Information Engineering, Qiongzhou University, Sanya,572022, China
[3] Depart .of Math .s and Stat., Hainan Normal University, Haikou, 571158, China
chengyizh@hainnu.edu.cn

Abstract. The transforming criterions and definitions from the interval valued data to the vague valued data are presented and the two transforming formulas from the interval valued data to the vague valued data are presented. The examples make known that the distinct formulas have distinct characters.

Keywords: interval valued data, vague valued data, transforming formulas, normal transforming formulas, unusual transforming formulas.

1 Introduction

In 1993, vague sets theory was given by W .L. Gau and D. J. Buehrer in [1].Since vague sets is an improvement of the fuzzy sets theory, specially, it has powerful presentation function, then vague sets theory had been applied successfully in many ways only ten years. When applying the vague set theory, the people resolve firstly the problem of transforming the original data into the vague data .Only after transform the original data into the vague data, then build the vague environment, thus, it is possible that applying the application methods of vague set method to research, analyze and solve practical problems. The problem has become one of the bottleneck of the application about vague set theory. In a general way, the original data have two types, one is single valued data, the other is interval valued data. In this paper, the problem of transforming interval valued data into the vague valued data is researched.

Transforming interval valued data $[x_{ij}, y_{ij}], (x_{ij} \le y_{ij})$ of the index $x_j (j=1,2,\cdots,n)$ in set $A_i (i=1,2,\cdots,m)$ into vague valued data $A_i(x_j) = [t_{A_i}(x_j), 1 - f_{A_i}(x_j)]$ is called interval valued data $[x_{ij}, y_{ij}], (x_{ij} \le y_{ij})$ responding to the vague valued data of vague sets A_i .In [2-9], the vague valued data for the discussed problem are given directly, then how to transform the original data into a vague valued data wasn't introduced, but this process just is key for the application of vague set method. This paper will introduce two methods for transforming directly interval valued data into vague valued data.

G. Lee (Ed.): Advances in Automation and Robotics, Vol.2, LNEE 123, pp. 417–422.
Springerlink.com

2 Transforming Rule and Definition of Transforming Formulas

Definition 1. Let $X=\{x_1,x_2,\cdots,x_n\}$ be index set, interval valued data $[x_{ij},y_{ij}],(x_{ij}\le y_{ij})$ be the index $x_j(j=1,2,\cdots,n)$ of $A_i(i=1,2,\cdots,m)$ .Let vague valued data $A_i(x_j)=[t_{A_i}(x_j),1-f_{A_i}(x_j)]$ transform from the interval valued data $[x_{ij},y_{ij}],(x_{ij}\le y_{ij})$ .If $A_i(x_j)=[t_{A_i}(x_j),1-f_{A_i}(x_j)]$ satisfies that:

(1) bounded principle : $t_{A_i}(x_j),1-f_{A_i}(x_j)\in[0,1]$;

(2) interval principle : $t_{A_i}(x_j)\le 1-f_{A_i}(x_j)$;

(3) positive sequence principle: let interval valued data $[w_i,w_i]$ and $[w_j,w_j]$ are transformed into vague valued data $A_i(x_j)=[t_{ij},1-f_{ij}]$ and $A_h(x_j)=[t_{hj},1-f_{hj}]$ respectively. If $w_i\ge w_h\ge 0$,then $t_{ij}>t_{hj},1-f_{ij}>1-f_{hj}$.

Then this conversion formula from interval valued data to vague valued data is called normal transforming formulas. And the vague valued data is called normal vague valued data, then is called vague valued data for short.

If $A_i(x_j)=[t_{A_i}(x_j),1-f_{A_i}(x_j)]$ satisfies that:

(1) bounded principle : $t_{A_i}(x_j),1-f_{A_i}(x_j)\in[0,1]$;

(2) interval principle : $t_{A_i}(x_j)\le 1-f_{A_i}(x_j)$;

(3)antitone principle: let interval valued data $[w_i,w_i]$ and $[w_j,w_j]$ are transformed into vague valued data $A_i(x_j)=[t_{ij},1-f_{ij}]$ and $A_h(x_j)=[t_{hj},1-f_{hj}]$ espectively. If $w_i>w_h\ge 0$, then $t_{ij}<t_{hj},1-f_{ij}<1-f_{hj}$. Then this conversion formula from interval valued data to vague valued data is called unusual transforming formulas. And the vague valued data is called normal vague valued data.

Note: bounded principle in the definition also is called normal principle. And vague value should satisfy the interval principle. Positive sequence is normal order standards, antitone principle is abnormal order standards.

3 Normal Transforming Formulas from Interval Valued Data to Vague Valued Data

Theorem 1. Let $X=\{x_1,x_2,\cdots,x_n\}$ be index set, interval valued data $[x_{ij},y_{ij}],(x_{ij}\le y_{ij})$ be the index $x_j(j=1,2,\cdots,n)$ of $A_i(i=1,2,\cdots,m)$, $x_{j\max}=\max\{x_{1j},y_{1j},x_{2j},y_{2j},\cdots,x_{mj},y_{mj}\}$ .Then $\forall k=1,2,\cdots$,

$$A_i(x_j)=[t_{ij},1-f_{ij}]=[\frac{x_{ij}^k}{x_{j\max}^k}(1+\frac{y_{ij}^k-x_{ij}^k}{x_{j\max}^k}),1-(1-\frac{y_{ij}^k}{x_{j\max}^k})(1+\frac{y_{ij}^k-x_{ij}^k}{x_{j\max}^k})] \tag{1}$$

is normal transforming formulas from interval valued data to vague valued data.

Proof: Since $t_{ij}=\frac{1}{x_{j\max}^{2k}}(x_{j\max}^k x_{ij}^k+x_{ij}^k x_{ij}^k-x_{ij}^{2k})$,

$$f_{ij}=(1-\frac{y_{ij}^k}{x_{j\max}^k})(1+\frac{y_{ij}^k-x_{ij}^k}{x_{j\max}^k})=1+\frac{y_{ij}^k-x_{ij}^k}{x_{j\max}^k}--\frac{y_{ij}^k}{x_{j\max}^k}-\frac{y_{ij}^k(y_{ij}^k-x_{ij}^k)}{x_{j\max}^{2k}}$$

$$1-f_{ij}-t_{ij}=\frac{1}{x_{j\max}^{2k}}(y_{ij}^{2k}-2x_{ij}^k y_{ij}^k+x_{ij}^{2k})=\left(\frac{y_{ij}^k-x_{ij}^k}{x_{j\max}^{2k}}\right)^2\in[0,1],$$

then $t_{ij}\in[0,1],1-f_{ij}\in[0,1]$, thus $t_{ij}\le 1-f_{ij}$ .Hence,(1) satisfies bounded principle and interval principle. let interval valued data $[w_i,w_i]$ and $[w_j,w_j]$ are transformed into vague valued data $A_i(x_j)=[t_{ij},1-f_{ij}]$ and $A_h(x_j)=[t_{hj},1-f_{hj}]$ respectively. If $w_i>w_h\ge 0$, then

$$t_{ij}=\frac{w_i^k}{x_{j\max}^k}(1+\frac{w_i^k-w_i^k}{x_{j\max}^k})=\frac{w_i^k}{x_{j\max}^k}>\frac{w_h^k}{x_{j\max}^k}=\frac{w_h^k}{x_{j\max}^k}(1+\frac{w_h^k-w_h^k}{x_{j\max}^k})=t_{hj},$$

$$1-f_{ij}=1-(1-\frac{w_i^k}{x_{j\max}^k})(1+\frac{w_i^k-w_i^k}{x_{j\max}^k})=\frac{w_i^k}{x_{j\max}^k}$$

$$>\frac{w_h^k}{x_{j\max}^k}=1-(1-\frac{w_h^k}{x_{j\max}^k})(1+\frac{w_h^k-w_h^k}{x_{j\max}^k})=1-f_{hj},$$

thus , (1) also satisfies positive sequence principle. Therefore, is normal transforming formulas from interval valued data to vague valued data.

4 Unusual Transforming Formulas from Interval Valued Data to Vague Valued Data

Theorem 2. Let $X=\{x_1,x_2,\cdots,x_n\}$ be index set, interval valued data $[x_{ij},y_{ij}],(x_{ij}\le y_{ij})$ be the index $x_j(j=1,2,\cdots,n)$ of $A_i(i=1,2,\cdots,m)$, $x_{j\max}=\max\{x_{1j},y_{1j},x_{2j},y_{2j},\cdots,x_{mj},y_{mj}\}$ .Then $\forall k=1,2,\cdots,$

$$A_i(x_j)=[t_{ij},1-f_{ij}]=[(1-\frac{y_{ij}^k}{x_{j\max}^k})(1+\frac{y_{ij}^k-x_{ij}^k}{x_{j\max}^k}),1-\frac{x_{ij}^k}{x_{j\max}^k}(1+\frac{y_{ij}^k-x_{ij}^k}{x_{j\max}^k})] \quad (2)$$

is unusual transforming formulas from interval valued data to vague valued data.
Proof: This is similar to theorem 1,(2) satisfies bounded principle and interval principle.Let interval valued data $[w_i, w_i]$ and $[w_j, w_j]$ are transformed into vague valued data $A_i(x_j)=[t_{ij},1-f_{ij}]$ and $A_h(x_j)=[t_{hj},1-f_{hj}]$ respectively. If $w_i > w_h \geq 0$, then

$$t_{ij}=(1-\frac{w_i^k}{x_{j\max}^k})(1+\frac{w_i^k-w_i^k}{x_{j\max}^k})=1-\frac{w_i^k}{x_{j\max}^k}$$

$$<1-\frac{w_h^k}{x_{j\max}^k}=(1-\frac{w_h^k}{x_{j\max}^k})(1+\frac{w_h^k-w_h^k}{x_{j\max}^k})=t_{hj}.$$

Similarly, we can prove $1-f_{ij}<1-f_{hj}$ .Then, (2) satisfies antitone principle. Hence, (2) is unusual transforming formulas from interval valued data to vague valued data.

5 Features of Two Types of Data Transforming Method

By the example as follows, the features of transforming interval valued data into vague valued data by normal transforming formulas and unusual transforming formulas are showed intuitively.

Example 1 Suppose the original data be in table 1 as follows. By theorem 1, take k = 2, then, transform data in the table 1 from interval valued data into the vague valued data in table 1.

Table 1. Three types of data

scheme	original data		vague valued data		normal vague valued data	
	x_1	x_2	x_1	x_2	x_1	x_2
A_1	[20,20]	[0,0]	[1,1]	[0,0]	[0,0]	[0.1,1]
A_2	[8,10]	[7,7]	[0.17,0.18]	[0.49,0.49]	[0.82,0.83]	[0.51,0.51]
A_3	[3,8]	[4,4]	[0.03,0.04]	[0.16,0.16]	[0.96,0.97]	[0.84,0.84]
A_4	[8,12]	[10,10]	[0.23,0.26]	[1,1]	[0.74,0.77]	[0,0]
A_5	[15,18]	[5,5]	[0.7,0.76]	[0.25,0.25]	[0.24,0.3]	[0.75,0.75]

Then, by theorem 2, take k = 2, then, transform data in the table 1 from interval valued data into the vague valued data in table 1.Application theorem 2 take k = 2, criterion but the table 1 of interval value data table into normal vague valued data in table 1.

By the example 1, the features of transforming interval valued data into vague valued data by normal transforming formulas and unusual transforming formulas are showed more intuitively.

By the normal transforming formulas in theorem 1, the smaller x_{ij} in interval valued data $[x_{ij}, y_{ij}], (0 \le x_{ij} \le y_{ij})$ is transformed into the smaller t_{ij} in vague valued data $A_i(x_j) = [t_{ij}, 1 - f_{ij}]$, and the smaller y_{ij} in interval valued data $[x_{ij}, y_{ij}], (0 \le x_{ij} \le y_{ij})$ is transformed into the smaller $1 - f_{ij}$ in vague valued data $A_i(x_j) = [t_{ij}, 1 - f_{ij}]$ .And the bigger x_{ij} in interval valued data $[x_{ij}, y_{ij}]$, $(0 \le x_{ij} \le y_{ij})$ is transformed into the bigger t_{ij} in vague valued data $A_i(x_j) = [t_{ij}, 1 - f_{ij}]$, and the bigger y_{ij} in interval valued data $[x_{ij}, y_{ij}]$, $(0 \le x_{ij} \le y_{ij})$ is transformed into the bigger $1 - f_{ij}$ in vague valued data $A_i(x_j) = [t_{ij}, 1 - f_{ij}]$ .Specially, the interval valued data $[x_{j\max}, x_{j\max}]$, $(x_{j\max} = \{x_{1j}, x_{2j}, \cdots, y_{mj}\})$ can be transformed into vague valued data $[1,1]$.

Hence, normal transforming formulas which transforms interval valued data $[x_{ij}, y_{ij}], (0 \le x_{ij} \le y_{ij})$ into vague valued data suits people's intuitive consciousness, and it is normal transforming order and the transforming method used frequently in application.

On the contrary, By the unusual transforming formulas in theorem 2, the smaller x_{ij} in interval valued data $[x_{ij}, y_{ij}], (0 \le x_{ij} \le y_{ij})$ is transformed into the bigger t_{ij} in vague valued data $A_i(x_j) = [t_{ij}, 1 - f_{ij}]$, and the smaller y_{ij} in interval valued data $[x_{ij}, y_{ij}], (0 \le x_{ij} \le y_{ij})$ is transformed into the bigger $1 - f_{ij}$ in vague valued data $A_i(x_j) = [t_{ij}, 1 - f_{ij}]$ .And the bigger x_{ij} in interval valued data $[x_{ij}, y_{ij}], (0 \le x_{ij} \le y_{ij})$ is transformed into the smaller t_{ij} in vague valued data $A_i(x_j) = [t_{ij}, 1 - f_{ij}]$, and the bigger y_{ij} in interval valued data $[x_{ij}, y_{ij}], (0 \le x_{ij} \le y_{ij})$ is transformed into the smaller $1 - f_{ij}$ in vague valued data $A_i(x_j) = [t_{ij}, 1 - f_{ij}]$. Specially, the interval valued data $[x_{j\max}, x_{j\max}]$, $(x_{j\max} = \{x_{1j}, x_{2j}, \cdots, y_{mj}\})$ can be transformed into vague valued data $[0,0]$.

Hence, unusual transforming formulas which transforms interval valued data $[x_{ij}, y_{ij}], (0 \le x_{ij} \le y_{ij})$ into vague valued data doesn't suit people's intuitive consciousness, and it is abnormal transforming order and the transforming method isn't used frequently in application, then it should be treated with caution in the applications.

6 Conclusion

When applying the vague set theory, the people should transform the original data into the vague data, and build the vague environment. In this paper, we introduce the transforming rule and definition of transforming formulas, and give two transforming formulas for the application. But we should notice, normal transforming formulas suits people's intuitive consciousness, and it is normal transforming order and the transforming method used frequently in application. And unusual transforming formulas doesn't suit people's intuitive consciousness, and it is abnormal transforming order and the transforming method isn't used frequently in application, then it should be treated with caution in the applications.

Acknowledgments. This work is supported by NSF of China (70940007), KSTF of Hainan (090802 &ZDXM20110047), KSTF of Haikou (2010072), Social Development Technology Fund of Hainan (2011SF003). Corresponding author: Chengyi Zhang. E-mail: chengyizh@hainnu.edu.cn.

References

1. Gau, W.L., Buehrer, D.J.: Vague sets. IEEE Trans. Systems Man Cybernet. 23, 610–614 (1993)
2. Peng, A.H.: Application of Similarity Measurement Analysis of Vague Sets to the Material Choice. Coal Mine Machinery 27, 891–893 (2006)
3. Ye, J., Xu, Y.C.: A Decision Method of Schemes Using Similarity Measures between Vague Sets. Computer Simulation 23, 90–93 (2006)
4. Lou, J.G.: Similarity Measurement Between Vague Sets and their Application in Scheme Decision. Journal of Engineering Design 12, 325–328 (2005)
5. Ye, J.: Research on Fault Diagnosis of Turbine Based on Similarity Measures Between Vague Sets. Proceedings of the CSEE 26, 16–20 (2006)
6. Zhao, Z.W.: Fault Diagnosis of Wear Type in Diesel Engine Based on Similarity Measures between Vague Sets. Journal of Shaoxing University 26, 68–72 (2006)
7. Ye, J., Lou, J.G., Li, W.B.: Application of Similarity Measurement Analysis of Vague sets to the Policy Decision of Mechanism Schemes. Journal of Maching Design 22, 10–12 (2005)
8. Lu, J.Y., Han, F.C.: Evaluating the Condition of over Head Transmission Lines Based on Similarity Measures between Vague Sets. Electrical Engineering 3, 58–61 (2008)
9. Zhang, J.F., Liang, L.: Evaluation Method for Green Packaging Design based on Vague Set. Packaging Engineering 28, 110–112 (2007)
10. Zheng, C.H., Li, T.F., Gui, J.Z.: Study on Aeroengine Fault Diagnosis Based on Similarity Measures between Vague Sets. Aeronautical Computing Technique 38, 34–36 (2008)

Ultrasonic Scanner for Heat Treatment Qualification in CNG Cylinder Fabrication

Asad Babakhani[1], Fereidoun A. Mianji[1], Ariya Parish[2],
Mahdi Mozaffarian[2], and Abdollah Dadkhah[2]

[1] National Radiation Protection Department of Iran
[2] Tabesh Pardaz Pegah Co. Tehran, Iran

Abstract. Designing and manufacturing of an automatic ultrasonic testing system (scanner) for flaw detection of compressed natural gas (CNG) cylinders, particularly in heat treatment quality assessment, is explained in this paper. Ultrasonic scanners have been installed on 5 different CNG production lines in the territory providing a fast and reliable test of cylinders with the capability of logging results according to customer demands. The paper explains both the machine specifications and the challenges faced during the manufacturing and starting up stages. A questioner was designed to collect the opinions of operators in different sites about handling easiness and performance of the scanners. It was proved that the ultrasonic scanners efficiently identify defects caused in different stages of CNG production line especially in heat treatment furnace. Besides the innovations applied to overcome the practical matters in machine manufacturing, the knowledge of ultrasonic interpreting has been transferred to operators and they earned enough proficiency to distinguish between flaws and noise signals.

Keywords: CNG production line, Heat treatment, ISO 11439 standard, Ultrasonic scanner.

1 Introduction

Iran was increasingly intending to substitute gasoline with gas in vehicles for last 5 years. Compressed natural gas (CNG) cylinders are mainly considered in this approach. It is essential to assure on the quality of CNG cylinders since the gas pressure is high enough (almost 200 bar) to cause an explosion and severe accident in case of any abnormality. Moreover, monitoring of used cylinders should be repeated every 5 years. ISO 114391 has been approved as the mandatory standard for the CNG cylinder production process in Iran2. Other complementary standard, ISO 190783 (with the national code: ISIRI-9426) is also considered applicable since it has the main concepts of inspecting, monitoring, and destructive and nondestructive testing (DT & NDT)4.

Steel, aluminum, and composite materials are the main materials used in the body of CNG cylinders which are divided into four categories as Table 1.

Three different types of testing including the prototype testing, batch testing, and whole testing are applicable in CNG production lines. Prototype testing shall be conducted on newly designed cylinders. Finished cylinders representing the normal

G. Lee (Ed.): Advances in Automation and Robotics, Vol.2, LNEE 123, pp. 423–430.
Springerlink.com

product of this type are tested and engraved with appropriate identification marks. Batch testing shall be conducted on finished cylinders which are representative of normal product of an approved type. The cylinders required for testing shall be randomly selected from each batch. All the cylinders of the bach receive the related identification marks provided that the representatives pass the test statistically. In whole test, the test shall be carried out on all cylinders of a batch. In all of these three types of testing, nondestructive examinations shall be carried out in accordance with a standard acceptable to the regulatory authority.

Table 1. Different types of compressed cylinders

Type 1	Metal	Complete Steel, 34CrMo4 alloy
Type 2	Hoop wrapped	Metal liner reinforced with resin impregnated continuous filament
Type 3	Fully Wrapped	Metal liner reinforced with resin impregnated continuous filament
Type 4	All Composite	Resin impregnated continuous filament with a nonmetallic liner

Table 2 shows a summary of different test types offered by ISO 11439.Ultrasonic test is one of the important examinations recommended by ISO 11439 to be implemented on every cylinder for determining the position and depth of flaws. An ultrasonic wave is sent into the metallic body and the echoes are received. In order to assure the preciseness of the test, four different artificial flaws (reference notches) made on internal and external surfaces of a cylinder's sidewall in accordance to annex B of ISO 11439 (Figure 1) are used as the calibration reference.

Table 2. Different test types for CNG production line

Type of examination		Examples
Test on every cylinders (Whole test)		Hardness test, Ultrasonic test, Hydrostatic test
Prototype and Batch test	Tolerance strain examination	Bonfire test, impact tests
	Environmental examination	hydrostatic pressure burst, tensile test, coating batch test, Penetration test, Sulfide stress cracking resistance test, material tests
	Cyclic examination	Leak before break test(LBB), Burst test, Ambient temperature pressure cycling test, Corrosion tests

The ultrasonic system shall detect internal and external flaws with equal amplitude and along with flaw length clearly. Cylinders with indications which are equal to or greater than the indications from the reference notches shall be withdrawn. The thickness of whole sidewall is investigated and shall not be less than the minimum acceptable figure according to the standard.

Both contact and immersion probes may be used, therefore, surface should be clean without any tiny pieces stuck on it. The procedures for ultrasonic examination are valid for sidewall zone and the neck and bottom areas cannot use ultrasonic examination due to thickness alteration in transition between sidewall to neck/bottom. Operators shall have ultrasonic level 2 certificate in accordance to the IRISI-ISO-9712 training method 5,6.

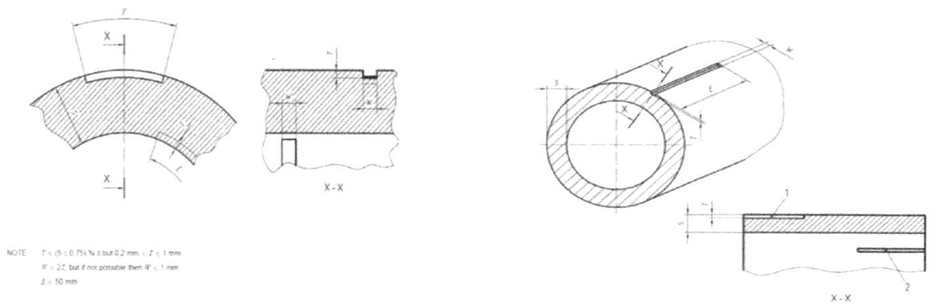

Fig. 1. Calibration block for checking the performance of ultrasonic testing system

2 System Description

2.1 Mechanics and Electrics

Cylinders are put horizontally on two parallel shafts equipped with 10 plastic wheels. A three-jaw chuck holds the neck of cylinder and keeps it in horizontal position during the rotation. Once the rotation starts, probe holder is descended on the initial position of the sidewall and starts the horizontal movement up to the end of the sidewall while the cylinder is rotating7 .The probe holder consists of 5 probes; 2 longitudinal, 2 transversal, 1 thickness and a plexiglas shoe. A suspension system with spring and shock absorber holds the probe holder and keeps a fixed distance between the holder and cylinder's outer surface. This mechanism is important to keep the examination reliable and repeatable. Probe holder has a space for water reservoir to supply the necessary couplant for the test. Used water is gathered in a repository tank to be used again after a filtration stage. The shoe of probe holder is made of a wear resistant material in different radiuses of convexes for different cylinders.

The scanning start/stop point and rotational and horizontal speeds are controlled by an industrial PC to ensure that not only all surfaces are scanned but also a 10% overlap between every two adjacent scans is guaranteed. The total testing time is about 3.5 minutes for a 40 litter water capacity cylinder; from which 2 minutes is for scanning and 1.5 minutes is for the load/unload process. The load and unload mechanism can be designed differently in accordance to the production line requirements. The mechanical and rotational system is installed in a stainless steel tank which serves as used-water collecting system. This tank is equipped with a filtration system for cyclic usage of the water too. Figures 2,3,4 show 3 different types of mechanical designs.

Two electric cabinets contain automation and control parts. They include all necessary tools for controlling the horizontal and rotational speeds and provide signal conditioning and analyses. An operator panel is the interface between operator and the machine (for manual commands). An industrial PC captures the ultrasonic signals from the used multichannel analyzer and presents them in several informative forms.

Fig. 2. The mechanism of type 'a' for ultrasonic testing system

Fig. 3. The mechanism of type 'b' for ultrasonic testing system

Fig. 4. The mechanism of type 'c' for ultrasonic testing system

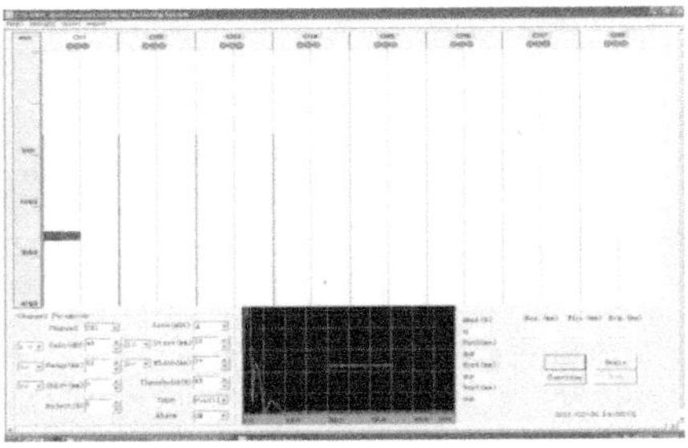

Fig. 5. Software overview

2.2 Software

The software of the system, written by visual C, enables the real time showing of the probes' signals, determining the flaw position, flaw length, and thickness measures (Figure 5). Information of each test is saved into a database for reporting purposes. Any specific request by the user can be considered in the software as well.

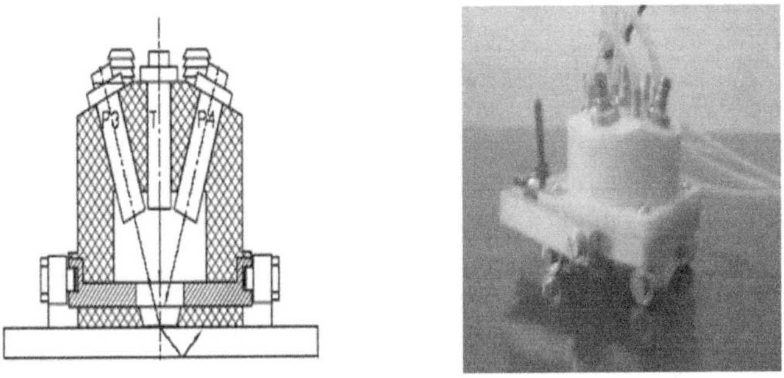

Fig. 6. Probe Holder design and sample

2.3 Immersion Probes

Four normal probes immersed in water send longitudinal waves to the cylinder surface obliquely. Mode conversion phenomena occurs on the boundary of water and steel and both longitudinal and transverse waves are produced. Since the incident angle is bigger than the first water/steel critical angle, only the transverse wave is propagated into the cylinder body. The fifth normal probe sends the ultrasonic wave normally into the cylinder body to measure its wall thickness. The probe holder is de

signed so that it keeps one pair of mirror probes parallel with the longitudinal axis of the cylinder for testing the transverse flaws. The probe holder also keeps the other pair of mirror probes perpendicular to the longitudinal axis of the cylinder for testing the longitudinal flaws as showed in Figure 6.

2.4 Manufacturing Challenges

2.4.1 Cylinder Unbalances

CNG cylinders in CNG production lines are usually made by forming billet, sheet or long length tubes. In these production methods, the section of cylinders is not perfectly round. Moreover, the metal forming and heat treatment process deform the shape of cylinders into asymmetrical shapes and degrades their tomographical circularity. The hot spinning machine which make the neck of cylinders do not have enough accuracy to perfectly center the neck on the longitudinal axis due to the chuck's (which holds the cylinder) vibration and bearings' forces during the rotational twisting8.

In automatic testing of moving parts, the stability of the pieces under examination and smoothness of the movement are the key factors. As it was explained above, the asymmetries caused in the production process of CNG cylinders cause considerable rotational unbalances during the ultrasonic testing of the cylinders. This in turn varys the distance between the cylinder wall and probe holder causing many error signals on the ultrasonic device. It also causes asymmetrical frictions on the probe holder shoe and depreciation of mechanical parts. In order to mitigate the unbalances problem, two extra supporting arms are designed that position on the top surface of the cylinder during the scanning process pushing it down. The end of each supporting arm has one highly abrasion resistant plastic wheel to avoid corrosion while compensating the swings.

2.4.2 Noise

The distance between the probes and ultrasonic multichannel analyzer (Reader) is about 10m owing to the cylinder length and the need to move the probe holder horizontally on the surface of cylinder. Ultrasonic signals are inherently weak, i.e., in the range of mV. They may thus be influenced by high voltage switching on/off of motors and pneumatic jacks. It is essential to consider noise reduction techniques such as using the shield wires, physical separation between low signals and power signals, separating digital and analogue grounds, establishing a proper earthing well, and using high quality probes.

2.4.3 Probes Calibration

Immersion probes must be calibrated for angle, sensitivity and beam index carefully. For every probe in the angular position, receipt of one extra echo from the opposite probe shall be considered. The ultrasonic wave passes through both the water and steel mediums. The multichannel ultrasonic device uses sound velocity in steel as reference for sound path calculation. As an example, if the ultrasonic wave runs 30 mm in water and then incidents the steel with an angle of 24 degree, the echo in steel would be seen at 73mm in the screen for a flaw in 3 mm depth (figure 7 and formulas 1to 3).

$$\frac{\sin 24^\circ}{1473} = \frac{\mathrm{Sin}\,\theta_{st}}{3251} \rightarrow \theta_{st} = 63.8^\circ \qquad BP_{2st} = \frac{3mm}{\cos 63.8^\circ} = 6.8mm \tag{1}$$

$$\frac{BP_{1w}}{V_w} = \frac{BP_{1i}}{V_{sst}} \rightarrow BP_{1st} = \frac{3251}{1473} \times 30 = 66.2mm \tag{2}$$

$$BP_{total} = 66.2mm + 6.8 = 73mm \tag{3}$$

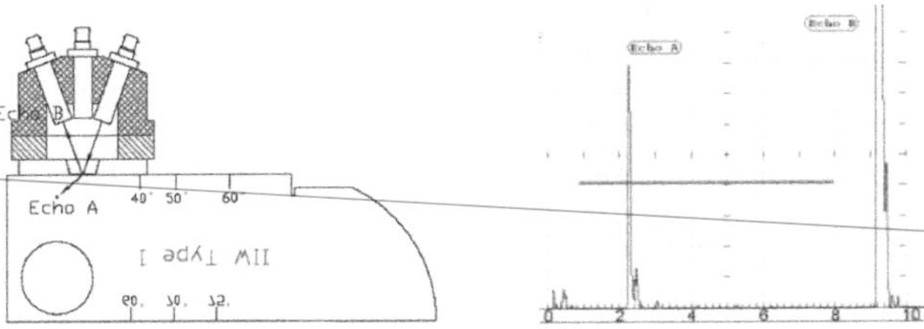

Fig. 7. Echoes of hole in Standard Block and mirror probe

3 Results

Ultrasonic nondestructive testing is a kind of NDT methods which needs experienced operators. The operators must be able to distinguish between noise and flaw pulses and shall be familiar with the standards and accept/reject criteria. That is why it is the only test which requires level II certified operators among all the other mentioned tests for CNG cylinder fabrication according to ISO 11439.

After three years of operation of the first ultrasonic scanner made by this group, a questionnaire was developed and distributed among the operators of 3 different workshops. The purpose of the questionnaire was collecting feedbacks from the users and technicians in technical and operational terms. The results are presented in Table 3. The result shows that all operators have enough knowledge for working with the scanner.

In this paper the applicability of automatic ultrasonic scanner in detection of flaws arising from ironing works (all the processes for making a steel pipe or plate) was validated. It was also proved that any damage caused by poor heat treatment is detectable by the ultrasonic test station immediately. These damages are often cracks arising from elongation and propagation of very small discontinuities. It clearly validates the good efficiency and performance of the scanner.

It was shown that the immersion technique is practically efficient for detecting the longitudinal and transverse flaws and for measuring the thickness of sidewalls. The system is able to operate in compliance with ISO11439 standard and to run a complete ultrasonic test within 2 minutes (excluded load/unload time) for a medium size CNG cylinder. Fortunately, with solving the hardware and software problems the scanners are successfully working in 3 cylinder manufacturing sites. Using of phased array probes9 and the C-scan presentation10 is considered for the future research.

Table 3. Views and comments of the users of the ultrasonic testing scanners collected from 3 different workshops.

Q1: How a flaw pulse is distinguishable from a noise pulse?	Answer 1: Flaw pulse is sharp and has a fix position. In contrast to the noise, it appears almost same by repeating the test. Answer 2: It is distinguishable in accordance with its place and operator's experience. Answer 3 :The noise pulse is happens on all 4 probes almost at the same position but the flaw pulse may not.
Q2: Where is often the flaws positions?	Answer 1: All the cylinder surface. Answer 2: All surface may have. Answer 3: On the central and near to neck area.
Q3: Where is the main place of noises?	Answer 1:No fixed position Answer 2:In middle and near to neck of cylinder Answer 3:No fixed position
Q4: How often is it necessary to change the water?	Answer 1: 2-3 days Answer 2: Every 3 months Answer 3: Each 3 weeks
Q5: To announce a flaw: both mirror probes indications are noticed OR one probe indication is enough.	Answer 1: The pulses of each probe is surveyed individually. Answer 2: Two pulses from mirror probes are surveyed simultaneously. Answer 3: Two pulses from mirror probes are surveyed simultaneously.
Q6: Quality of the ultrasoinc scanner's response to heat treatment furnace malfunction.	Answer 1: In case of heat treatment malfunction the rate of flaws increases considerably. Answer 2: The number of flaws increases whenever the heat treatment is not working properly. Answer 3: In case of bad operation in furnace, the cracks are expanded on body of cylinders and these are shown by the scanner.
Q7: Verification of flaws reported by the ultrasonic scanner with other NDT methods.	Answer 1: Verification of the flaws by magnetic particle testing (MT) and/or hydrostatic test approves them Answer 2: By MT and hydrostatic test is rechecked. They approve each other. Answer 3: By visual and Magnetic test is rechecked. They approve each other.

References

1. International Standard Organization, ISO-11439 Gas cylinders - High pressure cylind-ers for the on-board storage of natural gas as a fuel for automotive vehicles (2000)
2. Iranian National standard for inspection, installation on CNG cylinders for vehicles (2008)
3. International Standard Organization, 2006 ISO-19078, Gas cylinders - Inspection of the cylinder installation, and requalification of high pressure cylinders for the on-board storage of natural gas as a fuel for automotive vehicles, International Standard Organization (2006)
4. More, P.E.: Non-destructive Testing Handbook. In: Ultrasonic Testing, vol. 7. American Society for Nondestructive Testing, Inc., USA (2007)
5. International Atomic Energy Agency, IAEA TECDOC 462, Ultrasonic Testing of Materials at Level 2 (1988)
6. ISIRI-ISO-9712, Iranian national standards for NDT training condition certification (2009)
7. CNG 87 (Shirali, The investigation of safety standards on CNG cylinders (2008))
8. CNG 88 (Torabi, J. Arab, Estimation of CNG cylinder life time on vehicles (2009))
9. TOFD CNG 88 (Ghandchi, Ghasemi, Introduction of TOFD application for testing of CNG cylinders (2009))
10. Willcox, M.: Ultrasonic Inspection Equipment for Seamless Gas Cylinders. Insight NDT equipment Co. (2009), http://www.insight-ndt.com

The Theta Model: An Essential Forecasting Tool for Supply Chain Planning

Konstantinos Nikolopoulos[1], Vassilios Assimakopoulos[2], Nikolaos Bougioukos[2], Akrivi Litsa[2], and Fotios Petropoulos[2]

[1] The Business School, Bangor University, Bangor, Gwynedd LL57 2DG, U.K.
[2] Forecasting & Strategy Unit, School of Electrical and Computer Engineering, National Technical University of Athens, Athens, Greece
k.nikolopoulos@bangor.ac.uk, vassim@fsu,gr, nbougiou@gmail.com, alitsa@fsu.gr, fotis@fsu.gr

Abstract. The Theta model created a lot of interest in academic circles due to its surprising performance in the M3-competition, the biggest ever time series forecasting competition. As a result in the subsequent years it became a benchmark in any empirical forecasting exercise and an essential tool for efficient Supply Chain Management ad planning as it provides very accurate point forecasts. The present study focuses on if the Theta model is a special case of Simple Exponential Smoothing with drift (SES-d). The Theta model outperforms SES-d in the Quarterly-M3 and Other-M3 subsets by 0.30% and 0.36%.

Keywords: Theta model, Exponential Smoothing, M3-Competition, Supply Chain forecasting.

1 Introduction

The Theta model [1] is a times series forecasting model derived from the idea that: *"An extrapolative method is practically incapable of capturing efficiently all the available information hidden in a time series"*. On the one hand there are models that are too simple to catch all the available information. On the other hand there are methods with more parameters employed in order to cope with more demanding underlying patterns; unfortunately, while optimizing all these parameters usually these complex methods end up actually over-fitting the actual data. So, this approach aims to help the models capture the data. This is achieved by breaking the data down into several simpler series, each one of which captures *part* of the information included in the original series. Thus, In essence, a decomposition approach is employed. As a result of this process simpler models can adapt to these simpler series. For example, Instead of trying to adapt Holt Exponential Smoothing to an initial set of data we could alternatively create two series - one that captures the short term information and one that captures the long or the medium trend. Then we could try fit Naïve, Simple Exponential Smoothing (SES) or even a Linear Regression line (LRL) to each of those two series.

G. Lee (Ed.): Advances in Automation and Robotics, Vol.2, LNEE 123, pp. 431–437.
Springerlink.com

2 The Theta Model

The model is based on the concept of modifying the local curvatures of the time series. This change is obtained from a coefficient, called Θ which is applied directly to the second differences of the time series. Following this procedure, a set of new time series, the so-called Theta-lines are constructed noted as L(Θ). The initial time series is decomposed into two or more Theta-lines L(Θ). Each of the Theta-lines is extrapolated separately and the forecasts are combined either equally weighted or through a weight optimization procedure. Any forecasting method can be used for the extrapolation of L(Θ). In the M3-competition [2] exponential smoothing was used for the extrapolation of L(Θ).

If we consider one of the simplest cases in which the initial time series $X=\{X_1,\ldots,X_n\}$ is decomposed into two L(Θ), L(Θ=0) and L(Θ=2), then the algebra can be significantly simplified:

$$X_t = \tfrac{1}{2}\,(L_t(\Theta=0) + L(_t\Theta=2)),\ \forall\ t=1..n\ \ [1,3]$$

$$\Rightarrow \text{Data} = \tfrac{1}{2}\,(LRL_t + L_t(\Theta=2)) \Rightarrow$$

$$\Rightarrow L_t(\Theta=2) = 2\,X_t - LRL_t \tag{1}$$

Thus, in practice, the model can be easily implemented in a Microsoft Excel Worksheet via the following steps:

- Step 1: Apply Simple Linear Regression to non-seasonal data and prepare the LRL line and forecasts
- Step 2: Prepare the values for L(Θ=2) with formula [1], that is subtracting the LRL values from the actual data multiplied by two.
- Step 3: Extrapolate L(Θ=2) with either SES (optimized with Microsoft Solver) or with a simpler method, such as a Moving average or even a Naïve forecast [4,5]
- Combine with equal weights the forecasts from SES and LRL.

3 SES with Drift

Hyndman and Billah (H&B, [3]) consider the Theta model similar to Simple Exponential Smoothing with drift (SES-d) and in their analysis focused on the special case of the method where the decomposition is to L(Θ=0) and L(Θ=2). Formula-wise, Theta model extrapolations in this specific case may look like SES-d, but for two fundamental differences:

- The drift is predefined as being equal to half of the regression slope.
- The smoothing parameter is optimized on L(Θ=2) and not the original data as anybody would expect in any SES-with-drift approach

If we employ equally weighted extrapolations of L(0) and L(2) we end up with the following formula for Theta model point forecasts:

$$F_{t+1} = aY_t + (1-a)F_t + \frac{\hat{b}}{2} \tag{2}$$

where Ft is the forecast for period t prepared one period earlier, *a* the level smoothing parameter and $\hat{b}$ the linear regression slope. This is indeed the formula of SES-d with a drift equal to the half of the linear regression slope. However in any ES approach the smoothing parameter must be optimized on the actual data. Here is where the differences start. No matter what optimization technique employed the SES-d smoothing parameter will be identical to the one obtained in the Theta model only as a result of chance. This is simply because, in the case of Theta model, the smoothing parameter is calculated via a Mean Square Error (MSE) minimization procedure on a different time series: L(Θ=2). In any SES-d approach both the parameter and the drift are either set or optimized simultaneously in the original time series.

That is exactly the fundamental approach of Theta model. "*Do not use the original data; use instead series containing reduced amounts of information*"... such as L(Θ=2) in our case, so a series with doubled local curvatures that emphasizes the short term trend of the series. In our approach If you want to capture the short term behavior you just need to give emphasis to that (through L(Θ=2)) and at the same time use a method that gives more weight to the most recent observations, such as SES or Naive.

H&B also try to optimize this drift for SES-d instead of just using an a-priori value equal to half the regression slope. Some promising results are presented in their study through a Maximum Likelihood optimization, however it seems that the optimal value is always close to this "magical" constant drift – equal to half of the regression slope. This particular value for the drift was never given any attention before the introduction of the Theta model or the corroboration of the H&B later study. It seems that this SES-d approach should at least be used as a benchmark in any forthcoming forecasting competitions (The overall performance of SES-d in the 3003 M3 competition is 13.04% in terms of Average Symmetric MAPE, ranking marginally second behind Theta model, Table 1).

4 The H&B Optimized Version of SES-d

H&B provided a recalculated version of the Theta model. The published results of the H&B model could not match the performance of the Theta model in the M3-competition. H&B tried their approach in the Yearly-M3series with results that do significantly differ from the published ones (see table 1). In fact not only do their results differ from Theta but they also differ from each other (Recalculated Theta versus SES with drift). Theta resulted in a difference of 0.28% from the recalculated version in terms of the Symmetric Mean Absolute Percentage Error (SMAPE). This caused some doubts about the replication of the model. H&B blamed this on the SES initialization on L(Θ=2) in Assimakopoulos and Nikolopoulos' (A&N,[1]) study.

Table 1. Performance of ES approaches relative to the Theta model

M3-Competition (Average symmetric MAPE)						
Source	**Method**	**Yearly**	**Other**	**Monthly**	**Quart**	**All-M3**
Makridakis & Hibon (2000)	Single ES	17.82%	6.29%	15.32%	9.72%	14.32%
Makridakis & Hibon (2000)	**Theta model**	**16.90%**	**4.44%**	**13.85%**	**8.96%**	**13.01%**
Hyndman & Billah(2003)	Recalulated Theta	16.62%	---	---	---	---
Hyndman & Billah(2003)	SES-Drift-Optimized	16.55%	---	---	---	---
Current study	SES-Drift (=regr slope/2)	16.69%	4.94%	13.83%	9.22%	13.04%
Current study	SES-Drift-Optimized	16.55%	4.80%	13.93%	9.26%	13.09%
Current study	Generalising the Theta model	---	---	13.67%	---	---
Makridakis & Hibon (2000)	Top-Performer M3	16.42%	4.38%	13.85%	8.96%	13.01%
Makridakis & Hibon (2000)	#Series	645	174	1428	756	3003
Current study	# Seasonal Series	---	---	780	555	1335
Makridakis & Hibon (2000)	Horizon	6	8	18	8	18
Makridakis & Hibon (2000)	#Errors used	3870	1392	25704	6048	37014
Current study	Theta Lines/Extrapolation Methods that were Used in M3	L(0),L(1.4) DES	L(0),L(2) DES	**L(0),L(2) SES**	L(0),L(2), L(1.9),L(1.7) DES	
Current study	Type of Optimisation in M3	Second Theta Line	Extrapolation method	**None**	More Theta Lines	

So where does this difference come from? Can it solely be blamed on the difference in the smoothing parameter, as A&N have demonstrated, or in the initialization procedure as H&B suggest? In fact, neither of these two factors could lead to such a difference in the forecasting performance of the two versions of the model.

In the published results in 2000 and In the case of the 645 yearly M3 series, we used an equal weighted combination of LRL and L(Θ=1.4) for the extrapolation of the Theta model. Theta line L(Θ=2) was not used at all in this yearly subset and L(Θ=1.4) was used instead. As a result the SMAPE in that case went up to 16.90% instead of 16.69% that would be if L(Θ=2) had been used. The selection of L(Θ=1.4) was made empirically in order to reduce the out-of sample forecasting error of the model in the yearly subset; this decision was made as the equal weighted combination of LRL and L(Θ=2) was not performing well when we were holding out six periods. We tested combinations of LRL with L(Θ) for values of Θ=1.3, Step 0.1 and then chose one that performed best out of sample. This empirical optimization did not performed well as the Theta model had its worst performance in this subset. However we could not know that a-priori. Our aim was to test empirical optimization of the model at a real time competition where the holdout data were really unknown. We do believe that since the yearly series were the shortest ones it was extremely risky to use the full holdout of six periods in order to optimize our model. In others words we did try to empirically optimize the extrapolation part of our model in the yearly dataset, but this did not work well in that case. It did work however for the Quarterly-M3 and the Other-M3 datasets where we tried other types of optimizations, either by adding more Theta lines or changing the extrapolation method (details of the experimental optimizations are given in Table 1).

In the remaining three smaller datasets of the M3-competition (Quarterly, Other and Yearly) we tried various types of optimizations so as to test the limitations of our model. Some of them proved to increase accuracy performance, some of them did not. In the Other-M3 dataset, Dampen-trend Exponential Smoothing (DES) was used instead of SES while in the Quarterly–M3 dataset four Theta lines were employed (L(Θ=0), L(Θ=1.7), L(Θ=1.9), L(Θ=2)), and that was by far the best performance of the model.

The final suggestion of H&B was that a maximum likelihood optimization version of SES-d could produce better forecasts. This is based on a subset where Theta presented its worst performance: the Yearly-M3 645 series. We have prepared forecasts with SES-d (drift=half of the regression slope) as well as an optimized version of SES-d. The optimization has been done with an MSE minimization for values of drift up to 150% of the regression slope. We have managed to replicate exactly the performance of the H&B optimized version of SES-d (16.55% in the yearly data, Table 1) and used our approach to test it against Theta in the remaining datasets so as to have a clear view on the potential of the model SES-d.

H&B assertion is not supported by our findings. In both the Quarterly-M3 and Other-M3 data, the Theta model performs better with 0.30% and 0.36% respectively (Table 1). The statistical significance of these results is still open to question as Koning *et al.* have demonstrated [5].

5 Generalizing the Theta Model

In the last few years our research has been focused on using the model at its full potential, that is to use as many Theta lines as possible in each individual series that we analyze. Although there are clear biases associated with the testing of these enhancements on the M3 series, we have no option other than to continue using this dataset, at least until the next forecasting competition. A promising idea is to use all the Theta lines that seem to increase out-of-sample accuracy.We tested this approach on the monthly dataset of the M3 competition. We selected LRL and gave a 50% weight so as to ensure that our series regressed to the long term trend as we had to compete up to 18 periods ahead. We then select from a pool of five Theta lines { L(Θ=-1), L(Θ=0), L(Θ=1), L(Θ=2), L(Θ=3) } and kept those lines that, when added, improved accuracy on the selected error measure. Note that we even allowed negative values for the Theta coefficient (a negative value equal to -1 results to a symmetric line to the original data). The weights are also determined through this empirical optimization procedure. This approach is time consuming (it takes 20 times longer than running the basic version of the Theta model), however it achieved an average symmetric MAPE of 13.6% in the monthly-M3 subset; in other words managed to reduce error by almost .20 % versus the Theta model (equivalent to SES-d in this case). This is a very new idea that is currently in the initial stages of investigation.

6 Conclusions

Despite the interest created by the surprising performance of the Theta model (or method as referred by H&B) in the M3-competition only Hyndman and Billah in 2003 corroborated this research. The present study discussed the advances in the model and gave further insight on the research question: "Is the Theta model just a special case Simple Exponential Smoothing with drift?" It is obvious that the Theta model is far more generic than Exponential smoothing, basically because it is a data decomposition approach that can rely for extrapolation on any forecasting model [6]. If we do not use equally-weighted extrapolations of two specific Theta Lines, L(Θ=0) and L(Θ=2) in the Theta model, then we end up with a far more generic model than SES-d. It seems however that this SES-d approach performs very well on the M3 data, thus it should at least be used as a benchmark in the forthcoming forecasting competitions. The paper also examined an optimization version of SES-d so as to test the results of Hyndman and Billah. In contrast to their research results, the Theta model outperforms SES-d in the Quarterly-M3 and Other-M3 subsets with 0.30% and 0.36% respectively (SMAPEs).

Future research should focus on the generic formulation of the model. The two-line variant (Θ=0 and Θ=2) is only one of the several choices. The first extension is to use more than two Theta-lines. Another option is to use different Theta-lines combinations for each forecasting horizon. The last and most promising characteristic of the model is the utilization of different Theta-lines for each time series. A number of Theta-lines will correspond to each time series according to its qualitative and/or quantitative characteristics.

References

1. Assimakopoulos, V., Nikolopoulos, K.: The Theta Model: A Decomposition Approach to Forecasting. International Journal of Forecasting 16(4), 521–530 (2000)
2. Makridakis, S., Hibon, M.: The M3-Competition: results, conclusions and implications. International Journal of Forecasting 16(4), 451–476 (2000)

3. Hyndman, R.J., Billah, B.: Unmasking the Theta method. International Journal of Forecasting 19(2), 287–290 (2003)
4. Nikolopoulos, K., Maris, K., Tavanidou, E., Metaxiotis, K., Assimakopoulos, V.: Supporting Business Decisions through Forecasting: A Case Study in a Greek Tobacco Distribution Company. Journal of Information & Knowledge Management 2(3), 237–244 (2003)
5. Koning, A.J., Franses, P.H., Hibon, M., Stekler, H.O.: The M3 competition: Statistical tests of the results. International Journal of Forecasting 21(3), 397–409 (2005)
6. Nikolopoulos, K., Assimakopoulos, V.: Theta Intelligent Forecasting Information System. Industrial Management and Data Systems 103(9), 711–726 (2003)

Networked Control Systems Based on Modified Fuzzy Immune Control

Weidong Zhang, Feng Du, and Jinyu Li

College of Information Sciences & Technology Hainan University Haikou, Hainan, China
zhangwd106@163.com, dufeng488@sina.com, lijinyushmliy@163.com

Abstract. Because P type immune feedback controller can't well overcome dynamic interference and eliminate static error, based on P type controller, this paper designs a fuzzy-immune-pid controller, which uses fuzzy control rules and immune feedback control law to adjust the proportion, integral and differential parameters of controller on line, and it is applied to the wired networked control systems with software of truetime1.5 for simulation platform. Because the fuzzy control rules are more refined than immune, from theory analysis, the modified fuzzy immune controller has better control effect and robustness. The simulation results show that the improved control system has a quicker response speed and smaller overshoots and higher robustness.

Keywords: networked control systems (NCS), fuzzy control, feedback control, PID control.

1 Introduction

Networked control systems [1] (NCS) refers to the network which based on real-time close-loop feedback control system. Pablo [2] pointed out that NCS is provided with simple wiring, flexible structure, and can realize information resources sharing, etc. Due to the introduction of communication network, NCS produced some problems, such as communication protocol's choice, network delay [3], and packet loss [4] and so on. Du [5] pointed out that the existence of network delay will seriously affect the stability of system, even cause instability, and worse quality of performance (QoP) and quality of service (QoS), so the control methods applied to NCS is remained to be further studied.

In recent years, people have put forward many kinds of optimization algorithm. Hao etc [6] put forward that biological immune system can identify the external harmful substances and its own organization, and produce antibodies to remove pathogens. Wang etc [7] pointed out that the immune system is a highly parallel, and has the unique identification system, strong adaptive and the organization learning, identification and memory function of system. Wang [8] put forward artificial immune system is modeled on a natural immune function of intelligent method, so it provides a new method to solve the problems of NCS.

Liu etc [9] said the conventional PID control system mainly aimed at a precise model of linear process, and its PID parameters cannot adjust once it was determined. Si etc [10] combined PID control principle and immune feedback mechanism, and

G. Lee (Ed.): Advances in Automation and Robotics, Vol.2, LNEE 123, pp. 439–446.
Springerlink.com

reduce the contradiction of PID controller between quickness and overshoot by using the memory of immune system and the feedback principle. But Jiang etc [11] said it is difficult for P type immune controller to overcome the steady-state error, so they designed a kind of immune controller which combines P type controller with integral, differential controller. But the simulation of the controllers is only under the condition of without network (no occupied bandwidth and lost package), which exist some problems such as the reaction of slow speed and the larger overshoot. And it shows that while the larger occupying bandwidth and lost package rate existing.

Therefore, this paper designs a kind of modified fuzzy-immune-pid control system, the controller using fuzzy control rules alternative fuzzy immune control rules, and uses the immune feedback control law to adjust the proportion, integral and differential parameters of the controller on line and at the end it is applied to networked control systems with truetime1.5 as the simulation platform. The simulation results show that the modified control system has a faster response speed and higher robustness.

2 The Improvement of Fuzzy Immune Feedback Control

2.1 Adjustment Principle of Biological Immune System

Immune system is composed of antibody and lymphocyte, which is made up of T cells (auxiliary cell T_H and suppressor cell T_S) produced by thymocyte and B cells created by marrow. Humoral immune response in as an example, when antigen (Ag) invades organism and is digested by surrounding cells, messages are sent to T_H and T_S cells, and B cell is stimulated to create more antibody so as to eliminate antigen. If quantity of antigen is large, much more auxiliary cells T_H yield, but number of suppressor cell T_S reduces, which results in more B cell production; if antigen becomes less, number of T_S increases and that of T_H decreases, which results in the decrease of B cell. Synergism between suppressor mechanism and main feedback mechanism is realized through quick response of immune system to antigen and stabilizing immune system.

2.2 Normal Immune PID Control Algorithm

Intelligent behavior in creature information system provides various theory reference and technical methods for science and engineering field. Based on the immune feedback theory, assuming the number of kth generation of antigen is $\varepsilon(k)$, the output of auxiliary cell T_H is $T_H(K)$, impact of T_S, on B is $T_S(K)$, and all impetus that B cell received is

$$S(k)=T_H(k)-T_S(k) \tag{1}$$

Where $T_S(k)=k_2 f(S(k),\Delta S(k))\varepsilon(k)$, $T_H(k)=k_1\varepsilon(k)$. If number of antigen $\varepsilon(k)$ is supposed to be equivalent to the error $e(k)$, and overall impetus $S(k)$ that B received is equivalent to $u(k)$, then following control scheme is gained

$$u_1(k) = K_1(1-\eta_1 f_1(u_1(k), \Delta u_1(k))e(k) \tag{2}$$

Where $K_1 = k_1$ is used to control of response speed, and $\eta_1 = k_2/k_1$ is used to control the stabilization effect, and $f_1(u_1(k), \Delta u_1(k))$ is a selected nonlinear function. Considering control algorithm of regular P controller is

$$u_1(k) = k_{p1} e(k) \tag{3}$$

Where k_{p1} are the proportional gain, Comparing (2) to (3) we know that, the controller basing on immune feedback control principle is a nonlinear P controller actually, and its proportional gain is

$$k_{p1} = K_1(1-\eta_1 f_1(u_1(k), \Delta u_1(k)) \tag{4}$$

Following control rules are defined to determine $f_1(u_1(k), \Delta u_1(k))$:

① If u is P and Δu is P then $f_1(u, \Delta u)$ is $N(1)$.

② If u is P and Δu is N then $f_1(u, \Delta u)$ is $Z(1)$.

③ If u is N and Δu is P then $f_1(u, \Delta u)$ is $Z(1)$.

④ If u is N and Δu is N then $f_1(u, \Delta u)$ is $P(1)$.

The output of immune PID controller is

$$\begin{aligned} u(k) &= u(k-1) + k_{p1}((e(k) - e(k-1)) + k_i e(k) \\ &\quad + k_d(e(k) - 2e(k-1) + e(k-2)) \end{aligned} \tag{5}$$

2.3 Modified Immune PID Control Algorithm

Because P type controller is essentially a variable gain proportion of controller, therefore, the P type immune feedback controller cannot compensate for the error.

Based on the P type controller, which uses the fuzzy control rules, this paper modeled on the parameter control method of P type controller to control integral and differential parameters, in order to eliminate static deviation, improve the response speed, and improve the control quality.

The laws of the modified fuzzy immune controller feedback are as follows:

$$u_2(k) = K_2(1-\eta_2 f_2(m))e(k) = k_{p2} e_{21}(k) + k_{i2} e_{22}(k) + k_{d2} e_{23}(k) \tag{6}$$

Where $m = 1, 2, 3$, $k_{p2} = K_{21}(1-\eta_{21} f_2(1))$, $k_{i2} = K_{22}(1-\eta_{22} f_2(2))$, $k_{d2} = K_{23}(1-\eta_{23} f_2(3))$.

The control rules of $f_2(1)$ are shown in Table1.

Table 1. Fuzzy rules of k_{p2}

e \ △kp \ ec	NB	NM	NS	ZO	PS	PM	PB
NB	PB	PB	PM	PM	PS	ZO	ZO
NM	PB	PB	PM	PS	PS	ZO	NS
NS	PM	PM	PM	PS	ZO	NS	NS
ZO	PM	PM	PS	ZO	NS	NM	NM
PS	PS	PS	ZO	NS	NS	NM	NM
PM	PS	ZO	NS	NM	NM	NM	NB
PB	ZO	ZO	NM	NM	NM	NB	NB

The control rules of $f_2(2)$ are shown in Table2.

Table 2. Fuzzy rules of k_{i2}

e \ △ki \ Ec	NB	NM	NS	ZO	PS	PM	PB
NB	NB	NB	NM	NM	NS	ZO	ZO
NM	NB	NB	NM	NS	NS	ZO	ZO
NS	NB	NM	NS	NS	ZO	PS	PS
ZO	NM	NM	NS	ZO	PS	PM	PM
PS	NM	NS	ZO	PS	PS	PM	PB
PM	ZO	ZO	PS	PS	PM	PB	PB
PB	ZO	ZO	PS	PM	PM	PB	PB

The control rules of $f_2(3)$ are shown in Table3.

Table 3. Fuzzy rules of k_{d2}

e \ △kd \ Ec	NB	NM	NS	ZO	PS	PM	PB
NB	PS	NS	NB	NB	NB	NM	PS
NM	PS	NS	NB	NM	NM	NS	ZO
NS	ZO	NS	NM	NM	NS	NS	ZO
ZO	ZO	NS	NS	NS	NS	NS	ZO
PS	ZO	ZO	ZO	ZO	ZO	ZO	ZO
PM	PB	NS	PS	PS	PS	PS	PB
PB	PB	PM	PM	PM	PS	PS	PB

3 Simulation

The controlled plants are second order plus delay systems as follows

$$G(s)=\frac{1770}{s^2+60s+1770}e^{-0.02s} \tag{7}$$

The sampling time is 10ms, the reference signal is square wave signal, and the simulation time is 5s. In order to test the robustness of the controller, in the first 4.2 s, this paper adds an interference whose value is 0.25. The parameters of conventional fuzzy immune controller are $K=0.3,\ \eta=0.8, k_i^{'}=0.30, k_d^{'}=0.30$; the parameters of modified fuzzy immune controller are $K_{21}=0.001, \eta_{21}=0.01, K_{22}=0.08, \eta_{22}=0.01,\ K_{23}=0.10, \eta_{23}=0.60$. the simulation software is truetime1.5, this paper chooses CSMA/CD (Ethernet).

Case1: When occupy bandwidth of the interference is 0.45 and lost package is zero, the simulation results are shown in fig.1 to fig.3.

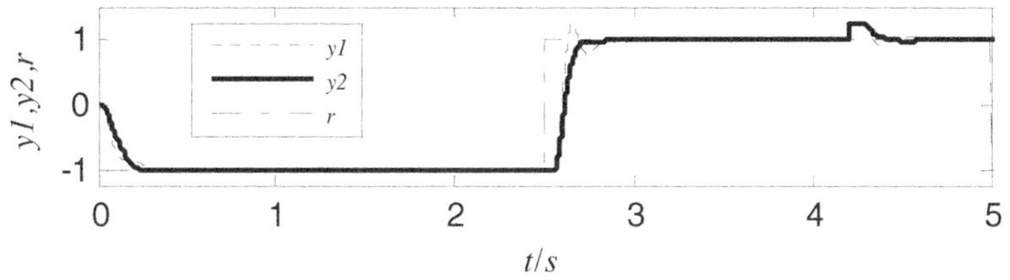

Fig. 1. Controlled plants output y_1 and y_2. The y_1 expresses conventional fuzzy immune, and y_2 is modified fuzzy immune PID control.

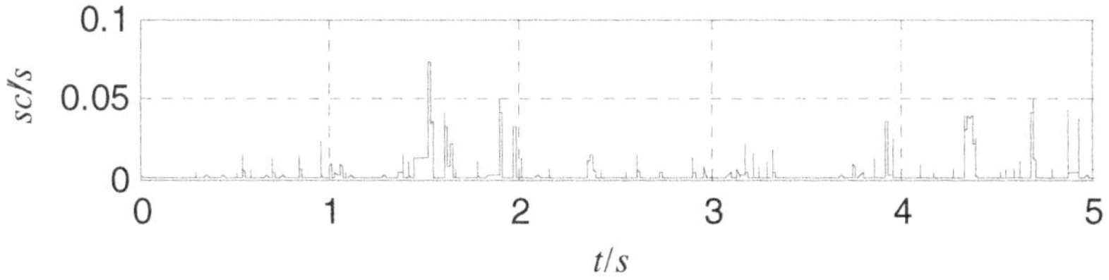

Fig. 2. The network delay t_{sc} from actuator(sensor) to controller.

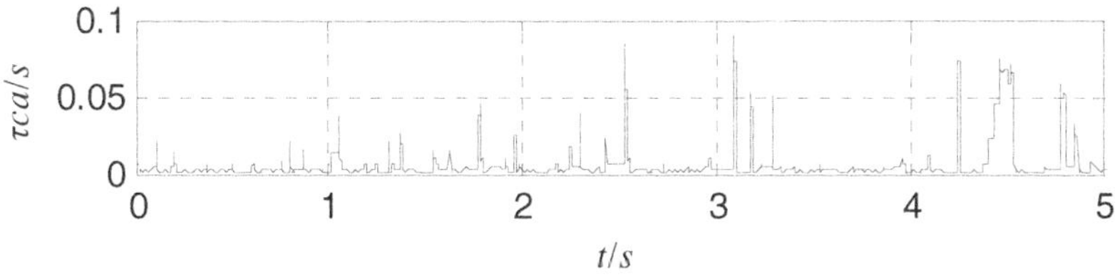

Fig. 3. The network delay t_{ca} from controller to actuator(sensor).

From fig.1 to fig.3, we can see that t_{sc} and t_{ca} are random, time-variant and uncertain, and their probability distributions are different, t_{sc} and t_{ca} maximum are respectively 0.072s and 0.09s (sampling period is 0.0ls). The $y2$ (in fig.1, thick line expression) track is timely, and the overshoot is none, it completely satisfies the performance requirements of the NCS. However, the $y1$ gives higher overshoot to begin at 2.6s, and it can't quickly reach stable state.

Case2: When occupy bandwidth is 0.45 and lost package is 0.4, the simulation results are shown in fig.4 to fig.8.

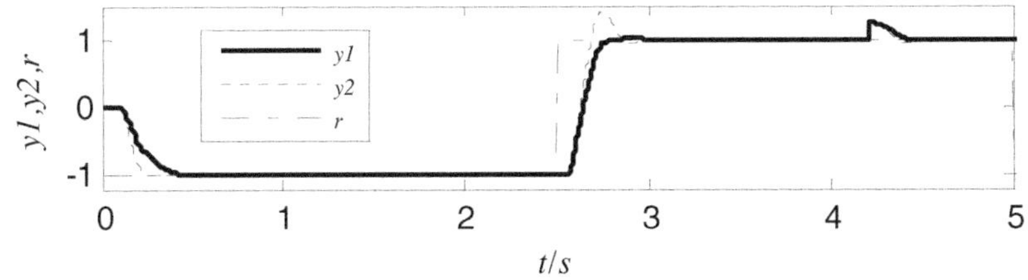

Fig. 4. Controlled plants output y_1 and y_2. The y_1 expresses conventional fuzzy immune, and y_2 is modified fuzzy immune PID control.

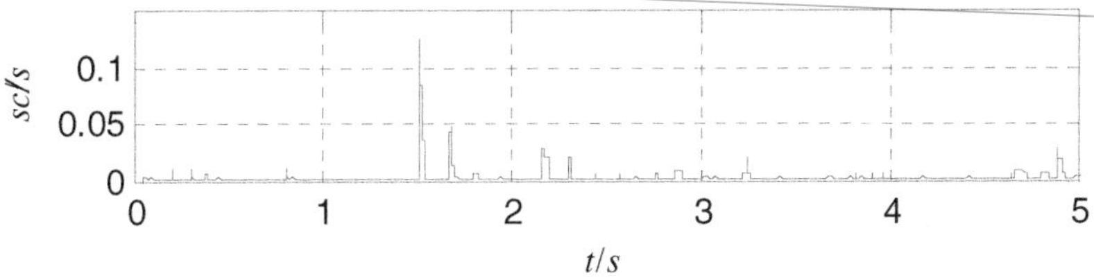

Fig. 5. The network delay t_{sc} from actuator(sensor) to controller

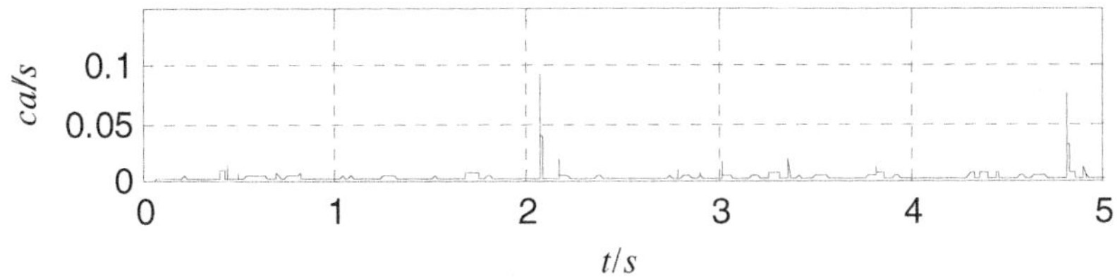

Fig. 6. The network delay t_{ca} from controller to actuator(sensor)

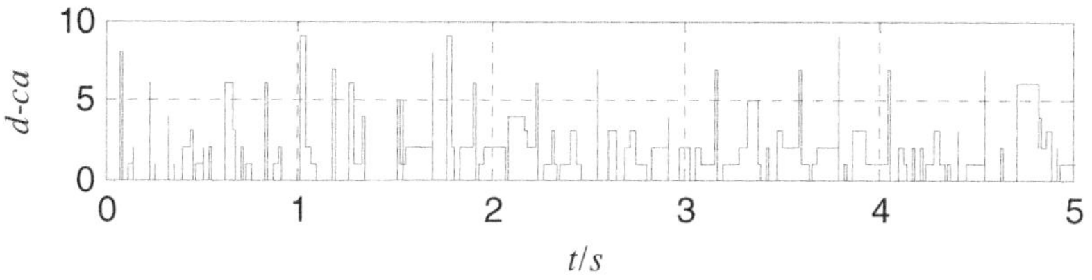

Fig. 7. The lost package d_{ca} of actuator (sensor)

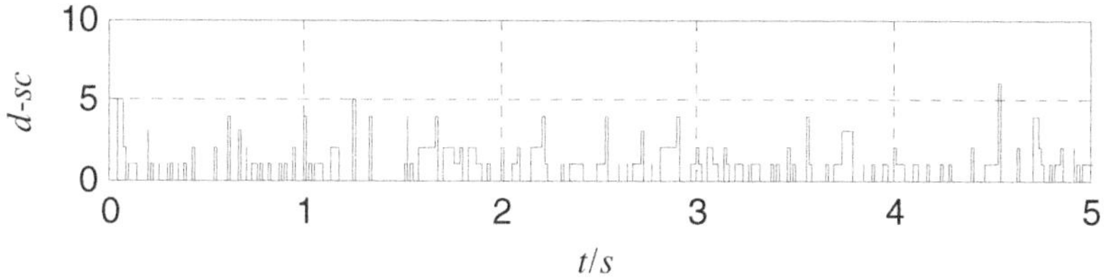

Fig. 8. The lost package d_{sc} of controller.

From fig.4 to fig.8, we can see that t_{sc} and t_{ca} are random, t_{sc} and t_{ca} maximum are respectively 0.12s and 0.1s (sampling period is 0.0ls). Also, we can see that d_{ca} and d_{sc} maximum are respectively 9 and 6.The $y2$ (in fig.4, thick line expression) track is timely, and the overshoot is none, it completely satisfies the performance requirements of the NCS. However, the $y1$ gives higher overshoot to begin at 2.6s. Apparently, $y1$ doesn't satisfy the performance requirements of the NCS.

4 Conclusion

This paper puts forward a modified fuzzy immune feedback PID controller which is based on fuzzy rules and immune feedback mechanism. The experimental results show that the modified controller has good dynamic performance and stable performance.

Corresponding author: Feng Du. This paper was supported by the 211 Project Central Special Fund of Hainan University, and the joint training funds of Hainan University & South China sea research station.

References

1. Wen, D.-L., Gao, Y.: Networked Control Systems in Multiple-packet Transmission. In: Chinese Control and Decision Conference, pp. 423–426 (2008)
2. Millan, P., Jurado, I., Vivas, C., Rubio, F.R.: Networked Predictive Control of Systems with Data Dropouts. In: Proceedings of the 47th IEEE Conference on Decision and Control, pp. 2704–2709 (2008)

3. Okano, R., Ohtani, T., Nagashima, A.: Networked control systems by PID controller Improvement of performance degradation caused by packet loss. IEEE Control Systems Magazine, 1126–1132 (2008)
4. Robinson, C.L., Kumar, P.R.: Networked control systems with packet delays and losses. In: Proceedings of the 47th IEEE Conference on Decision and Control Cancun., Mexico, pp. 4602–4607 (2008)
5. Du, F.: The research of Networked control systems base on Modified Smith Predictor. Ph.D. Dissertation of Xinan jiaotong university, pp.1-2 (2008)
6. Hao, W., Xie, K.: The design and simulation of fuzzy immune parameter self-setting PID control system. Journal of Taiyuan University of Technology 37, 274–276 (2006)
7. Wang, B., Li, A.: The optimization design of fuzzy immune nonlinear PID control. Control Engineering 14, 81–82 (2007)
8. Wang, Y.: The fuzzy immune optimization control of a kind of second order handstand pendulum, Ph.D. Dissertation of haerbin industrial university, pp. 8–10 (2007)
9. Liuliang, Wujie, Chenyuanrui: The application of greenhouse based on the genetic optimize fuzzy-immune-pid controller. In: Proceedings of the 24th Chinese Control Conference, pp. 1121–1126 (2005)
10. Si, T., Tian, L., Mao, Z.: The design and optimization of MIMO systems of nonlinear fuzzy-immune-pid. Computer Simulation 26, 204–206 (2009)
11. Jiang, Z., Cai, Z., Gong, T.: A modified fuzzy PID controller immune feedback. Control Engineering 15, 485–488 (2008)
12. Liu, J.: Advanced PID control Matlab simulation, pp. 129–130. Publishing house of electronics industry, beijing (2004)

The Research Networked Control Systems Based on Improved Fuzzy Immune Control

Weidong Zhang, Feng Du, and Jinyu Li

College of Information Sciences & Technology Hainan University Haikou, Hainan, China
{zhangwd106,lijinyushmliy}@163.com, dufeng488@sina.com

Abstract. Considering it is cockamamie to adjust the experiment data of the fuzzy control systems, and it's easy to get into the problem of local optimum, the researchers propose a fuzzy-immune-pid controller, which combines the fuzzy control and immune PID control, It does not only has the reliability of PID and the strong robustness of the fuzzy algorithm, but also has the ability of global optimizing of immune algorithm. Based on the conventional fuzzy immune controller, this paper improves the fuzzy-immune-pid controller, this improved controller which combined fuzzy immune rules and feedback control theory, which respectively controls the proportion, integral and differential parameters, it make more accurate parameters adjustment. At last, this paper uses the software of truetime1.5 as the simulation platform to apply in wired networked control systems. Simulation results show that the improved fuzzy-immune-pid controller has the characteristics of strong robustness and rapid response.

Keywords: networked control systems (NCS), fuzzy immune control, feedback control, PID control.

1 Introduction

The mode of traditional control system, has had a successful application in industry for decades, however, with the expansion of the physical equipment and system function, the point-to-point communication mode has reached the limitation of its application [1]. But the bus structure with universal network system, which is named the networked control systems (NCS), has integrated system structure, distributed operations, wiring simple and reliable signal. Du Feng etc [2] say the NCS not only can realize large complicated system with remote control, share resource, have higher ability of diagnosis, and good interaction, but also increase the system's reliability, easy installation and maintenance, etc. In addition, Kyung Chang Lee etc [3] put forward that through the use of wireless sensor network or other wireless network technology, which can realize the widely distributed sensors and remote controllers and actuators which constitute the wireless network control system (WNCS) with some special uses, but these are what a traditional control system can't realize.

Luca Schenato [4] pointed out that in the NCS, the data transfer between the sensor and controller, the controller and actuators usually exist randomized, time-varying and probabilistic network delay. The exists of the delay serious influence on the

G. Lee (Ed.): Advances in Automation and Robotics, Vol.2, LNEE 123, pp. 447–454.
Springerlink.com

control quality of the system, and even lead to the instability, and cause worse quality of performance (QoP) and quality of service (QoS). Although the study of the NCS has already made some achievements, but so far, it has not formed a set of mature and complete theory or method system. Du Feng [5] indicated that the NCS has network delay, packet timing disorder and packet loss, which make traditional control theory no longer fit for the analysis of NCS. I will do some researches on the control in this paper.

The conventional PID controller is simple in structure, and easy to be implemented [6], however Zhang Yi ect [7] pointed out that because of its parameters is fixed, it can't adapt the industrial control object which is nonlinear, time-varying uncertainties and complex characteristics. ZHANG WEN-JUAN etc [8] put forward that fuzzy controller don't need to know the mathematical model, so particularly suitable for controlling the complex controlled object of highly nonlinear and uncertainties. However A. Fadaei etc [9] pointed out the fuzzy control rules of the fuzzy controller are usually given by their own experience of experts or experienced operator, F.Du etc [10] pointed out there is not a normative analysis and design method for fuzzy controller, so it is difficult for the analysis of fuzzy controller. Along with the development of biological immune system, researchers use the powerful information processing mechanism of the biological immune system to solve the practical problems of science and engineering that it is to produce artificial immune system. Since the immune system has the characteristics such as immune discrimination, learning and optimization, the associative memory, self-organization and adaptive, distributed system, fault patience and diversity, therefore, Liao YiFan [11] pointed out to design a fuzzy immune PID controller connected the fuzzy controller with the immune PID controller, which has both the reliability of PID controller and the strong robustness of the fuzzy algorithm, Si TuYing etc [12] pointed out that the memory and the feedback principle of the immune system can reduce contradiction of quickness and overshoot of the PID controller, meanwhile Laihua Fang etc [13] put forward using fuzzy control method to approximate the nonlinear function of immune regulation controller, in order to replace the fixed forms of function.

Based on the conventional fuzzy-immune-pid controller, referring to the control method of proportional parameters, this paper controls integral and differential parameters by using the fuzzy control rules and immune feedback control law to adjust various parameters of fuzzy-immune-pid controller on line, which make more accurate control, and it is applied to the wired networked control systems with truetime1.5 as the simulation platform. The simulation results show that the improved fuzzy immune controller has stronger robustness and quicker response speed.

2 The Optimized Fuzzy Immune Feedback PID Control

2.1 Feedback Principle of Biological Immune System

Immune is a characteristic physiological reaction of biological body. Immune system of biology could produce relative antibody to resist invading anti-source from extraneous. After antibody combined with anti-source, a serial reaction will be brought to destroy antibody by swallowing effect or producing special enzyme. The most important part of biological immune system is lymph cell, and it concludes T cell generating by thymus (including assistant cell T_H and suppressive cell T_S) and B cell generating by marrow. When antibody invades the body and is absorbed by surrounding cell, it transmits the message to T cell, that is, transmits to T_H cell and T_S cell, and then stimulates B cell. B cell generates antibody to remove anti-source. When there is much anti-source, T_H cell is more and T_S cell is little in the body, thus better B cell is produced. As the decrease of anti-source, T_S cell increases, which restrain the creation of T_H cell, then B cell decreases accordingly. After some time-interval, immune feedback system trends to a balance, the cooperation between restrain principle and main feedback principle is done by the fast reaction of immune feedback principle to anti-source and steady immune system.

2.2 Immune *P* Control

The intelligent behaves of biology information system offer various theoretical consult and technical method. According to the feedback manipulating principle of immune system T cell, assuming the number of k generation of antigen is $\varepsilon(k)$, the output of auxiliary cell T_H is $T_H(k)$, the effect of B cell from T_S cell is $T_S(k)$ and all impetus that B cell received is

$$S(k)=T_H(k)-T_S(k) \tag{1}$$

Where $T_S(k)=k_2 f(S(k),\Delta S(k))\varepsilon(k)$, $T_H(k)=k_1\varepsilon(k)$. If number of antigen $\varepsilon(k)$ is supposed to be equivalent to the error $e(k)$, and overall impetus $S(k)$ that B received is equivalent to $u(k)$, then following control scheme is gained

$$u_1(k)=K_1(1-\eta_1 f_1(u_1(k),\Delta u_1(k))e(k) \tag{2}$$

Where $K_1=k_1$ is used to control of response speed, and $\eta_1=k_2/k_1$ is used to control the stabilization effect, and $f_1(u_1(k),\Delta u_1(k))$ is a selected nonlinear function. Considering control algorithm of regular P controller is

$$u_1(k)=k_{p1}e(k) \tag{3}$$

Where k_{p1} are the proportional gain, Comparing (2) to (3) we know that, the controller basing on immune feedback control principle is a nonlinear P controller actually, and its proportional gain is

$$k_{p1}=K_1(1-\eta_1 f_1(u_1(k),\Delta u_1(k)) \tag{4}$$

2.3 Improved Immune PID Control

Although the immune P controller has a fast speed of response, it could not effectively remove the noise or control deviation causing by disturbance, this paper put forward a kind of improved fuzzy-immune-pid control.

The improved fuzzy immune PID controller algorithm is as follows:

$$\begin{aligned} u_2(k) &= K_2(1-\eta_2 f_2(u_2(k),\Delta u_2(k))e_{2m}(k) \\ &= k_{p2}e_{21}(k)+k_{i2}e_{22}(k)+k_{d2}e_{23}(k) \end{aligned} \tag{5}$$

Where:

$m=1,2,3.k_{p2}=K_{21}(1-\eta_{21}f_2(u_2(k),\Delta u_2(k)),$

$k_{i2}=K_{22}(1-\eta_{22}f_2(u_2(k),\Delta u_2(k)),\ k_{d2}=K_{23}(1-\eta_{23}f_2(u_2(k),\Delta u_2(k)).$

Immune PID controller is a nonlinear P controller which is designed by learning from immune principle of biology system. Its nonlinear is realized mainly by $f_2(u_2(k),\Delta u_2(k))$. Assuming each input variable in the paper is confused by two fuzzy assemble, that is "positive"(P) and "negative"(N), each output variable in the paper is confused by three fuzzy assemble, that is "positive"(P),"zero" and "negative"(N). These servile function above are defined in the whole area $(+\infty,-\infty)$.

Fuzzy rational logical rule adopts 4 rules below:

① If u is P and Δu is P then $f_1(u,\Delta u)$ is $N(1)$.

② If u is P and Δu is N then $f_1(u,\Delta u)$ is $Z(1)$.

③ If u is N and Δu is P then $f_1(u,\Delta u)$ is $Z(1)$.

④ If u is N and Δu is N then $f_1(u,\Delta u)$ is $P(1)$.

In the rules above, we use fuzzy logical AND, OR operation of Zadeh, adopt common used anti-fuzzy way of mom to obtain the output $f_1(u_1(k),\Delta u_1(k))$ of fuzzy controller, so as to construct fuzzy immune PID controller.

3 Simulation Experiment

The controlled plants are first order plus delay systems as follows

$$G(s)=\frac{100}{s+100}e^{-0.02s} \tag{6}$$

Sampling time is 10ms, the reference signal is square wave signal, and the simulation time is 5s. In order to test the robustness of the controller, in the first 4.2 s, this paper adds an interference whose value is 0.25. The parameters of conventional fuzzy immune controller are $K = 0.3,\ \eta = 0.8, k_i' = 0.30, k_d' = 0.30$; the parameters of modified fuzzy immune controller are $K_{21} = 0.001, \eta_{21} = 0.01, K_{22} = 0.08, \eta_{22} = 0.01,\ K_{23} = 0.10, \eta_{23} = 0.60$. the simulation software is truetime1.5, this paper chooses CSMA/CD (Ethernet) network protocol.

Case1: When occupy bandwidth of the interference is zero and lost package is zero, the simulation results are shown in fig.1.

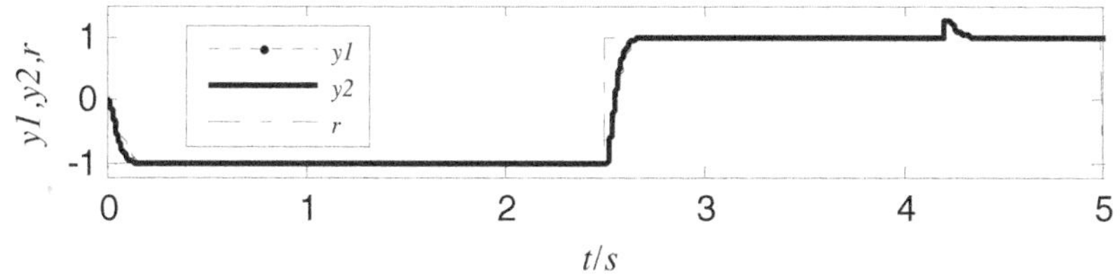

Fig. 1. Controlled plants output y_1 and y_2. The y_1 expresses conventional fuzzy immune, and y_2 is modified fuzzy immune PID control.

From fig.1, we can see that the two control effect is basically the same.

Case2: When occupy bandwidth of the interference is 0.45 and lost package is zero, the simulation results are shown in fig.2 to fig.4.

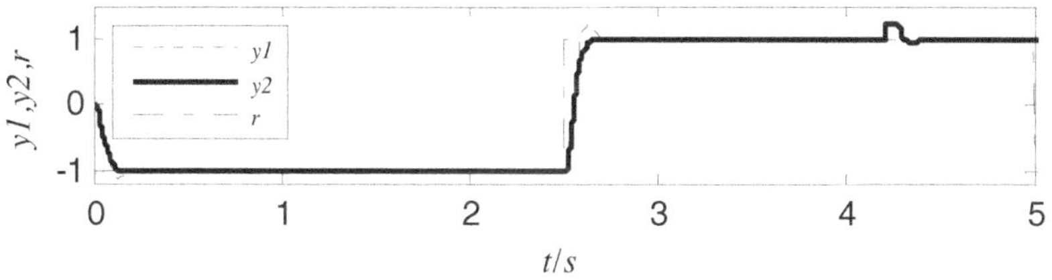

Fig. 2. Controlled plants output y_1 and y_2. The y_1 expresses conventional fuzzy immune, and y_2 is modified fuzzy immune PID control.

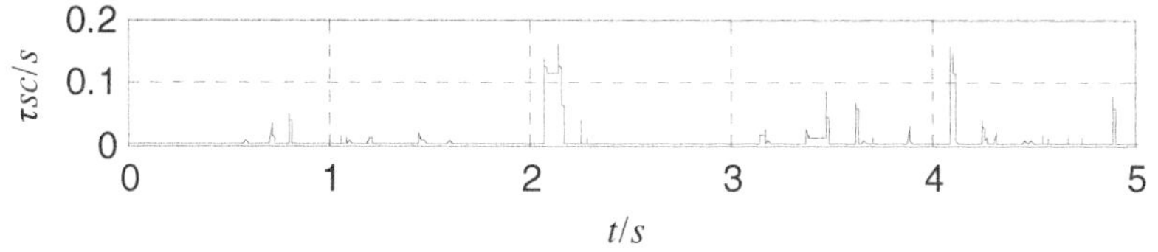

Fig. 3. The network delay t_{sc} from actuator(sensor) to controller

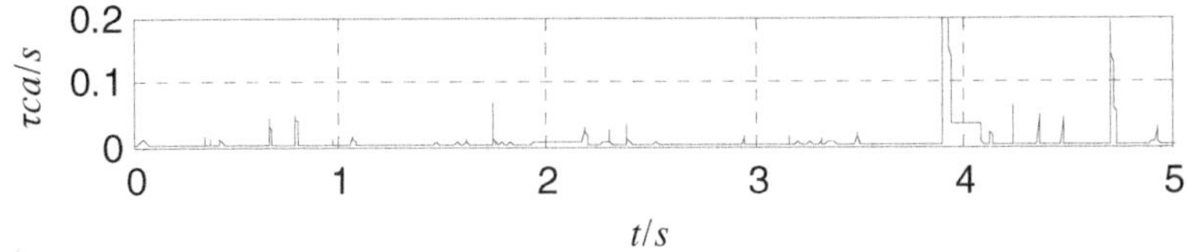

Fig. 4. The network delay t_{ca} from controller to actuator(sensor)

From fig.2 to fig.4, we can see that t_{sc} and t_{ca} are random, time-variant and uncertain, and their probability distributions are different, the maximum of t_{sc} is 0.15s which exceeds 15 sampling periods and the maximum of t_{ca} is 0.2s which exceeds 15 sampling periods (sampling period is 0.0ls). The $y2$ (in fig.2, thick line expression) track is timely, and the overshoot is none, it completely satisfies the performance requirements of the NCS. However, the $y1$ gives high overshoot to begin at 2.6s.

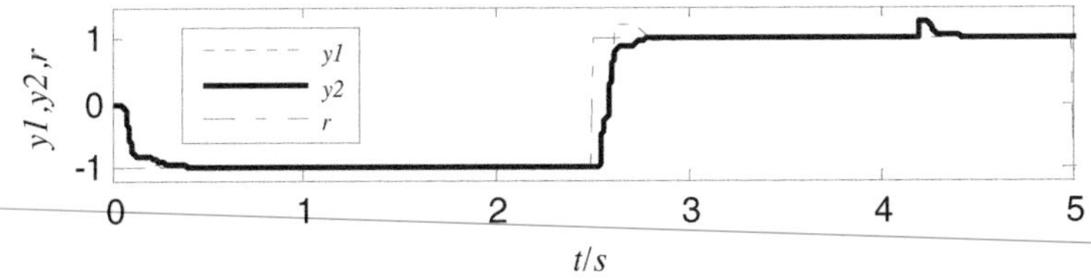

Fig. 5. Controlled plants output y_1 and y_2. The y_1 expresses conventional fuzzy immune, and y_2 is modified fuzzy immune PID control.

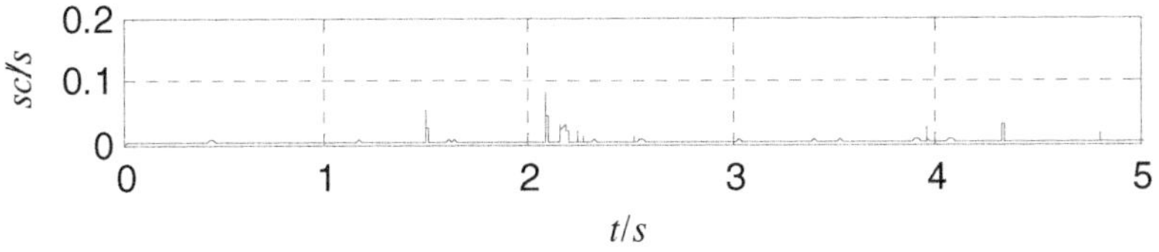

Fig. 6. The network delay t_{sc} from actuator(sensor) to controller.

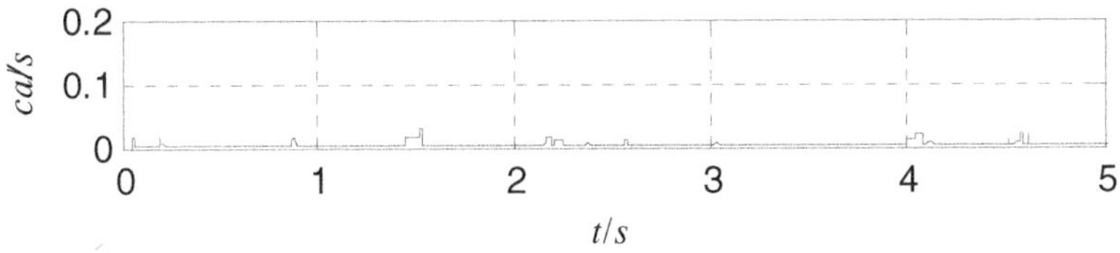

Fig. 7. The network delay t_{ca} from controller to actuator(sensor).

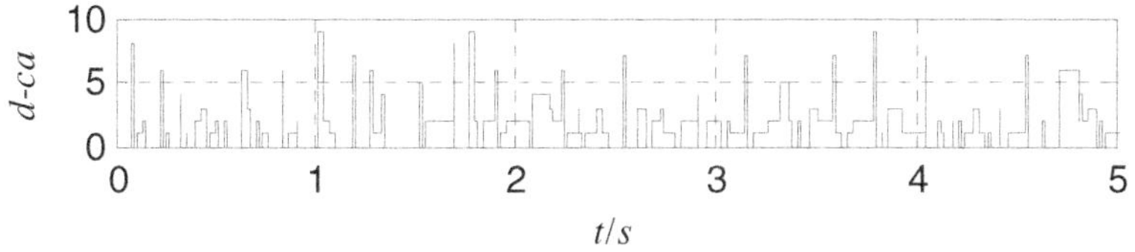

Fig. 8. The lost package d_{ca} of actuator (sensor).

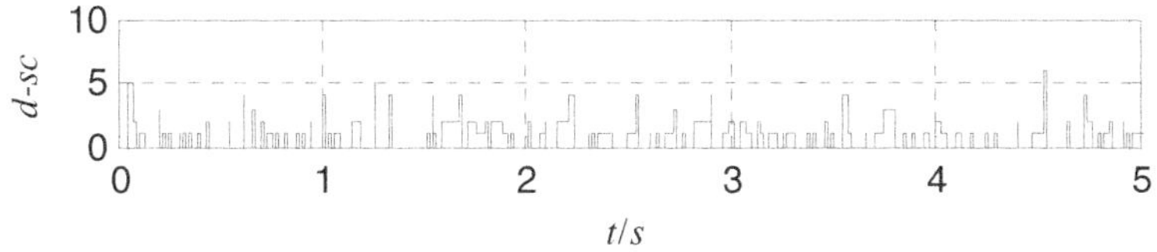

Fig. 9. The lost package d_{sc} of controller.

Case3: When occupy bandwidth of the interference is 0.45, and lost package is 0.4, the simulation results are shown in fig.5 to fig.9.

From fig.5 to fig.9, we can see that t_{sc} and t_{ca} are random, time-variant and uncertain, and their probability distributions are different, t_{sc} and t_{ca} maximum are respectively 0.08s and 0.03s (sampling period is 0.0ls). Also, we can see that d_{ca} and d_{sc} maximum are respectively 9 and 6.The $y2$ (in fig.5, thick line expression) track is timely, and the overshoot is none, it completely satisfies the performance requirements of the NCS. However, the $y1$ gives higher overshoot to begin at 2.6s. Apparently, $y1$ doesn't satisfy the performance requirements of the NCS.

3 Conclusion

Based on the conventional PID control cannot compensate for the control error caused by noise the nonlinear disturbance, this paper design a kind of fuzzy immune optimal controller, which can accurate control the parameters of proportion, integral and differential, so the system has better control effect and high robustness.

Corresponding author: Feng Du. This paper was supported by the 211 Project Central Special Fund of Hainan University, and the joint training funds of Hainan University & South China sea research station.

References

1. Zhang, Y., Zhu, X., Wang, J.: Networked Control Systems, pp. 1–2. The press of Chongqing University (2003)
2. Du, F., Qian, Q.: Networked Control Systems Based on Generalized Predictive Control and Modified Smith Predictor. Proceedings of the 7th World Congress on Intelligent Control and Automation, 7859–7863 (2008)

3. Lee, K.C., Lee, S.: Wireless Networked Control System Using NDIS-based Four-Layer Architecture for IEEE 802.11b. IEEE Transactions on Industrial Electronics (2008)
4. Schenato, L.: Optimal Estimation in Networked Control Systems Subject to Random Delay and Packet Drop. IEEE Transactions on Industrial Electronics 53, 1311–1317 (2008)
5. Du, F.: The research of Networked control systems base on Modified Smith Predictor, pp. 1–2. Ph.D. Dissertation of Xinan Jiaotong university (2008)
6. Liu, J.: Advanced PID control Matlab simulation, 2nd edn., pp. 129–130. Publishing house of electronics industry, beijing (2004)
7. Zhang, Y., Yang, X.: The design of fuzzy immune controller based on the genetic algorithm. The Journal of System Simulation 16, 1548–1557 (2004)
8. Zhang, W.-J., Deng, Y.-F., Wang, L.-M., Zhang, H.-W.: Fuzzy Pid Control Method for Networked Control System with Constant Delays. In: Proceedings of the Seventh International Conference on Machine Learning and Cybernetics, Kunming, pp. 1968–1973 (2008)
9. Fadaei, A., Salahshoor, K.: Improving the control performance of networked control systems using a new fuzzy PID. IEEE Transactions on Industrial Electronics, 2066–2071 (2008)
10. Du, F., Qian, Q.Q.: Fuzzy Immune Self-Regulating PID Control Based on Modified Smith Predictor for Networked Control Systems. IEEE Transactions on Industrial Electronics, 424–428
11. Liao, Y.: The design of drive system of soldiers arm hydraulic and the research on the control strategy of fuzzy immune PID [D].The master's degree thesis of Hunan normal university, pp:45-51 (2006)
12. Si, T., Tian, L., Mao, Z.: The design and optimization of the nonlinear fuzzy-immune-PID of the multiple-input multiple-output (MIMO) system. The Computer Simulation 26, 204–206 (2009)
13. Fang, L., Wu, Z., Wu, A., Zheng, A.: Fuzzy immune Self-regulating PID Control of Networked Control System. In: International Conference on Computational Intelligence for Modelling Control and Automation,and International Conference on Intelligent Agents, Web Technologies and Internet Commerce (2006)

Adaptive Background-Updating and Target Detection in Motion State

Zheng Mao[1], Anjie Gao[1], Wei Wei[2], Legong Sun[1], and Silin Chen[1]

[1] School of Electronic Information and Control Engineering, Beijing University of Technology,Beijing, 100124, China
maozheng@bjut.edu.cn
[2] Northwest Mechanical and Electrical Engineering Institute, Xianyang, Shaaxi, 712099

Abstract. A common method for moving target detection in image sequences involves a self-adaptive threshold for segmentation. This paper discusses each pixel modeling in the image as a mixture of Gaussian distributions and gives an improved method to update the model. First, the characteristics of the different frames method and a mixture of Gaussian distributions will be discussed. And then, the two methods, different frames method and a mixture of Gaussian distributions are combined to decide which learning rate is suitable for the updating method at this moment. If there is no moving target in the image sequences, the background model should be updated quickly to get the real background with low noise. The better foreground images will be gotten by this method with low computational complexity. The improved algorithm performs more robustly and powerfully than the classical Gaussian Mixture Model in moving target detecting.

Keywords: target detection, different frames method, Gaussian Mixture Model.

1 Introduction

Detecting moving target is a valuable field in computer vision. The accuracy of detection plays an important role in the next steps of target recognition and target tracking [1].

Locating the position and measuring the size and shape of the moving target are the aim for target detection. A common method for segmentation of moving regions in image sequences involves background subtraction, or finding the different part of the foreground between the two frames. The numerous approaches to target detection differ in the types of background model and the procedures used to update the model. Lipton shows the difference frames method [2], and Pfinder uses a multi-class statistical model for the tracked objects. But the background model is a single Gaussian per pixel [3]. Friedman and Russel have recently implemented a pixel-wise EM framework for vehicles detection [4].

This paper will discuss an approach that is adaptable and flexible enough to updating the background model. In section 2 and 3, we will discuss and demonstrate the different frame method and background model method, which are used widely in target detection and computer vision and are basic for our approach. In section 4, an improved approach will be put forward and validated.

G. Lee (Ed.): Advances in Automation and Robotics, Vol.2, LNEE 123, pp. 455–462.
Springerlink.com

2 Different Frames

The difference Computation between two sequence-frames is referred to as different frame method as (1).

$$D_t(x, y) = \left| f_t(x, y) - f_{t-1}(x, y) \right| \tag{1}$$

where $f_t(x,y)$ is the value of the point (x,y) at time t, $f_{t-1}(x,y)$ is the value at time t-1. And $D_t(x,y)$ is the image after different frame. Every pixel in $D_t(x,y)$ belongs to the target or background would be judged as follows:

$$R_k(x, y) = \begin{cases} foreground & D_t(x, y) > T \\ background & D_t(x, y) \le T \end{cases} \tag{2}$$

where $R_k(x,y)$ is the result of judgment and T is the threshold. Fig. 1 shows the processing of different frames method.

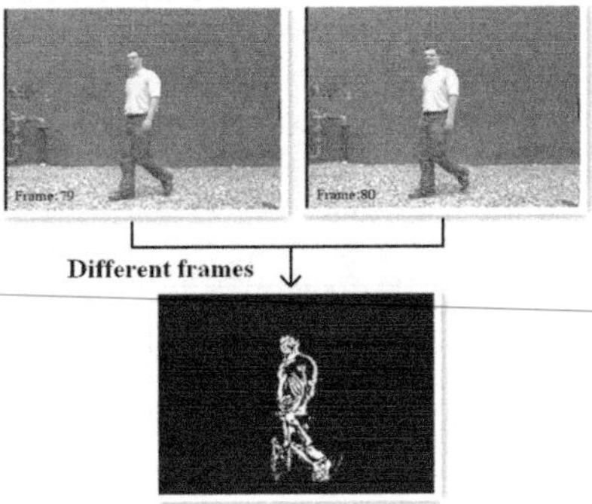

Fig. 1. Experiment of difference method based on frame

We differentiate Frame No.79 and Frame No.80, and the foreground image is judged by (2). The tailing area is also considered as the target, and the overlap is considered as the background. And the cavities are considered as the background mistakenly. So it is difficult to get the whole target perfectly.

Although the target image as in Fig.1 has low noise after segmenting by this method, it is imperfect. Because of the target moving, the tailing and overlap areas appear at the back of the target. And the tailing area is not only considered mistakenly as a part of the target, but also affects target recognition at the next stage. And when the two or more moving targets with physical contact, they may be processed into one target by the method based on difference. Traditionally, the disadvantage from this method disturbs the detecting, but the moving states of targets in the background sequences are detected easily, which is the useful information for us to solve the problem about updating background model.

3 Gaussian Mixture Model(GMM)

If lighting changed in the video sequences, it is necessary for the background model to update its parameters. If a moving target appears in the background, the corresponding pixels may be considered as foreground. These are important factors for us to build the model and study update method [5][6].

Gaussian Mixture model is a commonly used method. In this method, every pixel is modeled by a mixture of Gaussian distributions, and the GMM distribution of every pixel is measured in a time series. For a particular pixel(x_0,y_0), its history values is:

$$\{X_1,\ldots\ldots,X_t\}=\{I(x_0,y_0,i):1\le i\le t\} \tag{3}$$

where I is the image sequence. These history pixels are modeled by a mixture of K Gaussian distributions as:

$$P(X_t)=\sum_{i=1}^{K}\omega_{i,t}*\eta\left(X_t,\mu_{i,t},\Sigma_{i,t}\right) \tag{4}$$

$$\eta\left(X_t,\mu_{i,t},\Sigma_{i,t}\right)=\frac{1}{(2\pi)^{\frac{n}{2}}\left|\Sigma\right|^{\frac{1}{2}}}e^{-\frac{1}{2}(X_t-\mu_t)^T\Sigma^{-1}(X_t-\mu_t)} \tag{5}$$

Where K is the number of the Gaussian distributions, $\omega_{i,t}$ is the weight of the ith Gaussian in the mixture at time t, $\mu_{i,t}$ is the mean value of the ith Gaussian at time t, $\Sigma_{i,t}$ is the covariance matrix of the ith Gaussian at time t.η is a Gaussian function. K is usually determined from 3 to 5. The covariance matrix is defined as:

$$\Sigma_{k,t}=\sigma_k^2 I \tag{6}$$

This means that the RGB pixel values are independent and the variances are the same. Every new pixel value, X_t, is checked against the existing Gaussian distributions, until a match is found. The match is defined as a pixel value within 2.5 standard deviations of a distribution. The distribution which was unmatched will not update the$\mu_{K,t}$ and $\sigma_{K,t}$ parameters. And the weight $\omega_{K,t}$ is necessary. For the matched distributions, the $\mu_{K,t}$, $\sigma_{K,t}$ and$\omega_{K,t}$ are updated. If none of the Gaussian distributions match the new pixel value, a new distribution will be modeled with the current value as its mean value, an initially large variance, and low prior weight. And if the number of Gaussian distribution reaches the maximum K, the old distribution with the low prior weight should be replaced.

With a static background and unchanging lighting, the value of each pixel is relatively constant. However, if light changes in a static scene and a static object moves in the background image, the much changing values of these pixels will be considered as foreground. This may lead to the accumulated errors in the foreground estimation. So it is necessary that the parameters will be updated to match the new observation distribution as (7), (8) and (9).

$$\omega_{i,t}=(1-\rho)\omega_{i,t-1}+\rho M_{k,t} \tag{7}$$

$$\mu_t = (1-\rho)\mu_{t-1} + \rho X_t \tag{8}$$

$$\sigma_t^2 = (1-\rho)\sigma_{t-1}^2 + \rho(X_t - \mu_t)^T (X_t - \mu_t) \tag{9}$$

where the learning rate, ρ, is

$$\rho = \alpha\eta(X_t | \mu_{t-1}, \sigma_{t-1}) \tag{10}$$

One of the significant advantages of this method is that when the change of lighting and scene are allowed to become part of the background, it does not destroy the existing model of the background. The original background color remains in the GMM until it becomes the Kth Gaussian distribution. Therefore, if a static object in long time becomes part of the background and then it moves, the distribution describing the previous background still exists with the same $\mu$and σ, but a lower ω and will be quickly re-incorporated into the background.

4 Improved Method

Gauss model can draw the whole shape of the target which is not belonging to the background. But if this target stops in the background and it is necessary to consider this target as the background. This method can not update quickly and this may result in many mistake detections. So an approach is designed, which is based on both different-two-frames and gauss model to improve the method of updating.

If the target moves very slowly, it can not be easily detected by different-two-frames. As in Fig. 1, the target can not be wholly detected, some cavities appear, which are considered as the background. But there are still some useful target pixels left in the image, such as target contour. These pixels which leave in the images tell us that the moving target information appears in the view. So, no matter a target move fast or slow, the target pixels will be left in the different frame images, once the target stops and no target pixels are left in the different frame images. So we can take these characteristics to the gauss model method which can not be easily to detect target moving states.

Fig. 2 shows the flowchart of this improved method. There are two numbers which are the most important parameters. The num1 shows the information from the moving target and num2 shows the target which never appears in the background image. And then, the rate of updating background will be judged from comparing between the two numbers.

First, the two frames are differentiated to detect the moving target. And the number of the target pixels in the different image is counted named num1. If num1 is larger than the threshold, there is a moving target appearing in the background image.

Secondly, num2 is the quantity of the foreground pixels from Gaussian distributions. Num1 is from 0 to double num2. So the threshold may be determined by num2. By the experiments, 0.2*num2 is usually used as the threshold. If num1 is larger than the threshold, and smaller than double num2, the target is considered as a

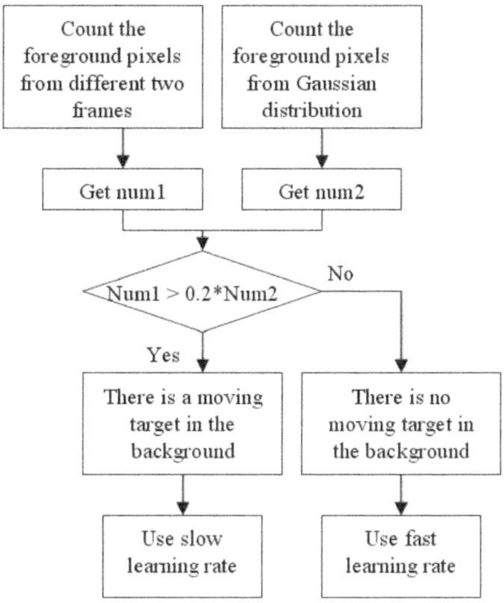

Fig. 2. Flow of this method we brought forward

moving target. At this time, the background model may be updated slowly. If num1 is nearly close to zero, even as equals to zero, this means that the target stops. The background model may be updated fast, which can make the target as the background model quickly. And then, we can get the new background model quickly. At same time, the disturbances from lighting changes and scene changes will be reduced perfectly.

Fig. 3 shows the results by the traditional method and the improved method. The first row is the original video sequences at frame 100,200,300,400. The target is moving fast at frame 100, and slowly at frame 200. It stops at frame 300 and 400. The second row shows the results by a mixture of Gaussian distributions with a low learning rate for the parameters updating. The shape of the target can be segmented from background perfectly. But when the target stops, there are still lots of error pixels left in the background images. And the third row is the GMM's experimental results with a higher learning rate. The shape of the target is eroded badly. The last row is the experiments with the improved method, which shows a better result of the adaptive learning rate.

The value of one pixel, at the point (150,150) from every image of the video sequences, is counted, which curves is drawn as Fig. 4. The black curve is the real-value. The mean value is marked in red and variance marked in blue. The target appears at the frame of No. 150, and stops at the frame of No 220. Fig. 4(a) shows that if the background model is updated fast, the mean value and variance of the background change quickly too.

The wrong judgment that the moving target is considered as background is made. If the background is updated slowly as before, the target which stops for a long time is considered as the moving target, shown as Fig. 4(b). A suitable convergence for the mean value and variance of the background distribution is shown as Fig. 4(c). At Frame 250, the target stops from the moving state, the convergence of the mean value and variance is robust because of the fast learning rate.

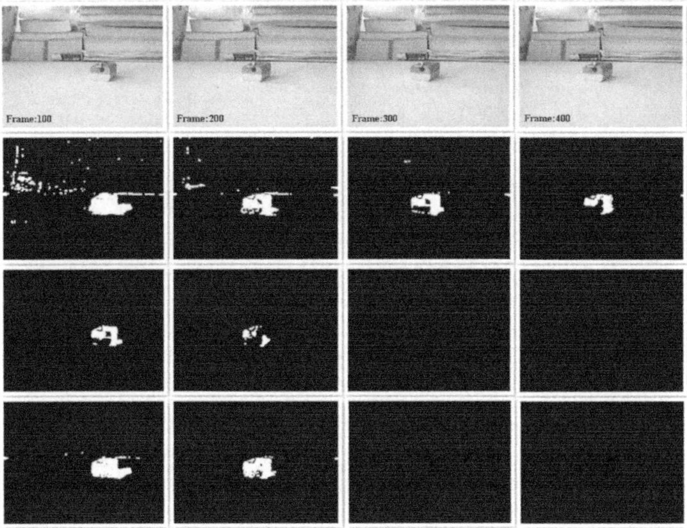

Fig. 3. Experiment of tranditional GMM and our improved method

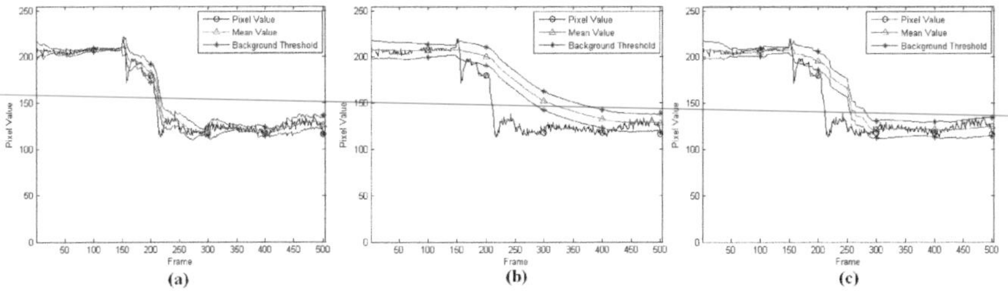

Fig. 4. Curve by the traditional method and our improved method

5 Post-processing

5.1 Shadow Removal in RGB Color Space

The existence of the target shadows by the motion or others may cause a series of problems in the process of moving target detection. As Fig. 5(b) shows, the shape of the moving target has been distorted when the shadow is adjacent to the target in the video images and the shadow may be recognized as a part of the target. This brings the significant error or mistake in the proceeding recognition or detection of the target. So it is necessary to remove this shadow.

A common method is to define the captured image regions into distinctly different classifications: foreground targets, background and shadows. The shadow pixels are detectable as well as target pixels since the both differ significantly from the background images. And the shadows have the same motion as the target which casts the shadows [7].

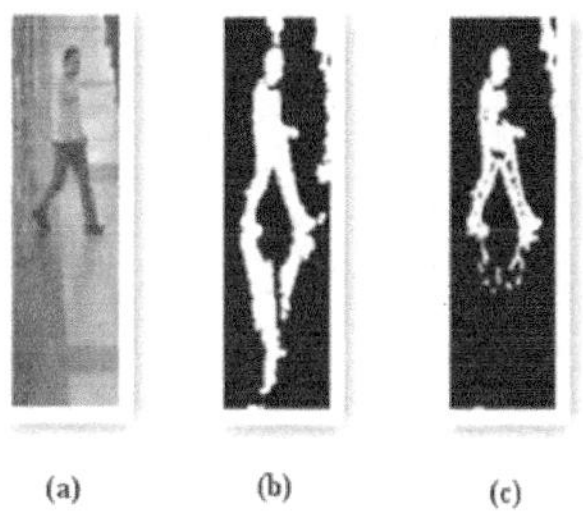

Fig. 5. Experiment of shadow removal

Many methods have been proposed in literatures on dealing with the problem of moving shadow detection and elimination recently. So the result of the experiment is shown without the specific discussion. Fig. 5(a) is the original image, in which the target is moving with a strong shadow because of the reflective from floor. Fig. 5(b) is the foreground after target detection. Fig. 5(c) shows the result from shadow removal in RGB color space. And a little part of the target is removed with the shadow removal.

5.2 Morphologic Operation and Components Connection

Although the foreground images are obtained by the improved method, there are some noise and cavities in the image. So it is necessary to solve these problems.

The better shape of the target can be segmented from the background by close operation of morphology, which can reduce the noise and fill the cavities. And then, the foreground pixels may be labeled and the components of one target will be connected. After these operations, the position, size and shape of one target are obtained and used to the next steps of target tracking and recognition. Fig. 6 shows the processing results.

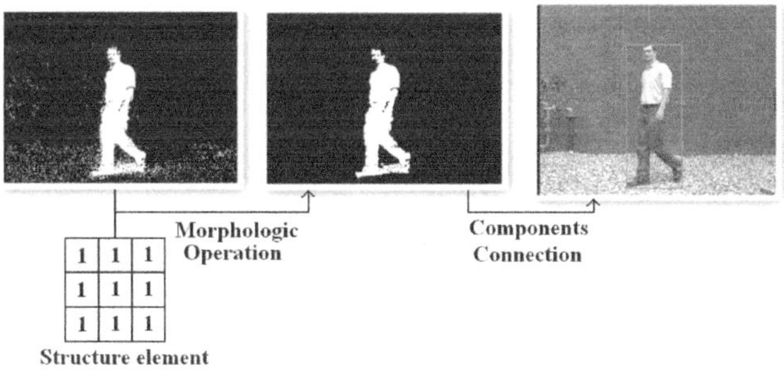

Fig. 6. Experiment of post-processing

6 Conclusions

In this paper, the different frames and Gaussian mixture model in target detection are discussed. The improved method combines the two traditional methods to improve the learning rate of the updating for GMM. The shape of target can be segmented from background easily and the new background model can be updated quickly when no moving target is left in the image sequences. The improved method has combined the advantages of different frames and GMM.

References

1. Li, S.: Research of hyperspectral target detection algorithms based on variance minimum. Guangxue Xuebao/Acta Optica Sinica 30, 2116–2122 (2010)
2. Lipton, A., Fujiyoshi, H., Patil, R.: Moving target classification and tracking from real-time video. In: Proceedings Fourth IEEE Workshop on Applications of Computer Vision, pp. 8–14 (1998)
3. Wren, C.R., Azarbayejani, A., Darrell, T., Pentland, A.: Pfinder: Real-Time Tracking of the Human Body. IEEE Transactions on Pattern Analysis and Machine Intelligence 19, 780–785 (1997)
4. Friedman, N., Russell, S.: Image segmentation in video sequences: A probabilistic approach. In: Proc. of the Thirteenth Conference on Uncertainty in ArtzJcial Intelligence (August 1997)
5. Bin, Z., Liu, Y.: Robust Moving Object Detection and Shadow Removing Based on Improved Gaussian Model and Gradient Information. In: 2010 International Conference on Multimedia Technology, ICMT (2010)
6. Stauffer, C., Grimson, W.E.L.: Adaptive background mixture models for real-time tracking. In: Proceedings of 1999 IEEE Computer Society Conference on Computer Vision and Pattern Recognition (Cat. No PR00149), vol. 2, pp. 246–252 (1999)
7. Dong, X., Wang, K., Jia, G.: Moving Object and Shadow Detection Based on. In: Proceedings of the 2009 2nd International Congress on Image and Signal Processing, CISP (2009)

Reliability Prediction for Components of Robotic Systems

Zdenek Vintr[1] and Michal Vintr[2]

[1] Faculty of Military Technology, University of Defence,
Kounicova 65, 662 10 Brno, Czech Republic
zdenek.vintr@unob.cz
[2] Faculty of Mechanical Engineering, Brno University of Technology,
Technicka 2896/2, 61669 Brno, Czech Republic
vintr@fme.vutbr.cz

Abstract. The article deals with the problem of reliability assessment of robotic systems and it is specifically focused on the analysis of the current situation in the field of component reliability prediction. The techniques of the component reliability prediction using the most credible methods (MIL-HDBK-217F, PRISM) and reliability databases (EPRD-97, NPRD-95) are presented in the article. Also the latest reliability prediction methods FIDES, 217Plus and reliability database SPIDR are presented.

Keywords: Reliability prediction, Reliability database, Robotic system.

1 Introduction

Nowadays the requirements on reliability of the robotic systems are constantly rising. Suppliers, who are able to "manage" reliability of developed and manufactured robotic systems, have significant competitive advantages. The basis of the mentioned "managing" is reliability assessment in the initial stages of product life.

The reliability of the whole robotic system can be assessed using the well-known methods (e.g. Reliability block diagrams, Fault tree analysis). The application of these methods requires to assess (or more precisely predict) the reliability of individual items. Reliability prediction can be carried out through various techniques based on the experience with similar items, expert's estimates, etc. However the most credible approach to prediction of item reliability is utilizing of internationally accepted reliability databases and reliability prediction methods.

Regarding specific types of elements (e.g. a screw with a hexagon head) the reliability databases contain numeric values of reliability measures. As for the individual groups of elements (e.g. resistors) the prediction methodologies contain calculation procedures which enable us to take into account a specific real situation by selecting different coefficients, and calculate the reliability measure value.

The article is focused on characterization of the most common tools in the field of item reliability prediction. The tools described in the article are usually used for prediction of the failure rate (λ) and Mean Time Between Failures ($MTBF$).

G. Lee (Ed.): Advances in Automation and Robotics, Vol.2, LNEE 123, pp. 463–470.
Springerlink.com

2 EPRD-97 and NPRD-95 Databases

The databases *EPRD-97 – Electronic Parts Reliability Data* [1] and *NPRD-95 – Nonelectronic Parts Reliability Data* [2] were developed by the System Reliability Centre (SRC). The SRC was called Reliability Analysis Centre (RAC) until June 2005 and it was the U.S. Department of Defense chartered Centre of Excellence.

The databases complement one another and do not contain duplicated data. The databases enable reliability prediction of most types of components used in robotic systems. The EPRD-97 database contains failure rate data on electronic components, namely capacitors, diodes, integrated circuits, optoelectronic devices, resistors, thyristors, transformers and transistors. The NPRD-95 database contains failure rate data on a wide variety of electrical, electromechanical and mechanical components.

Both databases contain data obtained by long-term monitoring of the components in the field. The collecting of the data was last from the early 1970's through 1994 (for NPRD-95) and through 1996 (for EPRD-97).

The purposes of the both databases are to provide failure rate data on commercial quality components, provide failure rates on state-of-the-art components in cases where data or analyses are not feasible or required and complement MIL-HDBK-217F by providing data on component types not addressed by it.

Both databases are sold in paper and electronic form and the prediction according to them is supported by most software products focused on reliability prediction.

3 SPIDR™

The database *SPIDR™ – System and Part Integrated Data Resource* is the most recent product in the field of reliability databases. The SPIDR™ was released at the beginning of 2006 by the System Reliability Centre (SRC).

The SPIDR™ is a complex product that replaces the NPRD-95 and EPRD-97 databases. It contains more than a double amount of data contained in the previous two databases. To be specific, it contains data on more than 6000 electronic, electric, electromechanical and mechanical component types. The database is based on nearly 40 years of experience and on the data collection completed by the RAC and the SRC.

The database is sold as the software product SPIDR™ produced by the SRC.

4 MIL-HDBK-217F

The *MIL-HDBK-217F – Military Handbook: Reliability Prediction of Electronic Equipment* [3] was developed by the U.S. Department of Defense in 1961 and it was revised several times. In 1995 the Department of Defense stopped supporting it.

This standard was primarily developed for reliability prediction of military electronic components. Nowadays the usage of the standard is common in many non-military areas and it is the most used reliability prediction method of electronic components.

The values included in the standard are based on statistical analysis of actual field failures and are used to calculate failure rates. The standard contains prediction about generic types of electronic components, namely microcircuits, semiconductors, tubes, lasers, resistors, capacitors, inductive devices, rotating devices, relays, switches, connectors, interconnection assemblies, meters, quartz crystals, lamps, electronics filters and fuses.

The standard contains two different prediction methods, a "parts count" method and a "parts stress" method.

The parts stress prediction method requires a greater amount of detailed information and is applicable during later design phase when stresses and other environmental and quality factors are known for each component.

The basic procedure in calculating the failure rate is by multiplying a base failure rate by operational and environmental stress factors. An example of a semiconductors component part stress model is as follows:

$$\lambda_p = \lambda_b \pi_T \pi_A \pi_R \pi_S \pi_C \pi_Q \pi_E \,. \tag{1}$$

where λ_p = predicted failure rate, λ_b = base failure rate, π_T = temperature factor, π_A = application factor, π_R = power rating factor, π_S = power stress factor, π_C = contact construction factor, π_Q = quality factor and π_E = environment factor.

The parts count prediction method is applicable in early stages of a design and development when little information about the design is known. The parts count method is a relatively simple prediction method using default stress values. The information needed to apply the method is a generic type of the component, the component quality level and the equipment environment.

The model for equipment failure rate with parts count method is as follows:

$$\lambda_{EQUIP} = \sum_{i=1}^{i=n} N_i \left(\lambda_g \pi_Q\right)_i \,. \tag{2}$$

where λ_{EQUIP} = total equipment failure rate, λ_g = generic failure rate for the i-th generic part, π_Q = quality factor for the i-th generic part, N_i = quantity of the i-th generic part and n = number of different generic part categories in the equipment.

The equation can be applied if the entire equipment is used in one environment. If the equipment consists of several units operating in different environments, the equation should be applied to the individual units separately. Specific values of a generic failure rate and a quality factor are included in the standard.

In general, the parts count method will usually result in more conservative estimation of a failure rate than parts stress method.

The prediction according to the MIL-HDBK-217F is supported by most software products focused on reliability prediction. The standard in the electronic form is freely available on the Internet.

5 PRISM©

The *PRISM© – Reliability Prediction and Database for Electronic and Non electronic Parts* is the reliability prediction method developed by the System Reliability Centre

(SRC). The method was released at the beginning of 1999. The main aim of the method was to replace the obsolete MIL-HDBK-217F.

The prediction method consists of two parts. First, the base failure rate of each component is predicted. Then these base failure rates are modified with system-level process assessment factors. The PRISM method failure rate model for a system is as follows:

$$\lambda_P = \lambda_{IA}\begin{pmatrix} \Pi_P \Pi_{IM} \Pi_E + \Pi_D \Pi_G + \Pi_M \Pi_{IM} \Pi_E \Pi_G + \\ + \Pi_S \Pi_G + \Pi_I + \Pi_N + \Pi_W \end{pmatrix} + \lambda_{SW} \,. \tag{3}$$

where λ_P = predicted failure rate of the system, λ_{IA} = initial assessment of the failure rate, Π_P = parts process multiplier, Π_{IM} = infant mortality factor, Π_E = environmental factor, Π_D = design process multiplier, Π_G = reliability growth factor, Π_M = manufacturing process multiplier, Π_S = system management process multiplier, Π_I = induced process multiplier, Π_N = no-defect process multiplier, Π_W = wear out process multiplier and λ_{SW} = software failure rate prediction.

The initial assessment of the failure rate λ_{IA} is the failure rate value which is obtained by using a combination of the RACRates model (it is a constituent of the PRISM method), the failure rate data contained in the reliability databases or the failure data defined by a user. The RACRates is a component reliability prediction model that uses a separate failure rate for each generic class of failure mechanisms for a component. Each of these failure rate terms is then accelerated by appropriate stress or component characteristic. This model form is as follows:

$$\lambda_{IA} = \lambda_o \pi_o + \lambda_e \pi_e + \lambda_c \pi_c + \lambda_i + \lambda_{sj} \pi_{sj} \,. \tag{4}$$

where λ_{IA} = predicted failure rate, λ_o = failure rate from operational stresses, π_o = product of failure rate multipliers for operational stresses, λ_e = failure rate from environmental stresses, π_e = product of failure rate multipliers for environmental stresses, λ_c = failure rate from power or temperature cycling stresses, π_c = product of failure rate multipliers for cycling stresses, λ_i = failure rate from induced stresses, including electrical overstress, λ_{sj} = failure rate from solder joints and π_{sj} = product of failure rate multipliers for solder joint stresses.

The RACRates model is available for capacitors, resistors, diodes, transistors, thyristors, integrated circuits and software. Specific values of failure rates, factors and multipliers are assessed according to the information on environment, operation, etc.

The method is sold as the software product PRISM© produced by the SRC and the prediction according to this method is supported by some software products focused on reliability prediction.

6 217Plus™

The *217Plus™ – Reliability Prediction Models* [4] is the most recent product in the field of a reliability prediction method that was developed by the Reliability

Information Analysis Centre (RIAC) which was established in June 2005 by a transformation of the Reliability Analysis Centre (RAC). The method was released in July 2006. The main aim of the method is to replace the obsolete MIL-HDBK-217F and the rival product PRISM©. Since the same producers who developed PRISM© worked on the 217Plus™, both methods are very much alike.

The method admits of reliability prediction of twelve types of electronic items, namely capacitors, connectors, diodes, inductors, integrated circuits, optoelectronic devices, relays, resistors, switches, thyristors, transistors and transformers.

The reliability prediction according to the *217Plus™* is carried out in two levels. First, the base failure rate of each component is predicted. Then these base failure rates are modified with various system-level factors. The mathematical models are similar to PRISM© models.

The method is sold in the paper and electronic form and as the software product 217Plus™ produced by the RIAC, and the prediction according to this method is supported by most software products focused on reliability prediction.

7 FIDES

The method FIDES is included in the *FIDES Guide 2009 – Reliability Methodology for Electronic Systems* [5]. This guide was developed by French defense and commercial aeronautical companies and it was published under the supervision of the French Ministry of Defense in 2009. It is a second edition of the guide, while the first edition appeared as the standard DGA-DM/STTC/CO/477-A [6] in 2004.

The method was developed using practical failure data from the aeronautical and military area and from manufacturers. The method takes into account the failures resulting from development and manufacturing, and overstress related to the application field, e.g. electrical, mechanical and thermal.

The method is focused on electric, electronic and electromechanical items, namely on integrated circuits, discrete semiconductors, capacitors, thermistors, resistors, potentiometers, inductors, transformers, relays, printed circuit boards, connectors and piezoelectric parts.

The general failure rate model is as follows:

$$\lambda = \lambda_{Physical} \Pi_{PM} \Pi_{Process} \,. \tag{5}$$

where λ = predicted item failure rate, $\lambda_{Physical}$ = physical contribution, Π_{PM} = quality and technical control of the item's manufacture (*PM* = Part Manufacturing) and $\Pi_{Process}$ = quality and technical control of the processes of development, manufacture and operation of the product containing the item.

The physical contribution can be expressed as follows:

$$\lambda_{Physical} = \left[\sum_{Physical\_Contributions} \left(\lambda_0 \cdot \Pi_{Acceleration} \right) \right] \cdot \Pi_{Induced} \,. \tag{6}$$

where λ_0 = base failure rate, $\Pi_{Acceleration}$ = acceleration factor indicating sensitivity to the conditions of use and $\Pi_{Induced}$ = contribution of induced factors (overstress) inherent to the field of application. The term in square brackets expresses contribution of the rated constraints.

The coefficient Π_{PM} specifying the item quality can be expressed by the following equation:

$$\Pi_{PM} = \exp[\delta_1(1 - Part\_Grade) - \alpha_1]. \tag{7}$$

where δ_1, α_1 = correlating factors that determine the extent of the effects of Π_{PM} on the item's reliability.

The coefficient Π_{PM} depends on the level of providing quality at a supplier, the level of providing element quality, the level of providing element reliability, and on the experience with an element supplier.

The factor $\Pi_{Process}$ specifying quality and technical management of the processes which influence product reliability during the life cycle can be expressed as follows:

$$\Pi_{Process} = \exp[\delta_2(1 - Process\_Grade)]. \tag{8}$$

where *Process_Grade* = grade indicating process control and δ_2 = correlation factor that determines the range of the $\Pi_{Process}$ factor.

Specific values of all factors, criteria and required inputs are assessed according to the tables, equations and recommendations included in the guide.

The prediction according to this method is supported by some software products focused on reliability prediction. The electronic form of the FIDES guide is freely available on the Internet.

8 RDF 2000 (IEC/TR 62380)

The RDF 2000 reliability prediction method is included in the technical report *IEC/TR 62380 – Reliability Data Handbook – A Universal Model for Reliability Prediction of Electronics Components, PCBs and Equipment* [7] published by IEC in August 2004. The technical report is based on the French telecommunication standard UTE C 80-810 [8] which was published in 2000. The methodology stated in both standards is known as RDF 2000.

This method provides models to calculate failure rate of electronic components, especially printed circuits boards (PCB) and their equipment. It makes reliability prediction easier to carry out, thanks to the introduction of influence factors.

The technical report IEC/TR 62380 is sold in the paper and electronic form and the prediction according to the RDF2000 method is supported by most software products focused on reliability prediction.

9 Telcordia SR-332

The standard *Telcordia SR-332 – Reliability Prediction Procedures for Electronic Equipment* was developed by the Telcordia Technologies Inc. (formerly Bell Communications Research Inc. or Bellcore). It originated from the Bellcore standard developed by the AT&T Bell Laboratories and sometimes it is called the Bellcore SR-332. The most recent version of this standard was released in January 2011.

The method is based on the MIL-HDBK-217F principles and reflects the Bellcore field experiences better. The prediction method is focused on equipment for the telecommunications industry and it is applicable to commercial electronic products.

The standard Telcordia SR-332 is sold in the electronic form and the prediction according to this standard is supported by most software products focused on reliability prediction.

10 GJB/z 299B

The standard *GJB/z 299B – Reliability Calculation Model for Electronic Equipment* was developed by the Chinese Army and it was translated into English in May 2001. Sometimes the standard is called China 299B.

The standard is based on the MIL-HDBK-217 and includes both a parts count and a parts stress prediction method for electronic components.
The prediction according to this standard is supported by most software products focused on reliability prediction.

11 NSWC-09

The *NSWC-09 – Handbook of Reliability Prediction Procedures for Mechanical Equipment* [9] was developed by the U.S. NAVY Naval Surface Warfare Centre. The most recent version of this standard was released in January 2009.

The standard contains models for various categories of mechanical components and enables to predict failure rates which are affected by temperature, stresses, flow rates and various other parameters. This methodology is the only widespread methodology developed for mechanical elements.

The prediction according to this standard is supported by most software products focused on reliability prediction.

12 Conclusion

The paper does not intend to offer a complete list of all databases and methodologies used for reliability prediction but it just introduces the most widespread and the most widely used present databases and methodologies applied in the area of robotic systems.

At present the most frequently used database applied to the prediction of non-electronic elements reliability is the database NPRD-95 which in course of time will be obviously replaced by more recent and much more extended database SPIDR™. These two databases are in the area of non-electronic elements without a serious rival.

Concerning the area of electronic elements reliability prediction, the situation is slightly more complicated. The database EPRD-97 is really widespread and it contains a great deal of real elements not included in other methodologies but it is not possible there to consider such extent of influencing factors. The reliability prediction based on MIL-HDBK-217F is at present the most widely used methodology when predicting electronic elements reliability but the standard has not been updated since 1995 and it got quite outdated. Therefore the methodologies PRISM©, FIDES and the most recent 217Plus™ provide different approaches which are intended to overcome the insufficiency of the standard MIL-HDBK-217F.

The decision on which method of reliability prediction to use is made after considering the reason why the prediction is performed, and above all, after taking into account the requirements of a customer for whom the prediction is made. Following practical experience of the authors, most of the component reliability prediction of robotic systems might be performed by using a combination of the databases NPRD-95, EPRD-97 and the standard MIL-HDBK-217F. In the future a combination of the database SPIDR™ and the methodology 217Plus™ seems to be the most convenient.

References

1. EPRD 1997: Electronic Parts Reliability Data. Reliability Analysis Center, Rome (1997)
2. NPRD 1995: Nonelectronic Parts Reliability Data. Reliability Analysis Center, Rome (1995)
3. MIL-HDBK-217F: Military Handbook: Reliability Prediction of Electronic Equipment. Department of Defense, Washington (1991)
4. Handbook of 217Plus™ Reliability Prediction Models. Reliability Information Analysis Center, Utica (2006)
5. FIDES Guide 2009 – Reliability Methodology for Electronic Systems. FIDES Group (2009)
6. DGA-DM/STTC/CO/477-A: FIDES Guide 2004 issue A – Reliability Methodology for Electronic Systems. FIDES Group (2004)
7. IEC/TR 62380: Reliability Data Handbook – A Universal Model for Reliability Prediction of Electronics Components, PCBs and Equipment. IEC, Geneva (2004)
8. UTE C 80-810: RDF 2000 – Reliability Data Handbook – A Universal Model for Reliability Prediction of Electronics Components, PCBs and Equipment. Union Technique de l'Electricité, Fontenay-aux-Roses (2000)
9. NSWC 2009: Handbook of Reliability Prediction Procedures for Mechanical Equipment. Naval Surface Warfare Center, West Bethesda (2009)

A Domain Upper Ontology for Smart Home

Moji Wei, Jianliang Xu, Xinjun Song, and Linlin Xu

Ocean University of China, Qingdao, China
weimoji@126.com, {xjl9898,songxinjun1226,gxl0216}@gmail.com

Abstract. Ontologies are essential for smart home, as they act as the semantics for describing the services provided by appliances. However there is no domain upper ontology to differentiate the concepts and explicate the relations among them. By analyzing the context of home service, we abstract six key concepts in the field and analyze the relations among them. As a result, we propose an upper ontology to illustrate the inter-relations and inter-dependency of concepts. Finally two scenarios are given to show the usage of our ontology.

Keywords: Smart Home, Service, Upper Domain Ontology.

1 Introduction

Smart home is different from the traditional home on its ability to perform function by integrating appropriate appliances automatically. To facilitate integrating services for households intelligently, ontologies are essential, as they act as the format explicit specification[1] for describing the services provided by networked devices, sensors, and other appliances. Thus the ontology-based semantic approaches could integrate services according to the specification automatically[2].

The designed ontologies focus mainly on context of service[3, 4] or QoS related service properties[5, 6]. Many domain ontologies have been constructed from above facets.

However to the best of our knowledge no attempts have been made to construct a domain upper ontology for smart home. Namely literatures independently construct various ontologies for the classification of concepts of smart home field without defining the concepts themselves and the relations among them formally. The absence of upper ontology would lead to concepts misuse. The current state of the literatures of home domain ontologies lacks the understanding of the correlation and inter-dependency of concepts within the field. In turn, this obscurity is hindering the advancement of this research field. As a result, the introduction of an organization of the domain knowledge and their relations (i.e. ontology) is necessary to help the research community achieve a better understanding of this field.

In this paper, we present an upper ontology for smart home, which would explicate the key concepts abstracted from home context and present the relations among the concepts formally. To explicate the definitions of concepts in smart home field, firstly we analyze the context of home service in order to abstract service-related concepts from the field. And then we construct home service ontology as domain upper ontology to formalize the relations among the concepts.

G. Lee (Ed.): Advances in Automation and Robotics, Vol.2, LNEE 123, pp. 471–477.
Springerlink.com

The rest of this paper is organized as follows. Section 2 analyzes the context of home service by differentiating the context into four worlds. Then in section 3 we abstract six key concepts of the field by analyzing the context of home service, and then constructs home service ontology to explicate the relations among the concepts. Section 4 takes two scenarios as examples to show the process from need motivation to service invocation with the upper ontology. Finally we recapitulate our work in section 5.

2 Context of Home Service

Home service is an algorithm that is implemented on the computer (here the computer is a broad sense computer including PC, SCM, etc.). In order to achieve specified function that people want, the algorithm would control home appliances to alter the states of context. Accordingly to extract service-related concepts from the field of smart home, we should analyze the context where home service runs first.

Rosenman and Gero[13,14] extend function-behavior- structure framework for improving intelligence of design system in mechanical manufacture, they analyze the relations among techno-physical environment, nature environment and socio-culture environment which have close relation with artifacts. For concept explanation they distinguish the context into three parts: external world, interpreted world and expected world. Home service is an algorithm and it is also an artifact, however unlike other artifacts home service is not substantial. It means that the realization of service needs device as its carrier. When analyzing home service environment the carrier environment should also be involved, so their classification can not satisfy our needs partially. Consulting predecessor research we introduce four different kinds of environments that interact with one another. Figure 1 depicts the relations among the four environments.

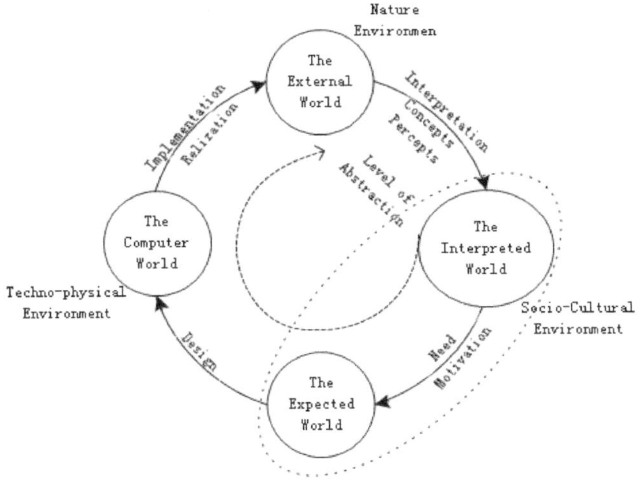

Fig. 1. Four Worlds of Home Service Environment

The external world pictures the current state of the physical world.

The interpreted world is the one built up inside the human beings in terms of sensory experiences, percepts and concepts. That is, the internal representation of the external world that human interacts with. It triggers lots of unsatisfied needs.

The expected world is the one being produced by using devices so as to satisfy the needs of human beings. The functions of devices can be imagined according to current goals of human and interpretations of the current state of the world.

The computer world is the one consisting of devices which are designed to satisfy human's needs. It is responsible for transforming the expected world into the external world.

The four worlds are recursively linked together by four classes of processes. The interpretation process transforms variables sensed in the external world into interpretations of sensory experiences, percepts and concepts which in turn become part of the interpreted world. This process is done by the interaction of sensation, perception and conception processes. Then the interpreted world is transformed into the expected world motivated by needs which are generated by human experiences. The needs are expressed as goals in the expected world. The devices in the computer world are designed to achieve the goals proposed in the expected world. The design process transforms the goals which are proposed in the expected world into the functions that could be provided by devices, and then the devices in the computer world would be designed according to the functions. Finally the process of implementation is an effect which brings about a change in the external world according to the function provided by the computer world. The four worlds from the interpreted world to the external world materialize conceptions of human beings gradually, and the level of abstraction is decreased as well. The home service which is a kind of artifact just sites in the computer world and would satisfy users' needs by controlling the device to realize specified function.

3 Home Service Ontology

By distinguishing home context into four worlds, we could abstract six service-related concepts including "Need", "Context", "Device", "Function", "Content" and "Service". The following would explicate these concepts respectively.

Home service would affect the external world by device operation. The effect of service is recognized as function which implies the goal of service. We say that the service satisfies need, when the effect plays a positive role for human.

In addition, the service as algorithm may operate some data which is defined as content in the paper. The service would be recognized as different applications when cooperates with different contents. Take the questionnaire service as example, when the content consists of simple choice questions of some course, the service becomes an examination; while the service becomes a psychological test if the content consists of choices to certain circumstances one meets in daily life.

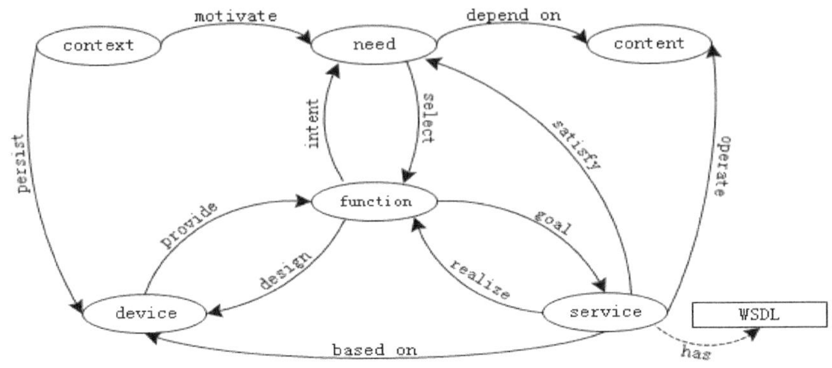

Fig. 2. Home Service Ontology

To clarify the concept relations among need, function, service, content etc, the paper constructs home service ontology as an upper ontology for smart home. The ontology is expressed in the form of RDF diagram in figure 2.

As shown in figure 2, human's need is motivated by context. Psychology researches [15, 16] that with the same need the goals may be such differences for kinds of people, and on the other side to the same goal the needs from different people are inconsistent as well. Such as the goal of purchasing house, someone purchases house for convenience need and another one may be for self-respect need. If we examine carefully the average needs that we have in daily life, we find that they have at least one important characteristic, i.e., that they are usually means to a goal rather than goals in themselves. Usually when a conscious need is analyzed we find that we can go behind it, so to speak, to other, more fundamental intentions of the individual. It is characteristic of this deeper analysis that it will always lead ultimately to certain intentions behind which we cannot go; that is, to certain intention-satisfactions that seem to be goals in themselves and seem not to need any further justification or demonstration. The fundamental or ultimate intentions of all human beings do not differ nearly as much as do their conscious everyday needs [17]. The intentions are always implied in the needs. Therefore to satisfy a need, user usually has to select a proper function which reflects the intention implied in the need. Moreover the function presents the goal of the service, and meanwhile it also guides the design of device, hence the device provides the function exactly. The realization of given function relies on service implementation, and the precondition is that the device which hosts the service should be persisted in the context. Additionally the service would be implemented as different applications when cooperates with different contents, so we can conclude that the need depends on specified content as well.

The home service ontology explicates our viewpoint and formally defines service-related concepts from the unified viewpoint. In the ontology we differentiate function from need, which are usually confused with each other in other literatures. For instance other literatures recognize "illumination" and "guard" as the functions of "turn-on light" service. However from our viewpoint described above, the intention of "turn-on light" service is to adjust the brightness state of the external world, accordingly the service affects human vision, and then the function of the service is vision-related. The service which realizes vision-related function would actually alter the state of external world, and in different contexts the altered state may lead to different effects which eventually satisfy various needs such as "illumination" or "guard". In our opinion, the function of a service is unique and fixed, while the needs which a service could satisfy are diverse with different contexts. Thus, annotating service with function could explicate the semantic of service.

The home service ontology defines six key concepts in the field, and formalizes relations among them. By formally clarifies the whole process from need motivation to service invocation, annotating the service with the ontology could promote the automation and intelligence of smart home. However, the concepts described above are highly abstracted, they can hardly be used to annotate services for exactly service searching. Thus the concepts in home service ontology could be combined with concrete ontologies which are explicated in other literatures.

4 Scenario

This section displays two scenarios from two classic applications of home services. One is for device operation application and the other is for user entertainment application. The scenarios show the entire process from need generation to service invocation to demonstrate the validation of the home service ontology.

4.1 Scenario for Device Operation

This section presents a scenario that displays the process of service invocation for device operation. The essential service that smart home should offer is secure service. The service should inform the insecurity when dangerous affairs occur. The following displays how to accomplish the application with home service ontology and the scenario shown in figure 3 simulates the solution.

It is obvious that insecure affairs would motivate "alarm notice" need. The model selects "visual" function or "hear" function for the intention. Then according to the

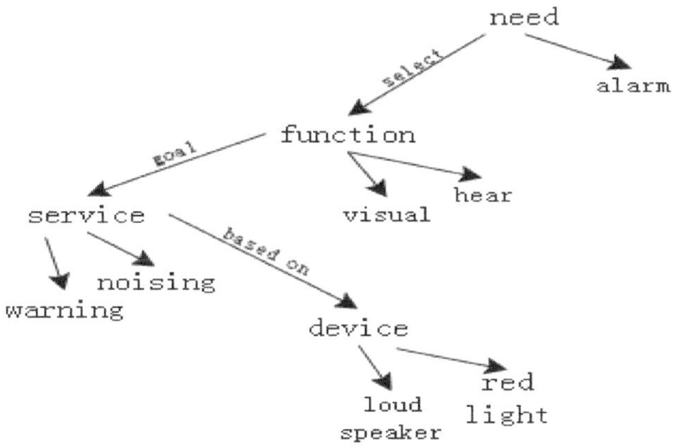

Fig. 3. Scenario for Device Operation

annotation of service, the "warning" service for "visual" function and "noising" service for "hear" function could be retrieved. Afterward the devices "loudspeaker" and "red light" which implements two services respectively would be checked. Finally the need could be satisfied, if the context holds at least one device.

4.2 Scenario for User Entertainment

This section presents a scenario that displays the process of service invocation for user entertainment. In smart home besides the essential guard services, the entertainment services should also be provided. After busy working, people may listen to the music for relaxation. Figure 4 displays the scenario to simulate the solution.

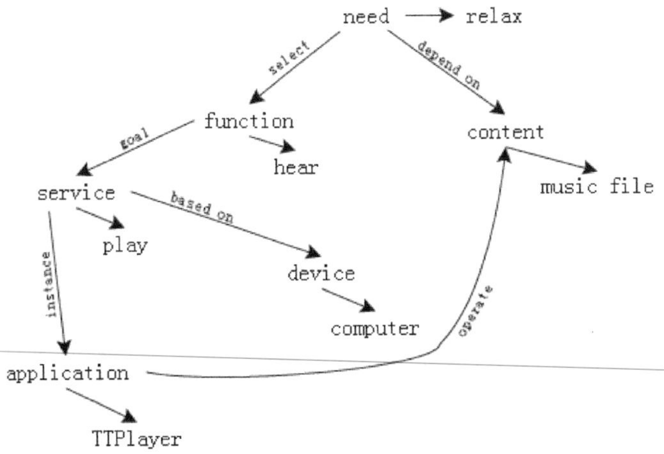

Fig. 4. Scenario for User Entertainment

It is obvious that the need in the scenario is relaxation. The intention could be achieved by "hear music". Thereby the selected function is "hear" and the depended content is "music file". The "play" service for the "hear" function may be retrieved according to the annotation of function concept. The model would locate the device "computer" which hosts the "play" service, and then check whether the required content is possessed in the device. If the device stores the content "music file", the model could use the application "TTPlayer" which is instance of service "play" to operate the content "music file".

5 Conclusion

The paper firstly analyzes the context of home service, and then presents home service ontology which is domain upper ontology for home field to explicate the relations among concepts. Finally two scenarios are used to show the process of ontology-based service retrieval.

Acknowledgment. This research is supported by National Natural Science Foundation of China (No.40806040) and Natural Science Foundation of Shandong Province (No.ZR2010FM002).

References

1. Studer, R., Benjamins, V.R., Fensel, D.: Knowledge Engineering: Principles and Methods. Data and Knowledge Engineering 25(1-2), 161–197 (1998)
2. Roussaki, I., Papaioannou, I., Tsesmetzis, D., Kantorovitch, J., Kalaoja, J., Poortinga, R.: Ontology Based Service Modelling for Composability in Smart Home Environments. Constructing Ambient Intelligence 11, 411–420 (2008)
3. Kim, E., Choi, J.: An Ontology-Based Context Model in a Smart Home. In: Workshop on Ubiquitous Web Systems and Intelligence (UWSI 2006), pp. 11–20 (2006)
4. Gu, T., et al.: An Ontology-based Context Model in Intelligent Environments. In: Communication Networks and Distributed Systems Modeling and Simulation Conference, pp. 270–275 (2004)
5. Maximilien, W.M., Singh, M.P.: A framework and ontology for dynamic web services selection. IEEE Internet Computing 8(5), 84–93 (2004)
6. Dobson, G., Lock, R., Sommerville, I.: Qosont: A qos ontology for service-centric systems. In: EUROMICRO-SEAA, pp. 80–87. IEEE Computer Society (2005)

Development of a Six-Torch Automatic Welding System for Aluminium Vacuum Chambers

C.L. Chen, C.C. Chang, C.K. Chan, S.N. Hsu, G.Y. Hsiung, and J.R. Chen

National Synchrotron Radiation Research Center, Hsinchu 30076, Taiwan

Abstract. A novel automatic gas-tungsten arc-welding (GTAW) system has been developed at NSRRC for welding the aluminium (A6061-T651) vacuum chambers to be installed in an Ultra-high vacuum (UHV) system. This welding system has a XY stage that is built and configured to provide high-performance positioning along multiple welding axes. The automatic system comprises six welding torches to implement simultaneously two longitudinal side welds of an aluminium chamber, and is innovative in using computer-based software to control the welding movements and the welding energy of the six-torch output. In traditional single-torch welding, the key success factors focus on eliminating as much as possible the distortions of structural assemblies. The six-torch welding and the clamp-free approach together address the issue of reducing distortion and minimizing residual stresses with a novel one-step welding process. This paper introduces the mechanical configuration and the motion-control program. Real-time measurements for thermal and deformation monitoring to analyze the temperature distribution along the chamber surface are discussed. The welded chambers are tested with a leakage-free hermeticity test <5×10-10 mbar·L/s; the maximum deformation is controlled under 0.3 mm.

Keywords: automatic welding systems, LabVIEW, TIG,GTAW.

1 Introduction

The 3-GeV Taiwan Photon Source (TPS) is a new facility at National Synchrotron Radiation Research Center (NSRRC) that will incorporate a storage ring of circumference 518.4 m. This synchrotron accelerator is designed to provide a beam of light with emittance <2 nmrad at a beam current 400 mA. The vacuum system for the storage ring provides a pressure less than 1 nTorr for the electron beam with a long beam lifetime and a small disturbance from trapped ions, as demonstrated (Hsiung et al., 2008). To achieve an ultra-high-vacuum system, the vacuum chambers are the most important components. Both Ishimaru (1984) and Chen et al. (1985) have shown that Aluminium alloy was chosen as the material for the storage-ring chamber because it machines easily and has a large thermal conductivity, a small rate of outgassing, and a small residual radioactivity. Combined with an appropriate manufacturing process, such as with a precise CNC machine, lubrication with pure alcohol and cleaning with ozonated water, the aluminium chambers have an oil-free interior surface finished for an ultra-high- vacuum environment before aluminium welding. Chan et al. (2008) indicates that ozonate water can be used to effectively clean aluminium chambers. The chamber consists of two symmetric plants that

G. Lee (Ed.): Advances in Automation and Robotics, Vol.2, LNEE 123, pp. 479–488.

are combined with an AC tungsten-inert-gas (TIG) welding method to form the vacuum chamber. The vacuum chambers to be welded have dimensions a length about 4 m and width 600 mm.

With manual welding applied for assembly, it is difficult to control the shrinkage deformation or thermal cracks because of the large energy input from the TIG torch. It is important to have an automatic welding system that can control the welding input current and the speed of torch movement, and control uniformly the welding seam path over a great welding length. With a real-time monitoring system, both thermal variation and shrinkage deformation can be measured. An automatic welding system has been investigated at NSRRC to develop a welding process that is free of leakage and has a high throughput.

2 Motivation for an Automatic TIG System

Figure 1 shows the three types of vacuum chambers and their dimensions. There are hence approximately 2700 aluminium welds for these TPS vacuum chambers, which would involve a heavy work load if these 2700 weld seams were completed by hand. A new and automatic system, which holds great promise as a replacement for manual arc welding, is required. As is generally known, arc welding is a continuous process in which the torch and filler are carried along the seam path to be welded. In the traditional manual single-torch welding method, the key success factors focus on eliminating, as much as practicable, the distortions of the structural assemblies. The six-torch welding process and a clamp-free method address the issues of reducing distortion and minimizing residual stresses with a new one-step welding process. This system is designed to use six welding torches to implement simultaneously two non-parallel welding seams on opposite sides of an aluminium chamber. Figure 2 and Figure 3 show the welding system with a unique welding configuration. This auto-welding system allows the welding torch to be manipulated in almost the same fashion as by a human being. Because the heat dissipation from torch during welding is large, too long a welding seam might cause the welder's hand to become burned. For an automatic welding system, this issue is avoided, and the welding seams can be much longer than with manual welding. The longer welding seam generates fewer overlapping junctions of two welding seams. As leakage is most likely leakage at those overlapping junctions, an automatic welding system can also decrease the possibility of vacuum leakage. The following section discusses the hardware of this auto-welding system in detail.

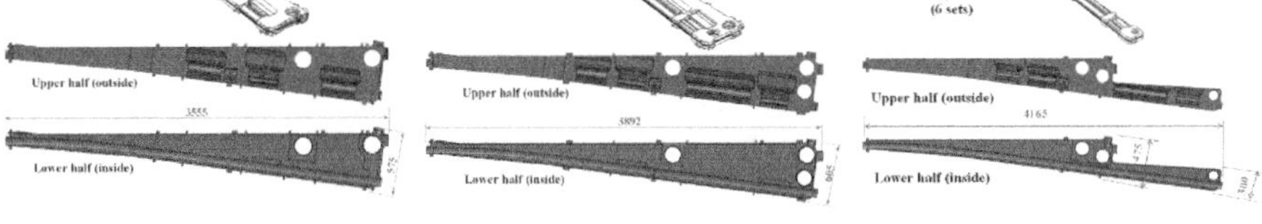

Fig. 1. Vacuum chamber of three types

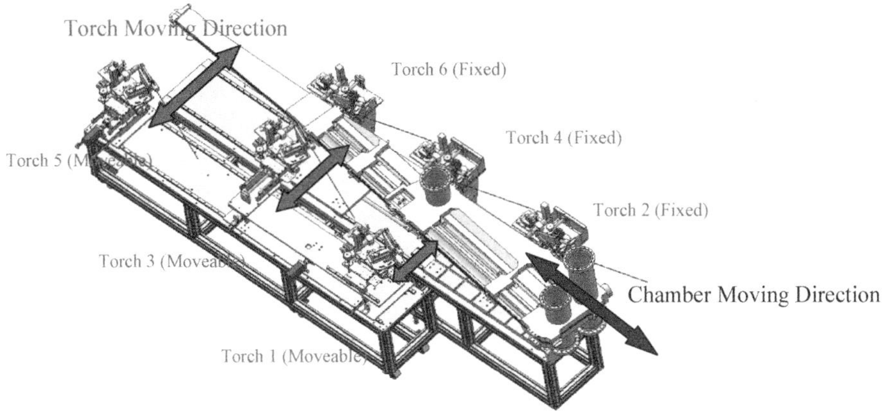

Fig. 2. Design of an auto-welding system with six TIG torches

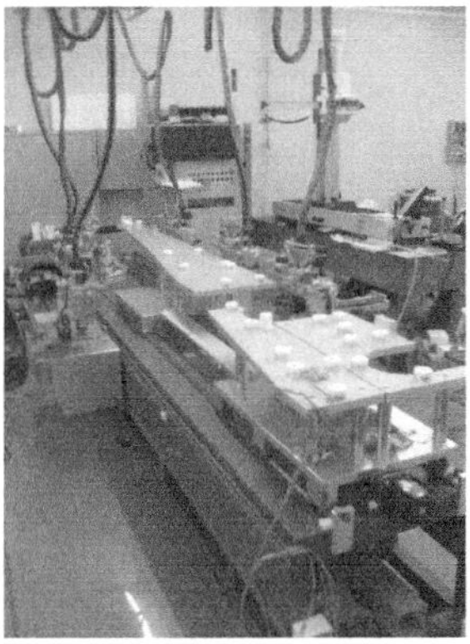

Fig. 3. Autucal auto-welding system with six TIG torches

3 Automatic Aluminium Welding System

In introducing this new automatic welding system, it is important to understand the material preparation before the TIG welding begins. To obtain ultra-high-vacuum chambers, the chamber surface must be clean and suffer no contamination from oil. Manufacturing procedures, from the selection of raw materials, mechanical machining, surface cleaning and drying, to TIG welding, must therefore proceed carefully.

3.1 Material Preparation

Aluminium alloy (A6061-T6) serves as the chamber material. Before welding aluminium, it is important to maintain the cleanliness of the surface of the aluminium chamber. Five factors that might influence the welding results are hydrocarbon solubility, aluminium oxides, thermal conductivity, thermal expansion and solidification shrinkage [5]. Hydrocarbon solubility and aluminium oxides must be taken into account to ensure the quality of the welds. To prevent hydrocarbon contamination and a thick layer of aluminium oxide, the chambers must be fabricated with oil-free computer- numerical-control (CNC) machining. All chambers must be cleaned properly after receipt from the machine shop. Ozonized water is used to clean the surface before welding, and the welding process must be completed within a definite period after cleaning. Moreover, Goeppner (1996) has recently shown that aluminium oxide can act as an insulator; this oxide layer is porous and tends to trap moisture. Such an oxide layer results in an erratic welding arc and causes poor welding quality. For this reason this layer must be scraped away manually before welding.

3.2 Welding Equipment

Gas-tungsten-arc welding (GTAW), also known as tungsten-inert-gas welding (TIG), was chosen for the aluminium welding. A commercial welding tool (OTC Daihen 500P) is used to weld all vacuum chambers. An AC-mode TIG welding process is selected in the automatic welding system. The head material of the welding electrode, of diameter 3.2 mm, is pure tungsten. Argon (15 L/min) serves as the shielding gas to protect the electrode and overflows inside the sample chamber. This process can provide welding seams of high quality. A4043 wire of diameter 3.2 mm is used as the filler material.

3.3 Auto-welding System

A system with X-Y stage motion is used to ensure the accuracy of the welding position. This system is sometimes called 'Cartesian' because the axes of motion are described with a X and Y coordinate system. This Cartesian design is claimed to produce the most accurate movements. According to figure 2, one axis is used to control the movement of the chamber in the X direction, and the other is to move the torch perpendicularly to the axis of chamber movement in the Y direction. A X-Y horizontal working plane is thus achieved. There are six welding torches with filler mechanisms; each side of the chamber has three torches. These six torches can be switched on to implement simultaneously two welds on each side of an aluminium chamber. Figure 4 illustrates the welding configuration of the torch and filler. The position between the torch and the filler in the auto-welding system is evaluated carefully, similarly to a human welding position. All motional axes are driven with servo motors. These motors are controlled with a LabVIEW motion program, which provides highly accurate control of the welding positioning.

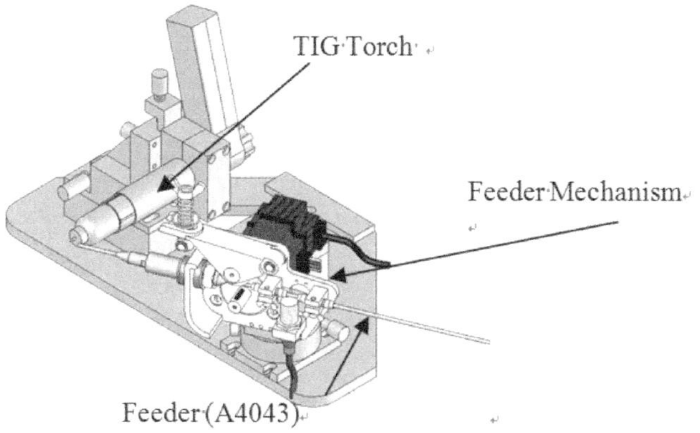

Fig. 4. Design of the welding configuration of the torch and the filler

4 Control of Motion with LabVIEW

Using a computer-based control program (LabVIEW) instead of a conventional programmable logic controller (PLC) to control welding procedures is an innovation. This program (LabVIEW) is a tool to provide a user interface, to determine inputs and outputs, and to control the motion and other related welding sequences. The entire program is separable into three parts:

1. graphical user interface (GUI),
2. digital input and output (DIO), and
3. motion and welding sequences.

These three parts are written into LabVIEW as three parallel 'While' loop processes. The data can thus be passed between parallel processes, which is advantageous for the synchronization of data transfer in the system. The system can be aborted when an irregular signal occurs during welding. These reasons are the most important for use of LabVIEW as the main controlling program. Figure 5 shows the programming flowchart for the auto-welding system.

First, the operator sets the welding parameters, such as torch current, travel speed, feeder motion modes, feeding speed, duration. Figure 6 shows the graphical user interface (GUI). These parameters can be saved as a welding recipe. Once these user parameters are saved into the system, the operator can then begin to teach the servo motors of the X- and Y- coordinators both the initial and final positions. After completing setting these points, the welding can begin. Welding commands are sent to the welder as a trigger signal to ignite or to extinguish the torch. According to the above description, these three parallel processes depend on each other.

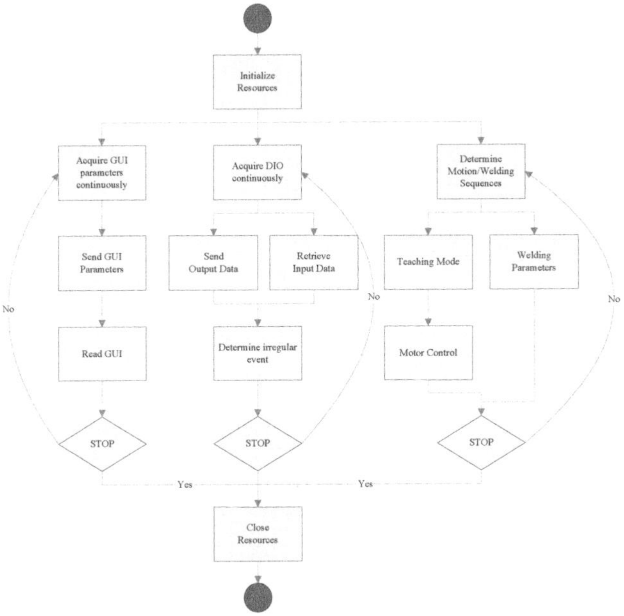

Fig. 5. Flow Chart for the Auto-welding System

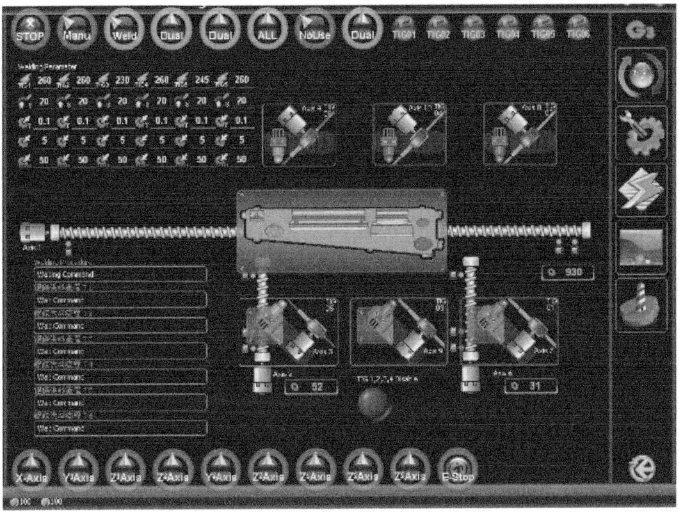

Fig. 6. Graphical user interface (GUI)

5 Real-Time Measurements for Thermal and Deformation Monitoring

A real-time system for thermal and deformation monitoring is established for the automatic welding system. To obtain the welding variations, the thermal and deformation changes of the vacuum chamber must be monitored. These thermal and deformation data are valuable for analysis of the quality of the welded seams, and are also useful to optimize the welding parameters. In the measuring system, K-type thermocouples are mounted near the welding seam to measure the temperature distribution of the chamber surface during and after the welding process. For deformation measurement, dial gauges measure vertical deformations while the welding is proceeding. The detailed setup is shown in Figure 7. Dial gauges are mounted on the top plates to measure the vertical deformation.

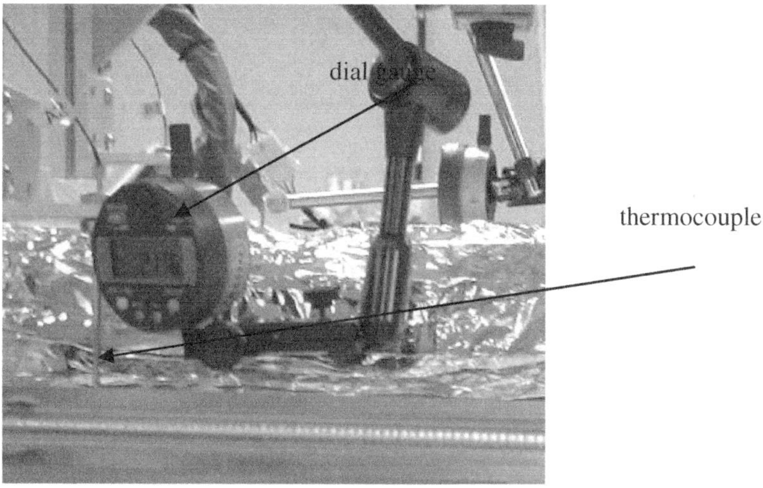

Fig. 7. Setup for real-time thermal and deformation measurements

6 Results and Discussion

A chamber of length 2 m is being tested. Figure 8 shows a sketch of the test chamber. This chamber follows the standard procedures for oil-free CNC machining, cleaning with ozonized water and auto-welding. All welding conditions are listed in Table 1. The chamber is placed on the welding stage and aligned carefully. Before the massive aluminium chamber is welded, it must be heated; heaters are therefore wrapped around the chamber. Dial gauges and thermocouples are then installed. After baking the chamber to 90°C, the welding process begins immediately. Figure 9 shows an actual welding seam on the chamber. A coordinate-measuring machine serves to measure the deformation of the test chamber after welding. Figure 8 also shows that the maximum deformation (0.112 mm) occurred at point B.

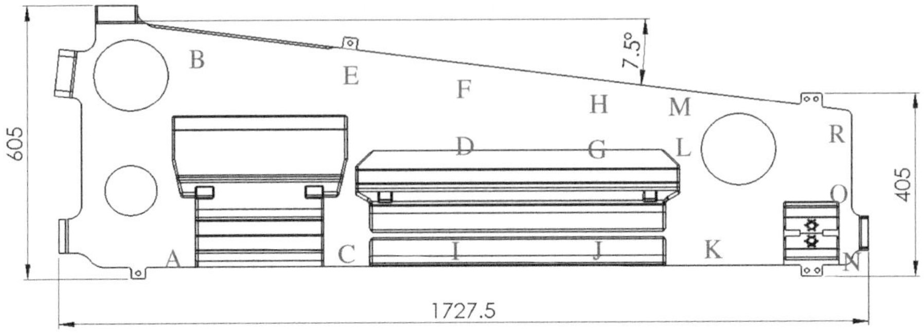

	A	B	C	D	E	F	G	H
△h	0.033	0.112	0.036	0.0005	0.0195	0.011	0.0105	0.0255

	I	J	K	L	M	N	O	R
△h	0.023	0.038	0.0345	0.001	0.040	0.0835	0.0595	0.026

Fig. 8. Deformation data, measured with a coordinate-measuring machine

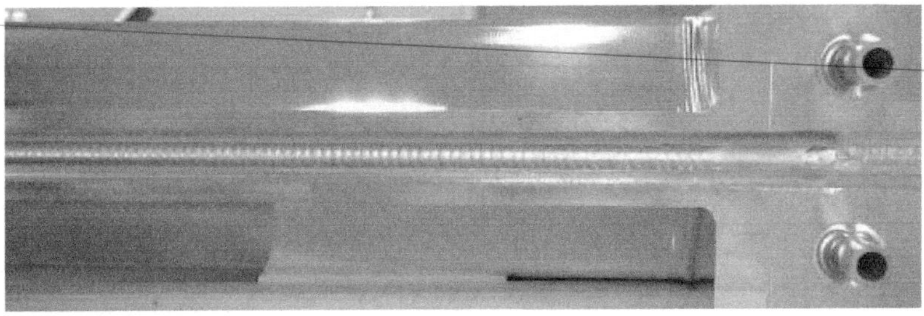

Fig. 9. Actual seam welded with an auto-welding system

Moreover, metallographic analysis of a welding seam has been made. According to photographs, with an optical microscope, of the cross section of the welded seam in Figure 10, several defects occur in a diameter range 18.4 – 107.8 μm. These defects are formed as the result of the contamination of the weld microstructure with hydrogen. When solidification occurs during or soon after the actual welding, hydrogen becomes trapped before escaping from the surface of the welding seam. Excluding hydrogen from a structure during welding is difficult. All possible contamination is eliminated before cleaning with ozonized water. These hydrogen-induced defects should be made as few as practicable. As long as these defects are not formed in a row in the welding seam, a source of leakage into the vacuum chamber might not develop.

For a pumping test, one sputtering-ion pump (150 L/s) and two lumped St-707 NEG pumps (500 L/s) are installed in the chamber. The system is baked at 150 °C for

12 h. Figure 11 show a pumping curve of the test chamber, corresponding to the baking temperature. From figure 11, an optimum pressure was achieved at 10 nPa after baking. The welded chambers are tested to be free of leakage (hermeticity test $<5\times10^{-10}$ mbar·L/s). According to both deformation data and the final pressure, these procedures are advantageous to establish ultra-high- vacuum chambers.

Table 1. Welding recipe for vacuum chamber

welding system	6-axis auto-welding system
initial current	280 A (for thicker portion) 240 A (for thinner portion)
time at start point	3 s
welding current	260 A (for thicker portion) 230 A (for thinner portion)
final current	200 A
time at end point	1.3 s
welded length	900 mm
temperature before welding	Baked to 90 °C
welding speed	180 mm/min
sample fixture type	horizontal free mounting
flow rate for Ar protection	10 L/min
humidity	50 %
room temperature	25 °C

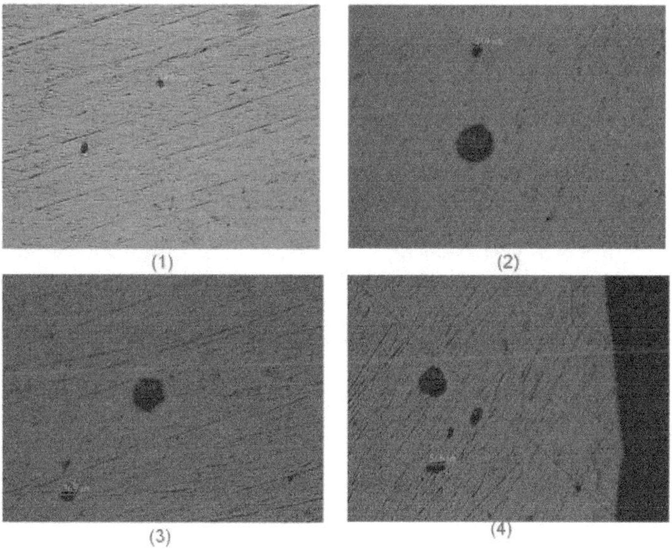

Fig. 10. Optical microstructure of welding seam

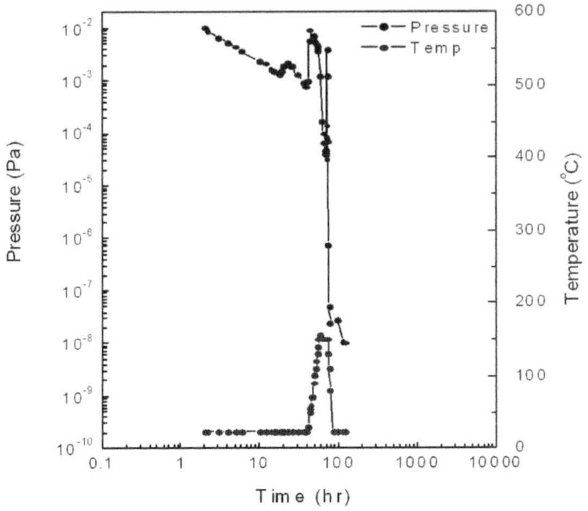

Fig. 11. Pumping curve of the test chamber

7 Conclusions

A six-torch auto-welding system and its computer-based control program are introduced. LabVIEW is used to provide a user interface, to determine inputs and outputs, and to control motion and other related welding sequences. A welded chamber, constructed with an auto-welding system, is verified to be free of leakage and the maximum deformation is controlled to be less than 0.15 mm. According to the pumping data and the prototype, the manufacturing process fulfills a promise of building a complete aluminium vacuum system with allowable deformation.

References

1. Chan, C.K., Hsiung, G.Y., Chang, C.C., Chen, R., Yang, C.Y., Chen, C.L., Hsueh, H.P., Hsu, S.N., Liu, I., Chen, J.R.: Cleaning of aluminium alloy chambers with ozonized water. Journal of Physics Conference Series 100 09025 (2008)
2. Chen, J.R., Narushima, K., Miyamoto, M., Ishimaru, H.: A large aluminum alloy molecular beam epitaxy chamber with an ultimate pressure of 7.5×10-12 Torr. J. Vac. Sci. Technol. A3 2200 (1985)
3. Goeppner, G.A.: TIG welding of aluminium alloys for the APS storage ring—A UHV application. APS LS Notes 254 (1996)
4. Hsiung, G.Y., Chan, C.K., Chang, C.C., Chen, Y.B., Hsueh, H.P., Chen, C.L., Yang, C.Y., Shu, S.N., Cheng, C.M., Kuan, C.K., Sheng, I.C., Chen, J.R.: Vacuum design for the 3 GeV TPS synchrotron light source. JOP: Conf. Series 100 092014 (2008)
5. Hajime, I.: All-aluminum-alloy ultrahigh vacuum system for a large-scale electron–positron collider. J. Vac. Sci. Technol. A2, 1170 (1984)

Application of Feedorward and Feedback Control Strategy in the Speech Acquisition and Production Model*

Shaobai Zhang and Liqin Gao

Computer College, Nanjing University of Posts&Telecommunications, Nanjing, 210003, China
adzsb@163.com

Abstract. The neural transmission delay of DIVA model was studied.Based on the detailed analysis of the speech transmission delay characteristics,the feedforward and feedback control strategy was proposed.It is proved that the delay can be controlled reasonably.Aiming at the network state disturbance caused by parameters changes during the neural network learning process, this paper does related emulation experiments. Experimental results show that this method is effective and gives good robustness.

Keywords: feedback control, feedforward control, perturbation, neural transmission delays.

1 Introduction

DIVA(Directions Into Velocities of Articulators) is a mathematical model of the processes behind speech acquisition and production, The model is an adaptive neural network that learns to control movements of a simulated vocal tract in order to produce words, syllables, or phonemes[1]. Since 1994 Guenther DIVA[2], there have been quite a few new version. Different versions of the DIVA model reflecs the relevance between the rough neural anatomy and brain areas[3].The model is relatively simple and is widely used to interpretate all kinds of the phenomenon,including motor equivalence, contextual variability, anticipatory and carryover coarticulation, velocity/distance relationships,speaking rate effects and speaking skill acquisition[4]. But at present, this model is not completely have nerve physiology function, despite of various aspects of improvement,but the sports system senses realistic neual processing delays problems don't getting good treatment[5].The present study shows that speech production is mainly nervous system network which is cortex to the cortex .In the cerebral cortex in processing mechanism on the basis of feedback control system is mainly using the error between action expectation and the actual sensory feedback.In this article we address the integration of feedback and feedforward control subsystems in the control of speech movements with realistic processing delays,and we provide model simulations of perturbation studies that probe the temporal response properties of feedback control mechanisms.

* This work is supported by the National Natural Science Foundation of China (No.61073115).

G. Lee (Ed.): Advances in Automation and Robotics, Vol.2, LNEE 123, pp. 489–494.
Springerlink.com

2 The DIVA Model Realistic Neural Processing Delays and The Processing Methods

2.1 Neural Processing Delays

Earlier versions of the DIVA model effectively assumed instantaneous transmission of neural signals. However the nervous system must cope with potentially destabilizing delays in the control of articulator movements.For example, sensory information from the articulators and cochlea are delayed by 10ms before they reach the primary sensory cortices. These transmission delays can be very problematic for a system that must control the rapid articulator movements underlying speech.

In addition, if the system suffers the influence of the external disturbance, such as the auditory feedback real-time distortion, so there had heard the mistakes voice, the system will activate error cells and try to correct disturbance. However, because of the neural transmission delay happened in the above mentioned muscle activation and the resulting activity, the system produce correct order's delay time is about 75-150 ms after suddenly disturbance.Therefore, in order to reduce the the impact of propagation delay , must control the rate of the vocal organs movement .

2.2 The Processing Methods

Because of the delay time may cause the deterioration of effect , and even make the system becomes unstable, so the process control system has been paying attention to the study of the delay, and put forward many control strategy.Such as, Time delay distinguish method based on linear system , but computational complexity is difficult to applicate in the practice; the chaos method based on time delay feedback control , the method is the difficulty to ensure parameters, such as delay time etc; Based on chaos optimization method to determine the delay feedback control delay time, and get a delay feedback control method of the parameters, then we can solve the problem that parameters is difficult to select in the delay feedback control ; Based on the combination of neural network identification and the delay of the compensation based on model system control strategy, this method can solve the model error problem,and have been applied in the delay system control that parameters are uncertainty.

Visible, the biggest advantage of neural network compensation is that it can overcome the change of time delay, this also is very important requirement in the control. Due to a purely based on feedback control system exists the transmission delay problems, and produce unstable speech rate. Therefore, in order to solve problems, this model using neural network control, namely compensation speech production system, it uses a strategy that feedforward control mechanism to supplement feedback control to deal with the delay change.

3 Feedback Control and Feedforward Control

3.1 Feedback Control Subsystem

The feedback control subsystem in the DIVA model carries out the following functions First, activation of the speech sound map cell corresponding to the sound in the model's premotor cortex leads to eadout of learned auditory and somatosensory targets for that sound.

The following paragraphs detail these processes, starting with descriptions of the auditory and somatosensory state maps, continuing with a treatment of the auditory and somatosensory targets for a speech sound, and concluding with a description of the circuitry involved in transforming auditory and somatosensory error signals into corrective motor commands.

3.1.1 Auditory State Map and Somatosensory State Map

In the model, the acoustic state(BA 41,42,22) is determined from the articulatory state as follows:

$$Au(t)=f_{AcAu}(Acoust(t-\tau_{AcAu})) \tag{1}$$

where $Au(t)$ is a vector of auditory state map cell activities, f_{AcAu} is a function that transforms an acoustic signal into the corresponding auditory cortical map representation, and τ_{AcAu} is the time it takes an acoustic signal transduced by the cochlea to make its way to the auditory cortical areas. Schroeder and Foxe (2002)[6] noted a response latency of approximately 10 ms in the posterior auditory cortex and 25 ms in STP.

The model also includes a somatosensory state map that corresponds to the representation of speech articulators in somatosensory cortical areas (BA 1,2,3,40,43):

$$S(t)=f_{Ars}(Artic(t-\tau_{Ars})) \tag{2}$$

where $S(t)$ is a 22-dimensional vector of somatosensory state map cell activities, f_{Ars} is a function that transforms the current state of the articulators into the corresponding somatosensory cortical map representation, and τ_{Ars} is the time it takes somatosensory feedback from the periphery to reach higher-order somatosensory cortical areas.

3.1.2 Auditory and Somatosensory Error Maps

Auditory and Somatosensory Error Maps The sensory target regions for the current sound are compared to incoming sensory information in the model's higher-order sensory cortices. The activity of the auditory error map cells is defined by the following equation:

$$\Delta Au(t)=Au(t)-P(t-\tau_{pAu})z_{pAu}(t) \tag{3}$$

where τ_{pAu} is the propagation delay for the signals from premotor cortex to auditory cortex ,and $z_{pAu}(t)$ are ynaptic weights that encode auditory expectations for the sound being produced. Houde, Nagarajan, Sekihara, and Merzenich (2002)[7] note that auditory evoked responses measured with MEG were smaller to self-produced speech than when the same speech was presented while the subject was not speaking, while response to a gated noise stimulus was the same in the presence or abscncc of self-produced speech.

3.1.3 Converting Sensory Errors into Corrective Motor Actions

In the model, production errors represented by activations in the auditory and/or somatosensory error maps get mapped into corrective motor commands through learned pathways projecting from the sensory cortical areas to the motor cortex. These projections form a feedback control signal that is governed by the following equation:

$$\dot{M}_{Feedback}(t) = \Delta Au(t-\tau_{AuM})Z_{AuM} + \Delta S(t-\tau_{SM})Z_{SM} \tag{4}$$

where Z_{AuM} and Z_{SM} are synaptic weights that transform directional sensory error signals into motor velocities that correct for these errors, and τ_{AuM} and τ_{SM} are cortico-cortical transmission delays (3ms).

Relatedly, Houde and Jordan (1998)[8] modified the auditory feedback of speakers and noted that the speakers compensated for the shifted auditory feedback over time. Tremblay, Shiller, and Ostry (2003)[9] performed an experiment in which jaw motion during syllable production was modified by application of a force to the jaw which did not measurably affect the acoustics of the syllable productions.

3.2 Feedforward Control Subsystem

According to the model, projections from premotor to primary motor cortex, supplemented by cerebellar projections, constitute feedforward motor commands.The primary motor and premotor cortices are well-known to be strongly interconnected. Furthermore, the cerebellum is known to receive input via the pontine nuclei from premotor cortical areas, as well as higher-order auditory and somatosensory areas that can provide state information important for choosing motor commands , and projects heavily to the motor cortex.

The feedforward motor command for production of a sound is represented in the model by the

$$\dot{M}_{Feedforward}(t) = P(t)Z_{PM}(t) - M(t) \tag{5}$$

The weights $Z_{PM}(t)$ encode the feedforward motor command for the speech sound being produced. This command is learned over time by incorporating the corrective motor commands from the feedback control subsystem on the previous attempt into the new feedforward command.

3.3 Feedforward Control Subsystem and Feedback Control Subsystem Combined

Feedforward and feedback composite control system can give play feedforward control timely adjust advantages, and feedback control has the strengths that restrain disturbance factors . According to the neurophysiology, the language of the human voice is a movement, it involves the movement of balance. And the cerebellum of people or animals have lots of function of motion control .Cerebellum was very good synchronization that provides the feedforward command to supplement the cortex based on feedback control command ,it also is a kind of very good solution. Therefore, the using of the cerebellum mechanism, speech production system uses feedforward control mechanism added feedback control strategy, deal with the transmission delay of speech motion control.

4 Simulation Examples

Aiming at the network state disturbance caused by parameters changes during the neural network learning process, the model adopts feedforward control .The following is simulation experiment of jaw disturbance compensationby using MATLAB and c + + .

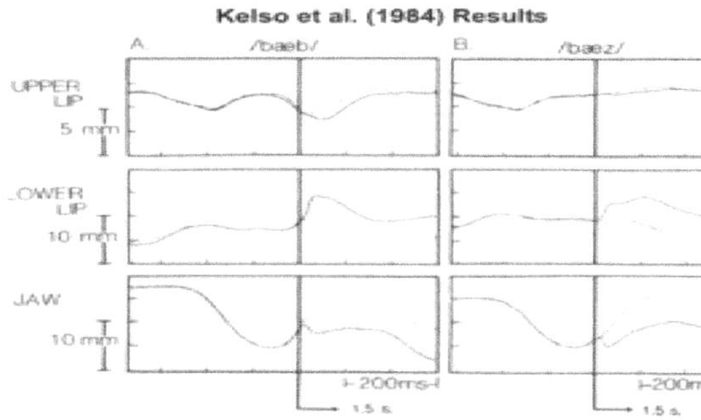

Fig. 1. Kelso et al. Results[1984]

Fig. 1. Results of Kelso et al. (1984)[10] jaw perturbation experiment. Solid lines indicate normal (unperturbed) trials, and dotted lines indicate perturbed trials. Vertical line indicates onset of perturbation. Lower lip position is measured relative to jaw.

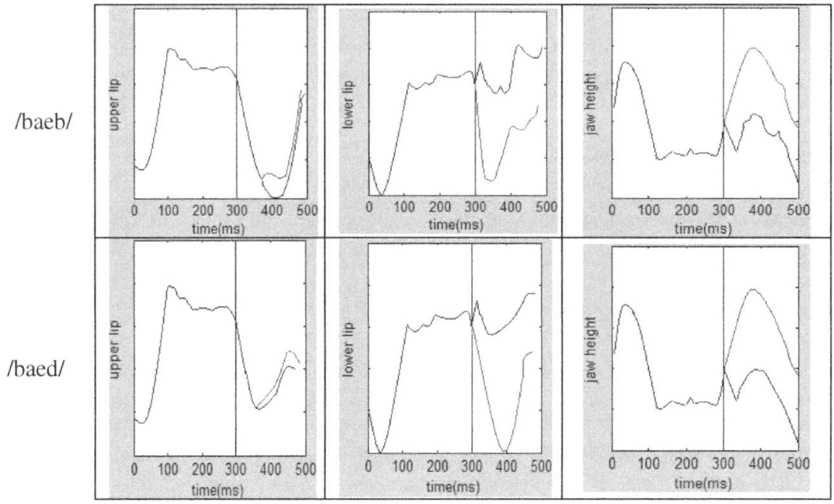

Fig. 2. DIVA simulation

Subjects produce compensatory downward movement of the upper lip for the bilabial stop /b/ but not for the alveolar stop /d/.

Fig.2. Corresponding DIVA simulation. The model produces a compensatory downward movement of the upper lip for the bilabial stop /b/ but not for the alveolar stop /d/.The onset of the perturbation is indicated by the vertical line in the simulation diagrams and was determined by the velocity and position of the jaw displacement. The red lines indicate the positions of the articulators in the normal (unperturbed) condition. The blue lines indicate the positions in the perturbed condition.

5 Conclusion

This article studies the question of nerve transmission delay in the DIVA model.By adopting feedforward control mechanism supplement feedback control strategy, processing speech control movement with realistic delays.Thus , in some of degree, the questions has been improved in related to the time delay synchronized question in organ feedback as well as the transmission delay speed question caused by each kind of nervous system. The simulation result indicates that the method enhances control system's robustness.

References

1. Zhang, S., Ruan, X., Cheng, X.: A new constructing method of cerebellum model applying to DIVA model. In: Control and Decision Conference, CCDC 2009, Guilin,Chinese (June 2009)
2. Guenther, F.H.: A neural network model of speech acquisition and motor equivalent speech production. Biological Cybernetics 72, 43–53 (1994)
3. Guenther, F.H., Ghosh, S.S., NietoCastanon, A.: A neural model of speech production. In: Proceedings of the 6th Internantional Seminar on Speech Production, Sydney, Australia, pp. 85–90 (2003)
4. Guenther, F.H., Ghosh, S.S., Tourville, J.A.: Neural modeling and imaging of the cortical interactions underlying syllable production. Brain and Language 96(3), 280–301 (2006)
5. Schroeder, C.E., Foxe, J.J.: The timing and laminar profile of convergeng inputs to multisensory areas of the macaque neocortex. Brain Research Cognitive Brain Research 14, 187–198 (2002)
6. Maeda, S.: Compensatory articulation during speech: Evidence from the analysis and synthesis of vocal tract shapes using an articulatory model. In: Hardcastle, W.J., Marchal, A. (eds.) Speech production and speech modeling, pp. 131–149. Kluwer Academic Publishers, Boston (1990)
7. Houde, J.F., Nagarajan, S.S., Sekihara, K., Merzenich, M.M.: Modulation of the auditory cortex during speech: an MEG study. Journal of Cognitive Neuroscience 14, 1125–1138 (2002)
8. Houde, J.F., Jordan, M.I.: Sensorimotor adaptation in speech production. Science 279, 1213–1216 (1998)
9. Tremblay, S., Shiller, D.M., Ostry, D.J.: Somatosensory basis of speech production. Nature 423, 866–869 (2003)
10. Kelso, J.A., Tuller, B., Vatikiotis-Bateson, E., Fowler, C.A.: Functionally specific articulatory cooperation following jaw perturbations during speech: evidence for coordinative structures. Journal of Experimental Psychology: Human Perception and Performance 10, 812–832 (1984)

An Improved Phonetic Learning Algorithm Based on the DIVA Model*

Shaobai Zhang and Xin Zhang

Computer College, Nanjing University of Posts and Telecommunications,
Nanjing 210003, China
adzsb@163.com

Abstract. DIVA (Directions Into Velocities of Artculators) is a mathematical model of the processes behind speech acquisition and production, it is also an adaptive neural network that learns to control movements of a simulated vocal tract in order to produce words, syllables or phonemes. But the research of recognition system in DIVA model is fuzzy. This article analyze the description of speech target based on the DIVA model, and the way of speech production is improved, so as to supplement the speech recognition system of DIVA model.

Keywords: DIVA model, speech acquisition and production, speech target, speech recognition system.

1 Introduction

In the past few decades, the hot-spot discussion in the speech production literature concerns the essence of the ‘‘targets’’ as specified to the production mechanism. Henke posited speech targets consisting of desired spatial positions of the articulators. The target for a flexible articulator such as the tongue consisted of a series of spatial target positions for small segments of the articulator. MacNeilage also proposed the control of spatial positions of articulators, suggesting that an articulator’s target could be specified as a set of desired muscle lengths. Muscle length targets have been proposed more recently by Cohen et al[1]. These spatial and muscle length target models suffer from the same shortcoming: they cannot account for compensatory movements of one articulator when another articulator cannot reach its ‘‘normal’’ position .

DIVA is a mathematical model of the processes behind speech acquisition and production, it is also an adaptive neural network that learns to control movements of a simulated vocal tract in order to produce words, syllables or phonemes [1][2].The model learns orosensory-to-articulatory mapping, which makes efficient use of redundancy in the articulator system to compensate for constraints or perturbations applied to the articulators automatically.

In the DIVA model presents a "convex region" theory, it specify the target for a speech sound as a high-dimensional orosensory space. This space includes not only the positions of the individual articulators, but also orosensory information related to

* This work is supported by the National Natural Science Foundation of China (No.61073115).

G. Lee (Ed.): Advances in Automation and Robotics, Vol.2, LNEE 123, pp. 495–500.
Springerlink.com

the speech target Each dimension of the orosensory target specifies a range of acceptable positions.

2 Speech Acquisition in the DIVA Model

Speech acquisition in the DIVA model consists of finding appropriate parameters, or synaptic weights, for the phonetic-to-orosensory and orosensory-to-articulatory mappings during a babbling phase. In the DIVA model speech recognition procedure is as follows [1][3][4]:

2.1 Randomly Activate an Articulator Velocity Vector

In the DIVA model, babbling is produced by superimposing random movements of the speech articulators on an oscillatory movement of the jaw. This corresponds to the phase in infant babbling known as variegated or nonreduplicated babbling which starts at an age of approximately 10 months; this phase has been hypothesized as the stage during which infants learn to produce the various phonemes of their native language. With the exception of the AVV cells coding jaw movement, each AVV cell is activated to a value of 1 with probability 0.1; otherwise, its value is 0.

2.2 For Each of 10 time Steps, Repeat the Following

(a) **Carry out learning in the phonetic-to-orosensory mapping.** The synaptic weights in the pathways projecting from a Speech Sound Map cell to the Orosensory Direction Vector cells represent a vocal tract target for the corresponding speech sound in orosensory space. When the changing vocal tract configuration is identified by the Speech Recognition system as producing a speech sound during babbling, the appropriate Speech Sound Map cell's activity is set to 1. This gates on learning in the synaptic weights of the phonetic-to-orosensory pathways projecting from that cell, resulting in a modification of the target associated with the active speech sound. This modification is appropriate because it expands the target to include the current configuration of the vocal tract, which is available through orosensory feedback at the Orosensory Direction Vector cells. See Section 4 for details of this learning process.

(b)**Carry out learning in the orosensory-to-articulatory mapping.** Random activation of AVV cells produces movements of the articulators which are transmitted through orosensory feedback to the Orosensory Direction Vector stage, resulting in changes of the ODV cell activities. These changes in activity gate on learning in the synaptic weights of the orosensory-to-articulatory pathways. If an ODV cell's activity is decreasing, the synaptic weights in pathways projecting from this cell to active Articulator Velocity Vector cells will increase; in this way, each ODV cell learns a set of articulator movements which will reduce the ODV cell's activity.

2.3 Go to (1)

After the above three steps to repeat the experiment, obtains the correct pronunciation the target area, complete the target sppech recognition. DIVA model used two mapping to complete the above steps, phonetic-orosensory mapping and orosensory-articulatory mapping. The advantages of the DIVA model is the simple mapping is effective. These mapping use of highly efficient pronunciation system redundant complete self compensate, reducing the compensation for redundancy noise and system overhead.

3 The Improved Algorithm

3.1 Backgroud

In the actual speech recognition system, which often need the certain scale adaptive data quantity to finish the speech recognition. For example, during a babbling phase in DIVA model, with the model simulation operating approximately in "real-time" the entire babbling sequence takes approximately 30 minutes,during which 5000 random movements of the articulators are carried out and the model learns to produce a set of 29 English phonemes[1][5]. Because a large number of random movement of the vocal organs, which makes the adaptability of the system rapidly expand, leading very big burden to the adaptive system .But people always hope to reduce adaptive quantity as few as possible, to achieve "quickly adapt ".

Phonemes is the minimum phonetic unit according to the natural attributes of voice. Phonemes generally with the international phonetic alphabet (IPA) mark. As is known to all, the English letters, [ei], [I:], [e], [ju:], [ai] phonemes occupied most of the vowels.

In DIVA model, during the mapping between articulators to orosensory, each letter pronunciation process is the individual treatment. Figure 1 schematizes the learning sequence for the vowel /i/ along two dimensions of orosensory space. Every time the model produces the vowel /i/ from any vocal tract configuration during babbling, the Speech Recognition System activates the cell for /i/ in the Speech Sound Map. The first time that the phoneme is produced during babbling, the corresponding cell in the Speech Sound Map learns the orosensory position that caused the phoneme. This corresponds to a point in orosensory position space, schematized in Figure 1(a). The next time the phoneme is babbled, the Speech Sound Map cell expands its learned target to be a convex region that encompasses both the previous orosensory position and the current orosensory position. During pronunciation stage, articulator according to the above areas to finish the action in order to produce the target speech. The final purpose of DIVA model is to set up the corresponding relations between the pronunciation position of the articulators and the speech, this relationship embodied in a sound happen for performance orosensory of a region of space[1][6]. When the articulator move to the designated area, through the collaborative of articulator generate pronunciation. During speech recognition, if the articulator reaches the corresponding areas, the system think successful identified the pronunciation.

Because inaccuracy of the pronunciation, when produce the target area or through orosensory space simulation of the speaker, the DIVA model must traverse all the mapping relationship to determine the pronunciation of the corresponding orosensory or space. This caused a large number of system cost, also make the speech access and produce become difficult and slow.

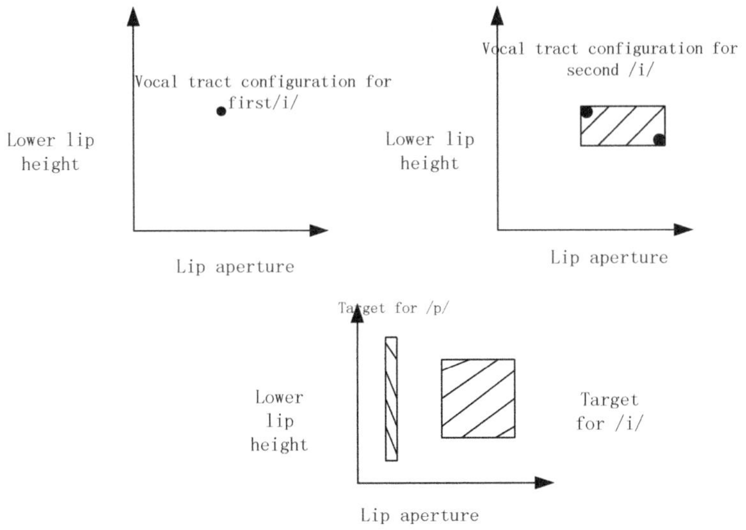

Fig. 1. Learning of the convex region target for the vowel /i/ along orosensory dimensions corresponding to lip aperture and lower lip height.

3.2 New Method of Speech Production and Acquisition

This paper put forward the English phonemes category method based on the classification, and the new search acoustic pronunciation methods of the acoustic model. According to the speecch acquisition phase as an example. This method divide the speech acquisition process into three stages:

1) General classification by the Phonemes identifier category, get the candidate acoustics model set of the speech, producing a simplified corresponding relationships;

2) The second stage respectively by the phonemes categories produce corresponding orosensory space mapping, here the mapping similar to the first stage which is the only part of the vowel phonemes mapping.

3) Because have the same speech phonemes, its corresponding to the target space is closely. Through the classification elements as the value or the middle, make each speech acquisition by simple mapping relation to determine. Thus, through the DIVA model to determine the corresponding mapping relation only the last step and classification standards to its target area mapping. Not every speech will go to identify or pass the multifarious formula, this reduced the time for speech acquisition. General ideas as shown in figure 2.

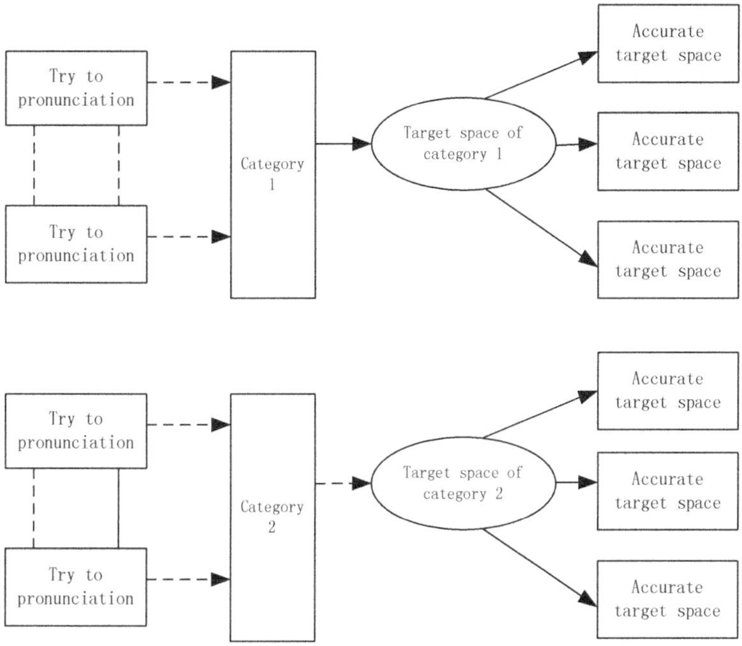

Fig. 2. The new method process of speech production and acquisition

The mainly implementation steps of new method:

1) Find the best classification of the rules is to establish phonemes preceded identifier. In the experiments ,using k-means clustering method and the Euclidean distance as metric distance, will be for speech in accordance with the number of categories set the clustering. Here we can use gaussian Mixture Model, Gaussion Mixture Model (GMM) as the phonemes identifier. The principle of GMM is which build and train a GMM for each phoneme. Then calculated the probability of this sample belongs to the category by each kind of GMM and the most probability will be as sample identification results.

2) Each category according to acquire the target area of DIVA model to get the ultimate mapping relationship. Due to set the number of categories, so in step 2 for speech learning only to classify study, its convenient and quick. When find the pronunciation area, and at the same time record study for mapping and the target area.

3) For another speech learning near the classification target space and finally confirmed the speech vocal tract target in the orosensory space.

3.3 The Algorithm Analysis

The improved algorithm made a speech classification and increased classification of the target area gain in the learning process, which increases the extra space system to store classification. But after such extension and the selection process, we get the candidate sound model set of the corresponding speech by speech phonemes category information of each speech paragraph. This set constitute the search space of the speech paragraph in decoding process. It is obvious that such phone-based search method, greatly reduce the search space of the acoustic model. The new algorithm is still have a lot of room for improvement, for example, the speech still operation many times, and the number of mapping in DIVA model also did not decrease, which made speech acquisition and production more convenient and accurate. The new algorithm is the main consideration to the habitual of articulator and the proximity position of corresponding similar pronunciation, but in the big speech samples and pronunciation similar cases, the algorithm can't play a very significant result.. The author considered this new algorithm only to the DIVA model and not extended to other speech model.

4 Concluding Remarks

This paper briefly introduces the basic structure of model DIVA, and introduced the speech production method of DIVA model. At the same time, the application of the "convex region" theory to the speech of the DIVA model to describe the target. In this paper the author put forward a new algorithm for speech acquisition and production based on the classification. The new algorithm, makes the speech acquisition and production more quickly and accurately. This method can be used widely in speech recognition system.

References

1. Guenther, F.H.: A neural network model of speech acquisition and motor equivalent speech production. Biological Cybernetics 72(1), 43–53 (1994)
2. Zhang, S., Ruan, X., Cheng, X.: A new constructing method of cerebellum model applying to DIVA model. In: Control and Decision Conference, Guilin, Chinese (Junuary 2009)
3. Guenther, F.H.: Speech sound acquisition, coarticulation, and rate effects in a neural network model of speech production. Psychological Review 102(1), 594–621 (1995)
4. Guenther, F.H., Hampson, M., Johnson, D.: A Theoretical Investigation of Reference Frames for the Planning of Speech Movements. Psychological Review 105, 611–633 (1998)
5. Guenther, F.H., Ghosh, S.S., Nieto-Castanon, A.: A neural model of speech production. In: Proceedings of the 6th International Seminar on Speech Production, Sydney, Australia, pp. 85–90 (2003)
6. Han, J., Wang, H.: Pronunciation Learning Technology Based on Speech Recognition. Voice Technology, 1002–8684 (2004)

Comparing Different Feature Extraction Methods of Pump Dynamograph Based on Support Vector Machine

Wei Wu and Yangyang Meng

School of Mechanical Engineering, Xian Shiyou University 710065

Abstract. The paper use two different methods of image moment invariants and wavelet packet-energy entropy (WP-EE) to extract the pump dynamograph's eigenvector, then making the collected date is training samples and test samples for the support vector machine (SVM) classifier separately, finally comparing the accuracy of two methods. The experimental results show that when choosing classifier based on support vector machine, the method of extracting image moment invariants has more higher accuracy than the other method.

Keywords: support vector machine, image moment invariants, wavelet packet-energy entropy.

1 Introduction

Pump dynamograph is a closed curve which reflects the change law of load-time and displacement-time of oil well pump, it contains abundant information about operation condition of oil pumping system. [1] Testing and analyzing pump dynamograph is an important method to understanding working conditions of pumping equipment and oil well performance, the success of fault classification and recognition are extracting the representative feature vector and selectting appropriate classifier.

In the description of pump dynamograph at present, extracting image moment invariants is a better method; meanwhile application of decomposition and reconstruction algorithm by wavelet packet in pump dynamograph achieve an ideal result. This paper will try to use the two methods to extract pump dynamograph's eigenvector, making the collected data is training and testing samples for the support vector machine classifier separately to study fault diagnosis technology about oil rod pumping system. The goal is searching an optimum composition method to improve the diagnosis accuracy of oil pumping system.

2 Support Vector Machine

Support vector machine is a machine learning algorithm based on the statistical learning theory, it's proposed by Vapnik in the 1990s,and obtained rapid development. It is a good way to solve the problems of small sample, nonlinear, high dimensional, local minimum point,etc. Because of based on structural risk minimization principle, SVM greatly improve the generalization of learning machine.It's widely used in the fields of pattern recognition, signal processing, regression analysis, etc.

G. Lee (Ed.): Advances in Automation and Robotics, Vol.2, LNEE 123, pp. 501–506.
Springerlink.com

Following picture explains the main idea of support vector machine: as Figure1.1 shows, with a straight line divides two kinds of linearly separable samples into two parts. Firstly, hypothesis the normal direction(w) of classification straight line(H) was given, hypothesis H_1 and H_2 singly through two kinds of samples which closest the H, at the moment the distance between two straight lines H1 and H2 is called largest interval (margin) corresponding with the method. Direct calculation shows that the distance between H_1 and H_2 is $2/\|\omega\|$. Obviously, we should choose the normal direction which making interval received maximum, the classification face is optimal classification face,also known as optimal classification hyperplane,the train samples were through by H1 and H2 called support vector.

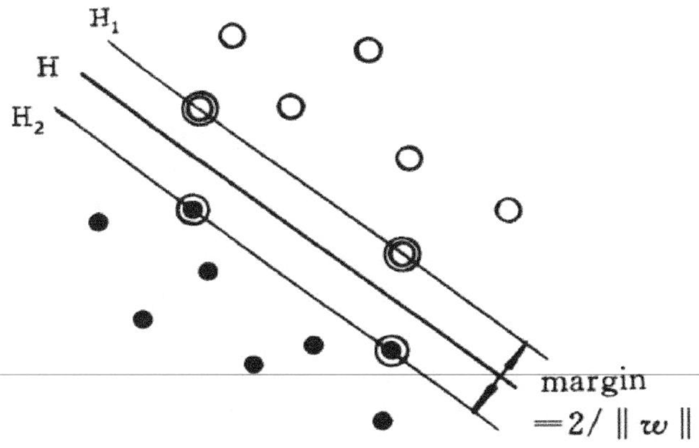

Fig. 1. The main idea of support vector machine

3 Extracting the Feature Vector of Pump Dynamograph

3.1 Preprocessing Image Date

The study based on MATLAB R2009a version. firstly, with the imread function reads and digitals the pump dynamograph, meanwhile binary and anti-color of the image; secondly, selecting imdilate function to linear expand of the image, to ensure coherence; thirdly, processing refining treatments by bwmorgh function, include extracting center framework and deburring; fourthly, selecting the minimum enclosed rectangle method to remove the excess part of the image, separating the target image; finally, with imresize function to scale the image 512 * 256 pixels, followed by imfill function to fill the target image. Thus far, the procedure of preprocessing image date is completed.

Figure 2 shows four common results of preprocessing pump dynamograph ,

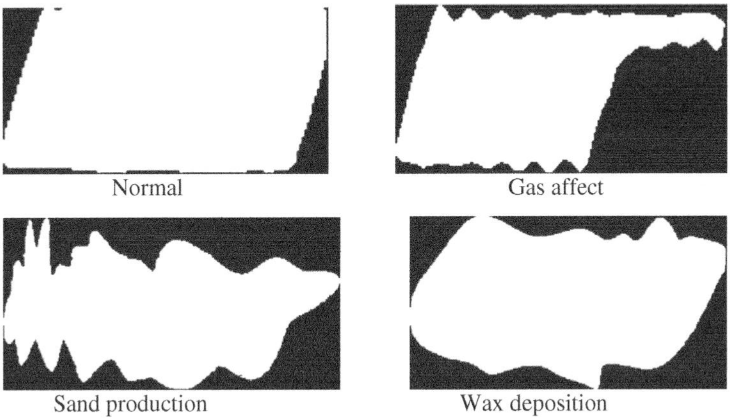

Fig. 2. Four common results of preprocessing pump dynamograph

3.2 Extracting Image Moment Invariants

Some moments of image region have invariant feature for geometric transformation, include translation, rotation, mirroring and scaling, etc. So the description of image moment invariants has an important significance in the image classification and recognition.

Define (j+k)-order moment of digital image as following,

$$m_{jk} = \sum_x \sum_y x^j y^k f(x,y) \qquad j,k=0,1,2 \tag{1}$$

M_{00} indicates the image area,m_{10} indicates the sum of x coordinate, m_{01} indicates the sum of y coordinate.

$$\text{Defined,} \qquad \bar{x} = \frac{m_{10}}{m_{00}} \qquad \bar{y} = \frac{m_{01}}{m_{00}} \tag{2}$$

$$\mu_{jk} = \sum_x \sum_y (x - \bar{x})^j (y - \bar{y})^k f(x,y) \tag{3}$$

Defined normalized (j+k)-order central moment as (4),

$$\eta_{jk} = \frac{\mu_{jk}}{(\mu_{00})^{\gamma}} \qquad j,\ k=0,1,2\cdots$$
$$\gamma = \left(\frac{(j+k)}{2} + 1\right) \qquad j+k=2,3, \tag{4}$$

Through the normalized central moment we can obtain seven moment invariants with invariant features,defined as following [4],

$$\varphi_1 = \eta_{20} + \eta_{02} \tag{5}$$

$$\varphi_2 = (\eta_{20} + \eta_{02})^2 + 4\eta_{11}^2 \tag{6}$$

$$\varphi_3 = (\eta_{30} + 3\eta_{12})^2 + (3\eta_{21} - \eta_{03})^2 \tag{7}$$

$$\varphi_4 = (\eta_{30} + \eta_{12})^2 + (\eta_{21} + \eta_{03})^2 \tag{8}$$

$$\begin{aligned}\varphi_5 = (\eta_{30} - 3\eta_{12})(\eta_{21} + \eta_{03})[(\eta_{30} + \eta_{12})^2 - 3(\eta_{21} + \eta_{03})^2] \\ + (3\eta_{21} - \eta_{03})(\eta_{21} + \eta_{03})[3(\eta_{30} + \eta_{12})^2 \\ - (\eta_{21} + \eta_{03})^2]\end{aligned} \tag{9}$$

$$\varphi_6 = (\eta_{20} - \eta_{02})[(\eta_{30} + \eta_{12})^2 - (\eta_{21} + \eta_{03})^2] + 4\eta_{11}^2(\eta_{30} +)(\eta_{21} + \eta_{03}) \tag{10}$$

$$\varphi_7 = (3\eta_{12} - \eta_{30})(\eta_{30} + \eta_{12})[(\eta_{30} + \eta_{12})^2 - 3(\eta_{21} + \eta_{03})^2] + (3\eta_{21} - \eta_{03})(\eta_{21} + \eta_{03})[3(\eta_{30} + \eta_{12})^2 - (\eta_{21} - \eta_{03})^2] \tag{11}$$

By the above formula, we can get the seven moment invariants dates of Figure2,

Table 1. Date of moment invariants

	φ_1	φ_2	φ_3	φ_4	φ_5	φ_6	φ_7
Normal	1.6028	4.3245	5.8685	8.1749	15.5180	10.3549	20.2004
Gas affect	1.4743	4.0012	5.9462	5.8558	13.1745	7.8664	14.2934
Sand production	1.4047	3.4962	6.8745	7.9243	16.9764	10.0090	16.7049
Wax deposition	1.4925	3.7600	6.1623	7.9527	15.7105	9.8756	17.8127

3.3 Extracting Image Wavelet Packet-Energy Entropy

The concept of wavelet analysis was first proposed by the French oil engineer J.Morlet in 1974, Wavelet analysis has made remarkable achievements in the field of information technology. Wavelet packet analysis of wavelet analysis is out of a signal to a more detailed analysis and reconstruction method, it is not only for the low frequency part decomposition, but also for the high frequency part decomposition [5].

Let $g_j^n(t) \in U_j^n$, g_j^n can be expressed as

$$g_j^n(t) = \sum_{\iota} d_{\iota}^{j,n} u_n(2^j t - \iota) \tag{12}$$

Algorithm of wavelet packet decomposition: according $\{d_l^{j+1,n}\}$ to calculate $\{d_l^{2n}\}$ and $\{d_l^{j,2n+1}\}$

$$d_l^{2n} = \sum_k \alpha_{k-2l} d_k^{j+1,n}$$
$$d_l^{j,2n+1} = \sum_k b_{k-2l} d_k^{j+1} \quad (13)$$

Where α_k and b_k are conjugate filter coefficients of wavelet decomposition.

Algorithm of wavelet packet reconstruction: according $\{d_l^{2n}\}$ to calculate$\{d_l^{j,2n+1}\}$ and $\{d_l^{j+1,n}\}$.

$$d_l^{j+1,n}=\sum_k[h_{l-2k} d_k^{j,2n} + g_{l-2k}d_k^{j,2n+1}] \quad (14)$$

Where $h_k$and g_k are conjugate filter coefficients of Wavelet reconstruction

As the square of wavelet transform coefficient $d^{j,k}$has the energy dimension, it can be used in the mechanical fault diagnosis [6].

Image are decomposed and reconstructed with three layers wavelet packet in this part, following calculates the energy entropy all nodes in the third layer with wentropy function, denote the results as $[E_{130},E_{131},\ldots\ldots E_{137}]$

$$E_{sum} = \sqrt{\sum_{i=0}^{7}(E_{13i}{}^2)} \text{ , where i=0,1......7} \quad (15)$$

In that way,

$$E_{3i}= E_{13i}/E_{sum} \quad (16)$$

Making the eight energy eigenvector which made up of E_{3i} is training and testing samples for the support vector machine classifier. By the algorithm of wavelet packet we can get the feature vectors of Figure 2.

Table 2. Date of wavelet packet

	E_{30}	E_{31}	E_{32}	E_{33}	E_{34}	E_{35}	E_{36}	E_{37}
Normal	0.3049	0.0618	0.0547	0.0168	0.0185	0.0178	0.0073	0.0082
Gas affect	0.4782	0.0636	0.0505	0.0254	0.0304	0.0274	0.0118	0.0145
Sand production	0.4748	0.0611	0.0622	0.0302	0.0286	0.0247	0.0143	0.0179
Wax deposition	0.4449	0.0657	0.0429	0.0253	0.0316	0.0282	0.0145	0.0174

4 Comparing Different Methods Based on SVM

The paper chosen 102 images consist of normal, gas effect, sand production, wax deposition, etc. as experiment samples, following extract the pump dynamograph's eigenvector with above two methods separately. Finally normalize the date to [-1,1] and using the toolbox of Format Data Libsvm to achieve the date conversion, making the date is standard format.

Based on DOS environment, the study takes RBF kernel function $K(x, x_i) = \exp\{-\frac{\|x-x_i\|^2}{\sigma^2}\}$ for classifier, Train procession chosen one-to-many algorithm while the parameters selected default by Libsvm toolbox. Combination with the production dates, we can get the following results,

Table 3. Comparison of experimental results

Name	Train samples	Test samples	Eigenvector	Error	Accuracy
Moment invariants	60	102	7	6	94.1176%
wavelet packet	60	102	8	11	89.2157%

The experimental results show that the method of extracting image moment invariants has more higher accuracy than the wavelet packet method when choosing classifier based on support vector machine; the success of fault classification and recognition are extracting the representative feature vector and selecting appropriate classifier.

Through seeking structural risk minimization to realize practical risk minimization, Support vector machine greatly improves the recognition accuracy. Under RBF kernel function, the paper choose nonlinear classify algorithms to compare two methods of image moment invariants and wavelet packet-energy entropy. To the diagnosis results, combining image moment invariants and SVM is an ideal method for oil fault diagnosing, it has good reference for practical engineering value.

References

1. Peng, Y.: Pattern recognition and its application in the petroleum industry. Publishing house of science and technology, Shaanxi (1998)
2. Yan, W., Zhao, H.: Application of medical diagnosis based on support vector machine. Computer Simulation 20(2), 69–71 (2003)
3. Zhang, X.: About statistical learning theory and support vector machine. Journal of Automation 26(1), 32–41 (2000)
4. Liu, W.: Applying the method of support vector machine to identifying the operation conditions of dyagragh of pumping unit. Xi'an University of Technology, Shaanxi (2009)
5. Zhang, D.: Digital image processing in MATLAB. China machine press (2009)
6. Wu, W., Sun, W., Wei, H.: A Fault Diagnosis of Suck Rod Pumping System Based On Wavelet Packet and RBF Network. Advanced Material Reserch 189-193, 2655–2669 (2011)

Analysis for Height Control-Loop of Cruise Missile to Different Kinds of Autopilot

Qiuqiu Wen, Qunli Xia, and Chuntao Cai

Department of Aerospace, Beijing Institute of technology
Beijing, People's Republic of China
wenqiuqiu82@gmail.com

Abstract. The models of two kinds of height control system which include the attitude autopilot inner loop and acceleration autopilot inner loop respectively are built. The response rate, stabilization and engineer realization of gravity compensation of the height control systems are analyzed and compared. The result shows that they have same performances and can be used both in missile altitude control. The gravity compensation can improve the transition quality of height control and decrease the overshoot. The gravity compensation command matching attitude autopilot depends on the flying parameter heavily, so the engineer reliability is lower than acceleration autopilot.

Keywords: Height control, attitude autopilot, acceleration autopilot, gravity compensation.

1 Introduction

The cruising height of the cruise missiles has become an important tactical and technological index in modern war attack and defense[3]. In order to improve the concealment quality of the penetration, the cruise missiles need to use sea-skimming or hedgehopping fly method. The high-performance height control system should be design to avoid the missile to have touchdown phenomenon.

The traditional height control system loops realize the height control by adopting the attitude autopilot to dominate the pitch angles of the missiles[1][4]. Compared with the attitude autopilot, the acceleration autopilot which is gradually applied in height control inner loops design has the high rate response and strong anti-interference characteristics.

This paper establishes the models of two kinds of height control system which include the attitude autopilot inner loop and acceleration autopilot inner loop. Through the theoretical analysis and the simulation calculation, the response rate, stabilization and engineer realization of gravity compensation of the height control systems are analyzed and compared. The analyzing result has great practical value for grasping exactly the height control loops design.

2 The Autopilot's Structural Height Control Loops

The principle of the typical cruise missiles height control loop is shown as the Fig.1 [2]. The height tracing error ΔH forms the height tracking command ΔH_c and then

G. Lee (Ed.): Advances in Automation and Robotics, Vol.2, LNEE 123, pp. 507–514.
Springerlink.com

accesses to the control system inner loop which includes command converter circuit, autopilot and missile body dynamics, the outputs is the missile' vertical speed $\dot{H}$. We can get the height of the missile by the kinematics integral.

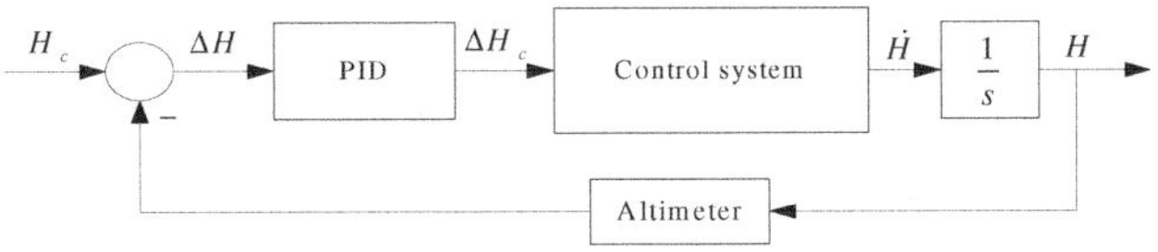

Fig. 1. The principle diagram of the height control loop

The height control loop can be composed by many different autopilots. The Fig.2 is the principle diagram of the height control loop using the attitude autopilot's loop. The height command changes into conversion gain $k_{\vartheta c}$ after the command and forms attitude angle command. And the attitude autopilot controls the missiles' attitude responding to the command, shown as Fig.3. The turning attitude angle forms the angle of attack to produce lift power so that it can change the speed direction to ensure the missile fly according to certain height. Based on the assumption of ignoring the missile damping and rudder lift, we simplified transfer function of the attitude angle changing to speed dip angle as a first-order lagging link which has time constant T_a.

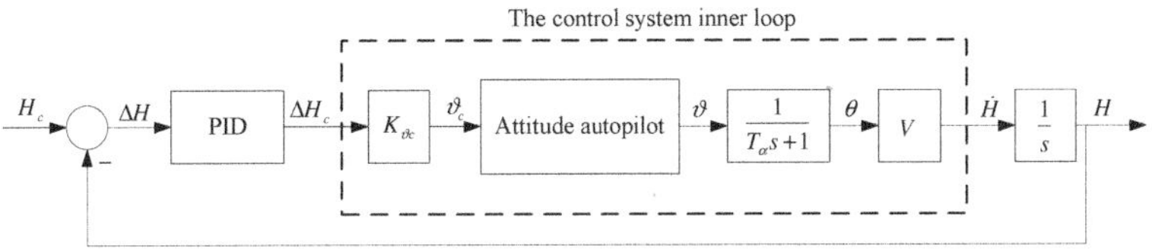

Fig. 2. Height control system with attitude autopilot

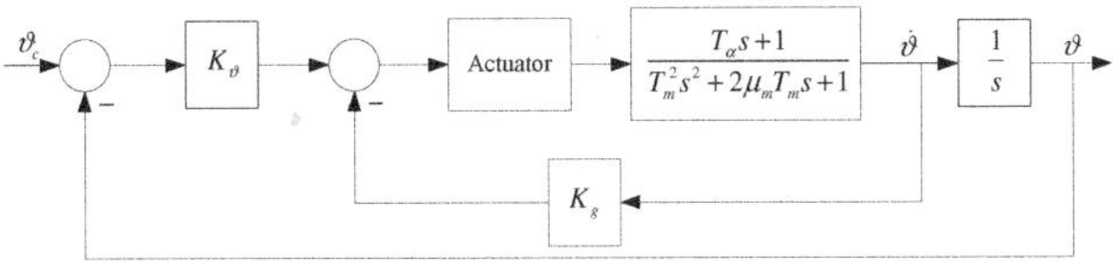

Fig. 3. Attitude autopilot block diagram

The principle diagram of the height control loop which is based on acceleration autopilot inner loop is shown as Fig.4. The height control command change into acceleration command through conversion gain k_{ac} and then enter into autopilot. After two integration of the missile overloaded output, it forms height so as to response to the height command. The acceleration autopilot uses the two-loop acceleration autopilot structure showing as Fig.5.

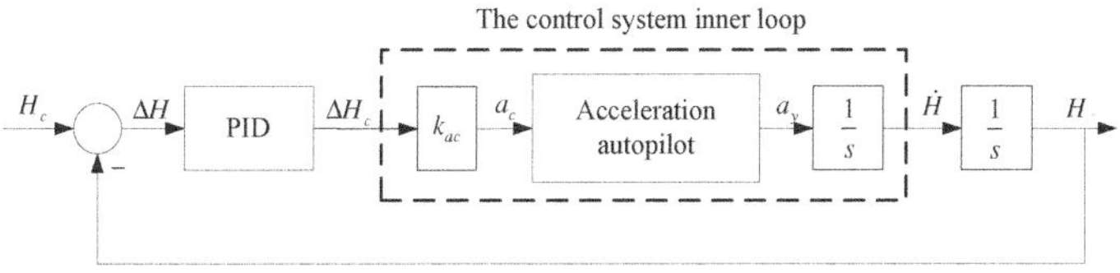

Fig. 4. Height control system with acceleration autopilot

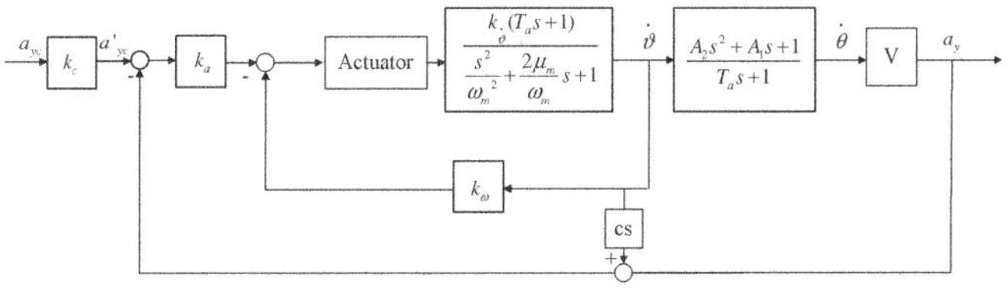

Fig. 5. Acceleration autopilot block diagram

3 Analysis and Comparison of the Height Control Loop

In order to analyze easily, we first ignored the autopilot actuator, angular velocity gyro, accelerometer, missile altimeter and etc hardware dynamic characteristic. At the meanwhile, the missile damping and lift of rudder is so small that we omit them.

According to Fig.3, we can easily calculate the closed-loop transfer function of the attitude autopilot:

$$G(\frac{\vartheta}{\vartheta_c}) = \frac{(T_a s+1)}{(T_\vartheta s+1)(\frac{s^2}{\omega^2}+2\frac{\mu}{\omega}s+1)} \tag{1}$$

Where: T_ϑ is the time constant of the attitude autopilot, its magnitude is decided by the response speed of the autopilot and has $T_\vartheta < T_a$.

Taking the attitude angle to speed velocity slope angle function:

$$G(\frac{\theta}{\vartheta}) = \frac{1}{T_a s+1} \tag{2}$$

The transfer function of the height control inner loop composed by the attitude autopilot is indicated as:

$$G(\frac{\dot{H}}{\Delta H_c}) = \frac{k_{\vartheta c}V}{(T_\vartheta s+1)(\frac{s^2}{\omega^2}+2\frac{\mu}{\omega}s+1)} \tag{3}$$

According to Fig.5, we can get the closed-loop transfer function of the acceleration autopilot

$$G(\frac{a_y}{a_{yc}}) = \frac{1}{(\frac{s^2}{\omega_{ay}^2} + 2\frac{\mu_{ay}}{\omega_{ay}}s + 1)} \quad (4)$$

Where ω_{ay} is the oscillation root frequency of the acceleration autopilot, its magnitude is decided by the respond speed of the autopilot.

So the transfer function of the height control inner loop composed by the acceleration autopilot is indicated as:

$$G(\frac{\dot{H}}{\Delta H_c}) = \frac{k_{ac}}{s(\frac{s^2}{\omega_{ay}^2} + 2\frac{\mu_{ay}}{\omega_{ay}}s + 1)} \quad (5)$$

Comparing the Equation (3) and (5), the closed-loop transfer function structure of height control with two kinds of autopilots as inner loop composed are basically same. The corresponding inner loop's main root of the attitude autopilot is its own first order root T_ϑ. Because of the delayed constant of the angle of attack is limited by the zero, $T_\vartheta > T_a$ [6]. Therefore the attitude autopilot inner loop's frequency band is impossible so high. Considering the phase advance ability of PID controller is limited, the height loop can not be designed so quickly in the context of ensuring the stability. But to the acceleration autopilot, the dominant pole of the closed-loop is the second order oscillation root which is not influenced by T_a. Its respond speed is bigger than the attitude autopilot. But from the autopilot overloading output to missile's longitudinal velocity contains a kinematic integral item. Considering of the stability, pure integral brings extra -90° phase lag which makes the height loop frequency band not be too wide so that it lowers the respond speed.

Through the above analysis, we can conclude that the height control loop, taking these two kinds of autopilots as its inner loop, is limited by the attitude autopilot itself and the kinematics integral link bring by the acceleration autopilot. Therefore its respond speed can not be designed so fast. Because the value of T_ϑ is generally higher, the break frequency of $1/T_\vartheta s+1$ is lower and its value in the open loop system is similar with the integral element. So in the same design requirement condition, the stability and rapidity of these two autopilot height loops is not very different.

4 Design Example

For further analysis through examples below, we take the missile dynamics parameters of a supersonic sea-skimming anti-ship missile in low flying section feature point, showing as the Table.1[1].

Table 1. Dynamic coefficients of missile

Flight Height/km	V/m/s	a_α /s-2	a_δ /s-2	b_α /s-1	b_δ /s-1
10	785	57.6	77.9	1.51	0.128

Design the acceleration autopilot and the attitude autopilot separately, the autopilot transmission gain is k_a =0.0011, k_ω =0.22,=1.6, k_g =0.15.The respond curve of the autopilot's step response is indicated by the fig.6. We can easily see that compared with the acceleration autopilot, the attitude autopilot's response speed is obviously slower. Especially, the velocity slop angle is more slowly than the attitude angle because it needs to go through a first-order lags.

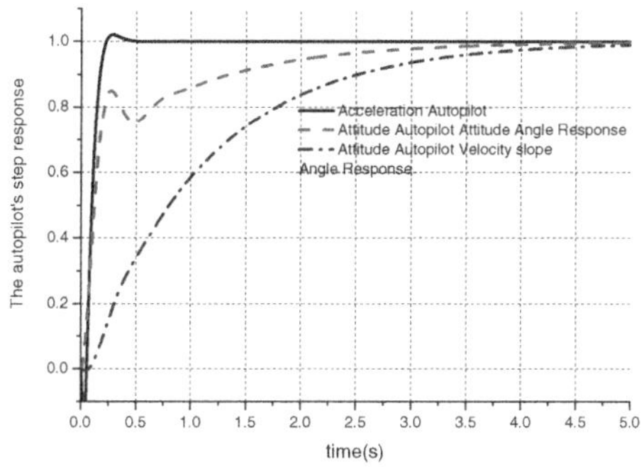

Fig. 6. Time responses of different type autopilots

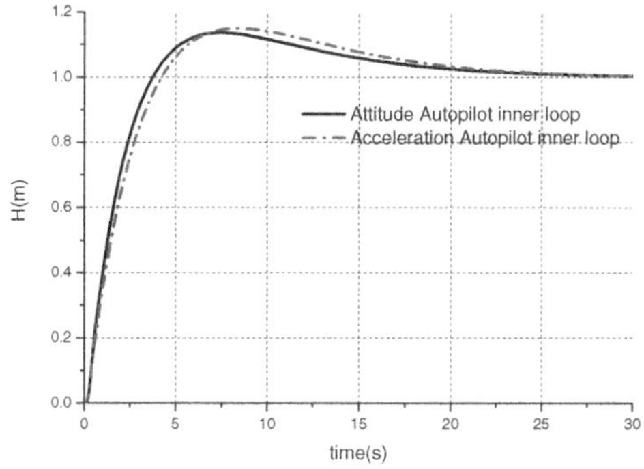

Fig. 7. Time responses of height control system

The closed-loop transfer functions of the acceleration is

$$G(\frac{a_y}{a_{yc}}) = \frac{(0.0341s-1)(0.0341s+1)}{(\frac{s^2}{14.0^2}+2\frac{0.75}{14.0}s+1)} \tag{6}$$

The attitude autopilot's closed-loop transfer function from attitude angle command to speed inclination angle:

$$G(\frac{\theta}{\vartheta_c}) = \frac{(0.0341s-1)(0.0341s+1)}{(1.06s+1)(\frac{s^2}{13.7^2}+2\frac{0.5}{13.7}s+1)} \tag{7}$$

Taking PID controller parameters of the height loop K_P=1.5, K_I=0.18, K_D=2.88 and taking separately the command conversion gains k_{ac}=0.96, $k_{\vartheta c}$=0.0026。

Through the compared curve of the height loop unit step response showing in Fig.7, we can see that by designing the parameters, the acceleration inner loop height control step response has responded with the attitude autopilot.

The calculating result of the open-loop system Bode Diagram and the stability margin indicated by Table.2 , we can conclude that in low-frequency place, the height control loop of the acceleration autopilot inner loop's phase lag is greater than the attitude autopilot. But near the cross-over frequency, the first order root of the attitude autopilot's phase lag is already close to-90°. Therefore the open-loop phase margin of the two autopilots is basically same. The attitude autopilot inner loop's magnitude margin is a litter lower but it is still can meet the requirement.

Table 2. Margin of stability for height control system

	Magnitude/dB	Phase/deg	Cross-over Frequency/rad/s
Acceleration Inner Loop	14.4	61.3	2.82
Attitude Inner Loop	5.57	64.1	6.53

Through the above comparative analysis, we can see that the performance of these two autopilots' inner loop acting as the height control loop is basically the same and both are feasible.

5 The Realization of Gravity Compensation

The gravity has much influence on the missile's practical flight. To height control loop, this accounts to lead in a constant interference item. It can let the height respond overshoot become larger. To low attitude skimming or hedgehopping cruise missile, in the process of dropping height control, we must avoid the larger height overshoot letting the missile hit the land. For the engineering aspect, we generally add a gravity compensation command at the autopilot command part and adopt the feed-forward way to eliminate the interference influence of the gravity.

To acceleration autopilot, the gravity compensation can take a constant value, $a_{cg} = g$; if we can get an exact speed dip angle information through a measure device in the missile, it can also be $a_{cg} = g\cos(\theta)$.

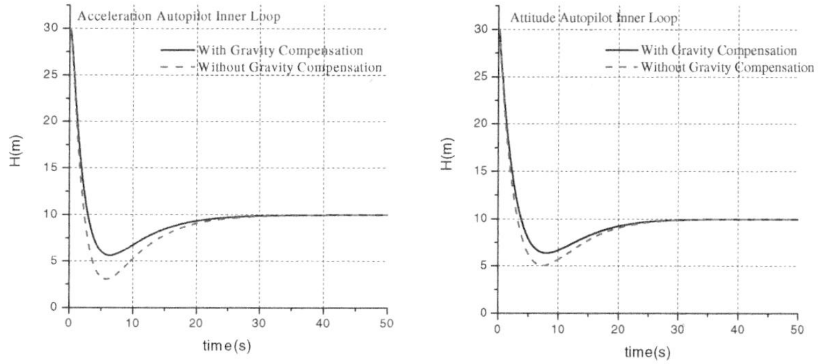

Fig. 8. Time responses of height control system with and without gravity compensation

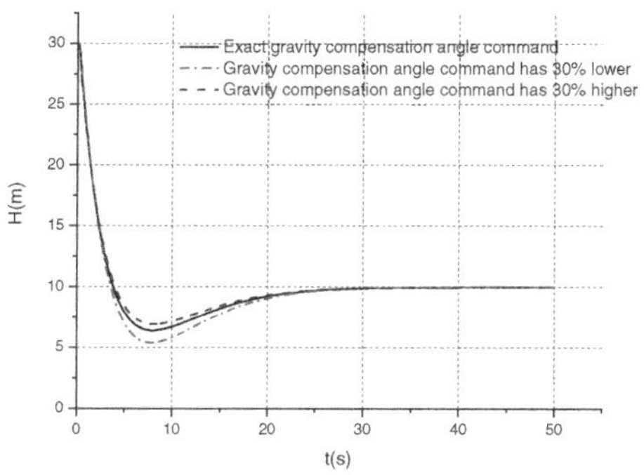

Fig. 9. Time responses of height control system with mistake of attitude command

To attitude autopilot, we must make the gravity acceleration change into corresponding attitude angle command [7]. According to the stress relationship while the missile is in the condition of level flight, we can take the angle of attack required by the balance gravity as the attitude gravity compensation command $\vartheta_{cg} = \alpha_B$.

Taking the missile initial height H_0=30m, the expect dropping cruising flight H_c=10m, the Fig. 8 shows the height changing curve under the imponderability compensation command condition, when using the gravity compensation, either the attitude autopilot or the acceleration autopilot inner loop, the height respond overstrike is both reduced. At the meanwhile, we can conclude that we can just improve the quality of the height control transition process by having the imponderability compensation command and it has no influence on the rapidity of the system.

Compared the corresponding gravity compensation command calculating equations of two autopilots, we can easily get that the numerical value of ϑ_{cg} relates to missile's current flight speed, atmospheric density, quality and aerodynamic characteristic. But all these above parameters can be gained exactly so it makes the gravity compensation command have some deviation. The diagram 10 is the height control curve under the assumption condition that the attitude gravity compensation has ±30% deviation. The transient process of the height control is obviously influenced by the gravity compensation. It also proves that compared with the acceleration autopilot, the calculation of the gravity compensation signal which is realized by the attitude autopilot inner loop has more dependence on the missile flight parameters and it is harder to realize.

6 Conclusion

Attitude autopilot is limited by the zero restriction of the angle of attack's lag constant so its respond speed is impossible fast. Although the acceleration autopilot's frequency band can be designed widely, but from the overload output to longitudinal velocity Kinematics integral items, this process has large influence on the stability of the height loop. In the same stability condition, on the rapidity of the acceleration autopilot inner loop's height control system has not obvious advantages. Therefore for the height control loop design, it can use these two kinds of autopilots.

The gravity compensation command can effectively improve the characteristic of the height control recovery curve. The acceleration autopilot use the literal gravity compensation and the command can easily calculate. But for the attitude autopilot, the gravity compensation command has much dependence on the missile flight parameters. And it can be influenced by the parameter so it can bring some errors.

References

1. Garnell, P.: Guided Weapon Control Systems. Beijing Institute of Technology, Beijing (2004); second revision by Qi, Z.-K., Xia, Q.-L
2. Zarchan, P.: Tactical and strategic missile guidance. American institute of Aeronautics and Astronautics, Washington, D C (1994)
3. Zhao, H.-C., Gu, W.-J., Peng, W.-H.: Design of the height control system for an anti-ship missile based on overload-control. Tactical Missile Control Technology 46(3) (2004)
4. Gormley, D.M.: Dealing With Cruise Missile Proliferation. In: AIAA 1998 Missile Sciences Conference The Emerging Threat and Nonproliferation Challenges (1998)
5. Li, Y., Huang, C., Hao, X.: Trajectory planning of the television-guided and air-to-groud missile. Journal of Ballistics 13(4) (2001)
6. Cai, L.: Guidance and control system of the cruise missiles. Tian Gong Press, Taibei (1989) (in Chinese)
7. Qian, X., Lin, R., Zhao, Y.: Missile flight mechanics. Beijing Institute of Technology press (2003)

An Optimized Scheduling Strategy Based on Task Type in Xen

Yuxin Wang[1], Xiaogang Wang[1], and He Guo[2]

[1] School of Computer Science and Technology
[2] School of Software Dalian University of Technology, Dalian, China, 116024
{Wyx,guohe}@dlut.edu.cn, xiaogangwang0417@qq.com

Abstract. Along with the development of virtualization technology, the computing tasks' type worked in virtual machine is more diverse and GPU intensive task has become a trend. But most current VMM scheduling algorithms pay little attention to the task types of diversification, but abide by "fair" principle and distribute CPU resources equally, which would be unable to meet the requirement for latency-sensitive tasks. Through theoretical analysis and a series of experiments, this paper confirme that the CPU resources sensitivity of different task type is different and put forward an optimized scheduling strategy based on task type. The VMM calculates the real-time scheduling priority according to CPU and GPU resource utilization of all active guest domains, and then redistributes the physical CPU resources. The experiments show that the optimized scheduling strategy can promote the performance of the GPU intensive tasks and improve user experience.

Keywords: CPU virtualization, GPU virtualization, scheduling strategy, task type, real-time priority.

1 Introduction

Virtualization, in computing, is the creation of a virtual version of something, such as a hardware platform, operating system, a storage device or network resources [1]. In the future, virtualization must be used by more and more normal users and the types of the tasks on virtual machines must be more diverse. However, as a developing technology, many problems still need to be solved urgently.

Currently, most prevalent processor scheduling strategies in virtual machine monitor consider the fairness as the first concern, but the diversity of computing tasks type is shown less regard, which deteriorate user experience seriously. Therefore, it is very necessary to improve existing scheduling strategy to optimize the system performance. Some relative works have been proposed. Chen et al. extend the Credit scheduler by using flexible time slice and real-time priority to meet the requirement for latency-sensitive tasks[2]. Ongaro et al. propose the priority of idle domains when they receive events by boosting[3]. Paper [4] presents a real-time scheduling mechanism based on messages. As mentioned above, most optimized schedulers are based on IO tasks and audio tasks, the GPU intensive tasks have not been considered in the related works, with this point, we design an optimized scheduling strategy based on task type in Xen[5].

G. Lee (Ed.): Advances in Automation and Robotics, Vol.2, LNEE 123, pp. 515–522.
Springerlink.com

The rest of this paper is organized as follows: In next section, we analyze and confirm that the CPU intensive and GPU intensive task have different CPU resource sensitivities. In the third section, data analysis of GPU intensive task performance in Xen environment is carried out to show that existing scheduler is not sufficient, and then we describe the framework of our optimized scheduling strategy. Performance evaluation is shown in the fourth section and the fifth section concludes the paper.

2 Tasks' Sensitivity of CPU Resource

The computing tasks are mainly divided into two kinds: the CPU intensive tasks and GPU intensive tasks. CPU intensive tasks make use of the computer's CPU resource for conventional logic operation and most floating-point arithmetic. GPU intensive tasks are usually CPU+GPU heterogeneous calculation mode, which are also divided into two types: one type is the traditional graphics rendering tasks, which mainly includes 3D image modeling and 3D gaming applications; The second one is based on General Purpose GPU(GPGPU), this general computing task makes use of GPU parallel characteristics to do massive parallel computing.

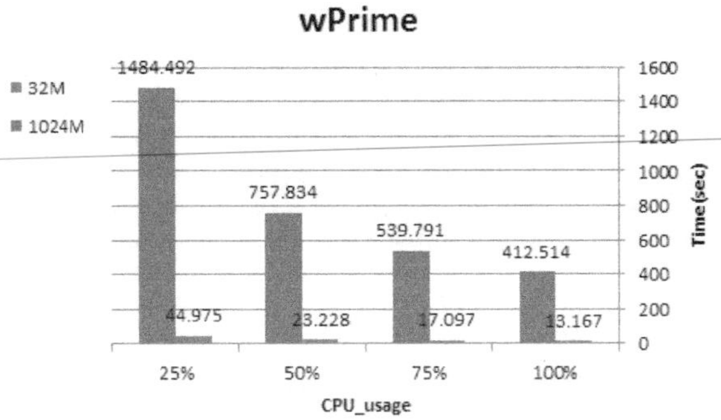

Fig. 1. Time of wPrime cost under different CPU utilization

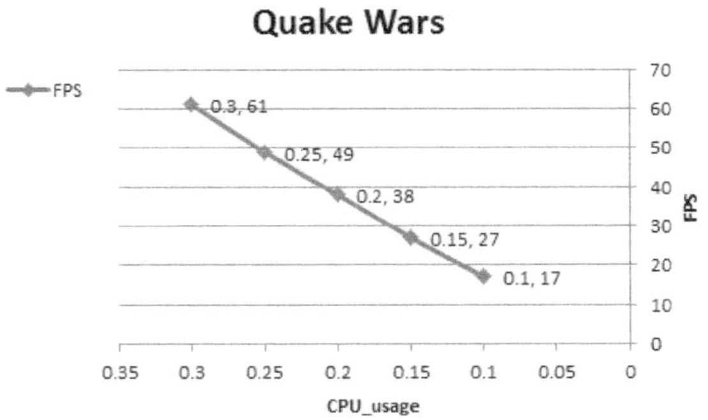

Fig. 2. FPS of Quake Wars under different CPU utilization

We suppose many scenarios, do many experiments and theoretical analysis, then draw the conclusion that the CPU resource sensitivity of GPU tasks are higher than that of CPU tasks in the situation that the CPU resource cannot meet their requirements. In order to verify it, three experiments with different types of tasks are shown here and the experiment results are shown in figure 1 and figure 2.

Experiment one is the execution of CPU intensive task wPrime. As shown in figure 1, with the CPU utilization reduces, the running time increases a little. This shows that the CPU resources sensitivity of CPU intensive task is not very high. Experiment two is the execution of GPU intensive task Quake Wars. As shown in figure 2, with CPU resource utilization rate descends 5%, the FPS declines about 20%, which shows that the CPU resources sensitivity of the GPU intensive task is high. Experiment three is the GPU intensive task MDGPU based on CUDA. Just like Quake Wars, its CPU resources sensitivity is also very high.

The traditional GPU intensive tasks mainly are the real-time rendering graphics tasks. In these applications, the CPU is responsible for logic strong affairs, and the GPU is responsible for calculation of high concentration calculated graphics rendering. When the GPU are rendering in current frame, CPU can calculate the content of the next frame needed to draw, i.e. the CPU computing and GPU computing are parallel. If the CPU resources cannot meet the needs of these tasks, then the CPU will not timely calculate the content of the next frame needed to draw and the GPU will wait for the CPU's calculation results, which will cause the phenomenon such as low FPS, freeze frame and affect the user experience seriously.

With the continuous improvement of GPU programmability, the research of using the GPU to do the general computing (called GPGPU) gradually becomes active. This kind of calculations usually use CPU+GPU heterogeneous mode. CUDA programming model takes CPU as the Host and takes GPU as a coprocessor. However, because the current GPU function is relatively limited, the serial part done by the CPU is still very impressive. Therefore, if the GPU-intensive tasks based on GPGPU could not get enough CPU resources, the CPU serial part of the task would become the bottleneck, which will directly impact on the GPU part of the task, resulting in lower GPGPU task parallelism and the whole system operational efficiency dropped significantly.

Through these experiments and theoretical analysis, it can be concluded that: the heterogeneous programming model of GPU-intensive tasks determines its sensitivity of the CPU resources higher than that of the CPU-intensive tasks.

3 Real-Time Priority Calculation of the Guest Domain

3.1 Credit Scheduling Algorithm

Credit scheduling algorithm[6] is used as the default scheduling algorithm since Xen 3.0, which is a fair share as a proportion and non-preemptive scheduling policy. The biggest advantage of credit algorithm is that it can globally manage multiple physical CPU, which will allocate the CPU resource to each virtual CPU fairly and efficiently. However, due to this algorithm follows the principle of fair allocation in any case, when multiple types of tasks are running on the virtual machine, the CPU resource may not meet the requirements of all tasks, especially the GPU-intensive tasks.

We also give an experiment, which has three stages, to verify this problem., and the results are shown in Figure 3 and Figure 4. Glxgears is chosen as GPU intensive task, which runs on Dom0 in all stages and requires 50% of the CPU resources normally. Benchmark cpu_scene is chosen as CPU-intensive task, which requires 100% of the CPU resources normally.

Stage 1: cpu_scene runs on guest domain VM1; domains VM2, VM3 are off.

Stage 2: cpu_scene runs on guest domain VM1, VM2; domain VM3 is off.

Stage 3: cpu_scene runs on guest domain VM1, VM2 and VM3.

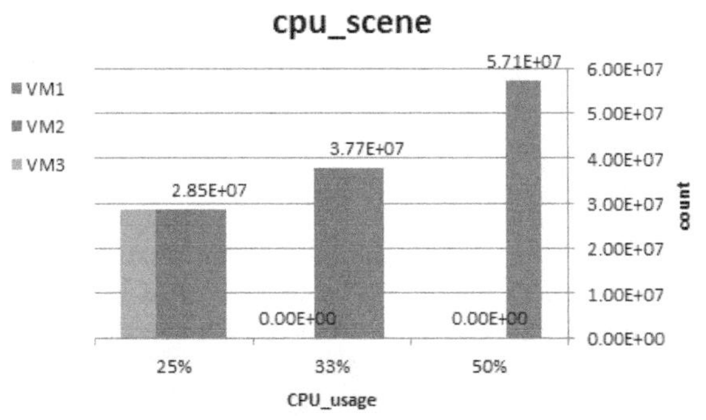

Fig. 3. Number of calculations of cpu_scene under different CPU utilization

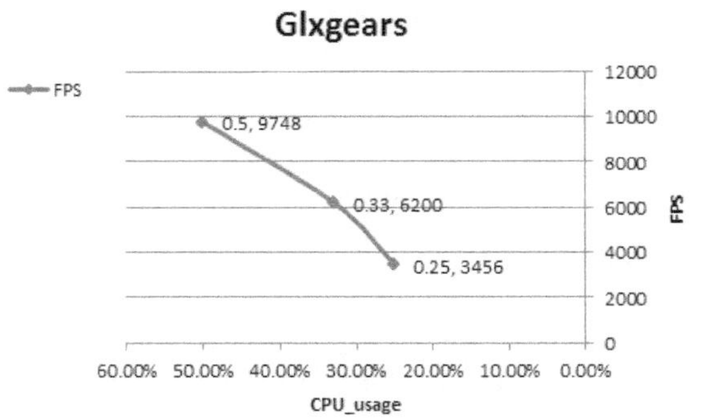

Fig. 4. FPS of Glxgears under different CPU utilization

The GPU-intensive tasks on are generally the foreground applications, which has to interact with the user frequently and need higher real-time requirements. If such applications have poor performance, it will seriously affect user experience. However, the CPU-intensive tasks are generally background calculation applications and the real-time requirements are lower, so the priority of such applications should be less than GPU intensive tasks. When the physical CPU resources cannot meet the requirements of the guest domain, VMM should allocate more CPU resources to the GPU-intensive tasks. According to this scheduling principle, we put forward the scheduling optimization system based on the types of the tasks.

3.2 The Framework of Optimized Credit Scheduling Strategy

The process of a complete improved scheduling strategy is described in Figure 5.

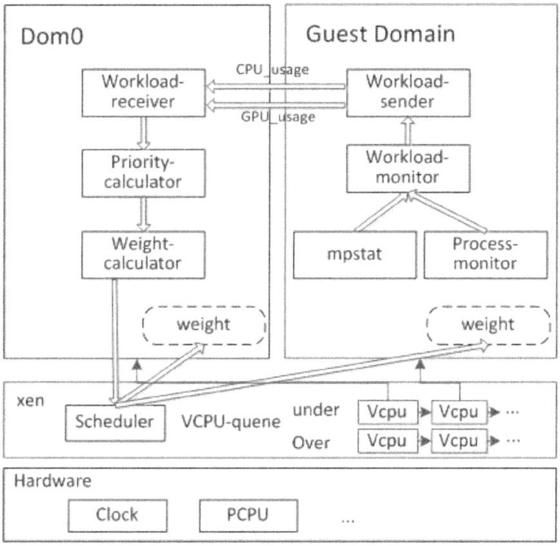

Fig. 5. Structure and workflow of optimized Xen scheduling strategy

The Workload-monitor that runs on the guest domain gets the CPU and GPU resources utilization every 10s and passes them on to Workload-sender, and then Workload-sender delivers the utilization information to Workload-receiver that runs on Dom0. Workload-receiver module is responsible for receiving load information from all the active guest domains and sending the information to the Priority-calculator that calculates each guest domains' real-time scheduling priority every 20s. Weight-calculator module recalculates domain weight value of each guest domain based on the Dom_priority and calls the character management tool xm redistribute the weight values of each guest domain.

3.3 Virtual Machine Workload Monitor

The Workload-monitor is run on the guest domain, which is responsible for obtaining the real-time CPU and GPU resource utilization. This process is the most important part in the scheduling system and premise of the scheduling optimization. We use Linux real-time system tool mpstat to obtain the CPU resource utilization of the guest domain system. With the continuous development of the hardware detection tool, monitoring the usage of the GPU has been very convenient in the native system, such as GPU tools Everest or GPU-Z in Windows OS. In Linux system, the system tools nvidia-smi provided by CUDA toolkit can be used to monitor the GPU utilization. But in current Xen virtual machine, all above methods can not work. At the same time, although running 3D applications in a virtual environment has become a direction, but for Xen, the 3D acceleration has not been supported very well. So we use OpenGL rendering acceleration solution VMGL[7], which is an open source virtual machine graphic library. We did a series of experiments, which mainly run OpenGL-based 3D applications in a non-virtualized Linux system and found that there is a nearly linear proportional relationship between GPU utilization and the called frequency of the OpenGL library functions. We divide the GPU utilization into six levels (0~5) according to the called frequency of the OpenGL library functions as formula (1). If the gpu_usage>5, we set gpu_usage =5.

$$\text{gpu\_usage} = \text{lib\_frequency} / 4000 \tag{1}$$

Generally there should be three types of computing tasks run on the guest domain: CPU-intensive tasks, GPU-intensive tasks, both CPU-intensive and GPU-intensive tasks. In order to distinguish GPU and hybrid tasks in the scheduling process, the Workload-monitor pass three required parameters to the Workload-sender scheduling, which are gpu_usage, cpu_usage and mixcpu_usage. The gpu_usage represents GPU resource utilization of the customer domain and cpu_usage represents CPU resource utilization of the GPU-intensive tasks. The mixcpu_usage represents the CPU resource utilization of the CPU-intensive tasks.

3.4 Real-Time Priority Calculation of the Guest Domain

The Workload-receiver module passes the CPU and GPU resource utilization of all active guest domains to the priority-calculator module every 20 seconds, and the latter transforms the utilization information into the required five parameters by formula (2) to formula (4) and two principles. The five parameters are the gpu_level of GPU-intensive tasks, the cpu_level of GPU-intensive tasks, the mixcpu_level of hybrid tasks, parameters k1 and k2. The two principles are: if gpu_usage = 0, then k1=0 and k2=-1; if gpu_usage > 0, then k1=1 and k2=1. Then Workload-receiver calculates the real-time scheduling priority of each guest domain by formula (5).

$$\text{gpu\_level} = \text{gpu\_usage} \tag{2}$$

$$\text{cpu\_level} = \text{cpu\_usage}*10+1 \tag{3}$$

$$\text{mixcpu\_level} = \text{mixcpu\_usage}*10+1 \tag{4}$$

$$\text{dom\_priority} = 0.5*((\text{gpu\_level}+1))+\text{k1}/\text{cpu\_level}-0.05*\text{mixcpu\_level}*\text{k2} \tag{5}$$

Formula (5) mainly considers following four factors:

First, the parameters dom_priority and gpu_level are proportional. An increased gpu_level means the task have higher GPU-intensive and the rendering cores involved in task are much more, thus the task has higher sensitivity of CPU resources, which should be given higher domain priority.

Second, dom_priority and cpu_level are inversely proportional. A very high cpu_level means the GPU-intensive task has gotten many CPU resources. Even the task is redistributed more CPU resources, its performance will not improve obviously.

Third, mixcpu_level of the CPU-intensive tasks will impact the real-time priority of the guest domain. If CPU+GPU hybrid tasks are running on a guest domain and the proportion of the CPU-intensive tasks is too large, the scheduling system should not assign excessive CPU resources to the guest domain.

Fourth, if different guest domains are running different types of tasks, the real-time priority among them should not be different too much; otherwise it will lead to very unbalanced distribution of CPU resources and low performance.

4 Performance Evaluation

The experimental platform is: Intel Core i5 750 CPU, NVIDIA 9800 GTX+ GPU, 2G RAM. The experiments mentioned above are all running on this platform. OS is Xen 4.0.1 + Ubuntu 9.10 64bit server. Benchmark cpu_scene is used to simulate the CPU-intensive tasks and records the task performance data; Glxgears is used to simulate the GPU-intensive tasks and records the FPS of the task. This experiment is also divided into three stages and the experimental comparison results are shown in Figure 6 and Figure 7.

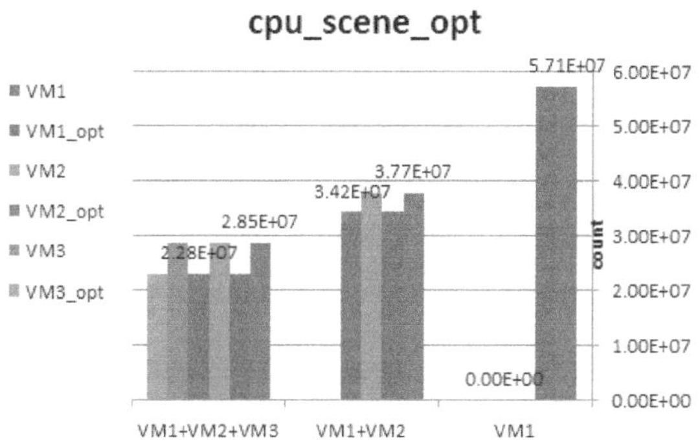

Fig. 6. Performance comparison of CPU-intensive tasks

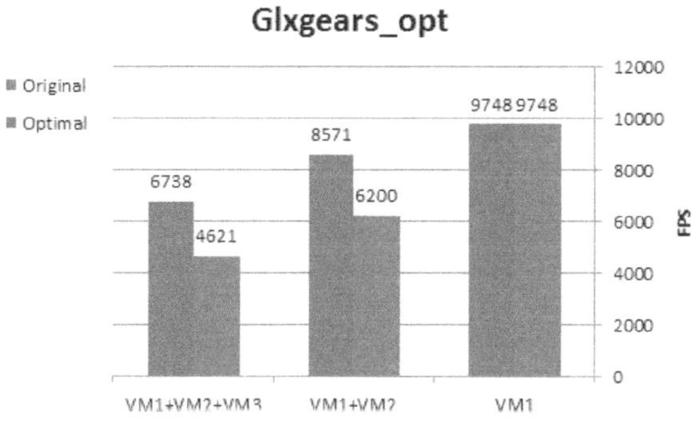

Fig. 7. Performance comparison of GPU-intensive tasks

From figure 6 we can see, after using the optimized strategy, the performance of cpu_scene has no change in the first stage; In second stage, the performance drop slightly and the computational efficiency reduce 9.3% than the previous; While in the third stage, the computational efficiency reduce 20%.

Figure 7 shows that, after using the optimized strategy, the performance of Glxgears has no change in the first stage for the CPU resource is enough and Glxgears can run at the highest performance; In second stage, the performance is improved significantly and in the third stage the performance improves 45%.

5 Conclusion

This paper presents an optimized scheduling strategy in Xen based on task type. The optimized scheduler monitors and calculates both the CPU resource utilization and the GPU resource utilization of all active guest domains, and recalculates the real-time scheduling priority, then makes the VMM redistribute the physical CPU resources. Compared with other scheduling algorithms, our optimized scheduler based on task type has following merits: 1)It can meet the diversity requirement of the tasks running on the guest domains; 2) The end-users would get better user experiences when they were running GPU intensive task; 3) The physical GPU resource utilization will be promoted to a certain extent.

References

1. Turban, E., King, D., Lee, J., Viehland, D.: Ch. 19: Building Ecommerce Applications and Infrastructure. In: Electronic Commerce A Managerial Perspective, 5th edn., p. 27. Prentice-Hall (2008)
2. Chen, H., Jin, H., Hu, K., Yuan, M.: Adaptive Audio-aware Scheduling in Xen Virtual Environment. In: The 8th ACS/IEEE International Conference on Computer Systems and Applications (AICCSA 2010), Hammamet, Tunisia, pp. 1–8 (May 16-19, 2010)
3. Ongaro, D., Cox, A.L., Rixner, S.: Scheduling I/O in virtual machine monitors. In: The Fourth ACM SIGPLAN/SIGOPS International Conference on Virtual Execution Environments (VEE 2008), New York, NY, USA, pp. 1–10 (2008)
4. Wang, Y., Zhang, J., Shang, L., Long, X., Jin, H.: Research of Real-time Task in Xen Virtualization Environment. In: The 2nd International Conference on Computer and Automation Engineering (ICCAE), Chongqing, China, pp. 496–500 (Febraury 26-28, 2010)
5. Barham, P., Dragovic, B., Fraser, K., Hand, S., Harris, T., Ho, A., Neugebauer, R., Pratt, I., Warfield, A.: Xen and the Art of Virtualization. In: The 19th ACM Symposium on Operating Systems Principles, New York, NY, USA, pp. 164–177 (October 2003)
6. Lee, M., Krishnakumar, A.S., Krishnan, P., Singh, N., Yajnik, S.: Supporting soft real-time tasks in the xen hypervisor. In: The 6th ACM SIGPLAN/SIGOPS International Conference on Virtual Execution Environments (VEE 2010), Pittsburgh, PA, USA, pp. 97–108 (March 17-19, 2010)
7. Lagar-Cavilla, H.A., Tolia, N., Satyanarayanan, M., de La-ra, E.: VMM-independent Graphics Acceleration. In: The 3rd International Conference on Virtual Execution Environments (VEE 2007), San Diego, California, USA, pp. 33–43 (June 13-15, 2007)

The Design of Machine Olfactory System Experiment Device and the Experiment on Meat Products Smell Recognition

Xiaojing Yang, Ju Yang, Jing Zhu, Zhenghu Yan, and Fujie Zhang

Faculty of Mordern Agricultural Engineering, Kunming University of Science and Technology, Kunming, 650500, China
xjyang@vip.sina.com

Abstract. With the improvement of living standard, people are paying more and more attention on the freshness of meat products. T he Machine olfactory system was established to detect meat freshness simply and quickly. In this paper the principle of electronic nose system was introduced, six types of gas sensors were selected according to the composition of smell of corrupted meat, and hardware system of a machine olfactory was built by sensor array and data acquisition card. In addition, LabVIEW and MATLAB were used to finish data acquisition, data analysis and waveform display. In order to find the relationship between corrupted meat odor and the freshness, a machine olfactory system experiment device was made to do identification experiments of meat sample that stored 1-5 days. The results of experiment showed that gas concentration of detected pork samples was related to the storage time. So, it is feasible to identify meat freshness with machine olfactory system experiment device.

Keywords: Machine olfactory, Gas sensor, Sensor array , Data acquisition, Freshness.

1 Introduction

Olfactory sensation is an indispensable and important part of human and animal's sensory system, and it plays a very important role for live. The machine olfactory system is an imitation of biological olfactory system, it is an electronic system which is able to recognize the gas and odor, it can also be consider as the simulation device of human and animals. It simulates the working principle of biological nose, and also be known as artificial olfactory system or electronic nose. In the recent years, many literatures point out that electronic nose is a very promising analysis method which can be applied in the meat detection[1]. It analyses the volatiles which are in the meat food so as to achieve the purpose of analysis and detection. we can detect smells with it very quickly. It operates simply and has a high sensitivity and good reproducibility, so it has been studied deeply and used widely in many fields, but there is a big gap between the current electronic nose and creature sensory system. In many studies of detecting the meat freshness by electronic nose, they still can't achieve content result on the sensitivity of detecting and recognize rate[2]. In this paper the project will imitate biological system, study the composition and theory of the machine olfactory system,

G. Lee (Ed.): Advances in Automation and Robotics, Vol.2, LNEE 123, pp. 523–528.
Springerlink.com

and design the experimental facility of the machine olfactory system which consists of sensor array. Then take the experiment which is about detecting the meat flavor with the machine olfactory system experiment device that is self-made, analysis the experiment result and summary a law.

2 The Component of Machine Olfactory System

The overall design scheme of the machine olfactory system is as follow: the sensor array is composed of different sensors which can detect diffident gas. The information of different sensors is transmitted to the computer through the data acquisition card. The data processing and analysis are took by computer software. So machine olfactory system is composed of sensor array, test cabinet, data acquisition card and PC. The type of the sensors which are chose to take experiment are as follow: TGS826, TGS813, TGS26220, TGS2602, TGS2201.These sensors are used to test hydrogen sulphide ammonia hydrocarbon, alcohols, air quality, nitric oxide and nitrogen dioxide. The type of the data acquisition card which is chose is YanHua PCI-1716 of advantech company. The test cabinet was self-made, and it is made of transparent organic glasses. These glasses are connected by aluminium alloy. There are two fans on both sides of the test cabinet, in the test cabinet, there is a platform to load test sample. The prototype of machine olfactory system experiment device is as shown in figure 1 and figure 2.

Fig. 1. Sensor array and test cabinet

Fig. 2. Machine olfactory system experiment device

3 The Experiment on Meat Products Smell Recognition

3.1 The Experiment Purpose

This experiment is used to test the smell of pork samples which are in different storage time. In the experiment, the machine olfactory system was designed to detect the pork freshness, the sensor array is important component of this system. The relationship between gas concentration and storage time will be find out, and the result of this experiment will be analyzed and discussed.

3.2 The Experiment Method

This experiment mainly use machine olfactory system experiment device to detect the relationship between pork freshness and the storage time. Its principle is as follow: the core of the machine-gas sensors transform the responses into different voltage signal through circuit, then analysis the voltage signal according to the experiment data that collected by the machine olfactory system. The specific process is as follow: pass a five volt voltage to the experiment device, so that all components of the experiment device can be in the normal work. Put the samples that are to be tested into test cabinet, then sensor array will acquire the signal of smells. After transforming the signal of the smells into voltage signal by circuit, the output signal will be present. The flow process of this experiment is as shown in figure 3.

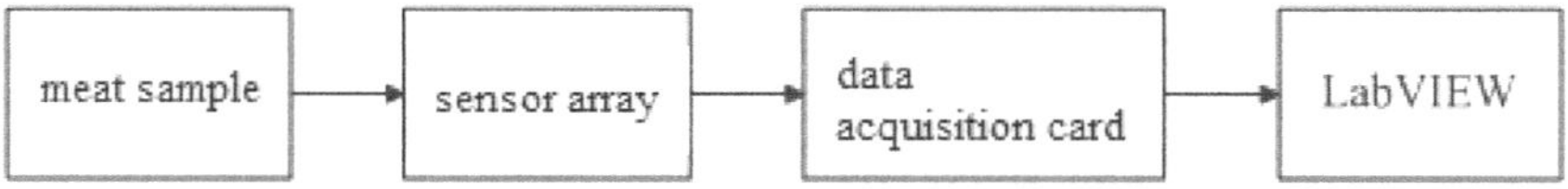

Fig. 3. Flow process of experiment on meat smell recognition

3.3 Experimental Data Collection

LabVIEW is a software about program development, it is developed by American national instrument(NI) company. LabVIEW is similar to C language and Basic development environment, but it uses graphical language to program, the program edited by G language in the form of block diagram, so that it makes program simple[3]. Due to LabVIEW is high reliability, professional configuration of the software, it collects with development software seamlessly, and it can collect any signal, and have a wide range of data collection hardware, so this project mainly use LabVIEW to collect experimental data. The LabVIEW software of machine olfactory system experiment device is as shown in figure 4 and figure 5.

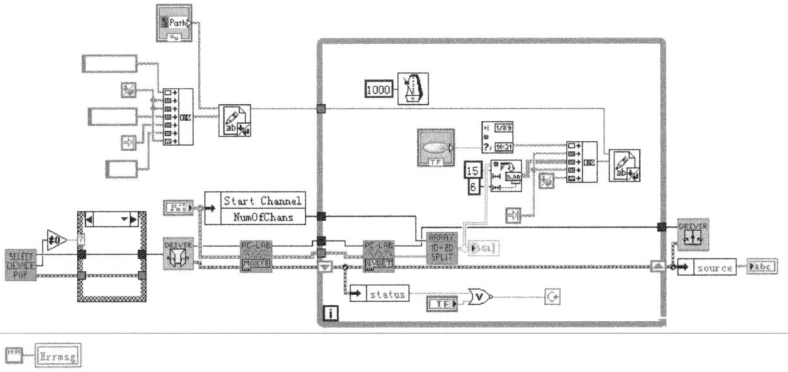

Fig. 4. LabVIEW program chart panel

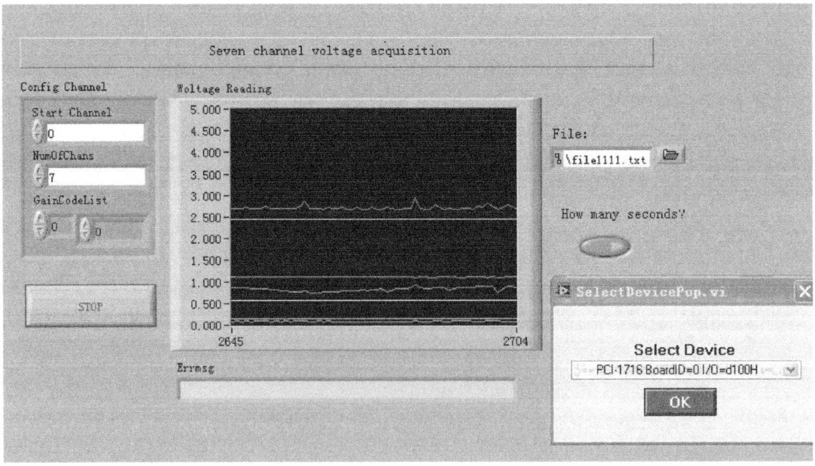

Fig. 5. LabVIEW front panel

3.4 The Data Analysis and Processing

MATLAB contains signal processing toolbox, it is designed for signal processing problems, and handle the conventional signal task about digital or analog systems, so this project mainly use MATLAB for data pretreatment. And then find out the relationship between pork freshness and storage time. The sample curve is as shown in figure 6. The ordinate of the graph stands for the voltage output of the gas sensors.

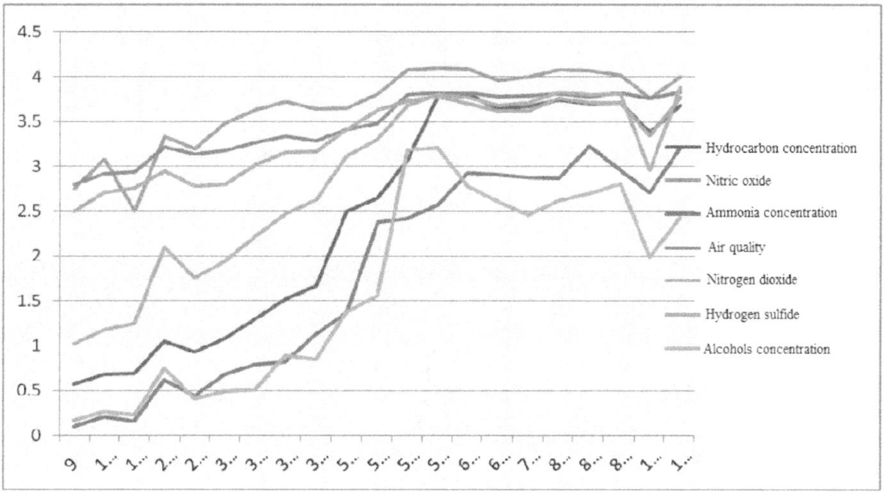

Fig. 6. Relationship between gas concentration and storage time

The abscissa is shown as the length of storage time, the unit is hour. The ordinate stands for the voltage value of the gas sensor, the unit is volt.

As the experiment graph shows, every smell concentration curves about the pork freshness has a rising trend with the extension of display time. However, because of some unpredictable factors is inevitable, such as defect of the sample, the mutations of the external environment influence on sensors, sensors themselves exist fatigue and incorrect operation etc, during the process of test sampling and operating, makes the experiment data produce some abnormal fluctuations. As far as the trend of the graph, a conclusion can be got: the concentration of the smells that produced by corrupt pork has a positive relationship with the display time. At the same time, it has practical value to choose the sensor array to measure the pork freshness. The sensor array composes of six kinds of sensors which are used to test hydrogen sulphide ammonia hydrocarbon, alcohols, air quality, nitric oxide and nitrogen dioxide.

4 Conclusion

The scheme and the detailed design about the machine olfactory system experiment device was put forward. The experiment device of the machine olfactory system is self-made. It divides into hardware and software components. Hardware part includes gas sensor array, data acquisition card, smell collection testing cabinet and relevant circuit. Software part includes data acquisition software which is LabVIEW, experiment data processing software which is MATLAB7.0. The software collects and processes experimental data of the pork smell, so as to judge the freshness of pork that are in different storage time. However some research works still need to continue. In the future, the research work can focus on every single gas that to be tested, consider the experiment environment requirements and other related condition requirements, choose suitable sensor for array, coupled with better signal acquisition equipment, method and signal processing method, to make the result more accurate and more practical. The subsequent work of the project will focus on data processing and the improvement of the ability to recognize smells.

Acknowledgements. The authors would like to acknowledge the support of nation college student innovation experimental plan project in china.

References

1. Peris, M., Escuder-Gilabert, L.: A 21st century technique for food control: Electronic noses. Analytica Chimica Acta 638, 1–15 (2009)
2. Zhang, Z., Tong, J., Chen, D.-H., Lan, Y.-B.: Electronic Nose with an Air Sensor Matrix for Detecting Beef Freshness. Journal of Bionic Engineering 5, 67–73 (2008)
3. Cheng, X., Zhang, Y.: Design of LabVIEW Software. Tsinghua university press, Beijing (2007)

Activity Analyzing with Multisensor Data Correlation

GuoQing Yin and Dietmar Bruckner

Institute of Computer Technology, Vienna University of Technology,
Gußhausstraße 27-29,
A-1040 Vienna, Austria
{Yin,Bruckner}@ict.tuwien.ac.at

Abstract. In an ambient assisted living project, a novel way to be proposed in order to protect privacy, increase comfort and safety: a system that with different kinds of sensors installed in the living environment and observe the daily activities of the elderly. Based on the daily activities of the user an activity model will be build. In case of unusual activities the system will send alarm signal to caregiver according the build activity model.

The huge amount data from sensors is a computational burden for the system and an obstacle for the system to get character parameters to build activity model. In the paper data correlation is used to deal with the data and detect the relationship between sensors. In predefined time interval the data from sensors is correlated. The huge amount data is translated to a few correlation parameters between sensors. Furthermore the values of correlation parameters between sensors are changing in different time interval. That means different activities of the user are detected by the system.

Keywords: Activity model, Ambient Assisted Living, Data Correlation, Sensor fusion.

1 Introduction

There are many papers about sensor fusion, sensor data correlation: in paper [1] sensor fusion architecture designed to bridge the gap between low-level sensor data and the high-level knowledge. The authors in paper [2] present a probabilistic approach to alert correlation, extending ideas from multisensor data fusion. Paper [3] shows how the spatial correlation can be exploited on the Medium Access Control (MAC) layer. The authors in paper [4] define the concept of sensor databases mixing stored data represented as relations and sensor data represented as time series. Techniques that exploit data correlations in sensor data to minimize communication costs were designed in paper [5]. The authors describe a system that provides a unified view of data handling in sensor networks, incorporating long-term storage, multi-resolution data access and spatio-temporal pattern mining in paper [6]. Some papers in ambient assisted living domain: in paper [7] a project was described that concern the ability to locate and track people within their homes, which based on a standard ceiling mounted camera, an on body accelerometer and simple yet robust image processing. Paper [8] provides a survey on Wireless sensor networks for healthcare. Paper [9] investigates the development of systems that involve different

G. Lee (Ed.): Advances in Automation and Robotics, Vol.2, LNEE 123, pp. 529–534.
Springerlink.com

types of sensors, monitoring activities and in order to detect emergency situations or deviations from desirable medical patterns, just like papers [10], [11], [12] from project ATTEND (AdapTive scenario recogniTion for Emergency and Need Detection).

2 Contribution and Innovation

This paper is a further research related earlier papers [10], [11], [12] based on the same idea: a system observes the activities of the user and builds the daily activities model of the user. According the model in case of unusual activities happened the system will send alarm signal to caregiver. But in papers [10], [11], [12] only motion detector is used to observe the activities of the user all the time. For a more flexible, robust, and accurate observation result many different types of sensors were installed in the living environment of user. For example passive infrared sensor (motion detector), door contactor, accelerometer, temperature sensor, light sensor. Each sensor sends signal to controller when it gets information from environment, such as: person movements, door closed or opened, vibration in the environment, temperature, and light. There are nearly 10,000 signals that the sensors send to controller every day in a real test experiment. How to deal with the huge data and furthermore find the relationship between these sensors are great challenges. In this paper data correlation is used to solve the problem.

Fig. 1. The experiment environment with different types of sensors

Don't like some other ambient assisted living projects in the project ATTEND will be without camera and microphone are used. This is because of privacy issues of the user. At the same time nothing should be wear on the body of the user and nothing should be activated by the user in case of emergency [10]. All of these measures make the user more comfortable but a great challenge to realization. In the following figure 1 the experiment environment and these different types of sensors will be introduced at first. There are sensors installed in different locations of the rooms: at the entrance is sensor 1, at WC is sensor 2, at bathroom is sensor 3, at kitchen are sensors 4 and 5, at living room are sensors 6, 7, 8, 9 and 12, at bedroom is sensor 13, and at restrooms are sensor 10 and 11. Each sensor is multisensor, which means each of them can detect some of the following information: movement, vibration, contactor of the door, temperature and light. For example sensor 1 at the entrance can get all the 5 physical parameters. Sensor 2 at WC has without the ability to detect movement of the user in WC. Sensor 6 in living room can only get information about light, temperature and vibration in the environment. The sensors with arrow indicate that motion detector included in the multisensor.

3 Data Correlation in Predefined Time Interval

Data correlation is used here to measure the relationship between different types of sensors. Because the user has activities in different areas in the rooms in different time of the day, for example at noon the user cooks for lunch, so perhaps activities from the user will be detected by sensor 4 and 5; in the evening the user watches TV in living room, so perhaps sensor 6, 7, 8, 9, and 12 detect activities from the user; at night the user sleeps at bedroom, so perhaps only sensor 13 detects activities from the user, but if the user goes from living room to WC, then goes to bathroom, then comes back to living room, so most of the sensors will be detect activities of the user. With sensor data correlation the relationship between these sensors will be find out.

Because of the user have activities in different areas in different time. So the relationship between sensors changes from time to time, in this paper we predefine a time interval (T_{int}), and try to find out the sensors relationship in the time interval. For example the sensor 1 has value $X=[x_1, x_2, \ldots x_t]$; the sensor 2 has value $Y=[y_1, y_2, \ldots y_t]$; Sensor 1 and 2 compose a set pair $[x_t, y_t]$ in time interval T_{int}.

1) The variance of X and Y

$$Var(X) = E[(X-\mu_x)]; \; Var(Y) = E[(Y-\mu_y)] \tag{1}$$

Here μ_x and μ_y are the mean value of X and Y.

2) The covariance of X and Y

$$Cov(X,Y) = E[(X-\mu_x)(Y-\mu_y)] \tag{2}$$

3) The correlation coefficient between X and Y

$$R_{xy} = Cov(X,Y) / \sqrt{Var(X)Var(Y)} \tag{3}$$

Because of Cauchy-Schwarz inequality $|Cov(X,Y)| \leq \sqrt{Var(X)Var(Y)}$, so $-1 \leq R_{xy} \leq 1$. If sensor data X and Y without relationship (they are independent) $Cov(X,Y)=0$, so $R_{xy} = 0$. If sensor data X and Y are the same values in time interval T_{int}, because of $Cov(X,Y)=E(XY)-E(X)E(Y)$, so $Cov(X,X)=E(XX)-E(X)E(X)=E(X^2)-(E(X))^2 = Var(X)$. So $R_{xy} = 1$.

In above figure 1 there are not only 2 sensors 1 and 2 but totally 13 sensors, so the correlations between sensors have to be done between each pair of the sensors. That means sensor 1 correlates with sensor 2, gets value $R_{1,2}$, then sensor 1 correlates with sensor 3 and gets value $R_{1,3}$, till all other sensors correlated with sensor 1. Then sensor 2 will be correlates with all other sensors and gets $R_{2,n}$. Here n means other different sensors. The correlation result shows in the following paragraph.

4 Result and Discussion

The figure 2 shows the sensors data correlation result. Here T_{int} is predefined as 15 minutes. The figure 2 displays the sensor data correlation result in the evening from 22:00 to 22:15. In the figure sensor 6 correlated with sensor 7, 8, and 13 with the same value 0.7008. Sensor 7 correlated with sensor 8 and 13 with value 1, with sensor 6 with value 0.7008. Sensor 8 correlated with sensor 7 and 13 with value 1, with sensor 6 with value 0.7008. Sensor 13 correlated with sensor 7 and 8 with value 1, with sensor 6 with value 0.7008. The result illustrated that the user has activities from 22:00 to 22:15 and the activities happened in living room and bedroom.

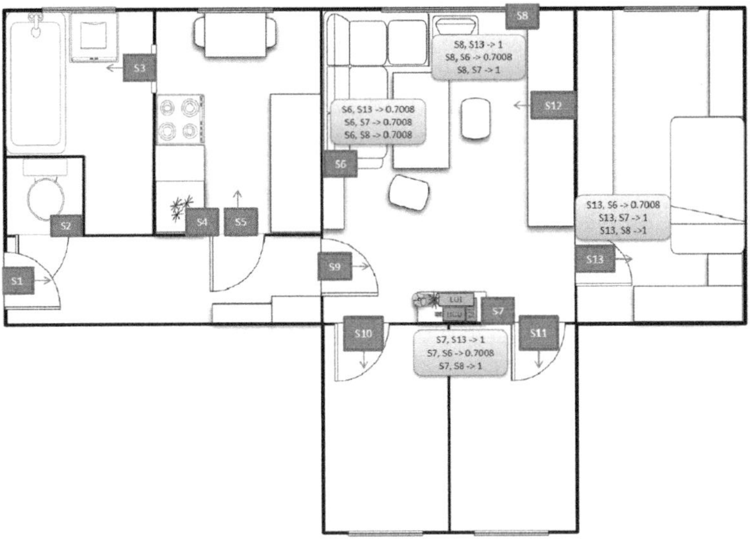

Fig. 2. The sensor data correlation result in the evening

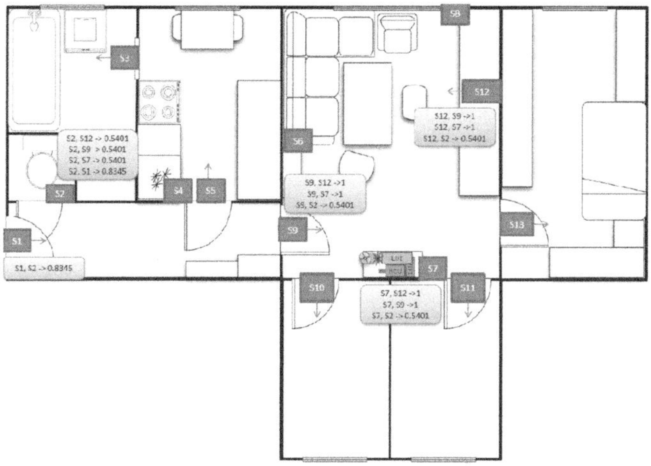

Fig. 3. The sensor data correlation result at night

The figure 3 shows the sensor data correlation result in another time interval from 1:30 to 1:45. Figure 3 demonstrates that the sensors 6, 7, and 8 in living room correlated well with value 1. At the same time interval sensor 1 and 2 correlated with value 0.8345. Furthermore the sensors in living room correlated with the sensor 2 at WC. So from the correlation result the system can judge that the user through living room goes to WC at night.

The above two examples illustrated that sensor data correlation cannot only detect the relationship between sensors but also when and where the activities happened by the user. Through comparing the result from figure 2 and 3 we can see that the correlation value changed in different time interval. This is because each sensor has limited observation area and the activities from the user are perhaps different in different time interval. If the activities happened at noon in kitchen so sensors 4 and 5 should be correlated. If the activities happened in the evening at living room so sensors 6, 7, 8, 9 and 12 should be correlated. But if the activities changed from time to time, even at the same location the correlation result should be not the same in different time interval.

5 Conclusion and Further Work

In this paper we illustrated sensor data correlation for ambient assisted living: sensor data from different types of sensors correlated in predefined time interval, and detected the relationship between sensors in the whole living environment of the user. According the correlation result the system knows when and where the activities happened. With the correlation result we can furthermore to search and build the activities model of the user. This is the work in the next step. Communicating sensor fusion and hidden Markov model together build flexible, robust, and as possible as accurate activities model of the user.

References

1. Sproull, T., Zuver, C.: Demo Abstract: Sensor Fusion and Correlation. In: SenSys 2005, San Diego, California, USA (November 2-4, 2005)
2. Valdes, A., Skinner, K.: Probabilistic Alert Correlation. Springer, Heidelberg (2001)
3. Vuran, M.C., Akyildiz, I.F.: Spatial Correlation-Based Collaborative Medium Access Control in Wireless Sensor Networks. In: IEEE/ACM Transactions On Networking, vol. 14(2) (April 2006)
4. Bonnet, P., Gehrke, J., Seshadri, P.: Towards Sensor Database Systems. In: Tan, K.-L., Franklin, M.J., Lui, J.C.-S. (eds.) MDM 2001. LNCS, vol. 1987, pp. 3–14. Springer, Heidelberg (2000)
5. Gupta, H., Navda, V., Das, S.R., Chowdhary, V.: Efficient Gathering of Correlated Data in Sensor Networks. In: MobiHoc 2005, Urbana-Champaign, Illinois, USA (May 25–27, 2005)
6. Ganesan, D., Estrin, D., Heidemann, J.: Dimensions: Why do we need a new Data Handling architecture for Sensor Networks? ACM SIGCOMM Computer Communications Review 33(1) (January 2003)
7. Bauer, G., Lukowicz, P.: Developing a Sub Room Level Indoor Location System for Wide Scale Deployment in Assisted Living Systems. In: Miesenberger, K., Klaus, J., Zagler, W.L., Karshmer, A.I. (eds.) ICCHP 2008. LNCS, vol. 5105, pp. 1057–1064. Springer, Heidelberg (2008)
8. Alemdar, H., Ersoy, C.: Wireless sensor networks for healthcare: A survey. Computer Networks 54, 2688–2710 (2010)
9. Rodrigues, G.N., Alves, V., Silveira, R., Laranjeira, L.A.: Dependability analysis in the Ambient Assisted Living Domain: An exploratory case study. The Journal of Systems and Software (2011)
10. Yin, G., Bruckner, D.: Daily Activity Model for Ambient Assisted Living. In: Technological Innovation for Sustainability, Portugal, pp. S. 197–S. 204 (2011)
11. Yin, G., Bruckner, D.: Data Analyzing and Daily Activity Learning with Hidden Markov Model. In: Proceedings of the 2010 International Conference on Computer Application and System Modeling, China, pp. S. 3–S. 7 (2010)
12. Yin, G., Bruckner, D.: Daily Activity Learning from Motion Detector Data for Ambient Assisted Living. In: Proceedings of the 3rd International Conference on Human System Interaction, Rzeszow, Poland, p. S. 6 (2010)

Enterprise Architecture and Business Intelligence: An Analysis of the Joint Application in the Sanitation Sector

Marcio Romero and Renato José Sassi

Production Engineering Program: Universidade Nove de Julho, São Paulo, Brazil
{mromero,sassi}@uninove.br

Abstract. The São Paulo sanitation sector generates annually great revenues, becoming a great attraction to investors. Improve results and expand the customer base is responsibility of the company's commercial area. Increase the sales has become a great challenge for the managers. One of the ways to increase the business is through the effective use of large amounts of information stored on corporate systems. In this context the Business Intelligence emerged as a key differentiator, since this architecture allow to access and explore the information contained in corporate databases. The Enterprise Architecture provides a set of best practices focused on improving the performance by structuring the information and integrated views of strategy, business processes, systems and technologies. The aim of this paper is to analyze the BI application in conjunction with Enterprise Architecture in the commercial area of a sanitation company to identify the contribution of the association of these two architectures.

Keywords: Business intelligence, decision making, enterprise architecture, sales, sanitation sector.

1 Introduction

In recent decades the sanitation sector has experienced significant changes where the private sector has increased considerably as the main source of investments required to lever investments in the sector, since it has experienced sharp reductions of budget resources by government investment in required to adequately provide services and also the universalisation of sanitation.

In order to enable the modernization and required expansion addressing the needs of Brazilian society, the municipal authorities, states and the federal government are seeking partnerships between the public and private sector as the main alternative to the implementation of investments necessary for the sanitation sector [1].

These partnerships allow a greater investment in the sector, but the sanitation companies have to universalize the services to the population and also to provide appropriate financial returns to the investors. Optimizing processes, improve performance and increase sales became a major challenge for companies in this market segment and the assertive decision making has become an important differential. But the decision making is only possible when there is access to the information organized and structured in this way to obtain important indicators to support the decisions of the company management.

G. Lee (Ed.): Advances in Automation and Robotics, Vol.2, LNEE 123, pp. 535–542.
Springerlink.com

Nowadays the companies use the databases as auxiliary tools for storing information and draw them through software applications, information and analysis of scenarios used in the enterprise management [2].

Analyze this information also becomes a complex and expensive task that requires specialized knowledge, as it requires analysts to manipulate multiple systems consult different types of data, files and reports to then compile the information, interpret them and provide results and reports that support the decision making.

In the case of a large company, the volume of data and the amount of systems is much larger and work this information (structured and organized) becomes a major challenge. Understand the behavior of the organization is the first step to define what information will be used as a guideline in decision making. One of the ways to access and explore the information contained in databases and through the architecture of Business Intelligence (BI), which is an important source of information for companies that need data and information work.

The BI systems include important artificial intelligence and analysis, opening a horizon of possibilities for small, medium and large enterprises, through the use of historical data from transactional legacy systems to support decision making from the strategic to the operational level [3].

The Enterprise Architecture describes how the business units of an organization and the infrastructure of Information Technology (IT) are logically organized to meet the requirements of integration and standardization, derived from operational models adopted by the organization [4].

In order to make use of this architecture of BI, companies must be clear what their processes, how they are organized and what are the information systems that support, thus adopt the concept of Enterprise Architecture added to the BI architecture can provide major contributions. The combination of these two architectures becomes an interesting object of study and application, since combined they can maximize the collection of data and leading indicators for decision making.

In this context emerges the question about which benefits from the association of architecture Business Intelligence (BI) in conjunction with Enterprise Architecture in the commercial area of a sanitation company that operates in the Metropolitan Region of São Paulo (MRSP)? The aim of this paper is to analyze the application of the association of BI with Enterprise Architecture in the commercial area and identify what the contribution obtained from the combination of these architectures.

Besides this introduction, the paper is organized as follows: section 2 presents an overview of the public and private partnership in the sanitation sector, the section 3 presents the Business Intelligence Architecture, while in the 4 section presents the Enterprise Architecture. In section 5 are describes the methodology adopted in this study and in section 6 the results and discussions. The work is concluded in section 7.

2 The Public and Private Partnership in the Sanitation Sector

The sanitation situation in Brazil is alarming. Only 44% of the population has access to the sewage system and 78.6% have access to safe drinking water. Just 29.4% of the total sewage generated is treated. More than 107 million of Brazilians have no access to the sewage system, 134 million do not have the treated sewage of their houses, 40 million have no access to drink water and 8 million do not even have toilets in home. The universalisation of basic sanitation services is therefore a long-term task. Its acceleration depends not only on greater investment effort, but also an increase in productivity, allowing more people are satisfied with each real invested [5].

One of the factors that explain the low level of investment in the sector is that, among the state companies, most do not have appropriate conditions to access long term financing. The sanitation sector has been facing some difficulties in obtaining resources for investment by public enterprises, as these are subject to tax rules such as limits on borrowing, credit and contingency surplus targets. The participation of the private sector was positive as it showed lower distribution losses and turnover, higher productivity rates, investment and quality of services. Therefore, it can be suggested that partnerships between public and private sector are viable alternatives to accelerate the goals of the sanitation sector [6].

But the private sector requires the positive results to continue investing, in this way the sanitation companies are seeking to improve their performance and achieve better financial performance. One way to improve the financial results of companies' sanitation is make better use of the customer base, which is made from low-income consumers to large businesses, trade and industry. Knowing the profile of these customers and offer customized solutions can generate more value to the company.

The company that serves as a base for this study operates in the country's largest state and the work presented here will focus on the Metropolitan Region of São Paulo (MRSP) that is one of the region's most complex of the country. This area centralizes the command of large private capital of Brazil and also where the most of the headquarters of the more important industrial complexes, commercial and financial, which control the economic activities in the country. At this company exist a Board which is responsible for this RMSP operations and there is a Superintendence who is responsible for managing business information and strategic information to subsidize the five Business Units responsible for the actions of sales and customer relationships. Explore the corporate information systems becomes a major challenge for the team in this area, since there are many scattered information in the enterprise systems in this context comes the BI as a tool that brings great contribution to the commercial area.

3 Business Intelligence

Currently, almost all business data and information are stored in databases, but not so classified, organized and orderly, so that when necessary, business managers can obtain important indicators to support decision making. Having access to quality information in a structured, organized, and speed has become a major challenge for professionals and areas of Information Technology (IT).

The concept of BI has been created by the Gartner Group in the 80's and refers to the process of collecting, organizing, analyzing, sharing and monitoring of information that supports the business management [7].

The BI architecture combines a set of tools for interrogation and exploitation of data with tools that allow you to generate reports to produce information that will be used later by the top management of organizations in support of decision making [8].

BI let's look at the organization as a whole, in search of points within the business processes that can be used as a competitive advantage. This is how executives are knowledgeable about the market, competition, customers, business processes and technology to anticipate changes and competitor actions [9].

The main objective of the BI architecture is to allow interactive access to data, provide for the manipulation of these data and provide managers with the ability to perform the proper analysis. When examining the data, situations, and historical and current performance, decision makers can better understand the data, which can serve as a basis for better and more informed decisions.

The BI process is based on transforming data into information. The main benefit to an organization's BI is the ability to provide accurate information when needed, including a real-time view of business performance. This information is required for all types of decisions, especially for strategic planning [3].

The BI architecture can include or make use of the following components: Transactional Systems, On-Line Transaction Processing (OLTP), Extraction, Transformation and Load, Data Warehouse (DW); Related Reports and On-Line Analytical Processing (OLAP), as Figure 1.

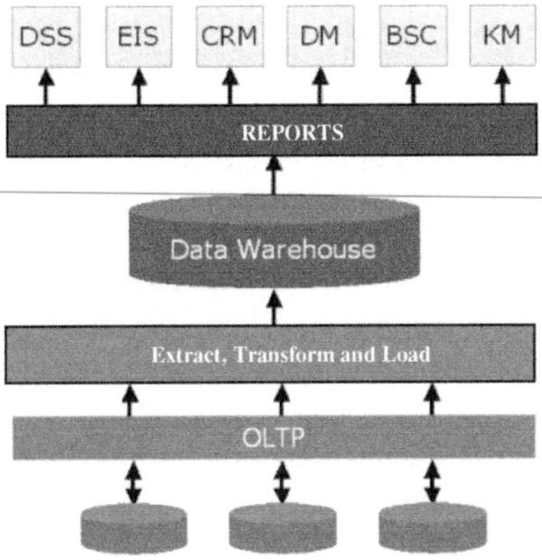

Fig. 1. Is possible to observe that the transactional systems capture and store operational data in its current state, thereby being the source of data for a BI solution [10]

Since the OLTP are software tools that implement access to transactional databases are needed in operational and BI systems directly connected to operational systems. The ETLs (Extract Transform Load) software are tools whose function is to extract data from multiple systems, processing of that data as business rules, and finally load the data into a data mart or data warehouse, and is considered one of the phases more critical than DW. The BI architecture gets more space within organizations, since it allows a look at the business as a whole and thus has a comprehensive view of the market, competition, business processes and especially their clients.

The company analyzed in this study implemented a DW in 2008 with a focus on structuring information in a single huge database using appropriate tools to work with extremely high volumes of data, which helps equip the Business Management with integrated information.

4 Enterprise Architecture

The breakdown of the company into manageable parts and the orchestration of the interaction between these parts, constitute the Enterprise Architecture [11].

Enterprise Architecture is a discipline on the IT architecture, which involves the organization's structure, systems, relationships between subsystems, their mechanisms of integration and connections with the outside world, the terminology known in the IT business, these technologies into its components, the accepted architectural principles for the design and development of the IT platform capabilities of current applications, as well as the deadlines, and resources available for investment decision-making involving IT an Enterprise Architecture requires an architect not only knowledge and technical skills, but also a business vision and a set of complementary expertise, often distant from the day-to-day software development only. Disciplines on people management, business strategies, product roadmaps, market knowledge, suppliers and technology trends are also involved [12].

It can be an important resource in helping an organization to find better ways to use technology to support their critical business processes.

The Enterprise Architecture consists of four areas that ensure the effectiveness and solutions that support the business strategy [13].

- Business Architecture - Definition of business and operational processes and functional characteristics, which are the basis for establishing the strategy of applications that support processes;
- Information Architecture (or data) - Are the data and the relationships with the goal of feeding strategies and optimize business decisions. It also serves as the basis for the design and delivery of applications;
- Application Architecture (or systems) - Identify which applications are required to support the business and leads the design, construction (or acquisition) and integration of these applications;
- Infrastructure Architecture (or technology) - covers all the support elements of the IT operation to be operated on a day-to-day, with tools, process monitoring and management.

The Enterprise Architecture enables change and manages the complexities inherent in large corporations; brings operational discipline to the organization and provides the necessary traceability of strategic intent with its productive capacity. These features include components of both business and technology. Enterprise Architecture does not happen at once it is an iterative process with maturity, which gives the administration an operational discipline to organize and engage business components and technology to fulfill the mission of the organization. This maturity model provides a scale or metric for understanding where the organization is and what measures are needed to take it to the next level of maturity [14].

One of the most common frameworks used today by the companies who adopt the practice of Enterprise Architecture is the TOGAF (The Open Group Architecture Framework), which consists of a standard architecture that can be used by any organization wishing to develop information systems architecture.

As defined in TOGAF a framework can be used as a tool to structure thinking and to ensure consistency and completeness. In other words, a framework is to "put the house in order" (content) and "teaching the ropes" (process), a framework developed and maintained by The Open Group, a nonprofit organization maintained by its members, who are mostly IT services companies such as IBM, SAP and HP [15].

The Enterprise Architecture maps the entire organization, it starts with strategy, continues with the mapping of business processes and how these processes are running (or fail to perform) the strategy, then moves to the Information Systems that automate (or not) these processes business (good or bad) and ends by identifying the technological infrastructure available to implement these systems. In addition, the process of Enterprise Architecture maps not only what exists today, but what is needed in the future for the organization's strategy can be implemented. Finally, the process also takes the architects to do the analysis between what exists today and what should exist.

According to the TOGAF Enterprise Architecture is twofold, depending on the context it uses, "A formal description of a system, or detailed plan for system-level components, like to guide its implementation" or "The structure of components, their interrelationships, the principles and guidelines governing their design and evolution over time. " [16].

The TOGAF provides a common sense, practical, prudent and effective method for developing enterprise architecture and is composed of three main parts [17]:

- The TOGAF Architecture Development Method (ADM) explains how to derive a specific enterprise architecture that meets business requirements and provides: A reliable and proven way of developing the architecture, the architecture vision that allow the architect to ensure that a complex set of conditions are adequately addressed; practical case studies, guidelines on tools for architecture development .
- The continuum Enterprise, which is a repository "virtual" of all architecture assets - models, patterns, architecture descriptions, - That exist both within the company as the IT industry in general that the company considers available for the development of architectures. In appropriate places throughout the TOGAF ADM, there are reminders to consider which architecture assets the architect should use.
- The TOGAF Resource Base, which is a set of resources - guidelines, templates, backgrounds, etc.

Using an Enterprise Architecture framework can simplify the mapping process and optimize the technological resources used in the business and thus allow growth aligned to business needs. The resulting architecture of the union of BI with Enterprise Architecture enables: the assessment of business impacts of changes caused by mergers, acquisitions or change of strategy, assessment of impacts of improvements in business processes, assessment of impacts and disaster recovery plans; management issues of safety and setting standards of architecture.

BI solutions can act as signals to market changes requiring review of strategy and architecture, since the Enterprise Architecture provides the information necessary to conduct properly change projects in BI solutions.

5 Methodology

The methodology used to support this study is grounded in a literature search of the references dealing with the topics covered in this study with the purpose of obtaining information regarding the Sanitation Sector, Enterprise Architecture and Business Intelligence. Experiments were performed with the database from the commercial sanitation company system's, was applied the Pentaho BI tools and the Enterprise Architecture Framework TOGAF - The Open Group Architecture Framework.

6 Results

The BI architecture has contributed to information gathering and correlation of external data with internal data of the company, but it is notable that there is also a need for systems and processes are mapped and organized in this way to optimize their use to the fullest. The results indicate that it is effective to use TOGAF to map the processes and systems this gap and act as a standard architecture is open source and can be used by any organization that is developing information systems architecture.

7 Conclusion

Improve processes, performance and increase sales of the companies operating in the sanitation sector is a key factor to ensure private investment in the sector and assertive decision-making is a determining factor for success in this regard.

Decision making is only possible when there is access to information in an organized and structured so that so obtained are important indicators to support the decisions of top management.

According Dias [18], it is essential to the application of information management to manage the informational chaos of the digital world. Often, these information are stored in databases are not integrated platforms and operating systems and database managers different. Thus, access to such information is difficult and consequently the process of decision making is also difficult, given an overview of the organization. Analyze business with another look has become a major challenge for business managers and sanitation in the area could not be different because the population has required increasingly corporate and otherwise positive results should be presented to shareholders. It is concluded that this study should be continued with the joint application of two architectures, thus to confirm the results. The BI architecture has contributed to information gathering and correlation of external data with internal data of the company, but it is notable that there is also a need for systems and processes are mapped and organized in this way to optimize their use to the fullest.

References

1. Petersen, O., Brancher, P.: The privatization of basic sanitation in Brazil, Jus Navigandi (2000), http://jus.uol.com.br/revista/texto/450 (access: February 19, 2011)
2. Arrivabene, A., Castro, J.P.S., Lima, J.A., Sassi, R.J.: Application of Zero Latency Data Mart in the Business Intelligence Architecture. In: 42º Simpósio Brasileiro de Pesquisa Operacional (SBPO), vol. 1, p. 150 (2010)
3. Turban, E., Sharda, R., Aronson, J., King, D.: Business Intelligence: a managerial approach to business intelligence. Artmed, p. 254 (2009)
4. Weill, P., Ross, J.W., Robertson, D.C.: Enterprise Architecture as Strategy: Creating a Foundation for Business Execution Havard Business (2006)
5. Oliveira, G., Marcato, F.S., Scazufca, A.P.: Scenarios and Requirements for Universal Sanitation in Brazil. In: FIPE (2011), http://www.fipe.org.br/publicacoes (access February 20, 2011)
6. Scriptore, J.S.: The public-private partnership in the Brazilian sanitation: a proposal for the development sector. Universidade de São Paulo (2010)
7. Primak, F.V.: Decisions with BI (Business Intelligence). Editora Moderna (2008)
8. Santos, M.Y., Ramos, I.: Business Intelligence: Information technology in knowledge management. FCA - Editora de Informática (2006)
9. Sassi, R.J.: Data Webhouse e Business Intelligence Operational: Revisiting the technology trends and analyzing warehouse data. In: XXX Encontro Nacional de Engenharia de Produção, São Carlos, SP, Brasil, 12 a15 de outubro de (2010)
10. Business Brain Services. Components and Architecture of a system of Business Intelligence, http://www.b2.adm.br (access April 17, 2011)
11. Cavalcanti, J.C.: Enterprise Architecture: A concept of interface enters the economy and corporate governance. Journal of Information Systems and Technology Management 6(3), 525–550 (2009)
12. Cambiucci, W.: Enterprise Architecture: A arquitetura corporativa e o papel do arquiteto de TI (December 2010) http://msdn.microsoft.com/pt-br/library/gg490650d=printer.aspx (access June 25, 2011)
13. Sávio, M.: Enterprise Architeture, http://www.slideshare.net/msavio/ (access May 01, 2011)
14. Scherkerman, J.: Extended Enterprise Architecture Maturity Model Support Guide. Institute For Enterprise Architecture Developments (2006)
15. Belloquim, A.: Frameworks de Arquitetura – Parte 1: Zachman (2009) http://blog.gnosisbr.com.br/sobre/serie-arquiteto-profissao-do-futuro/frameworks-de-arquitetura-%E2%80%93-parte-1-zachman/ (access May 10, 2011)
16. Rosa, F.A.J.: Method of Enterprise Architecture modeling. POLI São Paulo (2008)
17. TOGAF Version 9 Catalog number G091 UK ISBN 9789087532307 (2009)
18. Dias, C.: Corporate Portal: concepts and characteristics. Information Science 30(1), 50–60 (2001)

Research and Implementation of E-Commerce Workload

Qingsong Bao, Fang Yang, and Bing Su

Xi'an Communication Institute, xi'an, China
469759229@qq.com

Abstract. Nowadays, more and more people are shopping online. With the increase of user session, how to provide better e-commerce services and how to describe the network load as well as the advantages and the disadvantages of current services have become the urgent problems. Through the description of e-commerce workload, an intuitive and reliable result can be provided to the managers of website for the current runnability and the development trends, which can also provide the reference for the further improvement and the upgrade of website. And in this paper, the e-commerce workload has been analyzed from the aspects of theory and implementation.

Keywords: Customer Behavior Model, Customer Visit Model, e-commerce, workload, K-Means.

1 Introduction

With the rapid development of Internet in recent years, more and more people are shopping online. And the e-commerce websites are also vigorously developed, These websites have already focused on the studies of Customer Behavior Model Graph (CBMG) and the e-commerce workload and some appropriate products have been available in foreign countries.

At present, there are two kinds of products for analyzing the e-commerce websites. The first one is to directly insert the module which analyzes the workload and the Customer Behavior Model into the e-commerce websites and real-timely monitor the cared contents. However, as for this method, the burden of system servers will be increased. The site system will be more complex, the upgrade and the maintenance of which will become difficult. At the same time, the administrators should familiar with the system of websites. All of them have become the great burden for many small and medium websites. The second method refers that the websites and the analysis system should be separated and the web logs are used to discuss the results. This method is simple, which can be easily implemented and flexibly operated. But sometimes the subtle changes of websites can't be real-timely reflected. However, this method has been favored by most websites because of the simplicity and many products have been come out.

The existing CBMG of customers and the software for analyzing the e-commerce workload mainly include analog 6.0, visitors, rcnt2_engfree, weblog_files, Web Log Explorer and some others. But there are also some disadvantages, for example, they

G. Lee (Ed.): Advances in Automation and Robotics, Vol.2, LNEE 123, pp. 543–548.
Springerlink.com

just analyze the user session of a single web page, the clustering as well as the analysis of workload is not supported, the statistical information can't be selected, the interfaces are not friendly enough and they are all the command-line interfaces, etc. Therefore, based on the above software and the deep analysis of e-commerce, we should propose our own research and implementation of the e-commerce workload.

2 The Description of System

After the establishment of e-commerce websites, as time goes on, we should understand the proportion of site search, selection, purchase and payment of users in the time of visiting the websites. At the same time, we should also obtain whether the Web server has reached the full load movement and whether it is necessary to increase the investment of hardware or the improvement as well as the upgrade of software in the running process of e-commerce websites.

As for the existing method for describing the workload of e-commerce websites, the key issue refers to the description of customer behaviors. The description of customer behavior has two models: the Customer Behavior Model Graph (CBMG) and the Customer Visit Model (CVM). The first model is used to capture the navigation mode when the users access to the websites. The second mode does not involve the details, which just captures the frequency that the users perform each e-commerce function in one session. In order to describe the characteristics of e-commerce workload, it is required to get CBMG and CVM as well as the results from the HTTP logs. We should employ the clustering analysis to obtain a series of CBMG and CVM which can accurately describe the workload and get the parameters of resource model (such as the queuing network model) from the Customer Behavior Model.

The CBMG can pilot the customers who have patronized this e-commerce website and this section contains two parts: the transition and the temporal aspect. The former has determined the customers from one state to another state (such as an e-commerce functionality) and this is represented by a matrix which shows that the transition may be available. Cij refers to the transition times from state i to state j. As for the temporal aspect, the corresponding time from one state to another state should be recorded and this time is called server-preceived think time (abbr. think time). Wij refers to the time from state i to state j and this time is measured by the server.

Each session of customer can be represented by CBMG and we will explain how to represent the similar sessions with a CBMG in the following part. The purpose is: it is not required to deal with hundreds of thousands of CBMG and the typical CBMG with a relatively small number can be used to describe the load.

The specific steps can be shown as follows:

The first step: it is necessary to filter the irrelevant records of the HTTP logs in HTTP server of e-commerce site, such as the image requirements, errors and some others. Request log L has been produced in the first step and each line of assumption in this log has the following input:User ID (u): it is used to identify the customers and differentiate the different needs of customers who have carried on the same browsing in the session. Request Type (r): it is required to indicate the request types, such as returning to the home page, the browsing requirements (e.g. returning to another page), performing the search requirements, selecting a search result, adding a product to the shopping cart and checking out. Request Time (t): it refers to the time that the request gets to the site. ExceTime (x): it refers to the execution time of request. This value is not necessarily recorded in the HTTP logs and the server can record the information and make the appropriate changes.

The second step is called Get Sessions: it is required to input the request logs and a session log S will be produced. The Get Sessions algorithm is consisted of three steps. The first step: the request logs should be classified according to User ID and the requests with the same User ID should be put together and arranged according to the Request Time sequence. The second step: it is necessary to use the start time to select the session from the sub-sequence. Each sub-sequence may represent one or more sessions. For example, a customer may produce one request sequence and he will return to the site after an hour because of another session. Then the sub-sequence should employ an entry time T (e.g. 30 minutes) to divide the sessions. If a session exceeds the entry time T and the entry time T is just between the two consecutive requests R1 and R2, then R1 will be regarded as the last request in the session and R2 will be considered as the first request in the next session. The third step: it is required to scan the requests in the session, add the transition times of state to C and add the think time to W. The code can be shown as follows:

```
    C[i,j] ⇐0 for all i,j=1,…,n.
    W[i,j] ⇐0 for all i,j=1,…,n.
    for k=2 to Q do
Begin
    C[r k-1, r k]⇐ C[r k-1, r k]+1;
    w[r k-1, r k] ⇐ w[r k-1, r k]+(t k- t k-1 - x k-1);
End
    C[r Q, n] ⇐ 1;{the session is in the exit state }
```

The third step is named Get CBMG: CBMG will be produced after analyzing the session logs and these CBMGs can represent the session log S. After the production of session log S, we should carry out a cluster of analysis on S so as to get the integrated load composed by a small amount of CBMG. The centroid of this series has determined the characteristics of CBMG. And any type of clustering algorithm can be used such as the k-means clustering algorithm. At the beginning of this algorithm, k points will be selected in the point range which will be considered as the initial estimate centroids of k series. Then the rest of points will be distributed to the series which are closest to the centroid. The distribution process shouldn't be repeated for several times until there is no point for distribution or the repeated times have reached the maximum. The metric distance should be defined in the clustering algorithm so as to calculate the distance between the points and the centroid. Assuming that the session log is composed by M points Xm= (Cm, Wm), m=1,···, M and the distance between Xa and Xb in session log is d Xa, Xb: d Xa, Xb= $\sqrt{\sum_{i=1}^{n}\sum_{j=1}^{n}(Ca[i,j]-Cb[i,j])^2}$. In the implementation of k-means clustering algorithm, there are k centroids at any time. It is necessary to use s (k) to record the number of points contained by centroid k. When a new point is added to the series, there will be the new C value and W value. Assuming that the point Xm=(Cm, Wm) is added to the centroid k, then it will be represented as point (C, W) and the new centroid will be represented as point (C', W'), in which the calculation of C' and W' can be shown as follows:

C'[i,j]=(s(k)*C[i,j]+Cm[i,j])/(s(k)+1)
W'[i,j]=(s(k)*W[i,j]+Wm[i,j])/(s(k)+1)

How many series should we choose to accurately express the load during the clustering algorithm? We can solve this problem through testing the following two changes: the average distance between points in series and the average distance between centroids. It is hoped that the CV within the series should be small as possible and the CV between the series should be large as possible. This change can be illustrated by the variation of coefficient, for example, the deviation ratio between the average distance and the standard distance (CV) must be satisfied if the number of series and the number of points are the same. However, our purpose is to characterize the load, so we should select a relatively small series. The ratio between the internal CV and the external CV is expressed as Bcv. Then we can find that the changes of the number of series have little effect on the internal CV and the external CV will be increased with the increase of k. As for Bcv, Bcv is slowly reduced when k=3 and k=6, which indicates that we can choose k=6 in the process of testing the load.

3 The Implementation of Algorithm

Firstly, we should get the log files of e-commerce websites. But the log formats provided by the different Web servers in different systems are also different. For example, the log file of IIS service software in Windows operating system is under %system%\system32\ \Logfiles\W3SVC1 and the daily logging is considered as a file by default. At the same time, IIS allows to configure the information such as the location of logging, the size and the content of logging, etc. What's more, it can also choose the contents of logging such as the access to the user IP address, the response time and some others.

As for the Web servers which have employed the UNIX or the Linux system, the Apache service system has been widely used. We can get the log contents which need to be recorded through modifying the conf files of Apache system. Generally speaking, the logs of Apache system are stored under the /var directory of system with the text mode.

Secondly, in general, the contents of logging are more detailed and each line has recorded one access of the user. But not all of the information is needed. The User ID is the only information that we need, which can be used to identify the Request Type of customers and indicate the Request Type, such as returning to the home page, the browsing requirements (e.g. returning to another page), performing the search requirements, selecting a search result, adding a product to the shopping cart, checking out, the Request Time that the request gets to the site and the ExecTime.

Therefore, we should firstly read out the records in lines after getting the log files and then the useful information entry should be stored. At the same time, it is also necessary to filter the irrelevant records of the HTTP logs in HTTP server of e-commerce site, such as the images, the image requirements, errors and some others.

In order to sort out the Sessions from the filtered information according to the User ID and the time sequence, the useful filtered information should be stored in database and the structure of database should be established:

Logs Table:

Field Name	Field Type	Description
Seq	Automatic Numbering	Number
UserID	Text	Identification Mark of Customer
RequestType	Text	Request Type
RequestTime	Date/Time	Time that Request Gets to Site
ExecTime	Number	Execution Time of Request
WebPage	Text	File Name of Webpage

Userid Table :

UserID	Text	Identification Mark of Customer

Among them, the information from logs has been recorded in logs Table. After the processing of logs, the different User ID in logs table should be only saved in userid Table so as to make preparation for the classification of User ID in the next step. And it can be shown in following Figure.

After storing the information in database, the clustering analysis can be carried out. Firstly, the request logs should be classified according to the User ID and the requests with the same User ID should be put together. Then we need to take out a User ID (the identification mark of customer) from the Userid Table and filter the records with the same User ID value as the one we have obtained from logs Table. They will be arranged according to the Request Time sequence and then the session will be selected and recorded by the Request Time. Each sub-sequence may represent one or more sessions. For example, a customer may produce one request sequence and he will return to the site after an hour because of another session. Then the sub-sequence should employ an interval time T (e.g. 30 minutes) to divide the sessions. In our program, this interval time T can be manually adjusted and the default time is set to 30 minutes.

After setting the interval time of session, the k-means dynamic clustering algorithm can be used to analyze the results. The clustering algorithm can be divided into the static clustering algorithm and the dynamic clustering algorithm. The class center of static clustering algorithm is given in advance and the other nodes should be closer to the class center. However, in the load analysis, this algorithm can't reflect the real changes of CBMG and CVG, so we have employed the dynamic clustering rather than the static clustering algorithm. The class center of dynamic clustering algorithm will be dynamically changed with the adding of nodes, which can accurately reflect the changes of site load and access model.

The k-means algorithm is one of the popular dynamic clustering algorithms, which has been widely used in various areas of pattern recognition. It is flexible, which can select the number value k with different classes according to the request. Here we take k=6 for analysis and the reasons have been discussed above. The Sessions which are separated by the user and the interval time in database should be clustered into six categories and each category has two eigen-values: Cij refers to the transition times from state i to state j, and Wij refers to the time from state i to state j. After the algorithm, the results can be displayed and analyzed. At the same time, the final results of clustering analysis can be represented by the histogram and the results should be saved as files.

4 The Further Work

The further work include: firstly, this kind of workload analysis shouldn't be limited to the e-commerce sites, for the general websites we can also get the information of other non-ecommerce websites through modifying the contents of matrix. Cij can be regarded as the transition times from state i to state j, and Wij can be considered as the time from state i to state j. Secondly, we can also analyze the workload of network traffic for the servers and other equipments through expanding the tools. Thirdly, some websites will not provide the appropriate log messages because of the security or the commercial confidentiality. In this case, we can simulate the client access by combining the automated testing tools and the stress testing tools. Then we can also get the satisfied results of load analysis. Fourthly, it is required to add the daily or monthly contrast function for analyzing the performance of website, i.e. the previous performance load is saved as the reference for comparing with the changes of future load, which has provided the basis for the decision making.

5 Conclusion

Through the above analysis and implementation, an intuitive and reliable result can be provided to the managers of website for the current runnability and the development trends, which can also provide the reference for the further improvement and the upgrade of website. At the same time, the phenomenon that whether there exists the hacking and other events can be also found to some extent. If the daily analysis result can be jointly discussed, then more useful reference information will be provided. So for the websites, especially the e-commerce websites, the description of the characteristics of e-commerce workload is one of the important works.

References

1. Song, W.-G.: Practical E-commerce Tutorial. Higher Education Press (2002)
2. Menasce, D.A., Almeida, V.A.F.: Scalling for E-Business: Technologies, Models, Performance, and Capacity Planning. Prentice Hall (2000)
3. Menasce, D.A., Virgilio, A.F.: Capacity Planning for Web Services: Metrics, Models, and Methods, 2nd edn. Prentice Hall PTR

Knowledge Discovery from Noisy Datasets

Hong Li[1,2], Yu Zong[2], and Enhong Chen[2]

[1] Key Laboratory of Network and Intelligent Information Processing,
Department of Computer Science and Technology, Hefei University, 230001, China
[2] School of Computer Science and Technology,
University of Science and Technology of China, 230001, China
{xiaoke_93}@126.com

Abstract. It is a significant challenges to deal with the noise data in data mining and knowledge discovery applications. Most of previous works on data cleansing and correction have been focused on addressing class noise or attribute noise for the benefit of the subsequent mining process. In this paper, we propose an error-sensitive(ES) data mining framework, which makes use of noise knowledge to restore original data distributions and accommodates noise knowledge to enhance data classification accuracy. We materialize our main idea by constructing Attribute-Decision tree and measureing correlation among attributes. Experimental results show that ES data mining procedure has ability to significantly improve the quality of data mining results.

Keywords: Data mining, Noisy datasets, Noise handling, Classification.

1 Introduction

Real-world data are dirty, and therefore, building data mining models using unreliable or abnormal datasets presents a significant challenge to knowledge discovery. For most of the data mining applications, data imperfections still commonly exist in practice, where data often carry a significant amount of errors, which will have negative impact on the mining algorithms and result in sub-optimal performance. In a classification problem, for example, numerous studies have shown that the presence of errors in the training dataset lowers the predictive accuracy of a learner on test data[1-3]. Knowledge discovery from noisy datasets becomes a challenging and real issue for the data mining community.

There are many different dimensions of data quality, including class noise or labeling errors[4-5]and attribute noise [3,6]. The latter is the errors that are introduced in the attribute values or missing attribute value of the instances. Most of previous efforts on data cleansing and correction have been focused on addressing class noise[4,8-10] or attribute noise[12-15] for the benefit of the subsequent mining process. But there is little attention has been paid to address the challenge of unifying class noise and attribute noise to achieve an improved mining result. It is obvious that instance-based error information is difficult to get, but there are many cases in reality that statistical error information of the whole database is known a priori, which include information transformation errors, device errors and data perturbation errors

G. Lee (Ed.): Advances in Automation and Robotics, Vol.2, LNEE 123, pp. 549–556.
Springerlink.com

(for privacy preserving purposes). It leads to the question of how to integrate such information into the mining process.

In this paper, we report our recent research efforts toward this goal. In particularly, we, first, propose an ES(error-sensitive) data mining framework which accommodates noise knowledge to enhance data classification accuracy; and then, materialize our main idea by constructing Attribute-Decision tree and measureing correlation among attributes; at last, conduct serval experiments to show how efficicent of ES on improving the quality of data mining results.

The remainder of this paper is organized as follows. Related work on noise handling is provided in Section 2. Section 3 gives a detailed description of ES(error-sensitive) data mining framework. The experimental results are presented in Sections 4. Conclusions and future work are provided in Section 5.

2 Related Work

In supervised learning, noise usually takes two forms, namely, class noise and attribute noise. Due to the fact that class noise is caused by mislabeling, the most challenging task in identifying class noise is how to distinguish the mislabeled errors from the exceptions to general rules[7].

Guyon et al.[8] used an information criterion to measure an instance's typicality; and atypical instances are then presented to a human expert to determine whether they are mislabeled errors or exceptions. Brodley & Friedl[4] used multiple classifiers for noise identification where multiple classifiers are learned from a noise corrupted dataset. Gamberger et al.[9] suggested a Classification Filter approach, in which the training set E is partitioned into n subsets, a set of classifiers Hy trained from the aggregation of any n-1 subsets are used to classify the instances in the complementary (excluded) subset, and the instances that are incorrectly classified by Hy are identified as noise. Zhu et al.[10] presented a Partitioning Filter strategy to eliminate mislabeled instances from large or distributed datasets, in which a data set is partitioned into n subsets, the *good* rules constructed from each subset are used to evaluate the whole dataset. For a given instance I_k, two error count variables are used to count the number of times it has been identified as noise by all subsets, and the instance with higher error values will have a higher probability of being a mislabeled example.

An investigation of both class and attribute noise was conducted by Zhu and Wu[3] which concluded that the presence of both class and attribute noise can be harmful to a classifier, with the former having the more severe impact. However, attribute noise tends to happen more often in the real world. For example, if noise comes from entry mistakes, it is very likely that the class has fewer errors since the people involved know that it is the 'important' value and pay more attention to it[11].

For errors introduced by missing attribute values, how to distinguish this error becomes a challenging task, because such a noisy instance likely acts as a new training example with valuable information. Teng[12] proposed a *polishing* mechanism to correct noisy attribute values by training classifiers for each attribute. Yang et al.[13] presented a *sifting* approach to identifying, cleansing and measuring noise for predictive-but-unpredictable attributes. A recent study[14] proposes the concept of approximate weight for reducing the impacts of attribute noise on the system performance. The other reserches focus on reducing the impacts of attribute noise on the system performance. In data preprocessing, the suspicious instances are ranked based on user-specified attributes of interest[15] or their impacts on the system performance[16].

How to improve the performance of learners from the dataset that includes both class noise and attribute noise? Wu et al.[17] assumed that such noise knowledge is available in advance, and proposed error-aware naive Bayes classification algorithm which takes advantage of statistical error information to improve data mining results. Tao[18] regarded a reduced data set based on the rough set theory as a training data of BP neural network to reduce the influence of noise. However, the noise is random in real-world dataset so that it is difficult to be known in advance, and the reduced data set might lose a lot of useful information. In our system, we consider realistic cases where noises are randomly and independently introduced to each attribute or class label.

3 Error-Sensitive Data Mining from Noisy Datasets

For most of the datasets we used, the noise is random. In other words, given a dataset, it either contains some attribute noises or class noises, and the proportion of noise data is relatively small in datasets, otherwise, the dataset does not make sense in real-world. Typically, the attributes are assumed to be (somewhat) correlated to the class. It is obvious that some attributes have stronger correlations with the class than others, and in such scenarios, those attributes act more importantly in classification[3].

The flowchart of our proposed scheme (Error-Sensitive Data Mining) is depicted in Figure 1.

The crucial procedure in Fig. 1, the procedure *Adlearning* for distinguishing noise and outputtng good classification rule, is designed as follows:

```
Input: AD-Tree, x:noise laval, α: correlation
threshold;
Out: Classification rule;
  For any branch(A,Ci),(1.1)if sup(A,Ci)>x
    if exist branch(A,Cj)
        Using Ci instand of Cj;
        output rule A→Ci;
     else calculating correlation between A and Ci;
         if correlation(A,Ci)>α
               output rule A→Ci;
         else for j=1 to m
               if exist cor=max(correlation(A,Cj))
                  output rule A→Cj;
                else for i=1 to n
                      calculating correlation between
{A1^x,A2^x,...,An-1^x,Cj} and Ai^x;
                      if exist cor=max(correlation({A1^x,
A2^x,    ,An-1^x,Cj},Ai^x))
output rule{A1^x,A2^x,...,An-1^x,Ai^x}→Cj;
```

The notations of all the symbols are explained as follows: ①a learning set D which consists of a number of training instances, i.e., (A, C), is given in advance. ②each single instance (A, C)is characterized by a set of n attribute A={A_1, A_2, . . . , $_n$} and one class label $C \in \{C_1, C_1, \ldots, C_m\}$. ③$A_i^x$ is one of the attribute values of the ith attribute A_i, and C_j is the jth class label.

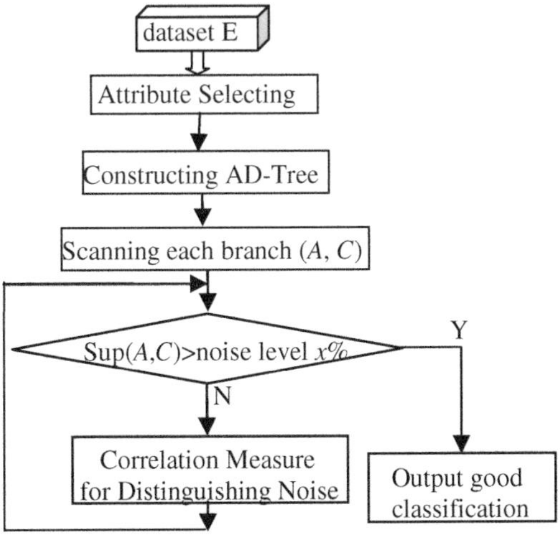

Fig. 1. An error-sensitive(ES) data mining framework

3.1 Attribute Selecting

The noise of different attribute has different impact with the system performance. The impact of the attribute noise critically depends on the correlation between the attribute and class[3]. However, we still have no idea about what types of attributes are sensitive to noise and why they are more sensitive than others. Therefore, we adopt the χ^2 test from statistics to analyze the correlations between each attribute and the class label[3]. The χ^2 value is defined as follows:

$$\chi^2 = \sum \frac{(f_0 - f)^2}{f} \tag{1}$$

Among them, f_0 is an observed frequency, and f is an expected frequency.

The higher the χ^2 value, the higher the correlation between the corresponding vectors. To execute the χ^2 test between an attribute A_i and the class label C, we take each of them as a vector, and calculate how many instances contain the corresponding values. The higher the correlation between A_i and C, the more impact could be found from this attribute A_i, if A_i includes noise. This implies that when handling attribute noise, it's not necessary to deal with all attributes, and we may select some noise sensitive attributes only.

3.2 Constructing Attribute-Decision Tree

Inspired by the construction of *FP*-tree[19], we construct a Attribute-Decision tree(*AD*-tree) where it only includes the selected attributes and class labels. Each layer of the *AD*-tree represents a attribute, and each branch of the tree describes some identical instances in dataset, and all class labels are represented as leaf nodes of the *AD*-tree. Each node of the *AD*-tree takes the following form $\{A_i^x| v_i\}$, where A_i^x is one of the attribute values of the ith attribute A_i, and v_i is the numbers of times that A_i^x has appeared in the branch. The model of *AD*-tree is expressed as Fig. 2.

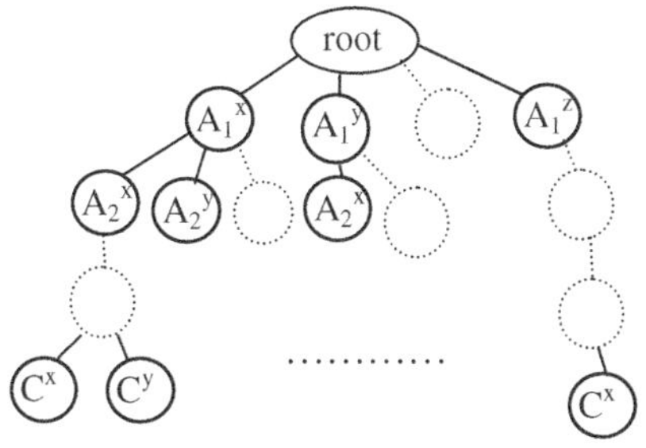

Fig. 2. The model of *AD*-tree

According to Fig. 2, any branch with two leaves must contains noise due to the same example with different class labels, and a branch may include noise if v_i is smaller than a given threshold in leaf node $\{C_i| v_i\}$ of the branch. This noise may be class noise or attribute noise. The procedure *ADlearning* describes the process of handling noise.

3.3 Correlation Measure for Distinguishing Noise

According to the above description, there is stronger correlation between the selected attributes and class. Then, for a set of n attribute value and a class label $S=\{A_1^x, A_2^x, \ldots, A_n^x, C_j\}$, noise may be introduced into A_i or C if the correlation between two vectors, $\{A_1^x, A_2^x, \ldots, A_n^x\}$ and C_j, or $\{A_1^x, A_2^x, \ldots, A_{n-1}^x, C_j\}$ and A_i^x, is very small (or not at all). In order to measure the correlation between two itemsets without the impact of unbalanced classification, we adopt the *cosine*(*A*, *B*) test from statistics to analyze the correlations between two item *A* and *B* [20]. The *cosine*(*A*, *B*) value is defined as follows:

$$\cos ine(A,B)=\frac{P(A\cup B)}{\sqrt{P(A)\times P(B)}}=\frac{\sup(A\cup B)}{\sqrt{\sup(A)\times \sup(B)}} \quad (2)$$

For a set of n attribute $A=\{A_1, A_2, \ldots, A_n\}$ and one class label $C\in\{C_1, C_1, \ldots, C_m\}$, A_i^x is one of the attribute values of the ith attribute A_i, the correlation measure includes two aspects as follows: ① the correlation between $\{A_1^x, A_2^x, \ldots, A_n^x\}$ and C_j. ②the correlation between $\{A_1^x, A_2^x, \ldots, A_n^x, C_j\}-\{A_i^x\}$ and A_i^x.

We first calculate the *cosine* values between $\{A_1^x, A_2^x, \ldots, A_n^x\}$ and C_j, for j=1,2, ..., m respectively, and then select the C_x with the largest *cosine* value of all. If the *cosine* value between $\{A_1^x, A_2^x, \ldots, A_n^x\}$ and C_x is larger than a given threshold α, then a classification rule $A_1^x, A_2^x, \ldots, A_n^x \rightarrow C_x$ is produced. Otherwise, we turn to the following procedure.

We calculate the *coisne* values between $\{A_1^x, A_2^x, \ldots, A_n^x, C_j\}-\{A_i^x\}$ and A_i^x, for each A_i^x of attribute A_i respectively, and if the largest *cosine* value is larger than the given threshold α, we produce a classification rule $A_1^x, A_2^x, \ldots, A_n^x \rightarrow C_j$ in which the value of A_i^x is fixed as the one with the largest *cosine* value. Otherwise, change A_i for other attribute A_j and repeat the above procedure until a classification rule has been produced.

4 Experimental Evaluations

To demonstrate the effectiveness of the proposed ES algorithm, we compare it with C4.5[21]. For each experiment, we use the classification accuracy as the measurement. We evaluate our approach on six benchmark datasets from the UCI database repository[22], where each numerical attribute is discretized with equal-width discretization approach. The main characteristics of our benchmark datasets are described in Table 1.

Table 1. Benchmark datasets used for evaluation

Dataset	# of Classes	# of Attributes		# of Instances
		Nominal	Continuous	
Adulte	2	8	6	48842
Krvskp	2	36	0	3196
Led24	10	24	0	1000
Nursery	5	8	0	12960
Wine	3	0	13	178
Zoo	7	15	2	101

The datasets in the UCI database repository have been carefully examined by domain experts; thus, they do not contain much noise. For comparative studies, we use manual mechanisms to add both class noise and attribute noise. With this scheme, the error values are introduced into each example with a level x×100%. For class noise, an instance with its label X might have a chance to be corrupted as other label Y, and for attribute noise, the error values might be introduced into each attribute. It must be emphasized that each instance only contains a noise.

Our experiments are designed to assess the performance of the proposed ES in noisy environments in comparison with the C4.5 classifiers trained from the same dataset. There are three types of noisy environments, which are only class noise, only attribute noise, and both class noise and attribute noise. In each experiment, the original dataset is randomly divided into a training set and a test set. The error corruption model was applied to the training set, and this corrupted dataset was used to build the ES and C4.5 classifiers. Both the classifiers are tested on the test set to evaluate their performance. When specifing a correlation threshold α=0.7, the summarized results are reported in Table 2.

Table 2. Classification accuracy comparison

Dataset	Noise Level x%	Original		Only class noise		Only attribute noise		Both two type noise	
		ES	C4.5	ES	C4.5	ES	C4.5	ES	C4.5
Adulte	10	85.61	84.77	83.12	83.94	**83.01**	80.56	**81.78**	79.76
	20			**81.23**	80.35	**82.35**	79.34	**80.32**	73.34
	30			**80.67**	75.46	**81.89**	80.56	**79.73**	70.85
Krvskp	10	98.35	99.64	96.12	96.53	**95.89**	94.76	**91.34**	90.23
	20			**93.42**	90.21	**92.45**	88.12	**89.43**	80.76
	30			**86.56**	82.63	**90.12**	79.56	**85.21**	75.21
Led24	10	99.78	100	94.32	94.41	**95.25**	90.35	**94.42**	85.32
	20			**94.12**	78.62	**93.67**	72.67	**90.38**	70.54
	30			**90.54**	69.27	**90.43**	60.29	**89.21**	57.32
Nursery	10	99.34	98.95	**92.42**	91.97	**91.67**	89.23	**90.78**	83.45
	20			**91.78**	85.34	**91.03**	78.67	**90.12**	70.54
	30			**90.38**	77.18	**90.14**	70.56	**88.45**	62.49
Wine	10	95.63	94.12	**96.54**	92.61	**94.23**	90.32	**91.54**	86.67
	20			**94.12**	91.35	**93.34**	88.54	**89.21**	82.32
	30			**92.25**	90.32	**92.03**	86.78	**88.05**	78.54
Zoo	10	93.43	92.16	**95.12**	92.01	**94.45**	90.23	**92.53**	87.13
	20			**90.45**	89.45	**90.12**	86.74	**89.36**	82.91
	30			**88.32**	87.23	**87.43**	80.21	**84.47**	73.28

5 Conclusions

In this paper, we have proposed an ES data mining framework for effective learning. The proposed effort has resulted in an improvement in comparison with the models built from the original noisy data and the noise-cleansed data. The algorithm takes advantage of the noise knowledge to polish the model trained from noisy data sources. Data mining from noisy information sources involves three essential tasks: noise identification, noise profiling, and noise-tolerant mining[3]. The ES data mining framework designed in this paper makes use of the noise knowledge for noise-tolerant mining. However, how to deal with different types of noise, namely, random or structured, for noise-tolerant mining, is still an open research issue.

Acknowledgement. The work was supported by the National Natural Science Foundation of China (Grant No. 60775037), and the Natural Science Foundation of Anhui Province,China(Grant No. KJ2011Z321), and the Key Natural Science Foundation of Hefei University,China(Grant No. 01KY03ZD).

References

1. Khoshgoftaar, T.M., Seliya, N.: The necessity of assuring quality in software measurement data. In: Pro. of 10th International Software Metrics Symposium, pp. 119–130 (2004)
2. Weiss, G.M.: Mining with rarity: a unifying framework. SIGKDD Explorations 6(1), 7–19 (2004)
3. Zhu, X., Wu, X.: Class noise vs. attribute noise: a quantitative study of their impacts. Artificial Intelligence Review 22(3-4), 177–210 (2004)
4. Brodley, C.E., Friedl, M.A.: Identifying mislabeled training data. Journal of Artificial Intelligence Research 11, 131–167 (1999)
5. Khoshgoftaar, T.M., Zhong, S., Joshi, V.: Enhancing software quality estimation using ensemble-classifier based noise filtering. Intelligent Data Analysis: An International Journal 9(1), 3–27 (2005)
6. Van Hulse, J., Khoshgoftaar, T.M., Huang, H.: The pairwise attribute noise detection algorithm. Knowledge and Information Systems Journal, Special Issue on Mining Low Quality Data 11(2), 171–190 (2007)
7. Srinivasan, A., Muggleton, S., Bain, M.: Distinguishing exception from noise in non-monotonic learning. In: Proc. of 2th ILP Workshop, pp. 97–107 (1992)
8. Guyon, I., Matic, N., Vapnik, V.: Discovering information patterns and data cleaning. Advances in Knowledge Discovery and Data Mining, 181–203 (1996)
9. Gamberger, D., Lavrac, N., Groselj, C.: Experiments with noise filtering in a medical domain. In: Proc. of 16th ICML, pp. 143–151 (1999)
10. Zhu, X., Wu, X., Chen., S.: Eliminating class noise in large datasets. In: Proc. of the 20th ICML, pp. 920–927 (2003)
11. Schwarm, S., Wolfman, S.: Cleaning data with Bayesian methods. Final project report for CSE574, University of Washington (2000)
12. Teng, C.M.: Correcting noisy data. In: Proc. of the 16th International Conf. on Machine Learning, pp. 239–248 (1999)
13. Yang, Y., Wu, X., Zhu, X.: Dealing with Predictive-but-Unpredictable Attributes in Noisy Data Sources. In: Boulicaut, J.-F., Esposito, F., Giannotti, F., Pedreschi, D. (eds.) PKDD 2004. LNCS (LNAI), vol. 3202, pp. 471–483. Springer, Heidelberg (2004)
14. Yun, U., Ryu, K.H.: Approximate weighted frequent pattern mining with/without noisy environments. Knowledge-Based Systems 24, 73–82 (2011)
15. Khoshgoftaar, T.M.: Identifying Noise in an Attribute of Interest. In: Proc. of ICMLA, pp. 124–131 (2005)
16. Zhu, X., Wu, X., Yang, Y.: Error detection and impact-sensitive instance ranking in noisy datasets. In: Proc. AAAI, pp. 378–384 (2004)
17. Wu, X., Zhu, X.: Mining With Noise Knowledge: Error-Aware Data Mining. IEEE Transactions on Systems, Man, and Cybernetics-part A: Systems and Humans 38(4), 917–932 (2008)
18. Tao, K.: A Novel Hybrid Data Mining Method Based on the RS and BP. In: Zhang, L., Lu, B.-L., Kwok, J. (eds.) ISNN 2010. Part II, LNCS, vol. 6064, pp. 346–352. Springer, Heidelberg (2010)
19. Han, J., Pei, J., Yin, Y.: Mining frequent patterns without candidate generation. In: Proc. of ACM SIGMOD, pp. 1–12 (2000)
20. Han, J., Kamber, M.: Data mining: concepts and techniques, 2nd edn., pp. 259–260. Elsevier Inc. (2006)
21. Quinlan, J.R.: C4.5: programs for machine learning. Morgan Kaufmann, San Mateo (1993)
22. Blake, C., Merz, C.: UCI Repository of Machine Learning Databases (1998)

Modeling and Simulation for Dynamics of Anti-HBV Infection Therapy

Xiao Chen[1], Lequan Min[1,2,*], Yongan Ye[3,*], and Yu Zheng[1]

[1] Automation School, University of Science and Technology Beijing
Beijing 100083, China
[2] Mathematics and Physics School, University of Science and Technology Beijing
Beijing 100083, China
[3] Traditional Chinese Internal Medicine Key Laboratory of China Education Ministry, Dongzhimen Hospital, Beijing University of Chinese Medicine
Beijing 100700, China
{chx717,minlequan}@sina.com

Abstract. Based on our previous researches, this paper introduces a new differential equation model to describe the dynamics of traditional Chinese medicines (TCM) + nucleoside analogues (NA) anti-HBV infection treatment for the chronic hepatitis B (CHB) patients. This model has five state variables. A treatment-activated specific term in this model is in charge of directly clearing patients' HBV rather than damaging patients' hepatocytes. As an application, we use this model to simulate the dynamics of TCM + NA anti-HBV infection personalized combination treatments. The numerical simulations and analysis show that the TCM + NA combination treatments are able to activate patients' abilities of cytokine-midiated noncytolytic HBV clearance.

Keywords: Chronic HBV infection, Mathematical modeling, Basic virus reproductive number, Global stability, Numerical simulation.

1 Introduction

Chronic hepatitis B virus (HBV) infection is a worldwide health problem. About two billion people worldwide have been infected with the HBV. An estimated 400 million people have chronic (long-term) liver infection [1]. Each year, about one million people die from HBV-related chronic liver disease worldwide [2].

Nucleotide analogues (NA) - Peginterferon, lamivudine, adefovir, entecavir and telbivudine are popular drugs to treat HBV infection. The treatments suppress HBV replication and result in improvement in liver architecture on microscopical evaluation during therapy [3], [4], [5], [6], [7]. The main roles of NA is to block the replication of HBV in patients' serums.

Some traditional Chinese medicines (TCM) have the functions to activate patients' immune function, especially reduce the levels of patients' serum HBeAg. TCM + NA anti-HBV infection therapies are able to increase proportion of

* Corresponding authors.

G. Lee (Ed.): Advances in Automation and Robotics, Vol.2, LNEE 123, pp. 557–566.
springerlink.com

patients achieving HBeAg loss [8]. Some evidences show that the main role of some TCMs is to activate patients' specific immune functions which clear HBV directly and almost does not damage patients' hepatocytes.

The basic viral infection mathematical model (BVIM) proposed by Nowak et al.([9], [10]) has been widely used in the studies of virus infection dynamics. The model (BVIM) takes the following form:

$$\begin{cases} \dot{x} = \lambda - dx - bvx, \\ \dot{y} = bvx - ay, \\ \dot{v} = ky - uv. \end{cases} \tag{1}$$

where x, y and v are the number of uninfected cells, virus-infected cells, and free virus, respectively. Uninfected cells are produced at constant rate λ, die at rate dx, and become infected at rate bvx. Virus-infected cells are thus produced at rate bvx and die at rate ay. Free virus are produced from virus-infected cells at rate ky and are removed at rate uv.

This model has a basic reproductive number $R_0 = \lambda bk/(adu)$. It has been proved that if $R_0 \leq 1$ then the infection free state of the model is globally attractive, otherwise the endemic state of the model is globally attractive (for example see [11]).

Since λ/d represents the total number of uninfected cells of a patient's organ, this implies that a person with small infected organ will be more difficult to be infected than a person with larger one. Recently, some amended basic viral infection models (ABVIM) are introduced ([12], [13], [14]). One of them has the following form [12]:

$$\begin{cases} \dot{x} = \lambda - dx - \dfrac{bvx}{x+y}, \\ \dot{y} = \dfrac{bvx}{x+y} - ay, \\ \dot{v} = ky - uv. \end{cases} \tag{2}$$

The ABVIM has a basic virus reproductive number $R_0 = bk/(au)$, which is independent on the total number of the patient's hepatocytes. In previous papers, it has been proved that if $R_0 \leq 1$, then the infection free equilibrium is globally attractive, otherwise the persistent infection equilibrium point is globally attractive ([12], [15]).

The rest of this paper is organize as follows. Section 2 presents a new model to describe the dynamics of TCM + NA anti-HBV infection treatment. An application to HBV infection dynamics is given in section 3. Finally, concluding remarks are addressed in section 4.

2 Modeling Anti-HBV Infection Treatment

For spontaneous recovered HBV infected patients, there are two mechanisms to clear patients' virus in vivo [16]. One is that the HBV is cleared via cytokine-mediated destruction of HBV-infected cells through CTL-activated perform, which causes patients' serum ALT level increasing. The other one is that the HBV is cleared via cytokine-midiated noncytolytic HBV clearance, which does not destruct HBV-infected cells and cause patients' serum ALT level increasing.

Therefore, we should add a suitable term in the new system which increases the rate of the virus death, and represents the activation of the immune function on noncytolytic HBV clearance. Furthermore, a variable representing ALT levels should be considered in the new system.

In summary, we propose the following improved system:

$$\begin{cases} \dot{x} = \lambda - dx - (1-m)\dfrac{bvx}{x+y}, \\ \dot{y} = (1-m)\dfrac{bvx}{x+y} - ay - \dfrac{k_1 ye}{x+y}, \\ \dot{v} = k(1-n)y - uv - \dfrac{k_4 ve}{x+y}, \\ \dot{e} = k_2(x+y) - k_3 e, \\ \dot{z} = k_5 + \dfrac{k_6 k_1 ye}{x+y} - k_7 z, \end{cases} \tag{3}$$

where the meanings of the variables x, y and v as well as the parameters λ, d, b, a, k and u are the same as those given in (1). The variable e represents the number of cytotoxic T lymphocytes(CTL) cells. The CTL cells (e) are produced at rate $k_2(x+y)$ and die at rate $k_3 e$. m, n $(0 < m, n < 100\%)$ are the efficacy variables of the treatment. $(k_1 ye)/(x+y)$ is the die rate of virus-infected cells generated by immune killing. $(k_4 ve)/(x+y)$ is the clearing rate of virus generated by some specific immune abilities activated via anti-virus infection treatment.

The new variable z represents the ALT levels. A liver without immune attacking produces ALT at rate k_5, and ALT dies at rate $k_7 z$. A chronic HBV patient's liver produces ALT at rate $k_5 + k_6 k_1 ye/(x+y)$.

All the 15 parameters are nonnegative. Among them, a, k, k_1, k_2, k_4, m and n may be changing during the treatment.

Before the treatment, we assume that all parameters are constant and $m = 0, n = 0, k_4 = 0$. System (3) has two equilibria Q_1 and Q_2, where

$$Q_1 = (\frac{\lambda}{d}, 0, 0, \frac{k_2\lambda}{k_3 d}, \frac{k_5\lambda}{k_7}) \tag{4}$$

is called the infection free equilibrium point, representing a patient's complete recovery.

$$Q_2 = (\bar{x}, \bar{y}, \bar{v}, \bar{e}, \bar{z},) \tag{5}$$

is called persistent infection equilibrium point, representing a patient's persistent HBV infection. Here

$$\begin{cases} \bar{x} = \dfrac{\lambda}{d + (bc_1c_2)/(1+c_2)}, \quad \bar{y} = c_2\bar{x}, \\ \bar{v} = c_1c_2\bar{x}, \quad \bar{e} = \dfrac{k_2(1+c_2)\bar{x}}{k_3}, \\ \bar{z} = \dfrac{k_5(a\bar{x} + b\bar{y})(\bar{x}+\bar{y}) + k_1k_6\bar{y}\bar{e}}{k_7(\bar{x}+\bar{y})}, \end{cases} \tag{6}$$

$$c_1 = \frac{kk_3}{uk_3 + k_2k_4}, \quad c_2 = \frac{bc_1k_3 - ak_3 - k_1k_2}{ak_3 + k_1k_2}. \tag{7}$$

Remark: The basic virus reproductive number of system (3) is

$$R_0 = \frac{kb(1-m)(1-n)}{au(1 + \frac{k_1k_2}{ak_3})(1 + \frac{k_2k_4}{uk_3})} < 1.$$

We can prove that if a basic virus reproductive number R_0 of system (3) satisfies $R_0 < \tilde{R}_0 = kb(1-m)(1-n)/(au) < 1$, then the infection free solution Q_1 of the model is globally attractive. This result suggests that if an anti-virus infection therapy makes an HBV infected patient's $R_0 < 1$, then the patient will eventually recover even if the patient's serum HBV DNA has very high levels.

3 Simulation for TCM and NA Combination Anti-HBV Infection Personalized Therapy

In this section, we will use a CHB patient's clinic data to simulate the evolution dynamics of the patient's HBV DNA, and ALT levels via system (3).

3.1 Personalized Therapy

A 57 years old male chronic HBeAg positive patient (nucleosides-native) with mild cirrhosis did not got any anti-HBV infection therapy in near 36 months before the combination treatment. During the February 2009 ~ the July 2011, the patient had received personalized TCM and NA combination therapy.

The patient's some characteristics on some weeks are listed in Table 1. The 500cps/ml is the HBV DNA test lower limit. The last HBV DNA test lower limit (at the week 131) is 12 IU/mL and the test result was undetectable. Figure.1 shows the evolutions of patient's HBV DNA (cps/mL), ALT (U/L), HBeAg (S/CO), and Anti-HBe (S/CO).

Table 1. The Patient's Clinic Characteristics.

Weeks	ALT*	HBeAg	Anti-HBe	HBV DNA†
0	45.7U/L	450.57S/CO	—	2.8e6 cps/mL
6	50.3U/L	279.72S/CO	14.68	7.5e7 cps/mL
13	63.4U/L	281.51S/CO	16.24S/CO	6.0e6 cps/ml
23	52.0U/L	7.61S/CO	1.131S/CO	6.24e4 cps/ml
32	32.9U/L	2.31S/CO	1.02S/CO	< 500 cps/mL
78	41.5U/L	1.40S/CO	1.09S/CO	< 500 cps/mL
131	42.8U/L	0.96S/CO	1.02S/CO	< 12 IU/mL ‡

∗ The upper limit of the normal range is 40U/L.
† The test lower limit is 500cps/mL.
‡ The test lower limit is 12 IU/mL.

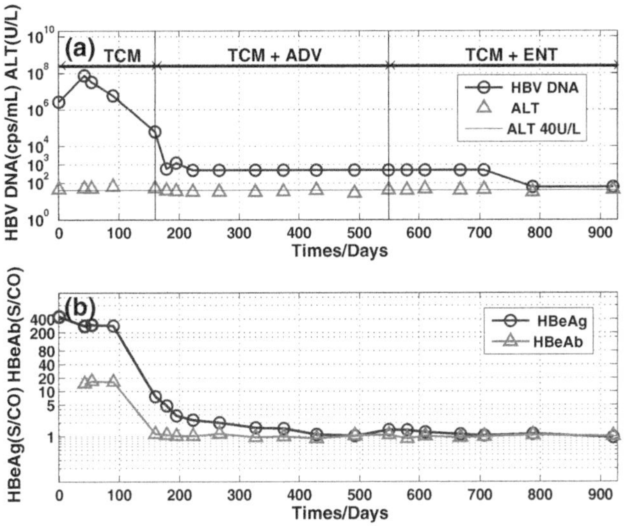

Fig. 1. Outcomes of the treatment of the patient. (a) Circles: HBV DNA levels, Triangles: ALT levels; (b) Circles: HBeAg levels, Triangles: HBeAb levels.

The patient received decocted TCM orally, twice daily. Each dose consists of 18∼24 ingredients, prescribed based on the patient's different symptom.

During the week 1∼ the week 23, he had received the TCM monotherapy. Then he had switched to TCM + ADV treatment until the the week 77. At the week 78, he had switched to TCM + ENT therapy for the following weeks. After the week 32, his serum HBV DNA has been kept to be less than unpredictable levels (500 cps/ml).

3.2 Numerical Simulations

The dynamic model of the patient's anti-HBV infection therapy is described by system (3).

Assume that before the treatment the "patient" was in a stable persistent infection state given by (5) and (6).

In (5). we only know $\bar{v}, \bar{z}$ (i.e. HBV DNA, ALT at the week 0, Table 1). We have to make some assumptions to determine approximately the other initial values and the 15 parameters in (3).

1) since a human liver contains around 2×10^{11} hepatocytes([9], p50), we choose

$$\frac{\lambda}{d} \approx 2 \times 10^{11}$$

2) Since the half-life of a hepatocyte is about half a year [17], we can choose

$$d = -\frac{ln(0.5)}{183}.$$

3) In a chronic HBV infection between $\delta = 5\%$ and $\delta = 40\%$ of all hepatocytes can be infected ([9], p50). Different chronic hepatitis B patient's serum HBV DNA load varies ranging from $\bar{v} = 10^3 cps/mL$ to $\bar{v} = 10^{12} cps/mL$. We assume that $\bar{v}$ and δ have the following relation

$$\delta = p + q\bar{v}.$$

We can determine p, q via solving equations: $5\% = p + q10^3, \quad 40\% = p + q10^{12}$. Consequently, we obtain $\delta = 0.05 + 3.5 \times 10^{-13}\bar{v}$.
4) In our example, $\bar{v} = 2.8 \times 10^6$ copies/ml since a human has around 3 liters of serum. Hence we obtain $\delta \approx 0.0500$.
5) Chronic HBV infection makes some infected hepatocytes undergo apoptosis and replaced by hepatic stellate cells [18]. we select a parameter δ_0, which represents that the patient's liver cells are reduced by $(1 - \delta_0) \times 100$ percent. Hence we obtain $\bar{x} + \bar{y} = \delta_0 \frac{\lambda}{d}$, where we choose $\delta_0 = 0.95$. Furthermore we obtain $\bar{x} = \delta_0 \delta \frac{\lambda}{d}$, $\bar{y} = (1 - \delta)\delta_0 \frac{\lambda}{d}$.
6) Select $u = 0.67$ ([9], p47).
7) Assuming that the baseline $k_4 = 0$, we obtain $k = u\bar{v}/\bar{y}$.
8) Since the half-life of CTL cells is about 77 days [19], we can assume that $k_3 = -\ln(0.5)/77$.
9) Since the half-life of ALT is about $2 \sim 3$ days [20], we select $k_7 = -\ln(0.5)/2.5$.
10) Since the upper limit of the normal range of ALT is 40U/L, and a human has about 3 liter serum, we assume that 30U/L is the mean normal ALT level. Form the 5th equation in system (3), we obtain $k_5 = 30 \times 3 \times k_7$.
11) Solving the equilibrium point equation gives

$$b = \frac{\lambda - d\bar{x}_0}{\bar{v}\delta}, \; k_1 = \frac{b\bar{v}\delta - a\bar{y}}{(1-\delta)\bar{e}}, \; k_2 = \frac{k_3\bar{e}d}{\delta_0\lambda}, \; k_6 = \frac{k_7\bar{z} - k_5(d\bar{x} + a\bar{y})(\bar{x} + \bar{y})}{k_1\bar{y}\bar{e}}.$$

Select the following initial condition

$$(x_0, y_0, v_0, e_0, z_0) = (\delta_0\delta\frac{\lambda}{d}, \; (1-\delta)\delta_0\frac{\lambda}{d}, \; 2.8 \times 10^6 \times 3 \times 10^3, \; 260 \times 3 \times 10^6, \; 45.7 \times 3)$$

where v_0, e_0, and z_0 represent the levels in 3 liter serum, respectively. Here we take $e_0 = 260 \times 3 \times 10^6$ because a healthy human has mean 618 CD8$^+$ T cells $/\mu$L.

The parameters $m, n, a, k, k_1, k_2, k_4, k_6$ will change during the treatment based on different combination schemes. Their values are listed in Table 2.

$*$ The patient had stopped receiving anti-HBV infection treatment during this period.

The evolutions of patient's HBV DNA, ALT, HBeAg, and anti-HBe are shown in Fig.2. Observe that on the days 398th, 518th, 579th, 636th, and 697th, the patient's ALT levels increased.

Table 2. Parameter values at different days.

Days	m	n	k_1	k_2	k_4	k	a
0∼42	0.4	0.2	$0.5k_1$	k_2	0	$14k$	$6.5a$
43∼54*	0	0	$0.5k_1$	k_2	0	k	a
55∼90	0.4	0.2	$0.7k_1$	$1.5k_2$	0	k	a
91∼160	0.4	0.2	$0.8k_1$	$1.7k_2$	$2000k_1$	k	$2a$
161∼549	0.99	0.99	$0.6k_1$	$1.5k_2$	$2000k_1$	k	$4a$
550∼921	0.999	0.999	$0.6k_1$	$1.5k_2$	$2000k_1$	k	$4a$

From 463th day to 752th day, the patient had been taking antihypertensive drugs Hyzaar (Losartan Potassium and Hydrochlorothiazide Tablets, 50mg /12.5mg daily) and Adalat (Nifedipine Controlled-release Tablets 30mg/daily). This two drugs may damage the patients's hepatocytes. We do not know the reason that the patient's ALTs increased on the 398th testing day. However we can assume that some unknown reason(s) caused additional damages of the patient's liver cells during the 342th day and the 398th day.

In summary during the above periods, we revise system (3) to the following form

$$\begin{cases} \dot{x} = \lambda - dx - (1-m)\dfrac{bvx}{x+y} - k_8 x, \\ \dot{y} = (1-m)\dfrac{bvx}{x+y} - ay - \dfrac{k_1 ye}{x+y} - k_8 y, \\ \dot{v} = k(1-n)y - uv - \dfrac{k_4 ve}{x+y}, \\ \dot{e} = k_2(x+y) - k_3 e, \\ \dot{z} = k_5 + \dfrac{k_6 k_1 ye}{x+y} - k_7 z + k_9 k_8 (x+y). \end{cases} \tag{8}$$

where $k_8 x$ and $k_8 y$ represent the damaging rates of the uninfected cells and the infected cells, respectively; $k_9 k_8(x+y)$ represents the increasing rate of ALT levels caused by the additional liver damages. The values of $k_8's$ and $k_9's$ are listed in Table 3.

Table 3. Selected values for $k_8's$ and $k_9's$.

days	344∼ 398	463 ∼752
k_8	1e-9	1.1e-9
k_9	6e-2	6.93e-2

The numerical simulations of systems (3) and (8) are shown in Figs 2 and 3. Observe that the simulated HBV DNA and ALT curves are agreement with the clinical testing data.

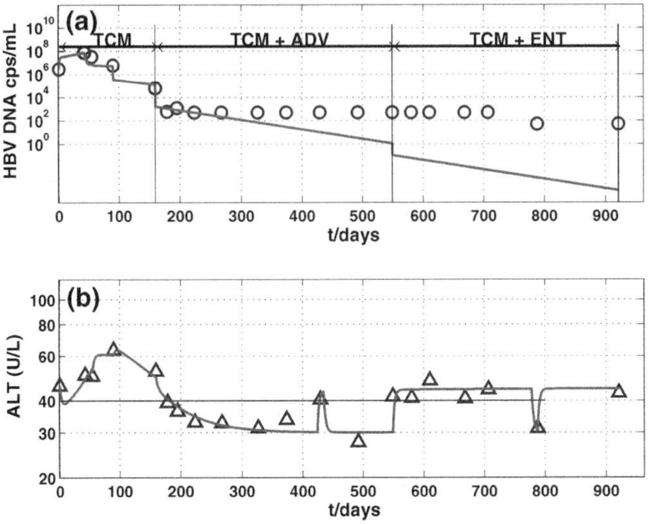

Fig. 2. Outcomes of the treatment efficacy of the patient. (a) Circles: the clinic HBV DNA levels. Solid line: numerical simulation of systems (3) and (8). (b) Triangles: the clinic ALT levels. Solid line: numerical simulation of systems (3) and (8).

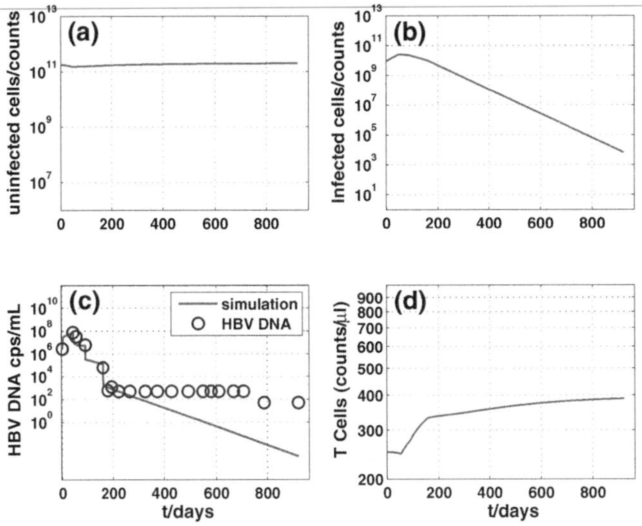

Fig. 3. Solid line: numerical simulation of system (3) and (8). Circles: clinic HBV DNA levels. (a) Uninfected cells x, (b) infected cells y, (c) free virus HBV DNA v, and (d) CTL cells e.

Remark: (1) A puzzling phenomenon is that during the first 1.5 months' treatment, the TCM made the patient's HBV DNA levels increased 27.7 times (see Table 1), and ALT levels almost no changed. As a result, we have to make virus replication parameter k increased 14 times, and the patient's infection cell death ("apoptosis") parameter a increased 6.5 times. Otherwise the numerical simulations showed that simulated patient's ALT levels would be much higher than the clinic testing data.

(2) From Tables 1, 2 and Figs.2 and 3, we can conclude that during the first 3 months, the main functions of the TCM are to regulate and activate the patients's immune system. The following 2.5 months TCM treatment have activated and increased two kinds of immune functions: cytokine-mediated destruction of HBV-infected cells (increasing the parameter k_1) and cytokine-mediated noncytolytic HBV clearance (increasing the parameter k_4). The following months' TCM + NA combination treatments have speeded up the blocks of the replications of HBV and the apoptosis of the HBV-infected cells.

4 Conclusions

In order to describe the dynamics of anti-HBV infection therapy, this paper proposes a differential equation model with five-variables (3). This model has two equilibrium points, Q_1 and Q_2, representing the patient's complete recovery, and persistent HBV infection, respectively.

Using the system (3) simulates the dynamics of the anti-HBV infection treatments with TCM + NA. We have found that we need to increase the values of the parameters a, k_1, k_4 in sometimes to make our simulations to be in agreement with the clinic HBV DNA and ALT levels. This means that the combination treatments increase the "apoptosis" of HBV-infected cells, and patients' immune abilities. It seems that the TCM can wake patients' immune functions via increasing the rate of the replications of the HBV in infected cells.

The combination treatment scheme may speed the patient's HBeAg seroconversion and HBV DNA reducing to undetectable levels. Chinese herbal medicine and nucleoside personalized combination therapy is promising.

Acknowledgments. This work is jointly supported by the 11th 5-Year Plan Key Research Project of China (No.2008ZX10005-006), and the National Nature Science Foundation of China (No.60674059).

References

1. WHO, Hepatitis B: Fact Sheet, World Health Organization (October 2000), `http://www.who.int/mediacentre/factsheets/fs204/en/` (accessed August 2011)
2. Maddret, W.C.: Hepatitis B: an important public health issue. J. Med. Virol. 61, 362–366 (2000)

3. Lau, G.K.K., Piratvisuth, T., Luo, K.X., et al.: Peginterferon Alfa-2a, Lamivudine, and the Combination for HBeAg-Positive Chronic Hepatitis B. New England Journal of Medicine 352, 2682–2695 (2005)
4. Dienstag, J.L., Schiff, E.R., Wright, T.L., et al.: Lamivudine as initial treatment for chronic hepatitis in the United States. N. Engl. J. Med. 34, 1256–1263 (1999)
5. Marcellin, P., Chang, T.T., Lim, S.: G., et al.: Long-term efficacy and safetyof adefovir dipivoxil in HBeAg + chronic hepatitis B patients: inceasing serologic, virologic, and biochemecal response over time (abstract). Hepatology. 40(suppl.1), 665A (2004)
6. Papathecodoridis, G.V., Hadziyannis, S.J.: Diagnosis and management of pre-core mutant chronic hepatitis B. J. Viral Hepat. 125, 562–567 (2001)
7. Hadziyannis, S.J., Vassilopoulos, D.: Hepatitis B e antigen-negative chronic hepatitis B. Hepatology 34, 617–624 (2001)
8. Ye, Y., Min, L., Zhang, Q.: Evaluation of 48 week adefovior dipvoxl (AD) and Chinese hebral medicine plus AD treatment in HBeAg(+) chronic hepatitis B Chinese patients: A double-bind randomized trial(Abstract). In: Hepatology (Supplement: The 62st Annual Meeting of the American Association for the Study of Liver Diseases: The Liver Meeting, 54 (S1) (2011) (in press)
9. Nowak, M.A., May, R.M.: Virus dynamics: Mathematical principles of immunology and virology. Oxford University, Oxford (2000)
10. Nowak, M.J., Bonhoeffer, S., Hill, A.M., et al.: Viral dynamics in hepatitis B virus infection. Proc. Natl. Acad. Sci. USA. 93, 4398–4402 (1996)
11. Leenbeer, P.D., Smith, H.L.: Virus dynamics: A global analysis. SIAM Journal on Applied Mathematics 63, 1313–1327 (2003)
12. Min, L., Su, Y., Kuang, Y.: Mathematical analysis of a basic model of virus infection with application to HBV infection. Rocky Mountain J. of Mathematics 38, 1573–1585 (2008)
13. Gourley, S.A., Kuang, Y., Nagy, J.D.: Dynamics of a delay differential model of hepatitis B virus infection. J. Biological Dynamics 2, 140–153 (2008)
14. Eikenberry, S., Hews, S., Nagy, J.D., Kuang, Y.: The dynamics of a delay model of HBV infection with logistic hepatocyte growth. Math. Biosc. and Eng. 6, 283–299 (2009)
15. Zheng, Y., Min, L., Ji, Y., Su, Y., Kuang, Y.: Global Stability of Endemic Equilibrium Point of Basic Virus Infection Model with Application to HBV Infection. J. Systems Science and Complexity 23, 1221–1230 (2010)
16. Feld, J., Lee, J., Locarnini, S.: New targets and possible new therapeutic approaches in the chemotherapy of chronic hepatitis B. Perspectives in Clinical Hepatology 38, 545–553 (2003)
17. Perelson, A.S.: Modelling viral and immune system dynamics. Nature Reviews Immunology 2, 28–36 (2002)
18. Bataller, R., Brenner, D.A.: Liver fibrosis. J. Clin. Invest. 115, 209–218 (2005)
19. Hellerstein, M., Hanley, M.B., Cesar, D., et al.: Directly measureed kinetics of circulating T lymphocyres in normal and HIV-1-infected humans. Nature Medicine 5, 83–89 (1999)
20. Stockham, S.L., Scott, M.A.: Fundamentals of Veterinary Clinical Pathology. Iowa State University Press (2002)

Signal Processing for Active Power Filters

Petr Simonik and Pavel Brandstetter

VSB – Technical University of Ostrava, FEI, Department of Electronics, 17.listopadu 15, 708 00 Ostrava, Czech Republic
{petr.simonik,pavel.brandstetter}@vsb.cz

Abstract. The paper describes special active power filter solution. There are given control algorithms and methods for the filter current regulation. Firstly, there is described power part structure of the filter with the resonant converter and general conception structure of the filter. Then there is description and calculations of the control algorithms and methods for the filter currents regulation. In the paper are given results of practical realization, which was, in this case, mainly orientated to control algorithm research. For this purpose was selected DSP TMS320F2812 and implemented to the control unit of the filter. Control unit was practically realized by use of development DSP kit. Sample of practically realized control unit was designed as universal control unit for power electronics applications. Reached results that are presented in the paper are simulation and practical realization results.

Keywords: Active Filter, Control Unit, Converter, Development Kit, DSP, Interface, Power Electronics.

1 Introduction

Power electronics and electric drives are a lot of years significant technical branch of world industry. High powered electric drives with power electronics applications are one of main equipments of modern industry manufacturing lines, mass transportation vehicles (as trains, buses, trams), power quality instruments in power supply systems and so on. There are many different transnational companies, research workplaces and educational institutions which invest financial resources to the development of new modern devices as power converters, inverters, pulse rectifiers, power active filters and so on. These applications require high powered microprocessor control systems (at present mainly DSP), modern semiconductor components (as FRED, GTO, RCT, IGCT, IGBT, MOSFET and so on), powered drivers for semiconductor switches, quality pulse capacitors, etc. We can say that just by usage of DSP can be realized modern control algorithms with specific requires to the microprocessor system. Traditional producers are developing DSP for specific application. Texas Instruments (TI) produce line of DSP named TMS320x2xxx which are especially intended for power electronics applications. This paper describes one practical application of TI DSP for branch of power electronics - active filtering, harmonics elimination.

As an alternative parallel harmonics correction technique can be use parallel active power filter (PAPF) with resonant converter that will be describe in the paper. The description will be mainly oriented on control method analyzing and practical

G. Lee (Ed.): Advances in Automation and Robotics, Vol.2, LNEE 123, pp. 567–574.
Springerlink.com

application (development and used techniques). Mentioned parallel active power filter is a device that is connected in parallel to compensated devices and cancels the reactive and harmonics currents from a group of these nonlinear loads so that the resulting total current drawn from the AC main is sinusoidal.

1.1 Structure of the Filter

Following picture shows block structure of an PAPF with resonant DC link.

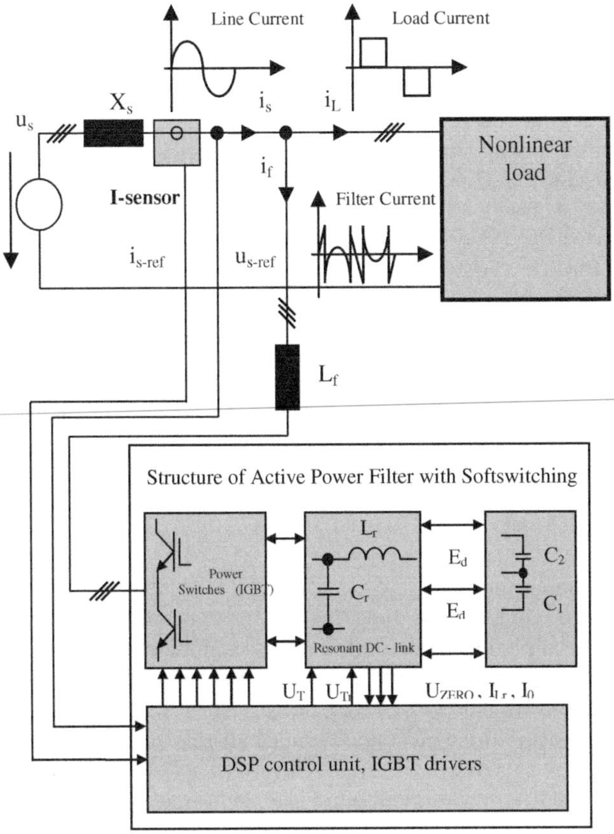

Fig. 1. Resonant active power filter and connection to the AC main.

PAPF is normally implemented with PWM voltage source inverters. In this case the PAPF operates as a voltage pulse source, so current is enforced by resonant voltage pulses. However the control unit with DSP is designed as universal control for application in structure of active power filters. Typical controls of PAPF are linear current control, digital deadbeat control, and hysteresis control. However, most methods for obtain right control require high-speed digital signal processor system with fast A/D.

Fig. 2 shows general conception of resonant PAPF with important control and measured signals. In the power part have to be implemented fast voltage and current sensors. Right calculation of all steps in method's diagram is based on knowledge of zero voltage time interval (in the case of PAPF with resonant DC link).

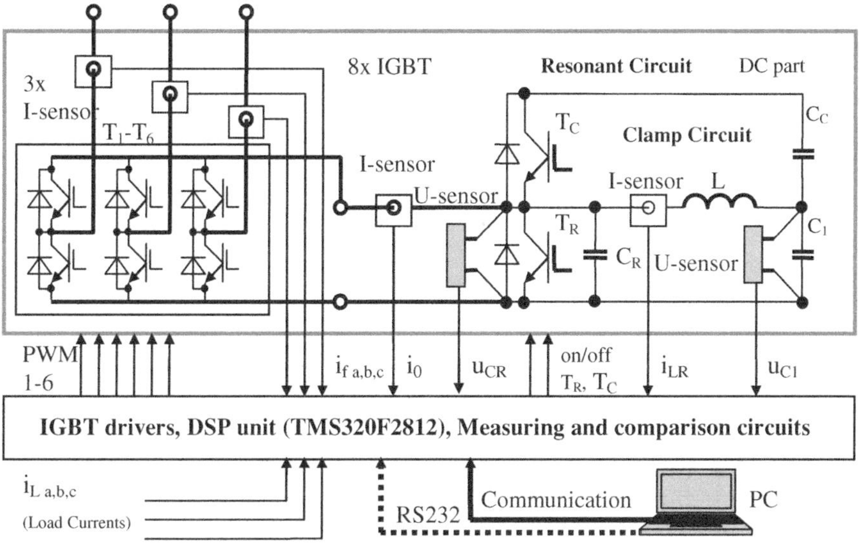

Fig. 2. Block structure of resonant PAPF, control unit signals.

Main circuit of the power part is resonant circuit with resonant capacitor CR and resonant inductor LR. The inductor was made by experimental testing on base of known value of capacitor CR. All experimental tests were realized on the base of computer simulation results. In the next chapter will men-tioned details of control circuit structure and used calculation and regulation techniques.

1.2 Realized Control Unit Description

More resonant process details have to be introduced as first. Fig. 3 shows theoretical voltage and current waveforms in resonant DC-link. Resonant frequency is defined as following:

$$\omega_0 = \frac{1}{\sqrt{L_R C_R}} \tag{1}$$

The resonant DC-link has implemented a limit voltage circuit, so called the active clamp circuit. Resonant frequency of the clamp circuit is defined as following:

$$\omega_C = \frac{1}{\sqrt{L_R C_C}} \tag{2}$$

The clamp factor k is related to the tank period T_k and resonant frequency.

$$\frac{f_0}{f_k} = T_k \varpi_0 = 2\left[\cos^{-1}(1-k) + \frac{\sqrt{k(2-k)}}{k-1}\right] \tag{3}$$

Voltage waveform of resonant DC-link on the Fig. 4 is divided by 10 for better visualization. Without an active clamp circuit are peaks of voltage resonant pulse slightly greater than twice the DC input voltage.

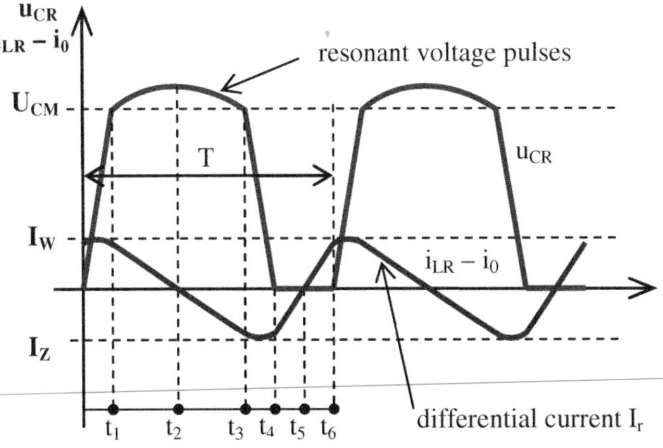

Fig. 3. Theoretical waveform of resonant voltage U_{CR} and differential current Ir ($I_{LR} - I_O$).

That is for fixed value of k, T_k can be determined for a given resonant circuit of the filter's converter.

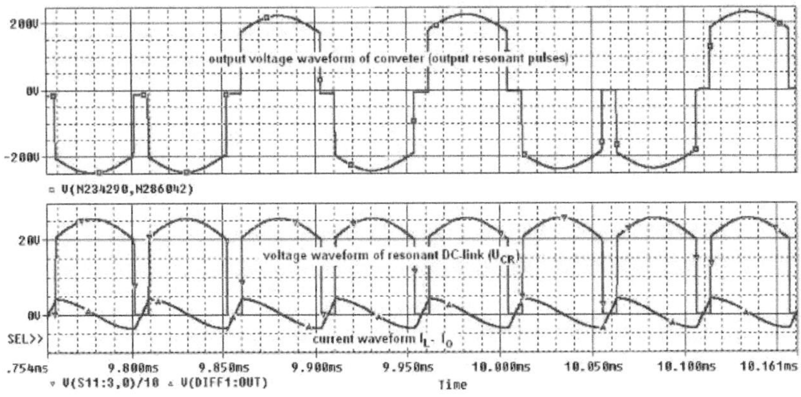

Fig. 4. Simulated waveform of resonant voltage U_{CR} and differential current Ir ($I_{LR} - I_O$).

2 Active Filter Control Methods and Algorithms

In the following will presented possibilities for acquisition of referential current value.

2.1 Calculation of Referential Current by Means of 1st Harmonics of Load Current

The method is based on calculation of required value of line current. However we can adjust it as following:

$$i_{a\_ref} = i_{a\_zátěž} - I_{1m} \,.\, \sin(\omega t) \tag{4}$$

$$i_{b\_ref} = i_{b\_zátěž} - I_{1m} \,.\, \sin(\omega t - 120^{o} - \varphi_1) \tag{5}$$

$$i_{c\_ref} = i_{c\_zátěž} - I_{1m} \,.\, \sin(\omega t + 120^{o} - \varphi_1) \tag{6}$$

For calculation of 1st harmonics load current can be used two ways. First of them is usage of Fast Fourier Transformation (good dynamic properties and can be used for symmetric and unsymmetric three phase distribution network, method is not dependant on value of line voltage and so on) and second is usage of mean value "p element" of instantenous power (simplicity and good dynamic properties, voltage dependence – needs sinusoidal voltage and knowledge of maximum voltage value). Last mentioned method also requires Phase Locked Loop – PLL and monitoring of line voltage and current at once).

2.2 Calculation of Referential Current by Means of Complete Clark-Park's Current Transformation (i_d, i_q Method)

The method can be used for PAPF compensative current harmonics and also zero equalizing current. Resulting transformed inharmonic current will be contain d, q, 0 elements and classical three phase system will be transformed into the rotary orthogonal system. Transformation equation can be defined as following:

$$\left[C_{\alpha\beta0}\right] = \sqrt{\frac{2}{3}} \cdot \begin{bmatrix} 1 & 0 & \frac{1}{\sqrt{2}} \\ -\frac{1}{2} & \frac{\sqrt{3}}{2} & \frac{1}{\sqrt{2}} \\ -\frac{1}{2} & -\frac{\sqrt{3}}{2} & \frac{1}{\sqrt{2}} \end{bmatrix} \tag{7}$$

Line currents and voltage are represented by u_{abc} a i_{abc}, then by means of Park's transformation can be transformed system converted into the d, q, 0 rotary orthogonal system. Transformation angle θ can be defined as θ = tan-1(uβ /uα).

$$\left[u_{\alpha\beta0}\right] = \left[C_{\alpha\beta0}\right]^{T} \cdot \left[u_{abc}\right] \tag{8}$$

$$\left[i_{\alpha\beta0}\right] = \left[C_{\alpha\beta0}\right]^{T} \cdot \left[i_{abc}\right] \tag{9}$$

Topology of current and voltage space vector lay-out shows Fig. 5. Size of transformation angle is not constant at the time, because this angle is very sensitive to voltage harmonics, voltage unbalance and zero component of voltage u_0. On the base of geometric relations on Fig. 5 we can write following:

$$\begin{bmatrix} i_d \\ i_q \\ i_0 \end{bmatrix} = \frac{1}{\sqrt{u_\alpha^2 + u_\beta^2}} \begin{bmatrix} u_\alpha & u_\beta & 0 \\ -u_\beta & u_\alpha & 0 \\ 0 & 0 & \sqrt{u_\alpha^2 + u_\beta^2} \end{bmatrix} \cdot \begin{bmatrix} i_a \\ i_b \\ i_0 \end{bmatrix} \tag{10}$$

Instantaneous real and imaginary component of currents i_d and i_q includes DC and AC component. First load current harmonics will be transformed into the DC components of currents i_d and i_q.

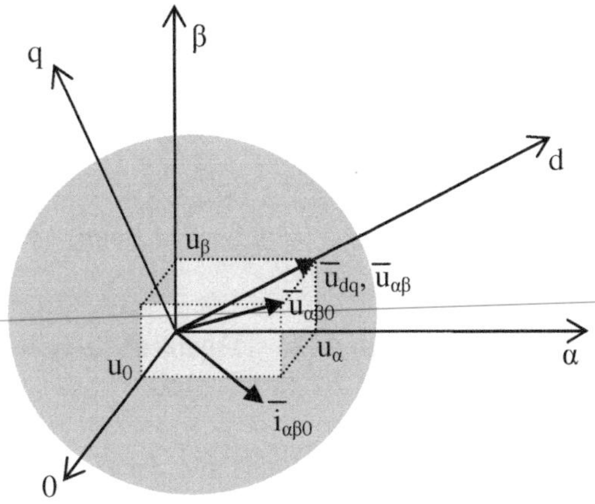

Fig. 5. Simulated waveform of resonant voltage U_{CR} and differential current Ir ($I_{LR} - I_O$).

Active components of first harmonics will be transformed to the DC component of current i_d and reactive component of first harmonics will be transformed to the DC component of current i_q. Referential current of the shunt active power filter is possible to calculate from an AC components. For filtration of first harmonics and also filtration of reactive components of first harmonics it is enough to have AC component of current i_d. Mean value can be separated by high pass filter or by direct numeric calculation per one period. There is lay-out in stationary orthogonal system α, β, 0 and rotary orthogonal system d, q, 0. For final shape of referential currents can be derived following relations:

$$\begin{bmatrix} i_{a\_ref} \\ i_{b\_ref} \\ i_{c\_ref} \end{bmatrix} = \frac{1}{\sqrt{u_\alpha^2 + u_\beta^2}} \cdot \begin{bmatrix} u_\alpha & -u_\beta & 0 \\ u_\beta & u_\alpha & 0 \\ 0 & 0 & \sqrt{u_\alpha^2 + u_\beta^2} \end{bmatrix} \cdot \begin{bmatrix} -\tilde{i}_d + i_{loss} \\ -\tilde{i}_q \\ -i_0 \end{bmatrix} \tag{11}$$

$$\left[i_{abc\_ref}\right] = \left[C_{\alpha\beta0}\right] \cdot \left[i_{\alpha\beta0\_ref}\right]$$

Components $\tilde{i}_d$ and $\tilde{i}_q$ are AC components of currents i_d and i_q. Components i_{loss} it's current, related with the losses in the power part of parallel active power filter. In the case of requirements to do just harmonics compensation we can write following:

$$\begin{bmatrix} i_{a\_ref} \\ i_{b\_ref} \\ i_{c\_ref} \end{bmatrix} = \frac{1}{\sqrt{u_\alpha^2 + u_\beta^2}} \cdot \begin{bmatrix} u_\alpha & -u_\beta & 0 \\ u_\beta & u_\alpha & 0 \\ 0 & 0 & \sqrt{u_\alpha^2 + u_\beta^2} \end{bmatrix} \cdot \begin{bmatrix} -\tilde{i}_d + i_{loss} \\ -i_q \\ -i_0 \end{bmatrix} \tag{12}$$

$$\left[i_{abc\_ref}\right] = \left[C_{\alpha\beta 0}\right] \cdot \left[i_{\alpha\beta 0\_ref}\right]$$

In the case of mentioned method has non-sinusoidal voltage of supply mains negative effect for right results of compensation. In the following will presented possibilities for acquisition of referential current value.

2.3 Method of Two-Value Hysteretic Regulation

Referential (i_{fa,b,c_ref}) and actual current's values (i_{fa,b,c_sk}) are compared in a hysteretic comparators with hysteresis Δi. Output signals from hysteretic comparators then determinate which one from the power switches couple will be turned on. Every single comparator is used for each phase of the filter. Two-value hysteretic regulator can be digital or analog. In case of digital regulator, comparison is realized with sampling frequency f_{VZ}. This is the reason why double amplitude (peak-to-peak) of the regulated current is higher. To undesirable double amplitude elimination, it should be used two-value regulator without hysteresis adjusting. Then the hysteresis is given by maximal value of sampling frequency and rate of current rise. In case of active filter with soft switching in the converter, the hysteresis is given by resonant frequency and resonant pulses steepness.

3 Control Unit Based on DSP and Practical Results

The system with TMS320F2812 was selected on the base of high speed calculation requirements. Developed control unit was designed as universal control unit and can be used also for other types of converters. System action is divided to both fast algorithm for control of resonant oscillation in resonant DC-link and compensating algorithm of own power active filter and own utility routines. For experimental testing of control unit was selected control algorithm based on Fast Fourier Transformation (FFT). In Fig. 6a (left side), there is repeating period of FFT coefficient calculation and time interval detail (7.2 μs) of one loop FFT calculation. This picture shows right behavior of digital block of Fourier analysis. Fig. 6b (right side) shows calculation results. All waveforms are inside signals of the control unit. These signals are watching by use of D/A converter which is for this cases placed in the control unit. There are four analog outputs for this case. The upper trace is input current i_{La} of device with non-linear load (load consist of resistor and inductor). The second trace is the computed 1st harmonic (required value) of line current. The third trace is computed line angle. The waveforms are computing results of FFT block. Fig. 7a (left side), shows right behavior of the filter for compensation of current harmonics.

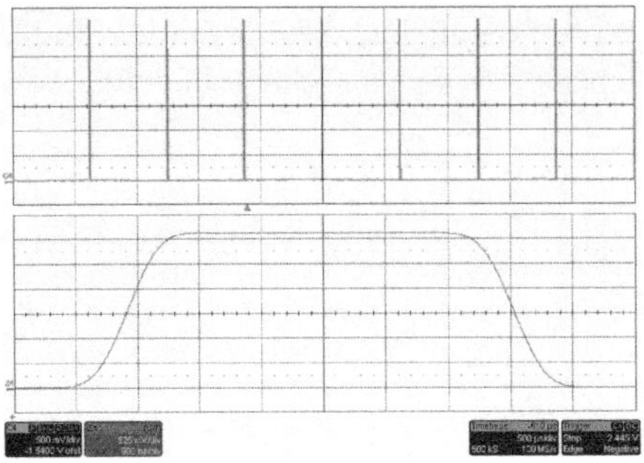

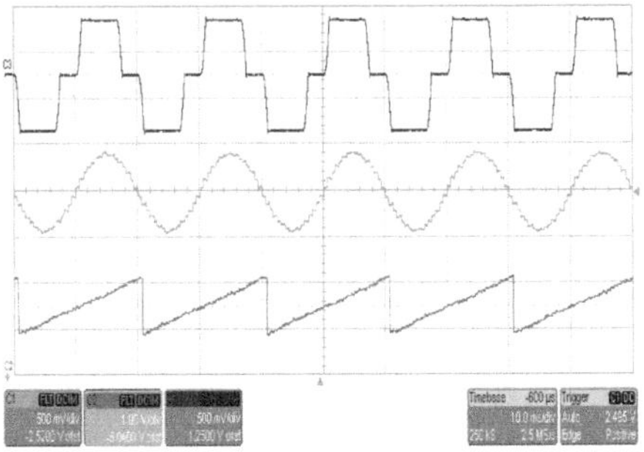

Fig. 6a. Repeating period of FFT coefficient calculation, time interval detail (7.2 μs) of one loop FFT calculation, **6b.** Calculated values of realized control unit.

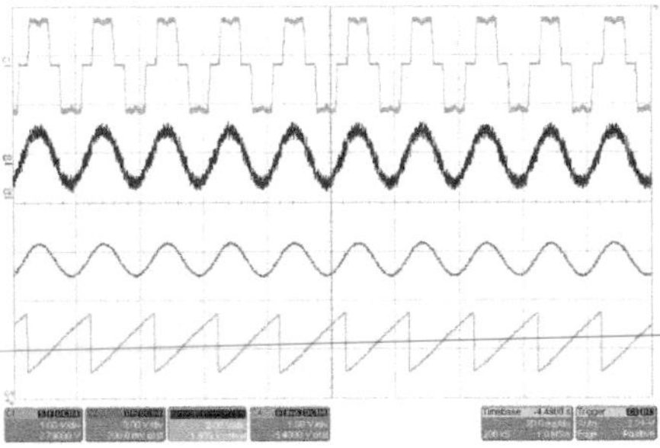

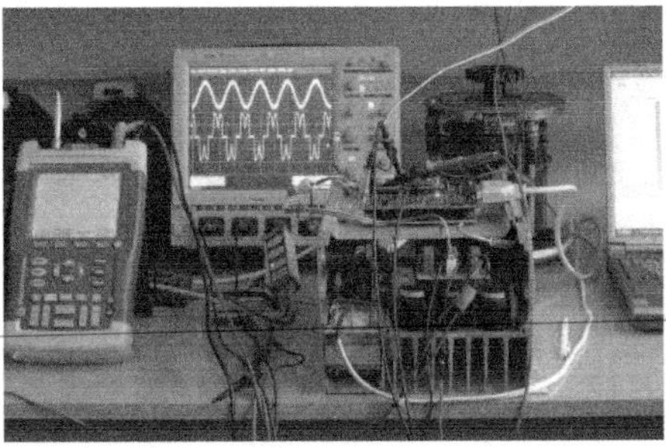

Fig. 7a. Measured and calculated waveforms, CH1: input current iLa of device with non-linear load, CH2: computed 1st harmonic of line current, CH3 : compensated current in the line, CH4: computed line angle., **7b.** Laboratory sample of the filter.

Acknowledgments. The project is supported by Grant Agency of Czech Republic (GA CR), project No. 102/09/P665, for which author express their sincere gratitude.

References

1. Brandstetter, P.: Research of Active Power Filters, final research report of project GA CR 102/99/0193, VSB-TU Ostrava, p. 74 (2001)
2. Qiao, C., Smedley, K.M.: Three-phase Active Power Filters with Unified Constant-frequency Integration control, `http://www.eng.uci.edu/~smedley/IPEMC-A185.PDF`

Steel Billet Recognition in the Heavy Rail Production Line

Hanyu Hong, Xiuhua Zhang, Yijun Yang, and Zhejun Yu

Laboratory for Image Processing and Intelligent Control, Wuhan Institute of Technology, Wuhan 430205, China
honghany@public.wh.hb.cn

Abstract. Steel billet recognition is an urgent requirement in the steel industry of heavy rail line. Due to high temperature and complex scene in the rolling line, the recognition at the end of billet is quite different from optical character recognition with simple background and vehicle license plate recognition. In order to solve the problem, this paper presents some new schemes in character location, segmentation and recognition in the complex scene. These proposed algorithms include the location algorithm of multilevel filter based on separability criterion, the segmentation algorithm based on intelligent multi-agent and the recognition algorithm of character features matching combining the template matching. The results show that the proposed algorithms are effective for steel billet character recognition in the complex scene.

Keywords: Steel billet character, character location, character segmentation, character recognition.

1 Introduction

Automatic recognition system of heavy rail billet is a high precision imaging system; it can complete character location and recognition at the end of billet on production line of heavy rail universal mill. The proposed system is shown in Fig. 1. Multi-subjects is adopted in this system to solve automatic detection and recognition problem at the end of billet on production line of heavy rail universal mill, its basic theory includes photoelectric imaging technology, modern digital image processing technology, artificial intelligence, pattern recognition, data communication and transmission technology.

The recognition of characters at the end of steel billet is quite different from optical character recognition with simple background and vehicle license plate recognition. The different scientific problems indicate as follow:(1)The steel code image has a different phenomenon such as uneven illumination in background and weak character information, even because of billet mutual friction and extrusion to cause roughness at the end of billet and character deformation or fracture.(2)Many factors influence the steel billet character recognition, such as the unknown existence of characters, the uncertainty of characters orientation, local characters pollution, missing, and adhesion. Thus the new requirement is proposed in the research of robustness and

G. Lee (Ed.): Advances in Automation and Robotics, Vol.2, LNEE 123, pp. 575–581.
Springerlink.com

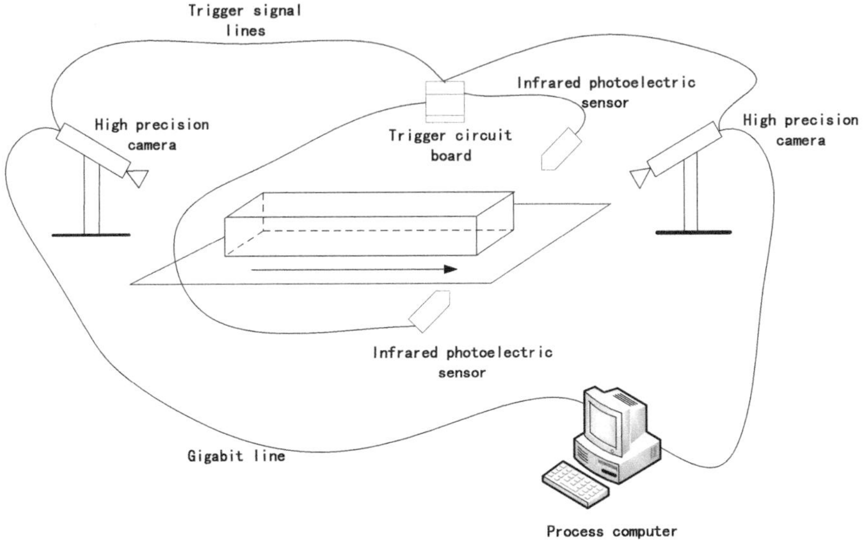

Fig. 1. The recognition system of steel billet

accuracy.(3)The flow of production line limits the time of the real-time detection location, there is a higher request in the recognition speed of characters.(4)Because of the complexity of target recognition, the traditional methods failed to find recognition at such situations in stability and reliability. This research helps to break through the key technology of artificial vision, and promote the development of the pattern recognition theory.

In order to solve these problems, this paper presents some new schemes in character location, segmentation and recognition to detect the billet characters in the complex scene. The results show that the proposed algorithms are effective in improving the adaptability, robustness and real-time capability of billet character recognition in the complex scene.

2 Steel Billet Character Recognition

2.1 Steel Billet Character Location

The location of character at the end of billet is quite different from optical character location with simple background and vehicle license plate location. The steel code image has a different phenomenon such as uneven illumination in background and weak character information, even because of billet mutual friction and extrusion to cause roughness at the end of billet and character deformation or fracture. The traditional methods failed to find location at such situations in stability and reliability. There are many factors to influence on the steel code location at the end of billet, such as the unknown existence of characters, the uncertainty of characters orientation, local characters pollution, missing, and adhesion. Thus the new requirement is proposed in the research of robustness and accuracy, and because the flow of production line limits the time of the real-time detection location, there is a higher request in the location speed of characters. The common character location methods include EunRyung etc[1, 2], who puts forward the location method which use the color components in image, Dai Ruwei etc[3], who proposed the location method based on Niblack binarization algorithm and adaptive boundary search algorithm, the location method based on character features or projection[4] and so on. The mentioned methods are mostly based on the characteristics of the license plate to study the license plate character location, and this research will have specific aim and limitations. The location result is not ideal in complex production line scene with unstable contrast ratio and great variation of lighting conditions, as well as dynamic geometric target.

In order to capture the target region from complex scene, this paper is adopted Mean Shift algorithm to restrain the background, and automatically distinguish region of interest and non-interest by gradually filtering and clustering method. Secondly, establish the quantitative model for the candidate character region, and join the statistical decision algorithm to complete screening. The results of the study show that this paper location method in complex production line scene has a much better stability and accuracy, and could solve the adaptability, robustness, real-time and applicability of the characters location, and other key scientific and technological problems in the complex conditions.

2.2 Steel Billet Character Segmentation

In the process of steel billet detection and recognition, how to accurately divide the characters at ends of steel in the complex scene is a highly complicated intelligence problem. In order to solve this complex problem, a segmentation algorithm based on intelligent multi-agent is proposed. The algorithm includes these main agents: the characters binarization and segmentation, the regions combination [5], the regions division and the features calculation. And then, these agents work in coordination with each other in the control of the master agent. At last, the billet characters are divided accurately through the proposed algorithm of intelligent multi-agent.

The algorithm structure of intelligent multi-agent is shown as follows: The master agent is responsible for the interaction and allocation of each agent. And the algorithm structure of intelligent multi-agent is shown as follows. Firstly, the master agent calls the agent of characters binarization and segmentation to divide the located billet characters roughly. Secondly, calling the agent of features calculation to mark and judge the features of each region, such as the width, height, center distance and so on. And then, call the agent of regions combination or regions division to combine the broken characters or to divide the connected characters. Finally, call the agent of features calculation once again to check the segmentation results. After the billet characters in the complex scene are divided accurately, the algorithm is stopped by the master agent. The algorithm structure of intelligent multi-agent is shown in **Fig**. 2.

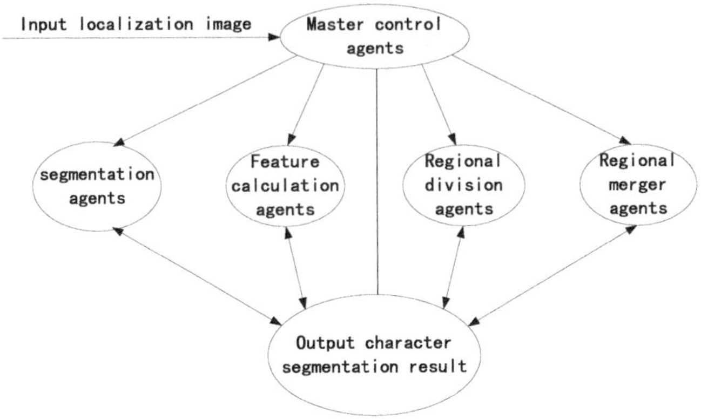

Fig. 2. The algorithm structure of intelligent multi-agent

2.3 Steel Billet Character Recognition

The information of steel billet characters is entirely spared after the intelligent segmentation, which provides the insurance for the character recognition. Though the accurate segmentation of steel billet character, the character recognition is also affected by the particularity of billet characters. Because of the printing of the billet characters, the characters have a lot of shortcomings such as the deformed characters, the lost characters, the blurry characters and so on. Therefore, during the process of recognition, the recognition algorithm is confronted with difficulty of the variational shape and the incomplete structure of the billet characters. And the much higher request is brought up for the adaptability and intelligence of the recognition algorithm. According to the diversity of the billet characters, it's badly needed to make a study of a recognition algorithm for the complex billet characters. Consequently, it's valuable to explore the recognition algorithm through extracting the structure feature of billet characters. Since the algorithm of template matching has simple structure and is easy to be realized [6]. A recognition algorithm of character features matching based on the template matching is proposed in this paper.

3 The Experiment Results and Analysis

The new character recognition method has been programmed on the computer (2.66 GHz, 0.98GB) by VC + + 6.0 in this paper. The character recognition tests were performed for billet images in the complex production line scene.

Test 1: Iterative segmentation filter. Fig. 3 is a grayscale image which shows the segmentation filtering process step by step after the restraining background, and the Tab. 1 shows the update growth of threshold. The example proves that the scheme of operation time had almost no growth compared to original algorithm, while the accuracy of extraction optimal threshold value is greatly increased.

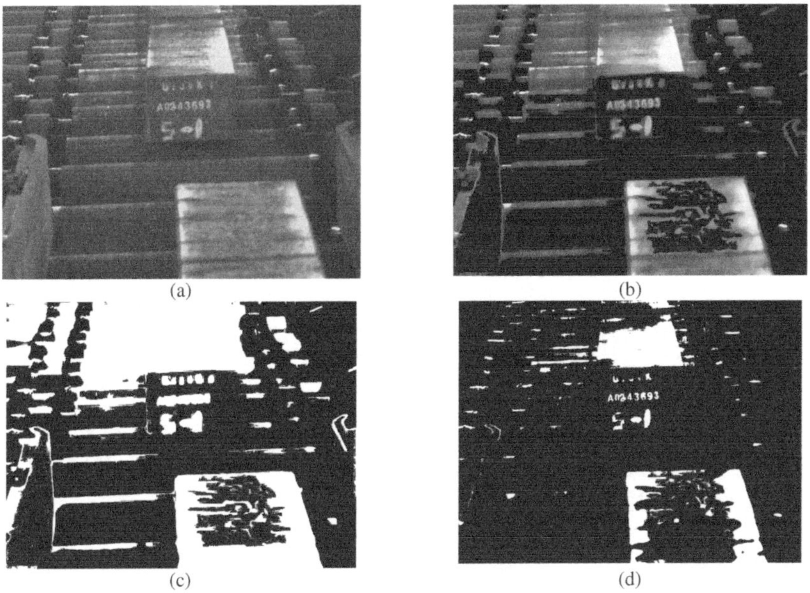

Fig. 3. Multistage filtering with our proposed method. (a)Original image. (b)Restraining background. (c) First stage filtering(threshold=60). (d) Second stage filtering(threshold=79).

Table 1. Comparison of threshold growth

	Threshold 1	Threshold 2	Threshold 3	Threshold 4	Threshold 5
updated threshold by traditional method	123	176	202	214	219
updated threshold by our method	60	69	79	106	145

Fig. 4. Comparison with result of location. (a)Location image by the traditional projector location method. (b)Location image by our proposed method. The located steel billet character images include the number of steel grade, the number of steel billet and the number of flow steel.

Test 2: Comparing the results of the billet image location. The location results of Fig. 3 are shown in Fig. 4. The traditional method of projection location relies more or less on the selection of the Coordinate datum, which is hard to get the exact location. Because the actual production line and the rest region of string are surrounded by complex and noisy environment, the direction of the string is random. The conventional coordinate datum might make projection information lost, leading to the wrong location. From Fig. 4(a) we could see that it could barely get the location, while Fig. 4(b) is a clear billet image location. This could demonstrate the advantages of the way in this paper, which is stable, and accurate

.

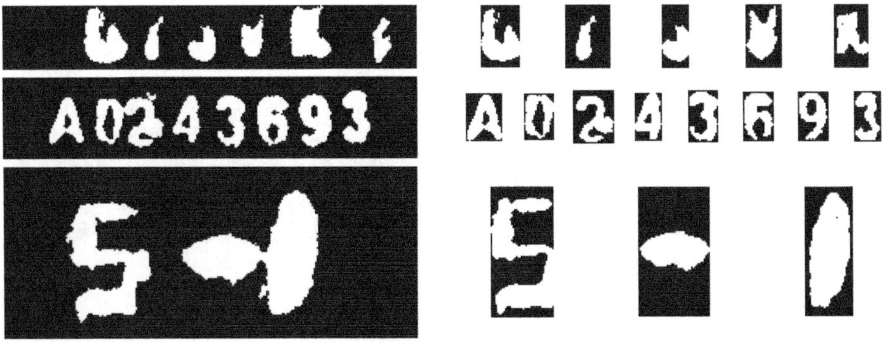

(a) (b)

Fig. 5. Character segmentation. (a)The binarization of located billet images. (b)The results of characters segmentation

Fig. 6. Interfance of the system and recognition results

Test 3: Character segmentation test. Fig. 5(a) indicates that the improved OTSU algorithm has good binarization effects. And a segmentation algorithm based on intelligent multi-agent is proposed this paper to solve the complex segmentation problem. As shown in Fig. 5(b), we can know that the billet characters are divided accurately through the proposed algorithm of intelligent multi-agent

Test 4: Chararcter recognition. A recognition algorithm of character features matching combining the template matching is proposed in this paper. As shown in the following Fig. 6, the result of recognition show that the algorithms proposed in this paper can recognize the billet characters in complex production line.

Acknowledgement. This work is supported by the National Science Foundation of China under grant 50975211 and 61175013, it is also supported by Science and Technology Brainstorm Project of Wuhan under grant 200810321164.

References

1. Lee, E.R., Kim, P.K.: Automatic recognition of a car license plate using color image processing. Journal of Korea Institute of Telematics and Electronics 24(2), 128–131 (1995)
2. Yen, J.C., Chang, F.J., Chang, S.: A New Criterion for Automatic Multilevel Thresholding. IEEE Trans. on Image processing 4(3), 370–377 (1995)
3. Dai, R., Liu, C., Xiao, B.: Chinese Character Recognition: History Status, and Prospects. Frontiers of Computer Science in China 1(2), 126–136 (2007)
4. Yang, W., Li, J., Shen, Z.: Technique of Automatic Orientation of Car License Plate Targets. Journal of Image and Graphics 7A(8), 835–839 (2002)
5. Zhang, Z., Huang, S., Li, D., Yan, G.: A Method Applied for Precise Segmentation of the Characters in the ID Card. Computer Engineering and Applications 13, 114–116 (2003)
6. Lu, X., Ling, X., Huang, W.: A Method of Vehicle Number Plate Recognition Based on Template Matching. Computer and Communications 6(22), 79–82 (2004)

Study on Vehicle Coupling Control for Automated Tracking Maneuver

Xinyu Liu[1,2], Yulin Ma[1,2], Qing Wu[3], and Zichao Huang[3]

[1] Intelligent Transport System Research Center, Wuhan University of Technology, Wuhan, 430063, Hubei, China
[2] Engineering Research Center for Transportation Safety (Ministry of Education), Wuhan University of Technology, Wuhan, 430063, Hubei, China
[3] School of Logistics Engineering, Wuhan University of Technology, Wuhan, 430063, Huibei, China
{xyliu,mayulin1983,wq}@hut.edu.cn, hzc_dragon@126.com

Abstract. When an autonomous vehicle (AV) performing automated tracking maneuver under complex road conditions such as driving with curves or with lane changing, the strong dynamic coupling effects exist in vehicle motion, thus it is very important to get ideal control effect by integrated lateral and longitudinal control system. In order to achieve automatic tracking control, assuming that a vehicle model based rigid body is make according to the simplified kinematic vehicle model. From this vehicle coupling model, the tracking control system of AV is derived. Based on Backstepping control method, the vehicle coupling variable structure control law has been designed. The stability of the tracking control system is analysed by using Lyapunov function method. Simulation results demonstrate that the control law designed could track the reference trajectory quickly, ang that it is effective and stable during AV's path tracking maneuver.

Keywords: Autonomous vehicle, automated tracking, rigid body, vehicle coupling, Backstepping.

1 Introduction

Over the past 10 years, automated vehicle has been a hot research at home and abroad. As the core content of it, vehicle longitudinal and lateral control is increasingly attracting researchers' attention[1,2]. Lateral vehicle control is mainly concerned with the road tracking ability, that is, how to control the vehicle along lane mark. In China, the lateral control is associated with vision navigation by image recognition. Domestic universities, such as National University of Defense Technology, Jilin University, and Tsinghua University, are most interested in it. The lateral vehicle controller has usually used optimal control, preview control and neural network control and other methods. Wuhan University of Technology, associated with National Center of ITS Engineering & Technology (ITSC) has carried out vehicle automated driving based on magnetic markers, and made a rapid progress by using fuzzy control method, some of which has been validated by a prototype car that

G. Lee (Ed.): Advances in Automation and Robotics, Vol.2, LNEE 123, pp. 583–591.
Springerlink.com

functions as removing snow in Xinjiang province[3]. The basic functions of longitudinal vehicle control are keeping the vehicle a safe distance behind another vehicle, maintaining a relatively constant speed. Recently, a variety of control methods, whether conventional PID control, or intelligent control methods including neural network control, fuzzy control, have been used for longitudinal vehicle control. [4] used a self-tuning fuzzy PID control method to design acceleration controller to keep spacing between two vehicles. By adjusting the three control parameters, the controller reduced the error about longitudinal spacing and velocity, and achieved the longitudinal control for two vehicles. [5] proposed a hierarchical switching control strategy based on robust control theory. The method is used to design velocity/acceleration system that realizes fast and accurate control for velocity/acceleration with unmodeled dynamics and parameter uncertainty. [6] proposed a new vehicle adaptive cruise control (ACC) algorithm based on driver optimal preview acceleration model. The algorithm was adapted to the strong nonlinearity of vehicle dynamic system.

However, in the above studies, the automated vehicle control has often progressed along two separate parts, without regard for the existing dynamic coupling between the lateral and longitudinal. Nevertheless, it is known that the motions are not completely independent. Consequently, unidirectional control designs invite unforeseen dangers arising from its neglect of the coupling dynamics. Moreover, the referencing systems utilized in these studies were often mutually exclusive, with little thought given to possible benefits derived from complementary coverage. So, when a controller is being designed, if the dynamic coupling effects are took into account, the accuracy and good tracking performance of the control system can be achieved, thus improve vehicles safety, mobility and collaboration[7-9].

In order to achieve automatic tracking control, according to the simplified kinematic vehicle model, a vehicle model based rigid body is made. Based on Backstepping control method, the vehicle coupling variable structure control law has been designed. The stability of the tracking control system is analysed by using Lyapunov function method. Finally, Simulation results demonstrate the effectiveness and stability of the control law designed.

2 Vehicle Model

The control object in this paper is an autonomous vehicle (AV). The rear wheels are regulated by motor, while the front wheels are regulated by steering engine. In order to simplify controller, assuming that: a) the entire vehicle is took as a rigid body, and the wheels as a rigid wheel, b) when velocity is low and the turning radius is large, all vehicle wheels are under pure rolling motion in the plane, without regard for the side slip between the tire and the road, c) without regard for the impact the uneven ground have on the trajectory[10-13]. Fig. 1 shows AV's kinematics diagram. According to the vehicle structure and the rigid body translation theory, AV is turning around the instantaneous center of vehicle body at any instant. Let o be the instantaneous center, and the turning radius ρ, angular velocity ω around o are got in the following equations respectively.

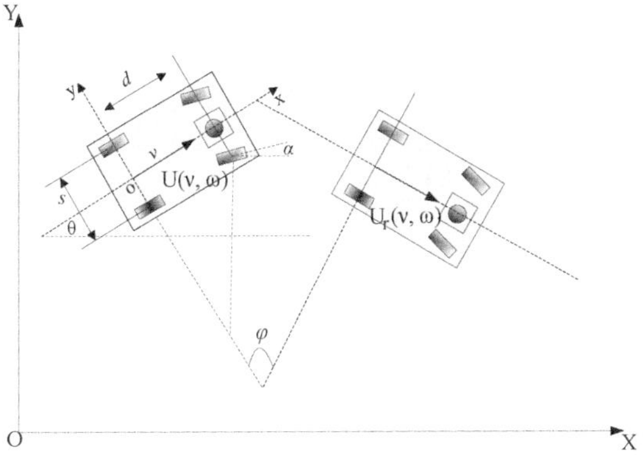

Fig. 1. The AV kinematics diagram. where OXY is the world coordinate system, d denotes the distance between rear wheels and front wheels, s denotes the distance between the two rear wheels, o is the center of the axis of the rear wheels, by which the vehicle coordinate system can be made. v denotes the center velocity, and is also AV's velocity, θ is the angle between the longitudinal vehicle body and the lateral axis of the world coordinate system, α is the corner of the front wheel.

$$\rho = \frac{d}{\tan\alpha} + \frac{s}{2}. \quad (1)$$

$$\omega = {}^{v}\!/_{\rho}. \quad (2)$$

At any time t, the relationship between the coordinate o(x, y) in the vehicle coordinate system and O(X, Y) in the world coordinate system is as follow

$$\begin{bmatrix} X \\ Y \\ 1 \end{bmatrix} = \Gamma(t)\begin{bmatrix} x \\ y \\ 1 \end{bmatrix} = \begin{bmatrix} \cos\theta & \sin\theta & x(\mathrm{o}) \\ -\sin\theta & \cos\theta & y(\mathrm{o}) \\ 0 & 0 & 1 \end{bmatrix}\begin{bmatrix} x \\ y \\ 1 \end{bmatrix}. \quad (3)$$

where $\mathbf{\Gamma}(t)$ is the transformation matrix from the vehicle coordinate system to the world coordinate system. (x(o), y(o)) is the coordinate of the origin in the vehicle coordinate system from the world coordinate system.

After the time interval Δt, the vehicle angle φ around o is

$$\varphi = \omega \cdot \Delta t. \quad (4)$$

So the transformation matrix $\mathbf{\Lambda}$ in vehicle coordinate system from t to (t+Δt) is

$$\Lambda = \begin{bmatrix} \cos\varphi & \sin\varphi & \rho\sin\varphi \\ -\sin\varphi & \cos\varphi & \rho(1-\cos\varphi) \\ 0 & 0 & 1 \end{bmatrix}. \quad (5)$$

and the transformation matrix **Γ** from the vehicle coordinate system to the world coordinate system at time (t+□t)

$$\Gamma(t+\Delta t)=\Gamma(t)\cdot\Lambda\,. \tag{6}$$

Let $p(x, y, \theta)$, $u(v, \omega)$ be AV's pose command, control command, respectively. From Eqs. (1), (2), (4), (6), we have that the increment in the world coordinate system, which is used for AV's actual trajectory.

$$\begin{cases}\Delta x=\rho\cdot\sin\varphi\cdot\cos\theta+\rho(1-\cos\varphi)\cdot\sin\theta\\ \Delta y=-\rho\cdot\sin\varphi\cdot\sin\theta+\rho(1-\cos\varphi)\cdot\cos\theta\,.\\ \Delta\theta=\varphi\end{cases} \tag{7}$$

where Δx is the incremental in X axis of the world coordinate system, Δy is the incremental in Y axis of the world coordinate system, $\Delta\theta$ is the incremental in vehicle angle.

Besides, according to the vehicle posture error in the vehicle coordinate system, we have the posture error equaiton[10], which is used for the control system.

$$\dot{\boldsymbol{p}}_{\mathrm{e}}=\begin{bmatrix}\dot{x}_e\\ \dot{y}_e\\ \dot{\theta}_e\end{bmatrix}=\begin{bmatrix}y_e\omega-v+v_{\mathrm{d}}\cos\theta_e\\ -x_e\omega+v_{\mathrm{d}}\sin\theta_e\\ \omega_{\mathrm{d}}-\omega\end{bmatrix}. \tag{8}$$

where v_{d}, ω_{d} is desired velocity, desired angular velocity of the vehicle ahead, respectively.

3 Coupling Variable Structure Control Law

As the existence of nonholonomic constraint of vehicle , the previous linear control theory is difficult to achieve the desired control effect. Backstepping method is developed as a stability theory of nonlinear system over the past decade. This method decomposes the complex nonlinear system into many subsystems, in each of which a basic controller has been designed, and state variables in each subsystem, that is the intermediate virtual control inputs, are associated with partial Lyapunov functions, so that the method simplifies the controller design by using virtual control inputs and partial Lyapunov functions[12,13]. In view of this, the backstepping method is used to design the coupling varialbe structure control law for AV. The objective of the control law is to control angular velocity and velocity of AV so that lateral and longitudinal tracking error converge to 0.

Assuming that the desired longitudinal displacement x_{d}, and desired velocity v_{d} are given. Define the longitudinal tracking error

$$\varepsilon_1=x-x_{\mathrm{d}}\,.$$

Choose a sliding surface as

$$s_1 = \lambda_1 \varepsilon_1 + \dot{\varepsilon}_1 . \tag{9}$$

Differentiating Eq. (9) and put Eq. (8) into it, yield

$$\dot{s}_1 = \dot{x} - \dot{x}_{\mathrm{d}} + \lambda_1 \varepsilon_1 = y_e \omega - v + v_{\mathrm{d}} \cos\theta_e - v_{\mathrm{d}} + \lambda_1 \varepsilon_1 . \tag{10}$$

It is clear that we choose the constant reaching law such that $\lim_{t\to\infty} |s_1(t)| = 0$, and the control law u_1 is

$$u_1 = v = y_e \omega + v_{\mathrm{d}} \cos\theta_e - v_{\mathrm{d}} + \lambda_1 \varepsilon_1 + k_1 \mathrm{sgn} s_1 . \tag{11}$$

Similarly, in the lateral, we also choose a sliding surface as

$$s_2 = \theta_e . \tag{12}$$

Differentiating Eq. (9) and put Eq. (8) into it, yield

$$\dot{s}_2 = \dot{\theta}_e = \omega_{\mathrm{d}} - \omega . \tag{13}$$

We have the control law u_2 from Eq. (13)

$$u_2 = \omega = \omega_{\mathrm{d}} + k_2 \mathrm{sgn} s_2 . \tag{14}$$

Next is stability analysis of the control system. We choose the following Lyapunov function candiate

$$V = \frac{1}{2} s_1^2 + \frac{1}{2} s_2^2 .$$

whose derivative along Eqs. (10), (13) is

$$\begin{aligned} \dot{V} &= s_1 \dot{s}_1 + s_2 \dot{s}_2 = s_1 (y_e \omega - v + v_{\mathrm{d}} \cos\theta_e - v_{\mathrm{d}} + \lambda_1 \varepsilon_1) + s_2 (\omega_{\mathrm{d}} - \omega) \\ &= -s_1 k_1 \mathrm{sgn} s_1 - s_2 k_2 \mathrm{sgn} s_2 = -k_1 |s_1| - k_2 |s_2| \le 0 \end{aligned} . \tag{15}$$

The Eq. (15) shows that when time $t \to \infty$, $\dot{V} \to 0$, thus we have that conditions like $\lim_{t\to\infty} |x_e| = 0$, $\lim_{t\to\infty} |y_e| = 0$ and $\lim_{t\to\infty} |\theta_e| = 0$ hold. By the Lyapunov stability criterion, using the designed control law, AV's tracking error $p_e(x_e, y_e, \theta_e)$ bounds and satisfies $\lim_{t\to\infty} \|[x_e, y_e, \theta_e]^{\mathrm{T}}\| = 0$ in the global scope. Therefore, the control system is stable.

To avoid AV skidding caused by too much changes in velocity or acceleration, limiting strategy has been added to the control law so as to ensure AV driving smooth.

$$u_1 = \begin{cases} v_{\min} & (|v| \le |v_{\min}|) \\ v_{\max} & (|v| \ge |v_{\max}|) \end{cases}.$$

$$u_2 = \begin{cases} \omega_{\min} & (|\omega| \le |\omega_{\min}|) \\ \omega_{\max} & (|\omega| \ge |\omega_{\max}|) \end{cases}.$$

4 Simulation Results

In order to test the proposed control law, we perform a series of simulation tests. Assuming that the sampling period of the control system is 10 ms, and the controller parameters and the desired vehicle state commands under various initial conditions are shown in Table. 1.

Table 1. Tracking control system parameters.

Initial conditions (x, y, θ)	Desired command (v_d, ω_d)	Control parameters (k_1, k_2, λ_1)
(1, -1, π/6)	(2, 0)	(6, 6, 2)
(10, -5, π/2)	(2, 0)	(6, 6, 2)
(-0.5, 0.5, π/8)	(2, 1)	(6, 6, 2)
(8, -6, -π/3)	(2, 1)	(6, 6, 2)
(3, 1, -π/3)	(2, sin*t*)	(6, 6, 2)

Fig. 2, Fig.3 shows AV line tracking simulation under different initial conditions of small error, large error, respectively.

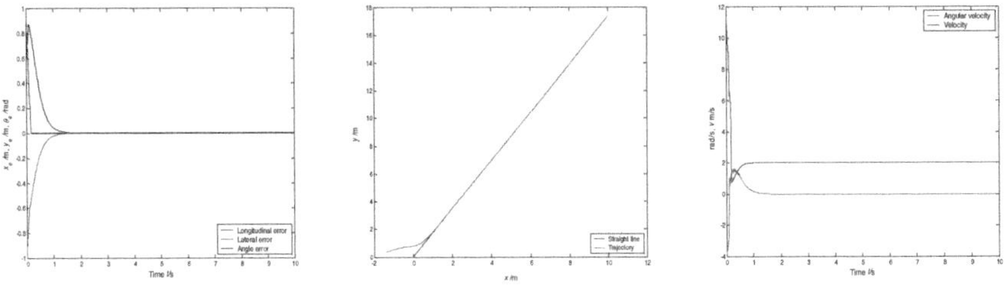

(a) (b) (c)

Fig. 2. AV line tracking simulation under the initial conditions of small error: (a) posture error profile, (b) trajectory profile, (c) control inputs profile.

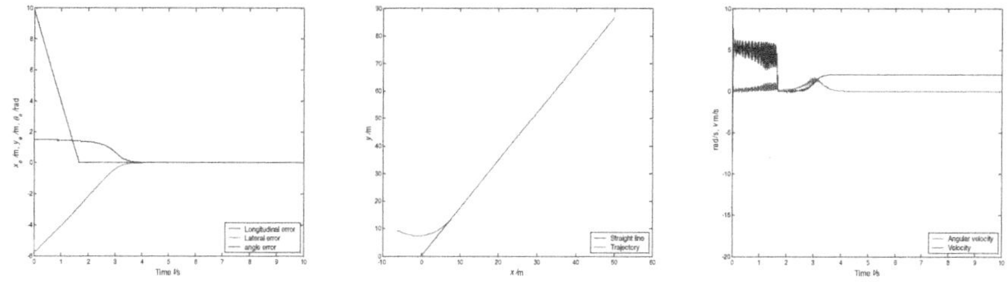

(a) (b) (c)

Fig. 3. AV line tracking simulation under the initial conditions of large error: (a) posture error profile, (b) trajectory profile, (c) control inputs profile.

Fig. 4, Fig.5 shows AV circle tracking simulation under different initial conditions of small error, large error, respectively.

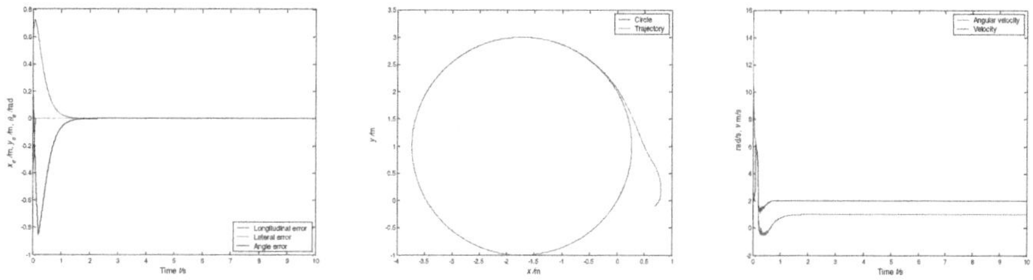

(a) (b) (c)

Fig. 4. AV circle tracking simulation under the initial conditions of small error: (a) posture error profile, (b) trajectory profile, (c) control inputs profile.

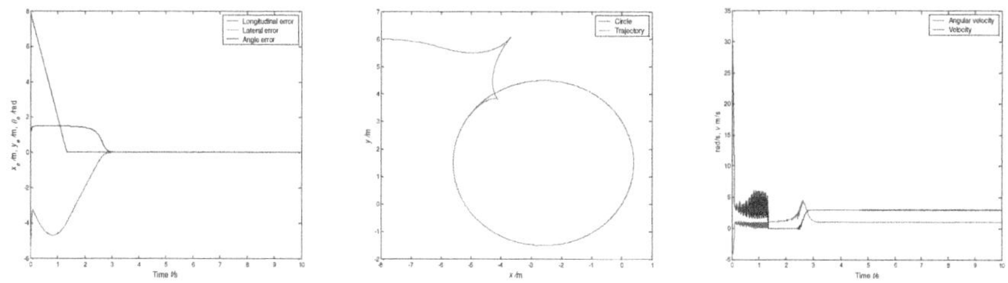

(a) (b) (c)

Fig. 5. AV circle tracking simulation under the initial conditions of large error: (a) posture error profile, (b) trajectory profile, (c) control inputs profile.

Fig. 6 shows AV curve tracking simulation.

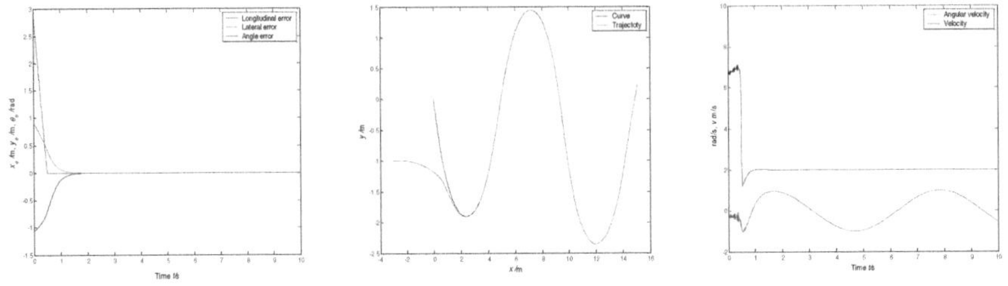

(a) (b) (c)

Fig. 6. AV curve tracking simulation: (a) posture error profile, (b) trajectory profile, (c) control inputs profile.

Seen from all figures, regardless of how to change the initial conditions, the control law designed has a satisfactory convergence for the tracking error. It is also seen that the control law track the reference trajectory quickly and stably. In a word, the tracking control system has a good tracking performance and global asymptotic stability properties.

5 Conclusion

In this paper, against "pure rolling without slipping" constraints, AV model based on rigid body is made, and the coupling variable structure control law for automatic tracking is proposed. The automatic tracking under complex road conditions can be achieved by the control for vehicle posture error. In consideration of vehicle coupling effects, the control law designed has a good tracking performance and global asymptotic stability properties.

Acknowledgments. The authors acknowledge the support of National Nature Science Foundation of China (60874081 and 61174173).

References

1. Nobe, S.A., Wang, F.Y.: An Overview of Recent Developments in Automated Lateral and Longitudinal Vehicle Controls. In: 2001 IEEE International Conference on Systems, Man, and Cybernetics, pp. 3447–3452. IEEE Press (2001)
2. Xu, Y.C., Wang, R.B., Li, B., et al.: A Summary of Worldwide Intelligent Vehicle. Automotive Engineering 23, 289–295 (2001)
3. Wu, C.Z., Wang, C.Y., Yang, L.B., et al.: Research on Fuzzy Control of Lateral Control in Intelligent Highway System. Journal of Highway and Transportation Research and Development 19, 131–133 (2002)
4. Li, Y.N., Zheng, L., Qiao, Y.J.: Fuzzy-PID Control Method on Vehicle Longitudinal Dynamics System. China Mechanical Engineering 17, 99–103 (2006)

5. Gao, F., Wang, J.Q., Li, K.Q.: Hierarchical Switching Control of Longitudinal Acceleration With Large Uncertainties. In: 2006 IEEE International Conference on Vehicular Electronics and Safety, pp. 270–275. IEEE Press, Shanghai (2006)
6. Guan, X., Wang, J.W., Gao, Z.H., et al.: An ACC Control Algorithm Based on Driver Behavior Model. Automotive Engineering 26, 205–209 (2004)
7. Mammar, S., Netto, M.: Integrated longitudinal and lateral control for vehicle low speed automation. In: 2004 IEEE International Conference on Control Applications, pp. 350–355. IEEE Press, Taipei (2004)
8. Li, Y.N., Lu, S.B., Zheng, L., et al.: Study on Lateral and Longitudinal Coupling Control when Vehicle Driving in Crooked Road with Variable Velocity. Journal of System Simulation 19, 5524–5528 (2007)
9. Espinosa, F., Awawdeh, A.M.H., Mazo, M., et al.: Reduction of lateral and longitudinal oscillations of vehicle's platooning by means of decentralized overlapping control. In: 46th IEEE Conference on Decision and Control, pp. 690–695. IEEE Press, New Orleans (2007)
10. d'Andréa-Nove, B., Campion, G., Bastin, G.: Control of Nonholonomic Wheeled Mobile Robots by State Feedback Linearization. The International Journal of Robotics Research 14, 543–559 (1995)
11. Ding, C.J., Zhang, M.L., Zhang, A.J.: Simulation of Mobile Robot's Road Recognition and Road Following. Journal of System Simulation 13, 133–135 (2001)
12. Wu, W.G., Chen, H.T., Wang, Y.J.: Global Trajectory Tracking Control of Mobile Robots. Acta Automatica Sinica 27, 326–331 (2001)
13. You, F., Wang, R.B., Zhang, R.H., et al.: Lane Changing and Overtaking Control Method for Intelligent Vehicle Based on Backstepping Algorithm. Transactions of the Chinese Society for Agricultural Machinery 39, 42–45 (2008)

Reviews on Planar Region Detection for Visual Navigation of Mobile Robot under Unknown Environment

Z.L. Wang, B.G. Cai, F.Z. Yi, and M. Li

School of Electronic and Information Engineering, Beijing Jiaotong University
Beijing 100044, China
{zlwang,bgcai,fzhyi,10125059}@bjtu.edu.cn

Abstract. Dominant plane detection is an essential task for an autonomous navigation of mobile robots equipped with a vision system, as we assume that the robot moves on the dominant plane. In this paper, a brief review about state-of-art method for planar region detection is presented. All these methods are classified into three types, that is stereo-based, monocular-based and multi-sensor fusion based method according to the sensor they used. And the monocular-based method is further divided into three kinds, projection invariance, homography based and optical flow based method. For each kind of the method, the correspondent references and its main process are summarized. Finally, the challenges for plane detection with vision are presented in authors view. We hope that this paper can be a start point to probe deep insight in this field.

Keywords: Planar region detection, homography matrix, multi-sensor fusion, visual navigation.

1 Introduction

Vision-based method for environmental perception and motion analysis has been a central research topic in the field of computer vision and robotics during last two decades. Particularly, some tasks, such as obstacle detection, traversable region detection, ego-motion estimation, localization etc, are the key issues for the mobile robot or vehicle working under unknown environment. In general, the moving platform with cameras is assumed to be moving along a planar surface, e.g. the road or floor. The detection of these regions in image space corresponding to this planar surface is a basic step towards the resolution of the aforementioned problems in robotics. Additionally, due to their abundance in man-made environments, as well as to their attractive geometric properties, planes are commonly employed in various vision tasks, such as feature matching [1, 2], grouping [3], camera self-calibration [4], obstacle detection [5], 3D reconstruction and scene analysis [6-10], camera relative positioning [11], object recognition [12], visual measurement [13], video stabilization, etc. Moreover, ground plane detection is the basis for the posterior detection of objects moving on that plane and tracking for mobile robot navigation.

G. Lee (Ed.): Advances in Automation and Robotics, Vol.2, LNEE 123, pp. 593–601.
Springerlink.com

The interest in finding planar regions in images is not new. In fact, due to its extensive application, this problem attracts many attentions, and a number of different approaches are proposed in the literatures. In this paper, for simplification, we classify these methods into thee types according to the sensors they used, that is, monocular-based, multi-camera based (including two or more cameras), and multi-sensor fusion-based method. The multi-camera based method is also called stereo-based method. Figure 1 illustrates the classification according to this paper.

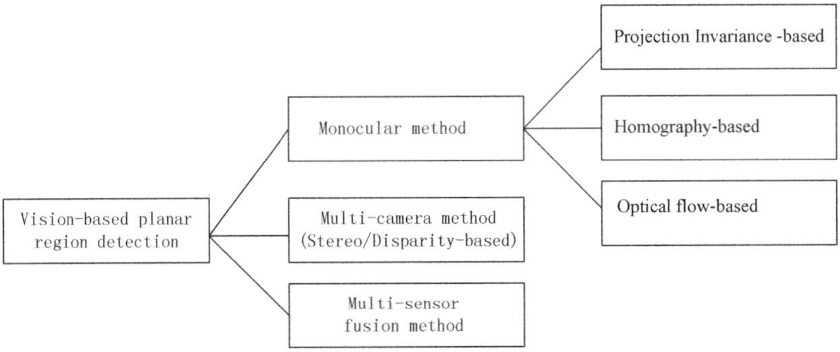

Fig. 1. The methods of planar region detection

In this paper, we give a brief review about the planar region detection method based on this taxonomy. For each kind of method, the principle and characteristics are discussed, as well as the reference in a separate section. Multi-camera or stereo-based method will be discussed in section 2. Monocular-based and multi-sensor fusion method will be discussed in section 3 and section 4 respectively. Section 5 is about some discussions and challenges on this issue.

2 Stereo-Based Planar Detection

Stereo-based planar detection method extracts the planar region in 3D space, so it is able to utilize three-dimensional structural information. Generally, plane detection constitutes a preprocessing step, 3D reconstruction, then applying the constraints imposed by planarity to extract the coplanar points among the reconstructed 3D points.

A randomized approach for planarity detection of an object and planar shape motion parameters is proposed by Imiya and Fermin [14], voting procedure is applied to increase the robustness of results. Some methods for plane detection that avoid 3D reconstruction are typically based on the extraction and matching of sparse geometric features from images. Sinclair and Blake [15], for example, detect the planes present in a scene by employing pairs of matched points extracted from a stereo pair and clustering them into coplanar sets using the two projective invariants defined by quintuples of coplanar points. Because a stereo pair was used, we class this reference into stereo-based method. As we will state in the section 3.1, the main shortcoming of this kind of approach is that the values assumed by these invariants are sensitive to errors in the localization of image points. Fornland and Schnorr [16] propose a two-step method for locating the dominant plane present in a scene by iteratively solving for both the plane homography and the stereo point correspondence. Lourakis and Orphanoudakis [5] solve a similar problem by employing a robust estimation technique to identify the dominant plane as the one the induced homography conforms to the motion of the majority of previously matched features. Both [16] and [5] share the common drawback of making implicit assumptions regarding the minimum fraction of point features belonging to the dominant plane, a factor that limits their applicability in certain cases.

During the recent years, some new kinds of methods are proposed. Piazzi and Prattichizzo [17] apply a merging strategy to group adjacent triangle, which is the triangulation of the set of features and its normal vector was the main factor to merge, by this way, all the points belonging to the same plane could be selected. Their method does not require any information about the camera calibration or the camera relative translation. Sappa [18] presents a histogram-based technique for detecting the points in disparity space which are relative to the environment's dominant surface area. An improved two stage histogram-based method, which is defined directly in 3D space, other than disparity space is proposed by Wang [19]. Simond [20] proposes a method consisting on extracting then tracking features (points, lines) from the road and estimate the homography induced by the plane between two poses.

Because most of the stereo-based methods extract the planar region in 3D space, it can make use of the 3D geometry constraints. But generally a 3D reconstruction preprocess is necessary and multiple camera were used, which make this method a number of disadvantages compared to monocular systems, especially in terms of cost and flexibility.

3 Monocular-Based Method

Monocular-based method extracts the planar region in an image sequence, which make use of the constraints between the adjacent images. As showed in figure 1, it can be subdivided into three types, that is,

- Perspective projection invariant parameter based method;
- Homography-based method;
- Optical flow based method.

3.1 Perspective Projection Invariant Parameter Based Method

Invariants are now a common tool in the computer vision community for model based recognition. Some perspective projection invariant parameters, such as the projection angle, the area ratio of two triangles, can be used for plane detection [15]. In reference [21], the basic transformation between planar points in image space, the planar homography is discussed. The properties of planar homography and invariants for homologies are probed into.

The main shortcoming of this method is that the values assumed by these invariants are sensitive to the localization errors of image features. And a few papers are published related to planar region detection and its application [25].

3.2 Homography-Based Method

The motion of the plane points is characterized by the planar homography between two consecutive views of the scene [21]. Some key steps for homography-based method include establishing point correspondences between consecutive image frames, feature tracking, estimation of the planar homography matrix or model parameter fitting, and the rules for distinguishing the point which is belong to the plane or not. Because this method is low cost and easy to implement, and moreover, the planar region detection results can be used to motion analysis and image understanding, it attracts many researchers in computer vision and robotics. During the recent years, many papers have been published based on planar homography.

To detect planar patches Kahler [22] first establish point correspondences between consecutive image frames in the sequence using KLT-tracking, then applying the random sampling and iterative dominant homography estimation to extract the planar patch, the results of extracted planar patches were refined with graph cut approach for motion segmentation. Reference [23] apply the SIFT algorithm to establish a correspondence between pixels from an image pair and then apply the homography constraint under a probabilistic framework as the criterion for the classification of the pixels as either ground plane or obstacle. Feature points extraction is a basic step for homography-based method. To improve the efficiency and stability of the homography estimation, reference [24] applys points and lines to estimate the homography matrix. Many other papers making use of the homography to identify plane region are published [25-31]. Reference [32] makes a very good summarization about the homography estimation.

Feature extraction and tracking in consecutive image is the basic step for homography-based method. But our experiments show that [53], either in indoor or outdoor environment, because of lacking of the texture on the ground plane region in the image, most of the extracted feature points is not located on the ground plane region, this make the homography-based plane detection more complicated in the application. Though some priori knowledge of the scenario, such as the line of the lane, some man-made markers [33], etc, can be used so that it is possible to obtain feature points that belong to the ground plane, but this limit these methods to a controlled environment application.

3.3 Optical Flow Based Method

Optical flow is the motion pattern of objects, surfaces, and edges in a visual scene caused by the relative motion between the observer (usually it is the camera) and the scene [37-40]. This motion pattern for the planar points is described by planar flow [47]. That is, for the planar points, the planar flow provides another geometry constraint for optical flow computing. Moreover, for robot navigation problem, optical flow is considered as fundamental information for the obstacle detection in the context of biological data processing [41]. Generally, they assume that there are two kinds of points in the image, ground plane points and obstacle points. From this point of view, the obstacle detection using optical flow is proposed in [42][43].

Reference [44] developed an algorithm for dominant plane detection using the optical flow, they model the dominant plane motion by affine transform when the camera displacement is small. All the points whose optical flow match the affine transformation model are taken as plane area points. The conception of planar flow was first proposed in this paper.

Samia et al [45] proposed the c-velocity, which is a Hough-like frame based on optical flow, based on a voting scheme, the approach are very efficient in making more robust the detection and coping with optical flow imprecision. Okamoto et al [46] apply neural network model for detecting a planar surface spatially from the optical flow. Marco et al [47] present a motion based segmentation algorithm to automatically detect multiple planes from sparse optical flow information.

Automatic detection of planar surfaces from flow fields belongs to the wider area of motion based segmentation, where the image is partitioned into regions of homogeneous 2D motion based on continuity or on fitting a parametric motion model. The advantages of this method are that, camera calibration is not necessary, the key motion between the mobile platform and the obstacle can be estimated. Moreover, the virtual plane can be avoided for density optical flow method. But optical flow is well-known for its large computing load, it's difficult to implement in real time. Some approaches can be employed to resolve this problem, such as a coarse-to-fine strategy [50]. Additionally, when the mobile platform is moving very slowly, even stops, this method will be failed or not robust.

4 Multi-sensor Fusion Method

Though so many methods are proposed for plane region detection as mentioned before, but many factors make the planar region detection task still very challenging. For example, many methods assume that the cameras are calibrated beforehand, which means the geometrical relation between the camera and the plane to be detected are fixed. This assumption is not practical in general case, e.g. the road may tilt, the mobile robot may incline due to its suspension, or the camera may vibrate. Another characteristic of the real application is that it will cause misdetection when there are large textureless areas in input image. A big problem in homography-based method is that the points for computing the homography matrix should be belong to the planar region, how to determine this is difficult. In many papers, prior knowledge about the environment is adopted, such as the pane line, the pavement pattern, et al. But for natural scene, looking for these marks may be failed. Additionally, virtual plane problem is encountered in many applications, it is one of the basic issue for plane detection [49], especially in feature based methods.

To compensate for the shortcomings of a single kind sensor, that is the camera(s), and to improve the robustness and accuracy of the planar region detection, some multi-sensor fusion methods which combine the vision sensor with inertial sensor, range sensor, et al., are proposed. Reference [33] applys Kalman filter to improve the robustness of homography estimation. The use of inertial sensors, such as speedmeters, accelerometers, gyroscopes, has been used in numerous works to facilitate ego-motion estimation [34, 35]. A probabilistic framework as the criterion for the classification of the pixels as either ground plane or obstacle is proposed by reference [23]. Christophe et al [36] proposed a method combines stereo-vision (involving geometric constraints), optical flow (involving photometric constraints) and additional model constraints. And these constraints are combined in the framework of direct method in order to deal with significant illumination drifts and obstacle detection on static objects or moving vehicle. Yang et al [54] combines the plane projective transform (PPT) and region segmentation (RS) to deal with the obstacle detection, the ground plane detection is the most important part in their algorithm.

Generally, there exist two kinds of fusions for plane detection, direct fusion and indirect fusion. Direct fusion is the fusion of sensor data from a set of homogeneous [36] or heterogeneous [34,35] sensors, and history values of sensor data [33], while indirect fusion uses information sources like a priori knowledge about the environment, even human input. Because of the challenges exist in plane detection for mobile robot (these challenges will be discussed in section V), multi-sensor fusion based method shows much more advantages than a single sensor system, for example, more robust, more accurate. But as the same time, the system complexity, the cost to implement such a system is a big problem.

5 Discussions

Because plane detection is an essential task for an autonomous navigation of mobile robot, and has many extensive applications, it attracts many researchers to work on it, and a lot of excellent works are presented here. But making a practical system for ground plane detection with vision sensor is still a very challenging task, the challenges mainly come from the complexity of the working site of the robots, in detail, these challenges are as follows

- In many cases, no matter in outdoor environment or indoor environment, there are always few features on the flat floor plane than on the road side. This phenomenon causes a lot of problems for the existing algorithm, just because they are feature extraction dependent. Though some prior knowledge about the environment maybe employed, for example, the line of lane, the road edge, the pavement mark, but in practical, these marks can be occluded by other vehicles on the road.
- The illumination variance brings great pressures for the hardware and the software of the system. Moreover, how to model the illumination variance is still a difficult work.
- There is not fixed pattern to make use for planar floor detection of mobile robot, because of the changing environment. The most important constraint which is maybe the only one is the planar homography for monocular image sequence. To estimate the H homography matrix, we first need know the planar points, it's a confliction.

There are many other factors which make this issue complex. To cope with these factors, some robust methods are necessary. We believe that the progresses in this field will greatly benefit the autonomy of the mobile robot under unknown environment.

Acknowledgment. This work was supported in part by the Natural Science Foundation of China under Grant No.61075079 and was supported in part by the State Key Laboratory of Robotics and System, Harbin Institute of Technology (SKLRS-2009-ZD-01).

References

1. Lourakis, M.I.A., Halkidis, S.T., Orphanoudakis, S.C.: Matching Disparate Views of Planar Surfaces Using Projective Invariants. IVC 18, 673–683 (2000)
2. Lourakis, M.I.A., Tzurbakis, S.V., Argyros, A.A., Orphanoudakis, S.C.: Using Geometric Constraints for Matching Disparate Stereo Views of 3D Scenes Containing Planes. In: Proc. of ICPR 2000, pp. 419–422 (2000)
3. Gool, L.V., Proesmans, M., Zisserman, A.: Planar Homologies as a Basis for Grouping and Recognition. IVC 16, 21–26 (1998)
4. Triggs, B.: Autocalibration from Planar Scenes. In: Burkhardt, H.-J., Neumann, B. (eds.) ECCV 1998. LNCS, vol. 1406, pp. 89–105. Springer, Heidelberg (1998)
5. Lourakis, M.I.A., Orphanoudakis, S.C.: Visual Detection of Obstacles Assuming a Locally Planar Ground. In: Proc. of ACCV 1998, vol. 2, pp. 527–534 (January 1998)
6. Irani, M., Anandan, P., Weinshall, D.: From Reference Frames to Reference Planes: Multi-View Parallax Geometry and Applications. In: Burkhardt, H., Neumann, B. (eds.) ECCV 1998. LNCS, vol. 1407, pp. 829–845. Springer, Heidelberg (1998)
7. Kaucic, R., Hartley, R., Dano, N.: Plane-based Projective Reconstruction. In: Proc. of ICCV 2001, vol. I, pp. 420–427 (2001)
8. Baillard, C., Zisserman, A.: Automatic Reconstruction of Piecewise Planar Models from Multiple Views. In: Proc. of CVPR 1999, pp. 559–565 (1999)
9. Bartoli, A., Sturm, P., Horaud, R.: A Projective Framework for Structure and Motion Recovery from Two Views of a Piecewise Planar Scene. Technical Report RR-4070, MOVI/INRIA (October 2000)
10. Criminisi, A., Reid, I.D., Zisserman, A.: Duality, Rigidity and Planar Parallax. In: Burkhardt, H., Neumann, B. (eds.) ECCV 1998. LNCS, vol. 1407, pp. 846–861. Springer, Heidelberg (1998)
11. Mohr, R., Morin, L., Grosso, E.: Relative Positioning with Uncalibrated Cameras. In: Mundy, J.L., Zisserman, A. (eds.) Geometric Invariance in Computer Vision, pp. 440–460. MIT Press, Cambridge (1992)
12. Rothwell, C.A., Zisserman, A., Forsyth, D.A., Mundy, J.L.: Planar Object Recognition Using Projective Shape Representation. IJCV 16(1), 57–99 (1995)
13. Criminisi, A., Reid, I., Zisserman, A.: A Plane Measuring Device. IVC 17, 625–634 (1999)
14. Imiya, A., Fermin, I.: Voting Method for Planarity and Motion Detection. IVC 17, 867–879 (1999)
15. Sinclair, D., Blake, A.: Quantitative Planar Region Detection. IJCV 18(1), 77–91 (1996)
16. Fornland, P., Schnörr, C.: A Robust and Convergent Iterative Approach for Determining the Dominant Plane from Two Views without Correspondence and Calibration. In: Proc. of CVPR 1997, pp. 508–513 (June 1997)
17. Piazzi, J., Prattichizzo, D.: Plane Detection with Stereo Images. In: Proceedings of the 2006 IEEE International Conference on Robotics and Automation, Orlando (May 2006)

18. Sappa, A.D., Dornaika, F., Ponsa, D., David, G., Antonio, L.: An Efficient Approach to Onboard Stereo Vision System Pose Estimation. IEEE Transactions on Intelligent Transportation Systems 9(3), 476–490 (2008)
19. Wang, Z., Zhang, B., Chen, L.: Motion Estimation for a Mobile Robot Based on Real-time Stereo Vision System. In: 2nd International Congress on Image and Signal Processing, pp. 1–5 (2009)
20. Nicolas, S.: Reconstruction of the road plane with an embedded stereo-rig in urban environments. In: 2006 IEEE Intelligent Vehicles Symposium, pp. 70–75 (2006)
21. Gool, L.V., Proesmans, L., Zisserman, A.: Grouping and Invariants Using Planar Homologies. In: Proc. Workshop Geometrical Modeling and Invariants for Computer Vision (1995)
22. Kahler, O., Denzler, J.: Detecting Coplanar Feature Points in Handheld Image Sequences. In: Proceedings Conference on Computer Vision Theory and Applications VISAPP (2007)
23. Conrad, D., DeSouza, G.N.: Homography-Based Ground Plane Detection for Mobile Robot Navigation Using a Modified EM Algorithm. In: IEEE International Conference on Robotics and Automation (ICRA 2010), pp. 910–915 (2010)
24. Lourakis, M.I.A., Argyros, A.A., Orphanoudakis, S.C.: Detecting Planes In An Uncalibrated Image Pair. In: BMVC 2002 (2002)
25. Gros, P., Bournez, O., Boyer, E.: Using Local Planar Geometric Invariants to Match and Model Images of Line Segments. Computer Vision and Image Understanding 69(2), 135–155 (1998)
26. Amirhasan, A., Boubakeur, B.: Homography-Based Plane Identification And Matching. In: 15th IEEE International Conference on Image Processing (ICIP 2008), pp. 297–300 (2008)
27. Benhimane, S., Malis, E.: Real-time image-based tracking of planes using Efficient Second-order Minimization. In: IEEE/RSJ International Conference on Intelligent Robots Systems, Japan (October 2004)
28. Aires, K.R.T., Araujo, H.d.J., de Medeiros, A.A.D.: Plane Detection Using Affine Homography, http://www.dca.ufrn.br/~adelardo/artigos/CBA08a.pdf
29. Aires, K.R.T., Araujo, H.d.J., de Medeiros, A.A.D.: Plane Detection From Monocular Image Sequences, http://www.dca.ufrn.br/~adelardo/artigos/IASTED08.pdf
30. Jozsef, N., Csaba, D., Zoltan, K.: Recovering Planar Homographies between 2D Shapes. In: ICCV 2009, pp. 2170–2176 (2009)
31. Silveira, G., Malis, E., Rives, P.: Real-time robust detection of planar regions in a pair of images. In: Proceedings of IEEE/RSJ International Conf. on Intelligent Robots and Systems, (IROS 2006), pp. 49–54.
32. He, Q., Chu, C.-H.H.: Planar surface detection in image pairs using homographic constraints. In: Proc of Int. Symp. on Visual Computing, Lake Tahoe, Nevada, USA, pp. 19–27 (2006)
33. Anubhav, A., Jawahar, C.V., Narayanan, P.J.: A Survey of Planar Homography Estimation Techniques. http://www.iiit.net/techreports/2005_12.pdf
34. Arrospide, J., Salgado, L., Nieto, M., Mohedano, R.: Homography-based ground plane detection using a single on-board camera. IET Intell. Transp. Syst. 4(2), 149–160 (2010)
35. Stein, G.P., Mano, O., Shashu, A.: A robust method for computing vehicle ego-motion. In: Proc. IEEE Intelligent Vehicles Symp., Dearborn, USA, pp. 362–36 (October 2000)
36. Blanco, J.L., Gonzale, J., Fernandez-Madrigal, J.-A.: Mobile Robot Ego-Motion Estimation By Proprioceptive Sensor Fusion. In: Proc. Int. Symp. on Signal Processing and its Applications, Sharjah, UAE, pp. 1–4 (February 2007)

37. Christophe Vincent, Y., Tjahjadi, T.: Planar Direct Method a New Framework for Stereo Vision Based Guidance and Obstacle Detection. In: IEEE International Conference on Image Processing (ICIP 2005), vol. 3, pp. III-381–4 (2005)
38. Barron, J.L., Fleet, D.J., Beauchemin, S.S.: Performance of optical flow techniques. International Journal of Computer Vision 12, 43–77 (1994)
39. Horn, B.K.P., Schunck, B.G.: Determining optical flow. Artificial Intelligence 17, 185–203 (1981)
40. Beauchemin, S.S., Barron, J.L.: The Computation Of Optical Flow. ACM, New York (1995)
41. Lucas, B., Kanade, T.: An Iterative Image Registration Technique with an Application to Stereo Vision. In: Proc. of 7th IJCAI, pp. 674–679 (1981)
42. Mallot, H.A., Bulthoff, H.H., Little, J.J., Bohrer, S.: Inverse Perspective Mapping Simplifies Optical Flow Computation And Obstacle Detection. Biological Cybernetics 64, 177–185 (1991)
43. Enkelmann, W.: Obstacle Detection By Evaluation Of Optical Flow Fields From Image Sequences. Image and Vision Computing 9, 160–168 (1991)
44. Santos-Victor, J., Sandini, G.: Uncalibrated Obstacle Detection Using Normal Flow. Machine Vision And Applications 9, 130–137 (1996)
45. Naoya, O., Atsuhi, I.: Dominant Plane Detection from Optical Flow For Robot Navigation. IPSJ SIG Technical Report, 2004-CVIM-145(7)
46. Naoya, O., Atsushi, I.: Dominant Plane Detection From Optical Flow For Robot Navigation. Pattern Recognition Letters 27(9), 1009–1021 (2006)
47. Samia, B., Antoine, P., Bertrand, Z.: Efficient Plane Detection from a Single Moving Camera. In: 16th IEEE International Conference on Image Processing (ICIP 2009), pp. 3493–3496 (2009)
48. Okamoto, H., Fujitsu, L.: A Neural Network Model for Detecting a Planar Surface Spatially from the Optical Flow in Area MST of the Visual Cortex. Systems and Computers in Japan 34(4), 46–59 (2003)
49. Marco, Z., Jose, S.V., Henrik, I.C.: Multiple Plane Segmentation Using Optical Flow. In: BMVC 2002, pp. 313–322 (2002)
50. Kanatani, K.: Geometric Computation for Machine Vision. Clarendon Press (1993)
51. Amintabar, A., Boufama, B.: Homography-Based Plane Identification And Matching. In: ICIP 2008, pp. 297–300 (2008)
52. Kim, Y.G., Kim, H.: Layered Ground Floor Detection for Vision-Based Mobile Robot Navigation. In: IEEE International Conference on Robotics and Automation (2004)
53. Wang, Z.L., Zhao, J.: Optical Flow Based Plane Detection for Mobile Robot Navigation, Published by WCICA 2011 (2011)
54. Yang, M., Qian, Y., Wang, H., Zhang, B.: Vision-Based Real-Time Obstacles Detection and Tracking for Autonomous Vehicle Guidance. In: Real-Time Imaging, vol. VI, pp. 65–74 (2002)

Research of Information Search Engine in Forestry Based on the Lucene

Yanling Cui, Yu Chen*, and Jingyu Li

Northeast Forestry University, Harbin, China
lg_chenyu@yahoo.com.cn

Abstract. This article based on the structure of Lucene indexing system , analyzes the common advantages of index-related technology and scalability. Finally, summed up some quality attributes , which need to focus on the index system, indicating that the distributed environment of the indexing system works. Then designed a full-text retrieval system based on Lucene model, and applied this model to the search engine system of the forestry information system to experiment. Experimental results show that, the indexing system based on Lucene is not only efficient but also faster retrieval speed.

Keywords: Lucene, Search, Forestry.

1 Introduction

With the rapid development of information technology, information on the Internet showed explosive growth, this trend allows the user to get more information at the same time, users will inevitably increase the difficulty of screening information. In order to make the user can quickly find information about forestry valid data form massive amounts of data , the development of high-performance information retrieval system of forestry information becomes increasingly important.

Lucene search consists mainly of two parts: First, parse the contents of the search and index; the second is indexed, that is based on user input query from the index database to find eligible documents, sort of the results through some order , then return to the user. This article describes the development and research of the forestry information search engine based on the Lucene , by theory, system structure, segmentation, indexing and other aspects of the establishment of a forestry information system dedicated search engine.

2 Lucene Introduction

2.1 Lucene Project

Lucene is a high-performance full-text search engine open source toolkit. It is not a complete full-text search engine, but a full-text search engine architecture, providing a

* Corresponding author.

G. Lee (Ed.): Advances in Automation and Robotics, Vol.2, LNEE 123, pp. 603–609.
Springerlink.com

complete query engine and index engine , providing data access and management provides a simple function call interface, can be easily embedded into applications to achieve for the application in full-text indexing. Lucene's API interface designed to be relatively common, input and output structures are like database tables, records and fields, many traditional applications, files, databases, etc., can be more easily mapped to the storage structure, or interface. Its open-source Lucene features, excellent index structure, a good system architecture, access to more and more applications. However, Lucene is not a full-featured search applications, the need for different types of data sources and application background for secondary development. There are many well-known projects as its background using Lucene full-text search engine, such as Web forums system Jive, mailing list system, Eyebrows, Institutional Repository DSpace, Java development platform Eclipse, XML-based Web publishing framework Cocoon, Apple's iTunes, Microsoft search plug-ins for Outlook. This article is based on the Lucene search engine dedicated forestry information systems.

2.2 Lucene Workflow

Lucene is something that for the class of text to retrieve. Assuming the computer's directory contains a lot of text documents, we need to find which documents contain certain keywords, like "pine", "willow" and so on. To achieve this function, we first Indexing using Lucene in this documents of directory, and search in the establishment of a good index of the documents you want to find.

The process of indexing and query process constitute the Lucene search process (Fig 1).

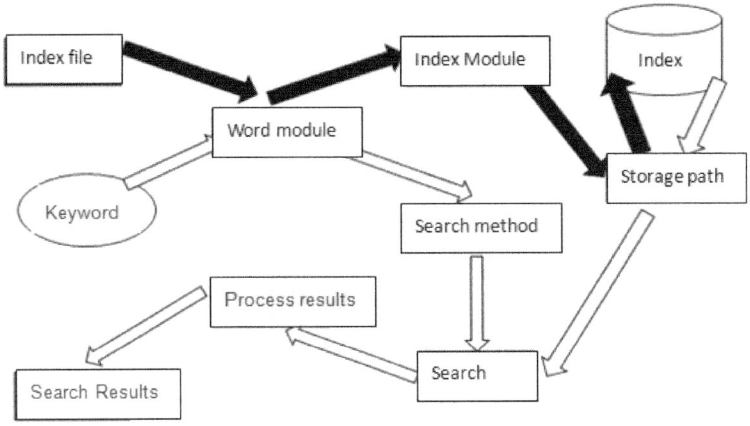

Fig. 1. Lucene indexing and search process

2.3 The Architecture of Lucene

The architecture of lucene build by layered approach. Each modules interact based on an agreement and formed a loosely coupled architecture of features. This greatly enhances the flexibility of the system. Lucene system consists of the infrastructure package, the core index and external interfaces.

Lucene's source code is divided into seven modules, and have three core class. They are analysis,index and search.Analysis is mainly used for word segmentation.Cut word is mainly finished by the expansion class of analyzer. Index provide the read and write interfaces of library. It can create library add and delete records and reading records. Search provide search interface. Through it we can input conditions and to get the query result and customize the query rule with the help of queryParser package.

2.4 The Key Technologies of Lucene

The key technologies of forestry information search is Mainly in the word segmentation, index management, and sort results.

Word segmentation:

Perform the word segmentation task is the object Analyzer in Lucene .And its most critical method is TokenStream.

Through the implementation of the method can return a set of token which is TokenStream object. TokenStream itself is similar to the iterator interface abstract class. And there two specific class,One is based on a Reader object as an input Tokenizer object, and another make another TokenStream object as input TokenFilter, Therefore in the forestry information system, word segmentation use IK_CAnalyzer to achieve. First it create Analyzer objects and Tokenizer, TokenFilter. Then word segmentation can be achieved by the help of these classes.

Index Management:

Basically, the index management includes two aspects. They are Indexing and retrieval based on index. Indexing is the segmentation of text words and index storage. After word segmentation it adds to the index database through the Lucene.index. Lucene.store responsible for data storage management, including some of the underlying I / O operations.

Sort Results:

The basic principle of sorting is to stand to the problem more valuable for users in the front without compromising performance as much as possible. Lucene sorts the results in accordance with their relevance algorithm (score) , In addition to matching score, you can also use the index to record the ID of the sort, Which is more efficient sorting method is: In the index, let the order into the Lucene full text of corresponds to certain rules (The evaluation value of forestry information users reflect higher, the higher the value of information), and invert file when search results according to the index when the record ID for (invert document).

3 Chinese Word

Need to contact the Chinese word for the specific context of the sentence cut to Chinese words into meaningful, and thus there is a higher degree of difficulty. Segmentation techniques are mainly three methods: 1) Based on the sub-string matching words; 2) based on understanding of the word; 3) is based on statistical word. This forestry information search engine system study IK_CAnalyzer word method. IK_CAnalyzer sub-word analysis is a dictionary word and the search for positive and negative way.

Test the core code as follow:

```
Public static void ik_CAnalyzer() throws Exception{
Analyzer analyzer = new IK_CAnalyzer();
Reader r = new StringReader("China is rich in forest resources. This is a the search engine dedicated forestry information system based on the Lucene.");
TokenStream ts = (TokenStream)analyzer.tokenStream("", r);
System.out.println("=====IK_CAnalyzer====");
Token t;
while ((t = ts.next()) != null) {
System.out.print(t.termText()+" | ");} }
```

Test results as follow:

=====IK_CAnalyzer====

China | Forest | Resources | ten | very | rich | This is | a | based on | lucene | Forestry | Information | Special | Search | Search Engine | Index | Engine | System |

It can be seen from the segmentation results, segmentation cutting out IK_C by Chinese word is specific and accurate.

4 Heritrix Web Crawler

Heritrix is a pure by Java developers, open source, Web crawlers, the user can use it to grab resources you want from the network. It comes from www.archive.org. Heritrix remarkable is that its scalability, developers can extend its various components, to achieve their capture logic.

After downloading the complete development Heritrix package, extract to a local directory. Which, Heritrix tools used by the library are present in the lib. In addition, there is a conf directory in Heritrix directory, which contains a very important document heritrix.properties. It configured to run a large number of closely related parameters with Heritrix in heritrix.properties. These parameters are mainly equipped with some of the default tools, WebUI startup parameters when Heritrix run, as well as Heritrix log formats(Fig 2).

```
# Below values are used if unspecified on the command line.
heritrix.cmdline.admin = admin:admin
heritrix.cmdline.port = 8080
heritrix.cmdline.run = false
heritrix.cmdline.nowui = false
heritrix.cmdline.order =heritrix.cmdline.jmxserver = false
heritrix.cmdline.jmxserver.port = 8081
```

Fig, 2. Heritrix configuration file

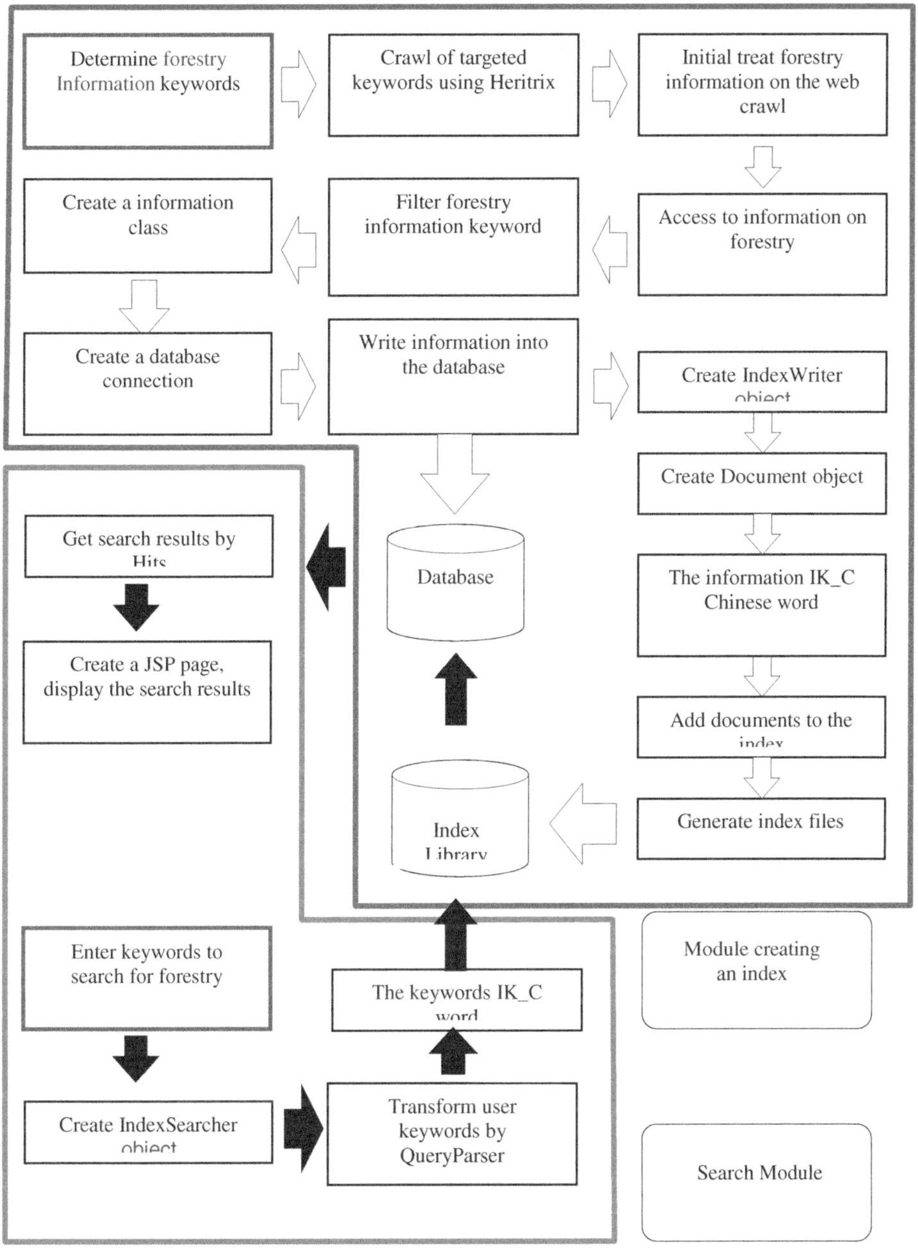

Fig. 3. System working model

5 Forestry Information Search Engine Systems Based on the Lucene

5.1 Working Model

\Crawl information by the Heritrix, processing information into the database and submitted to the Lucene indexer to index word and then available the formation of structured data to the indexer. Specific system architecture as shown below(Fig 3):

5.2 Module Creating an Index

Creating an index is the key to this system, if the index creation is not successful, the subsequent search will not be realized. Before creating the index, we must first ensure that the data exist in the database. Therefore, the index is created after the first screening on forestry words, to judge by forestjudge class. Then determine the data written to the database , if not written, not create the index, proceed to the next data.

Core code is as follows:

```
analyzer analyzer=new IK_CAnalyzer() ;
indexWriter indexWriter =new IndexWriter (indexDir ,  analyzer ,  true);
indexWriter.addDocument(document) ;
```

5.3 Search Module

Into the search page, enter the forest information keyword in the text box, Click on "search" button, to run a javascript function on the key words processing, then call SearchServiceImpl class SearchServiceImpl way to start the search.

Core code is as follows:

```
QueryParser queryParser =new QueryParser (field ,  new IK_CAnalyzer()) ;
Query luceneQuery=queryParser.parse(QueryString) ;
IndexSearcher indexSearcher =new IndexSearcher(IndexReader.open(indexDir)) ;
Hits hits=indexSearcher.search (luceneQuery) ;
```

6 Conclusion

This article describes the principles and the structure of the search engine dedicated forestry information system. Integrated use of Java, Lucene, SQL and other technologies to achieve the system. System uses the MVC design pattern to achieve a comprehensive Chinese word search function. It is a sense of the practical application of the search engine system.

Acknowledgement. This work is supported by Central Colleges basic scientific research projects special fund(DL10AB04),Northeast Forestry University Students Innovation Fund(111022540),Heilongjiang Provincial Department of Education Science and Technology Research Project(12513016),Postdoctoral Fund of Heilongjiang Province. corresponding author : Yu Chen.

References

1. Liu, Y., Huang, Y.: MYSearch full-text search engine based Lucence. Technology and Methods 29(18), 86–91 (2010)
2. Sui, l., Xu, C.: The design and implementation of a Chinese full-text search. Science and Technology Information 5(18), 29–34 (2007)
3. Man, Z., Qin: The research and design Personalized search engine. Computer Engineering and Design 30(2), 342–344 (2009)
4. Yong, Z., Jian, S., Rainsy, S.: The depth of mining information based on professional search engine. Spider Design and Implementation of Computer Engineering and Science 31(6), 18–20 (2009)

The Design of Broadband Electromagnetic Method System with GPS Simultaneous Localization

Xiaoliang Zheng[1] and Shengdong Liu[2,3]

[1] School of Electric and Information Engineering, Anhui University of Science and Technology, Anhui-huainan 232001

[2] State Key Laboratory of Deep Geamechanics & Underground Engineering, China University of Mining and Technology, Xuzhou 221008

[3] Shool of Resource and Earth Science, China University of Mining and Technology, Xuzhou 221116

zhengxl@aust.edu.cn

Abstract. Design a new type of broadband electromagnetic method system with GPS simultaneous localization, so as to meet the ever-increasing demand of geophysical prospecting. The system consists of three main parts, namely transmitter and receiver and GPS positioning. Magnetic field is stimulated by transmitter with broadband in a continuously controllable way, and three receiving coils with different spacing receive signals simultaneously, so that measurement data are more extensive and analysis results are more complete. Through simultaneously recording location information by GPS, a wide range of surveying tasks can be finished once. Compared to traditional electromagnetic method equipments, the system is more convenient and greatly reduces work intensity of technical personnel.

Keywords: Frequency domain electromagnetic method, global positioning system, broadband, synchronization, MOD5213.

1 Introduction

Electromagnetic prospecting is an effective means of geophysical exploration, being widely used in environment, subterranean cavitations, pipeline, groundwater pollution, archeology, explorations of ground water and mineral resources, as well as determinations of soil saline concentration and humidity. The main physical property basis of electromagnetic method lies in differences of electrical conductivity and magnetic conductivity of underground anomalies and surrounding medium. It achieves the purpose of finding underground anomalies through observing and studying electromagnetic spatial and temporal distribution according to the principle of electromagnetic induction. In actual measurements, alternating current is provided through transmitting coil so as to establish an alternating magnetic field (primary field) in the surrounding. Underground anomalies produce eddy current (induction current) under the effect of primary field, and then the eddy current produces secondary magnetic field (secondary field) in the surrounding. Electromagnetic method refers to detecting information of underground anomalies through receiving changes of primary field and secondary field.

G. Lee (Ed.): Advances in Automation and Robotics, Vol.2, LNEE 123, pp. 611–617.
Springerlink.com

Currently used frequency domain electromagnetic methods are mainly imported, including GEM-2 and GEM-3 developed by U.S. Geophex Company (multiple frequencies can be transmitted by one coil), EM31 and EM34-3 developed by Geonics Company of Canada (one coil transmits single frequency) [1], and EH-4 electrical conductivity imaging system which is jointly developed by U.S. EMI Company and Geometrics Company, etc. All these instruments are characterized by portable, easy-operating, high measuring efficiency and high-resolution. However, measuring range must be demarcated by people. The range of one manual measurement is quite small that only measurements of one transmitting frequency and one coil spacing can be done. In order to meet higher measuring requirements and the demand for rapid measurement of a wide range of plain areas, reduce work intensity of measuring crew, and improve the accuracy of measuring and positioning, a broadband electromagnetic system with GPS should be designed. The system can be moved by motor vehicles, achieving continuously changing measurements of transmitting frequency and simultaneous data-collecting of multiple receiving coils at different spacing. Meanwhile, GPS synchronous mark can be also done during measurement, finishing a wide range of positioning measurement tasks rapidly. Data processing is able to provide vertical subsurface information of different depths at the same measuring position, and shows directly measurement results of each position by combining map location.

2 System Design

Ordinary electromagnetic method usually generates one electromagnetic field by one transmitting coil and receives both primary and secondary signals by one receiving coil. The depth of exploration is decided by space between transmitting coil and receiving coil that are arranged in the same plane, as well as by transmitting frequency of electromagnetic field. In order to finish vertical exploration of different depths in one measurement, three receiving coils are used in the system design. The three coils at different spacing simultaneously receive signals of both primary and secondary fields when transmitting coil generates electromagnetic field. The alternating current frequency of transmitting coil is also designed as controllable and continuously variable, which realizes frequency-sweeping measurement of a whole frequency band in the same measuring area. Therefore, through processing groups of data measured in the same position, vertical information of different transmitting frequencies, different measuring intervals and different depths under measuring area can be acquired. Compared to other electromagnetic methods, this method detects more subsurface information, being directly used in drawing the geological structure chart of vertical direction.

Transmitting and receiving of the system are designed as two independent control modules, finishing tasks of transmitting and receiving electromagnetic field signals respectively. Both transmitting and receiving coils are placed inside a circular probe rod of 5 meters long. Intervals between the three receiving coils and transmitting coil are: 1m, 2m and 3.6m. Probe rod is made from PVC material, being portable and able to complete small-scale measurements manually. Measuring data are recorded in internal storage and then transferred into computer for processing after measurement.

In order to meet the demand for rapid positioning measurement of a wide range of plane areas, the circular probe rod is designed into a structure, on which a tire can be placed. The probe rod can be connected with a vehicle by fitting on two tires, and the measuring work of the whole system can be finished through the movement of vehicle. The moving speed reaches 5 to 10 kilometers per hour that is convenient for technical personnel to carry out long-distance measurement like river dike. Measuring data of that time are directly transferred to computer that records larger amount of data, greatly reducing work intensity and improving measuring speed.

In doing large-scale fast measurement, mark three-dimensional coordinate of each sampling location with GPS module, so as to make data and coordinates correspond with each other and provide essential data of geographical position for map drafting of large-scale fast measurement. The overall structure of the system is showed in Figure 1.

There are two measuring modes of the system: the vertical dipole mode, which is suitable for deeper probing depth; the horizontal dipole mode, which is appropriate for relatively shallow exploration.

Data that are collected can be directly analyzed not only by in-phase component and vertical component, but also by 1D or 2D alternatively. Besides, three-dimensional geological structure diagram can be formed through inverse model. The vertical component suggests that apparent resistivity of underground substances is largely decided by soil humidity and melting electrolyte concentration in soil gap, particle size, humidity and gap condition of soil, and surface chemical properties of clay particle. It can be used to determine various underground materials (such as sand, silt, clay, peat, etc.). The in-phase component indicates that magnetic conductivity of underground substances is mainly decided by types, oxidation degree and stability of ferromagnetic substances, continuous change of magnetic ions and the existence of ferromagnetic alloys. It can be used for determining different underground metallic (like mineral, cable, metal, etc.). Combining geographical location data collected by GPS module, the three-dimensional geological structure diagram, aerial photo and topographic map can all be used together to match geological state of each position with map.

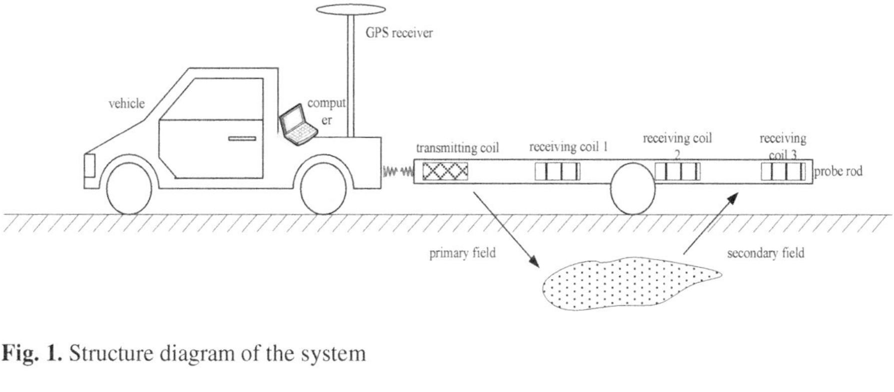

Fig. 1. Structure diagram of the system

2.1 Transmitting Control Module

The structure diagram of transmitting control module is as shown in Figure 2. The major function is to control alternating current of transmitting coil, generate alternating magnetic field and collect locating information of GPS module. The embedded processor module MOD5213 with ultra-low power of U.S. Net Burner Company is used as the control core, which receives control command from computer, sets frequency generator in the user-defined operating mode and controls the frequency of transmitting coil so as to get the required electromagnetic field. Meanwhile, it collects real-time locating information of GPS module.

Oscillator of SPPOSC series from U.S. Frequency Devices Company is used as the frequency generator. Based on the TMS320V5402 DSP processor, the generator can select and output programmable sine analog quantity with various frequencies through different digital input quantities of 8/9 bits. Frequency ranges from 1Hz to 102.4 kHz. MOD5213 sends 8-bit digital signal to SPPOSC through data port so as to change output frequency. According to different SPPOSC modules, changeable output frequencies of each time can be 1Hz, 10Hz, 100Hz and 200Hz. During the measurement of a certain position, the controllable output frequency continuously changes within the variation range in line with adjustment level, achieving the purpose of frequency scanning measurement.

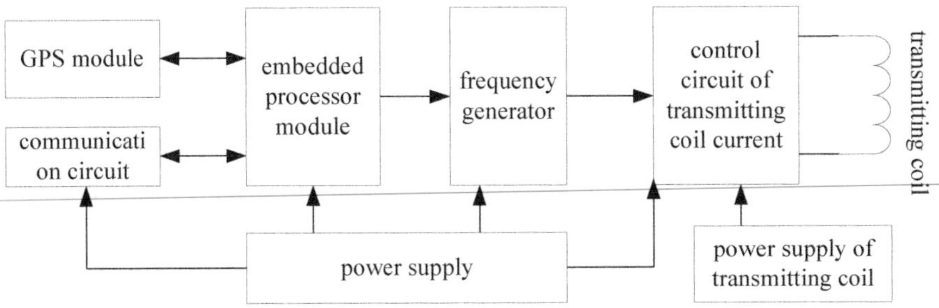

Fig. 2. Structure diagram of transmitting control module

The control circuit of transmitting coil current consists of digital adjustable potentiometer, operational amplifier and MOSFET switching tube, controlling the alternating current of transmitting coil. Chargeable and high-capacity lithium battery is used as the power supply of transmitting coil.

With the power supply of 3.3V, the communication circuit and the ST1480 RS-485 bus transceiver with ±15V electrostatic protection compose the 485 communication network, being in charge of data transmission between computers.

2.2 Receiving Control Module

The structure diagram of receiving control module is as shown in Figure 3. The major function is to collect data of primary field and secondary field, and then transfer data to computer through communication bus. Being the same as communication circuit and transmitting control module, the control core of receiving control module adopts MOD5213 as well. The communication circuit also uses ST1480 chip.

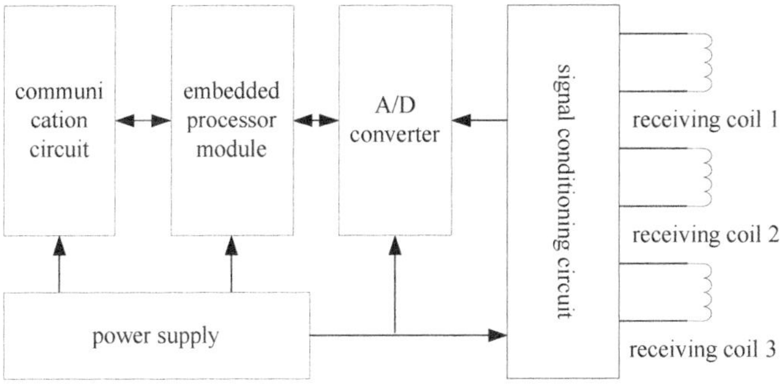

Fig. 3. Structure diagram of receiving control module

Power supplies of transmitting control module and receiving control module are basically the same. Levels of all powers that are used include ±12V, +5V, +3.3V and +1.8V, respectively supplied by DC-DC chip, PTK15-Q24-T512, PTK10-Q24-S3 and LM1117 chip.

3 System Software Design

System software mainly includes transmitting control module software and receiving control module software. Main program flowcharts are respectively shown in Figure 4 and Figure 5.

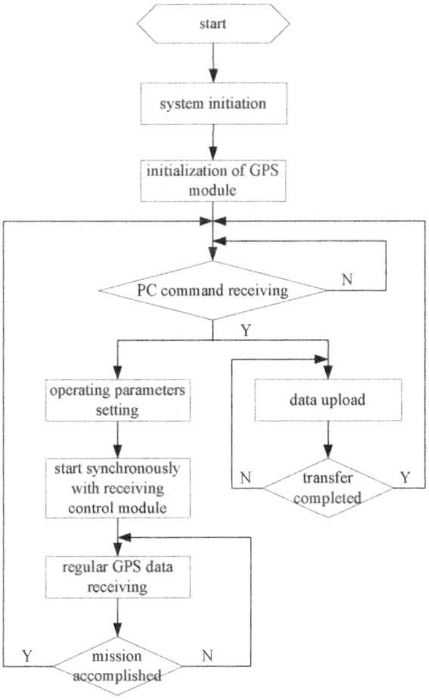

Fig. 4. Flowchart of transmitting control module

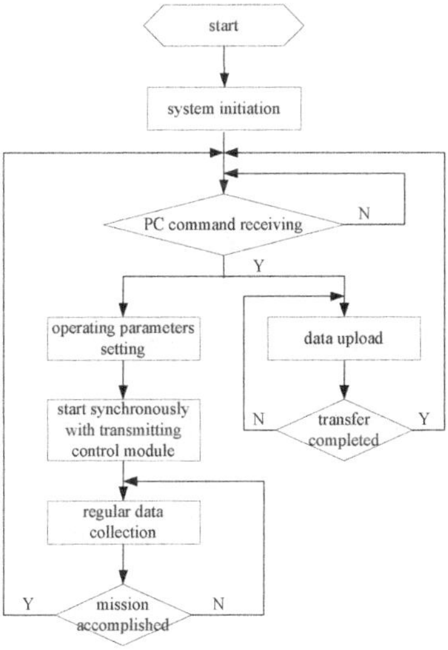

Fig. 5. Flowchart of receiving control module

4 Experiment Verification

Select a dam as the experimental site, and verify the experiment by making a comparison between electrical detection system and this system. The north side of the dam is the Huai River and the south side is the large stream channel. There is a bridge passing through the dam, facing water on both sides. It can be learned from cross-section diagrams of apparent resistivity of the two systems in Figure 6: there is an obvious high resistance anomaly（ρ_s>39.8$\Omega \bullet m$） in the range of 110<X<200 and Y<15. And ρ_s>55.9$\Omega \bullet m$ in the range of 130<X<140 and Y<15, corresponding to the location of drainage culvert. Dam in the west part of culvert has a perfect nature in terms of stratification and integrity. Yet during 190 to 220M in the X axis, electric resistance changes significantly and stratification and integrity are quite poor. Testing results of the two systems are similar to each other and also in line with the actual situation of the dam, indicating that the system performance meets the design requirements.

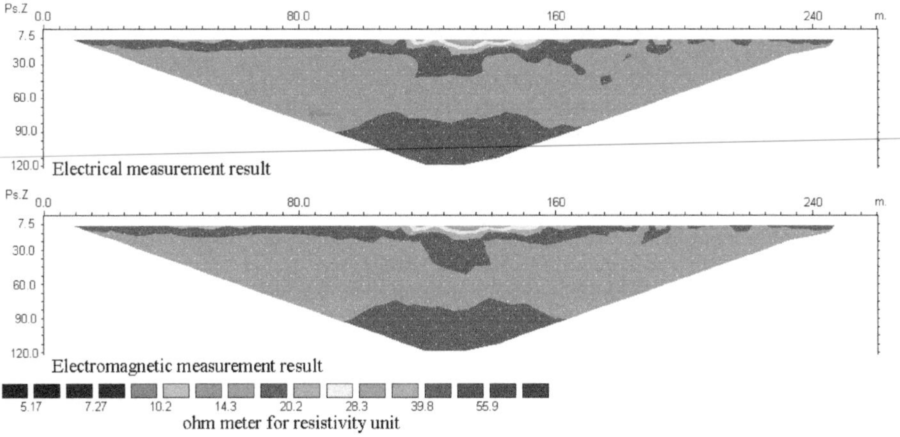

Fig. 6. Comparison chart of apparent resistivity cross-section diagrams of the two systems

5 Conclusion

The broadband electromagnetic method system with GPS simultaneous localization is able to complete measurement of multiple measuring lines conveniently and rapidly. With multiple receiving coils at various spacing and the continuously variable frequency transmitting mode, technical personnel can be provided with abundant underground vertical information. The system uses GPS to mark the three-dimensional coordinates of measured data, providing accurate geographic data analysis. Compared to other electromagnetic methods, this method is featured by rapider measurement and more plentiful measured information. It has a broad prospect of application in near-surface exploration, such as dam safety, geological disaster prediction, pollution investigation of underground water and soil, underground pipeline detection, underground cavity detection, unexploded ordinances identification and others.

References

1. Lina, P.: Research and Development of the Digital System of Electromagnetic Induction Detectors with Wideband Frequency Domain Based on DSK6713. Jilin University, Jilin (2006)
2. Net Burner, Inc. Mod5213 development2kit user manual [EB/ OL], `http://www.netburner.com/products/core_modules/mod5213.html2006205201`
3. Zhao, Y., Luan, J., Jiang, H.: Design of a Controllable High-resolution Data Acquisition System Based on ADS1274. International Electronic Elements

Intelligent Vision-Based Computerized Numerically Controlled (CNC) Machine

Ghassan Al-Kindi[1] and Hussien Zughaer[2]

[1] Faculty of Engineering, Sohar University, P.O. Box 44, PC-311, Sohar, Sultanate of Oman
[2] College of North Atlantic- Qatar, P.O. Box 24449, Doha, Qatar
gkindi@soharuni.edu.om

Abstract. In this article a system is developed to improve the performance of Computerized Numerically Controlled (CNC) machines by utilizing on-line vision based monitoring and control systems. Two different types of CNC machines are used in the investigation. Two cameras are mounted on the machine spindle of each of the two employed CNC machines to provide valid image data according to the cutting direction. Proper selection and activation of relative camera is achieved automatically by the developed system which analyze the most recent conducted tool path movement to decide on which camera is to be activated. In order to assess the machining surface quality and cutting tool status, image data are processed to evaluate resulting tool imprints on the machined surface. An indicating parameter to assess resulting tool imprints is proposed and used. Results show the validity of the approach and encourage further development to realize industrial scale intelligent vision-controlled-CNC machines.

Keywords: Vision-CNC machining, Surface roughness, Computer Vision, Intelligent machining.

1 Introduction

Although CNC machines are currently regarded as the heart of machining workshops, parts produced on these machines may not be as precise as expected. CNC machines cannot automatically monitor and judge the effectiveness of the applied machining parameters (i.e. cutting speed, feed rate, and depth of cut), tool wear and life status while machining, and the resulting accuracy and quality of machined surfaces. Providing CNC machines with abilities to monitor and assess the conduction of such tasks will open the way to introduce additional feedback control to these machines, hence improve their machining performance [1].

Surface roughness measurement is one crucial parameter that affects the preciseness of manufactured parts [2]. Therefore surface roughness could be used to decide on the quality of machining and tool condition.

Several studies on optimizing the cutting parameters to produce best surface roughness have been conducted, e.g. [3-4]. Although positive results are demonstrated in these studies to obtain best possible finish for a given tool shape and feed, satisfactorily implementation of these studies can only be approached if build up edge,

G. Lee (Ed.): Advances in Automation and Robotics, Vol.2, LNEE 123, pp. 619–628.
Springerlink.com

chatter, inaccuracies in machine-tool movement, etc., are eliminated [5], which cannot be avoided in real applications.

Roughness measurement is frequently achieved by stylus-based equipment, however using such technique to acquire on-line roughness measurement while machining is not possible due to the induced machine vibration and practical equipment setup. Recently vision based roughness measurement started to gain attention [6-8], therefore if such technique is introduced to CNC machines would enhance their performance and functionality by deciding whether to keep or alter the employed machining parameters, adjust the machining path, or to apply tool change.

In order to introduce vision-based roughness measurement technique to CNC machines a number of issues needs to be handled including scene visibility issues, effects of machine vibration on the quality of obtained roughness measurement, selection of a suitable roughness parameter to be employed in the analysis of vision data, machining parameters effects on acquired vision based roughness measurements.

The pioneer research to integrate CNC machines with vision systems is presented in [1] though only limited success was gained at that time due to hardware limitation. Several attempts were made later on to gain on-line assessment of machining tools [8-11]. However, conducted literature review has assured that "Vision controlled CNC machines" have not been realized yet.

2 Developed Experimental Rigs

In this research two separate experimental CNC-Vision rigs are established using two different types of CNC machines. These two different hardware setups aim to facilitate the generalization of the developed model to be applied to other types of CNC machines. In the first rig a Denford 'Novamill' CNC milling machine is employed, whereas in the second rig an Intelitek type 'Super-Pro-light 1000' CNC milling machine is used. Both machines can be characterised as bench type with PC-driven Fanuc-controller emulated machines. In order to introduce vision system to these two rigs, two high resolution Logitech C-910 cameras are used for each rig to acquire valid machined surface data. These four cameras are mounted on the two machine spindles to provide plan view of the machined part. Ambient lighting is used to minimise the effects of specular light reflection, [12], from the metallic work-parts that are machined [13]. Figure 1 provides a photograph of one of the developed rigs.

High speed steel (HSS) end milling cutters of $\Phi 10mm$ are used in the experiments of this research. Square size work-pieces, made of commercial aluminium, are used and their surfaces are machined using varying machining parameters. Depth of cut in the range of 1- 3 mm, feed rate of 50mm/min up to 300mm/min, and speed of cut of 1000 rpm are used in the investigation. More than 60 specimens were actually machined using varying values of machining parameters from the above ranges to enable the assessment of their surface finish.

In order to facilitate a comparison base for roughness measurement using vision system, surface roughness of these machined specimens are measured using SE1200 Kosaka-Lab stylus-based profile-meter. Seventeen different standard surface roughness parameters are computed for each specimen. However, analyses of acquired results of these different roughness parameters show that it is not always possible to predict the resulting surface roughness parameters even if only one machining variable is varied while all other machining variables are kept constant. Hence, based on acquired results, none of the employed surface roughness parameters could be used, with confidence, in this research to develop a reliable model that predicts and evaluates the performance of machining parameters used, especially that computer vision is planned to be employed as a source of surface data, thus adding more complications. Therefore, there is a need to introduce alternative surface roughness indicators that could provide direct and consistent relationship between resulting surface qualities and employed machining parameters.

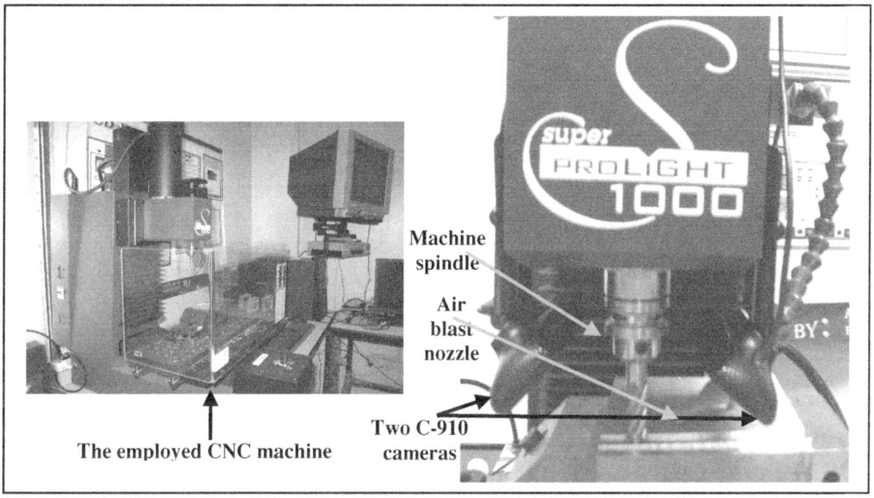

Fig. 1. Hardware setup utilizing "Super-Prolight1000" CNC milling machine and Two C910 cameras.

Nevertheless, if such novel surface roughness indicators are introduced they are not meant to replace the widely used and well known roughness parameters such as R_a but they will provide beneficial advantage to allow the evaluation of the performance of the cutting conditions in conjunction with the current existing roughness parameters.

In face milling operation, tool imprints on the machined surface are mainly influenced by feed per tooth. In other words, the tool imprints on the machined surface are affected by the cutting speed, the tool diameter, the number of cutting teeth in the tool, the feed rate used, and the overlap of cutting path (if any). Theoretically, tool movement on the main axis of the cutting direction is expected to generate tool imprints rate of I_t (imprints/mm):

$$I_t = \frac{60 \times V \times t_n}{\pi \times d \times f} \tag{1}$$

where V is the tangential cutting speed (mm/sec), t_n is the number of cutting teeth in the tool, d is the tool diameter (mm), and f is the cutting feed rate (mm/min).

It is worth stating that in face milling each tool tooth intersects the axis of the tool movement twice in each single rotation. Hence, this may alter the resulting pattern of the tool imprints on the machined surface if the second interception takes place in a deviated position from the already imprinted cuts, however from the theoretical point of view the added imprints will generate similar frequency of imprints though with certain value of phase shift. Such phase shift P_s (mm), if occurred, can be estimated by:

$$P_s = \left(\frac{f \times l_t}{d}\right) - int\left(\frac{f \times l_t}{d}\right) \tag{2}$$

where $int(x)$ is the integer value of x.

The monitoring of the resulting frequency of tool imprints on the machined surface provides a beneficial indication on the status of the tool condition, tool life, and correspondingly micro-scale surface roughness. For instant if one of the tool teeth (or more) has (have) broken or wear out, the resulting pattern of tool imprint on the machined surface is altered and therefore could be recognised if valid procedure for on line monitoring of such tool imprint is developed and applied. However, in practical applications there are a lot of other factors that affect the resulting pattern of tool imprints on the machined surface. The main of these are build up edge, chip trap in the tool/work-piece interface, chatter, inaccuracies in machine-tool movement, micro irregularities of tool cutting edges, and other reasons related to metal shear behaviour. All of these factors present a challenge to any developed model that aims to automatically monitor and control of resulting tool imprints.

3 Integration of Vision Systems with CNC Machines

In general, CNC machine controllers do not provide open platform to develop supervisory control programs by users [14], hence in order to facilitate the execution of this research a detailed technical design is proposed and employed to enable the integration of the vision system with employed CNC machine.

Figure 2 provides a flowchart of the developed system. The input to the system main computer is a normal G-code program that defines the required tool path, selected machining parameters as well as miscellaneous functions.

A dedicated program is developed by the authors of this article to enable data processing and system implementation. The program analyzes and processes the input G-codes in order to automatically segment the tool path and output a set of re-generated G-codes, thus allow the implementation of on-line monitoring and control. Each block of the re-generated G-codes facilitates the movement of one single segmented tool path movement of a predefined linear (or circular) length, where such movement is extracted from the originally programmed tool path. However, the execution of the newly generated G-code program assures to provide exactly the same tool path, machining parameters and miscellaneous functions, though it includes subroutine call functions and other control functions as will be discussed later. The re-generated G-codes are then stored in the common storage to enable their loading and execution by the CNC controller. Once the CNC controller starts to execute the newly generated G-code program, it will sequentially load and execute certain attached G-code subroutines that are also automatically created and attached to the main G-code program. These G-code subroutines include the spindle speed and feed rate as well as other added control codes; hence the values of cutting speed and feed rate can be replaced by other decided values without the interruption of the execution of the main G-code program. The instantaneous updating of either/both speed of cut or/and feed rate values is achieved by replacing relative G-code subroutine by a new version which include the new values of these machining parameters instead of the current values used. The new values of machining parameters are decided according to an automatic decision making action based on intelligent surface roughness feedback gained via the vision system.

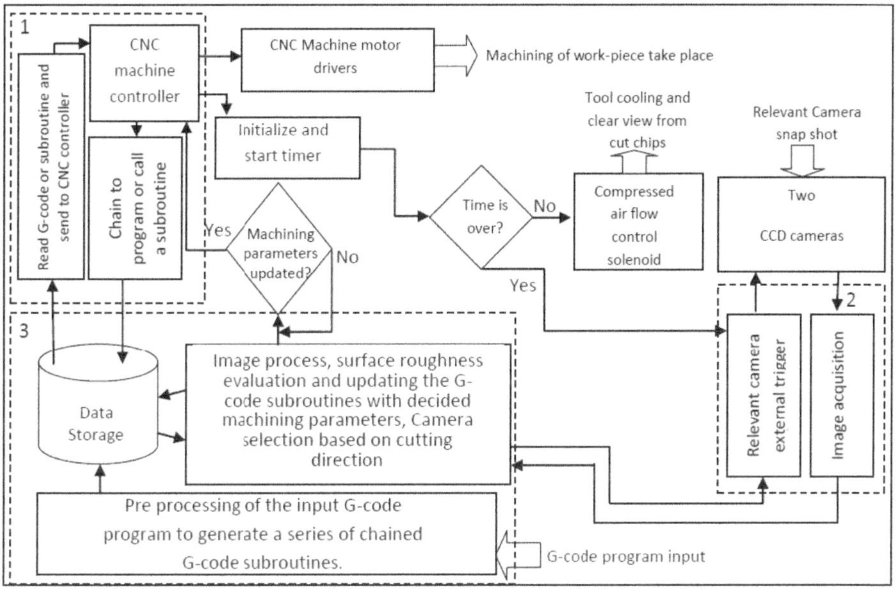

Fig. 2. Flowchart of the developed system.

In order to overcome the problem of coolant and cut chips obstructing the visual scene of the vision system, compressed air is used instead of liquid coolant. Activation and deactivation of compressed air is controlled via a solenoid valve that acquires its activation signal via the main G-code program. The compressed air, in this case, will also serve to remove the cut chips away from the camera field of view; however, to achieve this task, sequential control steps are used in the main G-code program. These steps were tested successfully on the two developed rigs. Figure 3 presents an example of a simple G-code program for the Super-Prolight 1000 CNC machine and the automatically generated output G-code subroutines from the system to run on the CNC machine controller.

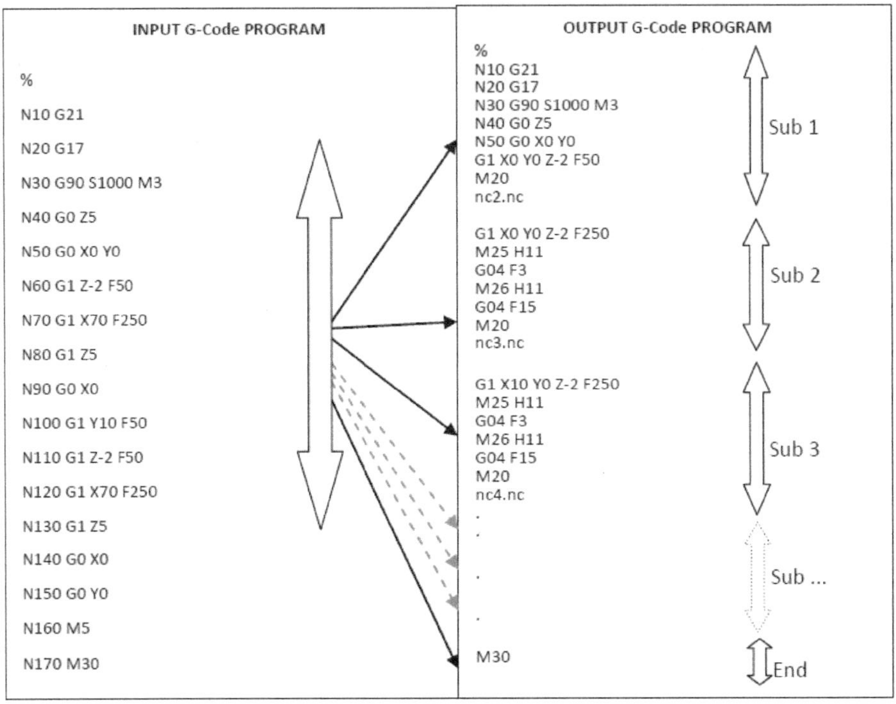

Fig. 3. Example of an input and output G-code programs.

4 Vision Data Processing and Tool Imprint Assessment

Roughness measurement using vision data is not a straight forward task. It requires a lot of attention and development of a methodology that works in a reliable way to provide on-line and semi real time results. Although a number of published articles assure that vision data provide valid information regarding surface roughness e.g. references [6-8], this task requires further investigation to suit the environment of each of the two developed CNC-Vision systems and ultimately the general case.

Experimental investigation is carried out by the authors of this article to analyse the images of the sixty machined specimens. Eight images of each specimen are acquired and processed to extract relative roughness data. Figure 4 gives examples of pixel intensities across single lines of acquired images. It could be noted from Figure 4 that although roughness information are inherited by vision data, the image data should undergo several processing to extract reliable roughness measurement from them.

One processing stage should be dedicated to remove localised lighting (reflection) variation from these images; a process which could be seen as an equivalent (but not identical) to the separation of waviness from roughness when employing stylus-based data.

In this research, we have developed a technique that aims to overcome problem areas associated with acquiring a valid measure of surface roughness assessment using vision data by employing moving average filter and introduction of a novel tool imprint indicating parameter C. The technique is designed to face the problems associated with irregular light reflection and colour variation of metal work parts.

The computation of the novel tool imprint indicating parameter C is achieved according to the following steps:

a. Set the value of Zero as an initial value to the tool imprints counter C.
b. Calculate the value of standard deviation (σ) of the filtered image data ($\bar{g}$).
c. Starting from the first element of resulting filtered image data, identify regions of neighbouring elements with values of less than M, where $M$is defined as:

$$M = \frac{\sum_{i=1}^{i=I} \bar{g}(i)}{I} \tag{3}$$

d. Record the resulting values as sub-sets.
e. Test the elements of each sub-set and increment C by 1 if at least one element in each sub-set has a value that is less than $(M - \sigma)$; this condition is meant to filter out intensity variations due to any unfiltered noise.

Example of computed reference surface based on the developed moving average method is imposed on Figure 4 graphs for completeness. It is clear from Figure 4 that resulting reference surface has successfully being gained from the vision-based roughness profile despite the several problem areas affecting the image data.

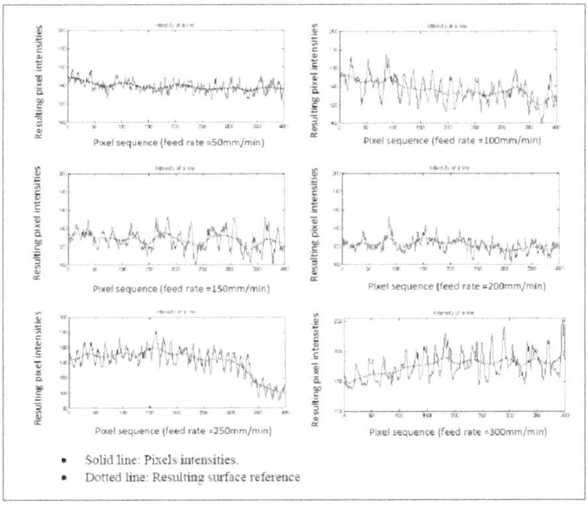

Fig. 4. Eexamples of pixel intensities across single lines of acquired images and resulting reference line computation using the developed moving average method.

In order to verify the validity of the proposed technique of reference surface calculation, the developed procedure is applied to a total of 480 images acquired for the sixty machined specimens. Satisfactory results are gained for all cases. Figure 5 shows a representative example of a set of acquired results which demonstrate the effectiveness of the developed procedure.

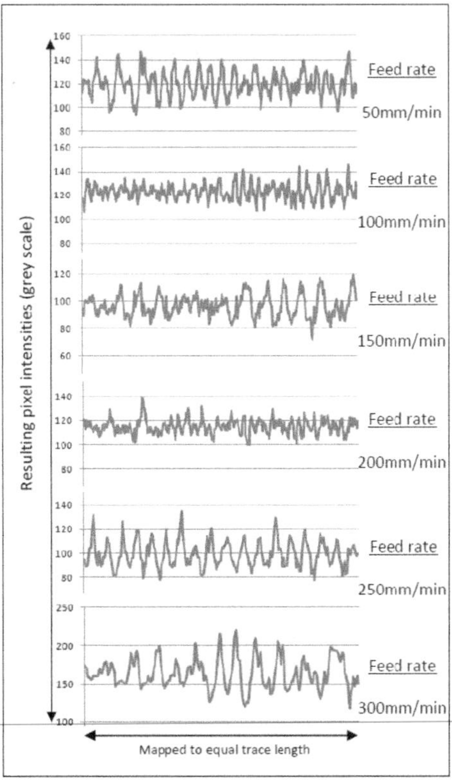

Fig. 5. Examples of processed and filtered image data of lines across a set of specimens (Depth of cut 2mm and speed of cut 1000 rpm).

Table 1. Example of acquired results of the resulting C values using different feedrate values.

Feed Rate mm/min	Resulting C value					
	Trial 1	Trial 2	Trial 3	Trial 4	Average	Max deviation from mean %
50	31	33	27	28	29.75	%10.9
100	32	25	25	26	27	%18.5
150	23	20	24	26	23.25	%14
200	18	16	17	22	18.25	%20.5
250	16	15	16	19	16.5	%15.2
300	12	12	15	13	13	%15.4

The proposed tool imprint indicating parameter $\boldsymbol{C}$ is also computed to verify its capability to handle the different cases provided by the acquired 480 images. Results are generally positive, however, further attenuation of the proposed model is required for more robustness. Table 1 show a representative example of acquired results where it can be seen that the developed model using the proposed tool imprint indicating parameter C has successfully provided consistent values of the tool imprints on the machined surface, except for few cases; these are mostly related to the lag in image resolution and quality.

5 Conclusions

In this work, considerable progress is achieved towards the challenging realization of intelligent vision-controlled CNC machines. The proposed frame work and the developed system show satisfactory results. The technical problems associated with visibility of the work piece scene are handled successfully through the designed and developed procedures. These procedures showed superb results to overcome effects of coolant blocking the image scene, discard cut chip from the image scene, and also reducing vibration effects on the acquired images.

The developed methodology of tool path segmentation to allow on-line monitoring and control of machining performance through adjusting or altering the machining parameters has showed to be very effective and provides the means to apply vision based feedback control to CNC machines to overcome limitations associated with CNC machine controllers which in general do not provide open platform to develop supervisory control programs.

The use of two cameras in the system hardware together with the developed procedure to activate relevant camera has enabled valid image data acquisition of the most recent cut of tool path segment.

Results showed that employing any of the standard surface roughness parameters in on-line assessment using vision system does not always provide reliable and consistent indicating values. Hence there is a real need to propose and develop alternative procedures to enable on-line in-situ assessment of machined surfaces to enable the integration of such developed techniques with CNC machines.

The developed technique to calculate reference surface for machined surfaces using image data showed to provide excellent results and overcome the problem areas associated with irregular light reflection and colour variation of metal work parts. Hence the authors strongly recommend employing of this technique in surface roughness evaluation using vision data.

The proposed tool imprint indicator C has showed to provide valid consistent indicating values that can be directly related to machining parameters employed. However more investigation and development of this parameter is needed for more robustness. In addition, further investigation of the use of this proposed indicator to assess tool condition is also required.

The outcome results gained from the practical execution of this research project encourage further investigation and development to beneficially realise the challenging aim of the development of intelligent computer vision controlled CNC machines. Further steps are already planed and under development by this research team to achieve more robust and reliable system that can be readily adopted and applied to industrial CNC machines.

Acknowledgements. The results presented in this article are part of an on-going research project funded by Qatar National Research Fund QNRF.

References

1. Al-Kindi, G., Gill, K., Baul, R.: Vision-Controlled CNC Machines. IEE Computing & Control Engineering Journal, UK, 92–96 (April 1993)
2. Hao, W., Wan, X.: Prediction surface topography in flank milling. In: Xie, M., Xiong, Y., Xiong, C., Liu, H., Hu, Z. (eds.) ICIRA 2009. LNCS (LNAI), vol. 5928, pp. 967–975. Springer, Heidelberg (2009)
3. Saravanan, R., Sankar, R., Asokan, P., Vijayakumar, K., Prabhaharan, G.: Optimization of Cutting Conditions during Continuous Finished Profile Machining using Non-Traditional Techniques. Int. J. of Adv. Manuf. Techn. 26(1-2), 30–40
4. Zhang, J., Chen, J.: Surface Roughness Optimization in a Drilling Operation using the Taguchi Design Method. Materials and Manuf. Proc. 4, 459–467 (2009)
5. El-Hossainy, T.: A New Technique for Enhancing Surface Roughness of Metals during Turning. Materials and Manufacturing Processes 25(12), 1505–1512 (2010)
6. Al-Kindi, G., Shirinzadeh, B.: An Evaluation of Surface Roughness Parameters Measurement using Vision-based Data. International Journal of Machine Tools & Manufacture 47, 697–708 (2007)
7. Al-Kindi, G., Shirinzadeh, B.: Feasibility Assessment of Vision-based Surface Roughness Parameters Acquisition for Different Types of Machined Specimens. Image and Vision Computing 27, 444–458 (2009)
8. Alegre, E., Barreiro, J., Castejón, M., Suarez, S.: Computer Vision and Classification Techniques on the Surface Finish Control in Machining Processes. In: Campilho, A., Kamel, M.S. (eds.) ICIAR 2008. LNCS, vol. 5112, pp. 1101–1110. Springer, Heidelberg (2008)
9. Castejon, M., Alegre, E., Barreiro, J., Hernandez, L.: On-line Tool Wear Monitoring using Geometric Descriptors from Digital Images. International Journal of Machine Tools & Manufacture 47, 1847–1853 (2007)
10. Eladawi, A., Gadelmawla, E., Elewa, I., Abdel-Shafy, A.: An Application of Computer Vision for Programming Computer Numerical Control Machines. Proc. Instn Mech. 217, 1315–1324 (2003)
11. Kerr, D., Pengilley, J., Garwood, R.: Assessment and Visualization of Machine Tool Wear using Computer Vision. I. J. of Adv. Manuf. Techn. 28, 781–791 (2006)
12. Mallick, P., Zickler, T., Belhumeur, P., Kriegman, D.: Specularity Removal in Images and Videos: A PDE Approach. In: Leonardis, A., Bischof, H., Pinz, A. (eds.) ECCV 2006. LNCS, vol. 3951, pp. 550–563. Springer, Heidelberg (2006)
13. Al-Kindi, G., Gill, K., Baul, R.: Experimental Evaluation of 'Shape from Shading' for Engineering Component Profile Measurement. Proc. Instn. Mech. Engrs. Part B: J. of Engg. Manuf. 203, 211–216 (1989)
14. Al-Kindi, G., Abdul Kareem, L.: An Application of Reverse Engineering using Vision–Based Data. Int. J. of Appl. Engg. Res (IJAER) 4(7), 1333–1346 (2009)

Robotic Arm, Wrist and Hand Using Pic Microcontroller

Nurul Huzaimi, Mohamad Basir, and Fuziah Sulaiman

Faculty of Electrical Engineering
Universiti Teknologi Mara, 40450 Shah Alam, Selangor, Malaysia
Fuziah807@salam.uitm.edu.my

Abstract. T A mobile robotic arm that can be controlled using computer is proposed in this paper. For PC based controller, a graphical user interface (GUI) software was used to provide control capability on the mobile robot developed using Visual Basic. The PC is connected to the mobile robot via serial port (RS232). The mobile robot also can be controlled using our portable control box. For hardware interfacing, Programmable Interface Controller (PIC) was used. The PIC receives the signal from the PC to provide motion control on the mobile robot. This PIC acts like analyzer that will analyze a serial signal from PC and produce a PWM (Pulse Width Modulation) combined with Electronic Speed Controller (ESC) that has high current spec to control high power motor. All of thees combinations were used to control motor speed. It has been shown that the designed robot can grip up to 500 grams load. The robotic arm can be used in multi-purpose applications such as in industrial manufacturing and servicing robot.

Keywords: Graphical user interface, programmable interface controller, pulse width modulation, and electronic speed controller.

1 Introduction

Nowadays many interesting events have been created and developed for the benefit of human being. Robot technology is widely recognized as one of the key technologies for the 21st century. In order to reduce human works robot has been created and the computer act as a brain. These phenomena can be clearly observed in the industries that mostly use robots in production, monitoring and others.Contrarily, human's arms are capable of adapting to various tasks due to a different mechanical structure. Thus, several works proposed human- like robotic manipulators for the service robots [7, 8]. First of all, the robotic arm is designed as a three degree-of-freedom mechanical structure. The robotic arm is composed of shoulder frame, a shoulder part, an elbow part, and a wrist part.

The direct current motor is one of the important parts in this project because the motor is used to control the speed and movement of the arm robot. It converts electrical energy into mechanical rotation [1]. A direct current motor is an electric motor that converts electrical energy (direct current) into mechanical motion. Direct current motor is the most often used in most application such as industrial, automobile and mobile robots since it is easy for speed control and torque regulation.

G. Lee (Ed.): Advances in Automation and Robotics, Vol.2, LNEE 123, pp. 629–640.
Springerlink.com

A serial interface controller will be used to control the robot arm power windows motor (direct current motor) by using Mosfet H-Bridges circuit via serial connection to a personal computer or directly from the onboard PIC microcontroller [2].

1.1 Objective of Project

The objective of this project is to develop a teleoperated robot using Microsoft Visual Basic. A program is created specially designed to control motor speed from computer. Most likely the true functions of this robot are to create human virtual interface software as a controller and for this project the robot will depend fully on human to give a signal from the computer.

MOSFET H-Bridges circuit was used to control the speed of the motor. A high current is needed to control the motor continuously. GUI software is develop using VB in order to control the arm robot or a controller box can be use to control the arm robot. The MOSFET H-Bridges act as a speed controller that will accept pulse width modulation (PWM) from PIC. A high current Mosfet is needed to control the power window motor continuously. Heat is one factor that has to be face in building this circuit. Mostly MOSFET, especially N type produce heat as pulse width modulation (PWM) is applied on one of its input.

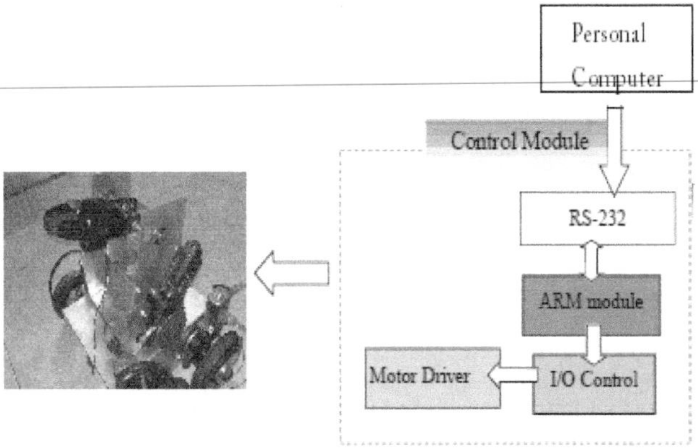

Fig. 1. System Architecture of Robotic Arm

Usually this H-bridge is a combination of two inverters of P and N channel. One act as source current circuit and one act as sinking current circuit. As sinking occur at NMOSFET, heat produced. This type of heat will damage the internal MOSFET circuitry. In order to overcome this problem several circuit that will act as a sinking circuit has to be added. The objective of this circuit is to protect the two NMOSFETs from damaging itself. The new circuit consists of N-Channel MOSFET, resistor and n-p-n transistor.

1.2 PIC Microcontroller

The PIC is compiled with a program to control the movement of the robot. The PIC is program to react for any incoming input from GUI interface or controller box. The PIC will constantly update the motor with required speed to drive the arm robot or wheel to the position required. A few external components needed to be checked and test whether their values correspond to the desired ones, and finally to see whether all the connections are done right.

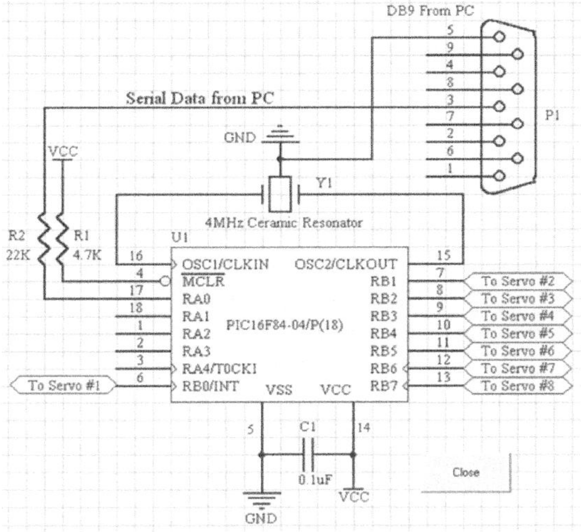

Fig. 2. Serial Interface Controller (The PIC16F84A is replaced with PIC16F877A)

The signal from the computer then will be analyzed by PIC Microcontroller. It will generate a Pulse Width Modulation (PWM) to control motor speed. There are several steps to complete the objective requirement and the primary goal is to design a simple and economical speed controller direct current motor using computer. The direct current motor will then be connected to the computer through a serial port.

2 Methodology

The main objective is to build a PIC microcontroller circuit as in Fig. 2 to listen for any incoming serial communication from the host computer, and then set (servos to the position received / Mosfet H-Bridges speed required). The PIC is programmed to constantly update the power windows with their positional speed required so that the power windows will hold their speed. These operations explain how remote control ESC work and how they are controlled will make a new understanding and gallant my knowledge. The R/C servo is geared, (power windows is direct from ESC), direct current (DC) motor with a built-in positional feedback control circuit. This makes it ideal for use with small robot (use servos) but ideal for medium robot if uses high power motor such as power windows.The PIC 16F877A software for the serial interface controller operates by receiving information in a serial format from a host computer (teleoperated robot). The PicBasic Pro compilers, by microEngineering Labs, include an instruction called *serin* command to receive data. *Serin* must receive the qualifier bytes in exact order before receiving any data items. If any of the bytes received do not match the bytes in the qualifier sequences, the qualification process starts over.

Another feature of the *serin* command is the ability to set a timeout and label. The timeout qualifier allows the program to continue execution at the point specified by label if a character is not received within a certain amount of time. Timeout is specified in 1 ms units. This is ideal because it gives the program the ability to constantly update the servos with positional information if nothing is happening on the communication port. For the program, the ASCII value of 255 is used as the qualifier to identify that data are being sent from the host computer. Once the ASCII value of 255 has been received, the next byte of information will be stored in a variable called slider. This variable determines which servo will be updated. The next byte of information is stored in variable called control and contains the position to which the servo will be set. The line code is shown below:

2.1 Serin com_in,baud,5,set_pos,[255],slider,control

The serial data are received on pin com_in (PORTC.0) at a baud rate of baud (2400). If nothing is received on pin 0 of port A within 5 ms, then program execution will continue at set_pos, where the servos are set. When the serial data that determine which servo and its position are received, the PIC will then transmit the data back to the computer pin com_out (PORTC.1). The serial servo control program running on the computer will use this information to verify that the PIC has received the correct information and will then send the next set of servo and position information. The program execution will remain in a tight loop af receiving servo position data and then setting the servos with the data received [4].This robot was designed with 3 degree-of-freedom mechanical structure as shown in Fig. 3.

Fig. 3. Robotic Arm Mechanical Structure

To communicate with the robot via serial interface circuit board, MSComm 6.0 control had been added. The MSComm control was one of the first custom controls designed for Visual Basic and is the original means of communications for my application by allowing the transmission and reception of data through the serial ports. The MSComm control provides serial communications for my application by allowing the transmissions and reception of data through the serial ports. The MSComm control provides two ways to handle communications. The first is event-driven communications, a very powerful method for handling serial port interactions. The second method of handling communications is by polling for events and errors by checking the value of the CommEvent property after each critical function of your program [6].

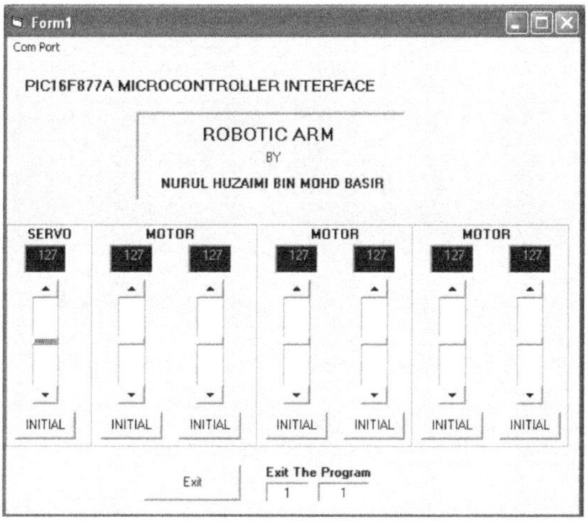

Fig. 4. Serial Interface Controller Program

The PIC16F877A microcontroller has a built in PWM command that is used to send a pulse width modulation signal on one of its output port. The code to send PWM signal through one of the output port is:

2.2 PWM Pin, Duty, Cycle

Outputs a pulse width modulated pulse train on Pin. Each cycle of PWM consists of 256 steps. The Duty cycle for each PWM cycle ranges from 0 (0%) to 255 (100%). This PWM cycle is repeated Cycle times. Pin may be a constant, 0 - 15, or a variable that contains a number 0 – 15 (e.g. B0) or a pin name (e.g. PORTA.0).

The Cycle time of PWM is dependent upon the oscillator frequency. If a 4MHz oscillator is used, each Cycle is about 5ms long. If a 20MHzoscillator is used, each Cycle is about 1ms in length. Defining an OSC value has no effect on PWM. The Cycle time always changes with the actual oscillator speed [1]-[4].

. The H-Bridges as in Fig. 5 uses MOSFETs for one main reason – to improve the efficiency of the bridge. When BJT transistors (normal transistors) were used, they had a saturation voltage of approximately 1V across the collector emitter junction when turned on. The power supply voltage was 10V and consuming 2V across the two transistor required to control the direction of the motor. 20% power was eaten up by the transistor.

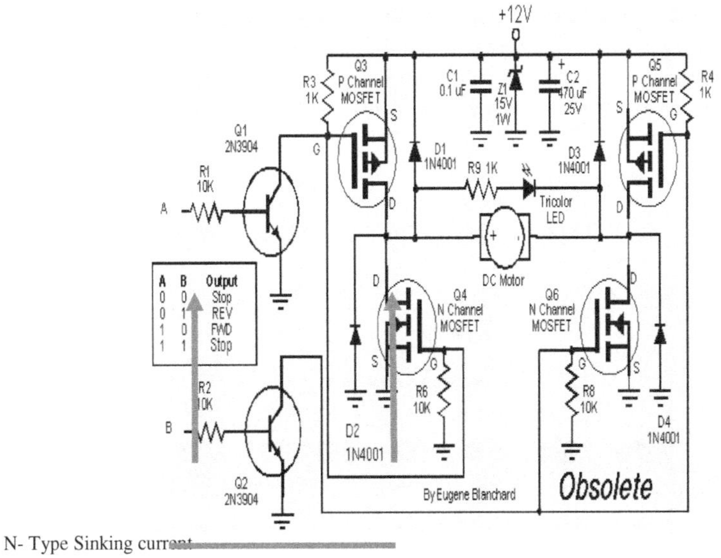

Fig. 5. Mosfet H- Bridge P-type and N-type sinking current.

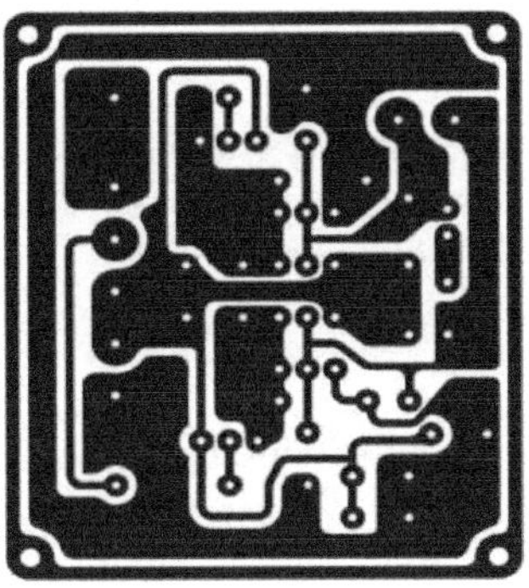

Fig. 6. PCB layout for MOSFET H-Bridge

MOSFETs were chosen because when they turn on they have an ON resistance called RDS (on). This is the resistance between the Drain and Sources when turned on. It is quite easy to buy MOSFETs that have very low RDS (on) ratings of less than 0.1 ohm. At 4 amps, this would mean that the voltage drop would be 0.4V per

MOSFET, a definite improvement. The MOSFETs chosen had (on) rating of 0.04 ohm which greatly improved the efficiency. Now, when MOSFETs has a low RDS (on) rating, it usually has quite a high current rating typically in the 10s of amps.

Naturally, the lower the RDS (on) rating, the more expensive the MOSFET Low RDS (on) P channel MOSFETs are more difficult to find than N channel. MOSFETs work by applying a voltage to the Gate. They called this transconductance. When a positive voltage greater than the Gate threshold voltage is applied, the MOSFET turns on (Q4 & Q6 – N channel only). The P channel works in reverse (See Q3 & Q5).

Relays circuit are used in this project. It is basically as same function as H-Bridge, which is to move the motor. But the difference is that relays cannot control the speed of the motor like H-Bridge circuits.\

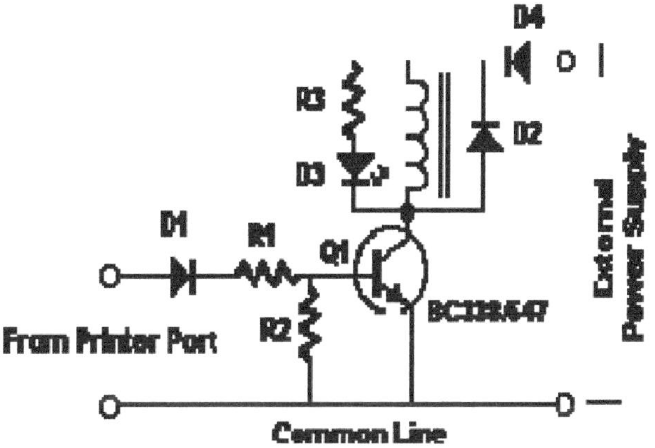

Fig.7. Relay Switching

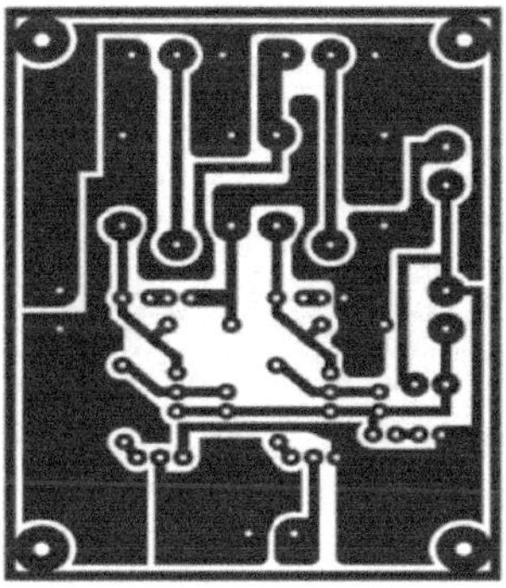

Fig. 8. PCB Layout for the 8x Relay switching

When one of the lines is taken HIGH (by a command from PIC output) 5 volts appears at the base of the transistor Q1. This switches on the transistor allowing a larger current to flow through the coil of relay RL1, turning it on. Taking PIC port line LOW will turns the relay off again Diode D1 prevents any 'reverse' potential from entering PIC and causing damage

3 Results and Discussion

A test was conducted by attaching the motors for mobility and movement. For the gripping system, a motor is needed for arm motion and one servo is needed for gripping.

The mobility system consists of 4 DC motors. Power window motors were chose for the mobility purpose. Each of the power windows controls one tire or one sprocket that produced enormous torque. The reasons why the power windows were proposed in this project are:

1. The mobile robot is slightly heavy. As a result it requires high torque for movement.
2. The environmental structure that unnoticed because of the earth environment such as slope that required high torque.
3. The application that this robot handle consists of different structure and different environments.

An experiment on burden measurement is conducted to measure how much weight the gripper can hold or raised.

Fig. 9. Robotic Arm with Burden of 500 Grams

The graph in Fig. 10 explains that the gripper can hold the amount of burden for about 500 grams. The amount exceed than that will cause the gripper loose the strength to hold.This is because of the servo that is used with the gripper. The servo produces low torque which results in low power to produce tight grip. The only solutions is that the servo is replace with direct current motor to have high output power thus have strong grip.

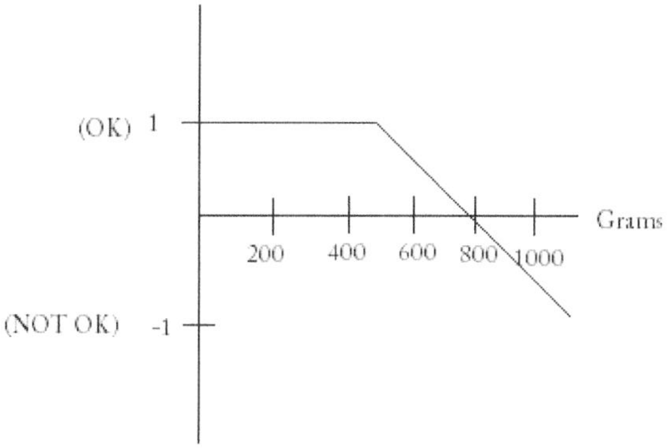

Fig. 10. Graph of Gripper Burden Measurement

	Gripper open	Gripper closed
Gripper		
Opening Distance	~ 40mm	0mm

Fig. 11. Robot Gripper

The gripper can hold thing which is not exceed 40mm of diameter. The gripper can open about 40mm only as shown in Fig. 10.

Gripper is used to pick up dangerous or suspicious items for the robot to carry, some can turn doorknobs, and others are designed to carry only very specific things like beer cans. Closing too tightly on an object and crushing it is a major problem with this gripper. There must be some way to tell how hard is enough to hold the object without dropping it or crushing it.

For these reasons, gripper design requires as much knowledge as possible of the range of items the gripper will be expected to handle. There are several basic types of gripper geometries. The most basic type which is used in this project has two simple jaws geared together so that turning the base of one turns the other. This pulls the two jaws together.

Actual Stresses Developed in the Robot Grip System: The effect of an external load on the gripper is similar to the effect of an external point load on a cantilever, fixed at one end, and free at the other. The effects of bending stresses due to external and internal loads and torsional stresses due to the motor are evaluated. Consider the situation in which an external load acts on the cantilever beam aside from the uniformly distributed self-weight of the beam. This results in a bending stress, σ calculated as:

$$\sigma = (FL + x\pi/4d^2 \rho L(L/2)/(\pi/32d^3), \quad \ldots \ldots \ldots (1)$$

where the length of the lead screw is 12 in, diameter is 0.5 in, and specific weight of mild steelis 0.283 lb/in3. Based on loading characteristics, the maximum bendingstress, _1 is found to be 5248 lbs/in2, while the minimum bendingstress, _2 is found to be 2786 lbs/in2, well within tolerance limits. The torsional stress induced in the mechanism because of motor torque may be calculated as:

$$T_{xy} = 16T = \pi d^3 \quad \ldots\ldots\ldots \quad (2)$$

where d is the diameter of the beam, and T is the torque generated by the motor. The value of torque (from motor specifications) is 2.187 lb-in. By substituting into the equation,

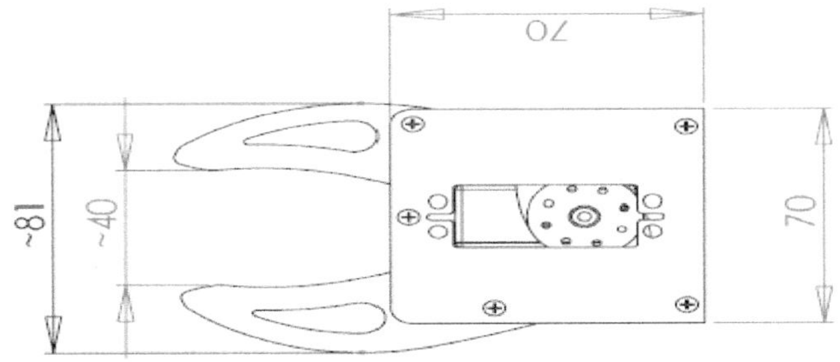

Fig. 12. Simple Direct Drive Swinging Jaw

In order for the gripper to open larger than before, a simple driven through a right angle drive gear motor could be used as shown in Fig. 13 which places the drive motor further out of the way of the gripper. This and similar designs have the drawback that the jaws are always at an angle to each other which tends to push the thing being grabbed out of the jaws[9-17].

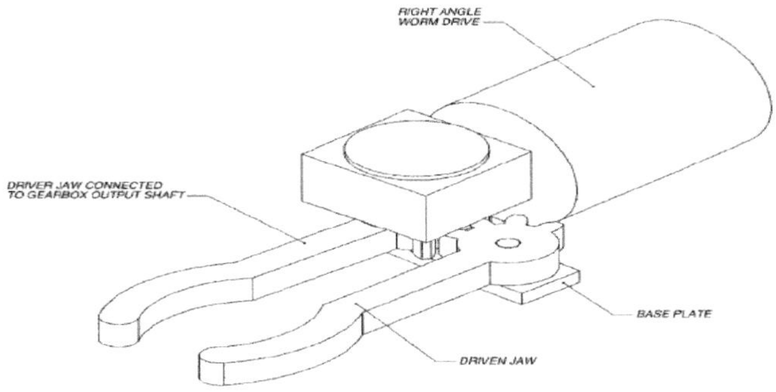

Fig. 13. Simple direct drive through right angle worm drive gearmotor

4 Conclusion

In this project, an economic mobile robot that can be controlled using computer via serial interface controller had been designed and developed. Basically the project can be divided into two parts, which is the hardware and mechanical development for mobile robot and software for designing an interfacing for controlling the mobile robot. For interfacing controlling method, Graphical User Interface (GUI) software was developed using Visual Basic. The instructions from the computer will be sent through the serial port communication to provide control capability on the mobile robot. The PC is connected to the mobile robot via serial port (RS232). For hardware interfacing, programmable interface controller (PIC) is used. The PIC receives the signal from the PC to provide motion control on the mobile robot. The fundamental of creating a serial communications project using Visual Basic was covered and it gives the ability to craft the robot control software and the ability to harness the computing power of the personal computer. To improve on the grapsing capability[18,19] the robotic arm in future will be equipped with tactile and force sensing sensors accompanying with visual servoing system.

5 Future Devepment

For future development is human-like robotic arm control based upon muscle tension effect. The muscles give birth to a tension effect where the muscles of a human are exerted gradually from neutral pose of each joint to the joint limits and they have trouble in terms of high tension in performing. Here, focus on such effects caused by tension of the muscles under the gravity. Also, the inertia gain C_0 of the second term in (7) was incorporated owing to consideration of muscle effects corresponding to individual joint muscles similar to a human as seen in Fig. 7. It is useful to keep the initial joint configuration of multi-DOF robotic systems such as a job position or set position. In consequence, this allows the human-like and energy efficient motion under the gravity including the revolute joints when the robotic arm leads the end position from the given initial $\boldsymbol{p}_0$ position to the target position $_d\boldsymbol{p}$ in task space. The overall control input designed by linear superposition is proposed in this work as follows:

$$_{\mathrm{T}}\boldsymbol{t} = -\boldsymbol{C}\,\boldsymbol{q} - x\boldsymbol{C}\,\mathrm{D}\boldsymbol{q} - \boldsymbol{J}\,k\mathrm{D}\boldsymbol{p} + z\,k\boldsymbol{p} + \boldsymbol{g}\,\boldsymbol{q} \quad (3)$$

where x is a torsional spring gain ($0 < x < 1$). Each desired angle $_n\boldsymbol{q}$ ÎÂ in (–) $_d\mathrm{D}\boldsymbol{q}$ $=\boldsymbol{q}\ \boldsymbol{q}$ for the joints of the robotic armis set to each joint angle referred to the kinematic configuration when the arm of a human is stretched along gravitational direction [20].

References

1. Williams, K.: Build Your Own Humanoid Robot, The McGraw-Hill Companies, Publication, (2004); ISBN: 0-07-142274-9
2. Williams, K.: Amphibionics, Build Your Own Reptilian Robot, The McGraw-Hill Companies, Publication (2003); ISBN: 0-07-142921-2
3. McComb, G.: Constructing Robot Bases, The McGraw-Hill Companies, Publication (2003); ISBN: 0-07-140852-5
4. Stevens, R.: Serial PIC'n: PIC Microcontroller Serial Communications, Square 1 Electronics, Publication (May 3, 1999); ISBN-10: 0965416224

5. Shircliff, D.R.: Build A Remote-Controlled Robot, The McGraw-Hill Companies (March 25, 2002); ISBN: 0-07-138543-6
6. Petrusha, R.: Visual Basic 2005: The Complete Reference, McGraw-Hill (July 27, 2006); ISBN-10: 0072260335
7. Lee, J.M., Park, B.S., Lee, Y.S., Ahn, J.S., Lee, S.H., Lim, S.J., Han, C.S.: The development of the robot manipulator for an intelligent service robot. In: International Joint Conference on SICE-ICASE, pp. 282–287 (2006)
8. Taipalus, T., Kazuhiro, K.: Development of service robot for fetching objects in home environment. In: IEEE International Symposium on Computational Intelligence in Robotics and Automation, pp. 451–456 (2005)
9. Sandin, P.E.: Robot Mechanisms and Mechanical Devices Illustrated, Publisher: McGraw-Hill (2003); ISBN: 0-07-142928-X
10. Tadakuma, R., Asahara, Y., Kajimoto, H., Kawakami, N., Tachi, S.: Development of Anthropomorphic Multi-D.O.F.Master-Slave Arm for Mutual Telexistence. IEEE Transactions on Visualization and Computer Graphics 11(6) (November-December 2005)
11. Atkeson, C.G., Hale, J.G., Pollick, F., Riley, M., Kotosaka, S., Schaal, S., Shibata, T., Tevatia, G., Ude, A., Vijayakumar, S., Kawato, M.: Using Humanoid Robots to Study Human Behavior. IEEE Intelligent Systems 15(4), 46–56 (2000)
12. Ambrose, R.O., Aldridge, H., Askew, R.S., Burridge, R., Bluethman, W., Diftler, M.A., Lovchik, C., Magruder, D., Rehnmark, F.: ROBONAUT: NASA's Space Humanoid. IEEE Intelligent Systems 15(4) (July-August 2000)
13. Diftler, M.A., Ambrose, R.O.: ROBONAUT, A Robotic Astronaut Assistant. In: Proc. Int'l Symp. Artificial Intelligence, Robotics, and Automation in Space (June 2001)
14. Bluethmann, W., Ambrose, R., Diftler, M., Askew, S., Huber, E., Goza, M., Rehnmark, F., Lovchik, C., Magruder, D.: Robonaut, aRobot Designed to Work with Humans in Space. Autonomous Robots 14, 179–197 (2003)
15. Tachi, S., Sakaki, T.: Impedance Controlled Master SlaveManipulation System Part I: Basic Concept and Application to the System with Time Delay. J. Robotics Soc. Japan 8(3), 241–252 (1990)
16. Winter, D.A.: Biomechanics and Motor Control of Human Movement, 2nd edn. Wiley, New York (1990)
17. Rosheim, M.: Robot Evolution: The Development of Anthrobotics. Wiley, NewYork (1994)
18. Zatsiorsky, V.M.: Kinematics of Human Motion. Human Kinetics, Urbana, Ill (1998)
19. Due, L.E., Syaifuddin, M., Toai, T.T., Tan, N.H., Saad, M.B., Wai, L.C.: Designing 8 Degrees of Freedom Humanoid Robotic Arm. In: IEEE International Conference on Intelligent and Advanced System (2007)
20. Bae, J.-H., Yang, W., Kim, D., Oh, Y., You, B.J., Oh, S.R.: Robotic Arm Control Inspired by Human Muscle Tension Effectunder the Gravity. In: IEEE International Conference on Robotics and Automation, Shanghai Chaina (2011)

A New Video Object Segmentation Algorithm by Fusion of Spatio-temporal Information Based on GMM Learning

Qingsong Zhu[1,2], Yaoqin Xie[1], Jia Gu[1], and Lei Wang[1]

[1] Shenzhen Institutes of Advanced Technology, Chinese Academy of Sciences, Shenzhen, China
[2] Chinese University of Hong Kong, Hongkong, China
qs.zhu@siat.ac.cn

Abstract. In the field of surveillance, Effective and rapid video object segmentation is a key technology for video analysis and processing. For the complex scene and noise that affect segmentation issue in the fixed occasion, on the base of classic Gaussian Mixture Background Model (GMM), a new algorithm named the fusion of Spatio-Temporal based on GMM is proposed for video object segmentation, which classifies for each pixel in Time and Space scales. Firstly, the algorithm constructs dynamically Gaussian Mixture Background Model for each pixel and segment foreground objects through background subtraction. Secondly, the algorithm detects synchronously the neighborhood statistic feature of each pixel through two lemmas. Finally, a result is produced using the spatial segmentation coupling with the temporal segmentation by "and" operator. Experiments show that our proposed algorithm can segment the moving object effectively and quickly from video sequences and has stronger robustness application prospect contrasted with other algorithms.

Keywords: Spatio-Temporal Information, Gaussian Mixture Model, background subtraction.

1 Introduction

Precise subtraction of moving object in video stream is a hot field of interdisciplinary study, which involves many fields such as image processing, artificial intelligence, pattern recognition, machine learning. It's also a key step to recognize and track objects in video surveillance, present multi-dimensional moving objects and so on. Background subtraction in video stream is a prerequisite and guarantee of accurate subtraction of moving objects. Fast, precise and robust background segmentation is becoming more difficulty in video object detection. Moving object detection and segmentation methods have been widely applied, and it mainly includes: background difference method [1], inter-frame difference method [2], method based on statistical model [3, 4], optical flow method [5] etc. Inter-frame method is not sensitive to changes of scene lighting. When grayscale in large area of the object surface is

G. Lee (Ed.): Advances in Automation and Robotics, Vol.2, LNEE 123, pp. 641–650.
Springerlink.com

uniform, "holes" will appear and segment the object into many regions. Background difference method detects moving area using difference of current video frame and background image. It can provide the most integrated information of moving object, and has a fast speed, a simple algorithm and a current detection to meet the real-time requirements of the system. But it is sensitive to changes of the moving scenes such as lighting and noise. So it needs to update background constantly. The method based on statistical model can subtract background from frame sequences well and segment a moving object, but it needs a former assumption of the background distribution feature density field. Optical flow method uses its physical properties and its advantage is that it can also segment the independent moving object under the condition of camera jitter. But the algorithm is particularly complex and time-consuming, therefore it is difficult to meet real-time requirements.

Classical adaptive Gaussian mixture background subtraction model [6, 7, 8] is a background modeling method based on pixel, proposed by Stauffer et al. It constructs a grayscale distribution model of each pixel based on the distribution information of each pixel in time domain and builds a background model of the pixels. Gaussian Mixture Background Model is a weighted sum of finite functions, which can describe the multi-peak state of pixels. It suits to model the complex background such as light gradients and trees swaying. Afterwards researchers do a variety of improvements on the Gaussian Mixture Model (GMM) background subtraction method [9, 10, 11, 12]. This method becomes a usual background subtraction method now. But this orthodox GMM models every pixel in every frame, which defaults that each pixel is strictly independent and irrelevant. From this analysis, it ignores the correlation rule of spatial adjacent domain among pixels in the same frame, and reduces the noise immunity and anti-interference. As a result, later researchers tend to fuse some spatial characteristic information of the video object, such as information of intensity, border, grayscale and texture, to segment the video object. For the above, this article proposes an algorithm for video object segmentation by fusion of spatio-temporal information based on GMM. This algorithm differs from orthodox algorithms based on spatio-temporal information [13, 14, 15]. Firstly it builds background GMM for each pixel in frame images in the time domain, and segments foreground object from the background by the background difference method in real time. Then by introducing two lemmas in the spatial domain, it synchronously detects the neighborhood statistic characteristics of each pixel on the video stream image sequence from the image grayscale, texture and other characteristics. After this processing, we can couple the results of the spatio-temporal segmentation of the same video frame through "and" operation, and then segment the accurate objects after morphological processing. The experimental result shows that this algorithm has high robustness, separates the foreground moving objects from the background well, has high adaptability to complex background environment, and obtains a good testing effect. At the same time, this algorithm effectively remedies the misjudged issues that the object that stagnates a long time changes from foreground to background in the Gaussian Mixture Background Model proposed by Stauffer et al.

2 Video Object Segmentation in Time Domain

2.1 Subtraction of Background in the Time Domain Based on GMM

We model each pixel in the frame images in the video stream using GMM. This GMM model uses $K(K=3,4,5)$ Gaussian distributions to count characteristics of the same pixel in each frame in the video stream. That is to set the pixel value in the latest t frames to $(X_1, X_2, ..., X_t)$, among them X_t is the pixel value at the time of t. The probability function of K mixture Gaussian distributions is:

$$p(X_t)=\sum_{k=1}^{K}\omega_{k,t}\cdot g(X_t;\mu_{k,t};\Sigma_{k,t}) \tag{1}$$

$$g(X_t;\mu_{k,t};\Sigma_{k,t})=\left[(2\pi)^{\frac{d}{2}}\left|Cov_{k,t}\right|^{\frac{1}{2}}\right]^{-1}\cdot\exp\left[-\frac{1}{2}\left(X_t-\mu_{k,t}\right)^T\cdot\Sigma_{k,t}^{-1}\left(X_t-\mu_{k,t}\right)\right] \tag{2}$$

The parameters of the probability function are:

$$(\omega_{1,t},...\omega_{K,t};\mu_{1,t},...\mu_{K,t};\Sigma_{1,t},...\Sigma_{K,t}) \tag{3}$$

Among them, $\omega_{k,t}, \mu_{k,t}, cov_{k.t}$ respectively are the weight, mean vector, and covariance matrix of the K th Gaussian distribution at time t. K is the number of Gaussian distributions. And $\omega_{1,t}, \omega_{2,t}, \ldots\omega_{K,t}$ meet the condition:

$$\sum_{k=1}^{K}\omega_{K,t}=1 \tag{4}$$

Assuming pixel values in the RGB color space are independent of one another and have the same variance, we have the following equation:

$$Cov_{k,t}=\sigma_k^2 I=\sigma_k^2\begin{pmatrix}1 & \cdots & 0\\ \vdots & \ddots & \vdots\\ 0 & \cdots & 1\end{pmatrix} \tag{5}$$

For the new sample point X_{t+1} in video stream frame image sequence, we get the current value $ImageData(X_{t+1})$ of pixel X_{t+1} and compare one by one with the mean value $\mu_{k,t}$ of K Gaussian model allocated by memory, and see which Gaussian model matches it. The matching principle is:

$$\left|ImageData(X_{t+1})-\mu_{k,t}\right|=\xi\cdot\sigma_{k,t} \tag{6}$$

Among it ξ more appropriately values 3 ± 0.5 in general. $\sigma_{k,t}$ is the standard deviation of the K th Gaussian distribution. According to the judging formula (5) we can find the matching Gaussian distribution. If the i th pixel exactly matches the $K(K = 1,2,3,4,5)$ th Gaussian distribution, the K th Gaussian distribution will be updated by X_{t+1}, while other Gaussian compositions remain the same. In actual software design, the structure body directed by the K th Gaussian distribution pointer will include four important member variables: number of matches, weight, variance, and mean value. Next we will update the weight, variance and mean value of the K th Gaussian distribution according to X_{t+1}. Update algorithm is as follows:

$$\omega_{k,t+1} = (1-\alpha)\cdot\omega_{k,t} + \alpha\cdot(M_{k,t}) \tag{7}$$

$$M_{k,t} = \begin{cases} 1 & match\ successes \\ 0 & match\ fails \end{cases} \tag{8}$$

Among them is the studying rate of Gaussian model.

$$\mu_{k,t+1} = (1-\rho)\cdot\mu_{k,t} + \rho\cdot X_{t+1} \tag{9}$$

$$(\sigma_{k,t+1})^2 = (1-\rho)\cdot(\sigma_{k,t})^2 + \rho\cdot(X_{t+1}-\mu_{k,t})^T\cdot(X_{t+1}-\mu_{k,t}) \tag{10}$$

$$\rho = \alpha\cdot g(X_{t+1}, \mu_{k,t}, \sigma_{k,t}^2) \tag{11}$$

After update parameters of each Gaussian distribution, we need to calculate the values of ω/σ in K Gaussian distributions and sort the Gaussian distributions by ω/σ from large to small. Determine the Gaussian distribution that can best characterize background characteristics according to sorted Gaussian distributions. The determine algorithm is: we only obtain B Gaussian distributions which have a greater former weight. That is:

$$B = argmin_b\left(\sum_{k=1}^{b}\omega_{k,t} > T\right) \tag{12}$$

Among them threshold T represents the prior probability value that the current pixel is judged as background. In the experiment, the T value is important because if T is too small, multi-Gauss model will deteriorate into single Gauss model, and if T is too large, the number of Gaussian distributions that describe the background increases and the system will regard moving objects as the background. Most literatures set T =0.8 or so. In this way, we accomplish the subtraction of background in the time domain based on GMM.

2.2 Foreground Segmentation Using Background Difference Method

Known from formula (11) above, if we use B Gaussian distributions that have bigger former weight to represent background distribution, the remaining $K-B$ Gaussian distributions will be positioned as foreground objects. That is:

$$F = K - B \tag{13}$$

Through a series of simulation experiments, we know that it appears a result of vague edge contour of segmentation objects to segment foreground objects using formula (13). It will greatly reduce the effect of segmentation. We know, background difference method received wide application for its simple algorithm, accurate detection, immediacy and effectiveness. The premise and key to obtain a fine effect of segmentation is to build a steady and reliable background model. So we can use the background difference method to segment the foreground based on background modeling by GMM. It will obtain a real-time robust preliminary segmentation result in the time domain.

3 Video Object Segmentation in the Spatial Domain

In the algorithm of video object segmentation in the spatial domain, since GMM is assumed independent among each pixel, it results in incomplete segmented foreground objects and existence of some small holes, at the same time a small part of the foreground objects will be misjudged as background. In response to these deficiencies, we do the second segmentation in the spatial domain based on the video object segmentation algorithm in the time domain. Its main purpose is to correct and compensate the background subtracted from the time domain and segmented foreground objects, that is, to secondarily cluster on the basis of the initial classification in the time domain. Correct those background points that are misjudged as foreground points, and those foreground points that are misjudged as background points. Its main segmentation idea is mainly based on the following two lemmas:

Lemma 1: If most pixels in a small neighborhood of a pixel belong to foreground in a foreground object, this pixel very likely belongs to foreground.

Lemma 1 illustrates: from the perspective of spatial domain of every frame in the video stream, the grayscale of all pixels in every small neighborhood affects the current pixel magnificently. So from this point of view, we construct a 3×3 filter to traverse all of the pixels in each frame. Consider the template of 8-neighborhood around the pixel (x,y), using the Bayesian decision theory in statistics, we model the priori probability $\mathrm{P}_{(x,y)}(fg)$ of pixel (x,y) as follows [16]:

$$P_{(x,y)}(fg) = \frac{1}{Z}\exp\{-[n_{hv}B + n_d C]\} \tag{14}$$

Among them, Z is an orthogonal constant, n_{hv} is the number of pixels on horizontal or vertical direction in the foreground template, n_d is the number of pixels on adjacent diagonal direction in the foreground template, B and C is corresponding penalty coefficient. Next we give out the likelihood probability $P_{(x,y)}(V_t \mid fg)$ (among them V_t is the foreground characteristic vector) of the foreground pixel (x, y). Then the pixel (x, y) clustered as foreground will meet the decision analysis formula as follows:

$$P_{(x,y)}(V_t \mid fg) \cdot P_{(x,y)}(fg) > T \tag{15}$$

Among them, T is a fixed threshold value and usually is set to 0.7. If formula (15) is satisfied, this pixel will be clustered as foreground. Conversely, this pixel will be divided into background.

Lemma 2: The pixels in a small neighborhood of a pixel in foreground segmentation objects have a local grayscale similarity.

Lemma 2 illustrates: in the foreground object neighborhood containing small holes obtained from video object segmentation algorithm in the time domain, by calculating local brightness similarity of foreground pixels, we can re-cluster those pixels misjudged as background to foreground objects. Assume the pixel grayscale value of foreground object P in the t th frame of the video images is: $f(x, y) = f(x, y \mid (t, p))$ Then the gradient vector of this point is:

$$gradf(x, y \mid (t, p)) = f_x(x, y \mid (t, p))i + f_y(x, y \mid (t, p))j \tag{16}$$

Where:

$$f_\delta(x, y \mid (t, p)) = \nabla_\delta f(x, y \mid (t, p)) \quad \delta = x \, or \, y \tag{17}$$

Assume two adjacent pixels in a small neighborhood of point P are p_i, p_j respectively, and assume the correspondent gradient vector similarity of the two pixels p_i, p_j is $\mathrm{Similar}(p_i, p_j)$, then:

$$Similar(p_i, p_j) = \frac{gradf(x_1, y_1 \mid (t, p_1)) \bullet gradf(x_2, y_2 \mid (t, p_2))}{\| gradf(x_1, y_1 \mid (t, p_1)) \| \cdot \| gradf(x_2, y_2 \mid (t, p_2)) \|} \tag{18}$$

By the definition of similarity, the following conclusions are obviously set up:

$$0 \le Similar(p_i, p_j) \le 1 \tag{19}$$

$$\mathrm{Similar}(p_i, p_i) + \mathrm{Similar}(p_j, p_j) \ge 2\mathrm{Similar}(p_i, p_j) \tag{20}$$

On the other hand, if p_i and p_j come from the surface of different objects, the local texture feature of p_i and p_j usually have big difference. The criterion above can be realized by introducing the discriminate function $\omega(p_i, p_j)$:

$$\omega(p_i, p_j) = 1 - \frac{2S(p_i, p_j)}{S(p_i, p_i) + S(p_j, p_j)} \tag{21}$$

The threshold of ω is usually set to 0.2 experiencely.

4 Spatio-temporal Fusion Based on GMM

Through the preliminary segmentation in the time domain and the following processing in the spatial domain, we obtain a general background distribution and foreground object of the video stream. That is through obtaining general regions of the video objects by segmentation in the time domain, and then through gain accurate information of video objects by secondary segmentation in the spatial domain, we clear and complete the foreground objects. After above processing, we can couple the results of spatio-temporal segmentations of the same video frame through pixel "and" operation. It successfully restores pixels misjudged as background to corresponding foreground pixels, and also makes those pixels belongs to foreground originally remain the same. Thus, we accomplish a perfect coupling of a preliminary segmentation based on the time domain and a secondary segmentation based on the spatial domain, and accurately segment the foreground objects in the video at last.

5 Morphological Procession

Due to the impact of noise, after the preliminary segmentation, secondary processing, and spatio-temporal coupling of video images based on GMM, there inevitably are some local small regions caused by several isolated points and noise. Here we use a 3×3 sliding window (20) to process these processed video images by morphology. The specific processing steps are as follows:

$$\mathbf{W}_3 = \begin{pmatrix} 0 & 1 & 0 \\ 1 & 1 & 1 \\ 0 & 1 & 0 \end{pmatrix} \tag{22}$$

1) Firstly do corrosion process twice, removing local small area and isolated points in the video images.
2) Secondly, making expansion process for those corroded video images again, making the edge of the corroded foreground objects smooth and recovered.

6 Experimental Results and Analysis

To verify the actual segmentation effect of the algorithm proposed by this article, we use the outdoor highway traffic video for detection. We did many groups of simulation experiment on the VC++6.0 platform on a Pentium 4, 1.8GHz dual core PC. Here in GMM, parameter K values 5, initial variance values is set to 100, α value is set to 0.06, ξ values is set to 3. Fig.1 is comparisons of the experimental results of the segmentation algorithm proposed by this article and Stauffer *et al.* The first and the second image series correspond to the processing results on frame 536 and frame 585 in the video sequence. Each row from left to right is followed by video image, background subtracted by the adaptive Gaussian Mixture Background Model, segmentation result using Stauffer algorithm, segmentation result using algorithm of this article. From the experimental results we can see: the foreground objects using the segmentation algorithm proposed by this article are more accurate and integrated than the segmentation result using Stauffer algorithm. This algorithm not only overcomes the deficiency of having vague edge profiles and the misjudging problem that the long stagnating objects change from foreground into background in the segmentation result of Stauffer, but also can segment the foreground objects accurately and completely under the condition that the road shakes up and down caused by the slight jitter of the camera, showing strong robustness.

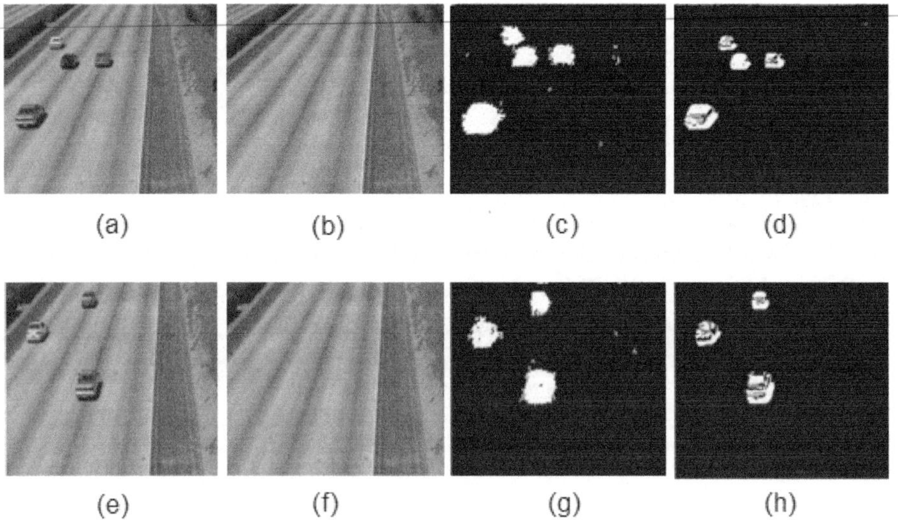

Fig. 1. Comparisons of the two segmentation algorithms. (a). video sequence (frame585). (b). Result of background modeling. (c) Video segmentation result of stauffer algorithm. (d) Video segmentation result of of our algorithm. (e). video sequence (frame602). (b). Result of background modeling. (c) Video segmentation result of stauffer algorithm. (d) Video segmentation result of of our algorithm.

7 Conclusion

This paper proposes a video object segmentation algorithm by coupling of spatio-temporal information based on GMM. This algorithm builds Gaussian Mixture Background Model dynamically in the time domain and preliminarily segments the foreground objects through background difference, then introduces two lemmas and secondarily clusters the preliminary segmentations in the spatial domain, at last couples the results of segmentations using the spatio-temporal information and thereby obtains an accurate segmentation result. The experiment verifies the effectiveness of this algorithm: not only improves the completeness and precision of the segmentation result, but also has a strong adaptability to the complex background environment. At the same time, this algorithm effectively makes up the misjudging problem like long stagnating objects change from foreground into background in the adaptive Gaussian Mixture Background Model proposed by Stauffer *et al.*

Acknowledgments. This paper is supported by National Natural Science Foundation of China (NSFC: 81171402), NSFC Joint Research Fund for Overseas Research Chinese, Hong Kong and Macao Young Scholars (30928030), grant number 6:National Basic Research Program 973 (2010CB732606) from Ministry of Science and Technology, China, Shenzhen Fundamental Research (JC201005270312A) and Shenzhen Science and Technology Program (JC200903160397A).

References

1. Palaniappan, V.J., Zhuang, X.H.: Automatic spatio-temporal video sequence segmentation. In: Proceeding of IEEE International Conference on Image Processing, pp. 958–962. IEEE Press, Chicago (1998)
2. Xuan, W.: Difference Image Based Multiple Motion Targets Detection and Tracking. J. of Image and Graphics
3. Lin, H.W., Tu, Y., Li, G.H.: Moving Objects Detection Method Based on Statistical Background Model. J. Computer Engineering 16, 125–132 (2003)
4. Liu, Y., Ai, H.Z., Xu, G.Y.: Moving object detection and tracking based on background subtraction. J. Information and Control 4, 315–319 (2002)
5. Li, J.Z., Yuan, L., Li, D.D.: Tracking method of movingtarget based on detection of feature optical flow. J. Systems Engineering and Electronics 3, 422–426 (2004)
6. Friedman, N., Russell, S.: Image segmentation in video sequences: Ap robabilistic approach. In: Proceedings of the 13th Annual Conference on Uncertainty in Artificial Intelligence, pp. 175–181. IEEE Press, Rhode Island (1997)
7. Stauffer, C., Grimson, W.: Adaptive background mixture models for real. time tracking. In: Proceedings of IEEE Conference on Computer Vision and Pattern Recognition, pp. 246–252. IEEE Press, Fort Collins Colorado (1999)
8. Stauffer, C., Grimson, W.E.L.: Learning patterns of activity using realtime tracking. IEEE Transactions on Pattern Analysis and Machine Intelligence 22(8), 747–757 (2000)
9. Kaewtrakulpong, P., Bowden, R.: An improved adaptive background mixture model for real time tracking with shadow detection. In: Proceedings of the 2nd European Workshop on Advanced Video Based Surveillance Systems, pp. 1–5. IEEE Press, Kluwer (2001)

10. Zivkovic, Z., Vander, H.F.: Effcient adaptive densityestimation per image pixel for the task of background subtraction. Pattern Recognition Letters 27(7), 773–780 (2006)
11. Lee, D.S.: Effective Gaussian mixture learning for videobackground subtraction. IEEE Transactions on Pattern Analysis and Machine Intelligence 27(5), 827–832 (2005)
12. Power, P.W., Schoonees, J.A.: Understanding background mixturemodels for foreground segmentation. In: Proceedings of Image and Vision Computing, New Zealand.Auckland, New Zealand, pp. 267–271 (2002)
13. Li, C.H., Zhang, Y.L., Yu, Y.L.: A Kind of Video Object Segmentation Based of Spatio-temporal Information. J. of Image and Graphics 8, 278–292 (2003)
14. Gao, L., Yang, S.Y., Li, H.Q.: New Efficient Automatic VOP Segmentation Based on Spatio-temporal Information. J. of Image and Graphics 10, 358–365 (2005)
15. Zhu, Z.J., Jiang, G.Y., Yu, M., Wu, X.W.: New Algorithm for Extracting Moving Object Based on Spatio-temporal Information. J. of Image and Graphics 8, 577–584 (2003)
16. Aach, T., Kaup, A.: Baysian Algorithms For Adaptive Change Detection In Image Sequences Using Markov Random Fields. Signal Processing: Image Communiation 7(2), 147–160 (1995)

Author Index

GPSR Compliance

The European Union's (EU) General Product Safety Regulation (GPSR) is a set of rules that requires consumer products to be safe and our obligations to ensure this.

If you have any concerns about our products, you can contact us on ProductSafety@springernature.com

In case Publisher is established outside the EU, the EU authorized representative is:

Springer Nature Customer Service Center GmbH
Europaplatz 3
69115 Heidelberg, Germany

Batch number: 10777868

Printed by Printforce, the Netherlands